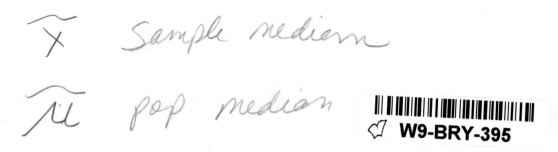

$\tilde{x}$ Sample median

$\tilde{\mu}$ pop median

f	frequency with which a value occurs	$z_{\alpha/2}$	critical value of z
Σ	capital sigma; summation	t	t distribution
Σx	sum of the scores	$t_{\alpha/2}$	ctitical value of t
Σx^2	sum of the squares of the scores	df	number of degrees of freedom
$(\Sigma x)^2$	square of the sum of all scores	F	F distribution
Σxy	sum of the products of each x score multiplied by the corresponding y score	χ^2	chi-square distribution
		χ^2_R	right-tailed critical value of chi-square
		χ^2_L	left-tailed critical value of chi-square
n	number of scores in a sample	p	probability of an event or the population proportion
$n!$	factorial		
N	number of scores in a finite population; also used as the size of all samples combined	q	probability or proportion equal to $1 - p$
k	number of samples or populations or categories	$\hat{p}$	sample proportion
		$\hat{q}$	sample proportion equal to $1 - \hat{p}$
$\overline{x}$	mean of the scores in a sample	$\overline{p}$	proportion obtained by pooling two samples
$\overline{R}$	mean of the sample ranges		
μ	mu; mean of all scores in a population	$\overline{q}$	proportion or probability equal to $1 - \overline{p}$
s	standard deviation of a set of sample values	$P(A)$	probability of event A
σ	lowercase sigma; standard deviation of all values in a population	$P(A\|B)$	probability of event A, assuming event B has occurred
s^2	variance of a set of sample values	${}_nP_r$	number of permutations of n items selected r at a time
σ^2	variance of all values in a population	${}_nC_r$	number of combinations of n items selected r at a time
z	standard score		

Sixth Edition

Elementary Statistics

Mario F. Triola

**Dutchess Community College
Poughkeepsie, New York**

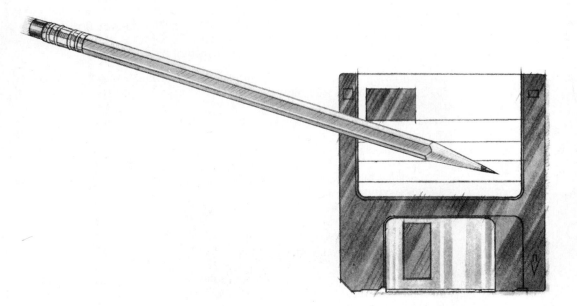

 ADDISON-WESLEY PUBLISHING COMPANY

*Reading, Massachusetts • Menlo Park, California • New York
Don Mills, Ontario • Workingham, England • Amsterdam • Bonn
Sydney • Singapore • Tokyo • Madrid • San Juan • Milan • Paris*

To Marc and Scott

Executive Editor: Michael Payne
Senior Sponsoring Editor: Julia G. Berrisford
Editorial Assistant: Maureen Lawson
Marketing Manager: Craig Bleyer
Senior Development Editor: Marilyn Freedman
Development Editor: Chere Bemelmans
Development Assistant: Susan Howard
Managing Editor: Kazia Navas
Production Supervisor: Kathy Diamond
Production Services: Lifland et al., Bookmakers
Text Designer: Janet Theurer

Cover Designer: Peter Blaiwas
Cover Illustrator: Terry Presnall
Art Buyer: Joe Vetere
Art Coordinator: Susan London-Payne
Technical Illustrator: Scientific Illustrators
Essay Illustrator: Terry Presnall
Prepress Services Manager: Sarah McCracken
Senior Manufacturing Manager: Roy Logan
Composition: Black Dot Graphics
Printer: R. R. Donnelley & Sons

Many of the designations used by manufacturers and sellers to distinguish their products are claimed as trademarks. Where those designations appear in this book, and Addison-Wesley was aware of a trademark claim, the designations have been printed in initial caps or all caps.

Photo credits: p. 32, Courtesy of Lester R. Curtin; p. 180, Courtesy of Anthony M. DiUglio, Jr., P.O. Box 3608, Poughkeepsie, NY 12603; p. 282, Courtesy of Drs. Lee Miringoff and Barbara Carvalho, Marist Institute for Public Opinion; p. 404, Courtesy of Jay Dean, Young & Rubicam San Francisco; p. 532, Courtesy of Barry Cook, A. C. Neilsen Co.; p. 648, Courtesy of Boeing Commercial Airplane Group.

Logos used with permission of The Philadelphia 76ers, Motorola Cellular, Motorola Inc., and The Procter & Gamble Company.

Library of Congress Cataloging-in-Publication Data
Triola, Mario F.
   Elementary statistics / Mario F. Triola.—6th ed.
     p.   cm.
     Includes bibliographical references and index.
     ISBN 0-201-57682-1
     1. Statistics. I. Title.
QA276.12.T76   1994
519.5--dc20   94-16049
Reprinted with corrections, May 1995.   CIP

ISBN 0-201-57682-1
4 5 6 7 8 9 10-DOC-979695

Preface

"Smoking is one of the leading causes of statistics," according to columnist Fletcher Knebel. Although somewhat amusing, that statement isn't very accurate. More realistic would be the claim that modern technology is a leading cause of statistics. Calculators and computers are making it possible for millions of people to process data painlessly, and statistical analysis of data is now occurring at an unprecedented rate.

Not very long ago, students who elected to take a statistics course enjoyed a competitive advantage in procuring jobs. Now students *without* any statistics courses are suffering from a real competitive disadvantage in the job market. Jay Dean, a Senior Vice President at Young & Rubicam Advertising, said in an interview with the author, "If I could go back to school, I would certainly study more math, statistics, and computer science." In another interview, David Hall, a Division Statistical Manager for the Boeing Commercial Airplane Group, told the author, "Right now, American industry is crying out for people with an understanding of statistics and the ability to communicate its use." And pollster Lee Miringoff said, "The study of statistics is important for understanding one aspect of knowledge and it's a key to opening up other avenues of pursuit; statistics cuts across disciplines." As employees, employers, and citizens, we must learn at least the elementary concepts of the field of statistics.

This book is designed to be a clear, readable, and even enjoyable introduction to the statistical concepts that have become such an important part of our everyday lives—the surveys and polls we hear and read about; research findings, from fields as diverse as medicine and marketing, that affect our outlooks and behaviors; the impact on each of us of statistical measures such as psychological and educational testing; and the increasingly important applications of quality control in business and engineering.

Audience

This book is an introduction to elementary statistics for students majoring in any field except mathematics. A strong mathematics background is not necessary, but students should have completed a high school or college elementary algebra course.

Although underlying theory is included, this book does not stress the mathematical rigor more suitable for mathematics majors.

Because the many examples and exercises cover a wide variety of different and interesting statistical applications, this book is appropriate for students pursuing majors in a wide variety of disciplines ranging from the social sciences of psychology and sociology to areas such as education, business and economics, engineering, the humanities, the physical sciences, and liberal arts.

Exercises

The Sixth Edition of **Elementary Statistics** includes more than 1400 exercises, many of them involving real-world data. In addition, many more exercises than in previous editions now involve *interpretation* of results. Because exercises are of such core importance to any statistics book, great care has been taken to ensure their usefulness, relevance, and accuracy.

Exercises are arranged in order of increasing difficulty. They are divided into two groups: Basic Skills and Concepts and Beyond the Basics. Beyond the Basics exercises address more difficult concepts or require a somewhat stronger mathematical background. In some cases, these exercises introduce a new concept.

Key Motivational Features

Real data sets are used throughout the book in examples and exercises. Data sets, found in Appendix B and on a separate data disk, include the following:

- The body temperatures of 107 healthy adults tested by University of Maryland researchers
- Weights of M&M candies
- Household garbage data from the famous study by University of Arizona archaeologist William Rathje

(A complete list of data sets can be found at the end of the Contents on page xiv.)

Marginal essays (117 in all) illustrate the uses and abuses of statistics in real practical applications. A small sample of these essays indicates their variety as well as their range of coverage:

- Biology: Were Mendel's experimental data manipulated?
- Business: How airlines save money by using sampling to determine revenues from split-ticket sales
- Ecology: How a Florida statistical study led to regulations that protect manatees
- Gambling: Why some lottery number combinations are better choices than others

- Sports: What happened to the 0.400 hitters in baseball?
- Public Policy: How a statistical analysis showed that the death penalty doesn't deter murders
- Entertainment: How it takes seven shuffles before a deck of cards is completely mixed
- Medicine: How the Salk vaccine was tested

Interviews with these six professionals focus on the importance and use of probability and statistics in their day-to-day work:

- (NEW) Lester Curtin, Chief of the Statistical Methods Staff at the National Center for Health Statistics
- (NEW) Anthony DiUglio, Nuclear Analyst in the Probabilistic Risk Assessment Department of the Consolidated Edison Company of New York, Indian Point Nuclear Power Plant
- (NEW) Barbara Carvalho, Director of the Marist College Poll, and Lee Miringoff, Director of the Marist College Institute for Public Opinion
- Jay Dean, Senior Vice President at Young & Rubicam Advertising
- Barry Cook, Senior Vice President at Nielsen Media Research
- David Hall, Division Statistical Manager at the Boeing Commercial Airplane Group

Flowcharts (19 in all) are used to visually depict and reinforce more complex procedures.

Key Pedagogical Features

Chapter-opening features

- A list of chapter sections with a brief description of their contents
- A chapter-opening problem which is solved later in the chapter
- A chapter overview which includes a statement of the chapter objectives

End-of-chapter features

- A Vocabulary List of important terms introduced in the chapter (A full glossary is found in Appendix C.)
- A Chapter Review
- A Summary of Important Formulas
- Review Exercises
- Computer Projects

- From Data to Decision, which is a capstone problem requiring the student to think critically and apply concepts and techniques found throughout the chapter (Many also contain a writing component that encourages students to improve both their critical thinking and their writing skills.)

Additional features

- Appendix C provides an expanded glossary of important terms.
- Appendix D is a bibliography of recommended and reference books.
- Appendix E contains answers to the odd-numbered exercises.
- A symbol table is included on the front inside cover for quick and easy reference to key symbols.
- Tables A-2 and A-3 are reprinted on the back inside cover pages for easy reference.
- A detachable formula/table card is enclosed for student use throughout the course.

Computer Usage

This text can be used without any reference to computers. However, for those who choose to supplement the course with computer usage, Computer Projects are included at the end of each chapter. Two different levels of software are available, and their use is discussed with sample displays throughout the book. The data sets in Appendix B (except Data Set 2) are available on disk for use with either Minitab or STATDISK.

STATDISK

STATDISK is an easy-to-use statistical software package that does not require any previous computer experience. Developed as a supplement specifically for this textbook, STATDISK is available for the IBM PC and Macintosh systems. This software is provided at no cost to colleges that adopt this text.

STATDISK Student Laboratory Manual and Workbook includes instructions on the use of the STATDISK software package. It also includes experiments to be conducted by students. The STATDISK software and the student manual/ workbook have been designed so that instructors can assign computer experiments without using valuable classroom time. STATDISK includes a wide variety of programs, which can be used throughout the course, and the experiments do more than number crunch or duplicate text exercises.

Minitab

For those who wish to use Minitab, Minitab displays have been included throughout the text.

Minitab Student Laboratory Manual and Workbook is designed specifically for this text and includes instructions on and examples of Minitab use. It also has experiments to be conducted by students.

Other Minitab software products available from the Publisher are as follows:

- **STAT101**

  An inexpensive software package based on Release 6 of Minitab, STAT101 provides an excellent foundation for progression to other, more powerful Minitab software products. It works with a spreadsheet-like Data Editor that allows up to 2000 data points.

- **Student Edition of Minitab for Windows**
  Student Edition of Minitab, Release 8 for DOS
  Student Edition of Minitab, Release 8 for the Macintosh

  The Student Editions of Minitab allow the user access to 3500 data points and to a full range of Minitab functionalities and graphics capabilities, as well as a wide range of data sets drawn from business, the social and physical sciences, and engineering. Accompanying the software for each product is a user's manual, which includes case studies and hands-on tutorials for using the software.

Supplements

The following student supplements are available from Addison-Wesley Publishing Company and can be purchased through your campus bookstore.

- *Student's Solutions Manual,* by Milton Loyer, which provides detailed worked-out solutions to odd-numbered exercises
- *STATDISK Student Laboratory Manual and Workbook,* by Mario Triola
- *Minitab Student Laboratory Manual and Workbook,* by Mario Triola
- **Data Disk,** prepared by Mario Triola, with Appendix B data files for use with Minitab software

It has been a genuine pleasure working with a truly exceptional publishing team, and I thank Marilyn Freedman, Maureen Lawson, Michael Payne, Julie Berrisford, Faith Sherlock, Kathy Diamond, Chere Bemelmans, Susan London-Payne, Susan Howard, Marcia Cole, and the entire Addison-Wesley staff.

I take great pride and pleasure in thanking my sons Marc and Scott for their help in proofreading and their suggestions for data sets and examples. Finally, I thank my wife Ginny for her support and encouragement.

M.F.T.
LaGrange, New York
April, 1994

Acknowledgments

Special thanks go to David Perkins, Emily Keaton, Marc Triola, David Lund, and Milton Loyer for their help in checking the accuracy of the answers. I also thank Lester Curtin, Anthony DiUglio, Barbara Carvalho, Lee Miringoff, Jay Dean, Barry Cook, and David Hall, who agreed to be interviewed for this book. I thank William Rathje and Masakazu Tani of the University of Arizona; Steve Wasserman, Philip Mackowiak, and Myron Levine of the University of Maryland; and the hundreds of researchers who did studies, conducted polls, and compiled data used in the examples and exercises of this book. I extend my sincere thanks for the suggestions made by the following colleagues:

Mary Abkemeier, Fontbonne College

Jules Albertini, Ulster County Community College

Mary Anne Anthony, Rancho Santiago College

James Baker, Jefferson Community College

James E. Beatty, Burlington County College

Philip M. Beckman, Black Hawk Community College

Michelle Benedict, Augusta College

Ronald Bensema, Joliet Junior College

David Bernklau, Long Island University

Maria Betkowski, Middlesex County College

Shirley Blatchley, Brookdale Community College

David Blaueuer, University of Findlay

John Buchl, John Wood Community College

Jerome J. Cardell, Brevard Community College

Don Chambless, Auburn University

Rodney Chase, Oakland Community College

Bob Chow, Grossmont College

Phillip S. Clarke, Los Angeles Valley College

Darrell Clevidence, Carl Sandburg College

Susan Cribelli, Aims Community College

Arthur Daniel, Macomb Community College

Tom E. Davis, III, Daytona Beach Community College

Richard Dilling, Grace College

Rose Dios, New Jersey Institute of Technology

Paul Duchow, Pasadena City College

Nadina Duran, Corpus Christi State University

Evelyn Dwyer, Walter State Community College

Jane Early, Manatee Community College

P. Teresa Farnum, Franklin Pierce College

Maggie Flint, Northeast State Technical Community College

Bob France, Edmonds Community College

Richard Fritz, Moraine Valley Community College

Mahmood Ghamsary, Long Beach City College

Tena Golding, Southeastern Louisiana University

Elizabeth Gray, Southeastern Louisiana University

Francis Hannick, Mankato State University

Sr. Joan Harnett, Molloy College

Leonard Heath, Pikes Peak Community College

Peter Herron, Suffolk County Community College

Larry Howe, Rowan College of New Jersey

Lloyd Jaisingh, Morehead State University

Herbert H. Jolliff, Oregon Institute of Technology

Martin Johnson, Gavilan College

Roger Johnson, Carleton College

Toni Kasper, Borough of Manhattan Community College

Alvin Kaumeyer, Pueblo Community College

William Keane, Boston College

Emily Keaton, Peabody, Massachusetts

Michael Kern, Bismarck State College

Marlene I. Kovaly, Florida Community College at Jacksonville

Richard Kulp, David Lipscomb University

Linda Kurz, SUNY College of Technology

Benny Lo, Ohlone College

Vincent Long, Gaston College

Phillip McGill, Illinois Central Community College

Marjorie McLean, University of Tennessee

Hossein Mansouri, Texas Technical College

Virgil Marco, Eastern New Mexico University

Joseph Mazonec, Delta College

Austen Meek, Canada College

Robert Mignone, College of Charleston

Kermit Miller, Florida Community College

Charlene Moeckel, Polk Community College

Gerald Mueller, Columbus State Community College

Sandra Murrell, Shelby State Community College

Faye Muse, Asheville-Buncombe Technical College

Gale Nash, Western State College

DeWayne Nymann, University of Tennessee

Patricia M. Odell, Bryant College

James O'Donnell, Bergen County Community College

Ron Pacheco, Harding University

Kwadwo Paku, Los Medanos College

S. A. Patil, Tennessee Technical University

Robin Pepper, Tri-County Tech College

David C. Perkins, Texas A&M University—Corpus Christi

Richard J. Pulskamp, Xavier University

Susan Riner, Oxford College

Sylvester Roebuck, Jr., Olive Harvey College

Kenneth Ross, Broward Community College

Charles M. Roy, Camden County College

Kara Ryan, College of Notre Dame

Jean Schrader, Jamestown Community College

Calvin Shad, Barstow College

Carole Shapero, Oakton Community College

Lewis Shoemaker, Millersville University

Joan Sholars, Mt. San Antonio College

Galen Shorack, University of Washington

Cheryl B. Slayden, Pellissippi State Technical Community College

Arthur Smith, Rhode Island College

Marty Smith, East Texas Baptist University

Sandra Spain, Thomas Nelson Community College

Carol Stanton, Contra Costa College

Richard Stephens, Western Carolina College

Terry Stephenson, Spartanburg Methodist College

Sr. Loretta Sullivan, University of Detroit—Mercy

Andrew Thomas, Triton College

Evan Thweatt, American River College

Judith A. Tully, Bunker Hill Community College

Randy Villa, Napa Valley College

Hugh Walker, Chattanooga State Technical Community College

Charles Wall, Trident Technical College

Glen Weber, Christopher Newport College

Roger Willig, Montgomery County Community College

Elyse Zois, Kean College of New Jersey

Contents

To the Student

When beginning a chapter, you should first read the overview carefully. Read each section quickly to get a general idea of the material, then reread it very carefully. Try the exercises. If you encounter difficulty, return to the section and work some of the examples in the text so that you can compare your solution to the ones given.

When working on assignments, first attempt the early odd-numbered exercises. Before moving on to the more difficult exercises, verify that you are correct by checking your answers against those given in Appendix E. Keep in mind that producing neat and well-organized written assignments tends to yield better results. When you finish a chapter, check the review section to make sure that you didn't miss any major topics. Before taking tests, do the review exercises at the ends of the chapters. In addition to helping you review, this will help you cope with a mixture of different problems.

You might consider purchasing the *Student's Solutions Manual* for this text. It gives detailed solutions to the odd-numbered text exercises.

About the Author

Mario F. Triola is a Professor of Mathematics at Dutchess Community College, where he has taught statistics, calculus, linear algebra, technical mathematics, programming, and other courses for 25 years. He is the author of *Mathematics and the Modern World* and *A Survey of Mathematics* and is a coauthor of *Introduction to Technical Mathematics* and *Business Statistics,* as well as the STATDISK computer software package. His consulting experience includes the mathematical design of casino slot machines, as well as work with attorneys in determining probabilities in paternity lawsuits and identifying salary inequities based on gender.

Sixth Edition

Elementary Statistics

1 Introduction to Statistics

1-1 Overview

The term *statistics* is defined, along with the terms *population*, *sample*, *parameter*, and *statistic*.

1-2 The Nature of Data

Different ways of arranging data are discussed. The four levels of measurement (nominal, ordinal, interval, ratio) are defined, along with discrete and continuous data.

1-3 Uses and Abuses of Statistics

Examples of beneficial uses of statistics are presented, along with some of the common ways in which statistics are used to deceive.

1-4 Methods of Sampling

The importance of good samples is discussed. Different sampling methods, including random sampling, stratified sampling, systematic sampling, cluster sampling, and convenience sampling, are presented.

1-5 Statistics and Computers

Computers and the statistical software packages of Minitab and STATDISK are briefly discussed. Some fundamentals of Minitab are introduced.

Chapter Problem:
Body temperatures, garbage, and movie stars

What do body temperatures, garbage, and movie stars have in common? We can consider each of them to be a *data set* (collection of information) that we can analyze, use to make predictions, and explore in a variety of ways.

Everyone knows that the average body temperature is 98.6°F, right? Not necessarily. By analyzing a list of 106 body temperatures collected by University of Maryland researchers, we will determine whether that value is correct.

Can we accurately determine the size of the population of a region by analyzing the amount of waste discarded there? In Chapter 9 we will use real data describing different amounts and types of waste discarded by a sample of households. We will consider household sizes and discarded amounts of metal, paper, plastic, glass, food, yard waste, textile waste, and other types of waste. The data we will use were collected as part of the Garbage Project at the University of Arizona.

Does a relationship exist among a movie's length, its rating (PG, PG-13, R), and its star rating (1 star, 1.5 stars, and so on)? Some examples and exercises in the text explore possible relationships.

In working with such real data sets, the particular statistical methods we use depend on the type of data we have and how the data are arranged. It makes sense to calculate the average amount of plastic discarded by households, but does it make sense to calculate an average star rating of movies? In this chapter we address such questions.

The State of Statistics

The word *statistics* is derived from the Latin word *status* (meaning "state"). Early uses of statistics involved compilations of data and graphs describing various aspects of a state or country. In 1662, John Graunt published statistical information about births and deaths. Graunt's work was followed by studies of mortality and disease rates, population sizes, incomes, and unemployment rates. Households, governments, and businesses rely heavily on statistical data for guidance. For example, unemployment rates, inflation rates, consumer indexes, and birth and death rates are carefully compiled on a regular basis, and the resulting data are used by business leaders to make decisions affecting future hiring, production levels, and expansion into new markets.

1-1 Overview

The word *statistics* has two basic meanings. We sometimes use this word when referring to actual numbers derived from data, such as the following:

- The highest baseball batting average for one season so far is 0.438, and it was achieved by Hugh Duffy in 1894.
- Last year's national homicide rate was 9.2 persons per 100,000 U.S. residents.
- Sixty-two percent of last year's high school graduates are now enrolled in a college or university.

A second meaning refers to statistics as a method of analysis.

DEFINITION

Statistics is a collection of methods for planning experiments, obtaining data, and then organizing, summarizing, presenting, analyzing, interpreting, and drawing conclusions based on the data.

Statistics involves much more than the simple collection, tabulation, and summarizing of data. In this introductory book you will learn how to develop general and meaningful conclusions that go beyond the original data.

In statistics, we commonly use the terms *population* and *sample*. These terms are at the very core of statistics, and we define them now.

DEFINITIONS

A **population** is the complete collection of elements (scores, people, measurements, and so on) to be studied. (The collection is complete in the sense that it includes all subjects to be studied.)

A **sample** is a subcollection of elements drawn from a population.

For example, a typical Nielsen television survey uses a *sample* of 4000 U.S. households, and the results are used to form conclusions about the *population* of all 91,947,410 U.S. households.

Closely related to the concepts of population and sample are the concepts of parameter and statistic.

DEFINITIONS

A **parameter** is a numerical measurement describing some characteristic of a *population*.

A **statistic** is a numerical measurement describing some characteristic of a *sample*.

Let's consider an example. Of 25 students in one particular statistics class, 21 have credit cards. Since 21 is 84% of 25, we can say that 84% have credit cards. The figure of 84% is a *parameter* (not a *statistic*) because it is based on the entire class. If we can somehow rationalize that this class is representative of all classes so that we can treat these 25 students as a sample drawn from a larger population, then the 84% becomes a statistic.

As you proceed with your study of statistics, you will learn how to extract pertinent data from samples and how to reach conclusions based on those samples. You will also learn how to assess the reliability of conclusions. You should always realize that, although the tools of statistics enable you to form generalizations about a population, you can never accurately predict the behavior of any one individual.

An important aspect of statistics is its obvious applicability to real and relevant situations. Although many branches of mathematics deal with abstractions that may initially appear to have little or no direct use in the real world, the elementary concepts of statistics have direct and practical applications. A wide variety of these applications will be found throughout this book.

1-2 The Nature of Data

When people think of collections of data, they tend to think of lists of numbers, such as the tuition fees of various colleges or the populations of different states. Yet data may be nonnumerical, and even numerical data may belong to different categories with different characteristics. For example, a pollster may compile nonnumerical data, such as the sex, race, and religion of each voter in a sample. Numerical data, instead of being in an unordered list, might be matched in pairs (discussed in Chapters 8, 9, and 13), as in the following two tables.

Pain before hypnosis	-5.5	-5.0	-6.6	-9.7	-4.0	-7.0	-7.0	-8.4
Pain after hypnosis	-1.4	-0.5	0.7	1.0	2.0	0.0	-0.6	-1.8

Weight of a bicyclist	167	191	112	129	140	173	119
Calories burned (1 min at 5.5 mi/h)	4.2	4.7	3.2	3.5	3.7	4.5	3.4

The Census

Every 10 years, the U.S. government undertakes a census intended to obtain information about each American. The last census cost about $10 per person, for a total of $2.5 billion. The results affect over $100 billion in government allocations, seats in Congress, and redistricting of state and local governments. Pollsters use census results for designing samples and weighting survey data. Businesses use the results in selecting plant and office locations. Governments make use of census data in planning new projects such as new schools and roads.

(The first table is based on data from "An Analysis of Factors That Contribute to the Efficacy of Hypnotic Analgesia," by Price and Barber, *Journal of Abnormal Psychology,* Vol. 96, No. 1. The second table is based on data from *Diet Free,* by Kuntzlemann.)

Another common arrangement for summarizing sample data is a contingency table (discussed in Section 10-3), such as the following table of survey refusal rates.

	Age		
	18–29	30–49	50 and over
Responded	328	381	340
Refused	31	49	76

Based on data from "I Hear You Knocking But You Can't Come In," by Fitzgerald and Fuller, *Sociological Methods and Research,* Vol. 11, No. 1.

In this table, the numbers are frequencies (counts) of sample results.

The nature and structure of the data can affect the nature of the relevant problem and the method used for analysis. With the paired before/after hypnosis data, the fundamental concern is whether there is a difference in the pain level before hypnosis and after hypnosis. Any analysis of these data should attempt to determine whether the before levels are substantially different from the after levels. With the paired weight/calorie data, the fundamental concern is whether a relationship exists between those two variables. This requires a different method of analysis. With the contingency table, the fundamental concern is whether a person's age is independent of whether the person being surveyed responds or refuses to respond. This requires yet another method of analysis. As we consider the topics of later chapters, we will see that the structure and nature of the data affect our choice of method.

Data may be categorized as qualitative or quantitative.

DEFINITIONS

Qualitative (or **categorical** or **attribute**) **data** can be separated into different categories that are distinguished by some nonnumerical characteristic.

Quantitative data consist of numbers representing counts or measurements.

One example of *qualitative data* is the *type* of product made by Motorola (pager, cellular telephone, two-way radio, and so on). An example of *quantitative data* is the *number* of pagers made by Motorola on different days.

We can further describe quantitative data by distinguishing between the discrete and continuous types.

DEFINITIONS

Discrete data result from either a finite number of possible values or a countable number of possible values. (That is, the number of possible values is 0, or 1, or 2, and so on.)

Continuous numerical data result from infinitely many possible values that can be associated with points on a continuous scale in such a way that there are no gaps or interruptions. *measurements over a continuum*

When data represent counts, they are discrete; when they represent measurements, they are continuous. For example, if we record the number of pagers (275, 281, 255, and so on) made by Motorola during each work day of last year, those values will be *discrete data* because they represent *counts*. If we measure and record the amounts of time that pager batteries last (such as 127.214 hours), those values will be *continuous data* because they are measurements that can assume any value *over a continuous time interval*.

Another common way to classify data is to use four levels of measurement: nominal, ordinal, interval, ratio.

DEFINITION

The **nominal level of measurement** is characterized by data that consist of names, labels, or categories only. The data cannot be arranged in an ordering scheme.

If you associate the term *nominal* with "name only," the meaning becomes easy to remember. An example of nominal data is the *gender* of each of the 107,000 Motorola employees. Data at the *nominal level of measurement* cannot be arranged according to some ordering scheme. That is, there is no criterion by which values can be identified as greater than or less than other values.

▶ EXAMPLE

The following are other examples of sample data at the nominal level of measurement:

1. Movies are listed according to their genre, such as comedy, adventure, and romance.
2. Responses to an exit poll are made by 45 Democrats, 80 Republicans, and 90 Independents.

Is Statistics Really Worth Anything?

With annual sales exceeding $17 billion, Motorola is one of the largest producers in the world of electronics equipment, including cellular telephones, cordless telephones, pagers, two-way radios, and modules that control car transmissions. In a recent five-year period, Motorola saved approximately $2.5 billion by implementing a quality-improvement plan that makes extensive use of statistical methods. Its pagers and cellular telephones are currently being produced with a target defect rate of only 0.00034%; Motorola is pursuing a goal popularly referred to as the "six sigma" level of quality, which corresponds to less than 3.4 defects per million parts produced. Motorola has found that the use of statistical methods is necessary for survival in an increasingly competitive market.

Because the categories lack any ordering or numerical significance, the preceding data cannot be used for calculations. We cannot, for example, average 12 comedies, 15 adventures, and 9 romances. Numbers are sometimes assigned to different categories, especially when the data are processed by computer. We might find that the Gallup Organization uses computerized survey data in which Democrats are assigned the number 0, Republicans are assigned 1, and Independents are assigned 2. Even though we now have number labels, the numbers lack any real computational significance, and an average calculated with those numbers would be a meaningless statistic.

DEFINITION

The **ordinal level of measurement** involves data that may be arranged in some order, but differences between data values either cannot be determined or are meaningless.

▶ EXAMPLE

The following are examples of data at the *ordinal level of measurement:*

1. In a sample of 36 stereo speakers, 12 were rated "good," 16 were rated "better," and 8 were rated "best."

2. The movie *Star Wars* was rated with 4 stars, while the movie *Godzilla* was rated with 1 star.

3. In considering employee promotions, a Motorola manager ranked Marilyn 3rd, Allyn 7th, and Michael 10th.

In the first example given, we cannot determine specific measured differences between such ratings as "good" and "better." In the second example, we can find a difference between 4 stars and 1 star, but the difference of 3 stars doesn't really mean anything. Similarly, in the third example, we can determine a difference between the rankings of 3 and 7, but the resulting value of 4 doesn't mean anything. That is, the difference of 4 between ranks of 3 and 7 isn't necessarily the same as the difference of 4 between ranks of 7 and 11. This ordinal level provides information about relative comparisons, but the degrees of difference are not available. Again, data at this level should not be used for calculations.

DEFINITION

The **interval level of measurement** is like the ordinal level, with the additional property that meaningful amounts of differences between data can be determined. However, there is no inherent (natural) zero starting point.

Body temperatures (such as those in Data Set 2 of Appendix B) of 98.2°F and 98.6°F are examples of data at this measurement level. Such values are ordered, and we can determine their difference (often called the *distance* between the two values). However, there is no natural starting point. The value of 0°F might seem like a starting point, but it is arbitrary. The value of 0°F does not indicate "no heat," and it is incorrect to say that 50°F is twice as hot as 25°F. For the same reasons, temperature readings on the Celsius scale are also at the *interval level of measurement*. (Temperature readings on the Kelvin scale are at the ratio level of measurement; that scale has an absolute zero.)

► **EXAMPLE**

The following are other examples of data at the interval level of measurement:

1. The years 1000, 2000, 1776, 1944, and 1995. (Time did not begin in the year 0, so the year 0 is arbitrary instead of being a natural zero starting point.)
2. Room temperatures (in degrees Celsius) of Motorola's sales offices.

DEFINITION

The **ratio level of measurement** is the interval level modified to include the inherent zero starting point. For values at this level, differences *and* ratios are meaningful.

► **EXAMPLE**

The following are examples of data at the *ratio level of measurement*:

1. Weights of plastic discarded by households
2. Lengths (in minutes) of movies
3. Distances (in miles) traveled by cars in a test of fuel consumption

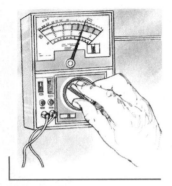

Measuring Disobedience

How are data collected about something that doesn't seem to be measurable, such as people's level of disobedience? Psychologist Stanley Milgram devised the following experiment: A researcher instructed a volunteer subject to operate a control board that gave increasingly painful "electrical shocks" to a third person. Actually, no real shocks were given, and the third person was an actor. The volunteer began with 15 volts and was instructed to increase the shocks by increments of 15 volts. The disobedience level was the point at which the subject refused to increase the voltage. Surprisingly, two-thirds of the subjects obeyed orders even though the actor screamed and faked a heart attack.

TABLE 1-1   Levels of Measurement of Data

Level	Summary	Example	
Nominal	Categories only. Data cannot be arranged in an ordering scheme.	Voter distribution: 45 Democrats 80 Republicans 90 Independents	Categories only.
Ordinal	Categories are ordered, but differences cannot be determined or they are meaningless.	Voter distribution: 45 low-income voters 80 middle-income voters 90 upper-income voters	An order is determined by "low, middle, upper."
Interval	Differences between values can be found, but there may be no inherent starting point. Ratios are meaningless.	Temperatures of steel rods: 45°F 80°F 90°F	90°F is not twice as hot as 45°F.
Ratio	Like interval, but with an inherent starting point. Ratios are meaningful.	Lengths of steel rods: 45 cm 80 cm 90 cm	90 cm is twice as long as 45 cm.

Values in each of these data collections can be arranged in order, differences can be computed, and there is an inherent zero starting point. *This level is called the ratio level because the starting point makes ratios meaningful.* Because a 200-lb weight is twice as heavy as a 100-lb weight, but 50°F is not twice as hot as 25°F, weights are at the ratio level while Fahrenheit temperatures are at the interval level. Table 1-1 summarizes the four levels of measurement and includes other examples.

In applying statistics to real problems, the level of measurement of the data is an important factor in determining which procedure to use. An understanding of the four levels of measurement should be supplemented with common sense—an indispensable tool in statistics. For example, does it make any sense to calculate the average of a list of social security numbers? No. Such numbers don't really *measure* or *count* anything; instead, they simply serve the function of identifying people. Social security numbers are different *names* for people and, as such, should not be used for calculations. In general, we should not calculate averages for data at the nominal or ordinal level of measurement.

1-2   Exercises A: Basic Concepts

In Exercises 1–8, identify each number as *discrete* or *continuous*.

1. Among 500,000 microcomputer chips made by Motorola, 2 are found to be defective.
2. Yesterday's records for Motorola's marketing department show that 25 employees were absent.
3. A Ford Taurus sedan weighs 3084 lb.
4. Among all SAT scores recorded last year, 23 were perfect.
5. Radar on Creek Road indicated that the driver was going 47.6 mi/h when ticketed for speeding.
6. The amount of time that a New York City cab driver spends yielding to individual pedestrians each year is 2.367 seconds.
7. Among 200 consumers surveyed, 186 recognize the Campbell's Soup brand name.
8. Upon completion of a diet and exercise program, Tony Hopkins weighed 12.37 lb less than when he started the program.

In Exercises 9–18, determine which of the four levels of measurement (nominal, ordinal, interval, ratio) is most appropriate.

9. Cars described as subcompact, compact, intermediate, or full-size
10. Weights of a sample of M&M candies
11. Colors of a sample of M&M candies
12. Years in which Republicans won presidential elections
13. Zip codes
14. Social security numbers
15. Total annual incomes for a sample of statistics students
16. Final course grades of A, B, C, D, F
17. Body temperatures (in degrees Fahrenheit) of a sample of bears captured in Wyoming
18. Instructors rated as superior, above average, average, below average, or poor

1-2   Exercises B: Beyond the Basics

19. Many people question what IQ scores actually measure. Consider this question: Is a person with an IQ score of 150 twice as intelligent as a person with an IQ score of 75?
    a.   What does an affirmative answer imply about the level of measurement corresponding to IQ scores?
    b.   What does a negative answer imply about the level of the data?
20. If a recipe requires cooking a meatloaf at 300°F for 2 hours, but you decide to cook it at 600°F for 1 hour, your result will be different. Explain.
21. U.S. presidents were assassinated in 1865, 1881, 1901, and 1963. Explain why those years form a collection of data at the interval level of measurement.

Is Army Food Good?

When *Miami Herald* humorist Dave Barry wrote that the U.S. Army's MREs (Meals Ready to Eat) are not very good, the *Register Guard*, Eugene, Oregon's newspaper, conducted a taste test. Using a 10-point rating scale, "the panel gave the MREs a rating of 8.1 on the taste scale. This is clearly a scientific result, because it contains a decimal point," Barry writes. Noting that the panel included the food service director of a local school district, Barry adds: "if anybody would recognize a delicious shelf-stable food substance, it would be the person responsible for the menu options at a public school."

Statistics Stops Poisoning

A study was made of 63 Nicaraguan mechanics who worked on airplanes used for crop dusting. It was found that 49% had been acutely poisoned through exposure to pesticides. As a result of this study, the Nicaraguan government now requires that crop duster planes be washed before mechanics work on them. Coveralls and special gloves are issued to mechanics who handle parts contaminated by pesticides. Also, the mixing and loading of pesticides are done with a closed system that reduces the exposure of the mechanics to the various chemicals.

1-3 Uses and Abuses of Statistics

Throughout this text, there are short essays in the margins that use real-world examples to illustrate uses and abuses of statistics. The many uses of statistics include applications in business, economics, psychology, biology, computer science, military intelligence, literature, physics, chemistry, medicine, sociology, political science, agriculture, and education.

Statistical theory applied to these diverse fields often results in changes that benefit humanity. Social reforms are sometimes initiated as a result of statistical analyses of factors such as crime rates and poverty levels. Large-scale population planning can result from projections devised by statisticians. Manufacturers can provide better products at lower cost through the effective use of statistics in quality control. Diseases can be controlled through analyses designed to anticipate epidemics. Endangered species of fish and other wildlife can be protected through regulations and laws that react to statistical estimates of small population sizes. Educators may implement innovative teaching techniques if statistical analyses show that they are more effective. By pointing to statistics that indicate lower fatality rates, legislators can better justify laws such as those governing air pollution, auto inspections, seat belt use, and drunk driving. Retired employees can benefit from financially stable pension plans that have been designed through statistical analysis.

Students choose a statistics course for many different reasons. Some take such a course because it's required; however, increasing numbers of students do so voluntarily because they recognize its value and application to whatever field they plan to pursue.

In addition to providing skills important in many jobs and disciplines, the study of statistics can help you become more critical in your analyses of information, making you less susceptible to misleading or deceptive claims, such as those commonly associated with polls, graphs, and averages. Everyone uses external data to make decisions, form conclusions, and build a warehouse of knowledge. If you want to build a sound knowledge base, make intelligent decisions, and form worthwhile opinions, you must be careful to filter out the incoming information that is erroneous or deceptive. As an educated and responsible member of society, you should sharpen your ability to recognize distorted statistical data; in addition, you should learn to interpret undistorted data intelligently.

Difficulties associated with statistical claims have been recognized for some time. For example, about a century ago, statesman Benjamin Disraeli said, "There are three kinds of lies: lies, damned lies, and statistics." It has also been said that "figures don't lie; liars figure." Historian Andrew Lang said that some used statistics "as a drunken man uses lampposts—for support rather than illumination." Economist Sir Josiah Stamp said, "The Government is very keen on amassing statistics. They collect them, add them, raise them to the nth power, take the cube root and prepare wonderful diagrams. But you must never forget that every one of these figures comes in the first instance from the village watchman, who just puts down what he damn well pleases."

The preceding statements refer to abuses of statistics in which data are presented in ways that may be misleading. Some abusers of statistics are simply ignorant or careless, whereas others have personal objectives and are willing to suppress unfavorable data while emphasizing supportive data. We will now present a few examples of the many ways in which data can be distorted.

Some use small sample results as a form of statistical "lying." The toothpaste preferences of only 10 dentists should not be used as a basis for a generalized claim such as "Covariant toothpaste is recommended by 7 out of 10 dentists." Even if the sample is large, it may not be unbiased and representative of the population from which it comes.

Sometimes the numbers themselves can be deceptive. A very precise figure, such as an annual salary of $27,735.29, might be used to instill a high degree of confidence in its accuracy. The figure of $27,700 doesn't convey that same sense of precision and accuracy. A statistic that is very precise with many decimal places, however, is not necessarily accurate.

Another source of statistical deception is estimates that are ultimately guesses and can therefore be in error by substantial amounts. We should consider the source of the estimate and how it was developed. When the Pope visited Miami, officials estimated the crowd size to be 250,000, but the *Miami Herald* used aerial photos and grids to come up with a better estimate of 150,000. As another example, the Associated Press ran an article in which the roach population of New York City was estimated to be 1 billion. That claim was made by a spokesperson for the Bliss Exterminator Company. The head of New York City's Bureau of Pest Control would not confirm that figure. He said, "I haven't the slightest idea if that's right. We do strictly rats here."

Misleading or unclear percentages are sometimes used. Continental Airlines ran full-page ads boasting about their better service. These ads claimed that dealing with lost baggage was "an area where we've already improved 100% in the last six months." In an editorial criticizing this statistic, the *New York Times* interpreted the 100% improvement figure to mean that no baggage is now being lost—an accomplishment not yet enjoyed by Continental Airlines.

"Ninety percent of all our cars sold in this country in the last 10 years are still on the road." Millions of consumers heard that commercial message and got the impression that the company's cars must be well built to last through those long years of driving. What the auto manufacturer failed to mention was that 90% of the cars it sold in this country were sold within the last three years. The claim was technically correct, but it was very misleading.

Many visual devices—such as bar graphs and pie charts—can be used to exaggerate or deemphasize the true nature of data. (Such devices will be discussed in Chapter 2.) The two graphs in Figure 1-1 depict the *same data* from the Bureau of Labor Statistics, but the graph in part b is designed to exaggerate the difference between the earnings of men and women. By not starting the horizontal axis at zero, part b tends to produce a misleading subjective impression. Too many of us look at a graph superficially and develop intuitive impressions based on the pattern we see. Instead, we should scrutinize the graph and search for distortions of the type illustrated in

Invisible Ink

The *National Observer* once hired a firm to conduct a confidential mail survey. Editor Henry Gemmill wrote in a cover letter that "each individual reply will be kept confidential, of course, but when your reply is combined with others from all over this land, we'll have a composite picture of our subscribers." One clever subscriber used an ultraviolet light to detect a code written on the survey in invisible ink. That code could be used to identify the respondent. Gemmill was not aware that this procedure was used, and he publicly apologized. Confidentiality was observed as promised, but anonymity was not directly promised, and it was not maintained.

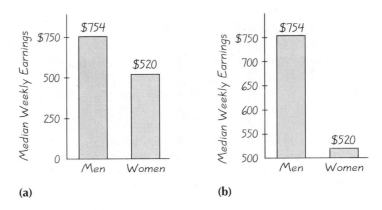

FIGURE 1-1   Earnings of Full-Time Professional Workers

Figure 1-1. By analyzing the *numerical* information given in the graph, we can avoid being misled by its general shape.

Drawings of objects may also be misleading. Some three-dimensional objects commonly used to depict data include moneybags, stacks of coins, army tanks (for military expenditures), cows (for dairy production), and houses (for home construction). When drawing such objects, artists can create false impressions that distort differences. For example, if an artist doubles each side of a cube, the volume doesn't double; it increases by a factor of eight. If taxes double over a decade, an artist can depict tax amounts with one moneybag for the first year and a second moneybag that is twice as deep, twice as tall, and twice as wide for the second year. Instead of appearing to double, taxes will appear to increase by a factor of eight, so the truth will be distorted by the drawing.

Another source of deceptive statistics is inappropriate methods of collecting data. It is common for a researcher to analyze data and form conclusions that are wrong because the method of collecting the data was poor. One typical example is the use of "900" telephone numbers that television viewers are asked to call in response to a survey question. The callers are charged a fee, usually around 50 cents or a dollar, and the viewers themselves decide whether to be included in the survey. It often happens that only those viewers with strong opinions participate. Consequently, the sample of people who respond is not representative of the whole population. This is only one way in which the method of collecting data can be seriously flawed. Because the method of sampling or collecting data is so important, we will devote the following section to it.

The examples in this section comprise a small sampling of the ways in which statistics can be used deceptively. Entire books have been devoted to this subject, including Darrell Huff's classic *How to Lie with Statistics* and Robert Reichard's *The Figure Finaglers*. Understanding these practices will be extremely helpful in evaluating the statistical data found in everyday situations.

1-3 Exercises A: Basic Skills and Concepts

1. A televised report by NBC News cited a study showing that eating walnuts lowers cholesterol by up to 10%. The report concluded with the comment that "You should also know that the study was sponsored by the walnut industry." Why did NBC News believe that you should know who sponsored the study?

2. To research recognition of the Estée Lauder brand name, you plan to conduct a telephone survey of 2000 consumers in the United States. What is wrong with using telephone directories as the population from which the sample is drawn?

3. A graph similar to the one shown appeared in *Car and Driver* magazine. What is wrong with it?

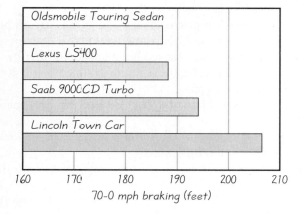

4. Seventy-two percent of Americans squeeze the toothpaste tube from the top. This and other not so serious findings are presented in *The First Really Important Survey of American Habits*. The results are based on 7000 responses to the 25,000 questionnaires that were mailed. What is wrong with this survey?

5. One study actually showed that smokers tend to get lower grades in college than nonsmokers. Does this mean that smoking *causes* lower grades? What other explanation is possible?

6. In a study on college campus crimes committed by students high on alcohol or drugs, a mail survey of 1875 students was conducted. A *USA Today* article noted, "Eight percent of the students responding anonymously say they've committed a campus crime. And 62% of that group say they did so under the influence of alcohol or drugs." Assuming that the number of students responding anonymously is 1875, how many actually committed a campus crime while under the influence of alcohol or drugs?

7. A study conducted by the Insurance Institute for Highway Safety found that the Chevrolet Corvette had the highest fatality rate—"5.2 deaths for every 10,000." The car with the lowest fatality rate was the Volvo, with only 0.6 death per 10,000. Does this mean that the Corvette is not as safe as a Volvo?

8. You plan to conduct a poll of students at your college. What is wrong with polling every 50th student who is leaving the cafeteria?

9. A college conducts a survey of its alumni in an attempt to determine their typical annual salary. Would alumni with very low salaries be likely to respond? How would this affect the result? Identify one other factor that might affect the result.

10. "According to a nationwide survey of 250 hiring professionals, scuffed shoes was the most common reason for a male job seeker's failure to make a good first impression." Newspapers carried this statement, based on a poll commissioned by Kiwi Brands, producers of shoe polish. Comment on why the results of the survey might be questionable.

11. An employee earning $400 per week was given a 20% cut in pay as part of her company's attempt to reduce labor costs. After a few weeks, this employee's dissatisfaction grew, and her threat to resign caused her manager to offer her a 20% raise. The employee accepted this offer because she assumed that a 20% raise would make up for the 20% cut in pay.
    a. What was the employee's weekly salary after she received the 20% cut in pay?
    b. Use the salary figure from part a to find the amount of the 20% increase, and determine the weekly salary after the raise.
    c. Did the 20% cut followed by the 20% raise get the employee back to the original salary of $400 per week?

12. The first edition of a textbook contained 1000 exercises. For the second edition, the author removed 100 of the original exercises and added 300 new exercises. Which of the following statements about the second edition are correct?
    a. There are 1200 exercises.
    b. There are 33% more exercises.
    c. There are 20% more exercises.
    d. Twenty-five percent of the exercises are new.

13. What differences are there between the following two statements, and which one do you believe is more accurate?
    a. Drunk drivers cause about half of all fatal car crashes.
    b. Of all fatal motor vehicle crashes, about 50% involve alcohol.

14. *Good Housekeeping* magazine reported on studies leading to the discovery that "the average person spends six years eating" (not all at once). Assuming an average life span of 75 years, develop your own estimate, and identify how it was obtained. Do you think the stated figure of 6 years is too low, too high, or about right?

15. In an advertising supplement inserted in *Time,* the increases in expenditures for pollution abatement were illustrated in a graph similar to the one shown. What is wrong with the figure?

$1864.8 Million

$1483.3 Million

$983.5 Million

$643.3 Million

1   2   3   4

Year

16. A *New York Times* article noted that the mean life span for 35 male symphony conductors was 73.4 years, in contrast to the mean of 69.5 years for males in the general population. The longer life span was attributed to such factors as fulfillment and motivation. There is a fundamental flaw in concluding that male symphony conductors live longer. What is it? (*Hint:* How old are males when they are identified as symphony conductors?)

1-3   Exercises B: Beyond the Basics

17. A researcher at the Sloan-Kettering Cancer Research Center was once criticized for falsifying data. Among his data were figures obtained from 6 groups of subjects, with 20 individual subjects in each group. These values were given for the percentage of successes in each group: 53%, 58%, 63%, 46%, 48%, 67%. What's wrong?

18. If an employee is given a cut in pay of x percent, find an expression for the percent raise that would return the salary to the original amount.

19. A *New York Times* editorial criticized a chart caption that described a dental rinse as one that "reduces plaque on teeth by over 300%."
   a.   If you remove 100% of some quantity, how much is left?
   b.   What does it mean to reduce plaque by over 300%?

20. Try to identify the four major flaws in the following paragraph:
   A daily newspaper ran a survey asking readers to call in with their responses to this question: "Do you support the development of atomic weapons that could kill millions of innocent people?" It was reported that 20 readers responded, and 87% said "no" while 13% said "yes."

1-4 Methods of Sampling

As noted in Section 1-3, a common mistake in using statistics to analyze data is to collect the data in a way that is inappropriate. Researchers often display disappointment, anger, and a wide variety of other emotions when they learn that the data they spent so much time collecting are essentially worthless. We cannot stress this very important point enough: *Data carelessly collected may be so completely useless that no amount of statistical torturing can salvage them.* Sampling and data collection usually require more time, effort, and money than the statistical analysis of the data does. Careful planning will minimize the waste of precious resources. The following are four important points to consider when collecting data.

First, ensure that the sample size is large enough for the required purposes. (Issues of sample size are discussed later in this text, especially in Chapter 6.) Many people incorrectly believe that large samples are good samples, but even large samples may be totally worthless if the data have been carelessly collected.

Second, if you are obtaining measurements of some characteristic from people (such as height), realize that you will get better results if you do the measuring instead of asking the subject for the value. Asking tends to yield a disproportionate number of rounded results. Thus the data are distorted, and the sample is flawed.

Third, when conducting a survey, consider the medium to be used. Mail surveys, telephone surveys, and personal interviews are most common, although other methods are used. Mail surveys tend to get lower response rates. Personal interviews are obviously more time consuming and expensive, but they may be necessary if detailed and complex data are required. Telephone interviews are relatively efficient and inexpensive.

Fourth, ensure that the method used to collect data actually results in a sample that is representative of the population. We now define and describe five of the more common methods of sampling.

DEFINITION

In **random sampling,** members of the population are selected in such a way that each has an equal chance of being selected.

Random sampling is also called *representative,* or *proportionate,* sampling because all groups of the population should be proportionately represented in the sample. Careless or haphazard sampling can easily result in a biased sample that has characteristics very different from those of the population from which it came. In contrast, random sampling requires much effort and planning to avoid any bias. For example, using telephone directories eliminates everyone with an unlisted number, and ignoring that segment of the population could easily yield misleading results. In Los Angeles, for example, 42.5% of the telephone numbers are unlisted (based on data from Survey Sampling, Inc.). Pollsters commonly circumvent this problem by using computers to generate phone numbers so that all numbers are possible. We must also be careful to include those who are initially unavailable or initially refuse to comment. Humphrey Taylor, president of the Harris polling company, states that the refusal rate for telephone interviews is generally at least 20%. Ignore those people who initially refuse and you run a real risk of having a biased sample.

DEFINITION

With **stratified sampling,** we subdivide the population into at least two different subpopulations (or strata) that share the same characteristics (such as gender), and then we draw a sample from each stratum.

In surveying views on the Equal Rights Amendment to the Constitution, we might use gender as a basis for creating two strata. After obtaining a list of men and a list of women (from census data), we use some suitable method (such as random sampling) to select a certain number of people from each list. If it should happen that some strata are not represented in the proper proportion, then the results can be adjusted or weighted accordingly. *Stratified sampling* is often the most efficient of the various sampling methods.

DEFINITION

In **systematic sampling,** we choose some starting point and then select every kth (such as every 50th) element in the population.

For example, if Motorola wanted to conduct a survey of its 107,000 employees, it could begin with a complete roster and then select every 100th employee to obtain a sample of size 1070. This method is simple and is often used.

DEFINITION

In **cluster sampling,** we divide the population area into sections (or clusters), randomly select a few of those sections, and then choose all the members from the selected sections.

Airlines Sample

Airline companies once used an expensive accounting system to split up income from tickets that involved two or more companies. They now use a sampling method in which a small percentage of these "split" tickets are randomly selected and used as a basis for dividing up all such revenues. The error created by this approach can cause some companies to receive slightly less than their fair share, but these losses are more than offset by the clerical savings accrued by dropping the 100% accounting method. This new system saves companies millions of dollars each year.

The *Literary Digest* Poll

In the 1936 presidential race, *Literary Digest* magazine ran a poll and predicted an Alf Landon victory, but Franklin D. Roosevelt won by a landslide. Maurice Bryson notes, "Ten million sample ballots were mailed to prospective voters, but only 2.3 million were returned. As everyone ought to know, such samples are practically always biased." He also states, "Voluntary response to mailed questionnaires is perhaps the most common method of social science data collection encountered by statisticians, and perhaps also the worst." (See Bryson's "The *Literary Digest* Poll: Making of a Statistical Myth," *The American Statistician,* Vol. 30, No. 4.)

For example, in conducting a preelection poll, we could randomly select 30 election precincts and survey all the people from each of those chosen precincts. This would be much faster and much less expensive than selecting one person from each of the many precincts in the population area. The results can be adjusted or weighted to correct for any disproportionate representations of groups. *Cluster sampling* is used extensively by government and private research organizations.

DEFINITION

With **convenience sampling,** we simply use results that are readily available.

In some cases results from *convenience sampling* may be quite good, but in other cases they may be seriously biased. It would be convenient for a teacher who is investigating the proportion of left-handed people to survey students. Even though the sample is not random, it will tend to be unbiased because left-handedness is not the type of characteristic that would be related to presence in class. Motorola's use of data collected by the Bureau of the Census is convenience sampling, because the company is using data already collected by another organization. Although these examples of convenience sampling are likely to yield good results, there are many other examples in which the results are not as good. When she wrote *Women and Love: A Cultural Revolution,* author Shere Hite based conclusions on 4500 responses from 100,000 questionnaires distributed to women. It was convenient to settle for the questionnaires that were returned, but it's very likely that the sample was biased. People are more likely to respond if they feel strongly about the survey topics. Even though the sample size of 4500 was large enough, Hite's book was widely criticized for its bias and lack of valid sampling methods.

Figure 1-2 illustrates the five common methods of sampling described above. The previous descriptions are intended to be brief and general. Thoroughly understanding these different methods so that you can successfully use them requires much more extensive study than is practical in a single introductory course of this type. As you read the frequent references in this text to "randomly selected" data, however, you should keep in mind that such data are carefully selected so that all members of the population have the same chance of being chosen. Although we will not make frequent reference to the other methods of sampling, you should understand that they exist and that the method of sampling requires careful planning and execution. The statistical methods we present throughout this text depend on samples that have been carefully obtained. Remember, data carelessly collected may be absolutely worthless, even if the sample is large.

Random Sampling:

Each member of the population has an equal chance of being selected. Computers are often used to generate random telephone numbers.

Stratified Sampling:

Classify the population into at least two strata, then draw a sample from each.

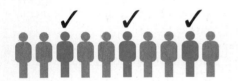

Systematic Sampling:

Select every kth member.

Cluster Sampling:

Divide the population area into sections, randomly select a few of those sections, and then choose all members in them.

Convenience Sampling:

Use results that are readily available.

Hey! Do you believe in the death penalty?

FIGURE 1-2   Common Sampling Methods

1-4   Exercises A: Basic Skills and Concepts

In Exercises 1–12, identify the type of sampling used.

1. Motorola selects every 50th pager from the assembly line for careful testing and analysis.
2. A reporter writes the name of each U.S. senator on a separate card, shuffles the cards, and then draws 5 names.
3. A dean at Ohio State University surveys all students from each of 12 randomly selected classes.
4. A dean at Menlo College selects 15 men and 15 women from each of 4 classes.
5. *Glamour* magazine obtains sample data from readers who decide to mail in a questionnaire printed in the latest issue.
6. An IRS auditor randomly selects 15 taxpayers with less than $25,000 in gross income and 15 taxpayers with gross income of at least $25,000.
7. CBS News polls 750 men and 750 women about their use of credit cards.
8. A market researcher for the Ford Motor Company interviews all drivers on each of 15 randomly selected city blocks.
9. A medical researcher from Johns Hopkins University interviews all leukemia patients in each of 20 randomly selected counties.
10. A reporter for *Business Week* magazine interviews every 50th chief executive officer in that magazine's listing of CEOs of the 1000 companies with the highest stock market values.
11. A reporter for *Business Week* magazine obtains a numbered listing of the 1000 companies with the highest stock market values, uses a computer to generate 20 random numbers between 1 and 1000, and then interviews the chief executive officers of companies corresponding to these numbers.
12. In conducting research for a psychology course, a student at Boston College interviews 40 students who are leaving the cafeteria.

1-4   Exercises B: Beyond the Basics

13. Assume that you are employed by General Motors to collect data on the waist sizes of drivers. Why is it better to obtain direct measurements than to ask people the sizes of their waists?
14. You plan to estimate the average weight of all passenger cars used in the United States.
    a. Is there universal agreement about what a "passenger car" is?
    b. Are there any factors that might lead to regional differences among the weights of passenger cars?
    c. How can you obtain a sample?
15. Two categories of survey questions are *open* and *closed*. An open question allows a free response, whereas a closed question allows only a fixed response. Here are examples based on Gallup surveys.   *(continued)*

Open question: What do you think can be done to reduce crime?
Closed question: Which of the following approaches would be most effective in reducing crime?

- Hire more police officers.
- Get parents to discipline children more.
- Correct social and economic conditions in slums.
- Improve rehabilitation efforts in jails.
- Give convicted criminals tougher sentences.
- Reform courts.

a. What are the advantages and disadvantages of open questions?
b. What are the advantages and disadvantages of closed questions?
c. Which type is easier to analyze with formal statistical procedures?

1-5   Statistics and Computers

Computers now play an important role in almost every aspect of statistical analysis. The widespread availability of computers and software packages has made the use of statistics possible for people with many different mathematical backgrounds, but there is also greater opportunity for misuses of statistics. It is important to recognize that statistical software and computers have a very serious limitation: They mindlessly follow instructions, even if those instructions are inappropriate or absurd. Computers don't do the necessary human reasoning, and they cannot exercise judgment. An understanding of the principles of statistics is an important prerequisite for correctly interpreting computer results. Even if you don't actually use computers in this course, you should try to develop some skill in interpreting computer displays of statistical analyses, such as those found throughout this text.

We will make frequent reference to two particular software packages: STATDISK and Minitab. STATDISK is easy-to-use software, designed specifically as a supplement to this textbook. STATDISK is available at no cost to colleges that use this textbook. Minitab is a higher-level statistical software package that requires knowledge of some important fundamentals, which are described below. (Separate workbook/manuals for STATDISK and Minitab are available as supplements to this book.)

 Minitab Fundamentals

The procedure for loading Minitab varies with different systems, so consult your instructor or computer lab assistant. Minitab can be run with commands, a menu-driven interface, or a Windows format. This text will use Minitab's commands for Minitab examples because they more clearly describe the

New Data Collection Technology

Measuring or coding products and manually entering results is one way of collecting data, but technology is now providing us with alternatives that are not so susceptible to human error. Supermarkets use bar code readers to set prices and analyze inventory and buying habits. Manufacturers are increasingly using "direct data-entry" devices such as electronic measuring gauges connected directly to a computer that records results. A third possibility is voice data entry, such as the Voice Data Logger. It consists of a headset connected to a small box that attaches to the speaker's belt. The box transmits the data to a computer.

operations performed. After Minitab has been successfully loaded, the screen should display MTB >. Everything immediately to the right of the symbol > is entered by you; everything else is displayed by Minitab. When you see an expression such as

$$\text{MTB} > \text{SET C1}$$

you should recognize that MTB > is displayed by Minitab and you enter the command SET C1.

We will now describe the fundamental operations of entering, displaying, saving, and retrieving data.

Entering Data   SET C1 is the Minitab command that allows you to enter data in a column that is designated by the label C1. Minitab can store different collections of data in different columns. Each column is represented by a number, such as C1, C2, or C3. For example, suppose that you want to enter the following values of SAT scores (obtained by Minitab from a northeastern university that wishes to remain anonymous):

1132	925	1343	1304	1429
1336	1236	1365	1306	1356

With the MTB > prompt displayed, begin by typing SET C1, then press the RETURN (or ENTER) key. Minitab will respond by displaying the prompt DATA >. Type the numbers, and press the RETURN key at the end of each line. Do not put commas between numbers or within numbers. A space is all that is necessary between numbers. When all of the data have been entered, type ENDOFDATA, and press the RETURN key. The following sequence of entries will cause the above values to be stored in a column identified as C1.

```
MTB > SET C1
DATA> 1132   925 1343 1304 1429
DATA> 1336 1236 1365 1306 1356
DATA> ENDOFDATA
```

Occasionally you will want to repeat the entry of a number several times. In the commands given below, column C2 will contain the number 1027.3 repeated five hundred times. Naturally, this is much better than typing 1027.3 five hundred times.

```
MTB > SET C2
DATA> 500(1027.3)
DATA> ENDOFDATA
```

READ C1 C2 is the Minitab command that allows you to enter matched data in two different columns. (You would use READ C1 C2 C3 for three columns, and so on.) With SET, you enter all of the data in one operation; with READ, you enter the data one row at a time. As an example, the following entries cause five pairs of data to be entered in columns C1 and C2. The first score of each entry is the SAT score of a college student, and the second score is his or her grade-point average.

```
MTB > READ C1 C2
DATA> 1132 2.6
DATA>  925 2.3
DATA> 1343 2.4
DATA> 1304 3.0
DATA> 1429 3.1
DATA> ENDOFDATA
```

Again, note that the process of entering the data is ended by the command ENDOFDATA (which can be abbreviated as END). READ tends to be slower than SET because you enter only one row at a time, but it is useful when you want to take the extra time to be sure that your data are matched correctly.

Displaying Data   PRINT C1 is the command used to display the data stored in column C1. This command is especially useful for verifying that data have been entered correctly. It's also useful when confusion reigns and you forget what data sets are stored where. Entry of the command PRINT C1 C2 will result in a display of the values stored in columns C1 and C2.

Saving Data   Data can be saved with the SAVE command used with a file name. For example, SAVE 'STATDATA' is the command that permanently stores all of the current information in a computer file identified by the name STATDATA. The name of the file must be enclosed within single quotes, and it must be unique (different from any name already used). To save the file on a disk in drive B, type SAVE 'B:STATDATA'.

Retrieving Data   RETRIEVE 'STATDATA' is the command that retrieves the data previously stored in a computer file identified by the name STATDATA. The name of the file must be enclosed within single quotes. (To retrieve from a disk in drive B, type RETRIEVE 'B:STATDATA'.)

Obtaining Printouts   At Minitab's prompt of MTB >, enter the command PAPER to have all of the subsequent commands and results sent to the printer connected to your microcomputer. The command NOPAPER discontinues the process of printing everything. Another method of obtaining printed results is to press the "Print Screen" key; that will initiate the printing of everything currently shown on the monitor's screen.

Logging Off   You can easily exit from the Minitab program by entering the command STOP.

Miscellaneous Notes   Below are several important rules that will help you work with Minitab successfully.

First, don't use commas in numbers. For example, enter 32156.50 instead of 32,156.50.

Second, if you know you have entered a wrong number, use one of the following methods to correct it. If you haven't yet hit the RETURN key, you can reenter the correct data set. Or, if you prefer to replace, delete, or insert a number, use the formats suggested by the following examples.

LET C3(7) = 9	The 7th entry of column C3 is *replaced* with the number 9.
DELETE 3 C5	The entry in the 3rd row of column C5 is *deleted*.
INSERT 5 6 C1 9 ENDOFDATA	The number 9 in column C1 is *inserted* between rows 5 and 6.

Third, to do arithmetic with data, use the LET command. The following examples demonstrate how to use LET.

LET C3 = C1 + C2	Column C3 is created by adding each entry of column C1 to the corresponding entry of column C2.
LET C2 = C2/5	Each entry of column C2 is divided by 5.
LET C5 = C2 - C1	Column C5 is created by subtracting the values in column C1 from the corresponding values in column C2.
LET C6 = C1*C2	Column C6 is created by multiplying the corresponding entries from columns C1 and C2.
LET C7 = C1**2	Column C7 is created by squaring the values in column C1. The symbol ** is used for exponents.

This section has presented only the fundamentals of the Minitab statistical software package. Future sections will include Minitab displays resulting from commands relevant to the statistical concepts being discussed. Brief discussions of those commands will also be included.

1-5 Exercises A: Basic Concepts

1. Load Minitab, use the SET C1 command to enter the following scores in column C1, and then enter PRINT C1 to display them. (The scores are from the U.S. National Center for Health Statistics and represent the numbers of motor vehicles owned or regularly used by a sample of families.)

2   4   1   2   3   2   3   1

2. Load Minitab and enter the following numbers in column C2, then enter PRINT C2 to display them. (The numbers are ages in years of U.S. commercial aircraft, based on data from Aviation Data Services.)

3.2   22.6   23.1   16.9   0.4   6.6   12.5   22.8

3. a.   Assume that Exercises 1 and 2 have been completed, so the given sets of numbers are stored in the columns designated as C1 and C2. Describe the results of entering the following Minitab commands.

```
MTB > LET C3 = C1 + C2
MTB > PRINT C3
```

   b.   Given that C1 consists of numbers of motor vehicles and C2 consists of ages of commercial aircraft, do the results from the above commands have any practical value? That is, are the resulting numbers meaningful?

4. Use Minitab's READ command to enter the following triplets of numbers in columns C4, C5, and C6. After entering ENDOFDATA, enter PRINT C4 C5 C6 to display those scores. (In each triplet, the first score is SAT verbal, the second score is SAT math, and the third score is grade-point average for a college student.)

623	509	2.6
454	471	2.3
643	700	2.4
585	719	3.0
719	710	3.1

   What Minitab command will result in the combined verbal and math SAT scores for the five students?

5. After completing Exercises 1–4, enter the command SAVE 'EXER', and then enter the command STOP to exit Minitab. Now load Minitab again, enter the following commands, and note the results.

```
MTB > RETRIEVE 'EXER'
MTB > PRINT C1-C6
```

6. Try to retrieve one of the data sets already installed with Minitab. If you are using Release 8 of the standard version, enter RETRIEVE 'TREES' and then PRINT C1 C2 C3. If you are using Release 8 of the Student Edition of Minitab, enter RETRIEVE 'NIELSEN' and then PRINT C1-C6. (If unsuccessful, consult your instructor or computer lab assistant.) Describe the results.

7. After loading Minitab, enter the following commands and describe the result.

```
MTB > SET C7
DATA> 25(12.345)
DATA> ENDOFDATA
MTB > PRINT C7
```

8. Enter the command SET C8, and proceed to enter

$$2 \quad 4 \quad 7 \quad 85 \quad 90 \quad 102.4$$

and then ENDOFDATA. Enter PRINT C8 to verify that the above numbers are correct.
   a. Enter LET C9 = C8/10, then display the numbers in C9 by entering PRINT C9. Describe the results.
   b. Enter LET C10 = C8**2, then display the numbers in C10 by entering PRINT C10. Describe the results.
   c. What Minitab command will create a column C11 consisting of the numbers in column C8 all multiplied by 50?

1-5   Exercises B: Beyond the Basics

9. Enter the scores given in Exercise 1 in Minitab's column C11. Enter each of the commands listed below and, based on the results and the suggestive name of the command, describe what each command does.

   a. MEAN C11   b. MAXIMUM C11
   c. MINIMUM C11   d. SUM C11
   e. SSQ C11

10. Repeat Exercise 9 after entering the Minitab command LET C11 = C11 * 100. How does this particular LET command affect the data stored in column C11? How are the results in parts a–e affected by that change?

VOCABULARY LIST

Define and give an example of each term.

statistics	nominal level of measurement
population	ordinal level of measurement
sample	interval level of measurement
parameter	ratio level of measurement
statistic	random sampling
qualitative data	stratified sampling
quantitative data	systematic sampling
discrete data	cluster sampling
continuous data	convenience sampling

REVIEW

This chapter described the general nature of statistics, along with some of its uses and abuses, and presented some very basic concepts dealing with the nature of data. Section 1-1 discussed statistics as a discipline and defined these very fundamental and important terms: *population, sample, parameter,* and *statistic.* Statistical analysis can involve all of the data in a population, or it can involve samples drawn from a population.

Section 1-2 discussed different ways in which data can be arranged, such as matched by pairs or listed in a table. We distinguished between qualitative and quantitative data and noted that some quantitative data are discrete whereas others are continuous. Data may also be categorized according to one of these levels of measurement: nominal, ordinal, interval, or ratio.

Section 1-3 presented uses of statistics as well as several examples of intentional or unintentional abuses. Section 1-4 discussed different methods of sampling, including the random, stratified, systematic, cluster, and convenience methods. It is extremely important to recognize that data collected carelessly may be absolutely worthless. Great care must be taken to ensure that samples are representative of the population from which they are drawn.

Section 1-5 discussed the STATDISK and Minitab software packages. The fundamental commands of the Minitab statistical software package were described.

REVIEW EXERCISES

1. In obtaining data on the following, determine which of the four levels of measurement (nominal, ordinal, interval, ratio) is most appropriate.
   a. Rankings (first, second, third, and so on) in order of quality for a sample of tire pressure gauges tested by *Consumer Reports* researchers
   b. The tire pressure readings (in psi or lb/in.²) for a sample of gauges tested by *Consumer Reports* researchers
   c. Ratings of tire pressure gauges as "recommended, acceptable, not acceptable"
   d. The college majors of researchers for *Consumer Reports*
   e. The internal air temperatures of a sample of inflated tires used in testing pressure gauges
2. A consumers' group measures the actual horsepower of a sample of lawn mowers labeled as 12 hp. The sample is obtained by selecting 3 lawn mowers from each manufacturer.
   a. Are the values obtained discrete or continuous?
   b. Identify the level of measurement (nominal, ordinal, interval, ratio) for the horsepower values.
   c. What type of sampling (random, stratified, systematic, cluster, convenience) is being used?

3. A news report states that the police seized forged record albums with a value of $1 million. How do you suppose the police computed the value of the forged albums, and in what other ways could that value have been estimated? Why might the police be inclined to exaggerate the value of the albums?

4. Identify each number as discrete or continuous.
   a. The Minolta Corporation surveyed 703 small business owners.
   b. The New York Metropolitan Transit Authority conducted a survey of commuting times, and the first result was 49 minutes.
   c. A consumer check of packaging revealed that a container of milk contained 30.4 ounces.

5. Identify the type of sampling (random, stratified, systematic, cluster, convenience) used in each of the following situations.
   a. A sample of products is obtained by selecting every 100th item on the assembly line.
   b. Random numbers generated by a computer are used to select the serial numbers of cars to be chosen for sample testing.
   c. An auto parts supplier obtains a sample of all stocked items from each of 12 different randomly selected retail stores.
   d. A car maker conducts a marketing study involving test drives performed by a sample of 10 men and 10 women in each of 4 different age brackets.
   e. A car maker conducts a marketing study by interviewing potential customers who happen to request test drives at a local dealership.

6. A newspaper article reports that a demonstration was attended by "1250 angry protesters." Comment.

7. Census takers have found that in obtaining people's ages, they get more people of age 50 than of age 49 or 51. Can you explain how this might occur?

8. It often happens that media reports dramatize the number of motor vehicle fatalities over a holiday weekend. In a typical year, about 46,000 deaths result from motor vehicle accidents, according to data from the National Safety Council.
   a. How many deaths would result from motor vehicle accidents in a typical day?
   b. How many deaths would result from motor vehicle accidents in a typical 4-day period?
   c. For the 4 days of the Memorial Day weekend (Friday through Monday), assume that driving increases by 25% and that there are 630 deaths resulting from motor vehicle accidents. Does it appear that driving is more dangerous over the Memorial Day weekend?

COMPUTER PROJECT

Chapter 2 uses data consisting of the body temperatures (in degrees Fahrenheit) of 106 people who participated in a research project. (University of Maryland researchers collected body temperature data and found that the average was not 98.6°F, which is the value that most of us assume to be the correct average.) The objective of this computer project is to enter the data and store them on a computer disk. This will allow you to have the data available for use in Chapter 2, and it will also allow you to develop the skill of entering and storing computer data—a skill that is critically important today.

Refer to the data in Table 2-1, near the beginning of the next chapter. Using STATDISK or Minitab, enter the 106 body temperatures, and save them under the name of TEMP. If STATDISK is used, select File from the menu bar at the top of the screen, then select New Sample Set to enter a new data set. When finished, select File and choose the option Save As and proceed to enter the file name of Temp. If Minitab is used, enter the data using the SET command as described in Section 1-5. After the data entry process is complete, enter the Minitab command SAVE 'TEMP' to store the data. If you wish to save the data on a separate disk in drive A, enter SAVE 'A:TEMP' instead. Save the data for use in Chapter 2.

FROM DATA TO DECISION: Misrepresented Data

Collect an example from a current newspaper or magazine in which data have been presented in a deceptive manner. Identify the source (including the publication date) from which the example is taken.

Explain the way in which the presentation is deceptive, and suggest how the data might be presented more fairly.

Interview

Lester Curtin

*Chief of the Statistical
Methods Staff in the Office of
Research and Methodology,
National Center for Health
Statistics, Centers for Disease
Control*

*Lester Curtin has a doctorate in
statistics and is temporarily the
Acting Director of the Division of
Vital Statistics. He regularly func-
tions as Chief of the Statistical
Methods Staff at the National
Center for Health Statistics,
which is part of the Centers for
Disease Control.*

What does the National Center for Health Statistics do?

The agency is designated by Congress to collect information
on the nation's health. We collect vital statistics, including
births, deaths, marriages, and divorces. We also do several
large-scale national health surveys, including the National
Health Interview Survey, which is a personal interview survey of about 120,000 people each year. We do the National
Health and Nutrition Examination Survey, which involves
mobile exam centers that go around the country taking
measurements on people. We get biological measurements
instead of just interview responses, which are subject to
measurement error. To determine calcium levels, in addition
to asking people what they ate, we can also draw blood to
get serum calcium levels. Our statistics are used to monitor
the nation's health and to determine what changes are occurring.

Could you cite an example of how your data are used?

The National Health and Nutrition Exam Survey was being
conducted in the field at about the time Congress banned
lead in gasoline. There was then a movement to put lead
back into gasoline. Because the surveyors were in the field
over time, we could do a temporal analysis, and we saw that
lead levels in people were dropping. This conclusion was
fought by the gasoline industry's experts. The government
had its experts, and the two groups went round and round.
We used data from the National Health and Nutrition Exam
Survey to present congressional testimony in favor of keeping
the lead out of gasoline. The validity of our data was upheld
and was very influential in saving the congressional law to
keep the lead out of gasoline.

It's a lot of fun dealing with health statistics. We're measuring serum cotinine as a way to see the effects of passive
smoking. We use data on monetary expenditures for mathematical modeling of health costs to be used for the health
care reform package that President Clinton is proposing.
We're looking for what might make a difference in terms of
dietary habits on future health.

In your field, is the use of statistics increasing, decreasing, or remaining about the same?

The National Center for Health Statistics once had a mandate that our group should simply collect and disseminate data. Over the last few years we've become much more involved with the analytic aspects of data. We are progressing more with the use of sophisticated statistical methods.

What statistical methods do you use?

We do just about everything. We do a lot of survey design work. We use a lot of quality control in coding and editing the data. We study the cognitive aspect of survey design and how people think when they answer a question. We might ask a question two or three different ways to see if that has an impact on how people are responding. We use confidence intervals, hypothesis testing, regression, time series analysis, ARIMA modeling, logistic regression on risk factors, and analysis of variance. The National Center for Health Statistics employs about 200 statisticians and another 50 people with general quantitative statistical backgrounds. We have employees with bachelor's degrees, master's degrees, and Ph.Ds.

How do you convince people of the validity of your data?

We have a basic policy of not presenting an estimate unless we present some measure of a sampling error. We also have a policy that when we publish a report from the National Center for Health Statistics and we say that one statistic is different from another, we have to back that up with a statistical test.

"The validity of our data was upheld and was very influential in saving the congressional law to keep the lead out of gasoline."

2 Descriptive Statistics

2-1 Overview

This chapter presents techniques for describing data sets. Tables, graphs, and measurements will be used to better understand important characteristics of the data. Later chapters will build on many of the concepts introduced in this chapter.

2-2 Summarizing Data

The construction of frequency tables, relative frequency tables, and cumulative frequency tables is described.

2-3 Pictures of Data

Methods for constructing histograms, relative frequency histograms, pie charts, and Pareto charts are presented.

2-4 Measures of Central Tendency

The following measures of central tendency are defined: mean, median, mode, midrange, and weighted mean. The concept of skewness is also considered.

2-5 Measures of Variation

The following measures of variation are defined: range, standard deviation, mean deviation, and variance.

2-6 Measures of Position

The standard score (or z score) is defined and used to illustrate how unusual values can be identified. Also defined are percentiles, quartiles, and deciles that are used to compare values within the same data set.

2-7 Exploratory Data Analysis

Techniques for exploring data with stem-and-leaf plots and boxplots are described.

Chapter Problem:
Is the 98.6°F body temperature a myth?

A person's body temperature is extremely important as a measure of health. Carl Wunderlich, a nineteenth-century medical researcher, analyzed measurements of the body temperatures of about 25,000 patients and concluded that, for healthy adults, the mean is 98.6°F. (We will formally define the term *mean* in Section 2-4; for now, simply think of it as an average.) However, more recent results suggest a different value. Dr. Philip Mackowiak, Dr. Steven Wasserman, and Dr. Myron Levine, University of Maryland researchers, conducted clinical tests, and we will analyze some of the results they obtained. Listed in Table 2-1 are the body temperatures (in degrees Fahrenheit) of 106 healthy adults taken at midnight on the second day of testing.

Table 2-1 Body Temperatures of 106 Healthy Adults

98.6	98.6	98.0	98.0	99.0	98.4	98.4	98.4	98.4	98.6	98.6
98.8	98.6	97.0	97.0	98.8	97.6	97.7	98.8	98.0	98.0	98.3
98.5	97.3	98.7	97.4	98.9	98.6	99.5	97.5	97.3	97.6	98.2
99.6	98.7	99.4	98.2	98.0	98.6	98.6	97.2	98.4	98.6	98.2
98.0	97.8	98.0	98.4	98.6	98.6	97.8	99.0	96.5	97.6	98.0
96.9	97.6	97.1	97.9	98.4	97.3	98.0	97.5	97.6	98.2	98.5
98.8	98.7	97.8	98.0	97.1	97.4	99.4	98.4	98.6	98.4	98.5
98.6	98.3	98.7	98.8	99.1	98.6	97.9	98.8	98.0	98.7	98.5
98.9	98.4	98.6	97.1	97.9	98.8	98.7	97.6	98.2	99.2	97.8
98.0	98.4	97.8	98.4	97.4	98.0	97.0				

We are interested in the mean (average) of the body temperatures, but there are other characteristics of the data that are also important. For example, how much do body temperatures vary? What temperatures are unusually low or high, suggesting that a person should be examined further? What is the cutoff for determining whether a patient has a fever? Although we may gain some insight into the important issues by examining Table 2-1, it is generally difficult to draw meaningful conclusions from a collection of raw data that are simply listed in no particular order. The major objective of this chapter is to develop a variety of methods that will provide more insight into data sets such as the one listed in Table 2-1. This particular data set will be used throughout the chapter, and we will eventually learn much about it. In later chapters we will address the question of whether the 98.6°F mean is correct.

2-1 Overview

In analyzing a data set, we should first determine whether we know all values for a complete population or whether we know only the values for some sample. (Data collected as the basis for making a general conclusion about a larger population are known as *sample* data.) That determination will affect both the methods we use and the conclusions we form.

We use methods of **descriptive statistics** to summarize or describe the important characteristics of a known set of population data. If the 106 values given in Table 2-1 represent the body temperatures of *everyone* living in a particular neighborhood, then we have known population data. We might then proceed to better understand this known population data by computing some average or by constructing a graph. Suppose we compute an average of the 106 temperatures in Table 2-1 and obtain a value of 98.20°F. In this case, the average describes and summarizes known population data and is therefore an example of descriptive statistics.

In contrast to descriptive statistics, **inferential statistics** goes beyond mere description; it involves the use of sample data to make inferences about a population. The 106 temperatures in Table 2-1 are the actual body temperatures of persons recruited from the larger population of all healthy adults. Treating those 106 temperatures as a sample drawn from a larger population, we might conclude that the average body temperature of all healthy adults is 98.20°F; in so doing, we have made an inference that goes beyond the known data.

Descriptive statistics and inferential statistics are the two basic divisions of the subject of statistics. This chapter deals with the basic concepts of descriptive statistics; after the introduction to probability theory in Chapter 3, subsequent chapters deal primarily with inferential statistics.

Important Characteristics of Data

We use the tools of descriptive statistics to understand an otherwise unintelligible collection of data. The following three characteristics of data are extremely important and can provide considerable insight:

1. Nature or shape of the distribution of the data, such as bell-shaped
2. Representative value, such as an average
3. Measure of scattering or variation

We can learn something about the nature or shape of the distribution by organizing the data and constructing graphs, as in Sections 2-2 and 2-3. In Section 2-4 we will learn how to obtain representative, or average, scores. We will measure the extent of scattering, or variation, among data as we use the tools found in Section 2-5. In Section 2-6 we will determine measures of position so that we can better analyze or compare various scores. And in Section 2-7 we will learn about methods for exploring data sets. As we proceed through this chapter, we will refer to the 106 scores given in Table 2-1 and increase our insight into that data set as its characteristics are revealed.

There is one last point that should be made in this overview. When collecting data, we must be extremely careful about the methods we use (common sense is often a critical requirement). If we plan our data collection with care and thoughtfulness, we can often learn much by using simple methods. If our data are collected without much thought, we may well end up with information that is misleading or worthless. As we consider the methods of descriptive statistics described in this chapter, remember that they will yield misleading results if the sample data are not representative of the population. (See Section 1-4.)

2-2   Summarizing Data

When beginning an analysis of a large set of values (such as the body temperatures listed in Table 2-1), we must often organize and summarize the data by developing tables and graphs. We begin with a frequency table.

Frequency Tables

DEFINITION

A **frequency table** lists categories (or classes) of scores, along with counts (or frequencies) of the number of scores that fall into each category.

Table 2-2 is a *frequency table* with 8 classes (or categories). The **frequency** for a particular class is the number of original scores that fall into that class. For example, the last class in Table 2-2 has a frequency of 4, indicating that there are 4 values between 99.3 and 99.6 inclusive.

TABLE 2-2   Frequency Table of Body Temperatures

Temperature (°F)	Tally Marks	Frequency
96.5–96.8	|	1
96.9–97.2	卌 |||	8
97.3–97.6	卌 卌 ||||	14
97.7–98.0	卌 卌 卌 卌 ||	22
98.1–98.4	卌 卌 卌 ||||	19
98.5–98.8	卌 卌 卌 卌 卌 卌 ||	32
98.9–99.2	卌 |	6
99.3–99.6	||||	4

The construction of a frequency table is not very difficult, and many statistics software packages can do it automatically. We will first present some standard terms used in discussing frequency tables, and then we will describe a procedure for constructing them.

DEFINITIONS

Lower class limits are the smallest numbers that can actually belong to the different classes. (Table 2-2 has lower class limits of 96.5, 96.9, . . . , 99.3.)

Upper class limits are the largest numbers that can actually belong to the different classes. (Table 2-2 has upper class limits of 96.8, 97.2, . . . , 99.6.)

Class boundaries are the numbers used to separate classes, but without the gaps created by class limits. They are obtained by increasing the upper class limits and decreasing the lower class limits by the same amount so that there are no gaps between consecutive classes. The amount to be added or subtracted is one-half the difference between the upper limit of one class and the lower limit of the following class. (Table 2-2 has class boundaries of 96.45, 96.85, 97.25, . . . , 99.65.)

Class marks are the midpoints of the classes. (Table 2-2 has class marks of 96.65, 97.05, . . . , 99.45.) They can be found by adding lower class limits to the corresponding upper class limits and dividing by 2.

Class width is the difference between two consecutive *lower class limits* or two consecutive *lower class boundaries*. (Table 2-2 uses a class width of 0.4.)

Two of the preceding definitions are a bit tricky. It's easy to make the mistake of determining the *class width* by finding the difference between the lower class limit and the corresponding upper class limit. Note that in Table 2-2 the class width is 0.4, not 0.3. However, students usually have the most difficulty with the concept of class boundaries. (We will use class boundaries later in the construction of some graphs.) Examine the class limits shown in Table 2-2, and note that there is a gap between 96.8 and 96.9, another gap between 97.2 and 97.3, and so on. *Class boundaries* basically split differences and fill in the gaps so that the construction of certain graphs will become easier. Carefully examine the definition of class boundaries, and spend some time working on understanding it.

The process of actually constructing a frequency table involves these steps:

Step 1: *Decide on the number of classes your frequency table will contain.* As a guideline, the number of classes should be between 5 and 20. The actual number of classes may be affected by the convenience of using round numbers or other subjective factors. With test grades, for example, it would be convenient to use these 10 classes: 50–54, 55–59, 60–64, . . . , 95–99.

Step 2: *Determine the class width* by dividing the number of classes into the range. (The *range* is the difference between the highest and lowest scores.) Round the result *up* to a convenient number. This rounding up (not off) not only is convenient, but also guarantees that all of the data will be included in the frequency table. (If the number of classes divides into the range evenly with no remainder, you will need to add another class for all of the data to be included.)

$$\text{class width} = \text{round } up \text{ of } \frac{\text{range}}{\text{number of classes}}$$

Step 3: *Select as the lower limit of the first class either the lowest score or a convenient value slightly less than the lowest score.* This value serves as the starting point.

Step 4: *Add the class width to the starting point to get the second lower class limit.* Add the class width to the second lower class limit to get the third, and so on.

Step 5: *List the lower class limits in a vertical column, and enter the upper class limits,* which can be easily identified at this stage.

Step 6: *Represent each score by a tally* in the appropriate class.

Step 7: *Replace the tally marks in each class with the total frequency count* for that class.

▶ **EXAMPLE**

Construct a frequency table for the 106 body temperatures given in Table 2-1.

● **SOLUTION**

We will list the steps that lead to the development of the frequency table shown in Table 2-2 (see page 37).

Step 1: We begin by selecting 10 as the number of desired classes. (Many statisticians recommend that we generally aim for about 10 classes.)

Step 2: With a minimum of 96.5 and a maximum of 99.6, the range is $99.6 - 96.5 = 3.1$, so

$$\text{class width} = \text{round } up \text{ of } \frac{3.1}{10}$$
$$= \text{round } up \text{ of } 0.31$$
$$= 0.4 \quad \text{(rounded up for the convenience of having only 1 decimal place instead of 2)}$$

(continued)

Buying Cars

Step 3:   The lowest value is 96.5. It becomes the starting point, as we use it for the lower limit of the first class.

Step 4:   Add the class width of 0.4 to the lower limit of 96.5 to get the next lower limit of 96.9. Continuing, we get the other lower class limits of 97.3, 97.7, and so on.

Step 5:   These lower class limits suggest these upper class limits.

96.5	96.8
96.9	97.2
97.3	97.6
.	.
.	.
.	.

Step 6:   The tally marks are shown in the middle column of Table 2-2. Each mark represents a value that falls within the corresponding class.

Step 7:   The frequency counts are shown in the extreme right column of Table 2-2. The final table should exclude the tally marks, because they are only a device for determining the frequencies.

Note that the resulting frequency table (as shown in Table 2-2) has only 8 classes instead of the 10 we began with. This is a result of rounding the class width up from 0.31 to 0.4. We could have forced the table to have exactly 10 classes, but that would have resulted in inconvenient limits such as 96.50–96.81. These limits are inconvenient because they have an extra decimal place.

Table 2-2 provides useful information by making the list of body temperatures more intelligible. Yet this information is not gained without some loss. In constructing frequency tables, we may lose the accuracy of the original data. To see how this loss occurs, consider the first class of 96.5–96.8. The table shows that there is one score in that class, but there is no way to determine from the table exactly what that score is. We cannot reconstruct the original 106 body temperatures; the exact values have been compromised for the sake of comprehension.

Summarizing data generally involves a compromise between accuracy and simplicity. A frequency table with too few classes is simple but not accurate. A frequency table with too many classes is more accurate but not as easy to understand. The best arrangement is arrived at subjectively. Some of these difficulties will be overcome in Section 2-7, when stem-and-leaf plots are discussed.

In constructing frequency tables, the following guidelines should be followed:

1. *Be sure that the classes are mutually exclusive.* That is, each of the original values must belong to exactly one class.

2. *Include all classes,* even if the frequency is zero.

3. *Try to use the same width for all classes,* although sometimes open-ended intervals such as "65 years or older" are impossible to avoid.

4. *Try to select convenient numbers for class limits.* Round up to use fewer decimal places or use numbers relevant to the situation.

5. *Try to use between 5 and 20 classes.*

 ## Using Computers to Generate Frequency Tables

Using the seven-step procedure and the five guidelines listed above, we can construct frequency tables. However, many statistical software packages allow us to obtain frequency tables with greater ease. In such software, frequency tables are often included with graphs called histograms (which will be described in the next section). With Minitab, for example, we can use the SET command to first enter the data and store them in column C1 (as described in Section 1-5). We can then use the command HISTOGRAM C1, and the display will include a frequency table. STATDISK also generates a histogram display that includes the frequencies for the different classes.

Relative Frequency Table

An important variation of the basic frequency table uses relative frequencies. The **relative frequency** for a particular class can be easily found by dividing the class frequency by the total of all frequencies. The **relative frequency table** includes the same class limits as a frequency table, but relative frequencies are used instead of actual frequencies.

$$\text{relative frequency} = \frac{\text{class frequency}}{\text{sum of all frequencies}}$$

Table 2-3 shows the relative frequencies for the 106 body temperatures summarized in Table 2-2. The first class has a relative frequency of $1/106 = 0.009$. (Relative frequencies can also be given as percentages; that is, 0.009 can be expressed as 0.9%.) The second class has a relative frequency of $8/106 = 0.075$, and so on. Relative frequency tables make it easier for us to compare different sets of data because we use comparable relative frequencies instead of original frequencies that might be very different in magnitude. (See Exercise 21.)

TABLE 2-3
Relative Frequency Table of Body Temperatures

Temperature (°F)	Relative Frequency
96.5–96.8	0.009
96.9–97.2	0.075
97.3–97.6	0.132
97.7–98.0	0.208
98.1–98.4	0.179
98.5–98.8	0.302
98.9–99.2	0.057
99.3–99.6	0.038

Authors Identified

In 1787–88 Alexander Hamilton, John Jay, and James Madison anonymously published the famous *Federalist* papers in an attempt to convince New Yorkers that they should ratify the Constitution. The identity of most of the papers' authors became known, but the authorship of 12 of the papers was contested. Through statistical analysis of the frequencies of various words, we can now conclude that James Madison is the *likely* author of these 12 papers. For many of the disputed papers, the evidence in favor of Madison's authorship is overwhelming to the degree that we can be almost certain of being correct.

TABLE 2-4   Cumulative Frequency Table of Body Temperatures

Temperature (°F)	Cumulative Frequency
Less than 96.9	1
Less than 97.3	9
Less than 97.7	23
Less than 98.1	45
Less than 98.5	64
Less than 98.9	96
Less than 99.3	102
Less than 99.7	106

Cumulative Frequency Table

Another variation of the standard frequency table is used when cumulative totals are desired. The **cumulative frequency** for a class is the sum of the frequencies for that class and all previous classes. Table 2-4 is an example of a **cumulative frequency table,** and it represents the same 106 body temperatures summarized in Table 2-2. A comparison of the frequency column of Table 2-2 and the cumulative frequency column of Table 2-4 reveals that the cumulative frequency values are obtained by starting with the frequency for the first class and adding successive frequencies for each class. For example, there is 1 value less than 96.9, there are 9 values (1 + 8) less than 97.3, and so on.

In the next section we will explore various graphic ways to depict data so that they are easy to understand. Frequency tables are necessary for constructing some of the graphs, which in turn are often necessary for considering the important characteristic of the nature or shape of the distribution of the data.

2-2   Exercises A: Basic Skills and Concepts

In Exercises 1–4, identify the class width, class marks, and class boundaries for the given frequency table.

1.

IQ	Frequency
80–87	16
88–95	37
96–103	50
104–111	29
112–119	14

2.

Time (hours)	Frequency
0.0–7.5	16
7.6–15.1	18
15.2–22.7	17
22.8–30.3	15
30.4–37.9	19

3.

Weight (kg)	Frequency
16.2–21.1	16
21.2–26.1	15
26.2–31.1	12
31.2–36.1	8
36.2–41.1	3

4.

Sales (dollars)	Frequency
0–21	2
22–43	5
44–65	8
66–87	12
88–109	14
110–131	20

In Exercises 5–8, construct the relative frequency table that corresponds to the frequency table in the exercise indicated.

5. Exercise 1   **6.** Exercise 2   **7.** Exercise 3   **8.** Exercise 4

In Exercises 9–12, construct the cumulative frequency table that corresponds to the frequency table in the exercise indicated.

9. Exercise 1   **10.** Exercise 2   **11.** Exercise 3   **12.** Exercise 4

In Exercises 13–16, use the given information to find upper and lower limits of the first class. (The data are in Appendix B, but there is no need to refer to the appendix for these exercises.)

13. A data set consists of weights of metal collected from households for one week, and those weights range from 0.26 lb to 4.95 lb. You wish to construct a frequency table with 10 classes.

14. A data set consists of surveyed people whose family incomes range from $3600 to $157,800. You wish to construct a frequency table with 16 classes.

15. A data set includes weights of cars (in pounds) with a minimum of 1650 lb and a maximum of 4367 lb. You wish to construct a frequency table with 7 classes.

16. A sample of M&M candies has weights that vary between 0.802 g and 1.027 g. You wish to construct a frequency table with 12 classes.

In Exercises 17–20, construct a frequency table, using the data given.

17. As part of the Garbage Project at the University of Arizona, weights of *metal* were collected from sample households. Those weights are given in Data Set 1 of Appendix B. Construct a frequency table with 10 classes.

18. A survey of 125 randomly selected persons includes their family incomes. Refer to the incomes listed in Data Set 3 in Appendix B, and construct a frequency table with 16 classes.

19. Weights are found for randomly selected cars, and they are listed in Data Set 4 of Appendix B. Construct a frequency table with 7 classes.

20. Refer to Data Set 6 in Appendix B, and use the weights of all 100 M&Ms to construct a frequency table with 12 classes.

2-2 Exercises B: Beyond the Basics

21. Given below are a frequency table of alcohol consumption prior to arrest for male inmates currently serving sentences for DWI and a corresponding table for women (based on data from the U.S. Department of Justice). First construct the corresponding relative frequency tables, and then use those results to compare the two samples. Note that it is difficult to compare the original frequencies, but it is much easier to compare the relative frequencies.

Ethanol Consumed by Men (oz)	Frequency	Ethanol Consumed by Women (oz)	Frequency
0.0–0.9	249	0.0–0.9	7
1.0–1.9	929	1.0–1.9	52
2.0–2.9	1545	2.0–2.9	125
3.0–3.9	2238	3.0–3.9	191
4.0–4.9	1139	4.0–4.9	30
5.0–9.9	3560	5.0–9.9	201
10.0–14.9	1849	10.0–14.9	43
15.0 or more	1546	15.0 or more	72

22. Listed below are two sets of scores that are supposed to be heights (in inches) of randomly selected adult males. One of the sets consists of heights actually obtained from randomly selected adult males, but the other set consists of numbers that were fabricated. Construct a frequency table for each set of heights. By examining the two frequency tables, identify the set of data that you believe to be false.

a. 70 73 70 72 71 73 71 67 68 72 67 72 71 73
    72 70 72 68 71 71 71 73 69 73 71 66 77 67

b. 70 73 70 72 71 66 74 76 68 75 67 68 71 77
    66 69 72 67 77 75 66 76 76 77 73 74 69 67

23. Data from the U.S. Bureau of the Census are summarized in the accompanying frequency table. Refer to the five guidelines for constructing frequency tables. Which of the guidelines are not followed?

Age	U.S. Population (millions)
Under 15	55
15–24	37
25–44	82
45 and older	79

24. In constructing a frequency table, **Sturges' guideline** suggests that the ideal number of classes can be approximated by the value of $1 + (\log n)/(\log 2)$, where n is the number of scores. Use this guideline to find the ideal number of classes (rounded off, not rounded up) corresponding to a data collection with the number of scores equal to

a. 50   b. 100   c. 150
d. 500   e. 1000   f. 50,000

2-3   Pictures of Data

In the preceding section we saw that frequency tables transform a disorganized collection of raw scores into an organized and understandable summary. In this section we will consider ways of representing data in pictorial form. This section has two major objectives:

1. To present methods for constructing a type of graph called a *histogram* in an attempt to depict the nature or shape of the distribution of the data.

2. To present methods for constructing other types of graphs, such as *pie charts* and *Pareto charts,* which may be much better in certain situations, depending on the type of data being considered and the characteristics of the data that are most important or relevant.

Histograms and the Shape of Data

A common graphic device for presenting data is the histogram, an example of which is shown in Figure 2-1. A **histogram** consists of a horizontal scale for values of the data being represented, a vertical scale for frequencies, and bars representing the frequency for each subdivision or class of values. We generally construct a histogram to represent a set of values after we have first completed a frequency table representing those values. The width of each bar extends

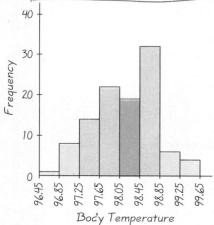

FIGURE 2-1   Histogram of Body Temperatures

Florence Nightingale

Florence Nightingale (1820–1910) is known to many as the founder of the nursing profession, but she also saved thousands of lives by using statistics. When she encountered an unsanitary and undersupplied hospital, she improved those conditions and then used statistics to convince others of the need for more widespread medical reform. She developed original graphs to illustrate that, during the Crimean War, more soldiers died as a result of unsanitary conditions than were killed in combat. Florence Nightingale pioneered the use of social statistics as well as graphics techniques.

from its lower class boundary to its upper class boundary, so we can mark the class boundaries on the horizontal scale. (Improved readability, however, is often achieved by using class limits or class marks instead of class boundaries.) The histogram in Figure 2-1 corresponds directly to the frequency table (Table 2-2) in the previous section.

Before constructing a histogram from a completed frequency table, we must give some consideration to the scales used on the vertical and horizontal axes. The maximum frequency (or the next highest convenient number) should suggest a value for the top of the vertical scale; 0 should be at the bottom. In Figure 2-1 we designed the vertical scale to run from 0 to 40. The horizontal scale should be designed to accommodate all the classes of the frequency table. Ideally, we should try to follow the rule of thumb that the vertical height of the histogram should be about three-fourths of the total width. Both axes should be clearly labeled.

A **relative frequency histogram** will have the same shape and horizontal scale as a histogram, but the vertical scale will be marked with *relative frequencies* instead of actual frequencies. Figure 2-1 can be modified to be a relative frequency histogram by labeling the vertical scale as "relative frequency" and by changing the values on that scale to range from 0 to 0.40 (see Figure 2-2). (The highest relative frequency for this data set is 0.302.) Just as the histogram in Figure 2-1 represents the frequency table in Table 2-2, the relative frequency histogram in Figure 2-2 represents the relative frequency table in Table 2-3.

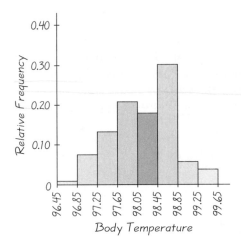

FIGURE 2-2   Relative Frequency Histogram of Body Temperatures

 Using Computers to Generate Histograms

Shown below are STATDISK and Minitab displays of histograms for the body temperature data we are considering in this chapter. The STATDISK display is obtained by first entering the data with the File and New options. Then proceed to use the View and Graph options. The Minitab display is obtained by using the HISTOGRAM command. First enter (or retrieve) the 106 temperatures in the column designated as C1, then enter the Minitab command HISTOGRAM C1 to get the display shown here. The nine values listed in the display under Midpoint are class marks.

STATDISK DISPLAY

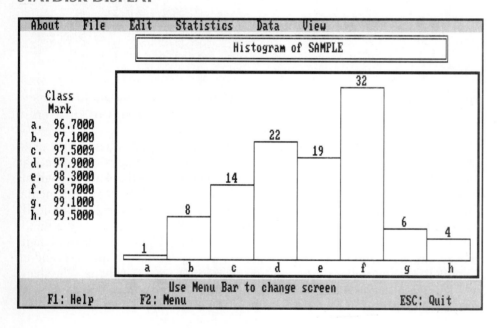

Best-Seller Lists

Listings of the best-selling books are based on sampling. The *New York Times*, for example, solicits book sales figures from about 3000 bookstores in seven regions of the country. *Publishers Weekly* polls about 2000 stores. One book that made the best-seller list is *Confessions of an S.O.B.* by Allen Neuharth; he heads a foundation that spent $40,000 to buy 2000 copies of the book from bookstores around the country.

MINITAB DISPLAY

```
MTB > HISTOGRAM C1

  Histogram of C1   N = 106

Midpoint   Count
    96.4       1    *
    96.8       1    *
    97.2      10    **********
    97.6      12    ************
    98.0      26    **************************
    98.4      33    *********************************
    98.8      15    ***************
    99.2       4    ****
    99.6       4    ****
```

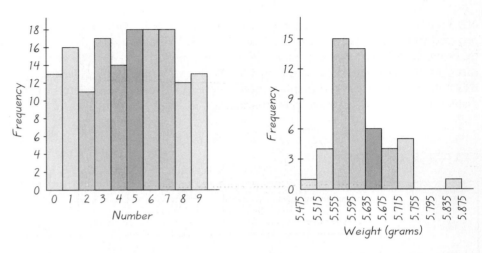

**FIGURE 2-3   Histogram of
Lottery Results**

**FIGURE 2-4   Histogram of
Weights of Quarters**

Frequency tables and graphs such as histograms make it possible for us to see the distribution of data. In any serious statistical analysis of data, the distribution is a critically important feature. Figures 2-3 and 2-4 are histograms of real data (see Data Sets 7 and 8) in Appendix B. Note the fundamental difference between the *shapes* of those two sets of data. Figure 2-3 is basically flat or uniform, whereas Figure 2-4 is roughly bell-shaped in the sense that it resembles the shape shown in the margin. Because Figure 2-3 depicts digits selected in Maryland's Pick Three Lottery, we expect that all digits should be equally likely and the histogram should therefore have a flat shape. The shape of Figure 2-3 suggests that the winning numbers seem to be selected as we expect. Any dramatic departure from a flat or uniform shape would raise serious questions about how that lottery is being run.

The bell shape of Figure 2-4 is typical of an incredibly wide variety of different circumstances, especially those in manufacturing. That bell shape shows the type of distribution we expect from errors that are naturally created when the same machine is used to manufacture the sample items. Because Figure 2-4 has such a good bell shape, those quarters seem to have been minted by the same machine. They were, however, randomly selected from the population of quarters minted by many different machines over a period of 20 years. This suggests that, over the years, quarters are being minted in a way that is very consistent. The mints of the United States government seem to have overcome the difficulties caused by using many different machines over long periods of time and seem to be doing a good job.

It is important to gain insight into the nature or shape of the distribution of data because that insight often reveals important information (as in Figures 2-3 and 2-4). Also, many procedures in statistics require a data set with a distribution that is approximately bell-shaped, and we must often construct frequency tables and/or graphs to see if that requirement is met. For these

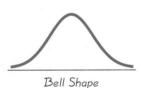

Bell Shape

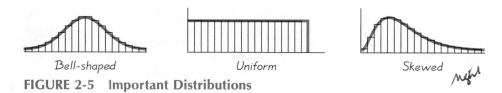

FIGURE 2-5   **Important Distributions**

reasons, the nature or shape of a distribution is a very important characteristic. The uniform shape of Figure 2-3 and the bell shape of Figure 2-4 illustrate two very important distributions. Figure 2-5 illustrates the bell-shaped, uniform, and skewed distributions. In a skewed distribution, there are many disproportionately large (or small) scores so the graph is lopsided.

Pareto Charts

Consider this statement: Among the 19,257 murders in the United States in a recent year, 11,381 were committed with firearms, 3957 with knives, 1310 with personal weapons (hands, feet, and so on), 1099 with blunt objects, and the remaining 1510 with a variety of other weapons. That written statement probably does a poor job of conveying the relationships among the different categories of murder weapons. We will now discuss a better way to convey relationships among data: Pareto charts.

Recall from Section 1-2 that quantitative data consist of numbers that are measures or counts, whereas qualitative data represent some nonnumeric characteristic. A histogram is a bar graph in which the horizontal scale represents a quantitative variable, such as the body temperatures of people. In contrast, a **Pareto chart** is a bar graph for *qualitative* data, with the bars arranged in order according to frequencies. As in histograms, in Pareto charts vertical scales can represent frequencies or relative frequencies. The tallest bar is at the left, and the smaller bars are farther to the right, as in Figure 2-6. This figure represents the FBI data described in the preceding paragraph, as listed in

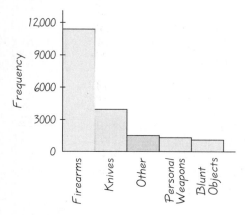

FIGURE 2-6   **Pareto Chart of Crime Data**

Uniform Crime Reports. By arranging the bars in order of frequency, the *Pareto chart* focuses attention on the more important categories. From Figure 2-6 we can see that the category of firearms is by far the largest category of murder weapons in the United States.

Pie Charts

Like Pareto charts, pie charts are used to depict qualitative data in a way that makes them much more understandable. Figure 2-7 shows an example of a **pie chart, which graphically depicts qualitative data as slices of a pie**. Construction of such a pie chart involves slicing up the pie into the proper proportions. If the category of firearms represents 59.1% of the total, then the wedge representing firearms should be 59.1% of the total. (The central angle should be $0.591 \times 360° = 213°$.)

The Pareto chart of Figure 2-6 and the pie chart of Figure 2-7 depict the same data in different ways. If you compare the Pareto chart and the pie chart, you will probably see that the Pareto chart does a better job of showing the relative sizes of the different components.

Numerous pictorial displays other than the ones just described can be used to represent data dramatically and effectively. Exercise 17 involves a frequency polygon, which is a variation of the histogram. Exercise 18 involves an ogive (or cumulative frequency polygon), which is based on a cumulative frequency table. *Pictographs* depict data by using pictures of objects, such as soldiers, tanks, airplanes, stacks of coins, or moneybags. Although there is almost no limit to the variety of different ways that data can be illustrated, the most common devices are histograms, relative frequency histograms, Pareto charts, pie charts, frequency polygons, and ogives.

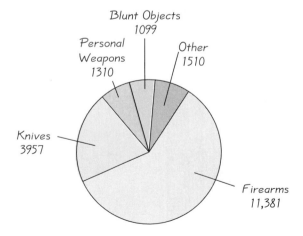

FIGURE 2-7   Pie Chart of Crime Data

In this section we have focused on the nature or shape of the distribution of data and methods of graphically depicting data. In the following sections we consider ways of measuring other characteristics of data.

2-3   Exercises A: Basic Skills and Concepts

1. The given frequency table describes the speeds of drivers ticketed by the Town of Poughkeepsie police. These drivers were traveling through a 30 mi/h speed zone on Creek Road, which passes the author's college. Construct a histogram corresponding to the given frequency table.

Speed	Frequency
42–43	14
44–45	11
46–47	8
48–49	6
50–51	4
52–53	3
54–55	1
56–57	2
58–59	0
60–61	1

2. Construct a histogram corresponding to the given frequency table that describes the age distribution of a randomly selected sample of United States residents (based on data from the U.S. Bureau of the Census).

Age	Frequency
0–9	37
10–19	35
20–29	39
30–39	46
40–49	33
50–59	22
60–69	20
70–79	17
80–89	9
90–99	3

3. Construct a relative frequency histogram that corresponds to the given frequency table. The data are based on information reported by Aviation Data Services.

Airplane Age (years)	Frequency
0.0–5.9	10
6.0–11.9	8
12.0–17.9	9
18.0–23.9	9
24.0–29.9	4

4. Insurance companies continually research ages at death and causes of death. Construct a relative frequency histogram that corresponds to the given frequency table. The data are based on a *Time* magazine study of people who died from gunfire in America during one week.

Age at Death	Frequency
16–25	22
26–35	10
36–45	6
46–55	2
56–65	4
66–75	5
76–85	1

5. A study was conducted to determine how people get jobs. The table lists data from 400 randomly selected subjects. The data are based on results from the National Center for Career Strategies. Construct a Pareto chart that corresponds to the given data.

Job Sources of Survey Respondents	Frequency
Help-wanted ads	56
Executive search firms	44
Networking	280
Mass mailing	20

6. Refer to the data given in Exercise 5, and construct a pie chart. Compare the pie chart to the Pareto chart, and determine which graph is more effective in showing the relative importance of job sources.

7. An analysis of train derailment incidents showed that 23 derailments were caused by bad track, 9 were due to faulty equipment, 12 were attributable to human error, and 6 had other causes (based on data from the Federal Railroad Administration). Construct a pie chart representing the given data.

8. Refer to the data given in Exercise 7, and construct a Pareto chart. Compare the Pareto chart to the pie chart, and determine which graph is more effective in showing the relative importance of the causes of train derailments.

In Exercises 9–16, refer to the data sets in Appendix B.
   a.   Construct a histogram.
   b.   Compare the result to the distributions illustrated in Figure 2-5, and categorize the given data set as bell-shaped, uniform, skewed, or a distribution different from those shown in the figure.

9. Data Set 5 in Appendix B: lengths (in minutes) of movies (Use 12 classes.)
10. Data Set 6 in Appendix B: weights of 100 M&Ms (Use 12 classes.)
11. Data Set 1 in Appendix B: weights of paper discarded by 62 households in one week (Use 10 classes.)
12. Data Set 1 in Appendix B: sizes of households (HHSIZE) (Use 11 classes.)
13. Data Set 7 in Appendix B: the 300 numbers selected in the Maryland Lotto (*not* the Maryland Pick Three lottery)
14. Data Set 2 in Appendix B: the ages of the 107 subjects who were measured for body temperatures
15. Data Set 3 in Appendix B: family incomes (Use 16 classes.)
16. Data Set 3 in Appendix B: heights of females

2-3   **Exercises B: Beyond the Basics**

17. A **frequency polygon** is a variation of a histogram that uses line segments connected to points (instead of using bars). Construct a frequency polygon for the data in Table 2-2 as follows. First construct a vertical scale for frequencies ranging from 0 to 40, and then construct a horizontal scale with class marks ranging from 96.65 to 99.45. Above each class mark, plot the point corresponding to the class frequency. After plotting the 8 points, extend the graph to the left and right so that it begins and ends with a frequency of 0. A portion of the graph is shown here; use the data in Table 2-2 to complete it.

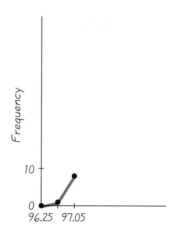

18. An **ogive** (also called a **cumulative frequency polygon**) is a common graphic display that is used when we want to see how many scores are above or below some value. Construct an ogive for the body temperature data of Table 2-2 as follows. First construct a vertical scale of cumulative frequencies ranging from 0 to 106. With class frequencies of 1, 8, 14, ..., 4, the cumulative frequencies of 1, 9, 23, ..., 106 are obtained by finding running totals as you go down the list of frequencies. For example, the third cumulative frequency of 23 is the sum of $1 + 8 + 14$. Use the class boundaries of 96.45, 96.85, 97.25, ..., 99.65 for the horizontal scale. Directly above each class boundary, plot a point corresponding to the cumulative frequency. After plotting the 8 points, connect them with straight line segments, and then extend the graph to the left so that it begins with a cumulative frequency of 0. A portion of the graph is shown here; use the body temperature data in Table 2-2 to complete it.

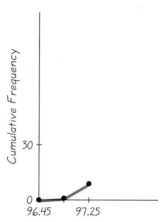

19. Frequency tables are given for the first 100 digits in the decimal representation of π and the first 100 digits in the decimal representation of 22/7.
   a.   Construct histograms representing the frequency tables, and note any differences.
   b.   The numbers π and 22/7 are both real numbers, but how are they fundamentally different?

	π		22/7
x	f	x	f
0	8	0	0
1	8	1	17
2	12	2	17
3	11	3	1
4	10	4	17
5	8	5	16
6	9	6	0
7	8	7	16
8	12	8	16
9	14	9	0

20. Using a collection of sample data, we construct a frequency table with 10 classes and then construct the corresponding histogram. How is the histogram affected if the number of classes is doubled but the same vertical scale is used?

21. In an insurance study of motor vehicle accidents in New York State, fatal crashes are categorized according to time of day, with the results given in the accompanying table (based on data from the New York State Department of Motor Vehicles).
    a. Complete the bar chart and the circular bar chart.
    b. Which is more effective in depicting the data?
    c. Because the time period of 4:00 AM to 6:00 AM has the lowest number of fatal crashes, is that time period the safest time to drive? Why or why not?

Time of Day	Number of Fatal Crashes
AM 12–2	194
2–4	149
4–6	100
6–8	131
8–10	119
10–12	160
PM 12–2	152
2–4	221
4–6	230
6–8	211
8–10	223
10–12	178

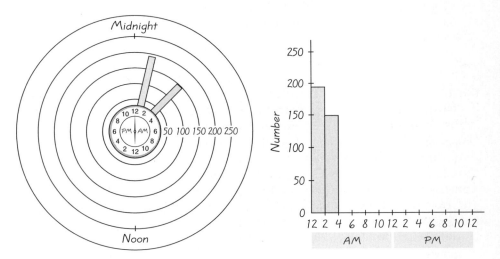

2-4   Measures of Central Tendency

In Sections 2-2 and 2-3 we considered frequency tables and graphs that reveal the nature or shape of the distribution of a data set. We generally also need information about a value that is typical or representative of the whole data set. That is, we need to identify a value that the data tend to center around. Because we seek a particular *value*, we refer to the statistic as a *measure*, and because we seek a value about which the data tend to center, we refer to the statistic as a measure of central tendency.

DEFINITION

A **measure of central tendency** is a value at the center or middle of a data set.

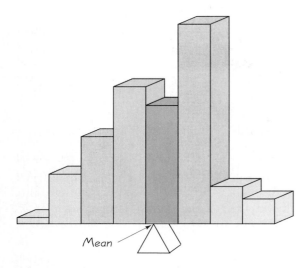

FIGURE 2-8   Mean as a Balance Point
If a fulcrum is placed at the position of the mean, it will
balance the histogram.

There are different criteria for determining the center, and so there are
different definitions of *measures of central tendency,* including the mean,
median, mode, and midrange. We begin our study of measures of central
tendency with the mean.

Mean

The (arithmetic) mean is the most important of all numerical descriptive
measurements, and it corresponds to what most people call an *average.* In
Figure 2-8 we illustrate the property that the mean is at the center of the data
set in the sense that it is a balance point for the data.

DEFINITION

The **arithmetic mean** of a list of scores is the value obtained by adding the
scores and dividing the total by the number of scores. This particular measure
of central tendency will be used frequently throughout the remainder of this
text, and it will be referred to simply as the **mean.**

What's a Degree Worth?

According to Jeff Thredgold, senior vice president and chief economist for KeyCorp, "In an increasingly sophisticated work environment, education, training and retraining rise in importance. The key to financial success and stability for many in the past was affiliation with a union. The key to financial success and stability for many more people in the future will be the rising importance and value of education." Based on census figures, the typical high school graduate will earn about $13,000 annually. The mean annual income for those with an associate degree is about $20,000, compared to $25,000 for a bachelor's degree, $34,000 for a master's degree, and $46,000 for those with a doctorate.

Many different formulas in statistics use the Greek letter Σ (uppercase sigma), which indicates summation of values. If we let x denote the value of a score, Σx represents the sum of all scores. The symbol n is used to represent the **sample size,** which is the number of scores being considered. Since the *mean* is the sum of all scores divided by the number of scores n, we have the following formula.

Formula 2-1
$$\text{mean} = \frac{\Sigma x}{n}$$

The mean can be denoted by $\overline{x}$ (pronounced "x-bar") if the available scores are samples from a larger population; if all scores of the population are available, then we can denote the computed mean by μ (lowercase Greek mu). (Sample statistics are usually represented with English letters, such as $\overline{x}$, whereas population parameters are usually represented with Greek letters, such as μ.) Many calculators can find the mean of a data set: simply enter the data, and press the key labeled $\overline{x}$. Entry of the data varies with different calculator models, so refer to the user manual for your calculator.

Notation	
Σ	denotes *summation* of a set of values.
x	is the *variable* usually used to represent the individual data values.
n	represents the *number* of values in a *sample*.
N	represents the *number* of values in a *population*.
$\overline{x} = \dfrac{\Sigma x}{n}$	is the *mean* of a set of *sample* values.
$\mu = \dfrac{\Sigma x}{N}$	denotes the *mean* of all values in some *population*.

▶ EXAMPLE

Methods of statistics are sometimes used to compare or identify authors of different works. Find the mean of the lengths of the first 10 words in Robert Frost's poem, *Stopping by Woods on a Snowy Evening.*

5   5   5   3   1   5   1   4   3   5   *(continued)*

● **SOLUTION**

The mean is computed by using Formula 2-1. First add the scores.

$$\Sigma x = 5 + 5 + 5 + 3 + 1 + 5 + 1 + 4 + 3 + 5 = 37$$

Now divide the total by the number of scores present. Because there are 10 scores, we have $n = 10$ and get

$$\overline{x} = \frac{37}{10} = 3.7$$

The mean value is therefore 3.7.

For the list of 10 values in the preceding example, 3.7 is at the center, according to the definition of the mean. Other definitions of a measure of central tendency involve different perceptions of how the center is determined. The median, for example, reflects another approach.

Median

DEFINITION

The **median** of a set of scores is the middle value when the scores are arranged in order of increasing (or decreasing) magnitude. The median is often denoted by $\tilde{x}$ (pronounced "x-tilde").

After first *ranking* the original scores (arranging them in increasing or decreasing order), we find the median in one of the following ways:

1. If the number of scores is *odd,* the median is the number that is exactly in the middle of the list.
2. If the number of scores is *even*, the median is found by computing the mean of the two middle numbers.

▶ **EXAMPLE**

Find the median of the scores

$$5 \quad 5 \quad 5 \quad 3 \quad 1 \quad 5 \quad 1 \quad 4 \quad 3 \quad 5 \quad 2 \qquad \textit{(continued)}$$

Leave Computer On

Some people turn off their computer whenever they finish an immediate task, whereas others leave it on until they know they won't be using it again until the next day. The computer's circuit board and chips do suffer from repeated on/off electrical power cycles. However, the monitor can be damaged when the same image is left on a screen for very long periods of time. The *mean time between failures* (MTBF) for hard disk drives was once around 5000 hours, but it's now up to about 30,000 hours. Considering the bad effects of on/off cycles on circuit boards and chips and the large MTBF for hard drives, it does make sense to leave computers on until the end of the day, provided the monitor's screen can be protected by making it blank or by using a screen-saver program. Many people use this strategy, which evolved in part from a statistical analysis of past events.

● **SOLUTION**

Begin by arranging the scores in increasing order.

$$1 \quad 1 \quad 2 \quad 3 \quad 3 \quad 4 \quad 5 \quad 5 \quad 5 \quad 5 \quad 5$$
$$\uparrow$$

With these 11 scores, the number 4 is at the exact middle, so 4 is the median.

▶ **EXAMPLE**

Find the median of the values

$$5 \quad 5 \quad 5 \quad 3 \quad 1 \quad 5 \quad 1 \quad 4 \quad 3 \quad 5$$

● **SOLUTION**

Again, begin by ranking the scores. That is, arrange them in increasing order.

$$1 \quad 1 \quad 3 \quad 3 \quad 4 \quad 5 \quad 5 \quad 5 \quad 5 \quad 5$$

With these 10 scores, no single score is at the exact middle. Instead, the two scores of 4 and 5 share the middle. We therefore find the mean of 4 and 5.

$$\frac{4 + 5}{2} = \frac{9}{2} = 4.5$$

The median is 4.5.

Mode

DEFINITION

The **mode** of a data set is the score that occurs *most frequently*. When two scores occur with the same greatest frequency, each one is a **mode** and the data set is **bimodal**. When more than two scores occur with the same greatest frequency, each is a mode and the data set is said to be **multimodal**. When no score is repeated, we stipulate that there is no mode.

► **EXAMPLE**

Find the modes of the following data sets.

a.   5   5   5   3   1   5   1   4   3   5
b.   1   2   2   2   3   4   5   6   6   6   7   9
c.   1   2   3   6   7   8   9   10

● **SOLUTION**

a.   The number 5 is the mode because it is the score that occurs most often.
b.   The numbers 2 and 6 are both modes because they occur with the same greatest frequency.
c.   There is no mode because no score is repeated.

Among the different measures of central tendency we are now considering, the mode is the only one that can be used with data at the nominal level of measurement, as illustrated in the next example.

► **EXAMPLE**

A town meeting is attended by 40 Republicans, 25 Democrats, and 20 Independents. Although we cannot numerically average these party affiliations, we can report that the mode is Republican, because that party had the greatest frequency of attendees.

Midrange

DEFINITION

The **midrange** is the value halfway between the highest and lowest scores. It is found by adding the highest score to the lowest score and then dividing the sum by 2.

$$\text{midrange} = \frac{\text{highest score} + \text{lowest score}}{2}$$

▶ **EXAMPLE**

Find the midrange of the scores

$$5 \quad 5 \quad 5 \quad 3 \quad 1 \quad 5 \quad 1 \quad 4 \quad 3 \quad 5$$

● **SOLUTION**

The midrange is found as follows:

$$\frac{\text{highest score} + \text{lowest score}}{2} = \frac{5 + 1}{2} = 3$$

The midrange is not used very often. We include it mainly to emphasize the point that there are several different ways to define the center of a set of data. (See also Exercises 26–28.)

Unfortunately, the term *average* is sometimes used for any measure of central tendency and is sometimes used for the mean. Because of this ambiguity, we should not use the term *average* when referring to a specific measure of central tendency. Instead, we should use the specific term, such as mean, median, mode, or midrange.

▶ **EXAMPLE**

Refer to Table 2-1, where we list the 106 body temperatures measured at the University of Maryland. We have constructed a frequency table (Table 2-2) for those values, and we have constructed a histogram (Figure 2-1) that reveals a bell-shaped distribution. Using the 106 temperatures in Table 2-1, find the (a) mean, (b) median, (c) mode, and (d) midrange.

● **SOLUTION**

a. Mean: The sum of the 106 temperatures is 10,409.2, so

$$\bar{x} = \frac{10,409.2}{106} = 98.20$$

b. Median: After arranging the scores in increasing order, we find that the 53rd and 54th scores are both 98.4, so the median is 98.40. (The scores can easily be arranged in increasing order by constructing a stem-and-leaf

(continued)

plot, as we will discuss in Section 2-7, or by using a computer program such as STATDISK or Minitab.) We express the result with an extra decimal place by using the round-off rule that follows this example.

c. Mode: The most frequent body temperature is 98.6, which occurs 15 times. It is therefore the mode.

d. Midrange: We use the following formula to find the midrange:

$$\text{midrange} = \frac{\text{highest score} + \text{lowest score}}{2} = \frac{99.6 + 96.5}{2} = 98.05$$

We now summarize these results.

mean: 98.20°F
median: 98.40°F
mode: 98.6°F
midrange: 98.05°F

Round-Off Rule

A simple rule for rounding answers is to *carry one more decimal place than is present in the original set of data*. We should round only the final answer and not intermediate values. For example, the mean of 2, 3, 5 is 3.33333333 . . . , and it can be rounded as 3.3. Because the original data were whole numbers, we rounded the answer to the nearest tenth. As another example, the mean of 2.1, 3.4, 5.7 is rounded to 3.73 with two decimal places (one more decimal place than was used for the original values).

Weighted Mean

In the preceding definition of the mean, every score is treated equally, but there are situations in which the scores vary in their degree of importance. In such cases, we can calculate the mean by applying different weights to different scores. A **weight** is a value corresponding to how much the score is counted. Given a list of scores $x_1, x_2, x_3, \ldots, x_n$ and a corresponding list of weights $w_1, w_2, w_3, \ldots, w_n$, the **weighted mean** is obtained by using Formula 2-2.

Formula 2-2

$$\text{weighted mean} = \frac{\Sigma(w \cdot x)}{\Sigma w}$$

where

$$w = \text{weight}$$

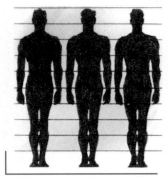

An Average Guy

The "average" American male is named Robert. He is 31 years old, 5 ft 9½ in. tall, weighs 172 lb, wears a size 40 suit, wears a size 9½ shoe, and has a 34-in. waist. Each year he eats 12.3 lb of pasta, 26 lb of bananas, 4 lb of potato chips, 18 lb of ice cream, and 79 lb of beef. Each year he also watches television for 2567 hours and gets 585 pieces of mail. After eating his share of potato chips, reading some of his mail, and watching some television, he ends the day with 7.7 hours of sleep. The next day begins with a 21-minute commute to a job at which he will work for 6.1 hours.

That is, first multiply each score by its corresponding weight; then find the total of the resulting products, thereby evaluating $\Sigma(wx)$. Finally, add the values of the weights to find Σw, and divide as indicated by Formula 2-2.

▶ **EXAMPLE**

A weighted mean is frequently used in the determination of a final average for a course that includes 4 tests plus a final examination. If the respective grades are 70, 80, 75, 85, and 90, the mean of 80 does not reflect the greater importance placed on the final exam. Suppose the instructor counts the respective tests as 15%, 15%, 15%, 15%, and 40%. Find the weighted mean.

● **SOLUTION**

Using the given scores and weights, we calculate the weighted mean by applying Formula 2-2.

$$\frac{(70 \cdot 15) + (80 \cdot 15) + (75 \cdot 15) + (85 \cdot 15) + (90 \cdot 40)}{100}$$

$$= \frac{1050 + 1200 + 1125 + 1275 + 3600}{100}$$

$$= \frac{8250}{100} = 82.5$$

Mean from a Frequency Table

Formula 2-2 can be modified so that we can approximate the mean from a frequency table. The body temperature data from Frequency Table 2-2 have been entered in Table 2-5, where we use the class marks as representative scores and the frequencies as weights. Then the formula for the weighted mean leads directly to Formula 2-3, which can be used to approximate the mean of a set of scores in a frequency table.

Formula 2-3

where

$$\bar{x} = \frac{\Sigma (f \cdot x)}{\Sigma f} \qquad \text{mean from frequency table}$$

$$x = \text{class mark}$$
$$f = \text{frequency}$$
$$\Sigma f = n$$

TABLE 2-5   Finding Σf and $\Sigma(f \cdot x)$

Temperature (°F)	Frequency	Class Mark x	$f \cdot x$
96.5–96.8	1	96.65	96.65
96.9–97.2	8	97.05	776.40
97.3–97.6	14	97.45	1364.30
97.7–98.0	22	97.85	2152.70
98.1–98.4	19	98.25	1866.75
98.5–98.8	32	98.65	3156.80
98.9–99.2	6	99.05	594.30
99.3–99.6	4	99.45	397.80
Total	106		10405.70

$$\uparrow \atop \Sigma f \qquad\qquad\qquad \uparrow \atop \Sigma(f \cdot x)$$

Class Size Paradox

There are at least two ways to obtain the mean class size, and they can have very different results. At one college, if we take the numbers of students in 737 classes, we get a mean of 40 students. But if we were to compile a list of the class sizes for each student and use this list, we would get a mean class size of 147. This large discrepancy is due to the fact that there are many students in large classes, while there are few students in small classes. Without changing the number of classes or faculty, we could reduce the mean class size experienced by students by making all classes about the same size. This would also improve attendance, which is better in smaller classes.

Formula 2-3 is really a variation of Formula 2-1, $\overline{x} = \Sigma x / n$. When data are summarized in a frequency table, Σf is the total number of scores, so $\Sigma f = n$. Also, $\Sigma(f \cdot x)$ is simply a quick way of adding up all of the scores. Formula 2-3 doesn't really involve a fundamentally different concept; it is simply a variation of Formula 2-1. We can now compute the weighted mean.

$$\overline{x} = \frac{\Sigma(f \cdot x)}{\Sigma f} = \frac{10405.7}{106} = 98.17$$

When we used the original collection of scores to calculate the mean directly, we obtained a mean of 98.20, so the value of the weighted mean obtained from the frequency table is just a little off. The procedure we use is justified by the fact that a class such as 96.5–96.8 can be represented by its class mark of 96.65, and the frequency number indicates that the representative score of 96.65 occurs one time. In essence, we are treating Table 2-2 as if it contained one score of 96.65, eight scores of 97.05, and so on.

The Best Measure of Central Tendency

We have found that for the data of Table 2-1, the mean, median, mode, and midrange have values of 98.20°F, 98.40°F, 98.60°F, and 98.05°F, respectively. Technically, someone could use any of the four preceding figures as the average

TABLE 2-6   Comparison of Mean, Median, Mode, and Midrange

Average	Definition	How Common?	Existence	Takes Every Score into Account?	Affected by Extreme Scores?	Advantages and Disadvantages
Mean	$\bar{x} = \dfrac{\Sigma x}{n}$	most familiar "average"	always exists	yes	yes	used throughout this book; works well with many statistical methods
Median	middle score	commonly used	always exists	no	no	often a good choice if there are some extreme scores
Mode	most frequent score	sometimes used	might not exist; may be more than one mode	no	no	appropriate for data at the nominal level
Midrange	$\dfrac{\text{high} + \text{low}}{2}$	rarely used	always exists	no	yes	very sensitive to extreme values

General comments:
* For a data collection that is approximately symmetric with one mode, the mean, median, mode, and midrange tend to be about the same.
* For a data collection that is obviously asymmetric, it would be good to report both the mean and median.
* The mean is relatively *reliable*. That is, when samples are drawn from the same population, the sample means tend to be more consistent than the other averages (consistent in the sense that the means of samples drawn from the same population don't vary as much as the other averages).

body temperature. The selection of the most representative measure is not always easy. The different measures of central tendency have different advantages and disadvantages, and there are no objective criteria that determine the most representative measure for all data sets. Some of the important advantages and disadvantages of the different measures are summarized in Table 2-6.

Skewness

A comparison of the mean, median, and mode can reveal information about the characteristic of skewness, as defined below and as illustrated in Figures 2-5 and 2-9.

DEFINITION

A distribution of data is **skewed** if it is not symmetric and extends more to one side than the other. (A distribution of data is **symmetric** if the left half of its histogram is roughly a mirror image of its right half.)

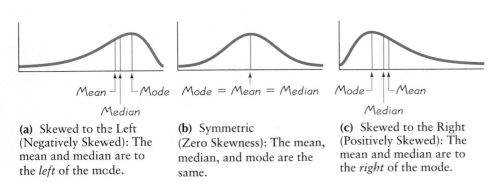

(a) Skewed to the Left (Negatively Skewed): The mean and median are to the *left* of the mode.

(b) Symmetric (Zero Skewness): The mean, median, and mode are the same.

(c) Skewed to the Right (Positively Skewed): The mean and median are to the *right* of the mode.

FIGURE 2-9 Skewness

Data skewed to the *left* are said to be **negatively skewed;** the mean and median are to the left of the mode. Although not always predictable, negatively skewed data generally have the mean to the left of the median. (See Figure 2-9a.) Data skewed to the *right* are said to be **positively skewed;** the mean and median are to the right of the mode. Again, although not always predictable, positively skewed data generally have the mean to the right of the median. (See Figure 2-9c.) If we examine the histogram in Figure 2-1 for the body temperature data we are considering in this chapter, we see a graph that is roughly bell-shaped without skewness, as in Figure 2-9b.

In reality, there are many such distributions of data that are symmetric and without skewness. Distributions skewed to the right are more common than those skewed to the left because it's often easier to get exceptionally large values than values that are exceptionally small. With annual incomes, for example, it's impossible to get values below the lower limit of zero, but there are a few people who earn millions of dollars in a year. Annual incomes therefore tend to be skewed to the right, as in Figure 2-9c.

If a data set is relatively small, the calculations required for finding measures of central tendency are not very difficult. For dealing with large data sets or with inconvenient numbers (numbers that are large or have many decimal places), it is helpful to use a statistical software package, such as STATDISK or Minitab. In Section 2-5 we will describe how to use statistical software to obtain descriptive statistics from data sets.

In this section we considered measures designed to identify a central or representative value. Specifically, we found that the mean body temperature for our sample data (Table 2-1) is 98.20°F, not the value of 98.6°F that is commonly used. But we also need a measure of the variation among the sample temperatures, and such a measure will be introduced in the following section. We are on our way to testing the researchers' claim that 98.6°F "should be abandoned as a concept relevant to clinical thermometry."

2-4   Exercises A: Basic Skills and Concepts

In Exercises 1–6, find the (a) mean, (b) median, (c) mode, and (d) midrange for the given sample data.

1. In Section 1 of an Introduction to Statistics course, 10 test scores were randomly selected, and the following results were obtained:

   74   73   77   77   71   68   65   77   67   66

2. In Section 2 of an Introduction to Statistics course, 10 test scores were randomly selected, and the following results were obtained:

   42   100   77   54   93   85   67   77   62   58

3. The blood alcohol concentrations of 15 drivers involved in fatal accidents and then convicted with jail sentences are given below (based on data from the U.S. Department of Justice):

   0.27   0.17   0.17   0.16   0.13   0.24   0.29   0.24
   0.14   0.16   0.12   0.16   0.21   0.17   0.18

4. The amounts of time (in hours) spent on paperwork in one day were obtained from a sample of office managers with the results given below (based on data from Adia Personnel Services):

   3.7   2.9   3.4   0.0   1.5   1.8   2.3   2.4   1.0   2.0
   4.4   2.0   4.5   0.0   1.7   4.4   3.3   2.4   2.1   2.1

5. The following are the ages of motorcyclists at the time they were fatally injured in traffic accidents (based on data from the U.S. Department of Transportation):

   17   38   27   14   18   34   16   42   28   24   40   20   23   31
   37   21   30   25   17   28   33   25   23   19   51   18   29

6. Given below are the ages of U.S. presidents when they were inaugurated:

   57   61   57   57   58   57   61   54   68   51   49   64   50   48
   65   52   56   46   54   49   51   47   55   55   54   42   51   56
   55   51   54   51   60   62   43   55   56   61   52   69   64   46

In Exercises 7–12, use the data randomly selected from the data sets in Appendix B, and find the (a) mean, (b) median, (c) mode, and (d) midrange.

7. Weights (in pounds) of plastic discarded by households in one week (data collected for the Garbage Project at the University of Arizona):

   2.19   2.10   1.41   0.63   0.92   1.40   1.74   2.87

8. Weights (in pounds) of paper discarded by households in one week (data collected for the Garbage Project at the University of Arizona):

$$11.42 \quad 15.09 \quad 13.61 \quad 16.39 \quad 8.08 \quad 9.46 \quad 10.58 \quad 11.03$$

9. Weights (in milligrams) of Bufferin tablets:

$$672.2 \quad 679.2 \quad 669.8 \quad 672.6 \quad 672.2 \quad 662.2$$
$$662.7 \quad 661.3 \quad 654.2 \quad 667.4 \quad 667.0 \quad 670.7$$

10. Weights (in milligrams) of randomly selected M&Ms:

$$0.957 \quad 0.912 \quad 0.842 \quad 0.925 \quad 0.939 \quad 0.886$$
$$0.914 \quad 0.913 \quad 0.958 \quad 0.947 \quad 0.920$$

11. Weights (in pounds) of new cars:

$$3536 \quad 3766 \quad 2526 \quad 2863 \quad 3315 \quad 2360 \quad 3253 \quad 3784 \quad 2992 \quad 2784$$

12. Family incomes (in dollars) (round answers to the nearest dollar):

10,400	67,000	26,400	7,400	6,400	23,600	7,800	34,800
26,800	17,200	47,400	22,000	25,400	15,000	40,000	

In Exercises 13–16, refer to the data set in Appendix B, and find the (a) mean, (b) median, (c) mode, and (d) midrange.

13. Data Set 2 in Appendix B: the body temperatures for 8:00 AM on day 1
14. Data Set 1 in Appendix B: the weights of discarded metal for all households
15. Data Set 1 in Appendix B: the weights of discarded paper for all households
16. Data Set 5 in Appendix B: the lengths of movies with a rating of R

In Exercises 17–20, find the mean of the data summarized in the given frequency table.

17. The given frequency table describes the speeds of drivers ticketed by the Town of Poughkeepsie police. These drivers were traveling through a 30 mi/h speed zone.

Speed	Frequency
42–43	14
44–45	11
46–47	8
48–49	6
50–51	4
52–53	3
54–55	1
56–57	2
58–59	0
60–61	1

18. The given frequency table describes the age distribution of a randomly selected sample of United States residents (based on data from the U.S. Bureau of the Census).

Age	Frequency
0–9	37
10–19	35
20–29	39
30–39	46
40–49	33
50–59	22
60–69	20
70–79	17
80–89	9
90–99	3

19. The given frequency table summarizes the ages of randomly selected U.S. commercial aircraft (based on information reported by Aviation Data Services).

Airplane Age (years)	Frequency
0.0–5.9	10
6.0–11.9	8
12.0–17.9	9
18.0–23.9	9
24.0–29.9	4

20. Insurance companies continually research ages at death and causes of death. The given frequency table summarizes ages at death for people who died from gunfire in America during a one-week period (based on a *Time* magazine study).

Age at Death	Frequency
16–25	22
26–35	10
36–45	6
46–55	2
56–65	4
66–75	5
76–85	1

2-4   Exercises B: Beyond the Basics

21. a. Find the mean, median, mode, and midrange for the following annual incomes (in dollars) of self-employed doctors (based on data from the American Medical Association):

       108,000   236,000   179,000   206,000   236,000

    b. Do part a after adding 20,000 to each income.
    c. In general, what is the effect of adding a constant k to every score? What is the effect of subtracting a constant k from every score?
    d. Find the mean, median, mode, and midrange of the given incomes after dividing each score by 1000.

(continued)

e. In general, if every score in a data set is divided or multiplied by some constant k, what is the effect on the mean, median, mode, and midrange?

22. a. A student receives quiz grades of 70, 65, and 90. The same student earns an 85 on the final examination. If each quiz constitutes 20% of the final grade and the final makes up 40%, find the weighted mean.

   b. A student earns the grades in the accompanying table. If grade points are assigned as A = 4, B = 3, C = 2, D = 1, F = 0 and the grade points are weighted according to the number of credit hours, find the weighted mean (grade-point average) rounded to three decimal places.

Course	Grade	Credit Hours
Math	A	4
English	C	3
Art	B	1
Physical education	A	2
Biology	C	4

23. Compare the averages that comprise the solutions to Exercises 1 and 2.
   a. Do these averages discriminate or differentiate between the two lists of scores?
   b. Is there any apparent difference between the two data sets that is not reflected in the averages?

24. Using the data from Exercise 1, change the first score (74) to 740, and find the mean, median, mode, and midrange of this modified data set. How does the extreme value affect these averages?

25. Data are sometimes transformed by the process of replacing each score x with $\log x$. Does

$$\log \overline{x} = \frac{\Sigma \log x}{n}$$

for a set of positive values? Explain.

26. The **harmonic mean** is often used as a measure of central tendency for data sets consisting of rates of change, such as speeds. To obtain the harmonic mean of a set of scores, divide the number of scores n by the sum of the reciprocals of all scores.

$$\text{harmonic mean} = \frac{n}{\Sigma \frac{1}{x}}$$

For example, the harmonic mean of 2, 3, 6, 7, 7, 8 is

$$\frac{6}{\frac{1}{2} + \frac{1}{3} + \frac{1}{6} + \frac{1}{7} + \frac{1}{7} + \frac{1}{8}} = \frac{6}{1.4} = 4.3$$

(Note that 0 cannot be included in the scores.)   *(continued)*

a. Find the harmonic mean of

$$2 \quad 3 \quad 6 \quad 7 \quad 7 \quad 8 \quad 9 \quad 9 \quad 9 \quad 10$$

b. Four students drive from New York to Florida (1200 miles) at a speed of 40 mi/h and return at a speed of 60 mi/h. What is their average speed for the round trip? (The harmonic mean is used in averaging speeds.)

c. A dispatcher of charter buses calculates the average round trip speed (in miles per hour) for a certain route. The results obtained for 14 different runs are listed below. Based on these values, what is the average speed of a bus assigned to this route?

$$
\begin{array}{ccccccc}
42.6 & 41.3 & 38.2 & 42.9 & 43.4 & 43.7 & 40.8 \\
34.2 & 40.1 & 41.2 & 40.5 & 41.7 & 39.8 & 39.6
\end{array}
$$

27. The **geometric mean** is often used in business and economics for finding average rates of change, average rates of growth, or average ratios. Given a collection of n scores (all of which are positive), the geometric mean is the nth root of their product. For example, to find the geometric mean of 2, 3, 6, 7, 7, 8, 9, 9, 9, 10, first multiply the scores.

$$2 \cdot 3 \cdot 6 \cdot 7 \cdot 7 \cdot 8 \cdot 9 \cdot 9 \cdot 9 \cdot 10 = 102,876,480$$

Then take the tenth root of the product, since there are 10 scores.

$$\sqrt[10]{102,876,480} = 6.3$$

The *average growth factor* for money compounded at annual interest rates of 10%, 8%, 9%, 12%, and 7% can be found by computing the geometric mean of 1.10, 1.08, 1.09, 1.12, and 1.07. Find that average growth factor.

28. The **quadratic mean** (or **root mean square,** or **R.M.S.**) is usually used in physical applications. In power distribution systems, for example, voltages and currents are usually referred to in terms of their R.M.S. values. The quadratic mean of a set of scores is obtained by squaring each score, adding the results, dividing by the number of scores n, and then taking the square root of that result.

$$\text{quadratic mean} = \sqrt{\frac{\Sigma x^2}{n}}$$

For example, the quadratic mean of 2, 3, 6, 7, 7, 8, 9, 9, 9, 10 is

$$\sqrt{\frac{4 + 9 + 36 + 49 + 49 + 64 + 81 + 81 + 81 + 100}{10}}$$

$$= \sqrt{\frac{554}{10}} = \sqrt{55.4}$$

$$= 7.4$$

Find the R.M.S. of the following power supplies (in volts): 151, 162, 0, 81, −68.

29. Frequency tables often have open-ended classes, such as the one given below. This table summarizes amounts of time spent studying by college freshmen (based on data from *The American Freshman* as reported in *USA Today*). Formula 2-3 cannot be directly applied because we can't determine a class mark for the class of "more than 20." Calculate the mean by assuming that this class is really (a) 21–25, (b) 21–30, and (c) 21–40. What can you conclude?

Hours Studying per Week	Frequency
0	5
1–5	96
6–10	57
11–15	25
16–20	11
More than 20	6

30. When data are summarized in a frequency table, the median can be found by first identifying the *median class* (the class that contains the median). We then assume that the scores in that class are evenly distributed and we can interpolate. This process can be described by

$$\text{(lower limit of median class)} + \text{(class width)}\left(\frac{\left(\dfrac{n+1}{2}\right) - (m+1)}{\text{frequency of median class}}\right)$$

   where n is the sum of all class frequencies and m is the sum of the class frequencies that *precede* the median class. Use this procedure and the data in Frequency Table 2-2 to find the median body temperature.

31. The mean is sensitive to extreme scores, but the **trimmed mean** is designed to be much less sensitive to them. To find the 10% trimmed mean for a data set, first arrange the data in order. Then delete the bottom 10% of the scores and the top 10% of the scores, and calculate the mean of the remaining scores.
   a.   Find the 10% trimmed mean for the data in Exercise 4.
   b.   Find the 20% trimmed mean for that data set.

32. Using an almanac, a researcher finds the average teacher's salary for each state. He adds those 50 values, then divides by 50 to obtain their mean. Is the result equal to the national average teacher's salary? Why or why not?

2-5   Measures of Variation

Sections 2-2 and 2-3 described frequency tables and histograms that are used to gain insight into the nature or shape of the distribution of data. Section 2-4 presented the basic measures of central tendency that are so important for obtaining a central or representative value. The present section is one of the most important sections in the book (the pressure is on) because it discusses the characteristic of variation (often referred to as *dispersion*), a concept that is critical to many methods of statistics.

Many banks once required that customers wait in separate lines at each teller's window, but they have now changed to one single main waiting line. Why did they make that change? The mean waiting time didn't change because the waiting line configuration doesn't affect the efficiency of the tellers. They changed to the single line because customers prefer waiting times that are more *consistent* with less variation. In this case, thousands of banks made a change that resulted in lower variation (and happier customers), even though the mean was not affected.

In this section we will begin with a brief discussion intended to illustrate the general concept of variation; then we will define important measures of variation, including the range, standard deviation, and variance. After describing the computation of these measures, we will proceed to consider ways of understanding or interpreting the values they produce.

First, consider the general concept of variation among scores by considering the waiting times (in minutes) for samples of 10 customers at the Jefferson Valley Bank and the neighboring Bank of Providence. Those times are listed below.

Jefferson Valley Bank	6.5 6.6 6.7 6.8 7.1 7.3 7.4 7.7 7.7 7.7
Bank of Providence	4.2 5.4 5.8 6.2 6.7 7.7 7.7 8.5 9.3 10.0

Customers at the Jefferson Valley Bank enter a single waiting line that feeds three teller windows. Customers at the Bank of Providence can enter any one of three different lines that have formed at three different teller windows. If we use the methods of Section 2-4, we get these results:

	Jefferson Valley Bank	Bank of Providence
Mean	7.15	7.15
Median	7.20	7.20
Mode	7.7	7.7
Midrange	7.10	7.10

The above results show that the two banks have the same measures of central tendency so, on average, customers wait the same time at both banks. Based on a comparison of the measures of central tendency alone, we can see no difference between the two banks. Yet scanning the original sample scores should reveal a fundamental difference: The Jefferson Valley Bank has waiting times with much less variation than the times for the Bank of Providence. If all other characteristics are equal, customers are likely to prefer the Jefferson Valley Bank, where they won't become annoyed by being caught in an individual line that is much slower than the others.

By intuitively comparing the differences in variation between the waiting times of the two banks, we get a general sense of the characteristic of variation. Let's now proceed to develop some specific ways of actually measuring variation. We begin with the range.

Range

One measure of variation is the range, which has the advantage of being very easy to compute. The **range** is simply the difference between the highest value and the lowest value. For the Jefferson Valley Bank customers, the range is the difference between 7.7 min and 6.5 min, which is 1.2 min. The range of the Bank of Providence waiting times is $10.0 - 4.2 = 5.8$ min. This much larger range suggests greater variation.

The range is very easy to compute, but, because it depends on only the highest and lowest scores, it's often inferior to other measures of variation that use the value of every score. For example, the Group A scores in Table 2-7 have a range of 19, whereas the Group B scores have a range of 16, suggesting that there is more variation in Group A. But the Group A scores are very close together, while those in Group B have more variation. The range may be misleading in this case and in many other cases because it uses only the maximum and minimum scores.

TABLE 2-7	Groups of Data
Group A	Group B
1	2
20	3
20	4
20	5
20	6
20	14
20	15
20	16
20	17
20	18
Range = 19	Range = 16

Standard Deviation and Variance

Another measure of variation is the standard deviation, which is generally the most important and useful such measure. It is defined below, but a complete understanding of this definition will require careful reading of the remainder of this section.

DEFINITION

The **standard deviation** of a set of *sample* scores is a measure of variation of scores about the mean. It is calculated by using Formula 2-4.

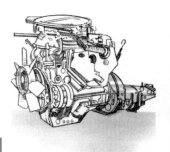

Lower Variation, Higher Quality

Ford and Mazda were producing similar transmissions that were supposed to be made with the same specifications. But the American-made transmissions required more warranty repairs than the Japanese-made transmissions. When investigators inspected samples of the Japanese transmission gearboxes, they first thought that their measuring instruments were defective because they weren't detecting any variability among the Mazda transmission gearboxes. They realized that although the American transmissions were within the specifications, the Mazda transmissions were not only within the specifications, but consistently close to the desired value. By reducing variability among transmission gearboxes, Mazda reduced the costs of inspection, scrap, rework, and warranty repair.

Formula 2-4 $s = \sqrt{\dfrac{\Sigma(x - \bar{x})^2}{n - 1}}$ sample standard deviation

Why define a measure of variation in the way described by Formula 2-4? In measuring variation in a set of sample data, one reasonable approach is to begin with the individual amounts by which scores deviate from the mean. For a particular score x, the amount of **deviation** is the amount of difference between the score and the mean, which can be expressed as $x - \bar{x}$. However, the sum of all such deviations is always zero. To get a measure that isn't always zero, we could take absolute values, as in $\Sigma|x - \bar{x}|$. If we find the mean of that sum, we get the **mean deviation** (or absolute deviation) described by the following expression:

$$\frac{\Sigma|x - \bar{x}|}{n}$$ mean deviation

Instead of using absolute values, we can obtain a better measure of variation by making all deviations $(x - \bar{x})$ nonnegative by squaring them. Finally, we take the square root to compensate for that squaring. As a result, *the standard deviation has the same units of measurement as the original scores.* For example, if customer waiting times are in minutes, the standard deviation of those times will also be in minutes. Based on the format of Formula 2-4, we can describe the procedure for calculating the standard deviation as follows.

Procedure for Finding the Standard Deviation with Formula 2-4

Step 1: Find the mean of the scores $(\bar{x})$.

Step 2: Subtract the mean from each individual score $(x - \bar{x})$.

Step 3: Square each of the differences obtained from Step 2. That is, multiply each value by itself. [This produces numbers of the form $(x - \bar{x})^2$.]

Step 4: Add all of the squares obtained from Step 3 to get $\Sigma(x - \bar{x})^2$.

Step 5: Divide the total from Step 4 by the number $(n - 1)$—that is, 1 less than the total number of scores present.

Step 6: Find the square root of the result of Step 5.

▶ **EXAMPLE**

Find the standard deviation of the Jefferson Valley Bank customer waiting times. Those times (in minutes) are reproduced below:

6.5   6.6   6.7   6.8   7.1   7.3   7.4   7.7   7.7   7.7

● **SOLUTION**

See Table 2-8, where the following steps are executed. It is often helpful to use the same vertical format shown in that table.

Step 1:   Obtain the mean of 7.15 by adding the scores and then dividing by the number of scores:

$$\bar{x} = \frac{\Sigma x}{n} = \frac{71.5}{10} = 7.15$$

Step 2:   Subtract the mean of 7.15 from each score to get -0.65, -0.55, ..., 0.55. These are the values of $x - \bar{x}$. (As a quick check, remember that these numbers must always total 0.)

Step 3:   Square each value obtained in Step 2 to get 0.4225, 0.3025, ..., 0.3025. These are the values of $(x - \bar{x})^2$.

Step 4:   Sum all of the preceding scores to get 2.045. That is,

$$\Sigma(x - \bar{x})^2 = 2.045$$

Step 5:   There are $n = 10$ scores, so divide by 1 less than 10:

$$2.045 \div 9 = 0.2272$$

Step 6:   Find the square root of 0.2272. The standard deviation is $\sqrt{0.2272} = 0.48$ min.   (*continued*)

TABLE 2-8   Calculating Standard Deviation for Jefferson Valley Bank Customers

x	$x - \bar{x}$	$(x - \bar{x})^2$
6.5	-0.65	0.4225
6.6	-0.55	0.3025
6.7	-0.45	0.2025
6.8	-0.35	0.1225
7.1	-0.05	0.0025
7.3	0.15	0.0225
7.4	0.25	0.0625
7.7	0.55	0.3025
7.7	0.55	0.3025
7.7	0.55	0.3025
Totals: 71.5		2.0450

$$\downarrow \qquad\qquad\qquad \downarrow$$

$$\bar{x} = \frac{71.5}{10} = 7.15 \qquad\qquad s = \sqrt{\frac{2.0450}{10 - 1}} = \sqrt{0.2272} = 0.48$$

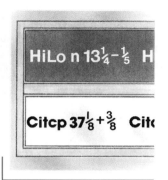

More Stocks, Less Risk

In their book *Investments,* authors Zvi Bodie, Alex Kane, and Alan Marcus state that "the average standard deviation for returns of portfolios composed of only one stock was 0.554. The average portfolio risk fell rapidly as the number of stocks included in the portfolio increased." They note that with 32 stocks, the standard deviation is 0.325, indicating much less variation and risk. They make the point that with only a few stocks, a portfolio has a high degree of "firm-specific" risk, meaning that the risk is attributable to the few stocks involved. With more than 30 stocks, there is very little firm-specific risk; instead, almost all of the risk is "market risk," attributable to the stock market as a whole. They note that these principles are "just an application of the well-known law of averages."

Ideally, we should interpret the meaning of that result, but such interpretations will be discussed a little later in this section. For now, practice calculating a standard deviation by using the customer waiting times given above for the Bank of Providence. Using those times, verify that the standard deviation is 1.82 min. Although the interpretations of those standard deviations will be discussed later, we can now compare them and note that the standard deviation of the times for the Jefferson Valley Bank (0.48 min) is much lower than the standard deviation for the Bank of Providence (1.82 min). This supports our intuitive observation that the waiting times at the Jefferson Valley Bank have much less variation than those at the Bank of Providence.

In our definition of standard deviation, we referred to the standard deviation of *sample* data. A slightly different formula is used to calculate the standard deviation σ of a population: Instead of dividing by $n - 1$, divide by the population size N, as in the following expression.

$$\sigma = \sqrt{\frac{\Sigma(x - \mu)^2}{N}} \qquad \text{population standard deviation}$$

For example, if the 10 scores in Table 2-8 constitute a *population,* the standard deviation is as follows:

$$\sigma = \sqrt{\frac{\Sigma(x - \mu)^2}{N}} = \sqrt{\frac{2.0450}{10}} = 0.45$$

Because we generally deal with sample data, we will usually use Formula 2-4, in which we divide by $n - 1$. Many calculators do standard deviations, with division by $n - 1$ corresponding to a key labeled σ_{n-1} or s, while the key labeled σ_n or σ corresponds to division by N. For some creative but strange reason, calculators use a variety of different notations; the following notations, however, are very standard in statistics. These notations include reference to the *variance* of a set of scores, and we will now proceed to describe that measure of variation.

Notation	
s	denotes the standard deviation of a set of *sample* data
σ	denotes the standard deviation of a set of *population* data
s^2	denotes the variance of a set of *sample* data
σ^2	denotes the variance of a set of *population* data

Note: Articles in professional journals and reports often use SD for standard deviation and Var for variance.

If we omit Step 6 in the procedure for calculating the standard deviation, we get the **variance**, defined in Formula 2-5.

Formula 2-5
$$s^2 = \frac{\Sigma(x - \overline{x})^2}{n - 1}$$
sample variance

Similarly, we can express the population variance as

$$\sigma^2 = \frac{\Sigma(x - \mu)^2}{N}$$
population variance

By comparing Formulas 2-4 and 2-5, we see that the variance is the square of the standard deviation. Although the variance will be used later in the book, we should concentrate first on the concept of standard deviation as we try to get a feeling for this statistic. A major difficulty with the variance is that it is not in the same units as the original data. For example, a data set might have a standard deviation of $3.00 and a variance of 9.00 *square* dollars. Since we can't relate well to square dollars, we find it difficult to understand variance.

Round-Off Rule

As in Section 2-4, we round off answers by carrying one more decimal place than was present in the original data. We should round only the final answer and not intermediate values. (If we must round intermediate results, we should carry at least twice as many decimal places as will be used in the final answer.)

Shortcut Formula and Grouped Data   We will now present two additional formulas for standard deviation, but these formulas do not involve a different concept; they are only different versions of Formula 2-4. First, Formula 2-4 can be expressed in the following equivalent form.

Formula 2-6
$$s = \sqrt{\frac{n(\Sigma x^2) - (\Sigma x)^2}{n(n - 1)}}$$

Formulas 2-4 and 2-6 are equivalent in the sense that they will always produce the same results. Algebra can be used to show that they are equal. Formula 2-6 is called the *shortcut* formula because it tends to be more convenient to use with messy numbers or with large sets of data. Also, Formula 2-6 is often used in calculators and computer programs because it requires only three memory registers (for n, Σx, and Σx^2), instead of a separate memory register for every individual score. However, many instructors prefer to use only Formula 2-4 for calculating standard deviations. They argue that Formula 2-4 reinforces the concept that the standard deviation is a type of average deviation, while Formula 2-6 obscures that idea. Other instructors have no objections to Formula 2-6. We have included the shortcut formula so that it is available for those who choose to use it. We have already presented an example illustrating the calculation of a standard deviation using Formula 2-4, and the following example illustrates the use of Formula 2-6.

Mail Consistency

A recent survey of 29,000 people who use the U.S. Postal Service revealed that they wanted better *consistency* in the time it takes to make a delivery. Now, a local letter could take one day or several days. *USA Today* reported a common complaint: "Just tell me how many days ahead I have to mail my mother's birthday card."

The level of consistency can be measured by the standard deviation of the delivery times. A lower standard deviation reflects more consistency. The standard deviation is often a critically important tool used to monitor and control the quality of goods and services.

▶ **EXAMPLE**

Find the standard deviation of the following Jefferson Valley Bank customer waiting times (in minutes), using Formula 2-6:

6.5   6.6   6.7   6.8   7.1   7.3   7.4   7.7   7.7   7.7

● **SOLUTION**

Formula 2-6 requires that we find the values of n, Σx, and Σx^2. Because there are 10 scores, we have $n = 10$. The sum of the 10 scores is 71.5, so $\Sigma x = 71.5$. The third required component is calculated as follows:

$$\Sigma x^2 = 6.5^2 + 6.6^2 + 6.7^2 + \cdots + 7.7^2$$
$$= 42.25 + 43.56 + 44.89 + \cdots + 59.29$$
$$= 513.27$$

Formula 2-6 can now be used to find the value of the standard deviation.

$$s = \sqrt{\frac{n(\Sigma x^2) - (\Sigma x)^2}{n(n-1)}}$$

$$= \sqrt{\frac{10(513.27) - (71.5)^2}{10(10-1)}}$$

$$= \sqrt{\frac{20.45}{90}}$$

$$= 0.4766783 = 0.48 \text{ min (rounded)}$$

We can develop a formula for standard deviation when the data are summarized in a frequency table. The result is as follows:

$$s = \sqrt{\frac{\Sigma f \cdot (x - \bar{x})^2}{n-1}}$$

We will express this formula in an equivalent expression that usually simplifies the actual calculations.

Formula 2-7 $s = \sqrt{\dfrac{n[\Sigma(f \cdot x^2)] - [\Sigma(f \cdot x)]^2}{n(n-1)}}$ standard deviation for frequency table

where x = class mark
f = frequency
n = sample size (or sum of the frequencies)

▶ **EXAMPLE**

Estimate the standard deviation of 106 body temperatures by using Frequency Table 2-2 and Formula 2-7.

● **SOLUTION**

Application of Formula 2-7 requires that we find the values of n, $\Sigma(f \cdot x)$, and $\Sigma(f \cdot x^2)$ (see Table 2-9). Formula 2-7 can now be used as follows:

$$s = \sqrt{\frac{n[\Sigma(f \cdot x^2)] - [\Sigma(f \cdot x)]^2}{n(n-1)}}$$

$$= \sqrt{\frac{106(1,021,536.72) - (10,405.70)^2}{106(106-1)}}$$

$$= \sqrt{\frac{4,299.83}{11,130}} = \sqrt{0.3863}$$

$$= 0.62°F$$

The 106 body temperatures have a standard deviation estimated to be 0.62°F.

TABLE 2-9   Calculating Standard Deviation from a Frequency Table

Score	Frequency	Class Mark	$f \cdot x$	$f \cdot x^2$
96.5–96.8	1	96.65	96.65	9,341.22
96.9–97.2	8	97.05	776.40	75,349.62
97.3–97.6	14	97.45	1364.30	132,951.04
97.7–98.0	22	97.85	2152.70	210,641.70
98.1–98.4	19	98.25	1866.75	183,408.19
98.5–98.8	32	98.65	3156.80	311,418.32
98.9–99.2	6	99.05	594.30	58,865.42
99.3–99.6	4	99.45	397.80	39,561.21
Total	106		10,405.70	1,021,536.72

$$n = \Sigma f = 106 \qquad \Sigma(f \cdot x) \qquad \Sigma(f \cdot x^2)$$

Understanding Standard Deviation   We will now attempt to make some intuitive sense of the standard deviation. First, we should clearly understand that the standard deviation measures the dispersion, or variation, among scores. Scores close together will yield a small standard deviation, whereas scores spread farther apart will yield a larger standard deviation. Figure 2-10 shows that as the data spread farther apart, the corresponding values of the standard deviation increase.

Because variation is such an important concept and because the standard deviation is such an important tool in measuring variation, we will consider three different ways of developing a sense for values of standard deviations. The first is the range rule of thumb, a rough estimate.

Range Rule of Thumb

The range of a set of data is approximately 4 standard deviations ($4s$) wide. This can be expressed as follows:

Samples	Populations
range $\approx 4s$	range $\approx 4\sigma$
or	or
$s \approx \dfrac{\text{range}}{4}$	$\sigma \approx \dfrac{\text{range}}{4}$

A consequence of the above estimates is that for data sets having a distribution with a symmetric shape, the maximum and minimum scores can be estimated as follows:

Samples	Populations
minimum $\approx \overline{x} - 2s$	minimum $\approx \mu - 2\sigma$
maximum $\approx \overline{x} + 2s$	maximum $\approx \mu + 2\sigma$

The above expressions for minimum and maximum scores make sense when we recognize that for data sets with a symmetric distribution, the mean should be at the center of values that span a range of $4s$; that is, the minimum value is the mean minus one-half of the range, or the mean minus 2 standard deviations. (In Figure 2-11 we illustrate this important principle for a data set with the common bell-shaped distribution, which is symmetric about a vertical line through the mean.) The maximum value can be estimated as the mean plus one-half of the range, or the mean plus 2 standard deviations. When calculating a standard deviation using Formula 2-4, you can use the range rule of thumb as a check on your result, but realize that although the approximation will get you in the general vicinity of the answer, it can be off by a fairly large amount.

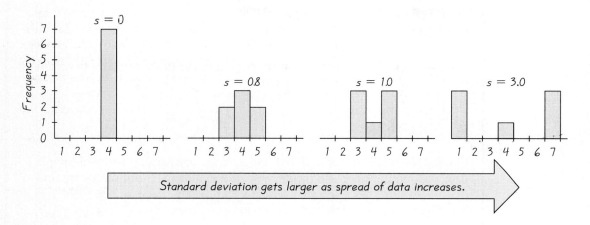

FIGURE 2-10   Same Means, Different Standard Deviations

For the data in the preceding example, the range is $7.7 - 6.5 = 1.2$, so we could use the range rule of thumb to get a rough estimate of s as follows:

$$s \approx \frac{\text{range}}{4} = \frac{1.2}{4} = 0.3$$

We found that the standard deviation s is actually 0.48, so the range rule of thumb gives us an estimate of 0.3 that is a bit too low here. However, our estimate does confirm that we are in the general ballpark, and we would know that a value such as 7 was probably not correct.

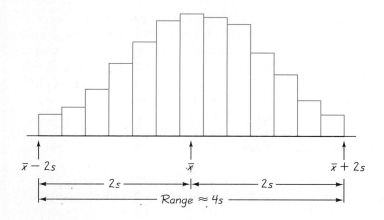

FIGURE 2-11   Range Rule of Thumb

Where Are the 0.400 Hitters?

The last baseball player to hit above 0.400 was Ted Williams, who hit 0.406 in 1941. There were averages above 0.400 in 1876, 1879, 1887, 1894, 1895, 1896, 1897, 1899, 1901, 1911, 1920, 1922, 1924, 1925, and 1930, but none since 1941. Are there no longer great hitters? Harvard's Stephen Jay Gould notes that the mean batting average has been steady at 0.260 for about 100 years, but the standard deviation has been decreasing from 0.049 in the 1870s to 0.031, where it is now. He argues that today's stars are as good as those from the past, but consistently better pitchers now keep averages below 0.400. Dr. Gould discusses this in Program 4 of the series *Against All Odds: Inside Statistics.*

▶ **EXAMPLE**

Use the range rule of thumb to find a rough estimate of the standard deviation of the sample of 106 body temperatures listed in Table 2-1.

● **SOLUTION**

In using the range rule of thumb to estimate the standard deviation of sample data, we find the range and divide by 4. By scanning the list of scores, we find that the lowest is 96.5 and the highest is 99.6, so the range is $99.6 - 96.5 = 3.1$. The standard deviation s is estimated as follows:

$$s \approx \frac{\text{range}}{4} = \frac{3.1}{4} = 0.78°\text{F}$$

This result is in the general ballpark of the correct value of 0.62 that is obtained by calculating the standard deviation with Formula 2-4.

The preceding example illustrated how we can use known information about the range to estimate the standard deviation. The following example is particularly important as an illustration of one way to *interpret* the value of a standard deviation.

▶ **EXAMPLE**

A sample of quarters minted by the United States government has a distribution that is bell-shaped, with a mean of 5.62 g and standard deviation of 0.68 g. Using the range rule of thumb, estimate the (a) range and (b) weights of the lightest and heaviest quarters.

● **SOLUTION**

a. We can estimate the range as follows:

$$\text{range} \approx 4s = 4(0.68) = 2.72 \text{ g}$$

b. The weights of the lightest (minimum) and heaviest (maximum) quarters can be estimated as follows:

$$\text{minimum} \approx \bar{x} - 2s = 5.62 - 2(0.68) = 4.26 \text{ g}$$
$$\text{maximum} \approx \bar{x} + 2s = 5.62 + 2(0.68) = 6.98 \text{ g}$$

Remember, these results are all rough estimates, but by knowing the mean and standard deviation, we are able to approximate the range of values as well as the lowest and highest values, thereby developing a much better sense of the data.

In attempting to understand or interpret a value of a standard deviation, we might also consider Chebyshev's theorem, which can be used with any set of data.

Chebyshev's Theorem

The *proportion* (or fraction) of any set of data lying within K standard deviations of the mean is always *at least* $1 - 1/K^2$, where K is any positive number greater than 1. For $K = 2$ and $K = 3$, we get the following two specific results.

- At least 3/4 (or 75%) of all scores will fall within the interval from 2 standard deviations below the mean to 2 standard deviations above the mean ($\bar{x} - 2s$ to $\bar{x} + 2s$).
- At least 8/9 (or 89%) of all scores will fall within 3 standard deviations of the mean ($\bar{x} - 3s$ to $\bar{x} + 3s$).

Another rule helpful in interpreting a value for a standard deviation is the **empirical rule,** which applies only to a data set having a distribution that is approximately bell-shaped, as in Figure 2-12. This figure shows how the mean and standard deviation of the data can be related to the proportion of data falling within certain limits. For example, data with a bell-shaped distribution will have about 0.34, or 34%, of its values in the interval between the mean ($\bar{x}$) and 1 standard deviation above the mean ($\bar{x} + s$). Similarly, we can expect only about 0.001, or 0.1%, of the values to be above the point identified by $\bar{x} + 3s$ (3 standard deviations above the mean). This rule is very important because statisticians often use measures based on standard deviations.

The empirical rule is often stated in an abbreviated form, sometimes called the *68-95-99 rule*.

- About 68% of all scores fall within 1 standard deviation of the mean.
- About 95% of all scores fall within 2 standard deviations of the mean.
- About 99.7% of all scores fall within 3 standard deviations of the mean.

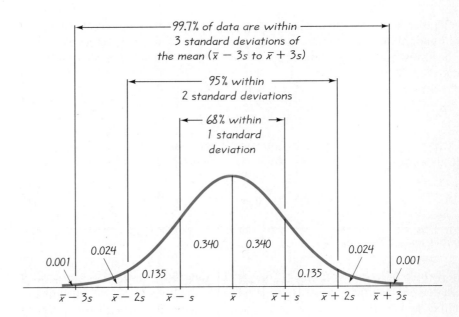

FIGURE 2-12   The Empirical Rule

▶ **EXAMPLE**

IQ scores have a distribution that is bell-shaped; the mean is 100, and the standard deviation is 15. Find the percentage of IQ scores between 70 and 130 by using (a) Chebyshev's theorem and (b) the empirical rule.

● **SOLUTION**

The key to solving parts a and b is to recognize that 70 and 130 are each 2 standard deviations away from the mean. Because the standard deviation is $s = 15$, it follows that $2s = 30$, so being within 2 standard deviations of the mean is equivalent to being within 30 IQ points of 100 (or between 70 and 130).

a. The IQ scores between 70 and 130 are within 2 standard deviations of the mean, and Chebyshev's theorem (for $K = 2$) states that at least 3/4 (or 75%) of all scores are within that interval.

b. The IQ scores between 70 and 130 are within 2 standard deviations of the mean, and the empirical rule states that 95% of the scores are within that interval.

In the preceding example, the empirical rule reveals that 95% of IQ scores fall between 70 and 130. One consequence of that result is that IQ scores less than 70 or more than 130 are relatively rare because they occur less than 5% of the time. An IQ of 147, for example, is considered unusually high. *In statistics, this ability to use the sample distribution, mean, and standard deviation to differentiate between usual and unusual scores is extremely important.* It will be used as the basis for some very important methods introduced in later chapters. For now, it is extremely important to develop an understanding of the standard deviation as a measure of variation.

After studying this section, you should understand that the standard deviation is a measure of variation among scores. Given sample data, you should be able to compute the value of the standard deviation. (For the body temperature data of Table 2-1, there are 106 scores so the calculation is messy; however, statistical software packages, such as STATDISK and Minitab, can be used to obtain the result of 0.62°F. The following section will include instructions and displays for these software packages.) You should be able to interpret the values of standard deviations that you compute. You should recognize that for many data sets, it is unusual for a score to differ from the mean by more than 2 or 3 standard deviations. [This suggests that a body temperature of 102.2°F is highly unusual because it differs from the mean of 98.20°F by 4.0°F, which is 6.45 standard deviations ($4.0 \div 0.62 = 6.45$) away from the mean. Because that temperature is so many standard deviations away from the mean, someone with a body temperature of 102.2°F is considered to be sick with a fever; it's very unlikely that such a high temperature will occur in a normal healthy person.] The mean gives us a central value for data, but the standard deviation gives us a measure of variation that we can use to differentiate between values that are usual and those that are unusual.

2-5 Exercises A: Basic Skills and Concepts

In Exercises 1–6, find the range, variance, and standard deviation for the given data. (The same data were used in Section 2-4 where we found measures of central tendency. Here we find measures of variation.)

1. In Section 1 of an Introduction to Statistics course, 10 test scores were randomly selected, and the following results were obtained:

    74 73 77 77 71 68 65 77 67 66

2. In Section 2 of an Introduction to Statistics course, 10 test scores were randomly selected, and the following results were obtained:

    42 100 77 54 93 85 67 77 62 58

3. The blood alcohol concentrations of 15 drivers involved in fatal accidents and

then convicted with jail sentences are given below (based on data from the U.S. Department of Justice):

0.27   0.17   0.17   0.16   0.13   0.24   0.29   0.24
0.14   0.16   0.12   0.16   0.21   0.17   0.18

4. The amounts of time (in hours) spent on paperwork in one day were obtained from a sample of office managers with the results given below (based on data from Adia Personnel Services):

3.7   2.9   3.4   0.0   1.5   1.8   2.3   2.4   1.0   2.0
4.4   2.0   4.5   0.0   1.7   4.4   3.3   2.4   2.1   2.1

5. The following are the ages of motorcyclists at the time they were fatally injured in traffic accidents (based on data from the U.S. Department of Transportation):

17   38   27   14   18   34   16   42   28
24   40   20   23   31   37   21   30   25
17   28   33   25   23   19   51   18   29

6. Given below are the ages of U.S. presidents when they were inaugurated:

57   61   57   57   58   57   61   54   68   51   49   64   50   48
65   52   56   46   54   49   51   47   55   55   54   42   51   56
55   51   54   51   60   62   43   55   56   61   52   69   64   46

In Exercises 7–12, use the data randomly selected from the data sets in Appendix B, and find the range, variance, and standard deviation.

7. Weights (in pounds) of plastic discarded by households in one week (data collected for the Garbage Project at the University of Arizona):

2.19   2.10   1.41   0.63   0.92   1.40   1.74   2.87

8. Weights (in pounds) of paper discarded by households in one week (data collected for the Garbage Project at the University of Arizona):

11.42   15.09   13.61   16.39   8.08   9.46   10.58   11.03

9. Weights (in milligrams) of Bufferin tablets:

672.2   679.2   669.8   672.6   672.2   662.2
662.7   661.3   654.2   667.4   667.0   670.7

10. Weights (in grams) of randomly selected M&Ms:

0.957   0.912   0.842   0.925   0.939   0.886   0.914   0.913   0.958
0.947   0.920

11. Weights (in pounds) of new cars:

3536   3766   2526   2863   3315   2360   3253   3784   2992   2784

12. Family incomes (in dollars) (round answers to nearest dollar):

10,400   67,000   26,400   7,400   6,400   23,600   7,800   34,800
26,800   17,200   47,400   22,000   25,400   15,000   40,000

In Exercises 13–16, refer to the data set in Appendix B, and find (a) the estimate of the standard deviation based on the range rule of thumb ($s \approx$ range$/4$) and (b) the standard deviation.

13. Data Set 2 in Appendix B: the body temperatures for 8:00 AM on day 1
14. Data Set 1 in Appendix B: the weights of discarded metal for all households
15. Data Set 1 in Appendix B: the weights of discarded paper for all households
16. Data Set 5 in Appendix B: the lengths of movies with a rating of R

In Exercises 17–20, find the standard deviation of the data summarized in the given frequency table.

17. The given frequency table describes the speeds of drivers ticketed by the Town of Poughkeepsie police. These drivers were traveling through a 30 mi/h speed zone.

Speed	Frequency
42–43	14
44–45	11
46–47	8
48–49	6
50–51	4
52–53	3
54–55	1
56–57	2
58–59	0
60–61	1

18. The frequency table describes the age distribution of a randomly selected sample of United States residents (based on data from the U.S. Bureau of the Census).

Age	Frequency
0–9	37
10–19	35
20–29	39
30–39	46
40–49	33
50–59	22
60–69	20
70–79	17
80–89	9
90–99	3

19. The frequency table summarizes ages of randomly selected U.S. commercial aircraft (based on information reported by Aviation Data Services).

Airplane Age (years)	Frequency
0.0– 5.9	10
6.0–11.9	8
12.0–17.9	9
18.0–23.9	9
24.0–29.9	4

20. Insurance companies continually research ages at death and causes of death. The given frequency table summarizes ages at death for people who died from gunfire in America during a one-week period (based on a *Time* magazine study).

Age at Death	Frequency
16–25	22
26–35	10
36–45	6
46–55	2
56–65	4
66–75	5
76–85	1

2-5   Exercises B: Beyond the Basics

21. a. Add 15 to each value given in Exercise 1, and then find the range, variance, and standard deviation. Compare the results to those of Exercise 1, and then form a general conclusion about the effect of adding a constant.

   b. Subtract 5 from each value given in Exercise 1, and then find the range, variance, and standard deviation. Compare the results to those of Exercise 1, and then form a general conclusion about the effect of subtracting a constant.

   c. Multiply each value given in Exercise 1 by 10, and then find the range, variance, and standard deviation. Compare the results to those of Exercise 1, and then form a general conclusion about the effect of multiplication by a constant.

   d. Halve each value in Exercise 1, and then find the range, variance, and standard deviation. Compare the results to those of Exercise 1, and then form a general conclusion about the effect of division by a constant.

22. a. Is it possible for a set of scores to have a standard deviation of zero? If so, how?

   b. Is it possible for a set of scores to have a negative standard deviation?

23. Test the effect of an *outlier* (an extreme value) by changing the 74 in Exercise 1 to 740. Calculate the range, variance, and standard deviation for the modified set, and compare the results to those originally obtained in Exercise 1.

24. Given a mean of 100 and a standard deviation of 15, what does Chebyshev's theorem say about the number of scores between 55 and 145? Between 85 and 115?

25. A large set of sample scores yields a mean and standard deviation of $\bar{x} = 56.0$ and $s = 4.0$, respectively. The distribution of the histogram is roughly bell-shaped. Use the empirical rule to answer the following:

   a. What percentage of the scores should fall between the values of 52.0 and 60.0?

   b. What percentage of the scores should fall within 8.0 of the mean?

(continued)

c.   About 99.7% of the scores should fall between what two values? (The mean of 56.0 should be midway between those two values.)

26. If we consider the values 1, 2, 3, . . . , n to be a population, the standard deviation can be calculated by the formula

$$\sigma = \sqrt{\frac{n^2 - 1}{12}}$$

This formula is equivalent to Formula 2-4 modified for division by n instead of $n - 1$, where the data set consists of the values 1, 2, 3, . . . , n.
   a.   Find the standard deviation of the population 1, 2, 3, . . . , 100.
   b.   Find an expression for calculating the *sample* standard deviation s for the sample values 1, 2, 3, . . . , n.

27. Find the mean and standard deviation for the scores 18, 19, 20, . . . , 182. Treat those values as a population (see Exercise 26).

28. Computers commonly use a random-number generator which produces values between 0.00000000 and 0.99999999. In the long run, all values occur with the same relative frequency. Find the mean and standard deviation for such numbers. (*Hint:* See Exercises 21c and 26.)

29. For the body temperature data listed in Table 2-1, the following statistics are calculated: $\bar{x} = 98.20°F$ and $s = 0.62°F$. Find the values of $\bar{x}$ and s for the same data set after each temperature has been converted to the Celsius scale. [*Hint:* $C = 5(F - 32)/9$.]

30. a.   The **coefficient of variation,** expressed as a percent, is used to describe the standard deviation relative to the mean. It allows us to compare variability of data sets with different measurement units (such as feet versus minutes) and it is calculated as follows:

$$\frac{s}{\bar{x}} \cdot 100 \qquad or \qquad \frac{\sigma}{\mu} \cdot 100$$

Find the coefficient of variation for the following sample scores:

2   2   2   3   5   8   12   19   22   30

   b.   Genichi Taguchi developed a method of improving quality and reducing manufacturing costs through a combination of engineering and statistics. A key tool in the Taguchi method is the **signal-to-noise ratio.** The simplest way to calculate this ratio is to divide the mean by the standard deviation. Find the *signal-to-noise* ratio for the sample data given in part a.

31. In Section 2-4 we introduced the general concept of skewness. Skewness can be measured by **Pearson's index of skewness:**

$$I = \frac{3(\bar{x} - \text{median})}{s}$$

If $I \geq 1.00$ or $I \leq -1.00$, the data can be considered to be *significantly skewed*. Find Pearson's index of skewness for the body temperature data given in Table 2-1, and then determine whether there is significant skewness.

2-6 Measures of Position

The preceding section ended with a discussion of how we can use standard deviations to better understand data. We noted, for example, that a body temperature of 102.2°F is extremely unusual for a normal healthy person because it is so many standard deviations (6.45) away from the mean. The standard score (or z score) is helpful in identifying such unusual data values. It is also helpful in comparing scores taken from different populations with different means and standard deviations. We begin this section with a discussion of z scores and then consider quartiles, deciles, and percentiles.

z Scores

In attempting to determine whether an individual score is within the realm of the typical or is unusual or rare, we might consider the difference between the score and the mean: $x - \overline{x}$. However, the size of such differences is relative to the scale being used. For example, an IQ score of 102 differs from the mean of 100 by 2 points, whereas an IQ score of 145 differs from the mean by 45 points. An IQ score of 102 is typical, whereas 145 is very rare. But the distinction between typical and rare scores requires an understanding of the IQ scale. It would be much better if we could develop a measure that was standard and did not require an understanding of the scale being used. With the standard score, we divide the difference $x - \overline{x}$ (or $x - \mu$) by the standard deviation to get such a result.

DEFINITION

The **standard score,** or **z score,** is the number of standard deviations that a given value x is above or below the mean. It is found by using

Sample		Population
$z = \dfrac{x - \overline{x}}{s}$	or	$z = \dfrac{x - \mu}{\sigma}$

(Round z to two decimal places.)

▶ **EXAMPLE**

For the body temperature data given in Table 2-1, we have found that the sample mean is $\overline{x} = 98.20°F$ and the standard deviation is $s = 0.62°F$. Find the z score corresponding to a body temperature of 102.2°F.

● SOLUTION

Because we are dealing with sample statistics, the z score is calculated as follows:

$$z = \frac{x - \bar{x}}{s} = \frac{102.2 - 98.20}{0.62} = 6.45$$

That is, the body temperature of 102.2°F is 6.45 standard deviations away from the mean.

The role of z scores in statistics is extremely important because they can be used to differentiate between ordinary values and unusual values. *Values with standard scores between $z = -2.00$ and $z = 2.00$ are ordinary, and values with z scores less than $z = -2.00$ or greater than $z = 2.00$ are unusual.* (See Figure 2-13.) This follows from the empirical rule and Chebyshev's theorem. Recall from the empirical rule that for data with a bell-shaped distribution, about 95% of the values are within 2 standard deviations of the mean. (See Figure 2-12 in the preceding section.) Also, Chebyshev's theorem states that for any data set, at least 75% of the values are within 2 standard deviations of the mean.

We noted earlier that z scores are also useful for comparing scores from different populations with different means and different standard deviations. The following example illustrates this use of z scores.

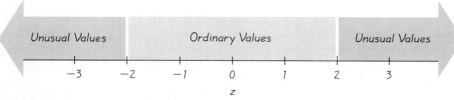

FIGURE 2-13   Interpreting z Scores
Unusual values are those with z scores less than $z = -2.00$ or greater than $z = 2.00$.

▶ EXAMPLE

Two equivalent intelligence tests are given to similar groups, but the tests are designed with different scales. The statistics for the tests are listed below. Which is better: a score of 130 on Test A or a score of 52 on Test B?

Test A	Test B
$\bar{x} = 100$	$\bar{x} = 40$
$s = 15$	$s = 5$

(continued)

● **SOLUTION**

For the score of 130 on Test A, we get a z score of 2.00, because

$$z = \frac{x - \overline{x}}{s} = \frac{130 - 100}{15} = 2.00$$

For the score of 52 on Test B, we get a z score of 2.40, because

$$z = \frac{x - \overline{x}}{s} = \frac{52 - 40}{5} = 2.40$$

That is, a score of 130 on Test A is 2.00 standard deviations above the mean, while a score of 52 on Test B is 2.40 standard deviations above the mean. This implies that the 52 on Test B is the better score. Although 52 is less than 130, it has a better *relative* position when considered in the context of the other test results. Later, we will make extensive use of these standard, or z, scores.

The preceding example illustrated that the z score provides a useful measurement for making comparisons between different sets of data. Likewise, quartiles, deciles, and percentiles are measures of position useful for comparing scores within one set of data or between different sets of data.

Quartiles, Deciles, and Percentiles

Just as the median divides the data into two equal parts, the three **quartiles,** denoted by Q_1, Q_2, and Q_3, divide the *ranked* scores into four equal parts. (Scores are ranked when they are arranged in order.) Roughly speaking, Q_1 separates the bottom 25% of the ranked scores from the top 75%, Q_2 is the median, and Q_3 separates the top 25% from the bottom 75%. To be more precise, at least 25% of the data will be less than or equal to Q_1, and at least 75% will be greater than or equal to Q_1. At least 75% of the data will be less than or equal to Q_3, while at least 25% will be equal to or greater than Q_3.

Similarly, there are nine **deciles,** denoted by D_1, D_2, D_3, . . . , D_9, which partition the data into 10 groups with about 10% of the data in each group. There are also **99 percentiles,** which partition the data into 100 groups with about 1% of the scores in each group. (Quartiles, deciles, and percentiles are examples of *fractiles*, which partition data into parts that are approximately equal.) A student taking a competitive college entrance examination might learn that he or she scored in the 92nd percentile. This does not mean that the student received a grade of 92% on the test; it indicates roughly that whatever score he or she did achieve was higher than 92% of the scores of those who took a similar test (and also lower than 8% of his or her colleagues').

Percentiles are useful for converting meaningless raw scores into meaningful comparative scores. For this reason, percentiles are used extensively in educational testing. A raw score of 750 on a college entrance exam means nothing to most people, but the corresponding percentile of 92% provides useful comparative information.

The process of finding the percentile that corresponds to a particular score x is fairly simple, as indicated in the following expression.

$$\text{percentile of score } x = \frac{\text{number of scores less than } x}{\text{total number of scores}} \cdot 100$$

▶ **EXAMPLE**

Table 2-10 lists the 106 body temperatures (in degrees Fahrenheit) ranked from lowest to highest. (These are the same body temperatures, from the University of Maryland study, that are listed in Table 2-1.) Find the percentile corresponding to 97.5°F.

● **SOLUTION**

From Table 2-10 we see that there are 15 temperatures less than 97.5°F, so

$$\text{percentile of } 97.5 = \frac{15}{106} \cdot 100 = 14 \text{ (rounded off)}$$

The body temperature of 97.5°F is the 14th percentile.

TABLE 2-10   *Ranked* Body Temperatures (in degrees Fahrenheit)

96.5	97.3	97.6	97.9	98.0	98.3	98.4	98.6	98.6	98.8	99.1
96.9	97.3	97.6	97.9	98.0	98.3	98.4	98.6	98.6	98.8	99.2
97.0	97.4	97.6	98.0	98.0	98.4	98.4	98.6	98.6	98.8	99.4
97.0	97.4	97.7	98.0	98.0	98.4	98.4	98.6	98.7	98.8	99.4
97.0	97.4	97.8	98.0	98.0	98.4	98.5	98.6	98.7	98.8	99.5
97.1	97.5	97.8	98.0	98.2	98.4	98.5	98.6	98.7	98.8	99.6
97.1	97.5	97.8	98.0	98.2	98.4	98.5	98.6	98.7	98.9	
97.1	97.6	97.8	98.0	98.2	98.4	98.5	98.6	98.7	98.9	
97.2	97.6	97.8	98.0	98.2	98.4	98.6	98.6	98.7	99.0	
97.3	97.6	97.9	98.0	98.2	98.4	98.6	98.6	98.8	99.0	

$100.00 ~~$100.00~~
$124.00

Index Numbers

Standard scores and percentiles allow us to compare different values, but they ignore any element of time. The Consumer Price Index (CPI) is an example of an *index number* that allows us to compare the value of some variable to its value at some base time period. The net effect is that we measure the change of the variable over some time span. We find the value of an index number by evaluating

$$\frac{\text{current value}}{\text{base value}} \times 100$$

The CPI is based on a weighted average of the costs of specific goods and services. When we use 1982–84 as a base with index 100, the CPI for a recent year is 124. Goods and services costing $100 in 1982–84 would cost $124 for this year.

The preceding example illustrated the procedure for finding the percentile corresponding to a given score. There are several different methods for the reverse procedure of finding the score corresponding to a particular percentile, but the one we will use is summarized in Figure 2-14, which uses the following notation.

Notation	
n	number of scores in the data set
k	percentile being used
L	locator that gives the *position* of a score
P_k	kth percentile

▶ EXAMPLE

Refer to the 106 body temperatures in Table 2-10, and find the score corresponding to the 25th percentile. That is, find the value of P_{25}.

● SOLUTION

We refer to Figure 2-14 and observe that the data are already ranked from lowest to highest. We now compute the locator L as follows:

$$L = \left(\frac{k}{100}\right) n = \left(\frac{25}{100}\right) \cdot 106 = 26.5$$

We answer no when asked in Figure 2-14 if 26.5 is a whole number, so we are directed to round L *up* (not off) to 27. (In this particular procedure we round L up to the next higher integer, but in most other situations in this book we generally follow the usual process for rounding.) The 25th percentile, denoted by P_{25}, is the 27th score, counting from the lowest. Beginning with the lowest score of 96.5, we count down the list to find the 27th score of 97.8, so $P_{25} =$ 97.8°F.

Verify that there are 24 scores below 97.8°F; be sure to count each individual score, including duplicates. Finding the percentile for 97.8°F therefore yields $(24/106) \cdot 100 = 23$ (rounded). There is a small discrepancy:

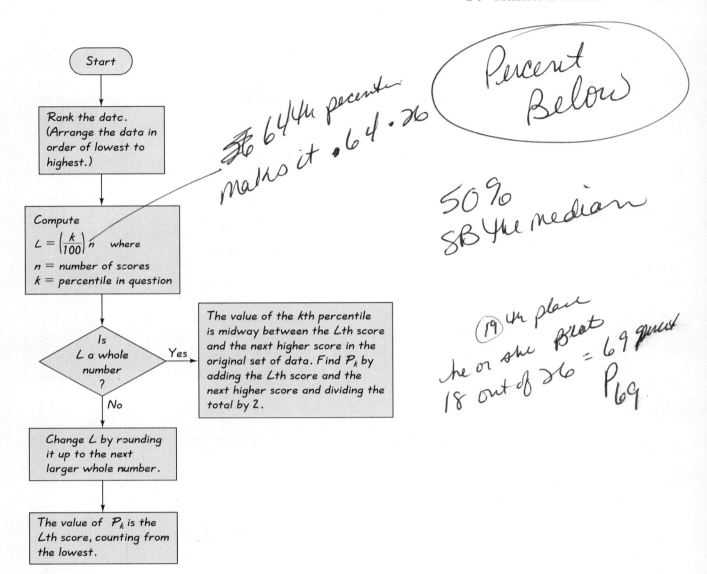

FIGURE 2-14   Finding the Value of the *k*th Percentile

In the preceding example we found the 25th percentile to be 97.8°F, but when we reverse the process we find that 97.8°F is the 23rd percentile. As the amount of data increases, such discrepancies become smaller.

Because of the sample size in the preceding example, the locator L first became 26.5, which was rounded to 27 because L was not originally a whole number. In the next example we illustrate a case in which L does begin as a whole number. This condition will cause us to branch to the right in Figure 2-14.

▶ **EXAMPLE**

Refer to the body temperature data listed in Table 2-10. Find P_{50}, which denotes the 50th percentile.

● **SOLUTION**

Following the procedure outlined in Figure 2-14 and noting that the data are already ranked from lowest to highest, we compute

$$L = \left(\frac{k}{100}\right) n$$

$$= \left(\frac{50}{100}\right) \cdot 106 = 53 \text{ (exactly)}$$

We note that 53 is a whole number and see that P_{50} is midway between the 53rd and 54th scores. Because the 53rd and 54th scores are both 98.4°F, we conclude that the 50th percentile is 98.4°F.

Once you have mastered these calculations with percentiles, similar calculations for quartiles and deciles can be performed with the same procedures by noting the following relationships.

Quartiles	Deciles
$Q_1 = P_{25}$	$D_1 = P_{10}$
$Q_2 = P_{50}$	$D_2 = P_{20}$
$Q_3 = P_{75}$	$D_3 = P_{30}$
	.
	.
	.
	$D_9 = P_{90}$

Using these relationships, we can see that finding Q_1 is equivalent to finding P_{25}. In an earlier example we found that $P_{25} = 97.8°F$, so it follows that the first quartile can be described by $Q_1 = 97.8°F$.

In addition to the measures of central tendency and the measures of variation already introduced, other statistics are sometimes defined using quartiles, deciles, or percentiles. For example, the *interquartile range* is a measure of dispersion obtained by evaluating $Q_3 - Q_1$. The *semi-interquartile range* is $(Q_3 - Q_1)/2$, the *midquartile* is $(Q_1 + Q_3)/2$, and the *10–90 percentile range* is defined to be $P_{90} - P_{10}$.

Using Computers for Descriptive Statistics

In dealing with large collections of data, reliable results are quickly obtained with greater ease when statistical software packages are used. The following STATDISK and Minitab computer displays for the 106 body temperatures listed in Table 2-1 show how the computer makes it easy to find the values of the various statistics that we have been discussing in this chapter. For the STATDISK display, first use the File option to enter (via New) or retrieve (via Open) the data file, then use the View option to select Descriptive Statistics. For the Minitab display, first enter the data in column C1 as described in Section 1-5 (or retrieve them), then enter the command DESCRIBE C1.

STATDISK DISPLAY FOR BODY TEMPERATURE DATA

```
About    File    Edit    Statistics    Data    View

                    Sample Size = 106

   Minimum ...... = 96.5          Sum of scores  =   10409.2
   Maximum ...... = 99.6          Sum of squares = 1022224.2

              MEASURES OF CENTRAL TENDENCY
   Mean ......... = 98.20          Geom. mean ... = 98.20
   Median ....... = 98.40          Harm. mean ... = 98.20
   Midrange ..... = 98.05          Quad. mean ... = 98.20

                MEASURES OF DISPERSION
   Samp. st. dev. =   .62          Samp. variance =  .39
   Pop.  st. dev. =   .62          Pop.  variance =  .38
   Range ........ =  3.10          Standard error =  .06

            MEASURES OF POSITION - Quartiles
     Q1 = 97.80          Q2 = 98.40          Q3 = 98.60

                Use Menu Bar to change screen
   F1: Help        F2: Menu                        ESC: Quit
```

MINITAB DISPLAY FOR BODY TEMPERATURE DATA

```
MTB > DESCRIBE C1
               N      MEAN    MEDIAN    TRMEAN    STDEV   SEMEAN
C1           106    98.200    98.400    98.205    0.623    0.061

              MIN       MAX        Q1        Q3
C1        96.500    99.600    97.800    98.600
```

2-6 Exercises A: Basic Skills and Concepts

In Exercises 1–4, express all z scores with two decimal places.

1. The body temperature data set in Table 2-1 has mean $\bar{x} = 98.20°F$ and standard deviation $s = 0.62°F$. Find the z score corresponding to

   a. 98.6°F   b. 98.2°F   c. 100.0°F

2. The typical IQ test is designed to yield a mean of 100 and a standard deviation of 15. Find the z score corresponding to IQ scores of

   a. 115   b. 90   c. 127

3. An investigation of the number of hours that college freshmen spend studying each week found that the mean is 7.06 h and the standard deviation is 5.32 h (based on data from *The American Freshman*). Find the z score corresponding to a freshman who studies 10.00 hours weekly.

4. A study of the amounts of time that high school students spend working at jobs each week found that the mean is 10.7 h and the standard deviation is 11.2 h (based on data from the National Federation of State High School Associations). Find the z score corresponding to a high school student who works 15.0 h each week.

In Exercises 5–8, express all z scores with two decimal places. Consider a score to be *unusual* if its z score is less than -2.00 or greater than 2.00.

5. The Beanstalk Club is limited to women and men who are very tall. The minimum height requirement for women is 70 in. Women's heights have a mean of 63.6 in. and a standard deviation of 2.5 in.
   a. Find the z score corresponding to a woman with a height of 70 in.
   b. Is that height unusual?

6. The heights of six-year-old girls have a mean of 117.8 cm and a standard deviation of 5.52 cm (based on data from the National Health Survey, USDHEW publication 73-1605).
   a. Find the z score corresponding to a six-year-old girl who is 106.8 cm tall.
   b. Is that height unusual?

7. For a certain population, scores on the Thematic Apperception Test have a mean of 22.83 and a standard deviation of 8.55 (based on "Relationships Between Achievement-Related Motives, Extrinsic Conditions, and Task Performance" by Schroth, *Journal of Social Psychology*, Vol. 127, No. 1).
   a. Find the z score corresponding to a member of this population who has a score of 10.00.
   b. Is that score unusually low?

8. For men aged between 18 and 24 years, serum cholesterol levels (in mg/100 mL) have a mean of 178.1 and a standard deviation of 40.7 (based on data from the National Health Survey, USDHEW publication 78-1652).

(continued)

    a. Find the z score corresponding to a male, aged 18–24 years, who has a
        serum cholesterol level of 275.2 mg/100 mL.
    b. Is this level unusually high?

9. Which of the following two scores has the better relative position?
    a. A score of 53 on a test for which $\bar{x} = 50$ and $s = 10$
    b. A score of 53 on a test for which $\bar{x} = 50$ and $s = 5$

10. Two similar groups of students took equivalent language facility tests. Which of
the following results indicates the higher relative level of language facility?
    a. A score of 60 on a test for which $\bar{x} = 70$ and $s = 10$
    b. A score of 480 on a test for which $\bar{x} = 500$ and $s = 50$

11. Three prospective employees take equivalent tests of communicative ability.
Which of the following scores corresponds to the highest relative position?
    a. A score of 60 on a test for which $\bar{x} = 50$ and $s = 5$
    b. A score of 230 on a test for which $\bar{x} = 200$ and $s = 10$
    c. A score of 540 on a test for which $\bar{x} = 500$ and $s = 15$

12. Three students take equivalent tests of neuroticism with the given results.
Which is the highest relative score?
    a. A score of 3.6 on a test for which $\bar{x} = 4.2$ and $s = 1.2$
    b. A score of 72 on a test for which $\bar{x} = 84$ and $s = 10$
    c. A score of 255 on a test for which $\bar{x} = 300$ and $s = 30$

In Exercises 13–16, use the 106 ranked body temperatures listed in Table 2-10. Find
the percentile corresponding to the given temperature.

13. 97.2°F   **14.** 98.6°F   **15.** 99.0°F   **16.** 97.6°F

In Exercises 17–24, use the 106 ranked body temperatures listed in Table 2-10. Find
the indicated percentile, quartile, or decile.

17. P_{80} **18.** P_{40} **19.** Q_3 **20.** Q_1
21. D_3 **22.** D_9 **23.** D_7 **24.** P_{37}

In Exercises 25–28, use the lengths (in minutes) of the movies listed in Data Set 5 of
Appendix B. Find the percentile corresponding to the given time.

25. 120 min   **26.** 90 min   **27.** 100 min   **28.** 123 min

In Exercises 29–36, use the lengths (in minutes) of the movies listed in Data Set 5 of
Appendix B. Find the indicated percentile, quartile, or decile.

29. P_{15} **30.** P_{30} **31.** P_{80} **32.** P_{10}
33. Q_1 **34.** Q_3 **35.** D_9 **36.** D_3

2-6 Exercises B: Beyond the Basics

37. Use the ranked body temperatures listed in Table 2-10.
    a. Find the interquartile range.
    b. Find the midquartile.
    c. Find the 10–90 percentile range.   *(continued)*

    d. Does $P_{50} = Q_2$? If so, does P_{50} *always* equal Q_2?

    e. Does $Q_2 = (Q_1 + Q_3)/2$? If so, does Q_2 *always* equal $(Q_1 + Q_3)/2$?

38. When finding percentiles using Figure 2-14, if the locator L is not a whole number, we round it up to the next larger whole number. An alternative to this procedure is to *interpolate* so that a locator of 23.75 leads to a value that is 0.75 (or 3/4) of the way between the 23rd and 24th scores. Use this method of interpolation to find Q_1, Q_3, and P_{33} for these scores:

$$16 \quad 49 \quad 53 \quad 58 \quad 60 \quad 63 \quad 63 \quad 65 \quad 72 \quad 80 \quad 84 \quad 89 \quad 92 \quad 98$$

39. For the 106 body temperatures given in Table 2-1, the mean is $\bar{x} = 98.20°F$ and the standard deviation is $0.62°F$. Find the two temperatures that are cutoff values separating ordinary values from unusual values.

40. Using the scores 2, 5, 8, 9, and 16, first find $\bar{x}$ and s, then replace each score by its corresponding z score. (Don't round the z scores; carry as many decimal places as your calculator can handle.) Now find the mean and standard deviation of the five z scores. Will these new values of the mean and standard deviation result from *every* set of z scores?

2-7 Exploratory Data Analysis

The techniques that were discussed in the previous sections can be used to summarize data, see their distribution, and find important measures such as the mean and standard deviation. In summarizing and describing data, we should be careful to avoid overlooking important information that might be lost in our summaries. About 30 years ago, statisticians began to use an approach that is now referred to as **exploratory data analysis,** or **EDA.** Many of the techniques used in this approach are introduced in John Tukey's book *Exploratory Data Analysis* (Addison-Wesley, 1977). *EDA* is more than simply a collection of new statistical techniques—it is a fundamentally different approach. With EDA we *explore* data rather than use a statistical analysis to *confirm* some claim or assumption made about the data. For data obtained from a carefully planned experiment with a very specific objective (such as comparing the mpg ratings of two different car models), traditional methods of statistics will probably be sufficient. But if you have a collection of data and a broad goal (such as trying to find out what the data reveal), then you might want to begin with exploratory data analysis. With EDA the emphasis is on original explorations with the goals of simplifying the way the data are described and gaining deeper insight into the nature of the data. Thus it is easier to identify relevant questions that might be addressed. The table on the next page compares EDA and traditional statistics in three major areas.

Exploratory Data Analysis	Traditional Statistics
Used to *explore* data at a preliminary level	Used to *confirm* final conclusions about data
Few or no assumptions are made about the data	Typically requires some very important assumptions about the data
Tends to involve relatively simple calculations and graphs	Calculations are often complex, and graphs are often unnecessary

In this section we consider two helpful devices used for exploratory data analysis: the stem-and-leaf plot and the boxplot. They are typical of EDA techniques because they involve relatively simple calculations and graphs.

Stem-and-Leaf Plots

Section 2-3 discussed histograms, which are extremely useful in graphically displaying the distribution of data. By using such graphic devices, we are able to learn something about the data that is not apparent while the data remain in a list of values. This additional insight is clearly an advantage. However, in constructing histograms, we suffer the disadvantage of losing some information in the process. Generally we cannot reconstruct the original data set from the histogram; this shows that some distortion has occurred. We will now introduce another device that enables us to see the distribution of data without losing information in the process.

In a **stem-and-leaf plot** we sort data according to a pattern that reveals the underlying distribution. The pattern involves separating a number (such as 96.5) into two parts, usually the first one or two digits (96) and the other digits (5). The *stem* consists of the leftmost digits (in this case, 96), and the *leaves* consist of the rightmost digits (in this case, 5). The method is illustrated in the following example.

▶ **EXAMPLE**

Use the body temperature data listed in Table 2-1, and construct a stem-and-leaf plot.

● **SOLUTION**

If we use the two leftmost digits for the stem, the stem consists of 96, 97, 98, and 99. (Because all of the values have 9 for their first digit, it wouldn't be wise to use a stem with that single digit; the stem-and-leaf plot would consist of only one row and would not reveal the information we seek.) We then draw a

(continued)

The Power of Your Vote

In the electoral college system, the power of a voter in a large state exceeds that of a voter in a small state. When voting power is measured as the ability to affect the outcome of an election, we see that a New Yorker has 3.312 times the voting power of a resident of the District of Columbia. This result is included in John Banzhof's article "One Man, 3.312 Votes."

As an example, the outcome of the 1916 presidential election could have been changed by shifting only 1,983 votes in California. If the same number of votes were changed in a much smaller state, the resulting change in electoral votes would not have been sufficient to alter the outcome.

vertical line and list the leaves as shown below. The first value in Table 2-1 is 98.6, and we include that value by entering a 6 in the third row (for 98). We continue to enter all 106 values, and then we arrange the leaves so that the numbers are arranged in increasing order.

The first row represents the numbers 96.5 and 96.9. The second row represents 97.0, 97.0, and so on.

```
Stem │ Leaves                   ┌─ 97.5
96.  │ 59                       │
97.  │ 00011123334445566666667888888999   ↓
98.  │ 00000000000002222233444444444445555666666666666666677777888888899
99.  │ 00124456
```

By turning the page on its side, we can see a distribution of these data, which, in this case, roughly approximates a bell shape. Here's the great advantage of the stem-and-leaf plot: We can see the distribution of the data and yet retain all the information in the original list; if necessary, we could reconstruct the original list of values.

You might notice that the rows of digits in a stem-and-leaf plot are similar in nature to the bars in a histogram. One of the guidelines for constructing histograms is that the number of classes should be between 5 and 20. Just as histograms with fewer than 5 classes might not depict the distribution of the data very well, stem-and-leaf plots with fewer than 5 rows might have that same flaw. Stem-and-leaf plots can be expanded to include more rows and can also be condensed to include fewer rows. The stem-and-leaf plot of the preceding example can be stretched out by subdividing rows into those with the digits 0 through 4 and those with digits 5 through 9, as shown below.

```
96. │
96. │ 59
97. │ 0001112333444
97. │ 55666666788888999
98. │ 00000000000002222233444444444444
98. │ 5555666666666666666677777888888899
99. │ 001244
99. │ 56
```

Using Computers for Stem-and-Leaf Plots

If we use Minitab to enter the 106 body temperatures in column C1 (as described in Section 1-5) and then enter the command STEM-AND-LEAF C1, we will obtain the following display of an expanded stem-and-leaf plot. In this

Minitab result, each two-digit stem is expanded into 5 rows that contain the digits 0–1, 2–3, 4–5, 6–7, and 8–9. Minitab also allows you to control (to some extent) the manner in which the expansion occurs.

In the display below, the leftmost column represents cumulative totals. For example, the 24 in the left column shows that there are 24 scores included from the top row to the row in which 24 occurs. The 38 in the left column indicates that there are 38 scores included from the bottom row to the row in which 38 occurs. The 16 enclosed within parentheses indicates that there are 16 scores in the row containing the median.

MINITAB DISPLAY OF STEM-AND-LEAF PLOT

```
   1    96   5
   1    96
   2    96   9
   8    97   000111
  12    97   2333
  17    97   44455
  24    97   6666667
  32    97   88888999
  45    98   0000000000000
  52    98   2222233
 <16>   98   4444444444445555
  38    98   66666666666666666777777
  17    98   888888899
   8    99   001
   5    99   2
   4    99   445
   1    99   6
```

When a stem-and-leaf plot is expanded or condensed, the apparent shape of the distribution may change. It wouldn't make much sense to condense the stem-and-leaf plot from the preceding example, so we will use a different data set in the next example to illustrate how adjacent rows can be combined to form a condensed stem-and-leaf plot. This data set is configured to illustrate how the shape of the distribution can change.

▶ **EXAMPLE**

The following 30 scores on the mathematics portion of the SAT were achieved by students who took a preparatory course before taking the test. First construct a stem-and-leaf plot, and then condense it to include fewer rows.

500	510	514	514	516	519	521	522	522	527
528	535	540	542	545	553	555	558	561	571
572	574	577	578	580	583	584	588	589	592

(continued)

● SOLUTION

The stem-and-leaf plot is shown on the left. In the condensed plot shown on the right, we separated digits in the leaves associated with the numbers in each stem by an asterisk. Every row in the condensed plot must include exactly one asterisk so that the shape of the plot is not distorted. Note that the condensed plot has a uniform distribution shape, whereas the original plot (on the left) does not appear to have a uniform distribution. With histograms and stem-and-leaf plots, we must be careful when forming conclusions about the nature or shape of the distribution because one graph could be misleading, as in this case.

50	0		50–51	0*04469
51	04469		52–53	12278*5
52	12278	condense	54–55	025*358
53	5	⟶	56–57	1*12478
54	025		58–59	03489*2
55	358			
56	1			↑ ↑
57	12478			583 592
58	03489			
59	2			

Another useful feature of stem-and-leaf plots is that their construction provides a fast and easy procedure for *ranking* data (arranging data in order). Data must be ranked for a variety of statistical procedures, such as the Wilcoxon rank-sum test (discussed in Chapter 13) and finding the median of a set of data.

Boxplots

When exploring a collection of numerical data, we should be sure to investigate (1) the value of a representative score (such as the mean or median), (2) a measure of scattering or variation (such as the standard deviation), and (3) the nature or shape of the distribution. The distribution of data should definitely be considered because it may strongly affect other statistical methods we use and the conclusions we draw. In the spirit of exploratory data analysis, we should not simply examine a histogram and think that we understand the nature of the distribution—we should *explore*. As an example, we show a STATDISK printout of a histogram of the 106 body temperatures listed in Table 2-1 with one change: In the histogram on the top, the first value of 98.6 was incorrectly entered without its decimal point as 986. The histogram on the bottom is correct. Note how one simple error in only one of 106 values has a dramatic effect on the shape of the histogram. In this case, the incorrect extreme value of 986 caused a severe distortion of the histogram. In other cases, such extreme values (often called outliers) may be correct but may disguise the true nature of the distribution when illustrated through a histogram. If we didn't further explore the data, we might draw conclusions from the histogram that are seriously wrong.

STATDISK DISPLAY OF INCORRECT DATA

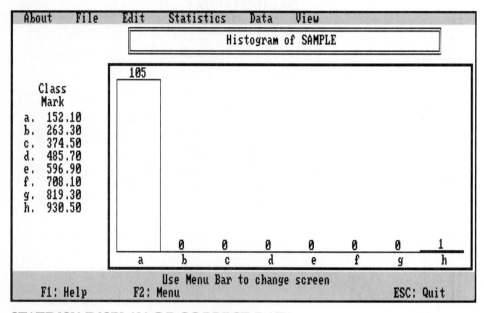

STATDISK DISPLAY OF CORRECT DATA

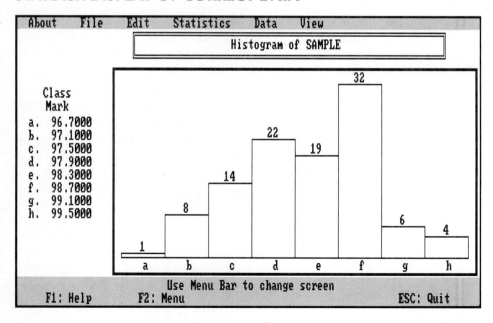

The stem-and-leaf plot of those body temperatures would be very helpful in revealing that the histogram on the top was incorrect. Boxplots are also helpful because they reveal information about how the data are spread out. The construction of a boxplot requires that we obtain the minimum score, the maximum score, the median, and two other values called *hinges*.

DEFINITIONS

After the data have been ranked from lowest to highest, the **lower hinge** is the median of the lower half of all scores (from the minimum score up to and including the original median).

The **upper hinge** is the median of the upper half of all scores (from the original median up to the maximum score).

The minimum score, the maximum score, the median, and the two hinges constitute a **5-number summary** of a set of data.

A **boxplot** is a graph of lines and a box that includes the minimum score, the maximum score, the median, and the two hinges (see Figure 2-15).

Hinges are very similar to quartiles. In fact, several texts and statistical software programs use quartiles instead of hinges for the construction of boxplots. The differences between hinges and quartiles are usually small, especially for larger data sets. The definitions of hinges given above are consistent with John Tukey's definitions. Hinges can be easily found by following these steps:

Step 1:   Arrange the data in increasing order. (That is, rank the data.)

Step 2:   Find the median. (With an odd number of scores, it's the middle score; with an even number of scores, the median is equal to the mean of the two middle scores.)

Step 3:   List the lower half of the data from the minimum score up to and including the median found in Step 2. The left hinge is the median of these scores.

Step 4:   List the upper half of the data, starting with the median and including the scores up to and including the maximum. The right hinge is the median of these scores.

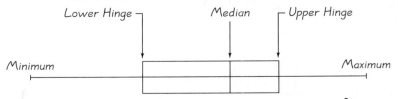

FIGURE 2-15   Boxplot

▶ **EXAMPLE**

The given values are 10 body temperatures selected from Table 2-1.

98.0   98.4   97.0   97.9   97.7   99.6   98.8   98.6   98.6   96.5

a. Find the values constituting the 5-number summary of these 10 temperatures.

b. Construct a boxplot for these 10 temperatures.

● **SOLUTION**

a. We begin by arranging the values in order from low to high.

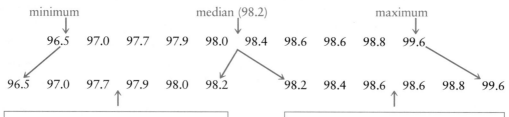

The *left hinge* of 97.8 is found by determining the median of the bottom half of the data up to and including the original median.

The *right hinge* of 98.6 is found by determining the median of the top half of the data from the original median up to the maximum.

The 5-number summary consists of the minimum (96.5), the left hinge (97.8), the median (98.2), the right hinge (98.6), and the maximum (99.6).

b. The values in the 5-number summary are included in the boxplot shown in Figure 2-16.

With the procedures used here, approximately one-fourth of the values fall between the minimum score and the left hinge, the two middle quarters are in the boxes, and approximately one-fourth of the values are between the right hinge and the maximum. A boxplot therefore shows how data are spread out.

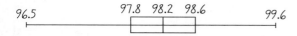

FIGURE 2-16   Boxplot of 10 Temperatures

Using Computers for Boxplots

Minitab can be used to create boxplots. Enter the data in column C1 (as described in Section 1-5), then enter the command BOXPLOT C1. The Minitab boxplot corresponding to the 10 scores in the preceding example follows.

MINITAB DISPLAY

```
MTB > SET C1
DATA> 96.5 97.0 97.7 97.9 98.0 98.4 98.6 98.6 98.8 99.6
DATA> ENDOFDATA
MTB > BOXPLOT C1
                                    ----------------
   --------------------I         +      I----------------
                                    ----------------
   ----+---------+---------+---------+---------+---------+--C1
   96.60      97.20      97.80      98.40      99.00      99.60
```

Figure 2-17(a) is the boxplot of the 106 body temperatures with the first entry of 98.6 incorrectly entered as 986; Figure 2-17(b) is a boxplot of the correct values. The uneven spread shown in Figure 2-17(a) is in strong contrast to the much more even spread shown in Figure 2-17(b). Suspecting that body temperatures have a symmetric bell-shaped distribution, we would expect to see a boxplot like the one in Figure 2-17(b), whereas the diagram in Figure 2-17(a) would raise suspicion and lead to further investigation. Figure 2-18 shows some common distributions along with the corresponding boxplots.

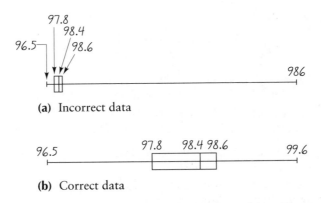

(a) Incorrect data

(b) Correct data

FIGURE 2-17   Boxplots for Incorrect and Correct Body Temperature Data

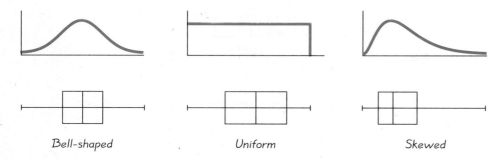

FIGURE 2-18   Boxplots Corresponding to Bell-Shaped, Uniform, and Skewed Distributions

2-7   Exercises A: Basic Skills and Concepts

In Exercises 1 and 2, list the original numbers in the data set represented by the given stem-and-leaf plots.

1.

Stem	Leaves
40	6678
41	09999
42	13466
43	088

2.

Stem	Leaves
68	45 45 47 86
69	33 38 89
70	52 59 93
71	27

In Exercises 3–6, construct the stem-and-leaf plots for the given data sets.

3. High temperatures (in degrees Fahrenheit) for the 31 days in July (recorded at Dutchess Community College):

```
80   68   84   86   85   77   64   81   93   94   97
93   89   82   76   75   83   90   83   84   92   94
90   92   91   84   81   84   79   80   80
```

4. Pulse rates (number of beats per minute) of 20 male statistics students:

```
82   74   77   62   78   58   85   74   66   71
58   80   65   60   54   75   71   74   73   82
```

5. Weights (in grams) of the 50 quarters listed in Data Set 8 of Appendix B. (Use an expanded stem-and-leaf plot with about 8 rows.)

6. Weights (in pounds) of plastic discarded by 62 households: Refer to Data Set 1 of Appendix B, and start by rounding the listed weights to the nearest tenth of a pound (or one decimal place). (Use an expanded stem-and-leaf plot with about 11 rows.)

In Exercises 7–12, use the given data to construct boxplots, and include the values of the minimum, maximum, median, and hinges.

7. Blood alcohol contents of drivers given breathalyzer tests:

    0.02   0.04   0.08   0.08   0.09   0.10   0.10   0.12   0.13   0.19

8. Time intervals (in months) between adjacent births for selected families:

    12.9   13.4   18.3   24.7   31.2   31.3   32.0   32.1
    33.4   33.8   34.1   36.2   41.7   41.9   52.5

9. Blood pressure levels (in mm of mercury) for patients who have taken 25 mg of the drug Captopril:

    198   180   142   157   181   183   162   130   170   164
    170   173   173   175   195   190   193   157   159   138

10. Number of words typed in a 5-minute civil service test taken by 25 different applicants:

    174   181   219   213   213   207   106   111   143   160   166   350   183
    198   193   190   190   185   220   221   229   257   243   281   308

11. Time (in hours of operation) between failures for prototypes of computer printers:

    34   22   4   9   27   36   12   40   29   32
    35   25   7   9   26   36   45   43   41   2
    31   31   30   14   15   18   10   27   38   21

12. Construct a boxplot for the data given in the stem-and-leaf plot.

Stem	Leaves
5	0
6	6678
7	02334489
8	33566677
9	025

2-7   Exercises B: Beyond the Basics

Stem-and-leaf plots and boxplots are often used to investigate data that don't behave as we might expect. In Exercises 13 and 14, there is a major problem with the data. Analyze the given information, and identify the problem.

13. An analyst claims to have collected data on the numbers selected in the Florida lottery, which are summarized in the given boxplot.

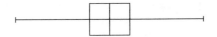

14. A researcher claims that he measured the
IQ scores of randomly selected adults,
and the results are summarized in the
given stem-and-leaf plot.

```
 9 | 035778999
10 | 012234899
11 | 22347789
12 | 013356778
```

15. In "Ages of Oscar-Winning Best Actors and Actresses" (*Mathematics Teacher*
magazine) by Richard Brown and Gretchen Davis, stem-and-leaf plots and
boxplots are used to compare the ages of actors and actresses at the time they
won Oscars. Here are the results for 30 recent and consecutive winners from
each category.

Actors:	32	51	33	61	35	45	55	39	76	37
	42	40	32	60	38	56	48	48	40	43
	62	43	42	44	41	56	39	46	31	47
Actresses:	80	26	41	21	61	38	49	33	74	30
	33	41	31	35	41	42	37	26	34	34
	35	26	61	60	34	24	30	37	31	27

a.   Construct a back-to-back stem-and-leaf plot for the above data. The first
two scores from each group have been entered.

Actors' Ages	Stem	Actresses' Ages
	2	6
2	3	
	4	
1	5	
	6	
	7	
	8	0

b.   Using the same scale, construct a boxplot for actors' ages and another
boxplot for actresses' ages.

c.   Using the results from parts a and b, compare the two different sets of data,
and try to explain any differences.

16. The boxplots discussed in this section are often called *skeletal* boxplots. In
investigating outliers, a useful variation is to construct boxplots as follows:

1.   Calculate the hinge difference:

$$D = \text{(upper hinge)} - \text{(lower hinge)}$$

2.   Draw the box with the median and hinges as usual, but when extending the
lines that branch out from the box, go only as far as the scores that are
within $1.5D$ of the box.

3.   **Mild outliers** are scores above the upper hinge by $1.5D$ to $3D$ or below the
lower hinge by $1.5D$ to $3D$. Plot them as solid dots.

4.   **Extreme outliers** are scores above the upper hinge by more than $3D$ or
below the lower hinge by more than $3D$. Plot them as small hollow circles.

(continued)

The accompanying figure is an example of the boxplot described here. Use this procedure to construct the boxplot for the given scores, and identify any mild outliers or extreme outliers.

3   15   17   18   21   21   22   25   27   30   38   49   68

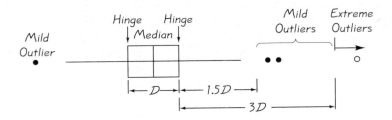

VOCABULARY LIST

Define and give an example of each term:

descriptive statistics	midrange
inferential statistics	weighted mean
frequency table	skewed
frequency	symmetric
lower class limits	negatively skewed
upper class limits	positively skewed
class boundaries	range
class marks	standard deviation
class width	deviation
relative frequency	mean deviation
relative frequency table	variance
cumulative frequency	range rule of thumb
cumulative frequency table	Chebyshev's theorem
histogram	empirical rule
relative frequency histogram	standard score
Pareto chart	z score
pie chart	quartiles
measure of central tendency	deciles
arithmetic mean	percentiles
mean	exploratory data analysis (EDA)
sample size	stem-and-leaf plot
median	lower hinge
mode	upper hinge
bimodal	5-number summary
multimodal	boxplot

Important Formulas

$$\overline{x} = \frac{\Sigma x}{n}$$

Mean

$$\overline{x} = \frac{\Sigma(f \cdot x)}{\Sigma f}$$

Computing the mean when the data are in a frequency table

$$s = \sqrt{\frac{\Sigma(x - \overline{x})^2}{n - 1}}$$

Standard deviation of sample data

$$s^2 = \frac{\Sigma(x - \overline{x})^2}{n - 1}$$

Variance of sample data

$$s = \sqrt{\frac{n(\Sigma x^2) - (\Sigma x)^2}{n(n - 1)}}$$

Shortcut formula for standard deviation of sample data

$$s = \sqrt{\frac{n[\Sigma(f \cdot x^2)] - [\Sigma(f \cdot x)]^2}{n(n - 1)}}$$

Computing the standard deviation when the sample data are in a frequency table

$$z = \frac{x - \overline{x}}{s} \text{ or } \frac{x - \mu}{\sigma}$$

Standard score or z score

REVIEW

Chapter 2 dealt mainly with the methods and techniques of descriptive statistics. It focused on developing the ability to organize, summarize, and illustrate data and to extract from data some meaningful measurements. Section 2-2 considered the frequency table as an excellent device for summarizing data, and then Section 2-3 dealt with graphic illustrations, such as histograms and relative frequency histograms, that help us determine the distribution of a set of data. We also considered Pareto charts and pie charts. Section 2-4 defined the common measures of central tendency. The mean, median, mode, and midrange represent different ways of characterizing the central value of a collection of data. The weighted mean is used to find the average of a set of scores that may vary in importance. The concept of skewness was also discussed. Section 2-5 presented the important measures of variation, including the range, standard deviation, and variance; these descriptive statistics are designed to measure the variability among a set of scores. We also presented the range rule of thumb, Chebyshev's theorem, and the empirical rule to show how the standard deviation relates to the actual variation of the data. The standard score, or z score, was introduced in Section 2-6 as a way of measuring the number of standard deviations by which a given score differs from the mean. That section also included these measures of position: quartiles, deciles, and percentiles.

Finally, Section 2-7 presented stem-and-leaf plots and boxplots and discussed their use in exploratory data analysis. Stem-and-leaf plots help us analyze the distribution of data, while boxplots are especially useful for depicting the spread of the data.

By now you should be able to organize, present, and describe collections of data composed of single scores. You should also be able to compute the key descriptive statistics that will be used in later applications.

The concepts developed in this chapter provided a better understanding of the 106 body temperatures listed in Table 2-1. The histogram and stem-and-leaf plot allowed us to see that the distribution is roughly bell-shaped. We know that the mean is 98.20°F, the median is 98.40°F, and the standard deviation is 0.62°F. The data were summarized in a frequency table and depicted in a histogram. Subsequent chapters will consider other ways of using sample statistics that go beyond summarizing them.

REVIEW EXERCISES

1. In "Determining Statistical Characteristics of a Vehicle Emissions Audit Procedure" (by Lorenzen, *Technometrics*, Vol. 22, No. 4), carbon monoxide emissions data (in g/m) are listed for vehicles. The values listed below are included. Find the (a) mean, (b) median, (c) mode, (d) midrange, (e) range, (f) variance, and (g) standard deviation.

5.01	14.67	8.60	4.42	4.95	7.24
7.51	12.30	14.59	7.98	11.53	4.10

2. The accompanying table lists times (in years) required to earn a bachelor's degree for a sample of undergraduate students (based on data from the National Center for Education Statistics). Use the table to find the mean and standard deviation.

Time (years)	Number
4	147
5	81
6	27
7	15
7.5–11.5	30

3. Using the frequency table given in Exercise 2, construct the corresponding relative frequency histogram.

4. A psychologist gave a subject 2 different tests designed to measure spatial perception. The subject obtained scores of 66 on the first test and 223 on the second test. The first test is known to have a mean of 75 and a standard deviation of 15, while the second test has a mean of 250 and a standard deviation of 25. Which result is better? Explain.

5. The NCAA was considering ways to speed up the end of college basketball games. The following values are the elapsed times (in seconds) that it took to play the last two minutes of regulation time in the 60 games of the first four rounds of a recent NCAA basketball tournament (based on data reported in *USA Today*). Using the minimum time as the lower class limit of the first class, construct a frequency table with 9 classes. (The data are on page 115.)

756	587	929	871	378	503	564	1128	693	748
448	670	1023	335	540	853	852	495	666	474
443	325	514	404	820	915	793	778	627	483
861	337	292	1070	625	457	676	494	420	862
991	615	609	723	794	447	704	396	235	552
626	688	506	700	240	363	860	670	396	345

6. Construct a relative frequency table (with 9 classes) for the data in Exercise 5.
7. Construct a histogram that corresponds to the frequency table from Exercise 5.
8. For the data in Exercise 5, find (a) Q_1, (b) P_{45}, and (c) the percentile corresponding to the time of 335 s.
9. Use the range rule of thumb to estimate the standard deviation of the data in Exercise 5.
10. Use the frequency table from Exercise 5 to find the mean and standard deviation for the times.
11. Use the data from Exercise 5 to construct a stem-and-leaf plot with 10 rows.
12. Use the data from Exercise 5 to construct a boxplot.
13. The values given below are snow depths (in centimeters) measured as part of a study of satellite observations and water resources (based on data in *Space Mathematics* published by NASA). Find the (a) mean, (b) median, (c) mode, (d) midrange, (e) range, (f) variance, and (g) standard deviation.

19   18   12   25   22   8   8   16

14. a. Scores on a statistics test have a mean of 67.3. What is the mean if 10 points are added to every score?
    b. Scores on a statistics test have a standard deviation of 14.2. What is the standard deviation if 10 points are added to each score?
    c. You just calculated the variance of a set of scores, and you got an answer of −5.5. What do you conclude?
    d. True or false: If set A has a range of 50 while set B has a range of 100, then the standard deviation for data set A must be less than the standard deviation for data set B.
    e. What do you know about a set of scores for which the standard deviation is zero?

15. The Telektronic Company records the number of computers sold each week, and the results are summarized in the accompanying stem-and-leaf plot. Find the (a) mean, (b) median, (c) mode, (d) midrange, (e) range, (f) variance, and (g) standard deviation.

0	244
1	02577
2	2578999
3	0122

16. The United States Coast Guard collected data on serious boating accidents and listed these categories, with their frequencies given in parentheses:

Colliding with another boat (2203)   Colliding with a fixed object (839)
Running aground (341)   Capsizing (458)
Person falling overboard (431)

Construct a Pareto chart summarizing the given data.

FROM DATA TO DECISION: Garbage In, Insight Out

Refer to Data Set 1 in Appendix B. That data set consists of weights of different categories of garbage discarded by a sample of 62 households. The data were collected as part of the Garbage Project at the University of Arizona. With such data sets, there are often several different issues that can be addressed. In Chapter 9 we will consider the issue of whether there is some relationship between household size and amount of waste discarded so that we might be able to predict the size of the population of a region by analyzing the garbage disposed there. For now, we will work with descriptive statistics based on the data.

a. Construct a Pareto chart and a pie chart depicting the relative amounts of the total weights of metal, paper, plastic, glass, food, yard waste, textile waste, and other waste. (Instead of frequencies, use the total weights.) Based on the results, which categories appear to be the largest components of the total amount of waste? Is there any single category that stands out as being the largest component?

b. A pie chart in *USA Today* depicted metal, paper, plastic, glass, food, yard waste, and other waste with the percents of 14%, 38%, 18%, 2%, 4%, 11%, and 13%, respectively. Do these percentages appear to be consistent with Data Set 1 in Appendix B?

c. For each category, find the mean and standard deviation, and construct a histogram of the 62 weights. Enter the results in the table below.

d. The amounts of garbage discarded are listed by weight. Many regions have household waste collected by commercial trucks that compress it, and charges at the destination are based on weight. Under these conditions, is the *volume* of the garbage relevant to the problem of community waste disposal? Are there any other factors that are relevant?

e. Based on the preceding results, if you had to institute conservation or recycling efforts because your region's waste facility was almost at full capacity, what would you do?

	Metal	Paper	Plastic	Glass	Food	Yard	Textile	Other
Mean								
Standard deviation								
Shape of distribution								

COMPUTER PROJECT

Refer to Data Set 7 in Appendix B, and use only the Pick Three data consisting of single digits drawn in the Maryland lottery.

a. Using STATDISK or Minitab, either enter the 150 digits as one combined data set or retrieve the data from the data disk available as a supplement to this book.

b. Use STATDISK or Minitab to find the mean, median, range, standard deviation, variance, first quartile Q_1, third quartile Q_3, minimum, and maximum. Also generate a histogram.

c. If the Maryland lottery is fair, we expect each digit to have an equal chance of being selected so that, in the long run, all digits occur with approximately the same frequency. Do your results suggest that the Maryland lottery is fair? Why or why not? (Be sure to include reference to the mean and histogram.)

d. Treat each of the 50 rows as a sample of 3 scores, and manually calculate the 50 sample means. Then enter the 50 means as a new sample, and proceed to find the mean, standard deviation, and shape of the distribution. Compare the results to those obtained in part b.

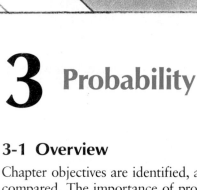

3 Probability

3-1 Overview

Chapter objectives are identified, and probability and statistics are compared. The importance of probability as a discipline in its own right is discussed, as well as its importance in statistics.

3-2 Fundamentals

Both relative frequency and classical definitions of probability are introduced and applied to simple events. The law of large numbers is described, and the complement of an event is defined and illustrated.

3-3 Addition Rule

The addition rule is described, and mutually exclusive events are defined. In applying the addition rule, we compensate for double counting of events that are not mutually exclusive. The rule of complementary events is also introduced.

3-4 Multiplication Rule

The multiplication rule is described, and independent events and conditional probability are defined. Applications to acceptance sampling and redundancy are discussed.

3-5 Probabilities Through Simulations

As an alternative to determining probabilities through theoretical calculations or through observed relative frequencies, simulations are often used to emulate experiments. Results from simulations are illustrated and then used to estimate actual probabilities.

3-6 Counting

Descriptions are provided of the roles played by the fundamental counting principle, factorial rule, permutations rule, and combinations rule in determining the total number of outcomes for a variety of different circumstances.

Chapter Problem:
You be the judge

Dr. Benjamin Spock was convicted of conspiracy to encourage resistance to the military draft during the Vietnam War. His defense attorney argued that he was handicapped by the fact that all 12 jurors were men. The defense argued that Spock was judged unfairly because women had been systematically eliminated in the jury selection process—women would have been more inclined to be sympathetic because of their greater opposition to the war and because Spock was a well-known baby doctor who wrote popular books used by many mothers.

What is the probability of getting an all-male jury? Does that probability actually support the claim that women were systematically eliminated? We will address these questions and reveal the final outcome of the trial.

3-1 Overview

In his excellent book *Innumeracy,* John Allen Paulos states that there is better than a 99% chance that if you take a deep breath, you will inhale a molecule that was exhaled in dying Caesar's last breath. In the same morbid spirit, consider this: If Socrates' fatal cup of hemlock was mostly water, then the next glass of water you drink will likely contain one of those same molecules.

In a class of 25 students, each is asked to identify the month and day of his or her birth. What are the chances that at least two students will share the same birthday? Although most of us tend to think that the chances of that event are quite small, it happens that at least two students will have the same birthday in more than half of all classes with 25 students.

In this country of 90 million voters, a pollster needs to survey only 1000 (or 0.001%) to get a good estimate of the number of voters who favor a particular candidate.

The preceding conclusions are based on simple principles of probability theory, which plays a critical role in statistical analysis. Probability theory is also playing an increasingly important role in a society that must attempt to measure uncertainties. Before firing up a nuclear power plant, we should have some knowledge about the probability of a meltdown. Before arming a nuclear warhead, we should have some knowledge about the probability of an accidental detonation. Before raising the speed limit on our nation's highways, we should have some knowledge of the probability of increased fatalities.

In Chapter 2 we stated that inferential statistics involves using sample evidence to formulate inferences or conclusions about a population. These inferential decisions are based on probabilities or likelihoods of events. As an example, suppose that a statistician plans to study the hiring practices of a large company. She finds that of the last 100 employees hired, all are men. Perhaps the company hires men and women at the same rate and the run of 100 men is an extremely rare chance event. Perhaps the hiring practices favor men. What should we infer? The statistician, along with most reasonable people, would conclude that men are favored. This decision is based on the very low probability of getting 100 consecutive men by chance alone. Subsequent chapters will develop methods of statistical inference that rely on this type of thinking.

In Figure 3-1 we illustrate a fundamental difference between probability and statistics. Figure 3-1(a) shows that we apply probability when we use known population data to determine the likelihood of getting a particular sample. Figure 3-1(b) shows that we apply statistics when we use known sample data to form a conclusion about the unknown population.

This chapter begins by introducing the fundamental concept of mathematical probability and then proceeds to investigate the basic rules of probability: the addition rule, the multiplication rule, and the rule of complements. It also considers techniques for counting the number of ways that an event can occur. The primary objective of this chapter is to develop a sound understanding of

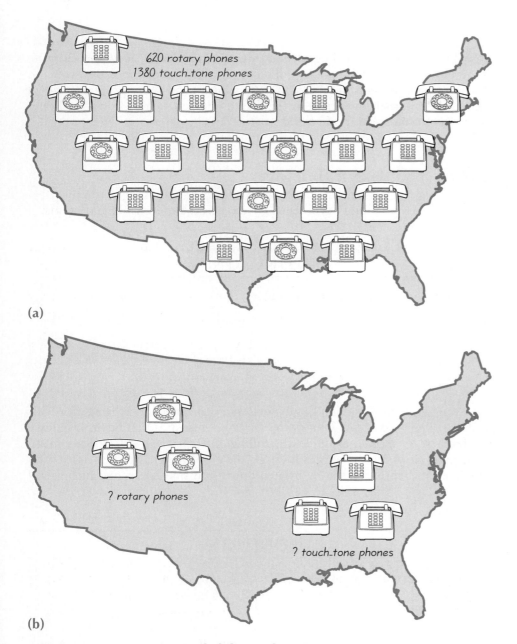

(a)

(b)

FIGURE 3-1 Comparing Probability and Statistics

(a) Probability: With a *known population*, we can see that if 1 telephone is randomly selected, there are 620 chances out of 2000 of getting a rotary phone. This corresponds to a probability of 620/2000, or 0.31. The population of all phones is known, and we are concerned with the likelihood of obtaining the particular sample consisting of a rotary phone. *We are making a conclusion about a sample based on our knowledge of the population.*

(b) Statistics: With an *unknown population,* we can obtain a fairly good idea of the proportion of rotary phones by assuming that this sample is representative of the whole population. After random selections have been made, our sample is known and we can use it to make inferences about the population of all phones in the region. *We are making a conclusion about the population based on our knowledge of the sample.*

probability values that will be used in subsequent chapters, as well as cultivate some intuitive feeling for what probabilities are. A secondary objective is to develop the basic skills necessary to solve simple probability problems, which are valuable in their own right as they are used to make decisions that better enable us to understand our world.

3-2 Fundamentals

In considering probability problems, we deal with experiments and events.

DEFINITIONS

An **experiment** is any process that allows researchers to obtain observations.

An **event** is any collection of results or outcomes of an experiment.

A **simple event** is an outcome or an event that cannot be broken down any further.

For example, the rolling of a single die is an *experiment,* and the result of 3 is an *event.* The outcome of 3 is a *simple event* because it cannot be broken down any further. As another example, the rolling of a pair of dice is an experiment, and the result of 7 is an event. However, the event of getting a 7 is not itself a simple event because it can be broken down into simpler events, such as 3-4 and 6-1.

DEFINITION

The **sample space** for an experiment consists of all possible simple events. That is, the sample space consists of all outcomes that cannot be broken down any further.

In the experiment of rolling a single die, the *sample space* consists of 6 simple events: 1, 2, 3, 4, 5, and 6. In the experiment of rolling a pair of dice, the sample space consists of 36 simple events: 1-1, 1-2, . . . , 6-6. In the experiment of randomly guessing the answer to a multiple-choice test question, the sample space can be listed as {a, b, c, d, e}.

There is no universal agreement on how to define the probability of an event, but among the various theories and schools of thought, two basic approaches emerge most often: the relative frequency approach and the classical approach. These approaches will be embodied in two rules for finding probabilities, and those rules use the following notation.

Notation for Finding Probabilities

P denotes a probability.

A, B, and C denote specific events.

$P(A)$ denotes the probability of event A occurring.

For example, A might represent the event of winning a million-dollar state lottery, and $P(A)$ would denote the probability of that event occurring.

Rule 1 Relative frequency approximation of probability

Conduct (or observe) an experiment a large number of times, and count the number of times that event A actually occurs. Then $P(A)$ is *estimated* as follows:

$$P(A) = \frac{\text{number of times } A \text{ occurred}}{\text{number of times experiment was repeated}}$$

Rule 2 Classical approach to probability

Assume that a given experiment has n different simple events, each of which has an *equal chance* of occurring. If event A can occur in s of these n ways, then

$$P(A) = \frac{s}{n}$$

The classical approach requires *equally likely* outcomes. If the outcomes are not equally likely, we must use the relative frequency estimate. Figure 3-2 illustrates this important distinction.

When determining probabilities by the relative frequency approach (Rule 1), we obtain an *approximation* instead of an exact value. As the total number of observations increases, the corresponding approximations tend to get closer to the actual true probability. This idea is stated as a theorem commonly referred to as the law of large numbers.

Coincidences?

John Adams and Thomas Jefferson (the second and third presidents) both died on July 4, 1826. President Lincoln was assassinated in Ford's Theater; President Kennedy was assassinated in a Lincoln car made by the Ford Motor Company. Lincoln and Kennedy were both succeeded by vice presidents named Johnson. Fourteen years *before* the sinking of the Titanic, a novel described the sinking of the Titan, a ship that hit an iceberg; see Martin Gardner's *The Wreck of the Titanic Foretold?* Gardner states, "In most cases of startling coincidences, it is impossible to make even a rough estimate of their probability."

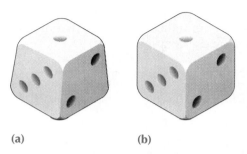

(a)   (b)

**FIGURE 3-2   Comparison of Relative Frequency and
                Classical Approaches**

(a) Relative Frequency Approach (Rule 1): When trying to
determine $P(2)$ on a "shaved" die, we must repeat the
experiment of rolling it many times and then form the ratio of
the number of times 2 occurred to the number of rolls. That
ratio is our *estimate* of $P(2)$.

(b) Classical Approach (Rule 2): With a balanced and fair die,
each of the six faces has an equal chance of occurring.

$$P(2) = \frac{\text{number of ways 2 can occur}}{\text{total number of simple events}} = \frac{1}{6}$$

Law of Large Numbers

As an experiment is repeated again and again, the relative frequency probability (from Rule 1) of an event tends to approach the actual probability.

The law of large numbers tells us that the relative frequency approximations
from Rule 1 tend to get better with more observations. This law reflects a
simple notion supported by common sense: A probability estimate based on
only a few trials can be off by substantial amounts, but with a very large
number of trials, the estimate tends to be much better. For example, if you
estimate the probability of a person being left-handed by randomly selecting
and testing only 5 subjects, you won't be close to the correct result of 0.1.
However, if you base the estimate on 1000 subjects, you will tend to be very
close to the correct probability of 0.1. Figure 3-3 illustrates the law of large
numbers by showing computer-simulated results. Note that as the number of
births increases, the proportion of girls approaches the 0.5 value.

It might seem that when dealing with experiments involving equally likely
outcomes, we should always use Rule 2 for finding probabilities. However,
many experiments involving equally likely outcomes are so complicated that
the classical approach of Rule 2 isn't practical. Instead, we can more easily get

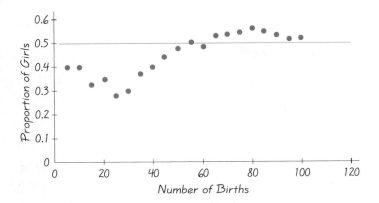

FIGURE 3-3   **Illustration of the Law of Large Numbers**

How Likely Is an Asteroid Strike?

NASA astronomer David Morrison says that there are about 2000 asteroids with orbits that cross Earth's orbit, yet we have found only 100 of them. It's therefore possible for an undetected asteroid to crash into our planet and cause a global catastrophe that could destroy most life. Morrison says that there is a 1/10,000 probability that within a person's lifetime, there will be an asteroid impact large enough to wipe out all agricultural crops for at least a year, with mass starvation following. Some astronomers recommend a 20-year program aimed at observing asteroids with the goal of early detection of those that are dangerous. Early detection could possibly allow us to alter an asteroid's course so that it isn't a fatal threat.

estimates of the desired probabilities by using the relative frequency approach of Rule 1. Simulations, often helpful in such cases, are discussed in Section 3-5.

The examples that follow are intended to illustrate the use of Rules 1 and 2. In some of these examples we use the term *random*.

DEFINITION

In a **random sample of one element** from a population, all elements available for selection have the same chance of being chosen.

The preceding definition applies to the selection of a single item, but throughout the text we often refer to a randomly selected *sample* of several items.

DEFINITION

A sample of *n* items is a **random sample** (or a *simple random sample*) if it is selected in such a way that every possible sample of *n* items from the population has the same chance of being chosen.

The general concept of randomness is extremely important in statistics. When making inferences based on samples, we must have a sampling process that is representative, impartial, and unbiased. Unlike a selection process that is haphazard, random selection usually requires great care and planning.

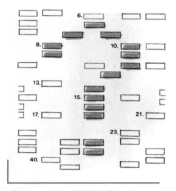

Guess on SATs?

Students preparing for multiple-choice test questions are often told not to guess, but that's not necessarily good advice. Standardized tests with multiple-choice questions typically *compensate* for guessing, but they don't *penalize* it. For questions with five answer choices, one-fourth of a point is usually subtracted for each incorrect response. Principles of probability show that in the long run, pure random guessing will neither raise nor lower the exam score. Definitely guess if you can eliminate at least one choice or if you have a sense for the right answer, but avoid tricky questions with attractive wrong answers. Also, don't waste too much time on such questions.

▶ **EXAMPLE**

On an ACT or SAT test, a typical question has 5 possible answers. If an examinee makes a random guess on one such question, what is the probability that the response is wrong?

● **SOLUTION**

There are 5 possible outcomes or answers, and there are 4 ways to answer incorrectly. Random guessing implies that the outcomes are equally likely, so we apply the classical approach (Rule 2) to get

$$P(\text{wrong answer}) = \frac{4}{5} = 0.8$$

In the next example, note that it is necessary to first determine the total number of cases before finding the desired probability.

▶ **EXAMPLE**

In a Bruskin-Goldring Research poll, respondents were asked how a fruitcake should be used. One hundred thirty-two respondents indicated that it should be used for a doorstop, and 880 other respondents cited other uses, including birdfeed, landfill, and a gift. If one of these respondents is randomly selected, what is the probability of getting someone who would use a fruitcake for a doorstop?

● **SOLUTION**

The total number of subjects in this study is $132 + 880 = 1012$. With random selection, the 1012 subjects are equally likely and Rule 2 applies.

$$P(\text{doorstop}) = \frac{\text{number of doorstop responses}}{\text{total number of responses}} = \frac{132}{1012} = 0.130$$

There is a 0.130 probability that when one of the survey respondents is randomly selected, the selected respondent will be someone who would use a fruitcake for a doorstop.

► **EXAMPLE**

Find the probability that a couple with 3 children will have exactly 2 boys. Assume that boys and girls are equally likely and that the sex of any child is not influenced by the sex of any other child.

● **SOLUTION**

We first list the sample space that identifies the 8 outcomes. Those outcomes are equally likely, so we use Rule 2. Of those 8 different possible outcomes, 3 correspond to exactly 2 boys, so

$$P(2 \text{ boys in 3 births}) = \frac{3}{8} = 0.375$$

There is a 0.375 probability that if a couple has 3 children, exactly 2 will be boys.

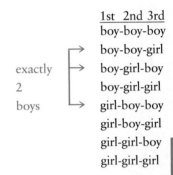

1st 2nd 3rd
boy-boy-boy
boy-boy-girl
boy-girl-boy
boy-girl-girl
girl-boy-boy
girl-boy-girl
girl-girl-boy
girl-girl-girl

exactly
2
boys

► **EXAMPLE**

Find the probability that a randomly selected college student owns a personal computer.

● **SOLUTION**

The two outcomes of owning a personal computer and not owning one aren't equally likely, so the relative frequency approximation must be used. This requires that we somehow observe a large number of college students and count those who own a personal computer. Let's assume that we survey 500 college students and find that 135 of them own a personal computer. (These are realistic figures based on data from the American Passage Media Corporation.) Then the relative frequency approximation becomes

$$P(\text{college student owning a personal computer}) = \frac{135}{500} = 0.27$$

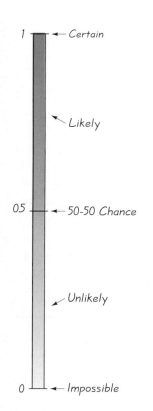

FIGURE 3-4   Possible Values for Probabilities

▶ **EXAMPLE**

If a year is selected at random, find the probability that Thanksgiving Day will be on a (a) Wednesday, (b) Thursday.

● **SOLUTION**

a.   Thanksgiving Day always falls on the fourth Thursday in November. It is therefore impossible for Thanksgiving to be on a Wednesday. When an event is impossible, we say that its probability is 0.

b.   It is certain that Thanksgiving will be on a Thursday. When an event is certain to occur, we say that its probability is 1.

The probability of any impossible event is 0.
The probability of any event that is certain to occur is 1.

Because any event imaginable is impossible, certain, or somewhere in between, it is reasonable to conclude that the mathematical probability of any event is 0, 1, or a number between 0 and 1 (see Figure 3-4). This property can be expressed as follows:

$$0 \le P(A) \le 1 \text{ for any event } A$$

In Figure 3-4, the scale of 0 through 1 is shown on the left, whereas the more familiar and common expressions of likelihood are shown on the right.

Complementary Events

Sometimes we need to find the probability that an event A does *not* occur.

DEFINITION

The **complement** of event A, denoted by $\overline{A}$, consists of all outcomes in which event A does not occur.

▶ **EXAMPLE**

A consumer test group consists of 80 college students, 30 of whom are women. If one person is randomly selected from this group, find the probability of *not* getting a woman.

● **SOLUTION**

Because 30 of the 80 consumers are women, it follows that 50 of them are *not* women, so

$$P(\text{not selecting a woman}) = P\,(\overline{\text{woman}}) = \frac{50}{80} = 0.625$$

Although it is difficult to develop a universal rule for rounding off probabilities, the following guide will apply to most problems in this text.

Rounding Off Probabilities

Either give the exact fraction or decimal representing a probability or round off final decimal results to three significant digits.

All the digits in a number are significant except for the zeros that are included for proper placement of the decimal point. The probability of 0.00128506 has six significant digits (128506), and it can be rounded to three significant digits as 0.00129. The probability of 1/3 can be left as a fraction or rounded in decimal form to 0.333, but not 0.3. The probability of heads in a coin toss can be expressed as 1/2 or 0.5; because 0.5 is *exact,* there's no need to express it as 0.500.

An important concept of this section is the mathematical expression of probability as a number between 0 and 1. This type of expression is fundamental and common in statistical procedures, and we will use it throughout the remainder of this text. A typical computer output, for example, may include a "*P*-value" expression such as "significance less than 0.001." We will discuss the meaning of *P*-values later, but they are essentially probabilities of the type discussed in this section. For now, you should recognize that a probability of 0.001 (equivalent to 1/1000) corresponds to an event so rare that it occurs an average of only once in a thousand trials.

Sensitive Surveys

Survey respondents are sometimes reluctant to honestly answer questions on a sensitive topic, such as employee theft or sex. Stanley Warner (York University, Ontario) devised a scheme that leads to more accurate results in such cases. As an example, ask employees if they stole within the past year and also ask them to flip a coin. The employees are instructed to answer no if they didn't steal and the coin turns up heads. Otherwise, they should answer yes. The employees are more likely to be honest because the coin flip helps protect their privacy. Probability theory can then be used to analyze responses so that more accurate results can be obtained.

Subjective Probabilities

This section presented the relative frequency approach and the classical approach as two formal methods for finding probabilities of events; another approach is to simply guess or estimate a probability. The technique of guessing is familiar to students who are sometimes not as well prepared for a test as they would like, but it is also used by professionals who set casino odds for sporting events. A guess or estimate based on knowledge of relevant circumstances is commonly referred to as a **subjective probability.** For example, a Las Vegas oddsmaker might estimate that there is a 0.05 probability that the New York Giants will win the Super Bowl next year. That estimate is based on knowledge of relevant factors, such as the quality of the coach and players.

How Probable?

How do we interpret such terms as *probable, improbable,* or *extremely improbable*? The FAA interprets these terms as follows. *Probable:* A probability on the order of 0.00001 or greater for each hour of flight. Such events are expected to occur several times during the operational life of each airplane. *Improbable:* A probability on the order of 0.00001 or less. Such events are not expected to occur during the total operational life of a single airplane of a particular type, but may occur during the total operational life of all airplanes of a particular type. *Extremely improbable:* A probability on the order of 0.000000001 or less. Such events are so unlikely that they need not be considered to ever occur.

3-2   Exercises A: Basic Skills and Concepts

1. Which of the following values *cannot* be probabilities?

$$4/3, \quad 0, \quad 0.999, \quad 1.000, \quad 1.001, \quad -0.2, \quad 2, \quad \sqrt{2}, \quad \sqrt{3/4}$$

2. a. What is $P(A)$ if event A is certain to occur?
   b. What is $P(A)$ if event A is impossible?
   c. A sample space consists of 200 separate events that are equally likely. What is the probability of each?
   d. On a college entrance exam, each question has 5 possible answers. If an examinee makes a random guess on the first question, what is the probability that the response is correct?

3. In a survey of 3630 college students, 1162 stated that they cheated on an exam (based on data from the Josephson Institute of Ethics). If one of these college students is randomly selected, find the probability that he or she cheated on an exam.

4. Among 750 taxpayers with incomes of less than $100,000, 20 are audited by the IRS (based on IRS data). Use this sample to estimate the probability of a tax return being audited if the income is less than $100,000.

5. Among 80 randomly selected blood donors, 36 were classified as group O (based on data from the Greater New York Blood Program). What is the approximate probability that a person will have group O blood?

6. If a person is randomly selected, find the probability that his or her birthday is October 18, which is National Statistics Day in Japan. Ignore leap years.

7. In a recent national election, there were 25,569,000 citizens in the 18–24 age bracket. Of these, 9,230,000 actually voted. Find the relative frequency estimate of the probability that a person randomly selected from this group did vote in that national election.

8. In a study of brand recognition, 831 consumers knew of Campbell's Soup, and 18 did not (based on data from Total Research Corporation). Use these results to estimate the probability that a randomly selected consumer will recognize Campbell's Soup.

9. In a survey of U.S. households, 288 had home computers, and 962 did not (based on data from Electronic Industries Association). Use this sample to estimate the probability of a household having a home computer.

10. In a study of train derailments, human error was found to be the cause in 36 cases, while other causes were cited in 114 different cases (based on data from the Federal Railroad Administration). Estimate the probability that in a train derailment, human error will be the cause.

11. A Bureau of the Census survey of 600 persons in the 18–25 age bracket found that 237 of them smoke. If a person in that age bracket is randomly selected, find the approximate probability that he or she smokes.

12. According to the U.S. Department of Transportation, American Airlines boarded 59,377,306 passengers in a recent year. In that same year, 82,796 passengers were voluntarily denied boarding, while 1664 other passengers were involuntarily denied boarding. Find the probability that a randomly selected passenger will be involuntarily denied boarding.

13. Among 400 randomly selected drivers in the 20–24 age bracket, 136 were in a car accident during the last year (based on data from the National Safety Council). If a driver in that age bracket is randomly selected, what is the approximate probability that he or she will be in a car accident during the next year?

14. The U.S. General Accounting Office recently tested the IRS for correctness of answers to taxpayers' questions. For 1733 trials, the IRS was correct 1107 times. Use these results to estimate the probability that a random taxpayer's question will be answered correctly.

15. When the allergy drug Seldane was clinically tested, 70 people experienced drowsiness, and 711 did not (based on data from Merrell Dow Pharmaceuticals, Inc.). Use this sample to estimate the probability of a Seldane user becoming drowsy.

16. Data provided by the Bureau of Justice Statistics revealed that for a representative sample of convicted burglars, 76,000 were jailed, 25,000 were put on probation, and 2000 received other sentences. Use these results to estimate the probability that a convicted burglar will serve jail time.

17. Blood groups were determined for a sample of people; the results are given in the accompanying table (based on data from the Greater New York Blood Program). If one person from this sample group is randomly selected, find the probability that the person has group AB blood.

Blood group	Frequency
O	90
A	80
B	20
AB	10

18. A study of credit card fraud was conducted by MasterCard International, and the accompanying table is based on the results. If one case of credit card fraud is randomly selected from the cases summarized in the table, find the probability that the fraud resulted from a counterfeit card.

Method of fraud	Number of cases
Stolen card	243
Counterfeit card	85
Mail or telephone order	52
Other	46

19. The accompanying table summarizes recent driver convictions for selected violations in two counties (the data are from the New York State Department of Motor Vehicles).

	Speeding	DWI
Dutchess County	10,589	636
Westchester County	22,551	963

If one of the convictions is randomly selected, find the probability that it is for DWI (driving while intoxicated).

20. A Gallup survey resulted in the sample data in the given table. If one of the survey respondents is randomly selected, find the probability of getting someone who brushes three times per day, as dentists recommend.

Number of tooth-brushings per day	Number of respondents
1	228
2	672
3	240

21. A couple plans to have 2 children.
   a. List the different outcomes according to the sex of each child. Assume that these outcomes are equally likely.
   b. Find the probability of getting 2 girls.
   c. Find the probability of getting exactly 1 child of each sex.

22. A couple plans to have 4 children.
   a. List the 16 different possible outcomes according to the sex of each child. Assume that these outcomes are equally likely.
   b. Find the probability of getting all girls.
   c. Find the probability of getting *at least* 1 child of each sex.
   d. Find the probability of getting *exactly* 2 children of each sex.

23. On a quiz consisting of 3 true/false questions, an unprepared student must guess at each one. The guesses will be random.
   a. List the different possible solutions.
   b. What is the probability of answering all 3 questions correctly?

(continued)

c. What is the probability of guessing incorrectly for all questions?

d. What is the probability of passing the quiz by guessing correctly for *at least* 2 questions?

24. Both parents have the brown/blue pair of eye-color genes, and each parent contributes one gene to a child. Assume that if the child has at least one brown gene, that color will dominate and the eyes will be brown. (Actually, the determination of eye color is somewhat more complex.)

   a. List the different possible outcomes. Assume that these outcomes are equally likely.

   b. What is the probability that a child of these parents will have the blue/blue pair of genes?

   c. What is the probability that the child will have brown eyes?

3-2   Exercises B: Beyond the Basics

25. The stem-and-leaf plot summarizes the time (in hours) that managers spend on paperwork in one day (based on data from Adia Personnel Services). Use this sample to estimate the probability that a randomly selected manager spends more than 2.0 hours per day on paperwork.

0.	00
1.	0578
2.	00113449
3.	347
4.	445

26. In Exercise 6 leap years were ignored in finding the probability that a randomly selected person will have a birthday on October 18.

   a. Recalculate this probability, assuming that a leap year occurs every 4 years. (Express your answer as an exact fraction.)

   b. Leap years occur in years evenly divisible by 4, except they are skipped in 3 of every 4 centesimal years (years ending in 00). The years 1700, 1800, and 1900 were not leap years, but 2000 will be a leap year. Find the exact probability for this case, and express it as a fraction.

27. If 2 flies land on an orange, find the probability that they are on points that are within the same hemisphere.

28. Two points along a straight stick are randomly selected. The stick is then broken at those 2 points. Find the probability that the 3 resulting pieces can be arranged to form a triangle. (This is a difficult problem.)

29. The **odds against** event A occurring are given by the ratio $P(\overline{A})/P(A)$, usually expressed in the form of $a{:}b$ (or "a to b"), where a and b are integers having no common factors. The **odds in favor** of event A are given by the reciprocal of the odds against that event. If the odds *against* event A are $a{:}b$, then the odds in favor are $b{:}a$.

   a. Find the odds against selecting the correct winning slot in roulette if you select 1 slot on the wheel among the 38 available slots that are all equally likely.

   b. Find the odds in favor of the event described in part a.

3-3 Addition Rule

The preceding section introduced the basic concept of probability and considered experiments and simple events. Many real situations involve compound events, such as the random selection of a consumer who is a woman *or* who is younger than age 40.

DEFINITION

Any event combining 2 or more simple events is called a **compound event.**

In this section we want to develop a rule for finding $P(A$ or $B)$—the probability that for a single outcome of an experiment, either event A or event B occurs or both events occur. (Throughout this text we use the inclusive *or*, which means either one or the other or both. We will *not* consider the exclusive *or*, which means either one or the other but not both.)

Notation for Addition Rule

$P(A$ or $B) = P$(event A occurs or event B occurs or they both occur)

Let's consider the sample survey results obtained from 2000 different crime victims, as summarized in Table 3-1 (based on data from the U.S. Department of Justice). In exploring data consisting of frequency counts for different categories, we find it helpful to arrange the data in such a table. (This type of table will be useful in solving several exercises.)

TABLE 3-1   Relationship of Criminal to Victim

	Homicide	Robbery	Assault	Totals
Stranger	12	379	727	1118
Acquaintance or relative	39	106	642	787
Unknown	18	20	57	95
Totals	69	505	1426	2000

The table lists frequencies of crime victims in different categories. By adding all of the individual cell frequencies, we see that 2000 crime victims are included in this experiment. We also see that 1118 of them were victimized by strangers, 787 were victimized by an acquaintance or relative, 95 were victimized by unknown criminals, 69 were homicide victims, 505 were robbery victims, and 1426 were assaulted. We will use two cases as a basis for developing our rule for determining $P(A \text{ or } B)$.

Case 1:   If 1 of the 2000 crime victims is randomly selected, the probability of getting someone who was assaulted or robbed is

$$\frac{1426}{2000} + \frac{505}{2000} = \frac{1931}{2000} = 0.966$$

That is, $P(\text{assault or robbery}) = P(\text{assault}) + P(\text{robbery})$. This suggests that $P(A \text{ or } B) = P(A) + P(B)$, but consider the next case before making that generalization.

Case 2:   Suppose that we are going to randomly select 1 of the 2000 crime victims, and we want the probability that we will get someone who was robbed or victimized by a stranger. Following the same pattern as in Case 1, we might write

$$P(\text{robbed}) + P(\text{stranger}) = \frac{505}{2000} + \frac{1118}{2000} = \frac{1623}{2000} = 0.812 \qquad (\textit{WRONG!})$$

This probability is wrong because 379 victims—those robbed by a stranger—were counted *twice*. One way to compensate for double counting is to subtract the 379 that was included twice. We get

$$P(\text{robbed or stranger}) = \frac{505}{2000} + \frac{1118}{2000} - \frac{379}{2000} = \frac{1244}{2000} = 0.622$$

This last calculation can be expressed as

$$P(\text{robbed or stranger}) = P(\text{robbed}) + P(\text{stranger}) - P(\text{both})$$

This approach is generalized in the following addition rule.

You Bet

In the typical state lottery, the "house" has a 65% to 70% advantage, since only 35% to 40% of the money bet is returned as prizes. The house advantage at racetracks is usually around 15%. In casinos, the house advantage is 5.26% for roulette, 5.9% for blackjack, 1.4% for craps, and 3% to 22% for slot machines. Some professional gamblers can systematically win at blackjack by using complicated card-counting techniques. They know when a deck has disproportionately more high cards, and this is when they place large bets. Many casinos react by ejecting card counters or by shuffling the decks more frequently.

Addition Rule

$$P(A \text{ or } B) = P(A) + P(B) - P(A \text{ and } B)$$

where $P(A \text{ and } B)$ denotes the probability that A and B both occur at the same time as an outcome in the experiment.

Total Area = 1

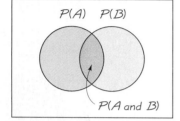

FIGURE 3-5   Venn Diagram Showing Overlapping Events

Total Area = 1

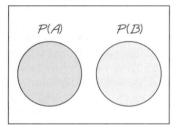

FIGURE 3-6   Venn Diagram Showing Nonoverlapping Events

The example in Case 2 addresses this key point: **When combining the number of ways event _A_ can occur with the number of ways _B_ can occur, we must avoid double counting of those outcomes in which _A_ and _B_ both happen together.**

Figure 3-5 shows a Venn diagram that provides a visual illustration of the addition rule. In this figure we can see that the probability of _A_ or _B_ equals the probability of _A_ (left circle) plus the probability of _B_ (right circle) minus the probability of _A_ and _B_ (football-shaped middle region). This figure shows that the addition of the areas of the two circles will cause double counting of the football-shaped middle region. This is the basic concept that underlies the addition rule. Because of the relationship between the addition rule and the Venn diagram shown in Figure 3-5, the notation $P(A \cup B)$ is often used in place of $P(A$ or $B)$. Similarly, the notation $P(A \cap B)$ is often used in place of $P(A$ and $B)$, so the addition rule can be expressed as

$$P(A \cup B) = P(A) + P(B) - P(A \cap B)$$

The addition rule is simplified whenever _A_ and _B_ cannot occur simultaneously, so $P(A$ and $B)$ becomes zero. Figure 3-6 illustrates that with no overlapping of _A_ and _B_, we have $P(A$ or $B) = P(A) + P(B)$. The following definition formalizes the lack of overlapping shown in Figure 3-6.

DEFINITION

Events _A_ and _B_ are **mutually exclusive** if they cannot occur simultaneously.

The flowchart of Figure 3-7 shows how mutually exclusive events affect the addition rule.

In the experiment of randomly selecting a crime victim, the event of getting someone who was victimized by a stranger and the event of getting someone who was victimized by an acquaintance are mutually exclusive, because no criminal can be both a stranger and an acquaintance of the victim. We can summarize the key points of this section as follows:

1.   To find $P(A$ or $B)$, begin by associating _or_ with addition.

2.   Consider whether events _A_ and _B_ are mutually exclusive. That is, can they happen at the same time? If they are not mutually exclusive (that is, if they can happen at the same time), be sure to avoid (or at least compensate for) double counting when adding the relevant probabilities. _Important hint:_ If you understand the importance of _not_ double counting when you find $P(A$ or $B)$, you don't necessarily have to formally calculate $P(A) + P(B) - P(A$ and $B)$. In finding $P(\text{robbed or stranger})$ from Table 3-1, you could sum the frequencies for the robbery column and the stranger row by being careful to count each cell exactly once. You would get $379 + 106 + 20 + 12 + 727 = 1244$, which,

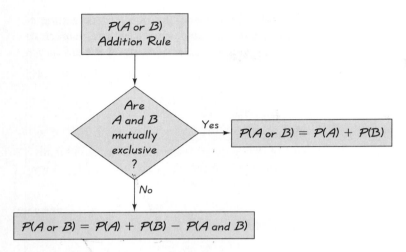

FIGURE 3-7 Applying the Addition Rule

Voltaire Beats Lottery

In 1729, the philosopher Voltaire became rich by devising a scheme to beat the Paris lottery. The government ran a lottery to repay municipal bonds that had lost some value. The city added large amounts of money with the net effect that the prize values totaled more than the cost of all tickets. Voltaire formed a group that bought all the tickets in the monthly lottery and won for more than a year. A bettor in the New York State Lottery tried to win a share of an exceptionally large prize that grew from a lack of previous winners. He wanted to write a $6,135,756 check that would cover all combinations, but the state declined and said that the nature of the lottery would have been changed.

when divided by the total number of subjects (2000), will yield the correct probability of 0.622. *It's much better to understand what you're doing than to blindly apply a formula.*

The following example further illustrates application of the addition rule.

▶ **EXAMPLE**

Survey subjects are often chosen by using computers to randomly select telephone numbers. Assume that a computer randomly generates the last digit of a telephone number. Find the probability that the outcome is (a) an 8 or 9, (b) odd or less than 4.

● **SOLUTION**

a. The outcome of 8 and the outcome of 9 are mutually exclusive events. This means that it is impossible for both 8 and 9 to occur together when 1 digit is selected, so $P(8 \text{ and } 9) = 0$ and the addition rule is applied as follows:

$$P(8 \text{ or } 9) = P(8) + P(9) - P(8 \text{ and } 9)$$
$$= \frac{1}{10} + \frac{1}{10} - 0$$
$$= \frac{2}{10} = \frac{1}{5} = 0.2$$

(continued)

b. The outcome of getting an odd number and the outcome of getting a number less than 4 are not mutually exclusive, since they both happen if the result is a 1 or a 3. Compensating for double counting, we get

$P(\text{odd or less than 4}) = P(\text{odd}) + P(\text{less than 4}) - P(\text{odd and less than 4})$

$$= \frac{5}{10} + \frac{4}{10} - \frac{2}{10} = \frac{7}{10}, \text{ or } 0.7$$

In this result, $P(\text{odd}) = 5/10$ as 5 of the 10 digits are odd (1, 3, 5, 7, 9). $P(\text{less than 4}) = 4/10$ because 4 of the digits are less than 4 (0, 1, 2, 3). Finally, $P(\text{odd and less than 4}) = 2/10$ because 2 of the digits are both odd and less than 4 (1 and 3). The correct answer is 7/10, or 0.7.

Errors made when applying the addition rule often involve double counting. That is, events that are not mutually exclusive are treated as if they were. One indication of such an error is a total probability that exceeds 1; however, errors involving the addition rule do not always cause the total probability to exceed 1.

Complementary Events

In Section 3-2 we defined the complement of event A and denoted it by $\overline{A}$. The definition of complementary events implies that they must be mutually exclusive because it is impossible for an event and its opposite to occur at the same time. Also, we can be absolutely certain that A either does or does not occur. That is, either A or $\overline{A}$ must occur. These observations enable us to apply the addition rule for mutually exclusive events as follows:

$$P(A \text{ or } \overline{A}) = P(A) + P(\overline{A}) = 1$$

We justify $P(A \text{ or } \overline{A}) = P(A) + P(\overline{A})$ by noting that A and $\overline{A}$ are mutually exclusive; we justify the total of 1 by our absolute certainty that A either does or does not occur. This result of the addition rule leads to the following three *equivalent* forms.

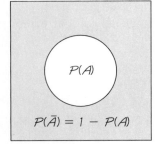

Total Area = 1

$P(A)$

$P(\overline{A}) = 1 - P(A)$

FIGURE 3-8   Venn Diagram for the Complement of Event A

Rule of Complementary Events
$P(A) + P(\overline{A}) = 1$
$P(\overline{A}) = 1 - P(A)$
$P(A) = 1 - P(\overline{A})$

The first form comes directly from our original result. The second (see Figure 3-8) and third variations involve very simple equation manipulations.

▶ **EXAMPLE**

If $P(A) = 0.4$, find $P(\overline{A})$.

● **SOLUTION**

Using the rule of complementary events, we get

$$P(\overline{A}) = 1 - P(A) = 1 - 0.4 = 0.6$$

A major advantage of the rule of complementary events is that it can be used to greatly simplify certain problems. We will illustrate this advantage in the following section.

3-3 Exercises A: Basic Skills and Concepts

In Exercises 1 and 2, for each pair of events given, determine whether the two events are mutually exclusive for a single experiment.

1. a. Selecting a student who regularly attends statistics class
   Selecting a student who owns a computer
   b. Selecting a person with blond hair (natural or otherwise)
   Selecting a person with brown eyes
   c. Selecting a required course
   Selecting an elective course
2. a. Selecting someone who colors his or her hair
   Selecting someone who has read *The Greening of America*
   b. Selecting an unmarried television viewer
   Selecting a television viewer with an employed spouse
   c. Selecting a new Corvette that is free of defects
   Selecting a car with inoperative headlights
3. a. If $P(A) = 0.45$, find $P(\overline{A})$.
   b. Based on recent data from the U.S. National Center for Health Statistics, the probability of a baby being a boy is 0.513. Find the probability of a baby being a girl.

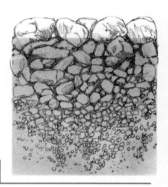

Growing Rocks

A *New York Times* article explained why the larger corn flakes seem to rise to the top of the box and why rocks rise to the surface of land that had been previously cleared of all surface rocks. A computer model was used to simulate mixtures of large and small objects. When such mixtures are shaken, gaps are created, but there are many more smaller gaps than larger ones. There is a greater probability that a smaller object will move downward into those more abundant smaller gaps. Larger objects tend to be forced upward, even if they are heavier. The same principle affects how we mix things such as cement and pharmaceuticals.

4. a. Find $P(\overline{A})$, given that $P(A) = 3/8$.
   b. Based on data from the National Conference of Bar Examiners, the probability of someone failing a bar exam is 0.43. Find the probability of someone passing the bar exam.

5. Pollsters are concerned about declining levels of cooperation among persons contacted in surveys. A pollster contacts 84 people in the 18–21 age bracket and finds that 73 of them respond and 11 refuse to respond. When 275 people in the 22–29 age bracket are contacted, 255 respond and 20 refuse to respond (based on data from "I Hear You Knocking but You Can't Come In," by Fitzgerald and Fuller, *Sociological Methods and Research,* Vol. 11, No. 1). Assume that 1 of the 359 people is randomly selected. Find the probability of getting someone in the 18–21 age bracket or someone who refused to respond.

6. Refer to the same data set from Exercise 5, and find the probability of getting someone who is in the 18–21 age bracket or someone who responded.

7. Problems of sexual harassment have been given much more attention in recent years. In one survey, 420 workers (240 of whom are men) considered a friendly pat on the shoulder to be a form of harassment, whereas 580 workers (380 of whom are men) did not consider that to be a form of harassment (based on data from Bruskin/Goldring Research). If one of the surveyed workers is randomly selected, find the probability of getting someone who does not consider a pat on the shoulder to be a form of harassment.

8. Refer to the same data set in Exercise 7, and find the probability of randomly selecting a man or someone who does not consider a pat on the shoulder to be a form of harassment.

9. Among 200 seats available on a British Airways flight, 40 are reserved for smokers (including 16 aisle seats) and 160 are reserved for nonsmokers (including 64 aisle seats). If a late passenger is randomly assigned a seat, find the probability of getting an aisle seat or one in the smoking section.

10. Refer to the crime data in Table 3-1. If one of the crime victims is randomly selected, find the probability of getting a homicide victim murdered by a stranger.

11. A study of consumer smoking habits includes 200 married people (54 of whom smoke), 100 divorced people (38 of whom smoke), and 50 adults who never married (11 of whom smoke) (based on data from the U.S. Department of Health and Human Services). If 1 subject is randomly selected from this sample, find the probability of getting someone who is divorced or smokes.

12. Refer to the data in Exercise 11, and find the probability of getting someone who was never married or does not smoke.

In Exercises 13–16, refer to the data in the accompanying table, which describes the age distribution of Americans who died by accident (based on data from the National Safety Council). In each case, assume that one person is randomly selected from this sample group.

13. Find the probability of selecting someone younger than 5 or older than 74.

Age	Number
0–4	3,843
5–14	4,226
15–24	19,975
25–44	27,201
45–64	14,733
65–74	8,499
75 and over	16,800

14. Find the probability of selecting someone between 15 and 64.
15. Find the probability of selecting someone younger than 45 or between 25 and 74.
16. Find the probability of selecting someone under 25 or between 15 and 44.

In Exercises 17–24, refer to the accompanying figure, which describes the blood groups and Rh types of 100 people (based on data from the Greater New York Blood Program). In each case, assume that 1 of the 100 subjects is randomly selected, and find the indicated probability.

17. P(group A or group B)
18. P(type Rh^+)
19. P(group A or type Rh^-)
20. P(group O or type Rh^+)
21. P(not group AB)
22. P(not type Rh^-)
23. P(group O or type Rh^-)
24. P(group A or group O or type Rh^-)

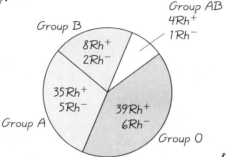

3-3 **Exercises B: Beyond the Basics**

25. a. If $P(A \text{ or } B) = 1/3$, $P(B) = 1/4$, and $P(A \text{ and } B) = 1/5$, find $P(A)$.
    b. If $P(A) = 0.4$ and $P(B) = 0.5$, what is known about $P(A \text{ or } B)$ if A and B are mutually exclusive events?
    c. If $P(A) = 0.4$ and $P(B) = 0.5$, what is known about $P(A \text{ or } B)$ if A and B are not mutually exclusive events?
26. If events A and B are mutually exclusive and events B and C are mutually exclusive, must events A and C be mutually exclusive? Give an example supporting your answer.
27. How is the addition rule changed if the exclusive *or* is used instead of the inclusive *or*? Recall that the exclusive *or* means either one or the other but not both.
28. Given that $P(A \text{ or } B) = P(A) + P(B) - P(A \text{ and } B)$, develop a rule for $P(A \text{ or } B \text{ or } C)$. (*Hint:* Draw a Venn diagram.)

3-4 **Multiplication Rule**

In Section 3-3 we developed a rule for finding the probability that event A or B will occur in a given experiment. We will now develop a rule for finding the probability that events A and B both occur. We begin with a simple example, which will suggest a preliminary multiplication rule. Then we use another example to develop a variation and ultimately obtain a generalized multiplication rule.

Independent Jet Engines

A three-engine jet departed from Miami International Airport en route to South America, but one engine failed immediately after takeoff. While the plane was turning back to the runway, the other two engines also failed, but the pilot was able to make a safe landing. With independent jet engines, the probability of all three failing is only 0.0001^3, or about one chance in a trillion. The FAA found that the same mechanic who replaced the oil in all three engines incorrectly positioned the oil plug sealing rings. A goal in using three separate engines is to increase safety with independent engines, but the use of a single mechanic caused their operation to become dependent. Maintenance procedures now require that the engines be serviced by different mechanics.

Probability theory is used extensively in the analysis and design of standardized tests. Practical considerations often require that such tests allow only those answers that can be corrected easily, such as true/false or multiple choice. Let's assume that the first question on a test is a true/false type, while the second question is multiple choice with 5 possible answers (a, b, c, d, e). We will use the following two questions. Try them!

1. True or false: At roughly 3/4 of U.S. large companies, some employees and job applicants are tested for drugs.
2. The father of Euclidean geometry is
    a. George Metry
    b. There was no father, only a mother.
    c. Euclid
    d. Triola
    e. None of the above

The grading of standardized tests often includes compensation for guessing, so there is a need to find the probabilities of guessing correctly. Suppose we want to determine the probability of getting both answers correct by making random guesses. We begin by listing the complete sample space of different possible answers.

T,a	T,b	T,c	T,d	T,e	1 case is correct
F,a	F,b	F,c	F,d	F,e	10 equally likely cases

If the answers are random guesses, then the 10 possible outcomes are equally likely. The correct answers are T and c, so

$$P(\text{both correct}) = P(\text{T and c}) = \frac{1}{10} = 0.1$$

Considering the component answers of T and c, respectively, we see that with random guesses we have $P(\text{T}) = 1/2$ and $P(\text{c}) = 1/5$. Recognizing that $1/10$ is the product of $1/2$ and $1/5$, we observe that $P(\text{T and c}) = P(\text{T}) \cdot P(\text{c})$. We use this observation as a basis for formulating the following preliminary multiplication rule: $P(A \text{ and } B) = P(A) \cdot P(B)$.

When the multiplication rule is used for finding probabilities in compound events, tree diagrams are sometimes helpful in determining the number of possible outcomes in a sample space. A **tree diagram** is a picture of the possible outcomes of an experiment, shown as line segments emanating from one starting point. These diagrams are helpful in counting the number of possible outcomes if the number of possibilities is not too large. The tree diagram shown in Figure 3-9 summarizes the outcomes of the true/false and multiple-choice questions.

Assuming that both answers are random guesses, all 10 branches are equally likely and the probability of getting the correct pair (T, c) is 1/10. For each response to the first question, there are 5 responses to the second. The total

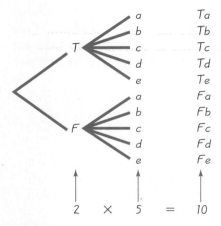

FIGURE 3-9   Tree Diagram of Test Answers

number of outcomes is, therefore, 5 taken 2 times, or 10. The tree diagram in Figure 3-9 illustrates the reason for the use of multiplication.

Probability theory and the multiplication rule are often used to test the quality of a group of products. When it's impractical or impossible to test every item in a group, samples are selected and tested. If all of the sampled items are good, the whole group is judged to be good. The next example is related to this procedure.

▶ **EXAMPLE**

Five hair dryers are produced; 4 of them are good and 1 is defective. If we randomly select 1 of them, there is a 4/5 probability that it is good. Suppose, however, that 2 hair dryers are randomly selected for testing, and the first is replaced before the second selection is made. Find the probability that both hair dryers selected are good.

● **SOLUTION**

Letting G represent the event of selecting a good hair dryer, we want $P(G$ and $G)$. With $P(G) = 4/5$, we apply the multiplication rule to get

$$P(G \text{ and } G) = P(G) \cdot P(G) = \frac{4}{5} \cdot \frac{4}{5} = \frac{16}{25} = 0.64$$

When 2 hair dryers are selected with replacement, there is a 0.64 probability that they are both good.

Probably Guilty?

A witness described a Los Angeles robber as a Caucasian woman with blond hair in a ponytail who escaped in a yellow car driven by a black male with a mustache and beard. Janet and Malcolm Collins fit this description, and they were convicted after a college math instructor testified that there is only about 1 chance in 12 million that any couple would have those characteristics. It was estimated that the probability of a car being yellow is 1/10; the other characteristics were also estimated. The convictions were later reversed when it was noted that no evidence was presented to support the estimated probabilities, and the independence of the characteristics was not established.

Although the preceding solution is mathematically correct, common sense suggests that product testing should be conducted without replacement of the items already tested. There is always the chance that you could test the same item twice. The next example uses an improved sampling procedure in which selected items are not replaced.

▶ **EXAMPLE**

Let's again assume that we have 5 hair dryers, 4 of which are good. We will again randomly select 2 of the hair dryers, but we will *not* replace the first selection. Find the probability of getting 2 good hair dryers with this improved sampling procedure.

● **SOLUTION**

To understand the sample space better, we represent the 4 good hair dryers by G_1, G_2, G_3, and G_4 and we represent the defective hair dryer by D. The sample space is listed below for the experiment of selecting 2 hair dryers without replacement.

$$
\begin{array}{cccc}
G_1G_2 & G_1G_3 & G_1G_4 & G_1D \\
G_2G_1 & G_2G_3 & G_2G_4 & G_2D \\
G_3G_1 & G_3G_2 & G_3G_4 & G_3D \\
G_4G_1 & G_4G_2 & G_4G_3 & G_4D \\
DG_1 & DG_2 & DG_3 & DG_4
\end{array}
$$

By examining this list of 20 equally likely outcomes, we see that 12 cases correspond to 2 good hair dryers, so $P(G \text{ and } G) = 12/20$, or 0.6. Without replacement of the first selection, there is a slightly smaller chance (0.6) of getting 2 good hair dryers than in the preceding example, which resulted in a probability of 0.64 with replacement. This is good, because there is a smaller chance of approving a group of items that contains too many defects.

Instead of constructing the sample space, we could have found the probability of 12/20 by reasoning as follows. On the first selection, the probability of getting a good hair dryer is 4/5, but the second selection would then begin with 4 hair dryers, 3 of which are good (assuming that a good hair dryer was obtained on the first selection), so there is a 3/4 probability that the second hair dryer is good. That is,

$$P(G \text{ and } G) = P(G \text{ on first selection}) \cdot P(G \text{ on second selection})$$
$$= \frac{4}{5} \cdot \frac{3}{4} = \frac{12}{20} = 0.6$$

Here is the key concept of this last example: **Without replacement of the first selection, the second probability is affected by the first result.** This principle is so important that we will introduce special notation and definitions.

Notation for the Multiplication Rule

$P(B|A)$ represents the probability of B occurring after it is assumed that A has already occurred. (We can read $B|A$ as "B given A.")

DEFINITIONS

Two events A and B are **independent** if the occurrence of one does not affect the probability of the occurrence of the other. (Several events are similarly independent if the occurrence of any does not affect the probabilities of the occurrence of the others.) If A and B are not independent, they are said to be **dependent.**

For example, flipping a coin and then tossing a die are *independent* events because the outcome of the coin has no effect on the probabilities of the outcomes of the die. However, the event of having your car start and the event of driving to class on time are *dependent,* because the outcome of trying to start your car does affect the probability of getting to class on time.

Using the preceding notation and definitions, along with the principles illustrated in the preceding two examples, we summarize the key concept of this section in Figure 3-10 and the following rule.

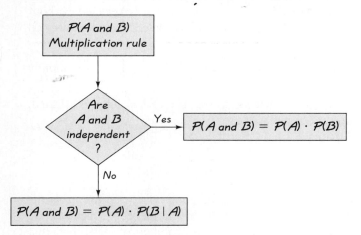

FIGURE 3-10   Applying the Multiplication Rule

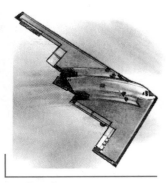

Redundancy

Reliability of systems can be greatly improved with redundancy of critical components. Airplanes have two independent electrical systems, and aircraft used for instrument flight typically have two separate radios. The following is from a *Popular Science* article about stealth aircraft: "One plane built largely of carbon fiber was the Lear Fan 2100 which had to carry two radar transponders. That's because if a single transponder failed, the plane was nearly invisible to radar." Such redundancy is an application of the multiplication rule in probability theory. If one component has a 0.001 probability of failure, the probability of two independent components both failing is only 0.000001.

Multiplication Rule

$P(A \text{ and } B) = P(A) \cdot P(B)$ if A and B are *independent*.
$P(A \text{ and } B) = P(A) \cdot P(B|A)$ if A and B are *dependent*.

In the last expression, it is easy to algebraically solve for $P(B|A)$; we simply divide both sides of the equation by $P(A)$. In so doing, we are finding an expression for $P(B|A)$, the probability of event B occurring after event A has already occurred. This result is formalized in the following definition.

DEFINITION

The **conditional probability** of B given A is the probability of event B occurring, given that event A has already occurred. It can be found by dividing the probability of events A and B both occurring by the probability of event A, as shown below.

$$P(B|A) = \frac{P(A \text{ and } B)}{P(A)}$$

▶ **EXAMPLE**

Refer to the preceding example involving 5 hair dryers, of which 4 are good and 1 is defective. Two hair dryers are to be randomly selected without replacement of the first result. Let events A and B be described as follows, and find the value of $P(B|A)$.

   A: A good hair dryer is obtained when the *first* hair dryer is randomly selected.

   B: A good hair dryer is obtained when the *second* hair dryer is selected. (The first selection is not replaced.)

● **SOLUTION**

$P(B|A)$ is the probability of selecting a second good hair dryer given that the first selection resulted in a good hair dryer. In the preceding example we already determined that this event has a probability of 3/4, but that probability can also be found as follows:

$$P(B|A) = \frac{P(A \text{ and } B)}{P(A)}$$

$$= \frac{12/20}{4/5} = \frac{0.6}{0.8} = \frac{3}{4}$$

In the multiplication rule for dependent events, if $P(B|A) = P(B)$, then the occurrence of event A has no effect on the probability of event B. This is often used as a test for independence. If $P(B|A) = P(B)$, then A and B are independent events; however, if $P(B|A) \neq P(B)$, then A and B are dependent events. Another test for independence involves checking for the equality of $P(A \text{ and } B)$ and $P(A) \cdot P(B)$. If they're equal, events A and B are independent. If $P(A \text{ and } B) \neq P(A) \cdot P(B)$, then A and B are dependent events. These results are summarized below.

Two Independent Events	Two Dependent Events
$P(B|A) = P(B)$	$P(B|A) \neq P(B)$
or	or
$P(A \text{ and } B) = P(A) \cdot P(B)$	$P(A \text{ and } B) \neq P(A) \cdot P(B)$

For example, if $P(B|A) = 0.2$ and $P(B) = 0.2$, then $P(B|A) = P(B)$ and we can conclude that A and B are independent events. However, if $P(B|A) = 0.5$ and $P(B) = 0.6$, then $P(B|A) \neq P(B)$ and we conclude that A and B are dependent events.

So far we have discussed two events, but the multiplication rule can be easily extended to several events. In general, **the probability of any sequence of independent events is simply the product of their corresponding probabilities.** The next example illustrates this extension of the multiplication rule.

▶ **EXAMPLE**

In the Chapter Problem we noted that Dr. Benjamin Spock was convicted by an all-male jury. Find the probability of getting such a jury if there is no discrimination based on gender. Assume that 50% of the potential jurors are men and that each selection is made in a way that is independent of other selections.

● **SOLUTION**

Because we are assuming that the events are independent, the probability of each individual juror being male is 0.5. Getting 12 males requires that the first juror be a male *and* the second juror be a male, and so on. The multiplication rule is applied as follows: *(continued)*

Component Failures

In a report on aircraft component failures, the FAA refers specifically to the multiplication rule as it relates to system independence and redundancy. The FAA states that the most common problem with quantitative analyses has been the improper treatment of events that are not independent. The FAA states, "The probability of occurrence of two events which are mutually independent may be multiplied to obtain the probability that both events occur using this formula: $P(A \text{ and } B) = P(A) \cdot P(B)$. This multiplication will produce an incorrect solution if A and B are not mutually independent."

$$P(12 \text{ males}) = P(\text{male}) \cdot P(\text{male}) \cdot \cdots \cdot P(\text{male}) \quad \text{(12 factors)}$$
$$= 0.5 \cdot 0.5 \cdot \cdots \cdot 0.5 = 0.5^{12}$$
$$= 0.000244$$

Based on this result, it appears extremely unlikely that an all-male jury would be selected by chance. It appears that the defense attorney has a valid claim of unfair treatment.

This trial took place in the Boston District Court, where more than 50% of the initial list of potential jurors were women. The first step in jury selection was taken when the court clerk was supposed to make random selections of 300 persons. Somehow, women made up only about 29% of those groups of 300. The next step involved selecting groups of 30. In these smaller groups of 30, Spock's judge somehow had an average of 14.6% women, whereas the other six judges in the same court had averages close to 29%. Spock's conviction was overturned for reasons other than the systematic elimination of women jurors, but to prevent such bias, we now randomly select federal court jurors.

The preceding example illustrates an extension of the multiplication rule for independent events; the next example illustrates a similar extension of the multiplication rule for dependent events.

▶ **EXAMPLE**

Find the probability of randomly selecting an all-male jury from a group of 30 potential jurors, 21 of whom are men. (See the discussion at the end of the preceding example.)

● **SOLUTION**

This example involves random selections made *without* replacement, because the 12 jurors must be 12 *different* persons. We are therefore dealing with *dependent* events, and we apply the multiplication rule as follows:

$$P(12 \text{ male jurors}) = P(\text{male}) \cdot P(\text{male, assuming first was male})$$
$$\cdot P(\text{male, assuming first 2 were male}), \text{ and so on}$$
$$= \frac{21}{30} \cdot \frac{20}{29} \cdot \frac{19}{28} \cdot \cdots \cdot \frac{10}{19} = 0.00340$$

The result indicates that it is highly unlikely that an all-male jury would be selected by chance.

In the last example we assumed that the events were dependent because the selections were made without replacement. However, it is a common practice to treat events as independent when *small* samples are drawn from *large* populations. (In such cases, it is rare to select the same item twice.) A common guideline is to assume independence whenever the sample size is at most 5% of the size of the population. When pollsters survey 1200 adults from a population of millions, they typically assume independence, even though they sample without replacement.

▶ **EXAMPLE**

As an insurance claims investigator, you are suspicious of 4 brothers who each reported a stolen car in a different region of Houston. If Houston has an annual car theft rate of 4.5%, find the probability that among 4 randomly selected cars, all are stolen in a given year. (There are 970,000 cars in Houston.)

● **SOLUTION**

We are selecting 4 cars without replacement; the events, therefore, seem dependent, but we can assume independence because the sample size of 4 is less than 5% of the population size of 970,000 cars. Treating the stolen cars as independent events and applying the extended multiplication rule, we get

$$P(\text{all 4 cars stolen}) = 0.045 \cdot 0.045 \cdot 0.045 \cdot 0.045$$
$$= 0.00000410$$

This corresponds to about 4 chances out of a million, so we are dealing with an event that is extremely rare.

The next example incorporates application of the addition rule and the multiplication rule. It brings together many of the important rules of probability.

▶ **EXAMPLE**

The preceding section used sample data in Table 3-1, which is reproduced on the next page as Table 3-2. Refer to Table 3-2, and find the following:

a. If 1 of the 2000 subjects is randomly selected, find the probability of getting someone who was assaulted, given that the person was victimized by an unknown assailant.

b. If 2 subjects are randomly selected with replacement, find the probability that they are both assault victims. *(continued)*

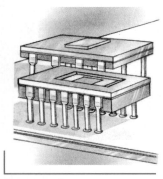

Bayes' Theorem

Thomas Bayes (1702–1761) said that probabilities should be revised when we learn more about an event. Here's one form of Bayes' theorem:

$$P(A|B) =$$
$$\frac{P(A) \cdot P(B|A)}{P(A) \cdot P(B|A) + P(\overline{A}) \cdot P(B|\overline{A})}$$

Suppose 60% of a company's computer chips are made in one factory (denoted by A) and 40% are made in its other factory (denoted by $\overline{A}$). For a randomly selected chip, the probability it came from factory A is 0.60. Suppose we learn that the chip is defective and the defect rates for the two factories are 35% (for A) and 25% (for $\overline{A}$). We can use the above formula to find that there is a 0.677 probability the defective chip came from factory A.

TABLE 3-2   Relationship of Criminal to Victim

	Homicide	Robbery	Assault	Totals
Stranger	12	379	727	1118
Acquaintance or relative	39	106	642	787
Unknown	18	20	57	95
Totals	69	505	1426	2000

c. If 2 different subjects are randomly selected, find the probability that they are both assault victims.

d. If 1 subject is randomly selected, find the probability of getting someone who was robbed or assaulted.

e. If 1 subject is randomly selected, find the probability of getting someone who was assaulted or victimized by an unknown assailant.

● SOLUTION

a. Use conditional probability, and note that among the 2000 subjects, 57 were assaulted by someone unknown to them. Overall, 95 subjects were victimized by unknown assailants.

$$P(\text{assault}|\text{unknown}) = \frac{P(\text{unknown and assault})}{P(\text{unknown})}$$
$$= \frac{57/2000}{95/2000} = 0.600$$

b. Because we are sampling with replacement, we use the multiplication rule for independent events. Overall, 1426 of the 2000 subjects were assaulted. We get

$$P(\text{assault and assault}) = P(\text{assault}) \cdot P(\text{assault})$$
$$= \frac{1426}{2000} \cdot \frac{1426}{2000} = 0.508$$

c. The requirement that 2 *different* subjects be randomly selected implies that we are sampling without replacement. We use the multiplication rule for dependent events.

$$P(\text{assault and assault}) = \frac{1426}{2000} \cdot \frac{1425}{1999} = 0.508$$

This result is slightly different from the result from part b, but the two answers agree when rounded to three significant digits. In other examples, the answers might be different when rounded to three significant digits.

(continued)

d. We use the addition rule for mutually exclusive events (because robbery and assault are separate categories of the Table 3-2 data that don't overlap).

$$P(\text{robbery or assault}) = P(\text{robbery}) + P(\text{assault}) = \frac{505}{2000} + \frac{1426}{2000}$$
$$= 0.966$$

e. We use the addition rule for events that are not mutually exclusive.

$$P(\text{assault or unknown}) = P(\text{assault}) + P(\text{unknown}) - P(\text{both})$$
$$= \frac{1426}{2000} + \frac{95}{2000} - \frac{57}{2000} = \frac{1464}{2000} = 0.732$$

The Probability of "At Least One"

The multiplication rule and the rule of complements can be used together to greatly simplify certain types of problems, such as those in which we want to find the probability that among several trials, *at least 1* will result in some specified outcome. In such cases, there are two issues of language that should be clearly understood. First, we should clearly understand that <u>at least 1 is equivalent to 1 or more</u>. Second, we should clearly understand that <u>the complement of getting at least 1 item of a particular type is getting no items of that type.</u> Suppose we're discussing a couple planning to have 7 children, and we want to determine the probability that they will have at least 1 girl. See the following interpretations.

    At least 1 girl = 1 or more girls
    Complement of "at least 1 girl" = no girls (or all boys)

The direct solution to this problem is complex, but a simple, indirect approach is illustrated in the following example.

▶ **EXAMPLE**

Find $P(\text{at least 1 girl in 7 births})$. Assume that the births are independent and that for each birth, $P(\text{boy}) = 1/2$. (Neither assumption is technically correct, but they will give us extremely good results.)

● **SOLUTION**

Step 1: Let G = at least 1 girl in 7 births.

Step 2: $\overline{G}$ = *not* getting at least 1 girl in 7 births
          = getting 7 boys in 7 births   *(continued)*

Step 3: $P(\overline{G}) = P(7 \text{ consecutive boys}) = \dfrac{1}{2} \cdot \dfrac{1}{2} \cdot \dfrac{1}{2} \cdot \dfrac{1}{2} \cdot \dfrac{1}{2} \cdot \dfrac{1}{2} \cdot \dfrac{1}{2} = \dfrac{1}{128}$

Step 4: $P(G) = 1 - P(\overline{G}) = 1 - \dfrac{1}{128} = \dfrac{127}{128}$

As difficult as this solution might seem, it is trivial in comparison to solutions using a direct approach.

Acceptance Sampling and Redundancy   We have just discussed a method for using the multiplication rule to find the probability of the occurrence of at least one of some event. An application of that method is found in the way many companies monitor quality by accepting or rejecting shipments of goods based on inspection of a sample. With the method known as **acceptance sampling,** a sample of *n* items is randomly selected without replacement, and the entire shipment is rejected if the number of defects is *at least* some predetermined number. In some cases, a shipment is rejected if there is at least one defect. The principles of this chapter can be used to determine the relevant probabilities, as in the preceding example.

Another important application of the multiplication rule involves the principle of **redundancy,** whereby critical components are duplicated so that their failure will not cause the failure of an entire system. Redundancy is used when backup components are installed. In a typical single-engine aircraft, for example, each cylinder is fired by a spark plug, but there is a second spark plug that acts as a backup so that if one plug fails, the cylinder can continue to operate with the backup. If the probability of a plug failing is 0.001, the probability of a cylinder not working because *both* plugs fail is $0.001 \cdot 0.001 = 0.000001$. With the backup, or redundant, spark plug, the probability of failure drops dramatically from one chance in a thousand to one chance in a million. This same principle is used in many different applications.

3-4   Exercises A: Basic Skills and Concepts

In Exercises 1 and 2, for each given pair of events, classify the two events as independent or dependent.

1. a. Randomly selecting a defective watch from a batch of 15 good and 5 defective watches
      Randomly selecting a second watch that is defective (assume that the same batch is used and the first selection was not replaced)
   b. Making a correct guess on the first question of a multiple-choice quiz
      Making a correct guess on the second question of the same multiple-choice quiz *(continued)*

c. Events A and B, where $P(A) = 0.40$, $P(B) = 0.60$, and $P(A \text{ and } B) = 0.20$

2. a. Finding your microwave oven inoperable
        Finding your battery-operated smoke detector inoperable
   b. Finding your kitchen light inoperable
        Finding your microwave oven inoperable
   c. Events A and B, where $P(A) = 0.90$, $P(B) = 0.80$, and $P(A \text{ and } B) = 0.72$

3. The Locust Tree Restaurant has found that 65% of its reservations are for nonsmoking tables, while 35% are for tables for smoking. Find the probability that 4 randomly selected reservations are all for nonsmoking tables.

4. A Pfaltzgraff survey by the ICR Research Group showed that 74% of those who date consider the worst turnoff to be a date who regularly uses vulgar language. If someone who regularly uses vulgar language randomly selects 2 different persons and asks them on dates, what is the probability that the 2 selected persons both consider vulgar language to be the worst turnoff?

5. The Federal Railroad Administration found that for a recent year, human error was cited as the cause in 24% of all derailments. A railroad safety specialist randomly selects records from 5 different derailment incidents. What is the probability that all 5 were caused by human error?

6. During one segment of ABC's television program *Nightline,* it was reported that there is a 60% success rate for those trying to stop smoking through hypnosis. Find the probability that 8 randomly selected smokers who undergo hypnosis all successfully stop smoking.

7. A quality control manager uses test equipment to detect defective computer disk drives. A sample of 4 *different* disk drives is to be randomly selected from a group consisting of 10 that are defective and 20 that have no defects. What is the probability that all 4 selected disk drives are defective?

8. A circuit requiring a 500-ohm resistance is designed with five 100-ohm resistors connected in series. The proper resistance is achieved only if all five resistors work successfully. There is a 0.992 probability that any individual resistor will work successfully. What is the probability that all five resistors will work successfully to provide the necessary resistance?

9. *Working Woman* magazine reports that currently 50% of new car buyers are women, but 60% will be women by the year 2000. Assuming that those figures are correct, find the probability of getting 4 women when randomly selecting 4 new car buyers if the selections are made
   a. now
   b. in the year 2000

10. The IRS reports that among all taxpayers audited, 70% end up owing more money. One new auditor randomly selected 8 tax returns, audited them, and boasted that he collected additional taxes from all of them. What is the probability of doing what he boasted about?

11. One couple attracted media attention when their 3 children, born in different years, were all born on July 4. Ignoring leap years, find the probability that 3 randomly selected people were all born on July 4.

12. Find the probability of answering the first 2 questions on a test correctly if random guesses are made.

   a. Assume that the first 2 questions are both true/false types.

   b. Assume that the first 2 questions are both multiple choice, each with 5 possible answers.

13. A Baltimore detective is suspicious about 5 deaths that were determined to be accidental. If a death is selected at random, there is a 0.0478 probability that it was caused by an accident, according to data from the *Statistical Abstract of the United States*. Find the probability that 5 randomly selected deaths were all accidental.

14. There are 6 defective fuses in a bin of 80 fuses. The entire bin is approved for shipping if no defects show up when 3 randomly selected fuses are tested.

   a. Find the probability of approval if the selected fuses are replaced.

   b. Find the probability of approval if the selected fuses are not replaced.

   c. Comparing the results of parts a and b, determine which procedure is more likely to reveal a defective fuse. Which procedure do you think is better?

15. From U.S. Census Bureau data, we know that 60% of those who are eligible to vote actually do vote. If a pollster surveys 10 people who are eligible to vote, what is the probability that they all vote?

16. The Providence Mint produces tokens for video games. Of 2 million tokens produced in one year, 5000 are defective. If 2 of these tokens are randomly selected and tested, find the probability that they are both good in each of the following cases.

   a. The first selected token is replaced.

   b. The first selected token is not replaced.

17. A sales manager must fly from New York to San Francisco, changing planes in St. Louis. For the manager to arrive on time, both legs of the flight must be on time. TWA Flight 25 from New York to St. Louis has an 80% on-time record, while TWA Flight 17 from St. Louis to San Francisco has a 60% on-time record (based on data from the EAASY SABRE reservation system). What is the probability of arriving on time? (Assume the 2 flights are independent, even though they may be dependent because of common factors, such as weather and strikes.)

18. A manager can identify employee theft by checking samples of shipments. Among 36 employees, 2 are stealing. If the manager checks on 4 different randomly selected employees, find the probability that neither of the thieves will be identified.

19. The Atlantic Delivery Company has 12 trucks. When 3 are inspected, it is found that all 3 have faulty brakes. The owner claims that all of the other trucks have good brakes and that it was just chance that led to selecting the trucks with faulty brakes. Find the probability of that event, assuming that the owner's claim is correct.

20. In a Riverhead, New York, case, 9 different crime victims listened to voice recordings of 5 different men. All 9 victims identified the same voice as that of the criminal. If the voice identifications were made by random guesses, find the probability that all 9 victims would select the same person.

21. If you bet on the number 7 for one spin of a roulette wheel, you have 1 chance in 38 of winning. If you continue to bet on 7, find the probability of losing 38 consecutive times.

22. The New England Life Insurance Company issues one-year policies to 12 men who are all 27 years of age. Based on data from the Department of Health and Human Services, each of these men has a 99.82% chance of living through the year. What is the probability that they all survive the year?

23. The Scott Computer Mail Order Company normally experiences a 20% reply rate on a coupon it mails. There is concern about the status of 1 batch of 6 coupons that resulted in no returns. What is the probability of this happening by chance if the coupons were delivered and the overall reply rate really is 20%?

24. An approved jury list contains 20 women and 20 men. Find the probability of randomly selecting 12 of these people and getting an all-male jury.

25. A blood testing procedure is made more efficient by combining samples of blood specimens. If samples from 5 people are combined and the mixture tests negative, we know that all 5 individual samples are negative. Find the probability of a positive result for 5 samples combined into 1 mixture, assuming the probability of an individual blood sample testing positive is 0.015.

26. An employee needs to call any 1 of 5 colleagues at home. Assume that the 5 colleagues are random selections from a population in which 28% have unlisted numbers. Find the probability that at least 1 of the 5 fellow workers will have a listed number.

27. A circuit is used to control the temperature in a museum room. It performs correctly if at least 1 of 4 identical and independent components does not fail. The probability of failure is 0.181. Find the probability that the critical function will be properly performed.

28. In a recent national election, 52% of the voters were women, according to a *New York Times*/CBS News Poll. For a random selection of 4 voters, find the probability of getting at least 1 woman.

In Exercises 29–32, use the data in Table 3-2.

29. a. Find the probability that when 1 of the 2000 subjects is randomly selected, the person chosen was victimized by a stranger, given that he or she was robbed.

   b. Find the probability that when 1 of the 2000 subjects is randomly selected, the person chosen was robbed by a stranger.

   c. Find the probability that when 1 of the 2000 subjects is randomly selected, the person chosen was robbed or was victimized by an unknown assailant.

30. a. Find the probability that when 1 of the 2000 subjects is randomly selected, the person chosen was victimized by an acquaintance or relative, given that he or she was a homicide victim.

   b. If 1 of the 2000 subjects is randomly selected, find the probability of getting someone who was murdered by an acquaintance or relative.

(continued)

c. If 1 of the 2000 subjects is randomly selected, find the probability of getting someone who was victimized by an acquaintance or relative or who was murdered.

31. a. If one of the subjects is randomly selected, find the probability of getting someone who was victimized by a stranger or who was assaulted.

b. If one of the subjects is randomly selected, find the probability of getting someone who was assaulted by a stranger.

c. If one of the crime victims represented in the table is randomly selected, find the probability of getting someone who was assaulted, given that the criminal was a stranger.

32. a. If one of the crime victims represented in the table is randomly selected, find the probability of getting someone who was victimized by someone unknown to the victim or who was murdered.

b. If one of the crime victims represented in the table is randomly selected, find the probability of getting someone who was murdered, given that he or she was victimized by a stranger.

c. If one of the crime victims represented in the table is randomly selected, find the probability of getting someone who was victimized by a stranger, given that he or she was murdered.

3-4 Exercises B: Beyond the Basics

33. Using a calculator, find the probability that of 25 randomly selected people,
    a. no 2 share the same birthday
    b. at least 2 share the same birthday

34. Let $P(A) = 3/4$ and $P(B) = 5/6$. Find $P(A \text{ and } B)$, given that
    a. A and B are independent events
    b. $P(B|A) = 1/2$
    c. $P(A|B) = 1/3$
    d. $P(A|B) = P(A)$

35. a. Develop a formula for the probability of not getting either A or B on a single trial. That is, find an expression for $P(\overline{A \text{ or } B})$.

b. Develop a formula for the probability of not getting A or not getting B on a single trial. That is, find an expression for $P(\overline{A} \text{ or } \overline{B})$.

c. Compare the results from parts a and b. Are they the same or are they different?

36. A poll of 500 carefully selected consumers included exactly 400 males. Of those polled, 200 preferred the cola labeled A, while the other 300 respondents preferred the cola labeled B.

a. If 1 of the 500 consumers is randomly selected, find the probability of getting a male.

b. If 1 of the 500 consumers is randomly selected, find the probability of getting someone who preferred cola A.   *(continued)*

c. Assuming that the gender of the consumer has no effect on cola preference, find the probability of randomly selecting 1 of the 500 consumers and getting a male who preferred cola A.

d. Using the probability from part c and the fact that the poll involves 500 consumers, what would you expect to be the number of males who prefer cola A?

37. Two cards are to be randomly selected without replacement from a shuffled deck. Find the probability of getting a 10 on the first card and a club on the second card.

3-5 Probabilities Through Simulations

You may have discovered that determining correct probabilities of events is sometimes quite difficult. Identifying the appropriate rule or principle is often hard to do. Sometimes the results, even though correct, just don't seem to be right. (For specific examples in which the correct probabilities are very different from what we might expect, see the first three paragraphs of Section 3-1.) Instead of relying solely on the abstract principles of probability theory, we can often benefit from the use of a simulation.

DEFINITION

A **simulation** of an experiment is a process that behaves the same way as the experiment, so that similar results are produced.

▶ **EXAMPLE**

A couple would prefer to have at least one child of each sex. They would like to know the average number of births it takes couples to get one child of each sex. Find that value.

● **SOLUTION**

Although the statement of the problem might seem simple, the theoretical calculation of the result isn't easy. One alternative is to conduct an experiment

(continued)

Maryland Pick Three: 50 consecutive drawings

1	0	0	0
2	7	1	3
3	3	6	4
4	6	8	6
5	2	4	7
6	7	6	9
7	6	2	5
8	6	7	7
9	6	1	1
10	3	3	3
11	8	2	2
12	1	9	7
13	7	6	9
14	8	7	4
15	7	1	5
16	1	0	3
17	6	4	5
18	8	0	0
19	6	5	0
20	9	5	1
21	5	4	9
22	2	6	2
23	1	2	1
24	5	6	0
25	0	3	6
26	3	8	2
27	9	2	6
28	9	5	3
29	0	3	9
30	7	4	4
31	7	8	2
32	5	0	4
33	8	5	5
34	5	5	1
35	4	3	8
36	9	4	4
37	1	1	8
38	1	7	1
39	0	6	6
40	3	5	6
41	3	4	5
42	9	9	9
43	4	4	1
44	8	2	9
45	7	3	0
46	7	5	7
47	7	5	7
48	5	3	3
49	8	7	3
50	8	1	6

with couples who plan to have children, but that would be very expensive and time consuming. Surveying a large number of families that already have children would be simpler, but it would still be expensive and time consuming. Instead, we will estimate the answer by developing a simulation. Refer to the table of results from the Maryland Pick Three lottery. We can simulate actual births by stipulating that the even digits (0, 2, 4, 6, 8) represent baby boys while the odd digits (1, 3, 5, 7, 9) represent baby girls. Just as boys and girls are (approximately) equally likely, the even digits and odd digits in the lottery are also equally likely. The first few results can be thought of as boys (B) and girls (G) as follows:

0	0	0	7	1	3	3	6	4	6	8	6	2	4	7	7	6
↓	↓	↓	↓	↓	↓	↓	↓	↓	↓	↓	↓	↓	↓	↓	↓	↓
B	B	B	G	G	G	G	B	B	B	B	B	B	B	G	G	B

Starting at the left, the first simulated "family" had 4 children (BBBG) before getting at least 1 of each sex. After we eliminate those results, the second simulated family had 4 children (GGGB) before getting at least 1 of each sex. The third simulated family had 7 children (BBBBBBG), and the fourth had only 2 children (GB). As we continue through the table, the numbers of children required are 4, 4, 7, 2, 2, 2, and so on. If we use all of the data in the table and compute the average number of children in these simulated families, we get 3.1, which is very close to the theoretical correct value of 3.

The simple simulation in the preceding example enabled us to get a very good result without much effort. When developing a simulation, we must be careful to create a process that imitates the actual process very well. The preceding example could be criticized because it assumes that boys and girls are equally likely, whereas in reality $P(\text{boy}) = 0.513$ and $P(\text{girl}) = 0.487$. Also, the preceding example assumes that the sex of a child is not affected by the sex of any siblings born earlier. It provides a reasonably close, but not perfect, simulation of couples actually having babies, at least insofar as determining the sexes of their children.

 ## Computer Simulations

The preceding example used random numbers obtained from the Maryland Pick Three lottery, but random numbers can be obtained from other sources. Tables of random numbers are included in some reference books, such as *CRC Standard Probability and Statistics Tables and Formulae*. Statistical software, such as STATDISK and Minitab, can be used to generate random numbers. The accompanying Minitab display shows the results from simulating 1000 births. Here, we can consider 0 to represent a baby girl and 1 to represent a baby boy. The command RANDOM 1000 C1 calls for 1000 random numbers to be stored in column C1, and the subcommand INTEGERS 0 1 stipulates that the random numbers must be integers from 0 to 1. The command HISTOGRAM C1 calls for a display of the histogram, which shows that there are 531 girls and 469 boys among the 1000 births.

MINITAB DISPLAY

```
MTB > RANDOM 1000 C1;
SUBC> INTEGERS 0 1.
MTB > HISTOGRAM C1

Histogram of C1    N = 1000
Each * represents 15 obs.

Midpoint     Count
       0       531    ***********************************
       1       469    *******************************
```

▶ EXAMPLE

Use Minitab to simulate rolling a pair of dice 500 times, and, based on the results, estimate $P(7)$.

● SOLUTION

The commands RANDOM 500 C1 and INTEGERS 1 6 simulate rolling a single die 500 times, with the results stored in column C1. Those same commands are then used for a second die, with the results stored in column C2. The command LET C3 = C1+C2 creates a column C3 consisting of the 500 sums of the two dice. The command HISTOGRAM C3 causes the histogram to be displayed (see the following page), and we can see the frequency for each total. Based on the displayed results, we can *estimate* that when a pair of dice is rolled, the most frequent outcome is 7 and $P(7) = 83/500 = 0.166$. (Using the rules of probability, we can determine that $P(7) = 6/36 = 0.167$.)

Monkey Typists

A classical claim is that a monkey randomly hitting a keyboard would eventually produce the complete works of Shakespeare if it could continue to type century after century. Dr. William Bennet used the rules of probability to develop a computer simulation that addressed this problem, and he concluded that it would take a monkey about 1,000,000,000, 000,000,000,000,000,000, 000,000,000 years to reproduce Shakespeare's works. In the same spirit, Sir Arthur Eddington wrote this poem: ''There once was a brainy baboon, who always breathed down a bassoon. For he said, 'It appears that in billions of years, I shall certainly hit on a tune.' ''

MINITAB DISPLAY

```
MTB > RANDOM 500 C1;
SUBC> INTEGERS 1 6.
MTB > RANDOM 500 C2;
SUBC> INTEGERS 1 6.
MTB > LET C3 = C1 + C2
MTB > HISTOGRAM C3

Histogram of C3    N = 500
Each * represents 2 obs.

Midpoint      Count
      2         16    ********
      3         24    ************
      4         39    *******************
      5         53    ***************************
      6         66    *********************************
      7         83    ******************************************
      8         65    ********************************
      9         51    **************************
     10         47    ************************
     11         35    *****************
     12         21    **********
```

Again, it's extremely important to carefully construct the simulation so that it closely imitates actual circumstances. A serious error would have been created in the preceding example if we had incorrectly believed that the dice totals of 2, 3, 4, . . . , 12 were equally likely and had used the Minitab command INTEGERS 2 12. The outcomes would have resembled dice totals in the sense that they would have been integers between 2 and 12 inclusive, but by failing to simulate each individual die and add their values, we would have failed to imitate real dice. That error would have produced terrible results.

▶ **EXAMPLE**

The Intel Corporation is one of the world's largest suppliers of computer chips. Suppose an Intel research and development team is working on manufacturing an expensive prototype for a new model, and it is estimated that there is a probability of 1/3 that an individual chip will be free of defects. The team would like some sense of how many chips must be produced to get one that is free of defects. Simulate this process.

(continued)

● SOLUTION

With P(defect free) $= 1/3$, we need to use a simulation that produces outcomes in such a way that 1 in 3 can be identified as a success. Let's refer to the same Maryland lottery results used earlier in this section. Because we want a probability of $1/3$, let's stipulate that the lottery results are as follows:

1, 2, 3 represent chips that are good (G) or free of defects.

4, 5, 6, 7, 8, 9 represent defective (D) chips.

0 doesn't mean anything; skip it and go to the next digit.

Having assigned meanings to the numbers, we can now refer to the Maryland lottery results. Skipping 0s and replacing all other digits by Gs and Ds, we get these results.

7	1	3	3	6	4	6	8	6	2	4	7
↓	↓	↓	↓	↓	↓	↓	↓	↓	↓	↓	↓
D	G	G	G	D	D	D	D	D	G	D	D

The first good chip was obtained in 2 tries, the next good chip was obtained on the first try, and so on. Recording the numbers of chips required to get a good one, we list 2, 1, 1, 6, 7, and so on. The mean of these numbers is 3.1. Based on the results, the Intel team could expect to produce about 3 chips before getting 1 that is defect free. It could get lucky and obtain a good chip on the first try, or it could get unlucky and produce several chips before getting a good one, but the mean number of chips is about 3.

In addition to solving some problems that might otherwise seem unsolvable, this simulation technique can be used for verifying results of probability calculations. One problem that gained much attention in recent years is the *Monty Hall problem*, based on the old television game show "Let's Make a Deal," hosted by Monty Hall. Suppose you are a contestant who has selected 1 of 3 doors after being told that 2 of them conceal nothing, but that a new red Corvette is behind 1 of the 3. Next, the host opens 1 of the doors you didn't select and shows that there is nothing behind it. He then offers you the choice of sticking with your first choice or switching to the other unopened door. Should you stick with your first choice or should you switch? The solution is far from obvious and we're not going to compute it here, but probability theory can be used to show that you should switch because your probability of winning then becomes $2/3$. An alternative to the theoretical calculation is to simulate this game by playing it with a friend. The simulation should verify that switching is better than sticking with your first choice because you will win $2/3$ of the time. According to *Chance* magazine, business schools at such institutions as Harvard and Stanford use this problem to help students deal with decision making.

Random Samples

In this section we have focused on the use of simulations to solve probability problems, but they can also be used to obtain random samples. Recall that a sample of size *n* is *random* if the probability of selecting it is the same as the probability of selecting every other sample of size *n*. Suppose that we want to select a random sample of 36 commuters from a population of 1000 commuters who take a train to work. There could be very serious problems of bias if we were to select the first 36 commuters who arrived at the train station; that would be convenience sampling, and those early birds might not reflect the population very well. Serious problems could also arise if we were to select those commuters who seemed approachable.

How do we decide which commuters to select? With a population size of 1000, we could randomly select 36 numbers between 1 and 1000. Representing 1000 by 000, we could also randomly select 36 numbers from the list 001, 002, 003, . . . , 999, 000. Where could we get such numbers? They could be generated by a computer program, such as STATDISK or Minitab, or they could be found in a list of random numbers, such as those obtained from the Maryland lottery. See the list of Maryland Pick Three results listed earlier in this section, and note that the first 36 results are 000, 713, 364, . . . , 438, 944. Replacing 000 by 1000, we can now proceed to select the 1000th commuter, the 713th commuter, and so on. We have established a method for obtaining a random sample.

3-5   Exercises A: Basic Skills and Concepts

1. The first example of this section illustrated a simulation for finding the average number of births necessary to get at least one child of each sex. That example used the Maryland lottery data included in this section. Use the same data set to estimate the average number of births necessary to get a girl.

2. Use the same Maryland lottery data to estimate the average number of births necessary to get at least 2 children of each sex. (Use odd numbers for one sex and even numbers for the other.)

3. Assume that the Intel Corporation is working on a prototype for a new model, and it is estimated that there is a probability of 1/10 that an individual chip will be free of defects. Use the Maryland lottery data from this section to estimate the average number of chips that must be produced to get one that is free of defects.

4. Repeat Exercise 3 after changing the probability from 1/10 to 1/5.

5. Repeat Exercise 3 by again using a probability of 1/10, but change the goal from getting 1 good chip to that of obtaining 2 good chips. (Let 0 represent a good chip.)

6. We know that when a single fair die is rolled, the probability of getting a 5 is 1/6, or 0.167. What is the estimated probability obtained when the rolling of a

die is simulated with the Maryland lottery data? (*Hint:* Skip any outcomes that are not 1, 2, 3, 4, 5, or 6.)

7. Use the Maryland lottery data to estimate the average number of rolls of a single die necessary to get a 6. (*Hint:* Skip any outcomes that are not 1, 2, 3, 4, 5, or 6.)

8. A student guesses answers to each of the 3 true/false questions on a quiz. Use the Maryland lottery data to estimate the mean number of correct responses for 50 such students.

3-5 Exercises B: Beyond the Basics

9. Use the Maryland lottery data to simulate 50 families with 3 children each. Let even numbers represent male children, and let odd numbers represent female children.
   a. Find the mean number of girls in a family
   b. Find the standard deviation of the numbers of girls.

10. a. What Minitab commands would generate 25 birthdays for 25 randomly selected people? (*Hint:* See the first Minitab display in this section, and note that instead of generating the actual birthdays, you could generate the birthdays coded as days of the year numbered from 1 to 365.)
   b. Use Minitab to generate 25 birthdays that are stored in column C1, and enter the command PRINT C1. Inspect the displayed values to determine whether any 2 birthdays are the same.
   c. Repeat part b until it has been done a total of 10 times. Based on the results, what is the probability that among 25 randomly selected people, at least 2 will share the same birthday?

3-6 Counting

In some probability problems the biggest obstacle is determining (counting) the total number of outcomes. In this section we examine some of the efficient ways this can be done. In addition to their use in probability problems, these counting techniques have also grown in importance because of their applicability to problems relating to computers and programming.

Suppose that we use a computer to randomly select 1 of the 2 Rh types (positive, negative) and 1 of the 4 blood groups A, O, B, or AB. Let's assume that we want the probability of getting a positive Rh type and group A blood. We can represent this probability as P(positive and A) and apply the multiplication rule to get

$$P(\text{positive and A}) = \frac{1}{2} \cdot \frac{1}{4} = \frac{1}{8}$$

The Number Crunch

Every so often telephone companies split regions with one area code into regions with two or more area codes because the increased number of telephones in the area has nearly exhausted the possible numbers that can be listed under a single code. A seven-digit telephone number cannot begin with a 0 or 1, but if we allow all other possibilities, we get $8 \cdot 10 \cdot 10 \cdot 10 \cdot 10 \cdot 10 \cdot 10 = 8,000,000$ different possible numbers! Even so, after surviving for 80 years with the single area code of 212, New York City was recently partitioned into the two area codes of 212 and 718. Los Angeles and Houston have also endured split area codes.

We could also arrive at the same probability by examining the tree diagram in Figure 3-11. The tree diagram has 8 branches that, by the random computer selection method, are all equally likely. Since only 1 branch corresponds to *P*(positive and A), we get a probability of 1/8. Apart from the calculation of the probability, this solution reveals another principle, which is a generalization of the following specific observation: With 2 Rh types and 4 blood groups, there are 8 different possibilities for the compound event of selecting a factor and type. We now state this generalized principle.

Fundamental Counting Rule

For a sequence of 2 events in which the first event can occur *m* ways and the second event can occur *n* ways, the events together can occur a total of $m \cdot n$ ways.

The fundamental counting rule easily extends to situations involving more than 2 events, as illustrated in the following example.

▶ **EXAMPLE**

In designing a computer, if a byte is defined to be a sequence of 8 bits and each bit must be a 0 or 1, how many different bytes are possible?

● **SOLUTION**

Since each bit can occur in 2 ways (0 or 1) and we have a sequence of 8 bits, the total number of different possibilities is given by

$$2 \cdot 2 \cdot 2 \cdot 2 \cdot 2 \cdot 2 \cdot 2 \cdot 2 = 256$$

The next three counting rules use the factorial symbol !, which denotes the product of decreasing whole numbers. For example, $5! = 5 \cdot 4 \cdot 3 \cdot 2 \cdot 1 = 120$. By special definition, $0! = 1$. (Many calculators have a factorial key.) Using the factorial symbol, we now present the factorial rule, which is actually a simple case of the permutations rule that will follow.

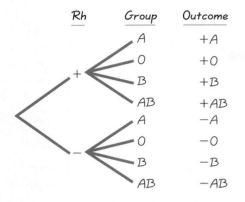

Rh	Group	Outcome
	A	+A
	O	+O
+	B	+B
	AB	+AB
	A	−A
	O	−O
−	B	−B
	AB	−AB

First City	Second City	Third City	Route
A	B	C	ABC
	C	B	ACB
B	A	C	BAC
	C	A	BCA
C	A	B	CAB
	B	A	CBA

3! = 6 different possible routes

FIGURE 3-11   Tree Diagram of Blood Types/Rh Factors

FIGURE 3-12   Tree Diagram of Routes

Factorial Rule

n different items can be arranged in order $n!$ different ways.

This factorial rule reflects the fact that the first item may be selected n different ways, the second item may be selected $n − 1$ ways, and so on.

▶ **EXAMPLE**

Routing problems are extremely important in many applications of the factorial rule. AT&T wants to route telephone calls through the shortest networks. Federal Express wants to find the shortest routes for its deliveries. Suppose a computer salesperson must visit 3 separate cities denoted by A, B, C. How many routes are possible?

● **SOLUTION**

Using the factorial rule, we see that the 3 different cities (A, B, C) can be arranged in $3! = 6$ different ways. In Figure 3-12 we can see that there are 3 choices for the first city and 2 choices for the second city. This leaves only 1 choice for the third city. The number of possible arrangements for the 3 cities is $3 \cdot 2 \cdot 1 = 6$.

▶ **EXAMPLE**

A presidential candidate plans to visit the capital of each of the 50 states. How many different routes are possible?

● **SOLUTION**

The 50 state capitals can be arranged 50! ways, so the number of different routes is 50!, or 30,414 *followed by 60 zeros;* that is an incredibly large number! Now we can see why the symbol ! is used for factorials!

The preceding example is a variation of a classical problem called the traveling salesman problem. It is especially interesting because the large number of possibilities precludes a direct computation for each route, even if computers are used. The time for the fastest computer to directly calculate the shortest possible route is about

$$1,000,000,000,000,000,000,000,000,000,000,000,000 \ centuries!$$

That's too long. Continuing efforts are aimed at finding efficient ways of solving such problems.

In the factorial counting rule, we determine the number of different possible ways we can arrange a number of items in some type of ordered sequence. The factorial rule tells us how many arrangements are possible when *all* of *n* different items are used. Sometimes, however, we want to select only *some* of the *n* items. Assuming that we have *n* different items available and we want to select *r* of them, how many different arrangements are possible? When we use the term *arrangements* or *sequences,* we imply that *order* is taken into account. Arrangements or sequences are commonly called *permutations,* which explains the use of the letter *P* in the following rule.

Permutations Rule

The number of **permutations** (or sequences) of *r* items selected from *n* available items (not allowing repetition) is

$$_nP_r = \frac{n!}{(n-r)!}$$

It must be emphasized that in applying the preceding permutations rule, we must have a total of n items available, we must select r of the n items, and we must consider rearrangements of the same items to be different. In the following example, we are asked to find the total number of different sequences that are possible. That suggests use of the permutations rule.

▶ **EXAMPLE**

A psychologist at the Industrial Testing Service is conducting research related to the effect of the order of test questions on test results. One experiment involves a computer test bank of 8 different questions. If 5 of the questions are to be selected for a test, how many different sequences are possible?

● **SOLUTION**

Here we want the number of sequences (or permutations) of $r = 5$ questions from $n = 8$ available questions, so the number of different possible sequences is

$$_8P_5 = \frac{8!}{(8-5)!} = \frac{8!}{3!} = 6720$$

The permutation rule can be thought of as an extension of the fundamental counting rule. We can also solve the preceding example by using the fundamental counting rule as follows: With 8 questions available and with a requirement that 5 are to be selected, we know that there are 8 choices for the first question, 7 choices for the second question, 6 choices for the third question, 5 choices for the fourth question, and 4 choices for the fifth question. The number of different possible arrangements is therefore $8 \cdot 7 \cdot 6 \cdot 5 \cdot 4 = 6720$, but $8 \cdot 7 \cdot 6 \cdot 5 \cdot 4$ is actually $8! \div 3!$. In general, whenever we select r items from n available items, the number of different possible arrangements is $n! \div (n - r)!$, and this is expressed in the permutations rule.

When we intend to select r items from n available items but *do not take order into account*, we are really concerned with possible combinations rather than permutations. That is, when **different orderings of the same items are to be counted separately, we have a permutation problem, but when different orderings are *not* to be counted separately, we have a combination problem** and may apply the following rule.

Promotion Contests

There have been many contests or games designed to promote or sell products, but some have encountered problems. The Beatrice Company ran a contest involving matching numbers on scratch cards with scores from Monday night football games. The game cards had too few permutations, and Frank Maggio was able to identify patterns and collect around 4000 winning cards worth $21 million. The contest was canceled, and lawsuits were filed. The Pepsi people ran another contest in which people had to spell out their names with letters found on bottle caps, but a surprisingly large number of people with short names like Ng forced the cancellation of this contest also.

Safety in Numbers

Some hotels have abandoned the traditional room key in favor of an electronic key with a number code. A central computer changes the access code to a room as soon as a guest checks out. A typical electronic key has 32 different positions that are either punched or left untouched. This configuration allows for 2^{32}, or 4,294,967,296, different possible codes, so it is impractical to develop a complete set of keys or try to make an illegal entry by trial and error.

Combinations Rule

The number of combinations of r items selected from n available items is

$$_nC_r = \frac{n!}{(n-r)!r!}$$

Because choosing between the permutations rule and the combinations rule can be confusing, we provide the following example, which is intended to emphasize the difference between them.

▶ **EXAMPLE**

Five students (Al, Bob, Carol, Donna, and Ed) have volunteered for service to the student government.

a. If 3 of the students are to be selected for a special committee, how many different committees are possible?

b. If 3 of the students are to be nominated for the offices of president, vice president, and secretary, how many different slates are possible?

● **SOLUTION**

a. In forming the committee, order of selection is irrelevant. The committee of Al, Donna, and Ed is the same as that of Donna, Al, and Ed. Therefore, we want the number of combinations of 5 students when 3 are selected. We get

$$_5C_3 = \frac{5!}{(5-3)!\,3!} = 10 \qquad \text{Use combinations when order is irrelevant.}$$

There are 10 different possible committees.

b. In forming slates of candidates, the order is relevant. The slate of Al for president, Donna for vice president, and Ed for secretary is different from the slate of Donna, Al, and Ed for president, vice president, and secretary, respectively. Here we want the number of permutations of 5 students when 3 are selected. We get

$$_5P_3 = \frac{5!}{(5-3)!} = 60 \qquad \text{Use permutations when order is relevant.}$$

There are 60 different possible slates.

We stated that the counting techniques presented in this section are sometimes used in probability problems. The following examples illustrate such applications.

▶ **EXAMPLE**

In the New York State lottery, a player wins first prize by selecting the correct 6-number combination when 6 different numbers from 1 through 54 are drawn. If a player selects 1 particular 6-number combination, find the probability of winning.

● **SOLUTION**

Since 6 different numbers are selected from 54 different possibilities, the total number of combinations is

$$_{54}C_6 = \frac{54!}{(54-6)!\,6!} = \frac{54!}{48!\,6!} = 25{,}827{,}165$$

With only 1 combination selected, the player's probability of winning is only 1/25,827,165.

▶ **EXAMPLE**

A home security device with 10 buttons is disarmed when 3 different buttons are pushed in the proper sequence. (No button can be pushed twice.) If the correct code is forgotten, what is the probability of disarming this device by randomly pushing 3 of the buttons?

● **SOLUTION**

The number of different possible 3-button sequences is

$$_{10}P_3 = \frac{10!}{(10-3)!} = 720$$

The probability of randomly selecting the correct 3-button sequence is therefore 1/720.

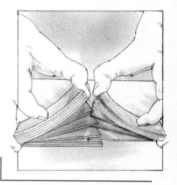

How Many Shuffles?

After conducting extensive research, Harvard mathematician Persi Diaconis found that it takes seven shuffles of a deck of cards to get a complete mixture. The mixture is complete in the sense that all possible arrangements are equally likely. More than seven shuffles will not have a significant effect, and fewer than seven are not enough. Casino dealers rarely shuffle as often as seven times, so the decks are not completely mixed. Some expert card players have been able to take advantage of the incomplete mixtures that result from fewer than seven shuffles.

The Random Secretary

One classic problem of probability goes like this: A secretary addresses 50 different letters and envelopes to 50 different people, but the letters are randomly mixed before he puts them into envelopes. What is the probability that at least one letter gets into the correct envelope? Although the probability might seem like it should be small, it's actually 0.632. Even with a million letters and a million envelopes, the probability is 0.632. The solution is beyond the scope of this text—way beyond.

▶ **EXAMPLE**

A UPS dispatcher sends a delivery truck to 8 different locations. If the order in which the deliveries are made is randomly determined, find the probability that the resulting route is the shortest possible route.

● **SOLUTION**

With 8 locations there are 8!, or 40,320, different possible routes. Among those 40,320 different possibilities, only 2 routes will be shortest (actually the same route in 2 different directions). Therefore, there is a probability of only 2/40,320, or 1/20,160, that the selected route will be the shortest possible route.

In this last example, application of the appropriate counting technique made the solution easily obtainable. If we had to determine the number of routes directly by listing them, we would labor for more than 11 hours while working at the rapid rate of 1 route per second! Clearly, these counting techniques are extremely valuable.

The concepts and rules of probability theory presented in this chapter consist of elementary and fundamental principles. A more complete study of probability is not necessary at this time, because our main objective is to study the elements of statistics and we have already covered the probability theory that we will need. We hope that this chapter generates some interest in probability for its own sake. The importance of probability is continuing to grow as it is used by more and more political scientists, economists, biologists, actuaries, business executives, and other professionals.

3-6 Exercises A: Basic Skills and Concepts

In Exercises 1–16, evaluate the given expressions.

1. $7!$

2. $9!$

3. $\dfrac{70!}{68!}$

4. $\dfrac{92!}{89!}$

5. $(9 - 3)!$

6. $(20 - 12)!$

7. $_6P_2$

8. $_6C_2$

9. $_{10}C_3$

10. $_{10}P_3$

11. $_{52}C_2$

12. $_{52}P_2$

13. $_nP_n$

14. $_nC_n$

15. $_nC_0$

16. $_nP_0$

17. Data are grouped according to sex (female, male) and income level (low, middle, high). How many different possible categories are there?

18. How many different ways can 5 cars be arranged on a carrier truck with room for 5 vehicles?

19. A computer operator must select 4 jobs from among 10 available jobs waiting to be completed. How many different sequences are possible?

20. A computer operator must select 4 jobs from 10 available jobs waiting to be completed. How many different combinations of 4 jobs are possible?

21. An IRS agent must audit 12 returns from a collection of 22 flagged returns. How many different combinations are possible?

22. A health inspector has time to visit 7 of the 20 restaurants on a list. How many different routes are possible?

23. Using a word processor, a pollster develops a market research survey of 10 questions. The pollster decides to rearrange the order of the questions so that any lead-in effect will be minimized. How many different versions of the survey are required if all possible arrangements are included?

24. How many different 7-digit telephone numbers are possible if the first digit cannot be 0 or 1?

25. A Federal Express delivery route must include stops at 7 cities.
    a. How many different routes are possible?
    b. If the route is randomly selected, what is the probability that the cities will be arranged in alphabetical order?

26. A 6-member FBI investigative team is to be formed from a list of 30 agents.
    a. How many different possible combinations can be formed?
    b. If the selections are random, what is the probability of getting the 6 agents with the most time in service?

27. Each Social Security number is a sequence of 9 digits. How many different Social Security numbers are possible?

28. A pollster must randomly select 3 of 12 available people. How many different groups of 3 are possible?

29. A union must elect 4 officers from 16 available candidates. How many different slates are possible if 1 candidate is nominated for each office?

30. a. How many different zip codes are possible if each code is a sequence of 5 digits?
    b. If a computer randomly generates 5 digits, what is the probability it will produce your zip code?

31. A typical combination lock is opened with the correct sequence of 3 numbers between 0 and 49 inclusive. How many different sequences are possible? (A number can be used more than once.) Are these sequences combinations or are they actually permutations?

32. One phase of an automobile assembly requires the attachment of 8 different parts, and they can be attached in any order. The manager decides to find the most efficient sequence by trying all possibilities. How many different sequences are possible?

33. A space shuttle crew has available 10 main dishes, 8 vegetable dishes, 13 desserts, and 3 appetizers. If the first meal includes 2 desserts and 1 item from each of the other categories, how many different combinations are possible?

34. In Denys Parsons' *Directory of Tunes and Musical Themes*, melodies for more than 14,000 songs are listed according to the following scheme: The first note of every song is represented by an asterisk (*), and successive notes are represented by R (for repeat the previous note), U (for a note that goes up), or D (for a note that goes down). Beethoven's "Fifth Symphony" begins as *RRD. Classical melodies are represented through the first 16 notes. With this scheme, how many different classical melodies are possible?

35. A television program director has 14 shows available for Monday night, and 5 shows must be chosen.
   a.   How many different possible combinations are there?
   b.   If 650 different combinations are judged to be incompatible, find the probability of randomly selecting 5 shows that are compatible.

36. A telephone company employee must collect the coins at 40 different locations.
   a.   How many different routes are possible?
   b.   If 2 of the routes are shortest, find the probability of randomly selecting a route and getting 1 of the 2 shortest routes.

37. A common lottery win requires that you pick the correct 6-number combination randomly selected from the numbers between 1 and 49 inclusive. Find the probability of such a win, and compare it to the probability of being struck by lightning this year, which is approximately 1/700,000.

38. A multiple-choice test consists of 10 questions with choices a, b, c, d, e.
   a.   How many different answer keys are possible?
   b.   If all 10 answers are random guesses, what is the probability of getting a perfect score?

39. The Bureau of Fisheries once asked Bell Laboratories for help in finding the shortest route for getting samples from locations in the Gulf of Mexico. How many different routes are possible if samples must be taken from 24 locations?

40. A manager must choose 5 secretaries from among 12 applicants and assign them to different stations.
   a.   How many different arrangements are possible?
   b.   If the selections are random, what is the probability of getting the 5 youngest secretaries selected in order of age?

3-6   Exercises B: Beyond the Basics

41. The number of permutations of n items when x of them are identical to each other and the remaining $n - x$ are identical to each other is given by

$$\frac{n!}{(n - x)! \, x!}$$

If a sequence of 10 trials results in only successes and failures, how many ways can 3 successes and 7 failures be arranged?

42. Among couples with 12 grandchildren, one couple is to be randomly selected.

(continued)

What is the probability that the 12 grandchildren will consist of 4 boys and 8 girls? (See Exercise 41.)

43. A common computer programming rule is that names of variables must be between 1 and 8 characters long. The first character can be any of the 26 letters, while successive characters can be any of the 26 letters or any of the 10 digits. For example, allowable variable names are A, AAA, and R2D2. How many different variable names are possible?

44. a.   Five managers gather for a meeting. If each manager shakes hands with each other manager exactly once, what is the total number of hand-shakes?

    b.   If n managers shake hands with each other exactly once, what is the total number of handshakes?

45. a.   How many different ways can 3 people be seated at a round table? (Assume that if everyone moves to the right, the seating arrangement is the same.)

    b.   How many different ways can n people be seated at a round table?

46. We say that a sample is *random* if all possible samples of the same size have the same probability of being selected.

    a.   If a random sample is to be drawn from a population of size 60, find the probability of selecting any one individual sample consisting of 5 members of the population.

    b.   Write a general expression for the probability of selecting a particular random sample of size n from a population of size N.

47. Many calculators or computers cannot directly calculate 70! or higher. When n is large, $n!$ can be *approximated* by

    $$n! = 10^K \text{ where } K = (n + 0.5)\log n + 0.39908993 - 0.43429448n$$

    Evaluate 50! using the factorial key on a calculator and also by using the approximation given here.

48. The Bureau of Fisheries once asked Bell Laboratories for help in finding the shortest route for getting samples from 300 locations in the Gulf of Mexico. There are 300! different possible routes. If 300! is evaluated, how many digits are used in the result? (See Exercise 47.)

49. Can computers "think"? According to the *Turing test*, a computer can be considered to think if, when a person communicates with it, the person believes he or she is communicating with another person instead of a computer. In an experiment at Boston's Computer Museum, each of 10 judges communicated with 4 computers and 4 other people and was asked to distinguish between them.

    a.   Assume that the first judge cannot distinguish between the 4 computers and the 4 people. If this judge makes random guesses, what is the probability of correctly identifying the 4 computers and the 4 people?

    b.   Assume that all 10 judges cannot distinguish between computers and people, so they make random guesses. Based on the result from part a, what is the probability that all 10 judges make all correct guesses? (That event would lead us to conclude that computers cannot "think" when, according to the Turing criterion, they can.)

VOCABULARY LIST

Define and give an example of each term.

experiment	mutually exclusive
event	rule of complementary events
simple event	tree diagram
sample space	independent events
relative frequency approximation of	dependent events
probability	multiplication rule
classical approach to probability	conditional probability
law of large numbers	acceptance sampling
random sample of one element	redundancy
random sample	simulation
complement	fundamental counting rule
subjective probability	factorial rule
compound event	permutations rule
addition rule	combinations rule

REVIEW

This chapter introduced the basic concept of probability. We began, in Section 3-2, with two rules for finding probabilities. Rule 1 represents the *relative frequency* approach, whereby the probability of an event is approximated by actually conducting or observing the experiment in question.

Rule 1

$$P(A) = \frac{\text{number of times } A \text{ occurred}}{\text{number of times experiment was repeated}}$$

Rule 2 is called the classical approach, and it applies only if all of the outcomes are equally likely.

Rule 2

$$P(A) = s/n = \frac{\text{number of ways } A \text{ can occur}}{\text{total number of different outcomes}}$$

Important Formulas

$0 \leq P(A) \leq 1$	for any event A
$P(A \text{ or } B) = P(A) + P(B)$	if A, B are mutually exclusive
$P(A \text{ or } B)$ $= P(A) + P(B) - P(A \text{ and } B)$	if A, B are not mutually exclusive
$P(\overline{A}) = 1 - P(A)$	
$P(A) = 1 - P(\overline{A})$	
$P(A) + P(\overline{A}) = 1$	
$P(A \text{ and } B) = P(A) \cdot P(B)$	if A, B are independent
$P(A \text{ and } B) = P(A) \cdot P(B|A)$	if A, B are dependent
$m \cdot n$	Total number of ways two events can occur, if the first can occur m ways while the second can occur n ways
$n!$	Number of ways n different items can be arranged
$_nP_r = \dfrac{n!}{(n-r)!}$	Number of *permutations* (arrangements) when r items are selected from n available items
$_nC_r = \dfrac{n!}{(n-r)!\, r!}$	Number of *combinations* when r items are selected from n available items

We noted that the probability of any impossible event is 0, while the probability of any certain event is 1, and for any event A,

$$0 \leq P(A) \leq 1$$

Also, $\overline{A}$ denotes the complement of event A. That is, $\overline{A}$ indicates that event A does not occur.

In Section 3-3 we consider the addition rule for finding the probability that A or B will occur. In evaluating $P(A \text{ or } B)$, it is important to consider whether the events are *mutually exclusive*—that is, whether they can both occur at the same time. We also introduced the rule of complementary events: $P(A) + P(\overline{A}) = 1$.

In Section 3-4 we considered the multiplication rule for finding the probability that A and B will occur. In evaluating $P(A \text{ and } B)$, it is important to consider whether the events are *independent*—that is, whether the occurrence of one event affects the probability of the other event.

In Section 3-5 we discussed the use of simulations as another approach to finding probabilities.

In Section 3-6 we considered techniques for determining the total number of different possibilities for various events. We presented the fundamental counting rule, the factorial rule, the permutations formula, and the combinations formula.

Most of the material in the following chapters deals with statistical inferences based on probabilities. As an example of the basic approach used, consider a test of someone's claim that a quarter used in a coin toss is fair. If we flip the quarter 10 times and get 10 consecutive heads, we can make one of two inferences from these sample results:

1. The coin is actually fair, and the string of 10 consecutive heads is a fluke.
2. The coin is not fair.

The statistician's decision as to which inference is correct is based on the probability of getting 10 consecutive heads, which, in this case, is so small (1/1024) that the inference of unfairness is the better choice. Here we can see the important role played by probability in the standard methods of statistical inference.

REVIEW EXERCISES

1. Ninety percent of all motorcycle drivers are male. If police stop 5 different motorcycle drivers at a license checkpoint, find the probability that
   a. all are males
   b. there is at least 1 female
2. A critical component in a circuit will work properly only if 3 other components all work properly. The probabilities of a failure for the 3 other components are 0.010, 0.005, and 0.012. Find the probability that at least 1 of these 3 components will fail.
3. The probability of a hang glider participant dying in a given year is 0.008 (based on the book *Acceptable Risks,* by Imperato and Mitchell). If 10 hang glider participants are randomly selected, find the probability that
   a. they all survive a year
   b. at least 1 does not survive a year
4. Among the respondents included in one survey, 98 drivers (29 of whom are males) said they used seat belts, while 52 drivers (21 of whom are males) said they did not use seat belts (based on data from Traffic Safety Now, Inc.). If one of these respondents is randomly selected, find the probability of getting a driver who said he or she used seat belts.
5. Refer to the survey respondents described in Exercise 4. If one of those respondents is randomly selected, find the probability of getting a female or someone who said he or she used seat belts.

6. In an attempt to gain access to a computer data bank, a computer is programmed to automatically dial every phone number with the prefix 478 followed by 4 digits. How many such telephone numbers are possible?

7. The accompanying table summarizes the ratings for 6665 films made before the NC 17 rating was introduced in 1990. (The table is based on data from the Motion Picture Association of America.) If one of these movies is randomly selected, find the probability that it has a rating of R.

Rating	Number
G	873
PG	2505
R	2945
X	342

8. Using the data from Exercise 7, find the probability of randomly selecting one of the movies and getting one with a rating of G or PG.

9. In the original New York State lottery, winning the grand prize required that you select the correct 6 numbers between 1 and 40 (in any order). Find the probability of winning if you select one combination of 6 numbers.

10. In one game of the New York lottery, your probability of winning by selecting the correct 6-number combination from the 54 possible numbers is 1/25,827,165. What is the probability if the rules are changed so that you must get the correct 6 numbers in the order in which they are selected?

11. Of 120 auto ignition circuits, 18 are defective. If 2 circuits are randomly selected, find the probability that they are both defective in each of the following cases.
   a. The first selection is replaced before the second selection is made.
   b. The first selection is not replaced.

12. In the televised NBC White Paper *Divorce,* it was reported that 85% of divorced women are not awarded any alimony. If 4 divorced women are randomly selected, what is the probability that at least 1 of them is not awarded alimony?

13. A pollster claims that 12 voters were randomly selected from a population of 200,000 voters (30% of whom are Republicans), and all 12 were Republicans. The pollster claims that this could easily happen. Find the probability of getting 12 Republicans when 12 voters are randomly selected from this population.

14. If a couple plans to have 5 children, find the probability that they will all be girls. Assume that boys and girls have the same chance of being born and that the sex of any child is not affected by the sex of any other children.

15. a. Find $P(A$ and $B)$ if $P(A) = 0.2$, $P(B) = 0.4$, and A and B are independent.
   b. Find $P(A$ or $B)$ if $P(A) = 0.2$, $P(B) = 0.4$, and A and B are mutually exclusive.
   c. Find $P(\overline{A})$ if $P(A) = 0.2$.

16. Evaluate the following:
   a. 0!   b. 8!   c. $_8P_6$ d. $_{10}C_8$ e. $_{80}C_{78}$

17. A question on a history test requires that 5 events be arranged in the proper chronological order. If a random arrangement is selected, what is the probability that it will be correct?

18. According to Bureau of the Census data, 52% of women aged 18 to 24 years do not live at home with their parents. If we randomly select 5 different women in that age bracket, what is the probability that none of them live at home with their parents?

19. The Allied Mail Express Company has a very large pool of thousands of job applicants, 50% of whom are women. If 4 employees are to be immediately hired and the selections are to be made randomly with respect to gender, find the probability that
    a. all 4 are women
    b. all 4 are men
    c. all 4 are of the same sex

20. In a poll of 1000 adult Americans conducted by the Family Violence Prevention Fund, 19% claim they have seen a robbery or mugging. Based on those results, estimate the probability that when 2 adult Americans are randomly selected, neither one has witnessed a robbery or mugging.

 COMPUTER PROJECT

In producing an experimental computer chip, a research group from the Clarion Computer Company experiences a 15% yield, meaning that there is a 0.15 probability of a chip being acceptable. Use a simulation approach (see Section 3-5) to "generate" chips until you get one that is acceptable. Use STATDISK or Minitab to generate 200 numbers between 1 and 100. Consider acceptable chips to be represented by 1, 2, 3, . . . , 15, and consider unacceptable chips to be represented by 16, 17, 18, . . . , 100. Examine the random numbers that are generated, and record the number of chips required to get one that is acceptable. Starting with the next computer-generated chip, begin another production run and record the number of chips required to get another good one. Repeat this process until the supply of the 200 random numbers is exhausted. Construct a frequency table, histogram, and stem-and-leaf plot, and find the mean, median, standard deviation, minimum, and maximum. Write a brief report describing important characteristics of the data.

FROM DATA TO DECISION: Drug testing of job applicants

According to the American Management Association, most U.S. companies now test at least some employees and job applicants for drug use. The U.S. National Institute on Drug Abuse claims that about 15% of people in the 18–25 age bracket use illegal drugs.

Allyn Clark, a 21-year-old college graduate, applied for a job at the Acton Paper Company, took a drug test, and was not offered a job. He suspected that he might have failed the drug test, even though he does not use drugs. In checking with the company's personnel department, he found that the drug test has 99% *sensitivity,* so only 1% of drug users test negative. Also, the test has 98% *specificity,* meaning that only 2% of non-users are incorrectly identified as drug users. Allyn felt relieved by these figures because he believed that they reflected a very reliable test that usually provides good results—but

should he be relieved? Can the company feel secure that drug users are not being hired?

The accompanying table shows data for Allyn and 1999 other job applicants. Based on those results, find the probability of a "false positive"; that is, find the probability of randomly selecting one of the subjects who tested positive and getting someone who does not use drugs. Also find the probability of a "false negative"; that is, find the probability of randomly selecting someone who tested negative and getting someone who does use drugs. Are the probabilities of these wrong results low enough so that job applicants and the Acton Paper Company need not be concerned?

	Drug users	Non-users
Positive test result	297	34
Negative test result	3	1666

Interview

Anthony DiUglio
Nuclear Analyst, Probabilistic
Risk Assessment
Consolidated Edison
Company of New York, Inc.

Anthony DiUglio works in the Probabilistic Risk Assessment (PRA) Group at Consolidated Edison's Indian Point Unit #2 nuclear generating facility in Buchanan, New York. In his work as a Nuclear Analyst, Tony develops probabilities that are used in quantifying various aspects of the plant-specific risk assessment. He is a former student of the author.

What is your job?

In PRA we are concerned with three basic questions about risk: what can happen, how likely is it to happen, and what are the consequences of it happening. We apply these questions about risk to the safe, reliable, and continuous operation of our power plant. When we quantify risk, we obtain numbers that are probabilities. If someone suggests a modification to a plant safety system, we analyze it from a risk perspective. Is the modification better for the system? Does it affect the operation of the plant or put the public health and safety at risk?

How do you use probability and/or statistics?

They're our primary tools. Our PRA requires that we quantify plant-specific repair rates for all safety-related components in our plant. In developing component-repair rates for pumps and valves, we look at industry-wide data (generic) and our plant-specific data. We combine this information together, under uncertainty, and end up with component-specific repair probabilities.

How do you use probability and/or statistics in other departments at Indian Point?

Our Performance Department measures various plant parameters, such as heat rate, megawatt generation, cost per kilowatt of generation, etc. These parameters are all done by use of statistics. The statistical tools they use are data trending, normal statistical curves, standard deviations, histograms, etc. Financial Planning makes extensive use of statistics in projecting budget needs and determining its constraints. Our corporate forecasters use probability theory to predict power demands at different times during the year (e.g., winter and summer, one, three, and five years down the road). We have so many people using statistics in their everyday work that statistics has now become a tool for engineers, planners, forecasters, and those of us in Risk Assessment.

In terms of statistics, what would you recommend for prospective employees?

They should have a good understanding of probability, statistics, and their applications. Because PRA is still a relatively new area, we often deal with problems that haven't been addressed before, so many of the problems we address require creative problem solving. Once you have the basic tools, your time is efficiently spent. You can't effectively communicate here unless you use a common language, and that common language happens to be statistics.

Has your work been helpful in convincing the public that your plant is safe?

Safety is always our first concern. In the early 1980s there was a series of public hearings conducted by the Nuclear Regulatory Commission (NRC) to discuss whether or not the plant should continue operation. Con Edison maintained that the plant was safe, and we were able to help justify the continued operation of our plant through the use of our PRA. At the conclusion of those hearings the NRC agreed with our position, and we continued to operate.

Who was your best math teacher?

Professor Mario Triola.

Is your use of probability and statistics increasing, decreasing, or remaining stable?

It's increasing all the time. We are very much involved with plant performance indicators as parameters for efficient plant operation. With PRA we now have a tool that allows us to focus our attention on the more important plant components and functions. In the case where three components all need maintenance, PRA allows us to identify which component should be returned to service first. In engineering, if we have several components that should be improved, PRA allows us to identify which component should be improved first. We can quantify the effects and thereby target our resources better, thus making the plant safer.

"You can't effectively communicate here unless you use a common language, and that common language happens to be statistics."

4 Probability Distributions

4-1 Overview

The objectives of this chapter are identified—developing probability distributions of discrete random variables and using various methods to find their means, standard deviations, and probability histograms.

4-2 Random Variables

Discrete and continuous random variables and probability distributions are described in this section. Methods of determining the mean, variance, and standard deviation for a given probability distribution are discussed. The expected value of a probability distribution is defined.

4-3 Binomial Experiments

Probabilities in binomial experiments are calculated using the binomial probability formula, a table of binomial probabilities, and statistical software. The Poisson distribution is discussed.

4-4 Mean, Variance, and Standard Deviation for the Binomial Distribution

The mean, variance, and standard deviation for a binomial distribution are calculated.

Chapter Problem:
What's new in pink and blue?

New techniques are making it possible for couples to select the gender of their children. Such techniques raise some serious ethical and religious questions. There is concern about the effects of changing our society's natural male/female ratio, and there is concern that infanticide and aborting of females will increase. (See also the essay "Big Brother" in Section 4-4.) Rational discussions of such issues must involve an understanding of the effectiveness of gender selection methods; we will focus on this aspect of the problem.

The Ericsson method of sperm separation (named for physiologist Ronald Ericsson) supposedly has a 75% success rate for couples wanting boys and a 69% success rate for couples wanting girls. (As of this writing, experimental data supporting those rates were unavailable.) Suppose that 100 couples prefer to have baby girls because the mothers are carriers of an X-linked recessive disorder that will occur in 50% of their sons but will not occur in their daughters. Let's assume that they use the Ericsson method, with the result that among 100 babies, there are 75 girls. Is the Ericsson method of gender selection effective in the sense that the birthrate of girls is greater than the 50% rate expected under normal conditions? If the Ericsson gender-choice method really has no effect, how likely is it that we will get a success rate of 75% for girls? We need some hard evidence to answer those questions, and the methods of this chapter will allow us to obtain that evidence. We will therefore address this issue later in the chapter.

4-1 Overview

In Chapter 2 we discussed the histogram as a device for showing the frequency distribution of a set of data; in Chapter 3 we discussed the basic principles of probability theory. In this chapter we combine those concepts to develop probability distributions that are basically theoretical models of the frequency distributions we produce when we collect sample data. We construct frequency tables and histograms using *observed* real scores, but we construct probability distributions by presenting possible outcomes along with the frequencies we expect, given an understanding of the relevant circumstances.

Suppose that a casino manager suspects cheating at a dice table. The manager can compare the frequency distribution of the actual sample outcomes to a theoretical model that describes the frequency distribution likely to occur with a fair die. (See Figure 4-1.) In this case the probability distribution serves as a model of a theoretically perfect population frequency distribution. In essence, we can determine what the frequency table and histogram would be like for a die rolled an infinite number of times. With this perception of the population of outcomes, we can then determine the values of important parameters such as the mean, variance, and standard deviation.

The concept of a probability distribution is not limited to casino management. In fact, the remainder of this book and the very core of inferential statistics depend on some knowledge of probability distributions. We begin by examining the concept of a random variable.

4-2 Random Variables

Many everyday situations can be used as experiments that yield outcomes corresponding to some value, such as the following:

- The number of parts replaced on Corvettes
- The number of correct answers on an ACT or SAT test
- The number of car crashes on the Golden Gate Bridge
- The number of passengers on American Airlines Flight 8088

We use the term *random variable* to describe the value that corresponds to the outcome from a given experiment. The word *random* is used to remind us that we don't usually know what that value is until after the experiment is conducted.

random not predetermined

DEFINITION

A **random variable** is a variable (typically represented by x) that has a single numerical value (determined by chance) for each outcome of an experiment.

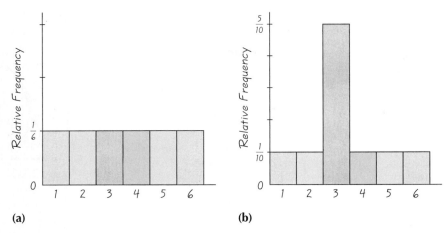

(a)   **(b)**

FIGURE 4-1   Histograms of Dice Outcomes for
(a) a Fair Die and (b) a Loaded Die

▶ **EXAMPLE**

An experiment consists of testing the mathematical skills of applicants for jobs at Taylor Advertising Company. The test is made up of 10 multiple-choice questions. If we let the random variable represent the number of correct answers for one applicant, this experiment has outcomes of 0, 1, 2, 3, 4, 5, 6, 7, 8, 9, 10. (Remember, 0 represents no correct answers, 1 represents 1 correct answer, and so on.) The variable is random in the sense that we don't know the values until after the test has been taken.

In Section 1-2 we made a distinction between discrete and continuous data. Random variables may also be discrete or continuous, and the following two definitions are consistent with those given in Section 1-2.

DEFINITIONS

A **discrete random variable** has either a finite number of values or a countable number of values.

A **continuous random variable** has infinitely many values, and those values can be associated with measurements on a continuous scale in such a way that there are no gaps or interruptions.

(a) Discrete Random Variable: Count of the number of movie patrons.

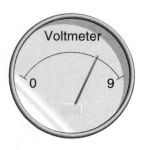

(b) Continuous Random Variable: The measured voltage of a smoke detector battery.

FIGURE 4-2   Discrete and Continuous Random Variables

▶ **EXAMPLE**

a. The count of the number of patrons viewing a movie is a finite number and is therefore a discrete random variable. See Figure 4-2(a).

b. The measure of voltage for a smoke detector battery can be any value between 0 volts and 9 volts. It is therefore a continuous random variable. See Figure 4-2(b).

In addition to identifying values of a random variable, we can often identify a probability for each of those values.

DEFINITION

A **probability distribution** gives the probability for each value of the random variable.

▶ **EXAMPLE**

Suppose the Telektronic Company plans to hire 10 new employees and there are hundreds of applicants, with equal numbers of men and women. Let the random variable x represent the number of women hired. If new employees are selected in such a way that men and women have an equal chance of being hired, then Table 4-1 describes the probability distribution for the random variable x. The probability that no women will be hired (among 10) is 0.001, the probability of exactly 1 woman (and 9 men) is 0.010, and so on. Later, we will see how the probabilities in Table 4-1 were found.

There are various ways to graph a probability distribution, but we present only the **probability histogram.** Figure 4-3 is a probability histogram that resembles the relative frequency histogram from Chapter 2, but the vertical scale delineates *probabilities* instead of actual relative frequencies.

In Figure 4-3, note that along the horizontal axis, the values of 0, 1, 2, . . . , 10 are located at the centers of the rectangles. This implies that the rectangles are each 1 unit wide, so the areas of the rectangles are 0.001, 0.010, and so on

x	P(x)
0	0.001
1	0.010
2	0.044
3	0.117
4	0.205
5	0.246
6	0.205
7	0.117
8	0.044
9	0.010
10	0.001

TABLE 4-1   Probability Distribution for Number of Women Hired

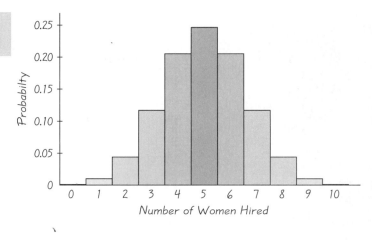

FIGURE 4-3   Probability Histogram for Number of Women Hired

(that is, the *probabilities* are equal to the corresponding rectangular *areas*). We will see in future chapters that this correspondence between area and probability is very useful in statistics.

If you hire 10 new employees, it is impossible to hire exactly 4 women and exactly 5 women at the same time, so the events of 4 women and 5 women are mutually exclusive. In general, if two events result in different values of a random variable, the events are mutually exclusive. Also, all the values of the random variable represent all events in the entire sample space. We can therefore use the addition rule for mutually exclusive events to conclude that the sum of P(x) for all values of x must be 1. Also, the probability rule stating that $0 \le P(A) \le 1$ for any event A implies that P(x) must be between 0 and 1 for any value of x. These two requirements for probability distributions are listed below.

Requirements for P(x) to Be a Probability Distribution

1. $\Sigma P(x) = 1$ where x assumes all possible values
2. $0 \le P(x) \le 1$ for every value of x

▶ EXAMPLE

Does $P(x) = x/5$ (where x can take on the values of 0, 1, 2, 3) determine a probability distribution?   *(continued)*

Is Parachuting Safe?

About 30 people die each year as more than 100,000 people make about 2.25 million parachute jumps. In comparison, a typical year includes about 200 scuba diving fatalities, 7000 drownings, 900 bicycle deaths, 800 lightning deaths, and 1150 deaths from bee stings. Of course, these figures don't necessarily mean that parachuting is safer than bike riding or swimming. A fair comparison should involve fatality *rates,* not just the total number of deaths.

The author, with much trepidation, made two parachute jumps but quit after missing the spacious drop zone both times. He has also flown in a hang glider, hot air balloon, and Goodyear blimp.

● **SOLUTION**

If a probability distribution is determined, it must conform to the preceding two requirements. But

$$\Sigma P(x) = P(0) + P(1) + P(2) + P(3)$$
$$= \frac{0}{5} + \frac{1}{5} + \frac{2}{5} + \frac{3}{5}$$
$$= \frac{6}{5} \qquad \Sigma P(x) \neq 1$$

Thus the first requirement is not satisfied, and a probability distribution is not determined.

▶ **EXAMPLE**

Does $P(x) = x/10$ (where x can be 0, 1, 2, 3, or 4) determine a probability distribution?

● **SOLUTION**

For the given function we conclude that

$$P(0) = \frac{0}{10} = 0$$

$$P(1) = \frac{1}{10}$$

$$P(2) = \frac{2}{10}$$

$$P(3) = \frac{3}{10}$$

$$P(4) = \frac{4}{10}$$

$$\frac{0}{10} + \frac{1}{10} + \frac{2}{10} + \frac{3}{10} + \frac{4}{10} = 1 \qquad \text{so } \Sigma P(x) = 1$$

The sum of these probabilities is 1, and each $P(x)$ is between 0 and 1, so both requirements are satisfied. Consequently, a probability distribution is determined.

Mean, Variance, and Standard Deviation

In Chapter 2 we saw that there are three extremely important characteristics of data:

1. Representative score, such as an average   *mean*
2. Measure of scattering or variation σ
3. Nature or shape of the distribution, such as bell shaped

The probability histogram can give us insight into the nature or shape of the distribution. Also, we can often find the mean, variance, and standard deviation of data, which provide insight into the other characteristics. The mean, variance, and standard deviation for a probability distribution can be found by applying Formulas 4-1, 4-2, 4-3, and 4-4.

Formula 4-1	$\mu = \Sigma x \cdot P(x)$	Mean for a probability distribution
Formula 4-2	$\sigma^2 = \Sigma[(x - \mu)^2 \cdot P(x)]$	Variance for a probability distribution
Formula 4-3	$\sigma^2 = [\Sigma x^2 \cdot P(x)] - \mu^2$	Variance for a probability distribution (shortcut version)
Formula 4-4	$\sigma = \sqrt{[\Sigma x^2 \cdot P(x)] - \mu^2}$	Standard deviation for a probability distribution

We should stress that $\Sigma x \cdot P(x)$ is the same as $\Sigma[x \cdot P(x)]$. That is, first multiply each value of x by its corresponding probability, and then add the resulting products.

Roundoff Rule for μ, σ^2, and σ

When using Formulas 4-1 through 4-4, round results by carrying one more decimal place than the number of decimal places used for the random variable x. If the values of x are integers, round μ, σ^2, and σ to one decimal place.

When we calculate the mean from a probability distribution, we get the average (mean) value that we would expect to get if the trials could be repeated indefinitely. We *don't* get the value we expect to occur most often. In fact, we often get a mean value that cannot occur in any one actual trial (such as 1.5 girls in 3 births). The standard deviation gives us a measure of how much the probability distribution is spread out around the mean. A large standard deviation reflects considerable spread, whereas a smaller standard deviation reflects lower variability with values relatively closer to the mean.

▶ **EXAMPLE**

Assuming that the Telektronic Company has hundreds of job applications from men and an equal number from women and assuming that 10 of the applicants will be selected in such a way that men and women have equal opportunities, find the mean number of women hired in such groups of 10, as well as the standard deviation and variance.

● **SOLUTION**

In Table 4-2, the two columns at the left describe the probability distribution given earlier, and we create the three columns at the right for the purposes of the calculations required.

TABLE 4-2   Calculating μ, σ^2, and σ for a Probability Distribution

x	$P(x)$	$x \cdot P(x)$	x^2	$x^2 \cdot P(x)$
0	0.001	0.000	0	0.000
1	0.010	0.010	1	0.010
2	0.044	0.088	4	0.176
3	0.117	0.351	9	1.053
4	0.205	0.820	16	3.280
5	0.246	1.230	25	6.150
6	0.205	1.230	36	7.380
7	0.117	0.819	49	5.733
8	0.044	0.352	64	2.816
9	0.010	0.090	81	0.810
10	0.001	0.010	100	0.100
Total		5.000		27.508
		↑		↑
		$\Sigma x \cdot P(x)$		$\Sigma x^2 \cdot P(x)$

Using Formulas 4-1 and 4-3 and the table results, we get

$$\mu = \Sigma x \cdot P(x) = 5.0$$

and

$$\sigma^2 = [\Sigma x^2 \cdot P(x)] - \mu^2$$
$$= 27.508 - 5.0^2$$
$$= 2.508 \text{ or } 2.5 \quad \text{(rounded)}$$

(continued)

The standard deviation is the square root of the variance, so

$$\sigma = \sqrt{2.508} = 1.6 \qquad \text{(rounded)}$$

We now know that among 10 new employees, the mean number of women is 5.0, the variance is 2.5 "women squared," and the standard deviation is 1.6 women.

Why do Formulas 4-1 through 4-4 work? A probability distribution is actually a model of a theoretically perfect population frequency distribution. The probability distribution is like a relative frequency distribution based on data that behave perfectly, without the usual imperfections of samples. Because the probability distribution allows us to perceive the population, we are able to determine the values of the mean, variance, and standard deviation. Formula 4-1 accomplishes the same task as the formula for the mean of a frequency table. (Recall that f represents class *frequency* and N represents population size.) Rewriting the formula for the mean of a frequency table so that it applies to a population and then changing its form, we get

$$\mu = \frac{\Sigma\,(f \cdot x)}{N} = \Sigma\,\frac{f \cdot x}{N} = \Sigma x \cdot \frac{f}{N} = \Sigma x \cdot P(x)$$

The fraction f/N is the relative frequency with which the value x occurs, and N is the population size, so f/N is the probability for the value of x.

Similar reasoning enables us to take the variance formula from Chapter 2 and apply it to a random variable for a probability distribution; the result is Formula 4-2. Formula 4-3 is a shortcut version that will always produce the same result as Formula 4-2. Although Formula 4-3 is usually easier to work with, Formula 4-2 is easier to understand directly. Based on Formula 4-2, we can express the standard deviation as

$$\sigma = \sqrt{\Sigma\,(x - \mu)^2 \cdot P(x)}$$

or as the equivalent form given in Formula 4-3. Some calculators are designed to give you values of the mean and standard deviation of probability distributions.

Expected Value

The mean of a discrete random variable is the theoretical mean outcome for infinitely many trials. We can think of that mean as the expected value in the sense that it is the average value that we would expect to get if the trials could continue indefinitely. The uses of expected value (also called expectation or mathematical expectation) are extensive and varied, and they play a very important role in an area of application called *decision theory*. (For a discussion of decision theory, see *Business Statistics,* by Triola and Franklin.)

Wage Gender Gap

Many articles note that, on average, full-time female workers earn about 70¢ for each $1 earned by full-time male workers. Researchers at the Institute for Social Research at the University of Michigan analyzed the effects of various key factors and found that about one-third of the discrepancy can be explained by differences in education, seniority, work interruptions, and job choices. The other two-thirds remains unexplained by such labor factors.

Expected Value is the same as the mean (handwritten margin note)

DEFINITION

The **expected value** of a discrete random variable is denoted by *E*, and it represents the average value of the outcomes. It is found by finding the value of $\Sigma x \cdot P(x)$.

$$E = \Sigma x \cdot P(x)$$

From Formula 4-1 we see that $E = \mu$. That is, the mean of a discrete random variable and its expected value are the same. Flip a coin 10 times and the *mean* number of heads is 5; flip a coin 10 times and the *expected value* of the number of heads is also 5.

▶ EXAMPLE

Consider the numbers game started many years ago by organized crime groups and now run legally by many organized governments. You can place a bet that the 3-digit number of your choice will be the winning number selected. The typical winning payoff is 499 to 1, meaning that for each winning $1 bet, you would be given $500; your net return is therefore $499. Suppose that you bet $1 on the number 327. What is your expected value of gain or loss?

● SOLUTION

For this bet there are two simple outcomes: You win or you lose. Because you have the number 327 and there are 1000 possibilities (from 000 to 999), your probability of winning is 1/1000 (or 0.001) and your probability of losing is 999/1000 (or 0.999). Table 4-3 summarizes this situation.

From Table 4-3 we can see that when we bet $1 in the numbers game, our expected value is

$$E = \Sigma x \cdot P(x) = -50¢$$

This means that in the long run, for each $1 bet, we can expect to lose an average of 50¢.

TABLE 4-3
The Numbers Game

Event	x	P(x)	x · P(x)
Win	$499	0.001	$0.499
Lose	−$1	0.999	−$0.999
Total			−$0.50 = −50¢

In the preceding example, a player will either lose $1 or win $499; there will never be a loss of 50¢, as the expected value of −50¢ might seem to suggest. The expected value of −50¢ is an *average* over a long run of bets placed. Even if we're thinking of placing only one bet, the expected value of −50¢ indicates that it's not a good bet. The potential gain is more than offset by the potential loss.

▶ EXAMPLE

The Dayton Machine Company is a defendant in a product liability case and must choose between settling out of court for a loss of $150,000 and going to trial and either losing nothing (if found not guilty) or losing $500,000 (if found guilty). The attorney estimates that the probability of a not-guilty verdict is 0.8. Assuming that the defendant decides to go to trial, find the expected value of the amount lost.

● SOLUTION

In Table 4-4 we summarize the values of the random variable x and the corresponding probabilities for the decision of going to trial.

TABLE 4-4   Product Liability Trial

Outcome	x	$P(x)$	$x \cdot P(x)$
Not guilty	$0	0.8	$0
Guilty	−$500,000	0.2	−$100,000
Total			−$100,000

The expected value is $\mu = \Sigma x \cdot P(x) = -\$100,000$. Because this isn't as bad as −$150,000, the defendant would be wise to go to trial.

In the preceding example, the defendant will lose either nothing, $150,000, or $500,000. The expected value of −$100,000 for a trial is an average for a long run of trials held under the same conditions. However, we have a situation that will occur only once. Yet based on the expected value, the best decision is to go to trial. Companies often use this type of decision-making strategy and have found it to be effective. It should, however, be tempered with common sense. If the Dayton Machine Company cannot absorb a $500,000 loss, it may be wise to settle the case, even though the expected value suggests that the case should be tried.

In Chapter 10 we use the concept of expected value to compare actual survey results to expected results. The degree of similarity or disparity will allow us to form some meaningful conclusions.

She Won the Lottery Twice!

Evelyn Marie Adams won the New Jersey Lottery twice in four months. This happy event was reported as an incredible coincidence with a likelihood of only 1 chance in 17 trillion. But Harvard mathematicians Persi Diaconis and Frederick Mosteller note that there is 1 chance in 17 trillion that a particular person with one ticket in each of two New Jersey lotteries will win both times. However, there is about 1 chance in 30 that someone in the United States will win a lottery twice in a four-month period. Diaconis and Mosteller analyzed coincidences and conclude that "with a large enough sample, any outrageous thing is apt to happen."

4-2 Exercises A: Basic Skills and Concepts

In Exercises 1–12, determine whether a probability distribution is given. In those cases where a probability distribution is not described, identify the requirement that is not satisfied. In those cases where a probability distribution is described, find its mean, variance, and standard deviation.

1. A vendor supplies refreshments at a baseball stadium and must plan for the possibility of a World Series contest. In the accompanying table (based on past results), x represents the number of baseball games required to complete a World Series contest.

x	$P(x)$
4	0.120
5	0.253
6	0.217
7	0.410

2. The Young Fun Company operates a national toy store chain that depends heavily on the size of the child population, and research is being conducted to learn more about the number of children in families. In the accompanying table, x represents the number of children (under 18 years of age) in families. (The table is based on data from the U.S. Census Bureau.)

x	$P(x)$
0	0.48
1	0.21
2	0.19
3	0.08

3. In the accompanying table, x represents the number of long-distance telephone calls made during a one-hour period in a telemarketing department.

x	$P(x)$
0	0.32
1	0.08
2	0.12
3	0.09
4	0.07
5	0.06
6	0.04

4. In the accompanying table, x represents the number of employees absent from the Flint Fabric Company on a given day.

x	$P(x)$
0	1/5
1	1/5
2	1/5
3	1/5
4	1/5

5. A drug supplied by the Medassist Pharmaceutical Company is administered to 3 patients, and the random variable x represents the number of cures that occur. The probabilities corresponding to 0 cures, 1 cure, 2 cures, and 3 cures are found to be 0.125, 0.375, 0.375, and 0.125, respectively.

6. If you randomly select a jail inmate convicted of DWI (driving while intoxicated), the probability distribution for the number x of prior DWI sentences is as described in the accompanying table (based on data from the U.S. Department of Justice).

x	$P(x)$
0	0.512
1	0.301
2	0.132
3	0.055

7. In assessing credit risks, Jefferson Valley Bank investigates the number of credit cards people have. With x representing the number of credit cards adults have, the accompanying table describes the probability distribution for a certain population (based on data from Maritz Marketing Research, Inc.).

x	$P(x)$
0	0.26
1	0.16
2	0.12
3	0.09
4	0.07
5	0.09
6	0.07
7	0.14

8. According to the Hertz Corporation, 69% of all workers commute in their own cars. For randomly selected groups of 8 workers, the accompanying table describes the probability distribution for the number of workers who commute in their own cars.

x	$P(x)$
0	0.000
1	0.002
2	0.012
3	0.053
4	0.147
5	0.261
6	0.290

9. To settle a paternity suit, two different people are given blood tests. If x is the number having group A blood, then x can be 0, 1, or 2 and the corresponding probabilities are 0.36, 0.48, and 0.16, respectively (based on data from the Greater New York Blood Program).

10. Car headlight manufacturers are concerned about failure rates. One headlight failure is an inconvenience, but if both lights fail, you can't drive at night. Assume that the probabilities of 0, 1, and 2 failures are 0.960, 0.036, and 0.004, respectively. (Round answers to 3 decimal places.)

11. United Airlines Flight 470 from Denver to St. Louis has an on-time performance described as follows: For 4 independent flights, the probabilities for 0, 1, 2, 3, and 4 on-time flights are 0.026, 0.345, 0.346, 0.154, and 0.129, respectively (based on data from the EAASY SABRE reservation system).

12. The Baltimore Computer House finds that the probabilities of selling 0, 1, 2, 3, and 4 microcomputers in one day are 0.245, 0.370, 0.210, 0.095, and 0.080, respectively.

13. If you have a 1/4 probability of gaining $500 and a 3/4 probability of gaining $200, what is your expected value of gain?

14. If you have a 1/10 probability of gaining $200, a 3/10 probability of losing $300, and a 6/10 probability of breaking even, what is your expected value?

15. The Kwik Klean Car Wash loses $30 on rainy days and gains $120 on days when it does not rain. If the probability of rain is 0.15, what is the expected value of net profit?

16. The Newman Construction Company bids on a job to construct a building. If the bid is won, there is a 0.7 probability of making a $175,000 profit and there is a probability of 0.3 that the contractor will break even. What is the expected value? Should the bid be submitted?

17. A 27-year-old woman decides to pay $156 for a one-year life insurance policy with coverage of $100,000. The probability of her living through the year is 0.9995 (based on data from the U.S. Department of Health and Human Services and AFT Group Life Insurance). What is her expected value for the insurance policy?

18. *Reader's Digest* recently ran a sweepstakes in which prizes were listed along with the chances of winning: $5,000,000 (1 chance in 201,000,000), $150,000 (1 chance in 201,000,000), $100,000 (1 chance in 201,000,000), $25,000 (1 chance in 100,500,000), $10,000 (1 chance in 50,250,000), $5000 (1 chance in 25,125,000), $200 (1 chance in 8,040,000), $125 (1 chance in 1,005,000), and a watch valued at $89 (1 chance in 3774).
    a. Compute the expected value of the amount won for one entry.
    b. If the only cost of entering this sweepstakes is a stamp, what is the mean of the net amount that is won or lost?

In Exercises 19–22, find the mean, variance, and standard deviation for the random variable.

19. The random variable x represents the number of girls in a family of 3 children. (*Hint:* Assuming that boys and girls are equally likely, we get $P(2) = 3/8$ by examining the sample space of bbb, bbg, bgb, bgg, gbb, gbg, ggb, ggg.)

20. The random variable x represents the number of boys in a family of 4 children. (See Exercise 19.)

21. The random variable x represents the number of foul shots a player makes when 2 shots are attempted. Assume that the player hits an average of 85% of his or her free throws and that the shots are independent. (*Hint:* Begin by using the multiplication rule to find $P(0)$ and $P(2)$.)

22. A batch of 20 computer chips contains exactly 6 that are defective. Two different chips are randomly selected without replacement, and the random variable x represents the number of defective chips selected. (*Hint:* Use the multiplication rule to first find $P(0)$ and $P(2)$.)

4-2   Exercises B: Beyond the Basics

23. a.   Let $P(x) = 0.4(0.6)^{x-1}$ where $x = 1, 2, 3, \ldots$. Is $P(x)$ a probability distribution?

b.   Let $P(x) = 1/2^x$ where $x = 1, 2, 3, \ldots$. Is $P(x)$ a probability distribution?

c.   Let $P(x) = 1/2x$ where $x = 1, 2, 3, \ldots$. Is $P(x)$ a probability distribution?

24. According to an analyst for Paine Webber, Intel's yield for computer chips is around 35%. Assume that we have 4 randomly selected chips and that the probability of an acceptable chip is 0.35. Let x represent the number of acceptable chips in groups of 4, and use the multiplication rule and the rule of complements (from Chapter 3) to complete the table so that a probability distribution is determined.

x	$P(x)$
0	
1	0.384
2	0.311
3	
4	

25. The variance for the discrete random variable x is 1.25.

a.   Find the variance of the random variable $5x$. (Each value of x is multiplied by 5.)

b.   Find the variance of the random variable $x/5$.

c.   Find the variance of the random variable $x + 5$.

d.   Find the variance of the random variable $x - 5$.

26. A discrete random variable can assume the values $1, 2, \ldots, n$, and those values are equally likely.

a.   Show that $\mu = (n + 1)/2$.

b.   Show that $\sigma^2 = (n^2 - 1)/12$.

(*Hint:* $1 + 2 + 3 + \cdots + n = n(n + 1)/2$.

$1^2 + 2^2 + 3^2 + \cdots + n^2 = n(n + 1)(2n + 1)/6$.)

27. Verify that $\sigma^2 = [\Sigma x^2 \cdot P(x)] - \mu^2$ is equivalent to $\sigma^2 = \Sigma(x - \mu)^2 \cdot P(x)$.

(*Hint:* For constant c, $\Sigma cx = c\Sigma x$. Also, $\mu = \Sigma x \cdot P(x)$.)

4-3   Binomial Experiments

On January 28, 1986, the space shuttle *Challenger* exploded, and its seven astronauts were killed. The disaster was apparently caused by the failure of a field-joint O-ring. The Rogers Commission investigated this disaster and concluded that when the *Challenger* was launched in 31°F weather, the chance of a catastrophic failure was at least 13%; if the launch had been delayed until the temperature reached 60°F, the chance of a catastrophic failure would have been around 2%. Methods' that we will discuss in this section were used to assess risks during the redesign of space shuttles following the *Challenger* tragedy.

Component failures are examples of a broad category of events that have an element of "twoness": In manufacturing, parts either fail or they don't. In medicine, a patient either survives a year or does not. In advertising, a consumer either recognizes a product or does not. These situations result in a special type of discrete probability distribution called the *binomial distribution,* which consists of a list of outcomes and probabilities for a binomial experiment. This section focuses on the binomial distribution, which is one of many such distributions that can be used to determine probabilities of outcomes for different situations.

DEFINITION

A **binomial experiment** is one that meets all the following requirements:

1. The experiment must have a *fixed number of trials*.
2. The trials must be *independent*. (The outcome of any individual trial doesn't affect the probabilities in the other trials.)
3. Each trial must have all outcomes classified into *two categories*.
4. The probabilities must remain *constant* for each trial.

▶ EXAMPLE

The Dean of Desks at a local college must furnish each seminar classroom with desks of two types: those made for right-handed students and those made for left-handed students. Ten percent of the students are left-handed, and the classrooms are being furnished with 15 desks each. If an experiment consists of finding the number of students in a class of 15 who are left-handed, does it meet the requirements for a binomial experiment?

● SOLUTION

This experiment does satisfy the requirements for a binomial experiment.

1. The number of trials (15) is fixed.
2. The trials are independent because the left-handedness or right-handedness of any one student doesn't affect the probability of any other student being left-handed.
3. Each trial has two categories of outcomes: the student either is or is not left-handed.
4. The probability (0.10) remains constant for the different students.

Notation for Binomial Distributions

S and F (success and failure) denote the two possible categories of all outcomes; p and q will denote the probabilities of S and F, respectively, so

$$P(S) = p$$
$$P(F) = 1 - p = q$$

n denotes the fixed number of trials.

x denotes a specific number of successes in n trials, so x can be any whole number between 0 and n, inclusive.

p denotes the probability of success in *one* of the n trials.

q denotes the probability of failure in *one* of the n trials.

$P(x)$ denotes the probability of getting exactly x successes among the n trials.

The word *success* as used here does not necessarily correspond to a desired result. Either of the two possible categories may be called the success S as long as the corresponding probability is identified as p. (The value of q can always be found by subtracting p from 1; if $p = 0.95$, then $q = 1 - 0.95 = 0.05$.) However, once a category has been designated as the success S, be sure that p is the probability of a success and x is the number of successes. That is, be sure that the values of p and x refer to the same category designated as a success.

▶ **EXAMPLE**

Again assume that 10% of students are left-handed and that there are 15 students in each class. Also assume that we want to find the probability of getting 3 left-handed students in a class. For now, identify the values of n, x, p, and q. (Later, we'll find the probabilities.)

● **SOLUTION**

1. With 15 students in each class, we have $n = 15$.
2. We want 3 left-handed students (successes), so $x = 3$.
3. The probability of a left-handed student (success) is 0.1, so $p = 0.1$.
4. The probability of failure (not left-handed) is 0.9, so $q = 0.9$.

In this section we present three methods for finding probabilities in a binomial experiment. The first method involves calculations using the *binomial probability formula* and is the basis for the other two methods. The second method involves the use of Table A-1, and the third method involves the use of statistical software. We will describe the three methods, illustrate them, and then provide a rationale for each one.

Method 1: Use the Binomial Probability Formula   After identifying the values of n, x, p, and q, calculate $P(x)$ by using the binomial probability formula (Formula 4-5).

Formula 4-5 $$P(x) = \frac{n!}{(n - x)!x!} \cdot p^x \cdot q^{n-x} \quad \text{for } x = 0, 1, 2, \ldots, n$$

Binomial probability formula

where n = number of trials

x = number of successes among n trials

p = probability of success in any one trial

q = probability of failure in any one trial ($q = 1 - p$)

The factorial symbol, introduced in Section 3-6, denotes the product of decreasing factors. Two examples of factorials are $3! = 3 \cdot 2 \cdot 1 = 6$ and $0! = 1$ (by definition). Many calculators have a factorial key, as well as a key labeled $_nC_r$ that can simplify the computations. Consult the manual for your particular calculator.

▶ **EXAMPLE**

Use the binomial probability formula to find the probability of having 3 left-handed students among 15 students when the probability of having a left-handed student is 0.1. That is, find $P(3)$ given that $n = 15$, $x = 3$, $p = 0.1$, and $q = 0.9$.

● **SOLUTION**

Using the given values of n, x, p, and q in Formula 4-5, we get

$$P(3) = \frac{15!}{(15 - 3)!3!} \cdot 0.1^3 \cdot 0.9^{15-3}$$

$$= \frac{15!}{12!3!} \cdot 0.001 \cdot 0.282429536$$

$$= (455)(0.001)(0.282429536) = 0.129$$

The probability that 3 of the 15 students will be left-handed is 0.129.

Method 2: Use Table A-1 in Appendix A   To use Table A-1 for finding probabilities in a binomial experiment, first locate n and the corresponding value of x that is desired. At this stage, one row of numbers should be isolated. Now align that row with the proper probability of p by using the column across the top. The isolated number represents the desired probability (missing its decimal point at the beginning). A very small probability, such as 0.000000345, is indicated by 0+.

Shown below is part of Table A-1. When $n = 15$ and $p = 0.10$ in a binomial experiment, the probabilities of 0, 1, 2, . . . , 15 successes are 0.206, 0.343, 0.267, . . . , 0+, respectively.

n	x	.10
15	0 . . .	206
	1 . . .	343
	2 . . .	267
	3 . . .	129
	4 . . .	043
	.	
	.	
	.	
	15	0+

x	$P(x)$
0	0.206
1	0.343
2	0.267
3	0.129
4	0.043
.	.
.	.
.	.
15	0+

We used the binomial probability formula to find the probability of 3 successes, given that $n = 15$, $x = 3$, $p = 0.1$, and $q = 0.9$. The display from Table A-1 confirms that when $n = 15$ and $p = 0.1$, $P(3) = 0.129$ is correct. In this case, the use of Table A-1 greatly simplifies the work required to find the probability we seek. To really appreciate the ease of using Table A-1, let $n = 15$ students and $p = 0.1$ for left-handed students, and find the probability of having at least 3 left-handed students by using Table A-1 and the binomial probability formula. Using Table A-1 involves looking up the 13 probabilities (for $x = 3$, 4, . . . , 15) and adding them. Using the binomial probability formula involves applying that formula 13 times, computing the 13 different probabilities, and then adding them. (A shortcut is to calculate $P(0)$, $P(1)$, and $P(2)$ and subtract their sum from 1, but even this shortcut would take longer than using Table A-1.) In this case, given a choice between formula and table, most people would choose the table. However, we must use the formula when the values of n or p do not allow us to use Table A-1 (because $n > 15$ or because the probability p is not included in the table).

Method 3: Use Computer Software   Many computer statistics packages include an option for generating binomial probabilities. The following are samples of output from STATDISK and Minitab obtained from a binomial experiment in which $n = 15$ and $p = 0.1$. With STATDISK, select Statistics from the main menu, then select the Binomial Probabilities option.

Enter the requested values for *n* and *p*, and the entire probability distribution is displayed. [The column labeled "Cum Prob" represents cumulative probabilities obtained by adding the values of $P(x)$ as you go down the column.] With Minitab, enter the command PDF (for probability density function), and then use the BINOMIAL command as shown in the accompanying Minitab display. The subcommand BINOMIAL N = 15 P = 0.1 specifies that binomial probabilities be given for $n = 15$ and $p = 0.1$.

STATDISK DISPLAY OF BINOMIAL PROBABILITIES

```
 About    File    Edit    Statistics    Data    View
                        Binomial Probabilities
       Mean = 1.5           St. Dev. = 1.161895        Variance = 1.35



                    X    P(X)  Cum Prob      X    P(X)  Cum Prob
                    0   .20589  .20589       8   .00003 1.00000
                    1   .34315  .54904       9   .00000 1.00000
                    2   .26690  .81594      10   .00000 1.00000
                    3   .12851  .94444      11   .00000 1.00000
                    4   .04284  .98728      12   .00000 1.00000
                    5   .01047  .99775      13   .00000 1.00000
                    6   .00194  .99969      14   .00000 1.00000
                    7   .00028  .99997      15   .00000 1.00000


                   Use Menu Bar to change screen
        F1: Help        F2: Menu                        ESC: Quit
```

MINITAB DISPLAY OF BINOMIAL PROBABILITIES

```
MTB > PDF;
SUBC > BINOMIAL N=15 P=0.1.

   BINOMIAL WITH N =  15  P = 0.100000
      K             P( X = K )                      .
      0               0.2059
      1               0.3432
      2               0.2669
      3               0.1285
      4               0.0428
      5               0.0105
      6               0.0019
      7               0.0003
      8               0.0000
```

If a computer and software are available, this third method of finding binomial probabilities is fast and easy. If a computer cannot be used, we recommend the use of Table A-1 if possible. If the probability cannot be found directly from that table, then use the binomial probability formula. We should note that Table A-1 was constructed by applying the binomial probability formula, and computer programs are also based on that formula.

In Section 4-2 we presented Table 4-1 (page 187) as an example of a probability distribution for the hiring of 10 employees. We stipulated that hundreds of applicants are available and the numbers of men and women are equal. If the hiring is done in such a way that men and women have equal chances, then we have a binomial distribution with $n = 10$ and $p = 0.5$. The probabilities in Table 4-1 can be found by referring to Table A-1, by using the binomial probability formula (but that formula would have to be used 11 times—once for $x = 0$, once for $x = 1$, and so on), or by using statistical software. In this particular case, the easiest method is to use Table A-1 or statistical software, whichever is more readily available.

But once we obtain those probabilities, how do we interpret them? For example, how can those probabilities be used to address the issue of whether the Telektronic Company is discriminating because it hired 9 men and only 1 woman? Is it possible that the company's hiring policies are unbiased and the selection of 9 men and 1 woman is simply a chance event? Assuming that the hiring policies have no gender bias, we need to find the probability of getting 9 *or more* men among 10 new employees. It might seem that we should focus on the probability of getting *exactly* 9 men, but that would not be correct. It's not the probability of any *individual* outcome that is relevant; it's the probability of getting a result *at least as extreme* as the one that was obtained.

Here's the key issue: When 10 new employees are hired with an unbiased policy, is the chance of getting 9 or more men so small that it suggests discrimination based on gender? By referring to Table 4-1 we see that

$$P(9 \text{ or more men}) = P(9) + P(10)$$
$$= 0.010 + 0.001$$
$$= 0.011$$

That is, if men and women have equal job opportunities, there is only a 0.011 probability (or 1.1% chance) that 9 or more men will be hired. Because this probability is so small, it appears that the assumption of equal opportunity (with a 0.5 probability for each gender) is probably not correct. It appears that the Telektronic Company does have a problem that should be immediately investigated and corrected. This conclusion is based on probabilities that can be found from the binomial probability formula, Table A-1, or statistical software.

In the next example, Table A-1 cannot be used, and we apply the binomial probability formula. We will provide a rationale for that formula following the example.

Picking Lottery Numbers

In a typical state lottery, you select six different numbers. After a random drawing, any entries with the correct combination share in the prize. Since the winning numbers are randomly selected, any choice of six numbers will have the same chance as any other choice, but some combinations are better than others. The combination of 1, 2, 3, 4, 5, 6 is a poor choice because many people tend to select it. In a Florida lottery with a $105 million prize, 52,000 tickets had 1, 2, 3, 4, 5, 6; if that combination had won, the prize would have been only $1000. It's wise to pick combinations not selected by many others. Avoid combinations that form a pattern on the entry card.

Prophets for Profits

You can spend a small fortune buying books, computer software, or magazine subscriptions that are supposed to help you select lottery numbers. These "aids" are apparently blind to the facts that lottery numbers are randomly selected and the outcomes are independent of previous results. These aids typically recommend some numbers as "hot" because they have been coming up often. Others are recommended as "due" because they haven't been coming up often. Other approaches involve numerology, astrology, dreams, and numbers that have "appeared or talked to" a seeress, as one book claims. It's all worthless in predicting winning lottery numbers.

▶ **EXAMPLE**

According to one study conducted at the University of Texas at Austin, 2/3 of all Americans can do routine computations. If an employer were to hire 8 randomly selected Americans, what is the probability that exactly 5 of them could do routine computations?

● **SOLUTION**

Let's stipulate that an employee being able to do routine calculations is a success. The experiment is binomial because of the following:

1. We have a fixed number of trials (8).

2. The trials are independent; the employees are randomly selected different people who have no effect on each other.

3. There are two categories; each employee either can or cannot do routine computations.

4. The probability of success is 2/3, and it remains constant from trial to trial. (The probability of 2/3 is not included in Table A-1, so we will use the binomial probability formula.)

We begin by identifying the values of n, p, q, and x so that we can apply the binomial probability formula. We have

$$n = 8 \qquad \text{Number of trials}$$

$$p = \frac{2}{3} \qquad \text{Probability of success}$$

$$q = \frac{1}{3} \qquad \text{Probability of failure}$$

$$x = 5 \qquad \text{Desired number of successes}$$

We should check for consistency by verifying that what we call a success, as counted by x, is the outcome with probability p. That is, we must be sure that x and p refer to the same concept of success. Using the values for n, p, q, and x in the binomial probability formula, we get

$$P(5) = \frac{8!}{(8-5)!5!} \cdot \left(\frac{2}{3}\right)^5 \cdot \left(\frac{1}{3}\right)^{8-5}$$

$$= \frac{40{,}320}{(6)(120)} \cdot \frac{32}{243} \cdot \frac{1}{27} = 0.273$$

There is a probability of 0.273 that 5 of the 8 employees can do routine computations.

In the last example we found P(5). We can also use the binomial probability formula to find P(0), P(1), P(2), P(3), P(4), P(6), P(7), and P(8) so that the complete probability distribution for this case will be known. The results are shown in Table 4-5, where x denotes the number of employees who can do routine calculations. We can depict Table 4-5 in the form of a probability histogram, as shown in Figure 4-4. The general nature of the distribution (its shape) can be seen from the probability histogram.

The entries in Table A-1 were calculated by using the binomial probability formula. A rationale for that formula follows.

In the preceding example we wanted the probability of getting 5 successes among the 8 trials, given a 2/3 probability of success in any 1 trial. It's correct to reason that for 5 successes among 8 trials, there must be 3 failures. A common error is to compute

$$\overbrace{\frac{2}{3} \cdot \frac{2}{3} \cdot \frac{2}{3} \cdot \frac{2}{3} \cdot \frac{2}{3}}^{5 \text{ successes}} \cdot \overbrace{\frac{1}{3} \cdot \frac{1}{3} \cdot \frac{1}{3}}^{3 \text{ failures}} = \frac{32}{6561} = 0.00488$$

The above calculation is wrong because it contains an implicit assumption that the *first* 5 employees are successes and the *last* 3 are failures. However, the 5 successes and 3 failures can occur in many different sequences, not only the one given above. In fact, there are 56 different sequences of 5 successes and 3 failures, each with a probability of 0.00488, so the correct probability is $56 \cdot 0.00488 = 0.273$.

TABLE 4-5 Probability Distribution for Number of Employees Who Can Do Routine Calculations

x	$P(x)$
0	0.000
1	0.002
2	0.017
3	0.068
4	0.171
5	0.273
6	0.273
7	0.156
8	0.039

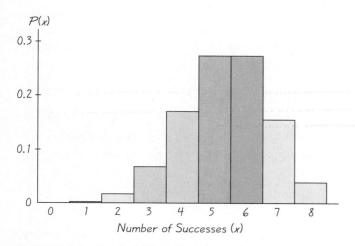

FIGURE 4-4 Probability Histogram for Number of Employees Who Can Do Routine Calculations

In general, the number of ways in which it is possible to arrange x successes and $n - x$ failures is shown in Formula 4-6.

Formula 4-6 $$\frac{n!}{(n - x)!x!}$$ Number of outcomes with exactly x successes among n trials

The expression given in Formula 4-6 does correspond to $_nC_r$ as introduced in Section 3-6. (Coverage of Section 3-6 is not required for this chapter.) We won't derive Formula 4-6, but its role should be clear: It counts the number of ways we can arrange x successes and $n - x$ failures. Combining this counting device (Formula 4-6) with the direct application of the multiplication rule for independent events results in the binomial probability formula.

The number of outcomes with exactly x successes among n trials

The probability of x successes among n trials for any 1 particular order

$$P(x) = \overbrace{\frac{n!}{(n - x)!x!}}^{} \cdot \overbrace{p^x \cdot q^{n-x}}^{}$$

To keep this section in perspective, remember that the binomial probability formula is only one of many probability formulas that can be used for different situations. It is often used in applications such as quality control, voter analysis, medical research, military intelligence, and advertising. Although the main focus of this section is the binomial distribution, other distributions can be found in Exercises 33–35, which deal with the geometric, hypergeometric, and multinomial distributions. Exercises 29–32 deal with the Poisson distribution, which we now discuss.

The Poisson Distribution

The Poisson distribution is often used as a mathematical model describing the probability distribution for the arrivals of entities requiring service, such as cars arriving at a gas station, planes arriving at an airport, or persons arriving at a ride in Disney World. It is defined as follows.

DEFINITION

The **Poisson distribution** is a discrete probability distribution that applies to occurrences of some event *over a specified interval*. The interval can be time, distance, area, volume, or some similar unit.

The random variable x is the number of occurrences of the event in an interval, such as the number of cars arriving at a gas station *during one minute,* the number of aircraft hijackings *in a day,* or the number of defective parts replaced on a new Corvette *during the first year* of warranty. The Poisson distribution requires that the occurrences be random and independent of each other and that they be uniformly distributed over an interval. The probability of the event occurring x times over an interval is given by

$$P(x) = \frac{\mu^x \cdot e^{-\mu}}{x!}$$

where $e \approx 2.71828$

The Poisson distribution has mean μ and standard deviation $\sqrt{\mu}$. The Poisson distribution differs from the binomial distribution in these important ways:

1. The binomial distribution is affected by the sample size n and the probability p, whereas the Poisson distribution is affected only by the mean μ.

2. In a binomial distribution, the possible values of the random variable x are 0, 1, ..., n, but a Poisson distribution has possible x values of 0, 1, 2, ... with no upper limit.

▶ EXAMPLE

For a recent year, there were 46 aircraft hijackings worldwide (based on data from the FAA). The mean number of hijackings per day is estimated as $\mu = 46/365 = 0.126$. If the United Nations is organizing a *single* international hijacking response team, there is a need to know about the chances of *multiple* hijackings in one day. Use $\mu = 0.126$ and find the probability that the number of hijackings (x) in one day is 0 or 1, so the single response team will be sufficient.

● SOLUTION

The Poisson distribution applies because we are dealing with the occurrences of hijacking events over a time interval of one day. The probability of 0 or 1 hijackings in one day is $P(0) + P(1)$, calculated as shown.

$$P(0) = \frac{0.126^0 \cdot 2.71828^{-0.126}}{0!} = \frac{1 \cdot 0.882}{1} = 0.882$$

$$P(1) = \frac{0.126^1 \cdot 2.71828^{-0.126}}{1!} = \frac{0.126 \cdot 0.882}{1} = 0.111$$

The probability of 0 or 1 hijacking in a day is therefore

$$P(0) + P(1) = 0.882 + 0.111 = 0.993$$

A single hijacking response team will be sufficient 99.3% of the days in the year. However, that leaves about 2 or 3 days each year when there will be multiple (2 or more) hijackings, so the single response team will not be sufficient.

BOARDING PASSES

Afraid to Fly?

Many of us tend to overestimate the probability of death from events that make headlines, such as jet crashes, murders, or terrorist attacks. Polls and fatality statistics show that although the past decade was the safest for air travel in the United States, people are now becoming more fearful of flying than they were in the past. In traveling from New York to California, your chances of dying are about 1 out of 11 million for a commercial jet, 1 out of 900,000 for a train, and 1 out of 14,000 for a car.

4-3 Exercises A: Basic Skills and Concepts

1. Which of the following can be treated as a binomial experiment?
   a. Testing a sample of 5 contact lenses (with replacement) from a population of 20 contact lenses, of which 40% are defective
   b. Testing a sample of 5 contact lenses (without replacement) from a population of 20 contact lenses, of which 40% are defective
   c. Tossing an unbiased coin 500 times
   d. Tossing a biased coin 500 times
   e. Surveying 1700 television viewers to determine whether they watched the Super Bowl on television
2. Which of the following can be treated as a binomial experiment?
   a. Surveying 1200 registered voters to determine their preferences for the next president   *no*
   b. Surveying 1200 registered voters to determine whether they would again vote for the current president   *yes*
   c. Sampling a randomly selected group of 500 prisoners to determine whether they have been in prison before   *yes*
   d. Sampling a randomly selected group of 500 prisoners to determine the lengths of their current sentences   *no*
   e. Testing 500 randomly selected drivers to determine whether their blood alcohol content levels are more than 0.10%   *yes*

In Exercises 3–6, assume that in a binomial experiment, a trial is repeated n times. Find the probability of x successes given the probability p of success on a given trial. (Use the given values of n, x, and p and Table A-1.)

3. $n = 12$, $x = 7$, $p = 0.4$ 4. $n = 6$, $x = 0$, $p = 0.05$
5. $n = 10$, $x = 10$, $p = 0.9$ 6. $n = 9$, $x = 2$, $p = 0.1$

In Exercises 7–10, assume that in a binomial experiment, a trial is repeated n times. Find the probability of x successes given the probability p of success on a single trial. Use the given values of n, x, and p and the binomial probability formula. Leave answers in the form of fractions.

7. $n = 5$, $x = 3$, $p = 1/4$ 8. $n = 4$, $x = 2$, $p = 1/3$
9. $n = 6$, $x = 2$, $p = 1/2$ 10. $n = 4$, $x = 4$, $p = 2/3$

In Exercises 11–27, identify the values of n, x, p, and q, and find the value requested.

11. Find the probability of getting exactly 4 girls in 10 births. (Assume that male and female births are equally likely and that the birth of any child does not affect the probability of the gender of any other children.)
12. A recent study by the A. C. Nielsen Company showed that 60% of all homes have cable TV. If 10 homes are randomly selected to test an experimental metering device, you would expect that about 6 of them would have cable. Find the probability that exactly 6 of these homes have cable TV.

13. In Table A-1 the probability corresponding to $n = 3$, $x = 2$, and $p = 0.01$ is shown as 0+. Find the exact probability represented by this 0+.

14. In Table A-1 the probability corresponding to $n = 5$, $x = 4$, and $p = 0.10$ is shown as 0+. Find the exact probability represented by this 0+.

15. There is a 90% chance that Domino's Pizza will make a delivery in less than 30 minutes (based on data from Domino's Pizza). An executive tests the operation of a franchise by ordering 10 pizzas at different times with delivery to different locations. If the 90% rate is correct, find the probability that 2 or more pizzas will be delivered in 30 minutes or more.

16. A new IRS employee learns that 70% of all audited taxpayers must pay more money (based on IRS data). During her first day, she must conduct 8 audits, and she hopes that at least 6 of the audited taxpayers must pay more money. Find the probability of that happening.

17. Air America has a policy of booking as many as 14 persons on an airplane that can seat only 12. (Past studies have revealed that only 90% of the booked passengers actually arrive for the flight.) Find the probability that if Air America books 14 persons, not enough seats will be available.

18. In a study of brand recognition, 95% of consumers recognized Coke (based on data from Total Research Corporation). Find the probability that among 12 randomly selected consumers, exactly 11 will recognize Coke.

19. According to *Discover* magazine, 95% of airline passengers will survive a crash under certain conditions. Given those conditions, what is the probability that exactly 18 of 20 passengers will survive a crash?

20. According to the U.S. Department of Justice, 5% of all U.S. households experienced at least one burglary last year. If you randomly select 30 households, what is the probability that fewer than 2 of them experienced at least one burglary last year?

21. According to the Labor Department, 40% of adult workers have a high school diploma but did not attend college. If 15 adult workers are randomly selected, find the probability that at least 10 of them have a high school diploma but did not attend college.

22. A Media General–AP poll showed that 20% of adult Americans are opposed to strict pollution controls on power plants that burn coal and oil. Assume that an environmental group launches a campaign to lower that number, and a post-campaign study begins with 15 randomly selected adults. If the 20% level of opposition hasn't changed, find the probability that among the 15 adults, fewer than 3 are opposed to the strict controls.

23. Fifty-three percent of those who live in the contiguous 48 states reside within 50 miles of a coastal shoreline (based on data from the U.S. Bureau of the Census). The Sandi Swimwear Company surveys randomly selected U.S. residents from the 48 contiguous states. Find the probability that among the first 20 residents selected, exactly 12 live within 50 miles of a coastal shoreline.

24. A quiz consists of 10 multiple-choice questions, each with 5 possible answers. For someone who makes random guesses for all of the answers, find the probability of passing if the minimum passing grade is 60%.

25. According to a Gannet News Service report on travel vouchers submitted by employees at the Department of Transportation's Cambridge branch, 98% of the travel vouchers contain errors and 51% of the errors result in wrong reimbursement amounts. An independent auditor is sent to investigate that report, and he begins by auditing 50 vouchers. If the 98% rate is correct, find the probability that
    a. all of the vouchers contain errors
    b. 49 of the vouchers contain errors

26. A *Computerworld* survey showed that 80% of top executives regularly use personal computers at work. The Telektronic Company plans to move 9 top executives to a new headquarters in Atlanta. If only 7 personal computers are available in Atlanta, find the probability that they will need more.

27. United Airlines Flight 470 from Denver to St. Louis has an on-time performance of 60% (based on data from the EAASY SABRE reservation system).
    a. Find the probability that among 12 such flights, at least 9 arrive on time.
    b. Find the probability that among 30 such flights, exactly 20 arrive on time.

28. Use the given assumption to find the probability of getting at least 4 girls when 5 babies are born.
    a. Assume that boys and girls are equally likely.
    b. Assume that the probability of a boy is 0.513 (based on past results).

The Poisson distribution is used to obtain probabilities for occurrences of some event *over a specified interval.* In Exercises 29–32, use the Poisson distribution to find the indicated probabilities.

29. A new hospital is being planned for Newtown, a community that does not yet have its own hospital. If Newtown averages 2.25 births per day, find the probability that in one day, the number of births is
    a. 0   b. 1   c. 4

30. The Scott Auto Park averages 0.5 car sale per day. Find the probability that for a randomly selected day, the number of cars sold is
    a. 0   b. 1   c. 2

31. Careful analysis of magnetic computer data tape shows that for each 500 ft of tape, the average number of defects is 2.0. Find the probability of more than 1 defect in a randomly selected length of 500 ft of tape.

32. The Townsend Manufacturing Company experiences a weekly average of 0.2 accident requiring medical attention. Find the probability that in a randomly selected week, the number of accidents requiring medical attention is
    a. 0   b. 1   c. 2

4-3   Exercises B: Beyond the Basics

33. Suppose that an experiment meets all conditions to be binomial except that the number of trials is not fixed. Then the **geometric distribution,** which gives us the probability of getting the first success on the xth trial, is described by

$P(x) = p(1 - p)^{x-1}$ where p is the probability of success on any one trial. Assume that the probability of a defective computer component is 0.2. Find the probability that the first defect is found in the seventh component tested.

34. If we sample from a small finite population without replacement, the binomial distribution should not be used because the events are not independent. If sampling is done without replacement and the outcomes belong to one of two types, we can use the **hypergeometric distribution.** If a population has A objects of one type, while the remaining B objects are of the other type, and if n objects are sampled without replacement, then the probability of getting x objects of type A and $n - x$ objects of type B is

$$P(x) = \frac{A!}{(A - x)!x!} \cdot \frac{B!}{(B - n + x)!(n - x)!} \div \frac{(A + B)!}{(A + B - n)!n!}$$

In Lotto 54, a bettor selects 6 numbers from 1 to 54 (without repetition), and a winning 6-number combination is later randomly selected. Find the probability of getting
   a. all 6 winning numbers
   b. exactly 5 of the winning numbers
   c. exactly 3 of the winning numbers
   d. no winning numbers

35. The binomial distribution applies only to cases involving 2 types of outcomes, whereas the **multinomial distribution** involves more than 2 categories. Suppose we have 3 types of mutually exclusive outcomes denoted by A, B, and C. Let $P(A) = p_1$, $P(B) = p_2$, and $P(C) = p_3$. In n independent trials, the probability of x_1 outcomes of type A, x_2 outcomes of type B, and x_3 outcomes of type C is given by

$$\frac{n!}{(x_1!)(x_2!)(x_3!)} \cdot p_1^{x_1} \cdot p_2^{x_2} \cdot p_3^{x_3}$$

   a. Extend the result to cover 6 types of outcomes.
   b. A genetics experiment involves 6 mutually exclusive genotypes identified as A, B, C, D, E, and F, and they are all equally likely. If 20 offspring are tested, find the probability of getting exactly 5 A's, 4 B's, 3 C's, 2 D's, 3 E's, and 3 F's.

36. Many companies monitor the quality of items supplied to them by using *acceptance sampling,* based on the binomial distribution. Suppose certain rubber parts are used to manufacture car brake components. The receiver gets large boxes of 5000 items and wants to have 99% (or more) of the items free of defects. There is a policy of randomly selecting 25 of the 5000 items and testing them; this testing process destroys those 25 items. The acceptance sampling policy is to accept the whole lot if the number of defective items is 0 or 1. Otherwise, the whole lot is rejected and returned to the supplier. Identify the values of n, x, p, and q, and find the value requested.
   a. If a lot of 5000 items has 1% defective, what is the probability the lot will be accepted? *(continued)*

b. If a lot of 5000 items has 20% defective, what is the probability the lot will be accepted? (*Hint:* Treat the selections as independent because the sample size is less than 5% of the large population size.)

37. If you bet on the number 7 for one spin of a roulette wheel, you have a 1/38 probability of winning. Assume that you bet on 7 for each of 20 spins.

a. Find the probability of exactly 1 win.

b. When $n \geq 20$ and $p \leq 0.05$, the Poisson distribution is sometimes used to approximate the binomial distribution. Use the Poisson distribution to approximate the probability for part a. (*Hint:* First identify the mean number of wins in groups of 20 spins.)

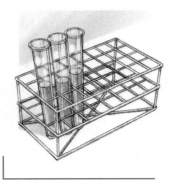

Composite Sampling

The U.S. Army once tested for syphilis by giving each inductee an individual blood test that was analyzed separately. One researcher suggested mixing pairs of blood samples. After the mixed pairs were tested, syphilitic inductees could be identified by retesting the few blood samples that were in the pairs that tested positive. The total number of analyses was reduced by pairing blood specimens, so why not put them in groups of three or four or more? Probability theory was used to find the most efficient group size, and a general theory was developed for detecting the defects in any population. This technique is known as *composite sampling.*

4-4   Mean, Variance, and Standard Deviation for the Binomial Distribution

The binomial distribution is a probability distribution, so the mean, variance, and standard deviation for the appropriate random variable can be found from the formulas presented in Section 4-2. The mean provides us with an important measure of central value, and the standard deviation provides us with an important measure of variation. Meaningful conclusions can often be obtained by using the mean and standard deviation with the range rule of thumb, the empirical rule, or Chebyshev's theorem. (See the second example in this section.)

Formula 4-1 $$\mu = \Sigma x \cdot P(x)$$

Formula 4-3 $$\sigma^2 = [\Sigma x^2 \cdot P(x)] - \mu^2$$

Formula 4-4 $$\sigma = \sqrt{[\Sigma x^2 \cdot P(x)] - \mu^2}$$

These formulas, which apply to all probability distributions, can be made much simpler for the special case of binomial distributions. Given the binomial probability formula and the above general formulas for μ, σ^2, and σ, we can pursue a series of somewhat complicated algebraic manipulations that ultimately lead to the following simple results. For a binomial experiment,

Formula 4-7 $$\mu = n \cdot p$$

Formula 4-8 $$\sigma^2 = n \cdot p \cdot q$$

Formula 4-9 $$\sigma = \sqrt{n \cdot p \cdot q}$$

The formula for the mean makes sense intuitively. If we were to analyze 100 births, we would expect to get about 50 girls, and np in this experiment becomes $100 \cdot 1/2$, or 50. In general, if we consider p to be the proportion of successes, then the product np will give us the actual number of expected successes among n trials.

The variance and standard deviation are not so easily justified, and we prefer to omit the complicated algebraic manipulations that lead to Formula 4-8. Instead, we will show that these simplified formulas (Formulas 4-7, 4-8, and 4-9) do lead to the same results as the more general formulas for binomial distributions (Formulas 4-1, 4-3, and 4-4).

▶ EXAMPLE

Like many companies, the Providence Electronics mail order company now has its telephone services configured so that callers can directly reach the department they want by pressing a number on their touch-tone phone. However, callers with rotary phones must listen to a long message before they can speak to an operator who can make the connection. Seventy percent of households have touch-tone phones (based on data from the Yankee Group and *USA Today* research). A manager monitors the first 4 calls to learn how the system is working. If x is the random variable representing the number of callers with touch-tone phones in a group of 4, find its mean, variance, and standard deviation.

● SOLUTION

In this binomial experiment we have $n = 4$ and $P(\text{touch-tone}) = 0.70$. It follows that $P(\text{rotary phone}) = q = 0.30$. We will find the mean and standard deviation by using two methods.

Method 1:   Use Formulas 4-7, 4-8, and 4-9, which apply to binomial experiments only.

$$\mu = n \cdot p = 4 \cdot 0.70 = 2.8 \text{ touch-tone phones}$$
$$\sigma^2 = n \cdot p \cdot q = 4 \cdot 0.70 \cdot 0.30 = 0.84$$
$$\sigma = \sqrt{n \cdot p \cdot q} = \sqrt{0.84} = 0.92 \text{ touch-tone phone}$$

Method 2:   Method 1 provides us with the solutions we seek, but we want to show that these same values will result from use of the more general formulas as well. Use Formulas 4-1, 4-3, and 4-4, which apply to all discrete probability distributions.

We begin by computing the mean using Formula 4-1: $\mu = \Sigma x \cdot P(x)$. The possible values of x are 0, 1, 2, 3, 4, but we also need the values of $P(0)$, $P(1)$, $P(2)$, $P(3)$, and $P(4)$. We can use the binomial probability formula or Table A-1 to find those values and then enter them in Table 4-6.   *(continued)*

Queues

Queuing theory is a branch of mathematics that uses probability and statistics. The study of queues, or waiting lines, is important to businesses such as supermarkets, banks, fast food restaurants, airlines, and amusement parks. Grand Union supermarkets try to keep checkout lines no longer than three shoppers. Wendy's introduced the "Express Pak" to expedite servicing its numerous drive-through customers. Disney conducts extensive studies of lines at its amusement parks so that it can keep patrons happy and plan for expansion. Bell Laboratories uses queuing theory to optimize telephone network usage, and factories use it to design efficient production lines.

TABLE 4-6   Calculating μ, σ^2, and σ

x	$P(x)$	$x \cdot P(x)$	x^2	$x^2 \cdot P(x)$
0	0.008	0.000	0	0.000
1	0.076	0.076	1	0.076
2	0.265	0.530	4	1.060
3	0.412	1.236	9	3.708
4	0.240	0.960	16	3.840
Total		2.802		8.684

We now use the results from Table 4-6 to apply the general formulas from Section 4-2 as follows:

$$\mu = \Sigma x \cdot P(x) = 2.8 \qquad \text{(rounded off)}$$
$$\sigma^2 = [\Sigma x^2 \cdot P(x)] - \mu^2$$
$$= 8.684 - 2.802^2 = 0.832796$$
$$= 0.83 \qquad \text{(rounded off)}$$
$$\sigma = \sqrt{0.832796} = 0.91 \qquad \text{(rounded off)}$$

The two methods produce the same results, except for minor discrepancies due to rounding.

We should recognize two important points. First, the simplified binomial formulas (Formulas 4-7, 4-8, and 4-9) do lead to the same results as the more general formulas that apply to all discrete probability distributions. Second, the binomial formulas are much simpler, provide fewer opportunities for arithmetic errors, and are generally more conducive to a positive outlook on life. If we know that an experiment is binomial, we should use the simplified formulas.

After finding values for the mean μ and standard deviation σ, we can use them to draw some conclusions about variation. In particular, we can often use the range rule of thumb, the empirical rule, and Chebyshev's theorem (see Section 2-5), as illustrated in the next example.

▶ **EXAMPLE**

Providence Electronics receives 400 calls in a typical day. Seventy percent of callers have touch-tone phones.

a. If x is the random variable representing the number of callers with touch-tone phones in a group of 400, find its mean and standard deviation.

b. According to the empirical rule (see Section 2-5), 95% of scores fall within 2 standard deviations of the mean. Apply this rule by using the result from part *a* and find the values that are 2 standard deviations away from the mean. (The empirical rule applies because this binomial distribution has a probability histogram that is approximately bell shaped.)

● SOLUTION

a. With $n = 400$ and P (touch-tone) = 0.70, we find the mean and standard deviation by using Formulas 4-7 and 4-9 as follows:

$$\mu = np$$
$$= 400 \cdot 0.70 = 280 \text{ touch-tone phones}$$
$$\sigma = \sqrt{npq}$$
$$= \sqrt{(400)(0.70)(0.30)} = 9.2 \text{ touch-tone phones}$$

b. Using the empirical rule, we know that 95% of the scores fall within 2 standard deviations (or $2 \cdot 9.2$) of the mean (280), so we get the following results.

$$\mu + 2\sigma = 280 + 2(9.2) = 298.4$$
$$\mu - 2\sigma = 280 - 2(9.2) = 261.6$$

That is, 95% of typical days with 400 calls will include between 261.6 and 298.4 callers with touch-tone phones. The same procedure reveals that 95% of such days will have between 101.6 and 138.4 callers with rotary phones. Such information is crucial to the manager who must provide enough operators to handle calls so that customers are not left on hold with the awful message "All of our operators are busy. Please stand by."

When n is large and p is close to 0.5 (as in the following example), the binomial distribution tends to resemble the smooth curve that approximates the probability histogram in Figure 4-5 on the following page. Note that the data tend to form a bell-shaped curve. In the following example we will use the empirical rule, which requires such a bell-shaped distribution.

▶ EXAMPLE

In the chapter opening problem we noted that 100 couples who were trying to have baby girls used the Ericsson method of gender selection. Seventy-five of the couples had girls and 25 had boys. Assuming that the Ericsson method has no effect and assuming that boys and girls are equally likely, find the mean and standard deviation for the number of girls in groups of 100 babies. Use the empirical rule (see Section 2-5) to determine whether these results support Dr. Ericsson's claim that his method is effective.   *(continued)*

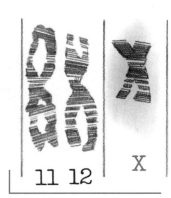

Big Brother

The Ericsson method of gender selection supposedly has a 75% success rate for couples wanting boys and a 69% success rate for couples wanting girls. Analysis of such experimental results typically requires use of a binomial distribution.

There are concerns that the use of gender-selection methods will change the natural male/female ratio of births. A Cleveland State University study of gender preferences found that among those who would use such methods, 94% of men and 81% of women want a boy first. Psychologist Roberta Steinbacher says that "if this technology were widely used, we could end up with a nation of older brothers and younger sisters. We [women] are in many cases second-class citizens now; this would mean we will be programmed from birth to be second."

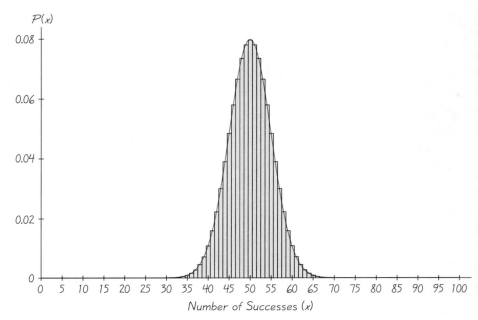

FIGURE 4-5   Probability Histogram for Binomial Experiment with $n = 100$ and $p = 0.5$

● SOLUTION

Let x represent the random variable for the number of girls in 100 births. Assuming that the Ericsson method has no effect and that girls and boys are equally likely, we have $n = 100$, $p = 0.5$, and $q = 0.5$. We can find the mean and standard deviation by using Formulas 4-7 and 4-9 as follows:

$$\mu = np = (100)(0.5) = 50$$
$$\sigma = \sqrt{npq} = \sqrt{(100)(0.5)(0.5)} = 5$$

For groups of 100 couples who each have a baby, the mean number of girls is 50 and the standard deviation is 5. Therefore, for the specific group that used the Ericsson method of gender selection, the 75 girls represent a value that is 5 standard deviations away from the mean. (In Section 2-6 we introduced the z score as a way to measure the number of standard deviations that a score is above or below the mean. Using $x = 75$, we get a z score of $z = (x - \mu)/\sigma = (75 - 50)/5 = 5$.) According to the empirical rule, about 99.7% of all scores fall within 3 standard deviations of the mean, but our sample is 5 standard deviations away from the mean. This suggests that under the assumption of equally likely girls and boys, our sample results are highly unusual. Because it's very unlikely that there would be 75 girls in 100 births, it appears that the Ericsson method is effective in increasing the likelihood of a baby being a girl.

4-4   Exercises A: Basic Skills and Concepts

In Exercises 1–4, find the mean μ, variance σ^2, and standard deviation σ for the given values of n and p. Assume the binomial conditions are satisfied in each case.

  1. $n = 49$, $p = 0.5$ **2.** $n = 64$, $p = 0.25$
  3. $n = 473$, $p = 0.855$ **4.** $n = 1067$, $p = 3/8$

In Exercises 5–12, find the indicated values.

  5. Several students are unprepared for a true/false test with 50 questions, and all of their answers are guesses. Find the mean, variance, and standard deviation for the number of correct answers for such students.

  6. On a multiple-choice test with 30 questions, each question has possible answers of a, b, c, and d, one of which is correct. For students who guess at all answers, find the mean, variance, and standard deviation for the number of correct answers.

  7. Among the 6665 films rated by the Motion Picture Association of America, 2945 have been rated R. Twenty rated films are randomly selected for a study. Find the mean, variance, and standard deviation for the number of R-rated films in randomly selected groups of 20.

  8. According to a survey by the Roper Organization, 64% of adults have money in regular savings accounts. If we plan to conduct a survey with groups of 50 randomly selected adults, find the mean, variance, and standard deviation for the number who have regular savings accounts.

  9. A study conducted by the National Transportation and Safety Board showed that among injured airline passengers, 47% of the injuries were caused by failure of the plane's seat. Two hundred different airline-passenger injuries are to be randomly selected for a study. Find the mean, variance, and standard deviation for the number of injuries caused by seat failure in such groups of 200.

  10. Among Americans aged 20 years or older, 12.5% sleep at least 9 hours each night. A study of dreams requires 36 volunteers who are at least 20 years old. Find the mean and standard deviation for the number of such people in groups of 36 who sleep at least 9 hours each night.

  11. Of all individual tax returns, 37% include errors made by the taxpayer. If IRS examiners are assigned randomly selected returns in batches of 12, find the mean and standard deviation for the number of erroneous returns per batch.

  12. Letter frequencies are analyzed in an attempt to decipher intercepted messages. In standard English text, the letter *e* occurs with a relative frequency of 0.130. Find the mean and standard deviation for the number of times the letter *e* will be found on standard pages of 2600 characters.

In Exercises 13–20, consider as *unusual* any result that differs from the mean by more than 2 standard deviations. That is, unusual values are either less than $\mu - 2\sigma$ or greater than $\mu + 2\sigma$.

  13. There is a 90% chance that a pizza from Domino's Pizza will be delivered in less than 30 minutes, and one franchise delivers 300 pizzas each day.

(continued)

    a. Find the mean and standard deviation for the number of on-time deliveries.

    b. Would it be unusual to have 244 pizzas delivered on time, with 56 late? Why or why not?

14. The Acton Paper Company receives 250 calls in a typical day, and 30% of the callers have rotary phones (based on data from the Yankee Group and *USA Today* research).

    a. Find the mean and standard deviation for the number of callers with rotary phones on such days.

    b. According to the empirical rule, 95% of all scores fall within 2 standard deviations of the mean. Apply that rule and find the values that are 2 standard deviations away from the mean.

    c. Is it unusual to get 50 callers with rotary phones?

15. One test of extrasensory perception involves the determination of a color. Fifty blindfolded subjects are asked to identify the one color selected from the possibilities of red, yellow, green, blue, black, and white.

    a. Assuming that all 50 subjects make random guesses, find the mean, variance, and standard deviation for the number of correct responses in such groups of 50.

    b. If 12 of 50 responses are correct, is this result within the scope of results likely to occur by chance?

16. A pathologist knows that 14.9% of all deaths are attributable to myocardial infarctions.

    a. Find the mean and standard deviation for the number of such deaths that will occur in a typical region with 5000 deaths.

    b. In one region, 5000 death certificates are examined, and it is found that 896 deaths were attributable to myocardial infarction. Is there cause for concern? Why or why not?

17. In New York State, 61.2% of motor vehicle accidents involve injuries (based on data from the N.Y. State Department of Motor Vehicles). When 80 accident reports for one road are examined, it is found that 72 of them involved injuries. Is the injury rate for this road unusually high?

18. Sixty-four couples plan to have children, and they all agree to test the Ericsson method of gender choice, which supposedly increases the probability of a baby being a girl.

    a. Assuming that the method has no effect, find the mean and standard deviation for the number of girls in groups of 64 babies. Assume that girls and boys are equally likely and that the gender of any baby is independent of the gender of any others.

    b. If each of the 64 couples has a baby and the results consist of 36 girls and 28 boys, does this suggest that the method is effective? Why or why not?

19. In a nationwide study of brand recognition, 95% of consumers recognized Coke (based on data from Total Research Corporation). In another study, 200 consumers were surveyed for brand recognition of Coke.

    a. For groups of 200 consumers randomly selected nationwide, find the mean and standard deviation for the number who recognize the Coke brand name.

(continued)

b. Would it be unusual to find that all members of a group of 200 consumers recognized the Coke brand name?

20. According to the U.S. Department of Justice, 5% of all U.S. households experienced at least one burglary last year.

a. For randomly selected groups of 150 households, find the mean and standard deviation for the number of households that experienced at least one burglary last year.

b. One hundred fifty homes are surveyed in one region, and it is found that 8% of them experienced at least one burglary last year. Is this unusual, or is it a result that is likely to occur by chance?

4-4 Exercises B: Beyond the Basics

21. A standard placement test consists of 100 multiple-choice questions, each with 5 possible answers. Some students are told that the results will be used for recommendations only and that no course changes will be mandated because of results on the test. Because the results aren't used for mandatory changes, some students guess at each answer to finish the test as quickly as possible.

a. For these students, find the mean and standard deviation for the number of correct answers.

b. Using Chebyshev's theorem and considering only those students who guess, what do you conclude about the proportion of tests having between 12 and 28 correct answers?

c. What is the z score that corresponds to 30 correct responses from a student who guessed a response to each question?

22. a. If a company makes a product with an 80% yield (meaning that 80% are good), what is the minimum number of items that must be produced to be at least 99% sure that the company produces at least 5 good items?

b. If the company produces batches of items, each with the minimum number determined in part a, find the mean and standard deviation for the number of good items in such batches.

VOCABULARY LIST

Define and give an example of each term.

random variable
discrete random variable
continuous random variable
probability distribution
probability histogram

expected value
binomial experiment
binomial probability formula
Poisson distribution

Important Formulas

Requirements for a discrete probability distribution:	1. $\Sigma P(x) = 1$	for all possible values of x
	2. $0 \leq P(x) \leq 1$	for any particular value of x
For *any* discrete probability distribution:	$\mu = \Sigma x \cdot P(x)$	Mean
	$\sigma^2 = \Sigma(x - \mu)^2 \cdot P(x)$	
	or	Variance
	$\sigma^2 = [\Sigma x^2 \cdot P(x)] - \mu^2$	
	$\sigma = \sqrt{[\Sigma x^2 \cdot P(x)] - \mu^2}$	Standard deviation
	$E = \Sigma x \cdot P(x)$	Expected value of discrete random variable
For *binomial* probability distributions:	$P(x) = \dfrac{n!}{(n - x)!x!} \cdot p^x \cdot q^{n-x}$	Binomial probability formula
	$\mu = n \cdot p$	Mean
	$\sigma^2 = n \cdot p \cdot q$	Variance
	$\sigma = \sqrt{n \cdot p \cdot q}$	Standard deviation

REVIEW

Central concerns of this chapter were the concepts of a random variable and a probability distribution. Here we dealt primarily with discrete probability distributions; successive chapters deal with continuous probability distributions.

In an experiment yielding numerical results, the random variable can take on those different numerical values. A probability distribution consists of all values of a random variable, along with their corresponding probabilities. By constructing a probability histogram, we can see a useful correspondence between those probabilities and the areas of the rectangles in the histogram.

Of the infinite number of different probability distributions, special attention was given to the important and useful binomial probability distribution, which is characterized by these properties:

1. There is a fixed number of trials (denoted by n).
2. The trials must be independent.
3. Each trial must have outcomes that can be classified in *two* categories.
4. The probabilities involved must remain constant for each trial.

We saw that probabilities for the binomial distribution can be computed by using the binomial probability formula, where n is the number of trials, x is the number of successes, p is the probability of a success, and q is the probability of a failure. The binomial probability formula was used to construct Table A-1, which lists binomial probabilities for select values of n, p, and x. That same formula is also used in many statistical software packages, such as STATDISK and Minitab.

We noted that the binomial probability distribution is one of many different discrete probability distributions. In Section 4-3 we briefly discussed the Poisson distribution, which applies to occurrences of some event over a specified interval of time, distance, area, volume, or similar unit.

For the special case of the binomial probability distribution, the mean, variance, and standard deviation of the random variable can be easily computed by using the formulas included in the summary of important formulas.

REVIEW EXERCISES

1. Thirty percent of college students own videocassette recorders (based on data from the America Passage Media Corporation). The Telektronic Company produced a videotape and sent copies to 10 randomly selected college students as part of a pilot sales program.
   a. Find the probability that exactly one-half of the 10 college students own videocassette recorders.
   b. Find the probability that at least one-half of the 10 college students own videocassette recorders.
   c. If videotapes are sent to randomly selected college students in many different groups of 10, find the mean and standard deviation for the number (among 10) who own videocassette recorders.
2. An experiment in parapsychology involves the selection of one of the numbers 1, 2, 3, 4, and 5, and they are all equally likely. The random variable x is the particular number selected.
   a. Find the mean, variance, and standard deviation for the random variable x.
   b. Fifty blindfolded subjects are asked to identify the number selected. Find the expected value for the number of correct identifications among 50.
   c. Fifty subjects attempt to identify the number selected. Would it be unusual to find that 15 of them made correct identifications?
3. The table of binomial probabilities (Table A-1) shows that for $n = 10$ trials, if the probability of success is $p = 0.05$, then the probability of $x = 6$ successes among the 10 trials is $0+$. Find the value represented by this $0+$, and express it as a decimal with three significant digits.
4. Does $P(x) = (x + 1)/5$ (for $x = -1$, 0, 1, 2) determine a probability distribution? Why or why not?
5. According to the National Highway Traffic Institute, seat belts are worn by 70% of California car passengers and 57% of Florida car passengers. If 5 car passengers are randomly selected in Florida and x is the random variable for the number of them who wear seat belts, the probability distribution can be described by the accompanying table. Complete the table to the right, and then find the mean and standard deviation for the random variable x.
6. Refer to the data given in Exercise 5 and construct a similar table that describes

x	$P(x)$
0	0.015
1	0.097
2	0.258
3	0.342
4	?
5	?

the probability distribution for the random variable x, where x is the number of randomly selected California passengers (among 5) who wear seat belts.

7. In Exercise 5 we noted that 70% of California car passengers wear seat belts.
   a. Find the mean, variance, and standard deviation for the number who wear seat belts in randomly selected groups of 500.
   b. Let's stipulate that an *ordinary* value is one that is within 2 standard deviations of the mean. Using the result from part a, find the lowest and highest ordinary values for the number of passengers who wear seat belts in randomly selected groups of 500.
   c. We expect that 350 of the 500 passengers wear seat belts. A media campaign is conducted to increase seat belt use, and a follow-up study reveals that among 500 randomly selected passengers, 375 wear seat belts. Does this result suggest that the campaign was effective? Why or why not?

8. An insurance association's study of home smoke detector use involves homes randomly selected in groups of 4. The accompanying table describes the probability distribution for x, the number of homes (in groups of 4) that have smoke detectors installed (based on data from the National Fire Protection Association). Find the mean, variance, and standard deviation for the random variable x.

x	$P(x)$
0	0.0004
1	0.0094
2	0.0870
3	0.3562
4	0.5470

9. Inability to get along with others is the reason cited in 17% of worker firings (based on data from Robert Half International, Inc.). Concerned about her company's working conditions, the personnel manager at the Flint Fabric Company plans to investigate the 5 employee firings that occurred over the past year. Assuming that the 17% rate applies, find the probability that among those 5 employees, the number fired because of an inability to get along with others is
   a. 0   b. 4   c. 5   d. at least 3
   (Once the actual reasons for the firings have been identified, such probabilities will be helpful in comparing Flint Fabric Company to other companies.)

10. Refer to the data given in Exercise 9. Let the random variable x represent the number of fired employees (among 5) who were let go because of an inability to get along with others.
   a. Find the mean value of x.
   b. Find the standard deviation of the random variable x.
   c. If we consider as unusual any values that are more than 2 standard deviations away from the mean, is it unusual to have 4 employees (among 5) fired because of an inability to get along with others?

COMPUTER PROJECT

The Boston School of Optometry conducts a study in which 150 workers are randomly selected from the population of white-collar workers aged 25–44. Among persons in that population, 54.9% wear corrective lenses (based on data from the U.S. Department of Health and Human Services). The sample size of 150 was selected because the school wanted to be reasonably sure of getting at least 70 persons who wear corrective lenses. Use STATDISK or Minitab to find the probability that among the 150 persons selected, the number who wear corrective lenses is (a) 70, (b) at least 70, (c) at least 82, and (d) at least 90. (This problem cannot be solved by using Table A-1 because that table stops at $n = 15$ and the probability of 0.549 is not included. It could theoretically be solved by using the binomial probability formula, but that approach would be extremely time consuming and painfully tedious. This problem is best solved by using statistical software.)

FROM DATA TO DECISION:

Is a Transatlantic Flight Safe with Two Engines?

Statistics is at its best when it is used to benefit humanity in some way. Companies use statistics to become more efficient, increase shareholders' profits, and lower prices. Regulatory agencies use statistics to ensure the safety of workers and clients. This exercise involves a situation in which cost effectiveness and passenger safety are both critical factors. With new aircraft designs and improved engine reliability, airline companies wanted to fly transatlantic routes with twin-engine jets, but the Federal Aviation Administration required at least three engines for transatlantic flights. Lowering this requirement was, of course, of great interest to manufacturers of twin-engine jets (such as the Boeing 767). Also, the two-engine jets use about half the fuel of jets with three or four engines. Obviously, the key issue in approving the lowered requirement is the probability of a twin-engine jet making a safe transatlantic crossing. This probability should be compared to that of three- and four-engine jets. Such a study should involve a thorough understanding of the related probabilities. A realistic estimate for the probability of an engine failing on a transatlantic flight is 1/14,000. Use this probability and the binomial probability formula to

find the probabilities of 0, 1, 2, and 3 engine failures for a three-engine jet and the probabilities of 0, 1, and 2 engine failures for a two-engine jet. Because of the numbers involved, carry all results to as many decimal places as your calculator will allow. Summarize your results by entering the probabilities in the two tables given below.

Three Engines		Two Engines	
x	$P(x)$	x	$P(x)$
0	?	0	?
1	?	1	?
2	?	2	?
3	?		

Use the results from the tables and assume that a flight will be completed if at least one engine works. Find the probability of a safe flight with a three-engine jet, and find the probability of a safe flight with a two-engine jet. Write a report for the Federal Aviation Administration that outlines the key issue, and include a recommendation. Support your recommendation with specific results.

5 Normal Probability Distributions

5-1 Overview

Chapter objectives are identified, and the normal distribution is defined.

5-2 The Standard Normal Distribution

The standard normal distribution is defined and methods are described for determining probabilities by using that distribution, as well as for determining standard scores corresponding to given probabilities.

5-3 Nonstandard Normal Distributions

The z score (or standard score) is used to work with normal distributions in which the mean is not 0, the standard deviation is not 1, or both. Methods are described for determining probabilities by using nonstandard normal distributions, as well as for determining scores corresponding to given probabilities.

5-4 The Central Limit Theorem

As the sample size increases, the sampling distribution of sample means is shown to approach a normal distribution with mean μ and standard deviation $\sigma/\sqrt{n}$, where n is the sample size and μ and σ represent the mean and standard deviation of the population, respectively.

5-5 Normal Distribution as Approximation to Binomial Distribution

Use of the normal distribution to determine probabilities in a binomial experiment is illustrated.

Chapter Problem:
Do military height restrictions affect men and women equally?

Height restrictions are used in many different situations, ranging from admittance to certain amusement park rides to enlistment in the United States Marine Corps. Table 5-1 lists some height restrictions in current use. For example, to be eligible for enlistment in the U.S. Army, men must be between 60 in. and 80 in. tall, whereas women's heights must be between 58 in. and 80 in.

Table 5-1 Height Restrictions (in inches)

	Men		Women	
	Minimum	Maximum	Minimum	Maximum
U.S. Army	60	80	58	80
U.S. Marine Corps	64	78	58	73
Rockette Dancer	–	–	65.5	68
Beanstalk Club	74	–	70	–

The height restrictions given in Table 5-1 raise some interesting questions. For instance, do the Army and Marine Corps treat both sexes equally by including the same percentages of men and women, or does one gender enjoy the advantage of having a greater percentage of eligible persons? Or suppose you plan to open a Beanstalk Club. Given the minimum height requirement of 70 in. for women, what percentage of women are eligible? In a metropolitan region of 500,000 adults, how many women actually satisfy that requirement? If women must be at least 70 in. tall and men must be at least 74 in. tall, are the percentages of eligible men and women approximately equal, or is one gender favored with a height requirement that is more liberal? Given the height requirement of 70 in. for women, what should be the corresponding height requirement for men if the percentage of eligible women is to be equal to the percentage of eligible men? Using the methods of this chapter, we will be able to answer such questions, and these particular questions will be addressed.

5-1 Overview

Chapter 4 introduced the concept of a probability distribution and considered only *discrete* types, such as those involving a finite number of possible values. The number of quarters made in a day at a U.S. mint is an example of a discrete distribution. There are also many different *continuous* probability distributions, such as the weights of the quarters produced at the mint. While distributions can be described as being discrete or continuous, they can also be described by their *shape*, such as a bell shape. This chapter focuses on normal distributions, which are extremely important because so many real applications involve data that are normally distributed. Heights of adult women, for example, are normally distributed.

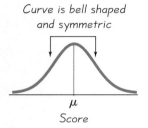

Curve is bell shaped and symmetric

μ

Score

FIGURE 5-1 The Normal Distribution

DEFINITION

A continuous random variable has a **normal distribution** if that distribution is symmetric and bell shaped, as in Figure 5-1, and the distribution fits the equation given as Formula 5-1.

Formula 5-1
$$y = \frac{e^{-(x-\mu)^2/2\sigma^2}}{\sigma\sqrt{2\pi}}$$

Fortunately, *it is not necessary for us to use Formula 5-1,* but we should note that it is an equation relating a horizontal scale of x values to a vertical scale of y values. The symbol μ represents the mean score of an entire population, σ is the standard deviation of the population, π is approximately 3.14159, and e is a constant with a value of approximately 2.71828. Formula 5-1 shows that any particular normal distribution is determined by two parameters: the mean μ and standard deviation σ.

This chapter presents the standard methods used to work with normally distributed scores, and it includes many real applications. Also, the methods of this chapter establish basic patterns and concepts that will apply to other continuous probability distributions. In later chapters, as we consider methods for using sample data to form conclusions about populations, we will make extensive use of the concepts presented here.

5-2 The Standard Normal Distribution

We noted in the overview that Chapter 4 discussed discrete probability distributions, whereas the normal probability distribution is continuous. Before formally considering the normal probability distribution, let's briefly examine a simpler continuous distribution called the uniform distribution, illustrated in Figure 5-2. This figure depicts the probability distribution for temperatures in a manufacturing process. Those temperatures are controlled

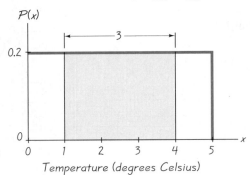

Shaded Area = $L \cdot W$
$= 3 \cdot 0.2 = 0.6$

FIGURE 5-2   Uniform Distribution of Temperatures

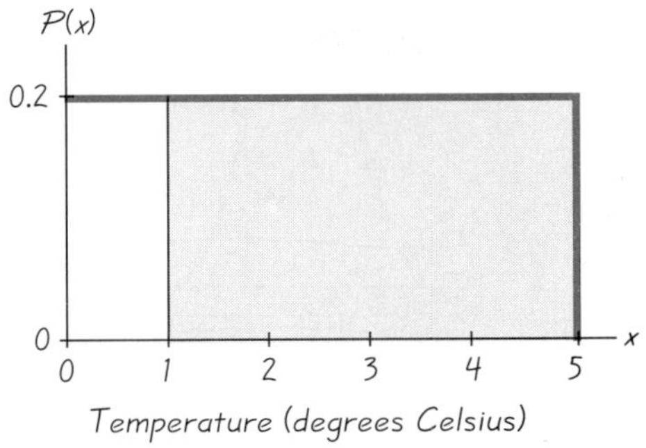

FIGURE 5-3   Temperatures Greater Than 1°C

so that they range between 0°C and 5°C, and every possible value is equally likely. A **uniform distribution** has equally likely values over the range of possible outcomes; therefore its graph (called a *density curve* for continuous random variables) is always a rectangle with an enclosed area equal to 1. (By setting the height of the rectangle in Figure 5-2 to be 0.2, we force the enclosed area to equal 1.) This property makes it very easy for us to solve probability problems. For example, given the uniform distribution of Figure 5-2, the probability of randomly selecting a temperature between 1°C and 4°C is 0.6, which is the area of the shaded region. Here's an important point: **For a density curve depicting the distribution of a continuous random variable, the area under the curve is 1 and there is a correspondence between area and probability.** In Figure 5-2, the probability of a value between 1°C and 4°C can be found by finding the area of the corresponding shaded rectangular region with dimensions 0.2 by 3.

▶ **EXAMPLE**

A uniform distribution of temperatures ranges from 0°C to 5°C. If one temperature is randomly selected, find the probability that it is greater than 1°C.

● **SOLUTION**

In Figure 5-3 the temperatures greater than 1°C are represented by the shaded rectangle that has an area of $4 \cdot 0.2 = 0.8$. Because there is a correspondence here between area and probability, we can conclude that there is a 0.8 probability that a randomly selected temperature will be greater than 1°C.

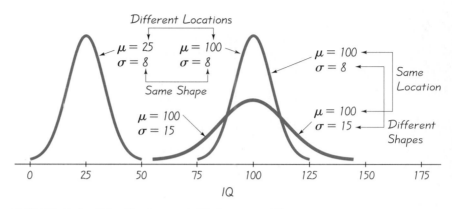

FIGURE 5-4   Distributions of IQ Scores with Different Means and Standard Deviations

Instead of the simple rectangle associated with uniform distributions, normal distributions have the more complicated bell shape shown in Figure 5-1, so it's more difficult to find areas. But the basic principle is the same: There is a correspondence between area and probability.

We saw in the overview that the normal distribution is a continuous probability distribution with a bell shape; it is defined by a particular equation that uses the population mean μ and standard deviation σ. There are actually many different normal probability distributions, each dependent on only the two parameters μ and σ. Figure 5-4 shows three different normal distributions of IQ scores, with the differences caused by changes in the mean and standard deviation. A change in the value of the population mean μ causes the curve to be shifted to the right or left. A change in the value of σ causes a change in the shape or variation of the curve; the basic bell shape remains, but the curve becomes broader or narrower, depending on σ. Among the infinite possibilities, one particular normal distribution is of special interest.

DEFINITION

The **standard normal distribution** is a normal probability distribution that has a mean of 0 and a standard deviation of 1. (See Figure 5-5.)

Suppose we had to perform calculations with Formula 5-1, such as finding heights y for given values of x or finding areas under the curve. If we could choose any values for μ and σ, we would soon recognize that $\mu = 0$ and $\sigma = 1$ lead to the simplest form of that equation, which is the *standard normal form*.

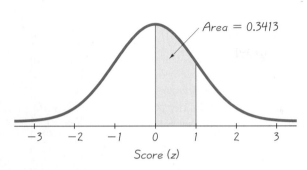

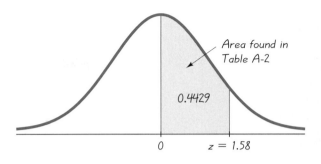

FIGURE 5-5   Standard Normal Distribution, with Mean $\mu = 0$ and Standard Deviation $\sigma = 1$

FIGURE 5-6   The Standard Normal Distribution
The area of the shaded region bounded by the mean of 0 and the positive number z can be found in Table A-2.

By letting $\mu = 0$ and $\sigma = 1$ in Formula 5-1, mathematicians are able to find areas under the curve. As shown in Figure 5-5, the area under the curve bounded by the mean of 0 and the score of 1 is 0.3413. The total area under the curve is always 1; this allows us to make the correspondence between area and probability, as we did in the preceding example with the uniform distribution.

Finding Probabilities When Given z Scores

Figure 5-5 shows that the area bounded by the curve, the horizontal axis, and the scores of 0 and 1 is an area of 0.3413. Although the figure shows only one area, Table A-2 (in Appendix A) includes areas (or probabilities) for many different regions.

Table A-2 gives the probability corresponding to the area under the curve bounded on the left by a vertical line above the mean of 0 and bounded on the right by a vertical line above any specific positive score denoted by z (see Figure 5-6). Note that when you use Table A-2, the part of the z score denoting hundredths is found across the top row. To find the probability associated with a score between 0 and 1.23, for example, begin with the z score of 1.23 by locating 1.2 in the left column. Then find the value in the adjoining row of probabilities that is directly below 0.03. The area (or probability) value of 0.3907 indicates that there is a probability of 0.3907 of randomly selecting a score between 0 and 1.23. It is essential to remember that this table is designed only for the standard normal distribution, which has a mean of 0 and a standard deviation of 1. Nonstandard cases will be considered in the next section.

Because theories concerning normal distributions originally resulted from studies of experimental errors, the examples in this section dealing with errors in measurements are particularly relevant.

▶ **EXAMPLE**

A manufacturer of scientific instruments produces thermometers that are supposed to give readings of 0°C at the freezing point of water. Tests on a large sample of these instruments reveal that at the freezing point of water, some thermometers give readings below 0° (denoted by negative numbers) and some give readings above 0° (denoted by positive numbers). Assume that the mean reading is 0°C and the standard deviation of the readings is 1.00°C. Also assume that the frequency distribution of errors closely resembles the normal distribution. If one thermometer is randomly selected, find the probability that, at the freezing point of water, the reading is between 0° and +1.58°.

● **SOLUTION**

We are dealing with a standard normal distribution and are looking for the area between 0 and z (the shaded region) in Figure 5-6 with $z = 1.58$. From Table A-2 we find that this area is 0.4429. The probability of randomly selecting a thermometer with an error between 0° and +1.58° is therefore 0.4429.

▶ **EXAMPLE**

Using the thermometers from the preceding example, find the probability of randomly selecting one thermometer that reads (at the freezing point of water) between −2.43° and 0°.

● **SOLUTION**

We are looking for the region shaded in Figure 5-7(a), but Table A-2 is designed to apply only to regions to the right of the mean (0) as in Figure 5-7(b). However, by observing that the normal probability distribution is symmetric about the vertical line through 0, we note that the shaded regions in parts (a) and (b) of Figure 5-7 have the same area. Referring to Table A-2, we can easily determine that the shaded area of Figure 5-7(b) is 0.4925, so the shaded area of Figure 5-7(a) must also be 0.4925. That is, the probability of randomly selecting a thermometer with an error between −2.43° and 0° is 0.4925.

The preceding solution illustrates an important principle: _Although a z score can be negative, the area under the curve (or the corresponding probability) can never be negative._

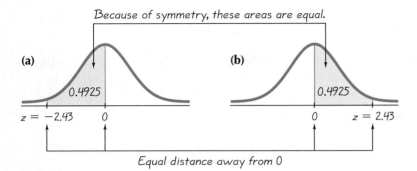

Because of symmetry, these areas are equal.

(a)

0.4925

$z = -2.43$ 0

(b)

0.4925

0    $z = 2.43$

Equal distance away from 0

FIGURE 5-7 **Using Symmetry to Find the Area
to the Left of the Mean**

Section 2-5 presented the empirical rule, which states that for bell-shaped distributions,

- about 68% of all scores fall within *1* standard deviation of the mean.
- about 95% of all scores fall within *2* standard deviations of the mean.
- about 99.7% of all scores fall within *3* standard deviations of the mean.

If we refer to Figure 5-5 with $z = 1$, Table A-2 shows us that the shaded area is 0.3413. It follows that the proportion of scores between $z = -1$ and $z = 1$ will be $0.3413 + 0.3413 = 0.6826$ (the size of the area from $z = -1$ to 0 is the same as that of the area from $z = 0$ to $z = 1$). That is, about 68% of all scores fall within 1 standard deviation of the mean. A similar calculation with $z = 2$ yields the values of $0.4772 + 0.4772 = 0.9544$ (or about 95%) as the proportion of scores between $z = -2$ and $z = 2$. Similarly, the proportion of scores between $z = -3$ and $z = 3$ is given by $0.4987 + 0.4987 = 0.9974$ (or about 99.7%). These exact values correspond very closely to those given in the empirical rule. In fact, the values of the empirical rule were found directly from Table A-2 and have been slightly rounded for convenience.

Because we are dealing with a probability distribution, the total area under the curve must be 1. Now refer to Figure 5-8 and see that a vertical line directly above the mean of 0 divides the area under the curve into two equal parts, each containing an area of 0.5. The following example uses this observation.

Misleading Norms

Dr. John Cannell, a West Virginia physician, observed that his state's high illiteracy rate was inconsistent with a state report claiming that West Virginia students were performing *above* the national average. Investigations showed that, among other problems, old norms were used as a basis for comparison. By comparing the new test results to those obtained many years ago, all 50 states achieved results that were above "the national average." Dr. Cannell charged, "The testing industry wants to sell lots of tests, and the school superintendents desperately need help in improving scores." The effect was to create misleading results.

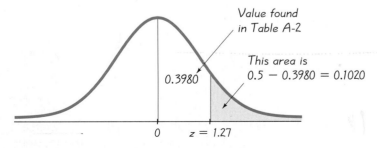

Value found
in Table A-2

This area is
$0.5 - 0.3980 = 0.1020$

0.3980

0    $z = 1.27$

FIGURE 5-8 **Finding the Area Above $z = 1.27$**

► **EXAMPLE**

Once again, make a random selection from this same sample of thermometers. Find the probability that the chosen thermometer reads (at the freezing point of water) greater than $+1.27°$.

● **SOLUTION**

We are again dealing with normally distributed values having a mean of $0°$ and a standard deviation of $1°$. The probability of selecting a thermometer that reads above $+1.27°$ corresponds to the shaded area of Figure 5-8. Table A-2 cannot be used to find that area directly, but we can use the table to find that $z = 1.27$ corresponds to the area of 0.3980, as shown in the figure. We now reason that because the total area to the right of zero is 0.5, the shaded area is $0.5 - 0.3980$, or 0.1020. We conclude that there is a probability of 0.1020 of randomly selecting one of the thermometers with a reading greater than $+1.27°$. Another way to interpret this result is to state that if many thermometers are selected and tested, then 0.1020 (or 10.20%) of them will read greater than $+1.27°$.

We are able to determine the area of the shaded region in Figure 5-8 by an *indirect* application of Table A-2. The following example illustrates yet another indirect use.

► **EXAMPLE**

Assuming that one thermometer in our sample is randomly selected, find the probability that it reads (at the freezing point of water) between $1.20°$ and $2.30°$.

● **SOLUTION**

The probability of selecting a thermometer that reads between $+1.20°$ and $+2.30°$ corresponds to the shaded area of Figure 5-9. However, Table A-2 is designed to provide only for regions bounded on the left by the vertical line above 0. We can use the table to find that $z = 1.20$ corresponds to an area of 0.3849 and that $z = 2.30$ corresponds to an area of 0.4893, as shown in the

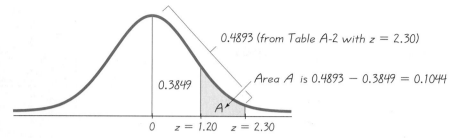

FIGURE 5-9   Finding the Area Between z = 1.20 and z = 2.30

figure. If we denote the area of the shaded region by A, we can see from Figure 5-9 that

$$0.3849 + A = 0.4893$$

so

$$A = 0.4893 - 0.3849 = 0.1044$$

The probability of a reading between 1.20° and 2.30° is therefore 0.1044.

The preceding example concluded with the statement that the probability of a reading between 1.20° and 2.30° is 0.1044. Such probabilities can also be expressed by using the following notation.

Notation	
$P(a < z < b)$	denotes the probability that the z score is between a and b.
$P(z > a)$	denotes the probability that the z score is greater than a.
$P(z < a)$	denotes the probability that the z score is less than a.

Using this notation, we can express the result of the last example as $P(1.20 < z < 2.30) = 0.1044$, which states in symbols that the probability that the z score falls between 1.20 and 2.30 is 0.1044. With a continuous probability distribution such as the normal distribution, $P(z = a) = 0$. With infinitely many different possible values, the probability of getting any one *exact* value is 0. As an example, consider the probability that $z = 1.33$ *exactly*. A single point such as $z = 1.33$ would be represented graphically by a vertical line above 1.33, but that vertical line contains no area, so $P(z = 1.33) = 0$. With any continuous random variable, the probability of any one exact value is 0, and it follows that $P(a \leq z \leq b) = P(a \leq z < b)$. It also follows that the probability of getting a z

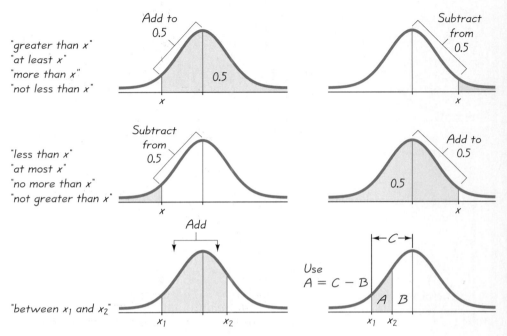

FIGURE 5-10   Interpreting Areas Correctly

score of *at most b* is equal to the probability of getting a *z score of less than b*. It is important to correctly interpret key phrases such as *at most, at least, more than, no more than*, and so on. The illustrations in Figure 5-10 provide an aid to interpreting several of the most common phrases.

Finding *z* Scores When Given Probabilities

So far, the examples of this section involving the standard normal distribution have all followed the same format: Given some value, we used Table A-2 to find a probability. Many other circumstances exist in which we know the probability and need to find the corresponding *z* score. In such cases, there are two important principles that we must know.

- In Table A-2 the numbers in the extreme left column and across the top are *z* scores, which are *distances* along the horizontal scale in Figure 5-6, but the numbers in the body of the table are *areas* (or probabilities).

- The *z* scores to the left of the centerline are always *negative* (as in Figure 5-7).

Based on the above principles, **if we know a probability and want to determine the corresponding z score by using Table A-2, we must be sure to look up the probability in the body of that table; if the z score is to the left of the centerline, we must be sure to make it a negative value.**

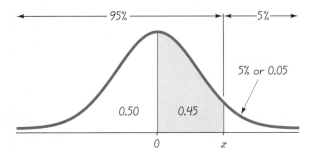

FIGURE 5-11   Finding the 95th Percentile

▶ **EXAMPLE**

Use the same thermometers with temperature readings that are normally distributed with a mean of 0°C and a standard deviation of 1°C. Find the temperature corresponding to P_{95}, the 95th percentile. That is, find the temperature separating the bottom 95% from the top 5%. See Figure 5-11.

● **SOLUTION**

Figure 5-11 shows the z score that is the 95th percentile, separating the top 5% from the bottom 95%. We must refer to Table A-2 to find that z score, and we must use a region bounded by the centerline ($\mu = 0$°C) on one side, such as the shaded region of 0.45 in Figure 5-11. (Remember, Table A-2 is designed to directly provide only those areas that are bounded on the left by the centerline and bounded on the right by the z score.) We first search for the area of 0.45 *in the body of the table* and then find the corresponding z score. In Table A-2 the area of 0.45 is between the table values of 0.4495 and 0.4505, but there's an asterisk with a special note indicating that 0.4500 corresponds to a z score of 1.645. We can now conclude that the z score in Figure 5-11 is 1.645, so the 95th percentile is the temperature reading of 1.645°C.

▶ **EXAMPLE**

Using the same thermometers, find P_{10}, the 10th percentile. That is, find the temperature reading separating the bottom 10% of all temperatures from the top 90%.   *(continued)*

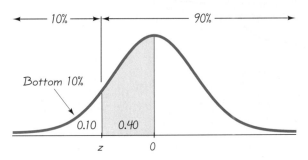

FIGURE 5-12 Finding the 10th Percentile

● **SOLUTION**

Refer to Figure 5-12, where the 10th percentile is shown as the z score separating the bottom 10% from the top 90%. Table A-2 is designed for areas bounded by the centerline, so we refer to the shaded area of 0.40. In the body of the table, we select the closest value of 0.3997 and find that it corresponds to $z = 1.28$. *However, because the z score is below the mean of 0, it must be negative.* The 10th percentile is therefore $-1.28°C$.

The examples of this section were contrived so that the mean of 0 and the standard deviation of 1 coincided exactly with the parameters of the standard normal distribution described in Table A-2. In reality, it is unusual to find such a nice relationship because typical normal distributions involve means different from 0 and standard deviations different from 1.

These nonstandard normal distributions introduce another problem. Since Table A-2 is designed around a mean of 0 and a standard deviation of 1, what table of probabilities can be used for such nonstandard normal distributions? For example, IQ scores are normally distributed with a mean of 100 and a standard deviation of 15. Scores in this range are far beyond the scope of Table A-2. Section 5-3 examines these nonstandard normal distributions and the methods used in dealing with them. We will see that Table A-2 can be used with nonstandard normal distributions after some very simple calculations are performed. Learning how to use Table A-2 will prepare you for the more practical and realistic situations found in later sections.

5-2 Exercises A: Basic Skills and Concepts

In Exercises 1–4, refer to the continuous *uniform* distribution depicted in Figure 5-2, assume that one temperature reading is randomly selected, and find the probability of each reading in degrees.

1. Between 3 and 5
2. Greater than 3
3. Less than 4.5
4. Between 1.5 and 4.5

In Exercises 5–24, assume that the readings on the thermometers are *normally* distributed with a mean of 0° and a standard deviation of 1.00°. A thermometer is randomly selected and tested. In each case, draw a sketch, and find the probability of each reading in degrees.

5. Between 0 and 2.00

6. Between −2.34 and 0

7. Greater than 1.05

8. Less than 0.82

9. Greater than 0.50

10. Less than −3.00

11. Greater than −1.09

12. Greater than −2.37

13. Between −1.00 and 1.00

14. Between −2.00 and 2.00

15. Between −1.15 and 2.60

16. Between −0.09 and 1.02

17. Between 1.20 and 1.80

18. Between 0.25 and 2.25

19. Between −1.05 and −2.30

20. Between −2.88 and −1.44

21. Greater than 0

22. Less than 0

23. Less than −1.96 or greater than 1.96

24. Less than −2.00 or greater than 1.00

In Exercises 25–28, assume that the readings on the thermometers are normally distributed with a mean of 0° and a standard deviation of 1.00°. Find the indicated probability, where z is the reading in degrees.

25. $P(z > 1.96)$

26. $P(-0.77 < z < 0.77)$

27. $P(z < 2.33)$

28. $P(1.11 < z < 2.22)$

In Exercises 29–36, assume that the readings on the thermometers are *normally* distributed with a mean of 0° and a standard deviation of 1.00°. A thermometer is randomly selected and tested. In each case, draw a sketch, and find the temperature reading corresponding to the given information.

29. Find P_{90}, the 90th percentile. This is the temperature reading separating the bottom 90% from the top 10%.

30. Find P_{40}, the 40th percentile.

31. If 15% of the thermometers are rejected because they have readings that are too high, but all other thermometers are acceptable, find the reading that separates the rejected thermometers from the others.

32. If 17% of the thermometers are rejected because they have readings that are too low, but all other thermometers are acceptable, find the reading that separates the rejected thermometers from the others.

33. A quality control analyst wants to examine thermometers that give readings in the bottom 1%. What is the highest reading in the bottom 1%?

34. Thermometers that give readings in the top 2.5% are to be recalibrated. Find the lowest temperature reading that is in the top 2.5%.

35. If 5% of the thermometers are rejected because they have readings that are too high and another 5% are rejected because they have readings that are too low, find the two readings that are cutoff values separating the rejected thermometers from the others.

36. If 15% of the thermometers are rejected because they have readings that are too high and another 20% are rejected because they have readings that are too low, find the two readings that are cutoff values separating the rejected thermometers from the others.

5-2 Exercises B: Beyond the Basics

37. Assume that z scores are normally distributed with a mean of 0 and a standard deviation of 1.
a. If $P(0 < z < a) = 0.4778$, find a.
b. If $P(-b < z < b) = 0.7814$, find b.
c. If $P(z > c) = 0.0329$, find c.
d. If $P(z > d) = 0.8508$, find d.
e. If $P(z < e) = 0.0062$, find e.

38. Assume that $\mu = 0$ and $\sigma = 1$ for a normally distributed population. Find the percentage of data that are
a. within 1 standard deviation of the mean
b. within 1.96 standard deviations of the mean
c. between $\mu - 3\sigma$ and $\mu + 3\sigma$
d. between 1 standard deviation below the mean and 2 standard deviations above the mean
e. More than 2 standard deviations away from the mean

39. In Formula 5-1, if we let $\mu = 0$ and $\sigma = 1$, we get

$$y = \frac{e^{-x^2/2}}{\sqrt{2\pi}}$$

which can be approximated by

$$y = \frac{2.7^{-x^2/2}}{2.5}$$

Graph the last equation after finding the y coordinates that correspond to the following x coordinates: $-4, -3, -2, -1, 0, 1, 2, 3$, and 4. (A calculator capable of dealing with exponents will be helpful.) Attempt to determine the approximate area bounded by the curve, the x-axis, the vertical line passing through 0 on the x-axis, and the vertical line passing through 1 on the x-axis. Compare this result to the value in Table A-2.

40. Assume that a random variable x has a probability distribution that is continuous with $\mu = 0$ and $\sigma = 1$ and that the distribution is uniform. Such a random variable has a minimum of $-\sqrt{3}$ and a maximum of $\sqrt{3}$.
a. Find $P(x > 1)$.
b. Find $P(x > 1)$ if you (incorrectly) assume that the distribution is normal instead of uniform.
c. Compare the correct result from part a to the incorrect result from part b. Does using the wrong distribution have much of an effect on the result?

5-3 Nonstandard Normal Distributions

This section extends to nonstandard normal distributions the same basic concepts applied previously to the standard normal distribution. This extension will greatly expand the variety of situations in which we can determine probabilities because, in reality, most normally distributed populations will have a nonzero mean, a standard deviation different from 1, or both.

In Section 5-2 we began with problems of finding probabilities (or areas) when given z scores; we then considered problems of finding z scores when given probabilities. We follow that same pattern in this section. We continue to use Table A-2, but we need to standardize these nonstandard cases. We convert the nonstandard case to the standard normal distribution by using Formula 5-2, in which *z is the number of standard deviations that a particular score x is away from the mean.* This *z score,* or *standard score,* is used in Table A-2 (see Figure 5-13). (This same definition was presented earlier in Section 2-6.)

Formula 5-2
$$z = \frac{x - \mu}{\sigma}$$

If all scores in a population are converted to z scores using Formula 5-2, the mean of the z scores is 0 and the standard deviation is 1. With a mean of 0 and a standard deviation of 1, we can use Table A-2, as we did in the preceding section.

Finding Probabilities When Given Scores

Suppose, for example, that we are considering a normally distributed collection of IQ scores known to have a mean of 100 and a standard deviation of 15. If we seek the probability of randomly selecting one IQ score that is between 100 and 130, we are concerned with the area shown in Figure 5-14. The difference between 130 and the mean of 100 is 30 IQ points, or exactly 2

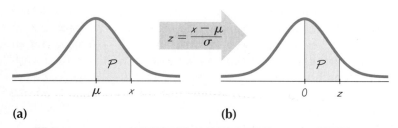

(a)   (b)

FIGURE 5-13   Converting from a Nonstandard Normal Distribution to the Standard Normal Distribution

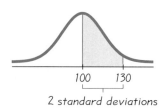

2 standard deviations

FIGURE 5-14   Normal Distribution of IQ Scores

standard deviations. We get $z = 2$ either by reasoning that 130 is 2 standard deviations above the mean of 100 or by computing

$$z = \frac{x - \mu}{\sigma} = \frac{130 - 100}{15} = \frac{30}{15} = 2$$

With $z = 2$, Table A-2 indicates that the shaded region we seek has an area of 0.4772, so the probability of randomly selecting an IQ score between 100 and 130 is 0.4772. Thus the table can be indirectly applied to any normal probability distribution if we use Formula 5-2 as the algebraic way of recognizing that the z score is actually the number of standard deviations that x is away from the mean. The following examples illustrate that observation.

▶ **EXAMPLE**

If IQ scores are normally distributed with a mean of 100 and a standard deviation of 15, find the probability of randomly selecting a person with an IQ score between 100 and 133.

● **SOLUTION**

Referring to Figure 5-15, we seek the probability associated with the shaded region. To use Table A-2, we must convert the nonstandard IQ scores (x) to standard z scores by applying Formula 5-2.

$$z = \frac{x - \mu}{\sigma} = \frac{133 - 100}{15} = \frac{33}{15} = 2.20$$

The score of 133 therefore differs from the mean of 100 by 2.20 standard deviations. Referring to Table A-2, we find that a probability of 0.4861 corresponds to $z = 2.20$. Thus there is a probability of 0.4861 of randomly selecting someone having an IQ score between 100 and 133.

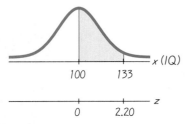

FIGURE 5-15   Probability of IQ Scores Between 100 and 133

Note that IQ scores and many other types of test scores are *discrete* whole numbers, whereas the normal distribution is *continuous*. We ignore that conflict in this section because the results are minimally affected, but in Section 5-4 we will introduce a *continuity correction factor* to deal with discrete data. At present, however, we should note that finding the probability of getting an IQ score of *exactly* 133 requires that the discrete value of 133 be represented by the continuous interval from 132.5 to 133.5; this illustrates how the continuity correction factor can be used to deal with discrete data.

In the preceding example we found that the probability of randomly selecting someone with an IQ score between 100 and 133 is 0.4861. Another way to present this result is to state that 48.61% of persons have IQ scores between 100 and 133. We could also express this result as $P(100 < x < 133) = 0.4861$ by using the same notation introduced in Section 5-2. Note that in this nonstandard normal distribution, we represent the score in its original units by the symbol x, not z. Note also that $P(100 < x < 133) = P(0 < z < 2.20) = 0.4861$.

▶ **EXAMPLE**

At the beginning of this chapter we identified the height requirements of several different organizations. The Beanstalk Club, a social organization for tall people, has a requirement that women must be at least 70 in. (or 5′ 10″) tall. Heights of adult women are normally distributed with a mean of 63.6 in. and a standard deviation of 2.5 in. (based on data from the National Health Survey). Suppose you are trying to decide whether to open a branch of the Beanstalk Club in a metropolitan area with 500,000 adult women.

a. Find the percentage of adult women who are eligible for membership because they meet the minimum height requirement of 70 in.

b. Among the 500,000 adult women living in this metropolitan area, how many are eligible for Beanstalk Club membership?

● **SOLUTION**

a. In Figure 5-16 the shaded region corresponds to the proportion of women who are eligible for membership in the Beanstalk Club because they are at least 70 in. tall. We cannot find that shaded area directly because Table A-2 is not designed for such cases, but we can find the shaded area indirectly by first finding the adjacent area A.

$$z = \frac{x - \mu}{\sigma} = \frac{70 - 63.6}{2.5} = 2.56$$

(continued)

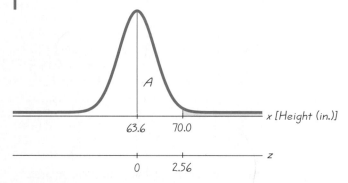

FIGURE 5-16 Women Eligible for the Beanstalk Club

Hottest Spot

A firefighter in Bullhead City, Arizona, provided daily weather statistics. One day, a Weather Service representative demanded that the thermometer be moved from the firehouse lawn to a more natural setting. The move to a drier area 100 yards away led to readings about 5° higher, which often made Bullhead City the hottest spot in the United States. As Bullhead City gained prominence in many television weather reports, some residents denounced the notoriety as a handicap to business, while others felt that it helped. Under more standardized conditions, measuring instruments, such as thermometers, tend to produce errors that are normally distributed.

We now use Table A-2 to find that $z = 2.56$ corresponds to an area of 0.4948. Because $A = 0.4948$, the shaded area must be $0.5 - 0.4948 = 0.0052$. That is, the *proportion* of women at least 70 in. tall is 0.0052. We can also state that the *percentage* of women at least 70 in. tall is 0.52%.

b. Among the 500,000 women in one metropolitan area, we expect that the actual number who are least 70 in. tall is

$$500,000 \cdot 0.0052 = 2600$$

With a maximum possible pool of 2600 women, there aren't too many tall women who would be likely candidates for a new branch of the Beanstalk Club. Based on these results, it would be wise to consider a different venture.

In the preceding problem we see that the Beanstalk Club's minimum height requirement of 70 in. for women results in a proportion of 0.0052 of women who are eligible and, in a region of 500,000 women, 2600 women who would be eligible. If we apply the same procedure to the minimum height requirement of 74 in. for men (using $\mu = 69.0$ in. and $\sigma = 2.8$ in.), we find that 0.0367 of them are eligible and, in a region of 500,000 men, 18,350 of them would be eligible. It appears that the height requirements do not treat both genders equally because a much greater proportion of men are eligible.

▶ **EXAMPLE**

At the beginning of the chapter, we noted that the United States Army requires that women's heights be between 58 in. and 80 in. Find the percentage of women satisfying that requirement. Again assume that women have heights that are normally distributed with a mean of 63.6 in. and a standard deviation of 2.5 in.

● **SOLUTION**

Figure 5-17 shows the normal distribution of women's heights, with the shaded region representing heights between 58 in. and 80 in., as required by the U.S. Army. The method for finding the area of the shaded region involves breaking it up into parts A and B as shown. We can use Formula 5-2 and Table A-2 to find the areas of those regions separately; then we can add the results.

For area A only:

$$z = \frac{x - \mu}{\sigma} = \frac{58.0 - 63.6}{2.5} = -2.24$$

We can use Table A-2 to find that $z = -2.24$ corresponds to 0.4875, so area A is 0.4875.

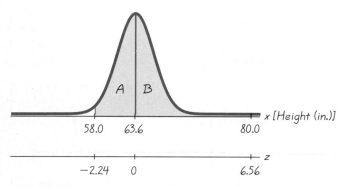

FIGURE 5-17   Women Eligible for the U.S. Army

For area B only:

$$z = \frac{80.0 - 63.6}{2.5} = 6.56$$

Table A-2 does not include z scores above 3.09, but it does include a note that for values of z above 3.09, we should use 0.4999 for the area. (If necessary, more accurate results can be obtained by using special tables or software.) Area B is 0.4999.

The shaded region consists of areas A and B combined, so

Area of regions A and B combined $= 0.4875 + 0.4999 = 0.9874$

That is, the proportion of women eligible for the U.S. Army is 0.9874. The statement of the problem indicated that a percentage should be found, so we express the result as 98.74%.

If we repeat the preceding example for men, we find that 99.98% of men are eligible. The percentage of eligible men exceeds the percentage of eligible women, but by a very small amount. Both percentages indicate that the U.S. Army rejects very few candidates because they fail to meet the height requirements.

▶ **EXAMPLE**

The heights of the Rockette dancers at New York City's Radio City Music Hall must be between 65.5 in. and 68.0 in. If a woman is randomly selected, find the probability that she meets the height requirement to be a Rockette. (Again assume that women's heights are normally distributed with $\mu = 63.6$ in. and $\sigma = 2.5$ in.)

(continued)

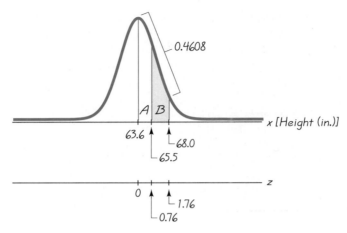

**FIGURE 5-18   Women Eligible to be
Rockette Dancers**

● SOLUTION

Figure 5-18 depicts the distribution of women's heights. The shaded region of that figure represents women who meet the height requirement of being between 65.5 in. tall and 68.0 in. tall. The design of Table A-2 requires that we work with vertical regions bounded by the mean, so the strategy for finding the shaded region B is to first find the area of regions A and B combined, then find the area of region A alone, and then find the difference between those two results. That is, $B = (A$ and B combined$) - A$.

For areas A and B combined:

$$z = \frac{68.0 - 63.6}{2.5} = 1.76$$

Table A-2 shows that $z = 1.76$ corresponds to an area of 0.4608, so regions A and B have a combined area of 0.4608, as shown in Figure 5-18.

For area A only:

$$z = \frac{65.5 - 63.6}{2.5} = 0.76$$

Referring to Table A-2, we now find that $z = 0.76$ corresponds to an area of 0.2764. Region A therefore has an area of 0.2764.

$$\text{Area } B = (\text{areas of } A \text{ and } B \text{ combined}) - (\text{area } A)$$
$$= 0.4608 - 0.2764 = 0.1844$$

If a woman is randomly selected, there is a 0.1844 probability that she meets the Rockettes' height requirement.

So far the problems we have considered in this section involved a normal distribution in this format: Given some score, we use Table A-2 to find a probability. We will now consider problems with this format: Given a probability, we find the score.

Finding Scores When Given Probabilities

In considering problems of finding scores when given probabilities, we must be careful to avoid confusion between areas and z scores. Areas (or probabilities) are found in the *body* of Table A-2. Once we have identified the probability or area that is known, we can use Table A-2 to find the corresponding z score and then use Formula 5-2 to convert the z score to an x score. The following example illustrates this method.

▶ **EXAMPLE**

A previous example of this section involved the Beanstalk Club, with a height requirement of at least 70 in. for women. We found that the proportion of women meeting that requirement is only 0.0052 and thus in a metropolitan area with 500,000 women, only 2600 are tall enough to be eligible. Suppose you plan to open a branch of the Beanstalk Club and you want the same proportion of men to be eligible. What should be the minimum height requirement for men if the proportion of them meeting that requirement is also 0.0052? Assume that the heights of men are normally distributed with a mean of 69.0 in. and a standard deviation of 2.8 in. (based on data from the National Health Survey).

● **SOLUTION**

Figure 5-19 on the next page shows the distribution of men's heights. Our objective is to find the height x that separates the shaded area from region A. (The shaded area represents the proportion of 0.0052 for the tallest men.) Because we will use Table A-2, we must work with a region bounded by the mean. Region A in Figure 5-19 must have an area of 0.4948, found by computing $0.5 - 0.0052$. (Remember, the total area under the curve is 1, and the combined area of A and the shaded region is 0.5.) Referring to the body of Table A-2, we find that the area of 0.4948 corresponds to $z = 2.56$. The score x is therefore 2.56 standard deviations away from the mean. The standard deviation is 2.8, so we conclude that 2.56 standard deviations is $2.56 \cdot 2.8 = 7.168$. Our x score is therefore 7.168 in. above 69.0 in., so $x = 69.0 + 7.168 = 76.2$ in. (rounded). That is, the height of 76.2 in. (6′ 4.2″) separates the tallest 0.0052 of men from the rest. We could have achieved the same result

(continued)

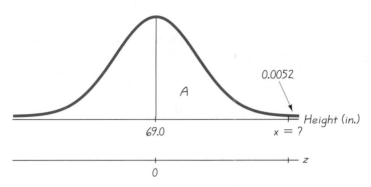

FIGURE 5-19   Men Eligible for the Beanstalk Club

by noting that

$$z = \frac{x - \mu}{\sigma} \quad \text{becomes} \quad 2.56 = \frac{x - 69.0}{2.8}$$

we substitute the given values for the mean μ, the standard deviation σ, and the z score corresponding to an area of 0.4948 to the right of the mean. We solve this equation by multiplying both sides by 2.8 and then adding 69.0 to both sides.

The Beanstalk Club currently has a 74 in. height requirement for men and a 70 in. height requirement for women, but if it were to establish a minimum height requirement of 76.2 in. for men and 70.0 in. for women, then the same proportion (0.0052) of men and women would be eligible.

In using Table A-2, it's very important to avoid confusion between areas found in the body of the table and the z scores listed at the left column and top row. It's also very important to *make z scores negative when they are located at the left half of the normal distribution.* The following example illustrates this point.

▶ **EXAMPLE**

A study compared the facial behavior of nonparanoid schizophrenic persons with that of a control group of normal persons. The control group was timed for eye contact in a 5-minute (300 s) period. The times were normally distributed with a mean of 184 and a standard deviation of 55 s (based on data from "Ethological Study of Facial Behavior in Nonparanoid and Paranoid Schizophrenic Patients," by Pitman, Kolb, Orr, and Singh, *Psychiatry*, 144:1).

Because results showed that nonparanoid schizophrenic patients had much lower eye contact times than did the control group, you have decided to further analyze people in the control group who are in the bottom 5%. For the control group, find P_5, the 5th percentile. That is, find the eye contact time separating the bottom 5% from the rest.

● SOLUTION

Constructing a graph is extremely helpful in solving such problems. Figure 5-20 shows the distribution relevant to this problem. We must first identify an area in the figure that can be located in Table A-2 so that the corresponding z score can be found. The table is designed for areas bounded by the mean, so we will work with region A, which must have an area of 0.45 (found by computing $0.5 - 0.05$). Referring to Table A-2, we find that an area of 0.45 corresponds to $z = 1.645$. *Because the z score is negative whenever it is below the mean, we set $z = -1.645$.* That is, the score x is 1.645 standard deviations *below* the mean of 184. Because the standard deviation is 55.0, we conclude that 1.645 standard deviations is $1.645 \cdot 55.0 = 90.475$. Our x score is therefore 90.475 *below* 184, and we get $x = 184 - 90.475 = 93.5$ s (rounded off). This same result can be found by noting that

$$z = \frac{x - \mu}{\sigma} \quad \text{becomes} \quad -1.645 = \frac{x - 184}{55.0}$$

We solve for x by multiplying both sides by 55.0 and then adding 184 to both sides.

The result indicates that $P_5 = 93.5$ s. That is, 5% of the times are less than 93.5 s. If you want to further analyze people in the bottom 5% of the control group, then 93.5 s is the cutoff.

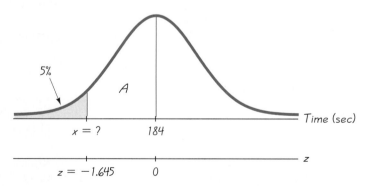

FIGURE 5-20 Finding the 5th Percentile for Eye Contact Times

Reliability and Validity

The reliability of data refers to the consistency with which results occur, whereas the validity of data refers to how well the data measure what they are supposed to measure. The reliability of an IQ test can be judged by comparing scores for the test given on one date to scores for the same test given at another time. To test the validity of an IQ test, we might compare the test scores to another indicator of intelligence, such as academic performance. Many critics charge that IQ tests are reliable, but not valid; they provide consistent results, but don't really measure intelligence.

In the preceding solution it would be very easy to make the mistake of forgetting the negative sign in -1.645. Omitting this negative sign would lead to an answer of 274.5 s, but Figure 5-20 shows that x can't possibly be 274.5 s. This illustrates the importance of drawing a graph when working with the normal distribution. *Always draw the graph with the relevant labels, and use common sense to check that the results are reasonable.*

Also note that in the preceding solution, Table A-2 led to a z score of 1.645, which is midway between 1.64 and 1.65. When using Table A-2, we can usually avoid interpolation by simply selecting the closest value. However, there are two special cases involving values that are important because they are used so often in a wide variety of applications (see the accompanying table). Except in these two special cases, we can select the closest value in the table. (If a desired value is midway between two table values, select the larger value.) Also, for z scores above 3.09, we can use 0.4999 as an approximation of the corresponding area.

z score	Area
1.645	0.4500
2.575	0.4950

5-3   Exercises A: Basic Skills and Concepts

In Exercises 1–4, assume that IQ scores are normally distributed with a mean of 100 and a standard deviation of 15. An IQ score is randomly selected from this population. Draw a graph, and find the indicated probability.

1. $P(100 < x < 127)$ 2. $P(x < 118)$
3. $P(x > 76)$ 4. $P(85 < x < 112)$

In Exercises 5–8, assume that IQ scores are normally distributed with a mean of 100 and a standard deviation of 15. Draw a graph, and find the IQ score for the indicated percentile.

5. P_{90} 6. P_{10} 7. P_{20} 8. P_{95}

9. Table 5-1 includes a height of 64 in. as the minimum for men in the U.S. Marine Corps. Find the percentage of men who have heights below that minimum height requirement. (Recall that heights of men are normally distributed with a mean of 69.0 in. and a standard deviation of 2.8 in.)

10. Table 5-1 indicates that to be eligible for the U.S. Marine Corps, a woman must have a height of between 58 in. and 73 in. Find the percentage of women who satisfy that requirement. (Recall that heights of women are normally distributed with a mean of 63.6 in. and a standard deviation of 2.5 in.)

11. Assume that the weights of paper discarded by households each week are normally distributed with a mean of 9.4 lb and a standard deviation of 4.2 lb (based on data from the Garbage Project at the University of Arizona). Find the weights that are the percentiles P_5 and P_{95}.

12. Referring to the data in Exercise 11, find the probability of randomly selecting a household and getting one that discards at least 5.0 lb of paper in a week.

13. In Chapter 2 we considered body temperature data collected by University of Maryland researchers. Based on those data, assume that body temperatures of

normal healthy persons are normally distributed with a mean of 98.2°F and a standard deviation of 0.62°F. If we define a fever to be a body temperature above 100°F, what percentage of normal and healthy persons would be considered to have a fever?

14. Refer to the same data given in Exercise 13. Instead of using 100°F, suppose we stipulate that a fever is any body temperature above a specific value and that value has the property that it is exceeded by only 0.005 of normal healthy persons. Find the minimum body temperature that qualifies as a fever.

15. In a study of employee stock ownership plans, satisfaction by employees is measured and found to be normally distributed with a mean of 4.89 and a standard deviation of 0.63 (based on "Employee Stock Ownership and Employee Attitudes: A Test of Three Models," by Klein, *Journal of Applied Psychology*, Vol. 72, No. 2). If employees in the bottom 8% are to be given interviews, find the score separating those who will receive interviews from those who will not.

16. On one measure of attractiveness, scores are normally distributed with a mean of 3.93 and a standard deviation of 0.75 (based on "Physical Attractiveness and Self Perception of Mental Disorder," by Burns and Farina, *Journal of Abnormal Psychology*, Vol. 96, No. 2). Find the probability of randomly selecting a subject with a measure of attractiveness that is greater than 2.75.

17. A study shows that Michigan teachers have measures of job dissatisfaction that are normally distributed with a mean of 3.80 and a standard deviation of 0.95 (based on "Stress and Strain from Family Roles and Work-Role Expectations," by Cooke and Rousseau, *Journal of Applied Psychology*, Vol. 69, No. 2). If subjects with scores above 4.00 are to be given additional tests, what percentage will fall into that category?

18. The serum cholesterol levels in men aged 18 to 24 are normally distributed with a mean of 178.1 and a standard deviation of 40.7. All units are in mg/100 mL, and the data are based on the National Health Survey (USDHEW publication 78-1654). If a man aged 18 to 24 is randomly selected, find the probability that his serum cholesterol level is between 100 and 200.

19. Scores on an anti-aircraft artillery exam are normally distributed with a mean of 99.56 and a standard deviation of 25.84 (based on "Routinization of Mental Training in Organizations: Effects on Performance and Well Being," by Larson, *Journal of Applied Psychology*, Vol. 72, No. 1). For a randomly selected subject, find the probability that a score will fall between 110.00 and 150.00.

20. On the Graduate Record Exam in economics, scores are normally distributed with a mean of 615 and a standard deviation of 107. If a college admissions office requires scores above the 70th percentile, find the cutoff point.

21. For a certain population, scores on the Miller Analogies Test are normally distributed with a mean of 58.84 and a standard deviation of 15.94 (based on "Equivalencing MAT and GRE Scores Using Simple Linear Transformation and Regression Methods," by Kagan and Stock, *Journal of Experimental Education*, Vol. 49, No. 1). If subjects who score below 27.00 are to be given special

training, what is the percentage of subjects who will be given the special training?

22. Scores on the numerical part of the Minnesota Clerical Test are normally distributed with a mean of 119.3 and a standard deviation of 32.4. This test is used for selecting clerical employees. (The data are based on "Modification of the Minnesota Clerical Test to Predict Performance on Video Display Terminals," by Silver and Bennett, *Journal of Applied Psychology*, Vol. 72, No. 1.) If a firm requires scores above 172, find the percentage of subjects who don't qualify.

23. For a certain population, scores on the Thematic Apperception Test are normally distributed with a mean of 22.83 and a standard deviation of 8.55 (based on "Relationships Between Achievement-Related Motives, Extrinsic Conditions, and Task Performance," by Schroth, *Journal of Social Psychology*, Vol. 127, No. 1). For a randomly selected subject, find the probability that the score is between 4.02 and 22.83.

24. For a certain group of students, scores on an algebra placement test are normally distributed with a mean of 18.4 and a standard deviation of 5.1 (based on data from "Factors Affecting Achievement in the First Course in Calculus," by Edge and Friedberg, *Journal of Experimental Education*, Vol. 52, No. 3). If 50 different students are randomly selected from this population, how many of them are expected to score above 16.0?

25. Scores on the biology portion of the Medical College Admissions Test are normally distributed with a mean of 8.0 and a standard deviation of 2.6. Among 600 individuals taking this test, how many are expected to score between 6.0 and 7.0?

26. The Chemco Company, which manufactures car tires, finds that the tires last distances that are normally distributed with a mean of 35,600 mi and a standard deviation of 4275 mi. The manufacturer wants to guarantee the tires so that only 3% will be replaced because of failure before the guaranteed number of miles. For how many miles should the tires be guaranteed?

27. One classic use of the normal distribution is inspired by a letter to *Dear Abby* in which a wife claimed to have given birth 308 days after a brief visit from her husband, who was serving in the Navy. The lengths of pregnancies are normally distributed with a mean of 268 days and a standard deviation of 15 days. Given this information, find the probability of a pregnancy lasting 308 days or longer. What does the result suggest?

28. You plan to open a men's clothing store. To minimize startup costs, you will not stock suits for the tallest 5% and the shortest 5% of men. Find the minimum and maximum heights of the men for whom suits will be stocked. (Men have normally distributed heights with a mean of 69.0 in. and a standard deviation of 2.8 in.)

29. Some vending machines are designed so that their owners can adjust the weights of the quarters that are accepted. If many counterfeit coins are found, adjustments are made to reject more coins, with the effect that most of the counterfeit coins are rejected along with many legal coins. Assume that quarters have weights that are normally distributed with a mean of 5.67 g and a standard

deviation of 0.0700 g. If a vending machine is adjusted to reject quarters weighing less than 5.50 g or more than 5.80 g, what is the percentage of legal quarters that are rejected?

30. Refer to the same data given in Exercise 29. If a vending machine is adjusted so that 15% of legal quarters are rejected because their weights are too low and another 8% are rejected because their weights are too high, find the minimum and maximum weights of legal quarters that are accepted.

5-3   Exercises B: Beyond the Basics

31. The following sample scores are times (in milliseconds) that it took the author's disk drive to make one revolution. The times were recorded by a diagnostic software program.

200.5	199.7	201.1	200.4	200.3	200.1	200.4	200.4	200.4	200.5
200.1	200.1	200.3	200.5	200.3	200.3	200.6	200.5	200.4	200.5
200.3	201.2	200.5	200.6	200.4	200.5	200.3	200.7	200.6	200.5
200.4	200.0	201.2	200.6	200.4	200.8	200.6	200.3	200.6	200.5

   a. Find the actual percentage of these sample scores that are greater than 201.0 milliseconds (ms).

   b. Find the mean $\bar{x}$ of this sample.

   c. Find the standard deviation s of this sample.

   d. Assuming a normal distribution, find the percentage of *population* scores greater than 201.0 ms. Use the sample values of $\bar{x}$ and s as estimates of μ and σ.

   e. The specifications require times between 198.0 ms and 202.0 ms. Based on these sample results, does the disk drive seem to be rotating at acceptable speeds?

32. A population of weights has a mean of 100 g, a standard deviation of 15 g, and a distribution that is uniform. Such a distribution has a minimum of 74 g and a maximum of 126 g. Find the probability of randomly selecting a value between 80 and 110 with this uniform distribution, and compare it to the area between 80 g and 110 g for a normal distribution with the same mean of 100 g and standard deviation of 15 g.

33. A teacher gives a test and gets normally distributed results with a mean of 50 and a standard deviation of 10. Grades are to be assigned according to the following scheme.

    A: Top 10%

    B: Scores above the bottom 70% and below the top 10%

    C: Scores above the bottom 30% and below the top 30%

    D: Scores above the bottom 10% and below the top 70%

    F: Bottom 10%

Find the numerical limits for each letter grade.

34. According to recent data from the College Entrance Examination Board, the mean math SAT score is 475, and 17.0% of the scores are above 600. Find the standard deviation, and then use that result to find the 99th percentile. (Assume that the scores are normally distributed.)

35. A city-sponsored cross-country race has 4830 applicants, but only 200 are allowed to run in the final race. A qualifying run was held two weeks before the final race, and the times were normally distributed with a mean of 36.2 min and a standard deviation of 3.8 min. If the 200 fastest times qualify, what is the cutoff time?

36. In constructing a boxplot for exploring a set of data, the median is found to be 650, while the hinges are 572 and 728. Given that the data appear to be normally distributed, use this information to estimate the standard deviation.

37. The College Entrance Examination Board writes that "for the SAT Achievement Tests, your score would fall in a range about 30 points above or below your actual ability about two-thirds of the time. This range is called the standard error of measurement (SEM)." Based on that statement, estimate the standard deviation for scores of an individual on an SAT Achievement Test. (Assume that the scores are normally distributed.)

5-4 The Central Limit Theorem

The central limit theorem is one of the most important and useful concepts in statistics. It forms a foundation for the important topics of estimating population parameters and hypothesis testing—topics discussed at length in the following chapters. Before considering this theorem, we will first try to develop an intuitive understanding of one of its most important consequences:

> **As the sample size increases, the sampling distribution of sample means approaches a normal distribution.**

We have previously considered uniform, normal, binomial, and other distributions, but they were all distributions of *individual scores,* such as distributions of body temperatures, airplane ages, and heights of women. We now want to consider the distribution of *sample means,* where the samples are all of the same size and are all drawn from the same population. For example, suppose that we collect many different samples, each consisting of 50 computer-generated random numbers from a uniform distribution, and then calculate the mean of each sample. What do we know about the distribution of those sample means? It will be a *normal* distribution.

In general, the **sampling distribution of sample means** is the distribution of the sample means obtained when we repeatedly draw samples of the same size from the same population. That is, if we collect samples of the same size from the same population, compute their means, and then develop a histogram of those means, it will tend to assume the bell shape of a normal distribution (see Figure 5-21). This is true regardless of the shape of the distribution of the

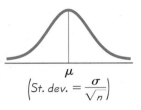

The distribution of these sample means will, as n increases, approach a *normal* distribution with mean μ and standard deviation $\sigma/\sqrt{n}$

μ

$\left(St.\ dev. = \dfrac{\sigma}{\sqrt{n}}\right)$

$\bar{x}_1$ $\bar{x}_2$ $\bar{x}_3$ etc.

Select samples of size n, and find $\bar{x}$ for each sample

Population:
Distribution can have *any* shape.
Mean is μ.
Standard deviation is σ.

FIGURE 5-21   The Central Limit Theorem

original population. The central limit theorem qualifies the preceding remarks and includes additional aspects, but let's try to understand the thrust of these remarks before continuing.

Let's begin with a concrete example based on a population of the numbers 1, 2, 3, 4, 5, and 6. Random samples can be drawn from this population by rolling a die. For the experiment of selecting samples of size $n = 1$ from this population, all of the outcomes are equally likely; the table in Figure 5-22(a) describes the probability distribution, and the graph illustrates that it's a uniform distribution. Now suppose that we select random samples of size $n = 2$ by drawing 2 members of the population of selected numbers *with replacement*. Drawing 2 numbers (with replacement) from the population {1, 2, 3, 4, 5, 6} is equivalent to rolling a pair of dice. If we repeatedly roll 2 dice and find their mean, we get the probability distribution described by the table in Figure 5-22(b) and illustrated in the graph. Note that whereas the original population in Figure 5-22(a) has a uniform distribution, the distribution of sample means in Figure 5-22(b) tends to be closer to a bell shape. It's a truly fascinating and intriguing phenomenon in statistics that by sampling from a uniform distribution, we can create a distribution that is normal or at least approximately normal.

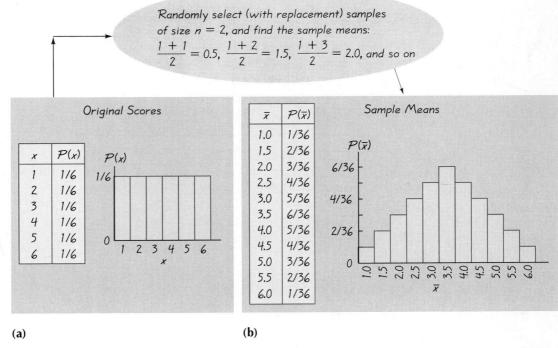

(a)   **(b)**

FIGURE 5-22   **Illustration of the Central Limit Theorem for (a) One Die and (b) Two Dice**

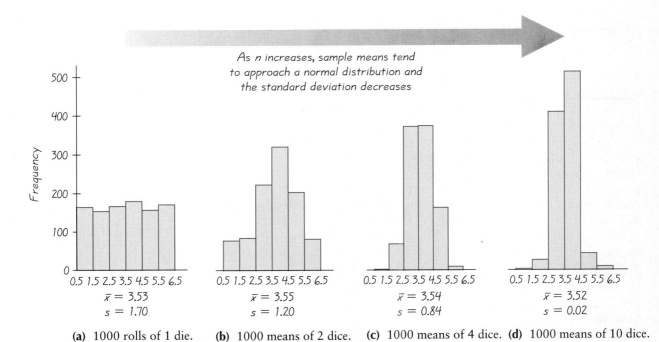

(a) 1000 rolls of 1 die.   **(b)** 1000 means of 2 dice.   **(c)** 1000 means of 4 dice.   **(d)** 1000 means of 10 dice.

FIGURE 5-23   **Central Limit Theorem Illustrated by Dice**

Figure 5-23 illustrates computer-simulated results of rolling a single die 1000 times, a pair of dice 1000 times, four dice 1000 times, and ten dice 1000 times. In each case the sample mean is calculated. Again note that the original population of {1, 2, 3, 4, 5, 6} has a uniform distribution, but when we select samples of size $n = 2$ (or roll two dice), the distribution shape tends to be closer to a bell shape. Figure 5-23 shows two very important trends that become evident as the sample size increases: the shape tends to become closer to a bell shape, and the spread of the data tends to decrease (indicated by a decrease in the standard deviation). Whereas Figure 5-23 illustrates particular results obtained from a uniform distribution, Figure 5-24 is a general illustration that also includes normal and skewed distributions. Observations exactly like these led to the formulation of the central limit theorem, which we will now discuss.

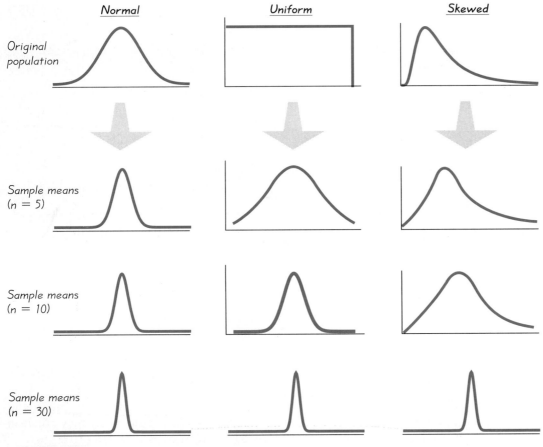

FIGURE 5-24   Normal, Uniform, and Skewed Distributions

Let's assume that the variable x represents scores that may or may not be normally distributed and that the mean of the x values is μ and the standard deviation is σ. Suppose that we collect samples of size n and calculate the sample means. What do we know about the collection of all sample means that we produce by repeating this experiment? The central limit theorem tells us that as the sample size n increases, the sampling distribution of sample means approaches a normal distribution with mean μ and standard deviation $\sigma/\sqrt{n}$. The distribution of sample means *approaches* a normal distribution in the sense that as n becomes larger, the distribution of sample means gets closer to a normal distribution. This conclusion is not intuitively obvious, and it was arrived at through extensive research and analysis. The formal rigorous proof requires advanced mathematics and is beyond the scope of this text. We now formally present the theorem and give examples of its use.

Central Limit Theorem

Given:

1. The random variable x has a distribution (which may or may not be normal) with mean μ and standard deviation σ.

2. Samples of size n are randomly selected from this population.

Conclusions:

1. The distribution of sample means $\overline{x}$ will, as the sample size increases, approach a normal distribution.

2. The mean of the sample means will be the population mean μ.

3. The standard deviation of the sample means will be $\sigma/\sqrt{n}$.

Practical Rules Commonly Used:

1. For samples of size n larger than 30, the distribution of the sample means can be approximated reasonably well by a normal distribution. The approximation gets better as the sample size n becomes larger.

2. If the original population is itself normally distributed, then the sample means will be normally distributed for *any* sample size n.

The central limit theorem involves two different distributions: the distribution of the original population and the distribution of the sample means. As in previous chapters, we use the symbols μ and σ to denote the mean and standard deviation of the original population. We now introduce new notation for the mean and standard deviation of the distribution of sample means.

Notation for the Central Limit Theorem

If all possible random samples of size n are selected from a population with mean μ and standard deviation σ, the mean of the sample means is denoted by $\mu_{\bar{x}}$, so

sample mean

$$\mu_{\bar{x}} = \mu$$

Also, the standard deviation of the sample means is denoted by $\sigma_{\bar{x}}$, so

std dev samp

$$\sigma_{\bar{x}} = \frac{\sigma}{\sqrt{n}}$$

$\sigma_{\bar{x}}$ is often called the *standard error of the mean*.

Many important and practical problems can be solved with the central limit theorem. In the following example we use the same basic methods introduced earlier in Section 5-3, but we make some important adjustments because our problem deals with the mean for a *sample* of scores (instead of an *individual* score). This example illustrates a method for dealing with the sampling distribution of means, rather than the distribution of individual scores.

▶ **EXAMPLE**

Assume that the population of human body temperatures has a mean of 98.6°F, as is commonly believed. Also assume that the population standard deviation is 0.62°F (based on data from University of Maryland researchers). If a sample of size $n = 106$ is randomly selected, find the probability of getting a mean of 98.2°F or lower. (The value of 98.2°F was actually obtained; see the midnight temperatures for Day 2 in Data Set 2 of Appendix B.)

● **SOLUTION**

We weren't given the distribution of the population, but because the sample size $n = 106$ exceeds 30, we use the central limit theorem and conclude that the distribution of sample means is the normal distribution with these parameters:

$$\mu_{\bar{x}} = \mu = 98.6$$

$$\sigma_{\bar{x}} = \frac{\sigma}{\sqrt{n}} = \frac{0.62}{\sqrt{106}} = 0.060$$

(continued)

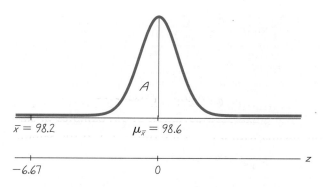

FIGURE 5-25   Distribution of Sample Mean Body Temperatures ($n = 106$)

Figure 5-25 shows the shaded area corresponding to the probability we seek. Having already found the parameters that apply to the distribution shown in Figure 5-25, we can now find the shaded area by using the same procedures developed in the preceding section. We first determine the value of the z score.

$$z = \frac{\overline{x} - \mu_{\overline{x}}}{\sigma_{\overline{x}}} = \frac{98.2 - 98.6}{0.060} = -6.67$$

Referring to Table A-2, we find that $z = -6.67$ is off the chart, but there is a notation that "for values of z above 3.09, use 0.4999 for the area." We therefore conclude that region A in Figure 5-25 is 0.4999 and the shaded region is $0.5 - 0.4999 = 0.0001$. (More precise tables indicate that the area of the shaded region is closer to 0.000000001.)

The result shows that if the mean of our body temperatures is really 98.6°F, then there is an extremely small probability of getting a sample mean of 98.2°F or lower when 106 subjects are randomly selected. University of Maryland researchers did obtain such a sample mean, and there are two possible explanations: Either the population mean really is 98.6°F and their sample represents a chance event that is extremely rare, or the population mean is actually lower than 98.6°F so their sample is typical. Because the probability is so low, it seems more reasonable to conclude that the population mean is lower than 98.6°F. This is the type of reasoning used in the formal methods of *hypothesis testing*, which are introduced in Chapter 7. For now, we should focus on the use of the central limit theorem for finding the probability of 0.0001, but we should also observe that this theorem will be used later in developing some very important concepts in statistics.

In the preceding example we justify use of the central limit theorem by noting that the sample size exceeds 30. If we know that the original population has a normal distribution, then samples of *any* size will yield means that are normally distributed, so the central limit theorem is applicable.

It is interesting to note that as the sample size increases, the sample means tend to vary less, because $\sigma_{\bar{x}} = \sigma/\sqrt{n}$ gets smaller as n gets larger. For example, IQ scores have a mean of 100 and a standard deviation of 15. Samples of size $n = 100$ will produce sample means with $\sigma_{\bar{x}} = 15/\sqrt{100} = 1.5$. If the sample size is increased to 400, $\sigma_{\bar{x}}$ decreases to $15/\sqrt{400} = 0.75$. By quadrupling the sample size, we halve the standard deviation. In general, as the sample size increases, the standard deviation of sample means decreases, so the amount of variation is less. This conclusion is supported by common sense: As the sample size increases, the corresponding sample mean will tend to be closer to the true population mean. The effect of an unusual or outstanding score tends to be diminished as it is averaged in as part of a larger sample.

In applying the central limit theorem, our use of $\sigma_{\bar{x}} = \sigma/\sqrt{n}$ assumes that the population has infinitely many members. When we sample with replacement (that is, put back each selected item before making the next selection), the population is effectively infinite. Yet many realistic applications involve sampling without replacement, so successive samples depend on previous outcomes. In manufacturing, quality control inspectors typically sample items from a finite production run without replacing them. For such a finite population, we may need to adjust $\sigma_{\bar{x}}$. Here is a common rule of thumb:

> When sampling without replacement and the sample size n is greater than 5% of the finite population size N (that is, $n > 0.05N$), adjust the standard deviation of sample means $\sigma_{\bar{x}}$ by multiplying it by the *finite population correction factor*

$$\sqrt{\frac{N - n}{N - 1}}$$

Except for Exercises 21 and 22, the examples and exercises of this section assume that the finite population correction factor does not apply, because the population is infinite or the sample size doesn't exceed 5% of the population size.

It's very important to note that the central limit theorem applies when we are dealing with the distribution of sample means and either the sample size is greater than 30 or the original population has a normal distribution. Once we have established that the central limit theorem applies, we can determine values for $\mu_{\bar{x}} = \mu$ and $\sigma_{\bar{x}} = \sigma/\sqrt{n}$ and then proceed to use the same methods presented in the preceding section. The following example illustrates two fundamentally different cases. Part a involves a simple and direct use of the normal distribution, and part b includes use of the central limit theorem.

Ethics in Experiments

Sample data can often be obtained by simply observing or surveying members selected from the population. Many other situations require that we somehow manipulate circumstances to obtain sample data. In both cases ethical questions may arise. Researchers in Tuskegee, Alabama, withheld the effective penicillin treatment to syphilis victims so that the disease could be studied. This experiment continued for a period of 27 years!

▶ **EXAMPLE**

In human engineering and product design, it is often important to consider the weights of people so that airplanes or elevators aren't overloaded, chairs don't break, and other such unpleasant happenings don't occur. Assume that the population of men has normally distributed weights, with a mean of 173 lb and a standard deviation of 30 pounds (based on data from the National Health Survey).

a. If 1 man is randomly selected, find the probability that his weight is greater than 180 lb.

b. If 36 different men are randomly selected from this population, find the probability that their mean weight is greater than 180 lb.

● **SOLUTION**

a. *Approach: Use the same methods presented in Section 5-3.* We seek the area of the shaded region in Figure 5-26(a).

$$z = \frac{x - \mu}{\sigma} = \frac{180 - 173}{30} = 0.23$$

We now refer to Table A-2 to find that region A is 0.0910. The shaded region is therefore $0.5 - 0.0910 = 0.4090$. The probability of the man weighing more than 180 lb is 0.4090.

b. *Approach: Use the central limit theorem.* Because we are now dealing with a distribution of sample means, we must use the parameters $\mu_{\bar{x}}$ and $\sigma_{\bar{x}}$. We want to determine the shaded area shown in Figure 5-26(b), and the relevant z score is calculated as follows:

$$z = \frac{\bar{x} - \mu_{\bar{x}}}{\sigma_{\bar{x}}} = \frac{180 - 173}{\dfrac{30}{\sqrt{36}}} = \frac{7}{5} = 1.40$$

Referring to Table A-2, we find that $z = 1.40$ corresponds to an area of 0.4192, so the shaded region is $0.5 - 0.4192 = 0.0808$. The probability that the 36 men have a mean weight greater than 180 lb is 0.0808.

The central limit theorem is one of the most important and useful concepts in statistics because it allows us to use the basic normal distribution methods in a wide variety of different circumstances. In Chapter 6, for example, we will

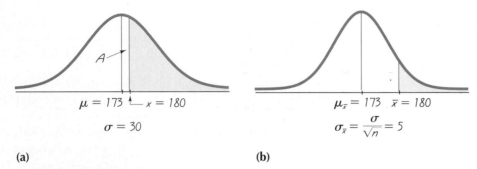

FIGURE 5-26   Distribution of (a) Individual Men's
Weights and (b) Means of Samples of
36 Men's Weights

apply the central limit theorem when we use sample data to estimate means of populations. In Chapter 7 we will apply the central limit theorem when we use sample data to test claims made about population means. Such applications of estimating population parameters and testing claims are extremely important uses of statistics, and the central limit theorem makes them possible.

5-4 Exercises A: Basic Skills and Concepts

1. IQ scores are normally distributed with a mean of 100 and a standard deviation of 15.
   a. If 1 person is randomly selected, find the probability that his or her IQ score is between 100 and 103.
   b. If 25 persons are randomly selected, find the probability that the mean of their IQ scores is between 100 and 103.
2. For women aged 18–24, systolic blood pressures (in mm Hg) are normally distributed with a mean of 114.8 and a standard deviation of 13.1 (based on data from the National Health Survey).
   a. If a woman between the ages of 18 and 24 is randomly selected, find the probability that her systolic blood pressure is above 120.
   b. If 30 women in that age bracket are randomly selected, find the probability that their mean systolic blood pressure is greater than 120.
3. Assume that the weights (in pounds) of paper discarded each week by households are normally distributed with a mean of 9.43 lb and a standard deviation of 4.17 lb (based on data from the Garbage Project at the University of Arizona). Find the probability that 12 randomly selected households have a mean between 10.0 lb and 12.0 lb.
4. In a study of work patterns in one population, the mean weekly family time devoted to child care is 23.08 h with a standard deviation of 15.58 h (based on

data from "Nonstandard Work Schedules," by Staines and Pleck, *Journal of Applied Psychology,* Vol. 69, No. 3). If 60 subjects are randomly selected, find the probability that the mean of this sample group is between 20.00 h and 24.00 h.

5. The Goodenough-Harris Drawing Test is used to measure the intellectual maturity of young people. For twelve-year-old girls, the scores are normally distributed with a mean of 34.8 and a standard deviation of 7.02. (The figures are based on data from the National Health Survey.) If 36 twelve-year-old girls are randomly selected, find the probability that their mean score is between 34.8 and 37.0.

6. A study of the amounts of time (in hours) college freshmen study each week found that the mean is 7.06 h and the standard deviation is 5.32 h (based on data from *The American Freshman*). If 55 freshmen are randomly selected, find the probability that their mean weekly study time exceeds 7.00 h.

7. A study of the amounts of time high school students spend working each week at a job found that the mean is 10.7 h and the standard deviation is 11.2 h (based on data from the National Federation of State High School Associations). If 42 high school students are randomly selected, find the probability that their mean weekly work time is less than 12.0 h.

8. The typical computer random-number generator yields numbers in a uniform distribution between 0 and 1 with a mean of 0.500 and a standard deviation of 0.289. If 45 random numbers are generated, find the probability that their mean is below 0.565.

9. The ages of U.S. commercial aircraft have a mean of 13.0 years and a standard deviation of 7.9 years (based on data from Aviation Data Services). If the Federal Aviation Administration randomly selects 35 commercial aircraft for special stress tests, find the probability that the mean age of this sample group is greater than 15.0 years.

10. The amounts of time that managers spend on paperwork each day have a mean of 2.7 h and a standard deviation of 1.4 h (data from Adia Personnel Services). If a computer sales team randomly selects 75 managers, find the probability that this sample group has a mean of less than 2.5 h.

11. A total of 630 students were measured for burnout. The resulting scores have a mean of 2.97 and a standard deviation of 0.60 (based on data from "Moderating Effects of Social Support on the Stress-Burnout Relationship," by Etzion, *Journal of Applied Psychology,* Vol. 69, No. 4). If 31 of these subjects are randomly selected, find the probability that their mean burnout score is between 3.00 and 3.10.

12. In a study of job training, the times required to learn how to use a word processing system are found to have a mean and a standard deviation of 462.1 min and 76.3 min, respectively. If 32 subjects are randomly selected, find the probability that their mean is between 447.0 min and 456.8 min.

13. A study of Reye's Syndrome (by Holtzhauer and others, *American Journal of Diseases of Children,* Vol. 140) found that children suffering from the disease

had a mean age of 8.5 years and a standard deviation of 3.96 years; their ages approximated a normal distribution. If 36 of those children are randomly selected, find the probability that their mean age is between 7.0 years and 10.0 years.

14. Refer to the population of children described in Exercise 13. What happens to the standard deviation of the sample means (also known as the standard error) if the sample size is quadrupled to 144 and all other values remain the same?

15. A study was made of the effectiveness of precautionary behavior as a deterrent to criminal victimization. For subjects who lock their cars when they are parked away from their homes, the mean crime prevention scale value was 3.54 and the standard deviation was 0.96 (based on data from "A Longitudinal Study of the Effects of Various Crime Prevention Strategies on Criminal Victimization, Fear of Crime, and Psychological Distress," by Norris and Kaniasty, *American Journal of Community Psychology*, Vol. 20, No. 5). If 80 such subjects are randomly selected, find the probability that their mean is between 3.50 and 3.60.

16. A study was made of seat belt use among children who were involved in car crashes that caused them to be hospitalized. It was found that children not wearing any restraints had hospital stays with a mean of 7.37 days and a standard deviation of 0.79 day (based on data from "Morbidity Among Pediatric Motor Vehicle Crash Victims: The Effectiveness of Seat Belts," by Osberg and Di Scala, *American Journal of Public Health*, Vol. 82, No. 3). If 40 such children are randomly selected, find the probability that their mean hospital stay is greater than 7.00 days.

17. SAT verbal scores are normally distributed with a mean of 430 and a standard deviation of 120 (based on data from the College Board ATP). Randomly selected SAT verbal scores are obtained from the population of students who took a test preparatory course from the Tillman Training School. Assume that this training course has no effect on test scores.
    a. If 1 of the students is randomly selected, find the probability that he or she obtained a score greater than 440.
    b. If 100 students are randomly selected, find the probability that their mean is greater than 440.
    c. If the 100 Tillman students achieve a sample mean of 440, does it seem reasonable to conclude that the course is effective because the students perform better on the SAT?

18. The lengths of pregnancies are normally distributed with a mean of 268 days and a standard deviation of 15 days.
    a. If 1 pregnant woman is randomly selected, find the probability that her length of pregnancy is less than 260 days.
    b. If 25 randomly selected women are put on a special diet just before they become pregnant, find the probability that their lengths of pregnancy have a mean that is less than 260 days (assuming that the diet has no effect). *(continued)*

c. If the 25 women do have a mean of less than 260 days, should the medical supervisors be concerned?

19. Using a standard measure of satisfaction with salaries, a study finds that college administrators have a mean of 38.9 and a standard deviation of 12.4 (based on data from "Job Satisfaction Among Academic Administrators," by Glick, *Research in Higher Education*, Vol. 33, No. 5). A pollster randomly selects 150 college administrators and measures their levels of satisfaction with their salaries.

a. Find the probability that the mean is greater than 42.0.

b. If a sample of 150 college administrators does yield a mean of 42.0 or greater, is there reason to believe that this sample came from a population with a mean that is higher than 38.9?

20. The population of weights of men is normally distributed with a mean of 173 lb and a standard deviation of 30 lb (based on data from the National Health Survey). An elevator in the Dallas Men's Club is limited to 32 occupants, but it will be overloaded if the 32 occupants have a mean weight in excess of 186 lb (yielding a total weight in excess of $32 \cdot 186 = 5952$ lb). If 32 male occupants are the result of a random selection, find the probability that their mean will exceed 186 lb, causing the elevator to be overloaded. Based on the value obtained, is there reason for concern?

5-4   Exercises B: Beyond the Basics

21. Repeat Exercise 20, assuming that the population size is $N = 500$ and all sampling is done without replacement. (*Hint:* See the discussion of the finite population correction factor.)

22. A population consists of these scores:

$$2 \quad 3 \quad 6 \quad 8 \quad 11 \quad 18$$

a. Find μ and σ.

b. List all samples of size $n = 2$ that are obtained without replacement.

c. Find the population of all values of $\bar{x}$ by finding the mean of each sample from part b.

d. Find the mean $\mu_{\bar{x}}$ and standard deviation $\sigma_{\bar{x}}$ for the population of sample means found in part c.

e. Verify that

$$\mu_{\bar{x}} = \mu \quad \text{and} \quad \sigma_{\bar{x}} = \frac{\sigma}{\sqrt{n}}\sqrt{\frac{N-n}{N-1}}$$

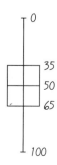

23. An education researcher develops an index of academic interest and obtains scores for a randomly selected sample of 350 college students. The results are summarized in the boxplot. If 15 of the students are randomly selected, find the probability that their mean is greater than 55.

24. Refer to Data Set 1 in Appendix B for the weights of plastic discarded by households. Using the sample mean and standard deviation as estimates of μ and σ, find the probability of randomly selecting 100 households and getting a mean weight greater than 1.75 lb.

5-5 Normal Distribution as Approximation to Binomial Distribution

Recall from Section 4-3 that a binomial probability distribution applies to a *discrete* random variable (rather than to a continuous random variable, as does the normal distribution). Recall also that outcomes belong to one of two categories, and that there is a fixed number n of independent trials, each having a probability p of success and a probability q of failure. A typical binomial probability problem involves finding $P(x)$, the probability of x successes among n trials. In Section 4-3 you learned (we hope) how to solve binomial problems by using Table A-1 or the binomial probability formula. However, it is impractical to solve many important binomial problems by such methods.

As an example, let's assume that the Telektronic Company has large and equal numbers of male and female job applicants who are all qualified. Telektronic claims that it hires employees without regard to gender, but records for the last year show that among the 100 newly hired employees, 64 are men. Does Telektronic really hire without regard to gender? To address that question, we need to find the probability of getting *at least* 64 men (among 100 new employees), assuming that men and women have equal chances of being hired. Because there are 100 trials, we have $n = 100$. Assuming that men and women have equal chances, the probability of hiring a man is $p = 0.5$ and the probability of hiring a woman is $q = 0.5$. The outcome of "at least 64 men" corresponds to $x = 64, 65, \ldots, 100$. (The probability of *exactly* 64 men doesn't really tell us anything, because with 100 trials, the probability of any exact number of men is fairly small. Instead, we need the probability of getting a result *at least* as extreme as the one obtained.)

Here's a condensed form of our binomial probability problem: Find $P(x \geq 64)$, given $n = 100$, $p = 0.5$, and $q = 0.5$. In Section 4-3 we used Table A-1 and the binomial probability formula, but those tools are not practical here. Table A-1 stops at $n = 15$ and therefore doesn't apply to this case. The binomial probability formula does apply, but it would be necessary to use it once for each value of x from 64 to 100, beginning with

$$P(64) = \frac{100!}{(100 - 64)!\ 64!} \cdot 0.5^{64} \cdot 0.5^{100 - 64}$$

The resulting 37 probabilities could then be added to produce the correct result. These calculations are of the type mathematicians often refer to as "tedious," meaning that they will seriously test your sanity. Fortunately, this section introduces a simple and practical alternative: Under certain circumstances, we can approximate the binomial probability distribution by the normal distribution. The following box summarizes the key point of this section.

Normal distribution as approximation to binomial distribution

If $np \geq 5$ and $nq \geq 5$, then the binomial random variable is approximately normally distributed with the mean and standard deviation given as

$$\mu = np$$
$$\sigma = \sqrt{npq}$$

Review Figure 4-5 (which applies to a binomial distribution with $n = 100$, $p = 0.5$, and $q = 0.5$) on page 216, and note that this particular binomial distribution does have a probability histogram that has roughly the same shape as that of a normal distribution. The justification that allows us to use the normal distribution as an approximation to the binomial distribution results from more advanced mathematics. Unfortunately (or fortunately, depending on your perspective), it is not practical to outline here the details of the formal proof of the above result. For now, try to accept the intuitive evidence of the strong resemblance between the binomial and normal distributions and take the more rigorous evidence on faith. For binomial experiments, past experience has shown that as long as $np \geq 5$ and $nq \geq 5$, the binomial distribution can be approximated reasonably well by the normal distribution.

We will now use the normal approximation approach to solve our Telektronic hiring problem. (For that binomial problem we have already ruled out the use of Table A-1 or the binomial probability formula.) We first verify that $np \geq 5$ and $nq \geq 5$.

$$np = 100 \cdot 0.5 = 50 \qquad \text{(Therefore } np \geq 5.)$$
$$nq = 100 \cdot 0.5 = 50 \qquad \text{(Therefore } nq \geq 5.)$$

As a result of satisfying these two requirements, we now know that it is reasonable to approximate the binomial distribution by a normal distribution, and we therefore proceed to find the values for μ and σ that are needed. We get the following:

$$\mu = np = 100 \cdot 0.5 = 50$$
$$\sigma = \sqrt{npq} = \sqrt{(100)(0.5)(0.5)} = 5$$

We want the probability of getting *at least* 64 men (among 100 new employees), so we include the probability of getting *exactly* 64. But the discrete value of 64 is approximated in the continuous normal distribution by the interval from

63.5 to 64.5. Such conversions from a discrete to a continuous distribution are called continuity corrections.

DEFINITION

> When we use the (continuous) normal distribution as an approximation to the (discrete) binomial distribution, a **continuity correction** is made to a discrete whole number x in the binomial distribution by representing the single value x by the *interval* from $x - 0.5$ to $x + 0.5$ (that is, add and subtract 0.5).

Because many people have some difficulty with application of the continuity correction, here are some suggestions.

1. When using the normal distribution as an approximation to the binomial distribution, *always* use the continuity correction.

2. In using the continuity correction, first identify the discrete whole number x that is relevant to the binomial probability problem. For example, if you're trying to find the probability of getting at least 64 men in 100 randomly selected people, the discrete whole number of concern is $x = 64$.

3. Draw a normal distribution centered about μ, then draw a *vertical strip area* centered over x. Mark the left side of the strip with the number $x - 0.5$, and mark the right side with $x + 0.5$. For example, if $x = 64$, draw a strip from 63.5 to 64.5. *Consider the entire area of the strip to represent the discrete number* x.

4. Now determine whether you want the probability of at least x, at most x, more than x, fewer than x, or exactly x. Also determine whether x itself should be included. Shade the area to the right or left of the strip, as appropriate; also shade the interior of the strip itself if and only if x is to be included. This total shaded region corresponds to the probability being sought.

In a binomial distribution, the following statements correspond to the indicated areas (see Figure 5-27).

Statement	Area
At least 64 (includes 64 and above)	To the *right* of 63.5
More than 64 (doesn't include 64)	To the *right* of 64.5
At most 64 (includes 64 and below)	To the *left* of 64.5
Fewer than 64 (doesn't include 64)	To the *left* of 63.5
Exactly 64	Between 63.5 and 64.5

Figure 5-28 illustrates how the discrete value of 64 is corrected for continuity when represented in the continuous normal distribution. In the Telektronic hiring problem we are considering, "at least 64" means that we include the entire interval (from 63.5 to 64.5) representing 64, along with

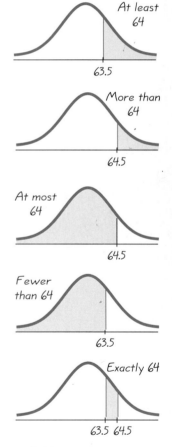

**FIGURE 5-27
Identifying the
Correct Area**

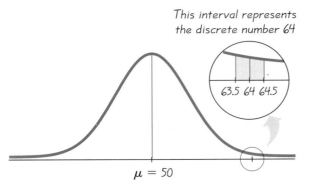

FIGURE 5-28   **Illustration of Continuity Correction**

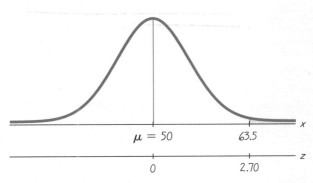

FIGURE 5-29   **Finding the Probability of "At Least" 64 Men Among 100 New Employees**

everything to the right of that interval; therefore 63.5 becomes the actual boundary that we use.

Although the continuity correction should always be used with a normal distribution to approximate a binomial distribution, if we ignore or forget the continuity correction, the additional error will be relatively small as long as n is large.

Now let's continue with our problem of finding the probability of getting at least 64 men among 100 newly hired employees. Figure 5-29 illustrates our problem and includes the continuity correction. We need to find the shaded area. We must first find the area bounded by 50 and 63.5; we get

$$z = \frac{63.5 - 50}{5} = 2.70$$

Using Table A-2, we find that $z = 2.70$ corresponds to an area of 0.4965, so the shaded region has an area of $0.5 - 0.4965 = 0.0035$. Assuming that men and women have equal chances of being hired, the probability of getting at least 64 men in 100 new employees is approximately 0.0035. (Using the software packages of STATDISK and Minitab, we get the answer of 0.0033, so the approximation is quite good here.) If we had neglected to correct for continuity by using 64 instead of 63.5, our answer would have been 0.0026. The answer of 0.0035 is a better result.

Because the probability of getting at least 64 men is so small, we conclude that either a very rare event has occurred or the assumption that men and women have the same chance is incorrect. It appears that the Telektronic Company discriminates based on gender. The reasoning here will be considered in more detail when we discuss formal methods of testing hypotheses in Chapter 7. For now, we should focus on the method of finding the probability by using the normal approximation technique.

Figure 5-30 summarizes the procedure for using the normal distribution as an approximation to the binomial distribution. The following example follows the procedure outlined in the figure.

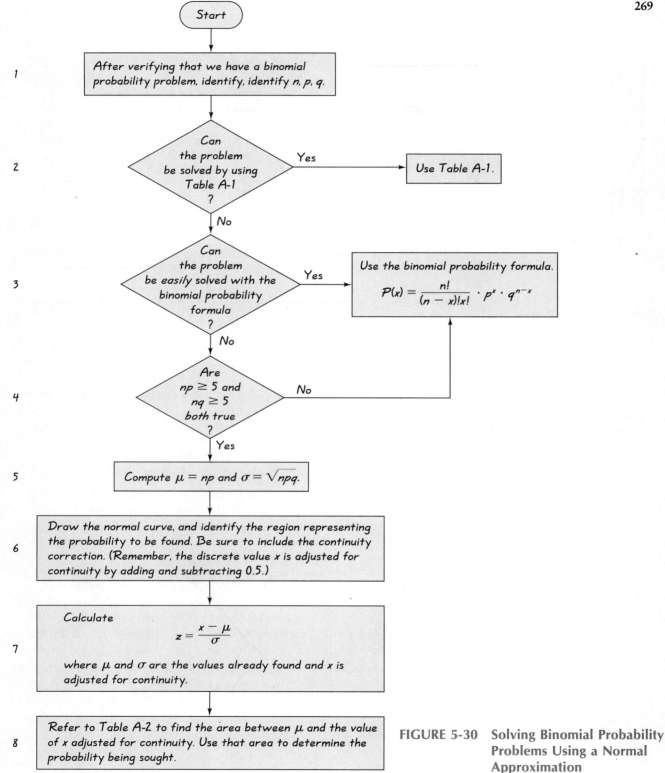

FIGURE 5-30   Solving Binomial Probability Problems Using a Normal Approximation

Survey Medium Can Affect Results

In a survey of Catholics in Boston, the subjects were asked if contraceptives should be made available to unmarried women. In personal interviews, 44% of the respondents said yes. But among a similar group contacted by mail or telephone, 75% of the respondents answered yes to the same question.

▶ EXAMPLE

There's an 80% chance that a prospective employer will check the educational background of a job applicant (based on data from the Bureau of National Affairs, Inc.). For 100 randomly selected job applicants, find the probability that *exactly* 85 have their educational backgrounds checked.

● SOLUTION

Refer to Figure 5-30.

Step 1:   The conditions described satisfy the criteria for the binomial distribution with $n = 100$, $p = 0.80$, $q = 0.20$, and $x = 85$.

Step 2:   Table A-1 cannot be used because n is too large. Table A-1 includes values of n up to 15, but in this problem the value of n is 100.

Step 3:   The binomial probability formula applies, but

$$P(85) = \frac{100!}{(100 - 85)! \; 85!} \cdot 0.80^{85} \cdot 0.20^{100 - 85}$$

is difficult to compute because many calculators cannot evaluate anything above 70!.

Step 4:   In this problem,

$$np = 100 \cdot 0.80 = 80 \qquad \text{(Therefore } np \geq 5.)$$
$$nq = 100 \cdot 0.20 = 20 \qquad \text{(Therefore } nq \geq 5.)$$

Because np and nq are both at least 5, we conclude that the normal approximation to the binomial distribution is satisfactory.

Step 5:   We obtain the values of μ and σ as follows:

$$\mu = np = 100 \cdot 0.80 = 80.0$$
$$\sigma = \sqrt{npq} = \sqrt{(100)(0.80)(0.20)} = 4.0$$

Step 6:   We draw the normal curve shown in Figure 5-31. The shaded region of that figure represents the probability we want. Use of the continuity correction results in the representation of 85 by the region extending from 84.5 to 85.5.

Step 7:   The format of Table A-2 requires that we first find the probability corresponding to the region bounded on the left by the vertical line through the mean of 80.0 and on the right by the vertical line through 85.5; therefore one of the calculations required in this step is as follows:

$$z_2 = \frac{85.5 - 80}{4} = 1.38$$

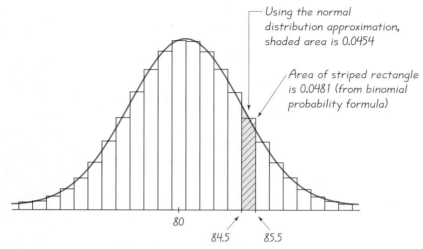

Using the normal distribution approximation, shaded area is 0.0454

Area of striped rectangle is 0.0481 (from binomial probability formula)

FIGURE 5-31   Distribution of Number of Job Applicants Whose Educational Backgrounds Are Checked

We also need the probability corresponding to the region bounded by 80.0 and 84.5, so we calculate

$$z_1 = \frac{84.5 - 80}{4} = 1.13$$

Step 8: We use Table A-2 to find that $z_2 = 1.38$ corresponds to a probability of 0.4162 and $z_1 = 1.13$ corresponds to a probability of 0.3708. Consequently, the entire shaded region of Figure 5-31 depicts a probability of $0.4162 - 0.3708 = 0.0454$. The probability of getting exactly 85 job applicants (out of 100) who have their educational backgrounds checked is approximately 0.0454.

Note on rounding: In the preceding example we found that $\mu = 80.0$ and $\sigma = 4.0$, but the results will not always be such round numbers. Be sure to carry at least six significant digits for each of μ and σ so that the resulting z scores will be accurate when rounded to two decimal places.

If we solve the preceding example using STATDISK or a calculator, we get a result of 0.0481 (rounded), but the normal approximation method resulted in a value of 0.0454. (See the following STATDISK display.) The discrepancy is very small; it occurs because we are finding the area of the shaded region in Figure 5-31, whereas the actual area is a rectangle centered above 85. (Figure 5-31 illustrates this discrepancy.) The area of the rectangle is 0.0481, but the area of the approximating shaded region is 0.0454.

Disease Clusters

Periodically, much media attention is given to a cluster of disease cases in a given community. One New Jersey community had 13 leukemia cases in 5 years, while the normal rate would have been 1 case in 10 years. Research of such clusters can be revealing. A study of a cluster of cancer cases near African asbestos mines led to the discovery that asbestos fibers can be carcinogenic.

When doing such an analysis, we should avoid the mistake of artificially creating a cluster by locating its outer boundary so that it just barely includes cases of disease or death. That would be like gerrymandering, even though it might be unintentional.

Drive to Survive Past Age 35

Avoid cars—motor vehicle crashes are the leading cause of death among Americans under 35 years of age. If you drive, don't drink—40% of fatally injured drivers are drunk. Drive in large cars—the death rate in the largest cars (1.3 per 10,000 vehicles) is less than half the rate in the smallest cars (3.0 per 10,000 vehicles). Wear seat belts—in a study of 1126 accidents, riders wearing seat belts had 86% fewer life-threatening injuries.

 Using Computers for Binomial Probabilities

A computer printout from the statistics package STATDISK follows. Our program in that package computes binomial probabilities, and the sample display shows one page of the output that results when we request the binomial probabilities corresponding to $n = 100$ and $p = 0.80$. In addition to the probability corresponding to each value of x, the display includes *cumulative probabilities*. For example, the probability that x is less than or equal to 85 is listed as 0.91956. For the problems of this section, STATDISK or Minitab is generally a good alternative to the normal approximation method, but every program will have some limitations that do not apply to the approximation method.

STATDISK DISPLAY OF BINOMIAL PROBABILITIES

```
About     File     Edit     Statistics     Data     View
                       Binomial Probabilities
    Mean = 80              St. Dev. = 4              Variance = 16

          X    P(X)   Cum Prob        X    P(X)   Cum Prob
          78   .08490  .34597         90   .00336  .99767
          79   .09457  .44054         91   .00148  .99914
          80   .09930  .53984         92   .00058  .99972
          81   .09807  .63791         93   .00020  .99992
          82   .09090  .72881         94   .00006  .99998
          83   .07885  .80766         95   .00001 1.00000
          84   .06383  .87149         96   .00000 1.00000
          85   .04806  .91956         97   .00000 1.00000
          86   .03353  .95309         98   .00000 1.00000
          87   .02158  .97467         99   .00000 1.00000
          88   .01275  .98743        100   .00000 1.00000
          89   .00688  .99430

                   Use Menu Bar to change screen        Page 3 of 3
    F1: Help       F2: Menu      PgUp/PgDn: More Data    ESC: Quit
```

5-5   Exercises A: Basic Skills and Concepts

In Exercises 1–4, do the following. (a) Find the indicated binomial probability by using Table A-1 in Appendix A. (b) If $np \geq 5$ and $nq \geq 5$, also find the indicated probability by using the normal distribution as an approximation to the binomial distribution; if $np < 5$ or $nq < 5$, then state that the normal approximation is not suitable.

1. With $n = 12$ and $p = 0.50$, find $P(8)$.

2. With $n = 15$ and $p = 0.40$, find $P(7)$.

3. With $n = 10$ and $p = 0.80$, find P(at least 8).

4. With $n = 14$ and $p = 0.60$, find P(fewer than 9).

5. Find the probability of getting at least 60 girls in 100 births.

6. Find the probability of getting exactly 40 boys in 80 births.

7. Find the probability of passing a true/false test of 50 questions if 60% (or 30 correct responses) represents a passing grade and all responses are random guesses.

8. A multiple-choice test consists of 40 questions with possible answers of a, b, c, d, and e. Find the probability of getting at most 30% correct if all answers are random guesses.

9. Susan Stein is the advertising sales director for a TV prime-time police drama, and she wants to analyze viewer use of videocassette recorders. She plans to persuade advertisers that the viewing audience is actually much larger than the number of people who watch the show at the time of broadcast. A Nielsen survey showed that when viewers use videocassette recorders, 66% of the shows taped are from the major networks. Find the probability that among 1000 randomly selected videotaped shows, at least 700 are from major networks, as the sales director claims.

10. The Telstar Appliance Company has developed a rotating tray designed for use in microwave ovens. In a trial sales program, expensive advertising packets are sent to 150 randomly selected households. A Roper poll showed that 75% of households have microwave ovens. Find the probability that fewer than 100 of the households receiving the packets have microwave ovens.

11. *Entertainment Report* magazine runs a sweepstakes as part of a campaign to acquire new subscribers. In the past, 26% of those who received sweepstakes entry materials have ended up entering the contest and subscribing to the magazine (based on data reported in *USA Today*). Find the probability that when sweepstakes entry materials are sent to 500 randomly selected households, the resulting number of new subscriptions is between 125 and 150 inclusive.

12. Among flight arrivals at Logan International Airport in Boston, 80% are on time (based on data from the U.S. Department of Transportation). The Boston Limousine Service plans to send vans to meet 220 flights over the next few months. Find the probability that the number of on-time arrivals is between 180 and 185 inclusive.

13. Among U.S. households, 24% have telephone answering machines (based on data from the U.S. Consumer Electronics Industry). If a telemarketing campaign involves 2500 households, find the probability that more than 650 have answering machines.

14. Based on U.S. Bureau of the Census data, 12% of the men in the United States have earned bachelor's degrees. If 140 U.S. men are randomly selected, find the probability that at least 20 of them have a bachelor's degree.

15. Among teenagers old enough to drive, 35% have their own cars (based on data from a Rand Youth Poll). If a marketing research team randomly selects 600 teenagers of driving age, find the probability that at least 210 of them have their own cars.

16. According to Bureau of the Census data, 60% of men aged 18 to 24 live at home with their parents. If 500 men aged 18 to 24 are randomly selected, find the probability that more than 325 of them live at home with their parents.

17. Among women aged 18 to 24, 75% are more than 159 cm tall (based on data from the National Health Survey, USDHEW publication 79-1659). If 320 women aged 18 to 24 are randomly selected, find the probability that more than 250 of them are taller than 159 cm.

18. Among workers aged 20 to 24, 26% work more than 40 hours per week (based on data from the U.S. Department of Labor). If we randomly select 350 workers aged 20 to 24, find the probability that the number who work more than 40 hours per week is between 80 and 90 inclusive.

19. According to a consumer affairs representative from Mars (the candy company, not the planet), 20% of all M&Ms produced are red. Data Set 6 in Appendix B shows that among 100 M&Ms chosen, 17 are red. Find the probability of randomly selecting 100 M&Ms and getting 17 or fewer that are red. Assume that the company's 20% red rate is correct.

20. Bob Taylor plans to place 200 bets of $1 each on the number 7 at roulette. On any one spin there is a probability of 1/38 that 7 will be the winning number. For Bob to end up with a profit, the number 7 must occur at least 6 times among the 200 trials. Find the probability that Bob finishes with a profit.

21. Based on U.S. Bureau of Justice data, 16% of those arrested are women. In one state, 400 arrest cases are randomly selected. Find the probability that the number of women is at least 100. If those arrest cases include at least 100 women, does it seem plausible that this state arrests women at the 16% rate?

22. Air America has been experiencing a 7% rate of no-shows on advance reservations. In a test project that requires passengers to confirm reservations, it is found that among 250 randomly selected advance reservations, there are 4 no-shows. Assuming that the confirmation requirement has no effect so the 7% rate applies, find the probability of 4 or fewer no-shows among 250 randomly selected reservations. Based on that result, does it appear that the confirmation requirement is effective?

23. A certain genetic characteristic is believed to appear in one-quarter of all offspring, yet in a randomly selected sample of 40 offspring, only 8 exhibit the characteristic. Find the probability that of 40 randomly selected offspring, 8 or fewer exhibit the characteristic in question. Based on that probability, does it seem that the one-quarter rate is correct?

24. The IRS finds that of all taxpayers whose returns are audited, 70% end up paying additional taxes. A review of 500 randomly selected returns audited by IRS agent Tom Clark shows that 340 resulted in payments of additional taxes. Find the probability that of 500 randomly selected returns, 340 or fewer result in additional tax payments. Based on that probability, does it appear that Tom's results are notably lower than the 70% rate for the IRS as a whole, or are Tom's results within the realm of what can be expected with a 70% rate?

25. According to the American Medical Association, 18.4% of college graduates smoke. A health study begins with the selection of 300 college graduates, but the number of smokers is 72, which is more than expected. The study is being questioned because the number of smokers does not seem to correspond to the overall rate of 18.4% for the population of college graduates. Find the probability of getting at least 72 smokers in a *random* sample of 300. Based on the result, does it appear that one could easily get 72 smokers by chance, or does it appear that there's something wrong with the sample?

26. The Life Trust Insurance Company, which underwrites medical malpractice insurance, is reviewing demographic data from its list of insured doctors. Among doctors in the United States, 25% percent are younger than 35 years of age (data from Health Care Market Research). Find the probability that among 40 randomly selected doctors, fewer than 5 are under 35 years of age.

27. Providence Memorial Hospital is conducting a blood drive because its supply of group O blood is low, and it needs 177 donors of group O blood. If 400 volunteers donate blood, find the probability that the number with group O blood is at least 177. Forty-five percent of us have group O blood, according to data provided by the Greater New York Blood Program.

28. We noted in Section 3-4 that some companies monitor quality by using a method of *acceptance sampling* whereby an entire batch of items is rejected if, in a random sample, the number of defects is at least some predetermined number. The Dayton Machine Company buys machine bolts in batches of 5000 and rejects a batch if, when 50 of them are sampled, at least 2 defects are found. Find the probability of rejecting a batch if the supplier is manufacturing the bolts with a defect rate of 10%.

5-5   **Exercises B: Beyond the Basics**

29. In a binomial experiment with $n = 15$ and $p = 0.4$, find P(at least 5) using the following, and then identify the result that is most accurate.
   a.   The table of binomial probabilities (Table A-1)
   b.   The binomial probability formula
   c.   The normal distribution approximation

30. Assume that a baseball player hits .350, so his probability of a hit is 0.350. Also assume that his hitting attempts are independent of each other.
   a.   Find the probability of at least 1 hit in 4 tries in 1 game.
   b.   Assuming that this batter gets up 4 times each game, find the probability of getting a total of at least 56 hits in 56 games.
   c.   Assuming that this batter gets up 4 times each game, find the probability of at least 1 hit in each of 56 consecutive games (Joe DiMaggio's 1941 record).
   d.   What minimum batting average would be required for the probability in part c to be greater than 0.1?

31. Air America works only with advance reservations and experiences a 7% rate of no-shows. How many reservations could be accepted for an airliner with a capacity of 250 if there is at least a 0.95 probability that all reservation holders who show will be accommodated?

32. The Telektronic Company manufactures integrated circuit chips with a 23% rate of defects. What is the minimum number of chips that must be manufactured if there must be at least a 90% chance that 5000 good chips can be supplied to General Motors?

33. Heights of women are normally distributed with a mean of 63.6 in. and a standard deviation of 2.5 in. Find the probability that when 200 women are randomly selected, at least 40 of them are taller than 66 in.

VOCABULARY LIST

Define and give an example of each term.

normal distribution   central limit theorem
uniform distribution   standard error of the mean
standard normal distribution   continuity correction
sampling distribution of sample
   means

REVIEW

The main topic of this chapter was the concept of a normal distribution, the most important of all continuous probability distributions. Many natural occurrences yield data that are normally distributed or can be approximated by a normal distribution. This particular distribution will be used extensively in the following chapters. The normal distribution, which appears bell shaped when graphed, can be described algebraically by an equation, but the complexity of that equation usually forces us to use a table of values (Table A-2) instead.

Table A-2 gives areas corresponding to specific regions under the standard normal distribution curve, which has a mean of 0 and a standard deviation of 1. This table relates deviations away from the mean with areas under the curve. Since the total area under the curve is 1, those areas correspond to probability values.

In this chapter we worked with the standard procedures used in applying Table A-2 to a variety of different situations. We saw that Table A-2 can be applied indirectly to normal distributions that are nonstandard. (That is, μ is not 0, σ is not

1, or both.) We were able to find the number of standard deviations that a score x is away from the mean μ by computing $z = (x - \mu)/\sigma$.

In Section 5-3 we considered practical examples as we converted from a nonstandard to a standard normal distribution. In Section 5-4 we looked at the distribution of sample means that come from normal or nonnormal populations. The central limit theorem asserts that the distribution of sample means $\bar{x}$ (based on random samples of size n drawn from the same population) will, as n increases, tend to approach a normal distribution with mean μ and standard deviation $\sigma/\sqrt{n}$. If the original population has a normal distribution or if samples are of size n where $n >$ 30, we can approximate the distribution of those sample means by a normal distribution. The standard error of the mean (or standard deviation of sample means) is $\sigma/\sqrt{n}$.

In Section 5-5 we noted that we can sometimes approximate a binomial probability distribution by a normal distribution. If both $np \geq 5$ and $nq \geq 5$, the binomial random variable x is approximately normally distributed with the mean and standard deviation given as $\mu = np$ and $\sigma = \sqrt{npq}$. Because the binomial probability distribution deals with discrete data and the normal distribution deals with continuous data, we applied the continuity correction, which should be used in normal approximations to binomial distributions.

As basic concepts of this chapter serve as critical prerequisites for the following chapters, it would be wise to master these ideas and methods now.

Important Formulas

Standard normal distribution has $\mu = 0$ and $\sigma = 1$.

$z = \dfrac{x - \mu}{\sigma}$	Standard score, or z score
$\mu_{\bar{x}} = \mu$	Population mean of sample means
$\sigma_{\bar{x}} = \dfrac{\sigma}{\sqrt{n}}$	Population standard deviation of sample means
$z = \dfrac{\bar{x} - \mu_{\bar{x}}}{\sigma_{\bar{x}}}$	z score used when applying the central limit theorem
$np \geq 5, \quad nq \geq 5$	Prerequisites for approximating binomial by normal
$\mu = np, \quad \sigma = \sqrt{npq}$	Parameters used in approximating binomial by normal

---

REVIEW EXERCISES

1. When an individual takes several versions of the same IQ test, his or her scores tend to be normally distributed with a standard deviation of 3 (based on data from *Probabilities in Everyday Life,* by McGervey). When Mike Taylor takes IQ tests, his scores are normally distributed with a mean of 106 and a standard deviation of 3. If Mike Taylor takes one IQ test, find the probability that his score is
   a. less than 100
   b. more than 110
   c. less than 112
   d. between 100 and 109
   e. between 110 and 115
2. Refer to the scores described in Exercise 1 and find
   a. P_{90}, the 90th percentile   b. P_{20}, the 20th percentile
3. The Gleason Supermarket uses a scale to weigh produce, and errors are normally distributed with a mean of 0 oz and a standard deviation of 1 oz. (The errors can be positive or negative.) One item is randomly selected and weighed. Find the probability that the error is
   a. between 0 and 1.25 oz
   b. greater than 0.50 oz
   c. greater than −1.08 oz
   d. between −0.50 oz and 1.50 oz
   e. between −1.00 oz and −0.25 oz
4. A study conducted by the International Council of Shopping Centers revealed that 4% of those who visit a mall or shopping center spend more than 3 hours there. If 850 people are randomly selected from those who visit malls or shopping centers, find the probability that at least 50 of them spend more than 3 hours there.
5. The Atlanta Video Store plans to target wealthy families in its membership campaign. Household incomes of VCR owners have a mean of $41,182 and a standard deviation of $19,990 (based on data from Nielsen Media Research). If 125 households with VCRs are randomly selected, find the probability that the mean income is above $40,000. Is it necessary to make any assumptions about the distribution of incomes? Why or why not?
6. The New England Insurance Company finds that the ages of motorcyclists killed in crashes are normally distributed with a mean of 26.9 years and a standard deviation of 8.4 years (based on data from the U.S. Department of Transportation).
   a. If we randomly select 1 such motorcyclist, find the probability that he or she is younger than 25 years of age.
   b. If we randomly select 40 such motorcyclists, find the probability that their mean age is less than 25 years.
   c. Find the age corresponding to P_{40}, the 40th percentile.

7. Among Americans aged 18 and older, 8% are now divorced (based on U.S. Bureau of the Census data). If 225 Americans aged 18 or older are randomly selected, find the probability that at least 20 of them are now divorced.

8. The weights (in pounds) of metal discarded in one week by households are normally distributed with a mean of 2.22 lb and a standard deviation of 1.09 lb (based on data from the Garbage Project at the University of Arizona).
   a. If 1 household is randomly selected, find the probability that it discards more than 2.00 lb of metal in a week.
   b. If 25 households are randomly selected, find the probability that the mean weight of metal discarded in a week is more than 2.50 lb.
   c. Find P_{30}, the weight separating the bottom 30% of all weights from the top 70%.

9. According to data from the American Medical Association, 10% of us are left-handed. In a freshman class of 200 students, find the probability that
   a. exactly 18 are left-handed
   b. fewer than 25 are left-handed

10. The mean IQ score of engineers is estimated to be 120. Among adults, IQ scores are normally distributed with a mean of 100 and a standard deviation of 15. If an adult is randomly selected, find the probability that his or her IQ score is above the mean IQ score of engineers.

 COMPUTER PROJECT

In this computer project we illustrate the important central limit theorem by developing computer simulations similar to those described in Section 5-4.

a. First simulate rolling 1 die 1000 times, and then find the mean, standard deviation, and shape of the histogram for the results.

b. Now simulate rolling 2 dice 1000 times, and find the mean, standard deviation, and shape of the histogram for the totals of the 2 dice.

c. Third, simulate rolling 3 dice 1000 times, and find the mean, standard deviation, and shape of the histogram for the totals of the 3 dice.

If you are using STATDISK or Minitab, parts a, b, and c can be completed as follows:

STATDISK: From the main menu item of `File`, select `New Generated Sample`, then select the option `Roll`, and choose 1, 2, or 3 dice.

Minitab: Enter the following commands, which simulate rolling 3 dice 1000 times each. Column C1 will store the results of the single die, column C2 will store the totals for 2 dice, and column C3 will store the totals for 3 dice. The means, standard deviations, and histograms will be displayed.

```
MTB > RANDOM 1000 C1 C2 C3;
SUBC> INTEGER 1 6.
MTB > LET C3 = C1 + C2 + C3
MTB > LET C2 = C1 + C2
MTB > DESCRIBE C1 C2 C3
MTB > HISTOGRAM C1
MTB > HISTOGRAM C2
MTB > HISTOGRAM C3
```

d. After obtaining the computer-generated results described in parts a, b, and c, identify and compare the shapes of the three distributions.

e. For rolling 2 dice, we generated *totals* instead of means. When 2 dice are rolled, what simple arithmetic operation is required to convert the 1000 sample *totals* to 1000 sample *means?* How does that operation affect the values of the mean and standard deviation from part b? Find the values of the mean and standard deviation for the 1000 sample means obtained when 2 dice are rolled.

f. For the rolling of three dice, we generated *totals* instead of means. When 3 dice are rolled, what simple arithmetic operation is required to convert the 1000 sample *totals* to 1000 sample *means?* How does that operation affect the values of the mean and standard deviation from part b? Find the values of the mean and standard deviation for the 1000 sample means obtained when 3 dice are rolled.

g. Use the results from parts a, e, and f to complete the following table, and then explain how the table illustrates the central limit theorem.

	Number of dice		
	1	2	3
Mean			
Standard deviation			

FROM DATA TO DECISION:

How can we outsmart users of counterfeit coins in vending machines?

Operators of vending machines are continually devising strategies to combat use of counterfeit coins and bills. One machine collected dollar bills in a tray and supplied change for use in other vending machines. A rod would pass through the center of the receiving tray and pull the dollar bill into a holding box. A creative but dishonest person found that if he used a dollar bill with a slit through the center, the machine would provide the change but the bill would not be removed. He could empty the machine of all its change, then move on to another machine.

The use of slugs also plagues vending machine operators. Let's develop a strategy for minimizing losses from slugs that are counterfeit quarters. In Data Set 8 of Appendix B, we have a random sample of weights (in grams) of legal and legitimate quarters. Based on that sample, let's assume that for the population of all such quarters, the distribution is normal, the mean weight is 5.622 g, and the standard deviation is 0.068 g. Let's also assume that a supply of counterfeit quarters has been collected from vending machines in downtown Dallas. Analysis of those counterfeit quarters shows that they have weights that are normally distributed with a mean of 5.45 g

and a standard deviation of 0.112 g. A vending machine is designed so that coins are accepted or rejected according to weight. The limits for acceptable coins can be changed, but the machine currently accepts any coin with a weight between 5.500 g and 5.744 g. Find the percentages of legitimate coins and counterfeit coins accepted by the machine.

Here's a dilemma: If we restrict the weight limits on the coins too much, we reduce the proportion of legitimate coins that are accepted and customers are lost because their legitimate coins won't work. If we don't restrict the weight limits on coins enough, too many counterfeit coins are accepted and losses are incurred. What are the percentages of legitimate and counterfeit coins accepted if the lower weight limit is changed to 5.550 g? Experiment with different weight settings and identify limits that seem reasonable in that the machine does not accept too many counterfeit coins but still does accept a reasonable proportion of legitimate coins. (There isn't necessarily a unique answer here; to some extent the choices depend on subjective judgments.) Write a brief report that includes the settings you recommend, along with specific reasons for your choices.

Interview

Barbara Carvalho
Director of the Marist College Poll

Lee Miringoff
Director of the Marist College Institute for Public Opinion

Barbara Carvalho and Lee Miringoff report on their poll results in many interviews for print and electronic media, including news programs for NBC, CBS, ABC, FOX, and public television. Lee Miringoff appears regularly on NBC's Today *show.*

What types of polls do you conduct?

We do public polling. We survey public issues, approval ratings of public officials in New York City, New York State, and nationwide. We don't do partisan polling for political parties, political candidates, or lobby groups. We are independently funded by Marist College, and we have no outside funding that in any way might suggest that we are doing research for any particular group on any one issue. Our program is really an educational program, but it has wide recognition because the results are released publicly. Reporters have come to depend on our results not only for their accuracy and professionalism, but also because they know that the poll is independent and is not commissioned by any one media source, as many polls are.

Who does your interviewing, and what are their backgrounds?

All of our interviewing is done by paid students who are trained in interviewing and the specific study they are working on. Students can take course work in survey research, public opinion, and data collection. Students in political science are a natural, but we also get many communication arts majors, as well as students interested in statistics, computer analysis, psychology, economics, sociology, business, and marketing.

Would you recommend statistics for students in fields like history, government, or social science?

Absolutely. They should have at least a research course that deals with the basics of statistical analysis and gives them a foundation by which they can deal with the numbers that they're going to be seeing—no matter what field they go into. The study of statistics is important for understanding one aspect of knowledge, and it's a key to opening up other avenues of pursuit. Statistics cuts across disciplines. Students will inevitably find it in their careers at some point. It might be an evaluation of their work or workplace, or it might involve some marketing aspect or promotional aspect. Surveys are now so pervasive throughout our culture that students are going to see them in their later lives, whether it's in their careers or just as citizens. People are now bombarded with survey information, so it's absolutely vital that as citizens they are in a position to evaluate survey accuracy and worth.

What concepts of statistics do you use?

Statistics comes into play in our sampling, even before we get to data analysis. We use statistics to determine our sample size and to develop an estimate of what would be statistically significant. In the data analysis most of our studies use basic descriptive statistics. Some of the academic studies get into regression analysis.

How do you select survey respondents?

For a statewide survey we select respondents in proportion to county voter registrations. Different counties have different refusal rates, and if we were to select people at random throughout the state, we would get an uneven model of what the state looks like. We stratify by county and use random digit dialing so that we get listed and unlisted numbers.

What's your typical sample size?

About four people, but they're very carefully selected. Seriously, it can be anywhere from 400 to 1200 or 1500. If we wanted to do an analysis of subgroups within our population group, we would increase our sample size so that we could look at subgroups such as men versus women, or different regional groups, or different income groups.

Is the political process actually influenced by poll results?

Although most polls that people see are public polls, the reality is that the political process is influenced by private polls that the public never sees. No one runs for high office today without using a private poll.

"The study of statistics is important for understanding one aspect of knowledge, and it's a key to opening up other avenues of pursuit."

6 Estimates and Sample Sizes

6-1 Overview

Objectives are identified, and fundamental methods for estimating values of population parameters are introduced. Methods for determining the sample sizes necessary to estimate those parameters are presented.

6-2 Estimating a Population Mean

The value of a population mean is approximated with a point estimate and confidence interval, and a method for determining the sample size is presented. The Student t distribution is introduced.

6-3 Estimating a Population Proportion

The value of a population proportion is approximated with a point estimate and confidence interval. Procedures are described for determining how large a sample must be to estimate a population proportion.

6-4 Estimating a Population Variance

The value of a population variance is approximated with a point estimate and confidence interval, and a method for determining the sample size is presented. The chi-square distribution is introduced.

Chapter Problem:
Is the mean body temperature really 98.6°F?

In Chapter 2 we used descriptive statistics to better understand the sample of 106 body temperatures (listed once again in Table 6-1). Our study of those temperatures revealed these important characteristics:

- The distribution is approximately bell shaped.
- The mean is $\bar{x} = 98.20°$ F.
- The standard deviation is $s = 0.62°$ F.
- The size of the sample is $n = 106$.

Currently, it is commonly believed that the mean body temperature is 98.6° F, but the Table 6-1 data seem to suggest that the mean is 98.20° F. We know that samples naturally vary, and we might explain the discrepancy by attributing it to typical sample variation; we could then continue to believe that 98.6° F is the correct value of the population mean. But should we? We can use the Table 6-1 sample data to develop estimates of the true population mean μ. Perhaps those estimates will indicate that 98.6° F is a possible value of μ; perhaps they will indicate that it's very unlikely that 98.6° F is correct. We shall see.

Table 6-1 Body Temperatures of 106 Healthy Adults

				99.0	98.4	98.4	98.4	98.4	98.6
98.6	98.6	98.0	98.0	97.0	98.8	97.6	97.7	98.8	98.0
98.6	98.8	98.6	97.0	98.7	97.4	98.9	98.6	99.5	97.5
98.0	98.3	98.5	97.3	98.7	99.4	98.2	98.0	98.6	98.6
97.3	97.6	98.2	99.6	98.0	97.8	98.0	98.4	98.6	98.6
97.2	98.4	98.6	98.2	98.0	96.9	97.6	97.1	97.9	98.4
97.8	99.0	96.5	97.6	98.2	98.5	98.8	98.7	97.8	98.0
97.3	98.0	97.5	97.6	98.6	98.4	98.5	98.6	98.3	98.7
97.1	97.4	99.4	98.4	98.8	98.0	98.7	98.5	98.9	98.4
98.8	99.1	98.6	97.9	98.7	97.6	98.2	99.2	97.8	98.0
98.6	97.1	97.9	98.8	98.7	97.0				
98.4	97.8	98.4	97.4	98.0					

Sample temperatures were obtained by Dr. Philip Mackowiak, Dr. Steven Wasserman, and Dr. Myron Levine, University of Maryland researchers.

6-1   Overview

In Chapter 1 we differentiated between descriptive and inferential statistics. We noted that with descriptive statistics we attempt to describe or understand known data, whereas with inferential statistics we use sample data to form inferences or conclusions about populations. In Chapter 2 we considered ways of describing population data by using graphs (such as histograms) and measures (such as means or standard deviations). In Chapter 3 we discussed the basic principles of probability. In Chapter 4 we looked at the concept of a probability distribution and proceeded to consider a variety of *discrete* probability distributions, including the important binomial probability distribution. In Chapter 5 we examined the normal distribution, an extremely important continuous probability distribution. It should be noted that with Chapter 5 we began the transition to inferential statistics that allows us to make inferences or form conclusions about populations. In this and the following chapters we begin to use methods of inferential statistics quite extensively.

As we proceed to consider methods of using sample data to form inferences about populations and processes, we should give some thought to the methods used to collect those samples in the first place. Sampling and data collection usually require more time, effort, and money than statistical analysis of the data. Careful planning can minimize the expenditure of those precious resources. In Chapter 1 we described different types of sampling procedures, including random sampling, stratified sampling, systematic sampling, cluster sampling, and convenience sampling. As we now proceed to make inferences about populations based on sample data, we should recall an extremely important point that was first made in Chapter 1: *Data collected carelessly can be absolutely worthless, even if the sample is quite large.*

One major topic of this chapter is the development of estimates of population parameters. In the problem described at the beginning of the chapter, for example, we are given the body temperatures of 106 persons and want to estimate the mean body temperature of *all* persons. As another example, Young and Rubicam Advertising runs tests of commercials by first carefully selecting a sample of consumers. The objective is to use the sample results to estimate the proportion of the entire consumer population that will react positively to a particular commercial. In this chapter we will see how such estimates are obtained for means, proportions, and variances. Methods for determining the accuracy of these estimates will be developed. We will also discuss how to determine optimum sample sizes. Section 6-2 begins with estimates of population means.

6-2   Estimating a Population Mean

Consider the 106 body temperatures given in Table 6-1 at the beginning of this chapter. Using only these sample values, we want to estimate the mean of *all* body temperatures. We could use a statistic such as the sample median, midrange, or mode as an estimate of this population mean μ, but the sample mean $\bar{x}$ usually provides the best estimate of a population mean. This is not simply an intuitive conclusion. It is based on careful study and analysis of the distributions of the different statistics that could be used as estimators.

DEFINITIONS

An **estimator** is a sample *statistic* (such as the sample mean $\bar{x}$) used to approximate a population parameter. An **estimate** is a specific *value or range of values* used to approximate some population parameter.

For example, based on the data in Table 6-1, we might use the *estimator* of $\bar{x}$ to conclude that the *estimate* of the mean body temperature of healthy adults is 98.20°F.

For many populations, the distribution of sample means $\bar{x}$ tends to be more consistent than the distributions of other sample statistics. That is, if you use the sample mean $\bar{x}$ to estimate the population mean μ, your errors are likely to be smaller than they would be with other sample statistics, such as the median or mode. For all populations, we say that the sample mean $\bar{x}$ is an *unbiased* estimator of the population mean μ, meaning that the distribution of sample means tends to center about the value of the population mean μ. For these reasons, we will use the sample mean $\bar{x}$ as the best estimate of the population mean μ. Because the sample mean $\bar{x}$ is a single value that corresponds to a point on the number scale, we call it a *point estimate*.

DEFINITION

A **point estimate** is a *single value* (or point) used to approximate a population parameter.

The sample mean $\bar{x}$ is the best point estimate of the population mean μ.

A Professional Speaks About Sampling Error

Daniel Yankelovich, in an essay for *Time*, commented on the sampling error often reported along with poll results. He stated that sampling error refers only to the inaccuracy created by using random sample data to make an inference about a population; the sampling error does not address issues of poorly stated, biased, or emotional questions. He said, "Most important of all, warning labels about sampling error say nothing about whether or not the public is conflict-ridden or has given a subject much thought. This is the most serious source of opinion poll misinterpretation."

▶ **EXAMPLE**

Use the sample data given in Table 6-1 to find the best point estimate of the population mean μ of all body temperatures.   *(continued)*

Captured Tank Serial Numbers Reveal Population Size

During World War II, Allied intelligence specialists wanted to determine the number of tanks Germany was producing. Traditional spy techniques provided unreliable results, but statisticians obtained accurate estimates by analyzing serial numbers on captured tanks. As one example, records show that Germany actually produced 271 tanks in June of 1941. The estimate based on serial numbers was 244, but traditional intelligence methods resulted in the extreme estimate of 1550. (See "An Empirical Approach to Economic Intelligence in World War II" by Ruggles and Brodie, *Journal of the American Statistical Association,* Vol. 42.)

● **SOLUTION**

The sample mean $\overline{x}$ is the best point estimate of the population mean μ, and for the data in Table 6-1 we have already found (in Chapter 2) that $\overline{x} = 98.20°F$. Based on the table's sample data, the best point estimate of the population mean μ of all body temperatures is therefore 98.20°F.

In the preceding example we see that 98.20°F is our *best* point estimate of the population mean μ, but we have no indication of just how good that estimate is. Suppose that we had only the first 4 temperatures of 98.6, 98.6, 98.0, and 98.0. Their mean of 98.3°F would be the best point estimate of the population mean μ, but we should not expect this point estimate to be very good because it is based on such a small sample. Statisticians have therefore developed another type of estimate that does reveal how good the point estimate is. This estimate, called a confidence interval or interval estimate, consists of a range (or an interval) of values instead of just a single value.

DEFINITION

A **confidence interval** (or **interval estimate**) is a range (or an interval) of values that is *likely* to contain the true value of the population parameter.

A *confidence interval* is associated with a degree of confidence, which is a measure of how certain we are that our interval contains the population parameter. The definition of degree of confidence uses the Greek letter α (alpha) to describe a probability that corresponds to an area. Refer to Figure 6-1, where the probability α is divided equally between two shaded extreme regions (often called *tails*) in the standard normal distribution. (We'll describe the role of $z_{\alpha/2}$ later; for now, simply note that α is divided equally between the two tails.)

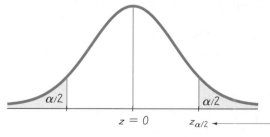

FIGURE 6-1   The Standard Normal Distribution

DEFINITION

The **degree of confidence** is the probability $1 - \alpha$ that the population parameter is contained in the confidence interval. This probability is often expressed as the equivalent percentage value. The degree of confidence is also referred to as the **level of confidence** or the **confidence coefficient.**

Common choices for the *degree of confidence* are 95% (with $\alpha = 0.05$), 99% (with $\alpha = 0.01$), and 90% (with $\alpha = 0.10$). The choice of 95% is most common because it provides a good balance between precision (as reflected in the width of the confidence interval) and reliability (as expressed by the degree of confidence).

Here's an example of a confidence interval based on the sample data of 106 body temperatures given in Table 6-1.

The 0.95 (or 95%) degree of confidence interval estimate of the population mean μ is 98.08°F $< \mu <$ 98.32°F.

Note that this estimate consists of a *range* of values and is associated with a degree of confidence. We interpret this confidence interval as follows: If we were to select many different samples of size $n = 106$ from the population of all healthy people and construct a similar confidence interval for each sample, in the long run 95% of those intervals would actually contain the value of the population mean μ.

We will now proceed to describe the process of constructing a confidence interval, such as the one just given. Recall from the central limit theorem that sample means $\overline{x}$ tend to be normally distributed, as in Figure 6-1. It's *likely* that sample means will fall within the unshaded region of such a figure, and it's *unlikely* that they will fall in the shaded area. From Figure 6-1 we can see that there is a total probability of α that a sample mean will fall in either of the two shaded tails. By the rule of complements from Chapter 3, it follows that there is a probability of $1 - \alpha$ that a sample mean will fall within the unshaded region of Figure 6-1. This is all somewhat abstract, so let's use some concrete numbers.

Let's stipulate that the degree of confidence is 95%, so $\alpha = 0.05$ (a very common choice). With $\alpha = 0.05$ in Figure 6-1, there is an area of $\alpha/2 = 0.025$ in each of the two shaded tails and an area of 0.95 in the unshaded middle region. If Figure 6-1 represents the distribution of sample means $\overline{x}$, then there is a probability of 0.95 that a sample mean will fall within the unshaded region; it's unlikely (with a probability of only 0.05) that a sample mean will fall within either of the shaded tails. The symbol $z_{\alpha/2}$ used in the figure is described as follows.

Notation for Critical Value

$z_{\alpha/2}$ is the positive z value that is at the vertical boundary for the area of $\alpha/2$ in the *right* tail of the standard normal distribution. (The value of $-z_{\alpha/2}$ is at the vertical boundary for the area of $\alpha/2$ in the *left* tail.) Because $z_{\alpha/2}$ is on the borderline separating sample means that are *likely* to occur from those that are *unlikely*, it is often called a *critical value*.

DEFINITION

A **critical value** is the number on the borderline separating sample statistics that are likely to occur from those that are unlikely to occur. The number $z_{\alpha/2}$ is a critical value with the property that the size of the area under the curve bounded by $-z_{\alpha/2}$ and $z_{\alpha/2}$ is $1 - \alpha$. [That is, $P(-z_{\alpha/2} < z < z_{\alpha/2}) = 1 - \alpha$.]

▶ **EXAMPLE**

Given a 95% degree of confidence, find the critical value $z_{\alpha/2}$.

● **SOLUTION**

A 95% degree of confidence corresponds to $\alpha = 0.05$. Refer to Figure 6-1 and stipulate that each of the shaded tails has an area of $\alpha/2 = 0.025$. We find $z_{\alpha/2} = 1.96$ by noting that the region to its left (and bounded by the mean of $z = 0$) must be $0.5 - 0.025$, or 0.475. In Table A-2 an area of 0.4750 corresponds exactly to a z value of 1.96. For a 95% degree of confidence, the critical value is therefore $z_{\alpha/2} = 1.96$.

When we collect a set of sample data, such as the set of 106 body temperatures listed in Table 6-1, we can calculate the sample mean $\bar{x}$, and that sample mean is typically different from the population mean μ. The difference between the sample mean and the population mean can be thought of as an *error*. In Section 5-4 we saw that $\sigma/\sqrt{n}$ is the standard deviation of sample means. Using $\sigma/\sqrt{n}$ and the $z_{\alpha/2}$ notation, we now define the margin of error E as follows.

DEFINITION

When sample data are used to estimate a population mean μ, the **margin of error**, denoted by E, is the maximum likely (with probability $1 - \alpha$) difference between the observed sample mean $\bar{x}$ and the true value of the population mean μ. The margin of error E is also called the **maximum error of the estimate** and can be found by multiplying the critical value and the standard deviation of sample means, as shown in Formula 6-1.

Formula 6-1
$$E = z_{\alpha/2} \cdot \frac{\sigma}{\sqrt{n}}$$

[handwritten margin note: $z_{\alpha/2}$ $95 = 1.96$ $90 = 1.645$]

Given the way that the margin of error E is defined, there is a probability of $1 - \alpha$ that a sample mean will be in error (different from the population mean μ) by no more than E, and there is a probability of α that the sample mean will be in error by more than E. The calculation of the margin of error E as given in Formula 6-1 requires that you know the population standard deviation σ, but in reality it's rare to know σ when the population mean μ is not known. The following method of calculation is common practice.

Calculating E When σ Is Unknown

If $n > 30$, we can replace σ in Formula 6-1 by the sample standard deviation s.

If $n \leq 30$, the population must have a normal distribution and we must know σ to use Formula 6-1. [An alternative method for calculating the margin of error E for small ($n \leq 30$) samples will be discussed more fully later in this section.]

Based on the definition of the margin of error E, we can now identify the confidence interval for the population mean μ.

Confidence Interval (or Interval Estimate) for the Population Mean μ

$$\bar{x} - E < \mu < \bar{x} + E$$

We will use this form for the confidence interval, but other equivalent forms are $\mu = \bar{x} \pm E$ and $(\bar{x} - E, \bar{x} + E)$.

Estimating Sugar in Oranges

In Florida, members of the citrus industry make extensive use of statistical methods. One particular application involves the way in which growers are paid for oranges used to make orange juice. An arriving truckload of oranges is first weighed at the receiving plant, then a sample of about a dozen oranges is randomly selected. The sample is weighed and then squeezed, and the amount of sugar in the juice is measured. Based on the sample results, an estimate is made of the total amount of sugar in the entire truckload. Payment for the load of oranges is based on the estimate of the amount of sugar because sweeter oranges are more valuable than those less sweet, even though the amounts of juice may be the same.

Once the sample mean $\bar{x}$ is known and the margin of error E has been found, we can calculate the values of $\bar{x} - E$ and $\bar{x} + E$, which are called **confidence interval limits.**

The basic idea underlying the construction of a confidence interval relates to the central limit theorem, which indicates that with large ($n > 30$) samples, the distribution of sample means is approximately normal with mean μ and standard deviation $\sigma/\sqrt{n}$. Let's consider the specific case of a 95% degree of confidence, so $\alpha = 0.05$ and $z_{\alpha/2} = 1.96$. For this case there is a probability of 0.05 that a sample mean will be more than 1.96 standard deviations (or $z_{\alpha/2}\sigma/\sqrt{n}$, which we denote by E) away from the population mean μ. Conversely, there is a 0.95 probability that a sample mean will be within 1.96 standard deviations (or $z_{\alpha/2}\sigma/\sqrt{n}$) of μ. (See Figure 6-2.) If the sample mean $\bar{x}$ is within $z_{\alpha/2}\sigma/\sqrt{n}$ of the population mean μ, then μ must be between $\bar{x} - z_{\alpha/2}\sigma/\sqrt{n}$ and $\bar{x} + z_{\alpha/2}\sigma/\sqrt{n}$; this is expressed in the general format of our confidence interval (with $z_{\alpha/2}\sigma/\sqrt{n}$ denoted as E):

$$\bar{x} - E < \mu < \bar{x} + E$$

In Chapter 2 statistics such as the sample mean were rounded to one more decimal place than was used in the original set of data. We follow that same guideline in the following rule for rounding in confidence intervals used to estimate the population mean μ.

Roundoff Rule for Confidence Intervals Used to Estimate μ

1. When using the original set of data to construct a confidence interval, round the confidence interval limits to one more decimal place than is used for the original set of data.
2. When the original set of data is unknown and only the summary statistics $(n, \bar{x}, s)$ are used, round the confidence interval limits to the same number of decimal places used for the sample mean.

The following example illustrates the construction of a confidence interval. The original data from Table 6-1 use one decimal place, so the confidence interval limits will be rounded to two decimal places.

▶ **EXAMPLE**

For the body temperature data given in Table 6-1, we have $n = 106$, $\bar{x} = 98.20$, and $s = 0.62$. For a 0.95 degree of confidence, use these statistics to find both of the following:

a. The margin of error E b. The confidence interval for μ

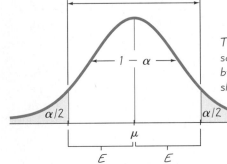

There is a $1 - \alpha$ probability that a sample mean will be in error by less than E or $z_{\alpha/2}\sigma/\sqrt{n}$

There is a probability of α that a sample mean will be in error by more than E (in one of the shaded tails)

$\alpha/2$

$\alpha/2$

$1 - \alpha$

μ

E E

FIGURE 6-2   Distribution of Sample Means

● **SOLUTION**

a.   The 0.95 degree of confidence implies that $\alpha = 0.05$, so $z_{\alpha/2} = 1.96$. (See the preceding example.)

$$E = z_{\alpha/2} \cdot \frac{\sigma}{\sqrt{n}} = 1.96 \cdot \frac{0.62}{\sqrt{106}} = 0.12$$

(Note that because σ is unknown but $n > 30$, we used $s = 0.62$ for the value of σ.)

b.   With $\bar{x} = 98.20$ and $E = 0.12$, we get

$$\bar{x} - E < \mu < \bar{x} + E$$
$$98.20 - 0.12 < \mu < 98.20 + 0.12$$
$$98.08 < \mu < 98.32$$

Based on the sample of 106 body temperatures listed in Table 6-1, the confidence interval for the population mean μ is $98.08°F < \mu < 98.32°F$, and this interval has a 0.95 degree of confidence. This means that if we were to select many different samples of size 106 and construct the confidence intervals as we did here, 95% of them would actually contain the value of the population mean μ. Note that this confidence interval does not contain 98.6°F, the value generally believed to be the mean body temperature. Based on the sample data in Table 6-1, we are now 95% confident that the true population mean μ is contained within the interval with limits of 98.08°F and 98.32°F. It seems very unlikely that 98.6°F is the correct value of μ.

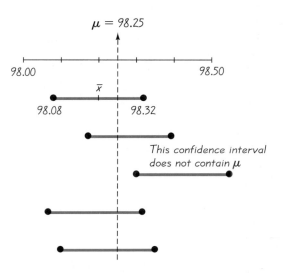

FIGURE 6-3   Confidence Intervals from Different Samples
The graph shows several confidence intervals, one of which
does not contain the population mean μ. For 95% confidence
intervals, we expect that among 100 such intervals, 5 will not
contain $\mu = 98.25$ while the other 95 will contain it.

We must be careful to correctly interpret confidence intervals. Once we use
sample data to find specific limits of $\bar{x} - E$ and $\bar{x} + E$, those limits either
enclose the population mean μ or do not, and we cannot determine if they do
or don't without knowing the true value of μ. It is incorrect to state that μ has
a 95% chance of falling within the specific limits of 98.08 and 98.32, because μ
is a constant, not a random variable, and either it will fall within these limits or
it won't; there's no probability involved. It is correct to say that in the long run
these methods will result in confidence intervals that will contain μ in 95% of
the cases.

Suppose that in the preceding example, body temperatures really come from
a population with a true mean of 98.25°F. Then the confidence interval
obtained from the given sample data does contain the population mean,
because 98.25 is between 98.08 and 98.32. This is illustrated in Figure 6-3. (See
the first confidence interval on the graph.)

Small Sample Cases and the Student *t* Distribution

Unfortunately, finding a confidence interval using Formula 6-1 requires either
a large ($n > 30$) sample or a normally distributed population with a known
value of σ. Factors such as cost and time often severely limit the size of a

sample, so the normal distribution may not be an appropriate approximation of the distribution of means from small samples. If we have a small ($n \leq 30$) sample and intend to construct a confidence interval but do not know σ, we can sometimes use the Student t distribution developed by William Gosset (1876–1937). Gosset was a Guinness Brewery employee who needed a distribution that could be used with small samples. The Irish brewery where he worked did not allow the publication of research results, so Gosset published under the pseudonym Student. As a result of his early experiments and studies of small samples, we can now use the Student t distribution.

Student t Distribution

If the distribution of a population is essentially normal (approximately bell shaped), then the distribution of

$$t = \frac{\bar{x} - \mu}{\frac{s}{\sqrt{n}}}$$

is essentially a **Student t distribution** for all samples of size n. The Student t distribution is often referred to as the ***t* distribution**.

Table A-3 lists t scores along with corresponding areas denoted by α. Values of $t_{\alpha/2}$ are obtained by locating the proper value for degrees of freedom in the left column and then proceeding across the corresponding row until reaching the number directly below the applicable value of α for two tails.

DEFINITION

The number of **degrees of freedom** for a data set corresponds to the number of scores that can vary after certain restrictions have been imposed on all scores.

For example, if 10 students have quiz scores with a mean of 80, we can freely assign values to the first 9 scores, but the 10th score is then determined because the mean of the 10 scores must be 80. Therefore, there are 9 degrees of freedom available. For the applications of this section, the number of degrees of freedom is simply the sample size minus 1.

$$\text{degrees of freedom} = n - 1$$

For the Student t distribution to be applicable, the distribution of the parent population must be *essentially* normal; it need not be exactly normal, but if it

Excerpts from a Department of Transportation Circular

The following excerpts from a Department of Transportation circular concern some of the accuracy requirements for navigation equipment used in aircraft. Note the use of the confidence interval.

"The total of the error contributions of the airborne equipment, when combined with the appropriate flight technical errors listed, should not exceed the following with a 95% confidence (2-sigma) over a period of time equal to the update cycle."

"The system of airways and routes in the United States has widths of route protection used on a VOR system use accuracy of ±4.5 degrees on a 95% probability basis."

has only one mode and is basically symmetric, we will generally get good results (such as accurate confidence intervals). If there is strong evidence that the population has a very nonnormal distribution, then nonparametric methods (see Chapter 13) or bootstrap resampling methods (see the Computer Project at the end of this chapter) should be used instead.

Important Properties of the Student *t* Distribution

1. The Student *t* distribution is different for different sample sizes. (See Figure 6-4 for the cases $n = 3$ and $n = 12$.)

2. The Student *t* distribution has the same general symmetric bell shape as the normal distribution, but it reflects the greater variability (with wider distributions) that is expected with small samples.

3. The Student *t* distribution has a mean of $t = 0$ (just as the standard normal distribution has a mean of $z = 0$).

4. The standard deviation of the Student *t* distribution varies with the sample size, but it is greater than 1 (unlike the standard normal distribution, which has $\sigma = 1$).

5. As the sample size *n* gets larger, the Student *t* distribution gets closer to the standard normal distribution. For values of $n > 30$, the differences are so small that we can use the critical *z* values instead of developing a much larger table of critical *t* values. (The values in the bottom row of Table A-3 are equal to the corresponding critical *z* values from the standard normal distribution.)

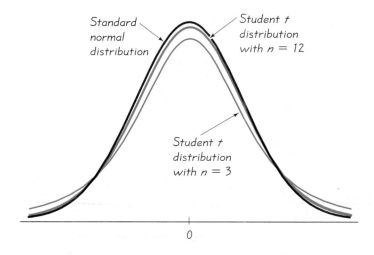

Standard normal distribution

Student *t* distribution with $n = 12$

Student *t* distribution with $n = 3$

0

FIGURE 6-4   Student *t* Distributions for $n = 3$ and $n = 12$
The Student *t* distribution has the same general shape and symmetry as the normal distribution, but it reflects the greater variability that is expected with small samples.

Following is a summary of the conditions indicating use of a t distribution instead of the standard normal distribution. (These same conditions will also apply in Chapter 7.)

Conditions for the Student t Distribution

1. The sample is small ($n \leq 30$); and
2. σ is unknown; and
3. The parent population has a distribution that is essentially normal. (Because the distribution of the parent population is often unknown, we often estimate it by constructing a histogram of sample data.)

We can now determine values for the maximum error E in estimating μ when a t distribution applies.

Margin of Error for the Estimate of μ

Formula 6-2
$$E = t_{\alpha/2} \frac{s}{\sqrt{n}}$$

when *all* of the conditions for the Student t distribution are met:

1. $n \leq 30$; and
2. σ is unknown; and
3. The population is normally distributed.

In the next example we use only the first 10 body temperatures from Table 6-1 so that you can better compare and contrast the large ($n > 30$) sample case and the small ($n \leq 30$) sample case.

▶ **EXAMPLE**

Suppose that we have only the following 10 randomly selected body temperatures.

> 98.6   98.6   98.0   98.0   99.0   98.4   98.4   98.4   98.4   98.6

For these scores, $n = 10$, $\bar{x} = 98.44$, and $s = 0.30$. Recognizing that this is a small sample ($n \leq 30$) and σ is unknown, construct the 95% confidence interval for the mean of all body temperatures. (Assume that a prior study has shown that body temperatures are normally distributed.)   *(continued)*

Large Sample Size Isn't Good Enough

Biased sample data should not be used for inferences, no matter how large the sample is. For example, in *Women and Love: A Cultural Revolution in Progress*, Shere Hite bases her conclusions on 4500 replies that she received after mailing 100,000 questionnaires to various women's groups. A *random* sample of 4500 subjects would usually provide good results, but Hite's sample is biased. It is criticized for overrepresenting women who join groups and women who feel strongly about the issues addressed. Because Hite's sample is biased, her inferences are not valid, even though the sample size of 4500 might seem to be sufficiently large.

● SOLUTION

With a small sample from a normally distributed population and unknown σ, we know that the Student t distribution applies, and we begin by calculating the margin of error E. In this calculation we use $t_{\alpha/2} = 2.262$, which is found in Table A-3 corresponding to $n - 1 = 9$ degrees of freedom and $\alpha = 0.05$ in two tails.

$$E = t_{\alpha/2} \cdot \frac{s}{\sqrt{n}} = 2.262 \cdot \frac{0.30}{\sqrt{10}} = 0.21$$

Because we have a smaller sample size and are using $t_{\alpha/2}$ instead of $z_{\alpha/2}$, this margin of error $E = 0.21$ is much larger than the value $E = 0.12$ found in the preceding example.

We now substitute $E = 0.21$ and $\bar{x} = 98.44$ to get the confidence interval shown below.

$$\bar{x} - E < \mu < \bar{x} + E$$
$$98.44 - 0.21 < \mu < 98.44 + 0.21$$
$$98.23 < \mu < 98.65$$

That is, based on the 10 sample temperatures, we are 95% confident that the limits of 98.23°F and 98.65°F do contain the true mean body temperature.

Confusion can sometimes arise in considering whether to use Formula 6-1 (and the standard normal distribution) or Formula 6-2 (and the Student t distribution). The flowchart in Figure 6-5 summarizes the key points to be considered when constructing confidence intervals for estimating μ, the true population mean.

Using Computers for Confidence Intervals

Once you understand the basic theory underlying the use of confidence intervals in estimating parameters and how to construct and interpret them, you can take advantage of software that simplifies the manual calculations. Following is the Minitab display for the data used in the preceding example, where $n = 10$ and σ is unknown. The first two lines show the entering of the sample data, and the third line shows the key command of TINTERVAL (for confidence interval using the t distribution). The display includes the sample size, sample mean, sample standard deviation, standard deviation of sample means (labeled SE MEAN for standard error of mean), and confidence interval limits enclosed within parentheses.

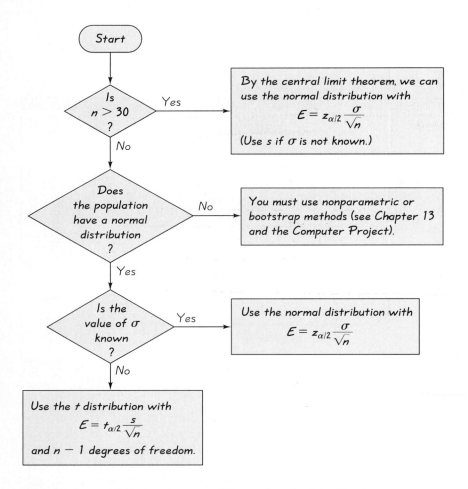

FIGURE 6-5   Choosing between Normal and *t* Distributions

MINITAB DISPLAY OF CONFIDENCE INTERVAL

```
MTB > SET C1
DATA> 98.6 98.6 98.0 98.0 99.0 98.4 98.4 98.4 98.4 98.6
DATA> ENDOFDATA
MTB > TINTERVAL 95 C1


              N       MEAN     STDEV   SE MEAN    95.0 PERCENT C.I.
C1           10     98.4400    0.2951   0.0933   < 98.2288, 98.6512>
```

STATDISK also calculates confidence intervals, and the following display corresponds to the data from the preceding example. Simply select Statistics from the main menu, then select Confidence Intervals and proceed to answer the questions that lead to the following result.

STATDISK DISPLAY OF CONFIDENCE INTERVAL

```
 About    File    Edit    Statistics    Data    View

                        Confidence Intervals

        Confidence Level ............................. =   .95
        Sample size ................................ n = 10
        Sample mean ................................. = 98.44
        Sample standard deviation ................... =   .3


              ┌─────────────────────────────────────────┐
              │         The Confidence Interval is        │
              │     ===================================   │
              │                                           │
              │   98.225229 < Population mean < 98.654771  │
              │                                           │
              │                                           │
              │                                           │
              └─────────────────────────────────────────┘

                    Use Menu Bar to change screen      Page 3 of 3
         F1: Help         F2: Menu                      ESC: Quit
```

We have described common methods of determining confidence interval estimates of a population mean μ, but these methods require that either the sample size exceeds 30 or the sample is drawn from a population with a normal distribution. If we have a small ($n \leq 30$) sample drawn from a very nonnormal distribution, we can't use the methods described. One alternative is to use nonparametric methods (see Chapter 13), and another alternative is to use the computer-oriented *bootstrap* method, which makes no assumptions about the original population. This method is described in the Computer Project at the end of this chapter.

Determining Sample Size

So far in this section we have discussed how to find point estimates and interval estimates of a population mean μ. We based our procedures on known sample data. But suppose that we haven't yet collected the sample. How do we know how many members of the population should be selected? For example, suppose we want to estimate the mean weight of plastic discarded by households in a week. How many households should be surveyed? Determining the size of a sample is a very important problem, because needlessly large samples waste time and money and samples of insufficient size may lead to poor results. In many cases we can determine the minimum sample size needed to estimate some parameter, such as the population mean μ.

If we begin with the expression for the margin of error E (Formula 6-1) and solve for the sample size n, we get

Sample Size for Estimating Mean μ

Formula 6-3	$n = \left[\dfrac{z_{\alpha/2}\sigma}{E} \right]^2$

This formula may be used to determine the sample size necessary to produce results accurate to a desired degree of confidence and margin of error. It should be used when we know the value of σ and want to determine the sample size necessary to establish, with a confidence level of $1 - \alpha$, the value of μ to within $\pm E$. (It's unusual to know σ without knowing μ, but σ might be known from a previous study or it might be estimated reasonably well from a pilot study.) The existence of such a formula is somewhat remarkable because it implies that *the sample size does not depend on the size (N) of the population.*

▶ **EXAMPLE**

We want to estimate the mean weight of plastic discarded by households in one week. How many households must we randomly select if we want to be 99% sure that the sample mean is within 0.250 lb of the true population mean μ? Assume that a previous study has shown that $\sigma = 1.100$ lb.

● **SOLUTION**

We seek the sample size n given that $\alpha = 0.01$ (from 99% confidence), so $z_{\alpha/2} = 2.575$, $E = 0.250$, and $\sigma = 1.100$. Using Formula 6-3, we get

$$n = \left[\frac{z_{\alpha/2}\sigma}{E} \right]^2 = \left[\frac{2.575 \cdot 1.100}{0.250} \right]^2 = 11.33^2 = 128.3689$$

$$= 129 \text{ (rounded up)}$$

We should therefore obtain a sample of at least 129 randomly selected households. With such a sample, we will be 99% confident that the sample mean $\overline{x}$ will be within 0.250 lb of the true population mean μ.

If we are willing to settle for less accurate results by using a larger margin of error such as 0.500 lb, the sample size drops to $n = [(2.575)(1.100)/0.500]^2 = 32.09$, which is rounded up to 33. We *always round up in sample size computations so that the required number in the sample is at least adequate instead of being slightly inadequate.*

Shakespeare's Vocabulary

According to Bradley Efron and Ronald Thisted, Shakespeare's writings included 31,534 different words. They used probability theory to conclude that Shakespeare probably knew at least another 35,000 words that he didn't use in his writings. The problem of estimating the size of a population is an important problem often encountered in ecology studies, but the result given here is another interesting application. (See "Estimating the Number of Unseen Species: How Many Words Did Shakespeare Know?" in *Biometrika*, Vol. 63, No. 3.)

Roundoff Rule for Sample Size

If the computed sample size is not a whole number, round it up to the next *higher* whole number.

Note that doubling the margin of error caused the required sample size to decrease to one-fourth of its original value. Conversely, halving the margin of error would quadruple the sample size. All of this implies that if you want more accurate results, then the sample size must be substantially increased. Because large samples generally require more time and money, there is often a need for a tradeoff between sample size and the margin of error E.

What If σ Is Unknown?   Formula 6-3 requires that we substitute some value for the population standard deviation σ. In the preceding example it was assumed that $\sigma = 1.100$ lb. In reality, the value of σ might be unknown, but we often have some prior information from a previous sample so we can use some value of s in place of σ. For example, if we want to solve the preceding example without the assumption that $\sigma = 1.100$, we can refer to the sample of 62 households given in Data Set 1 of Appendix B. From that sample, we can calculate the value of the standard deviation to get $s = 1.065$ lb so

$$ n = \left[\frac{z_{\alpha/2}\sigma}{E} \right]^2 = \left[\frac{2.575 \cdot 1.065}{0.250} \right]^2 = 120.3 = 121 \qquad \text{(rounded up)} $$

Based on the population standard deviation estimated from Data Set 1 in Appendix B, we must sample at least 121 randomly selected individuals to be 99% confident that the sample mean is within 0.250 lb of the true population mean.

This section dealt with the construction of point estimates and confidence interval estimates of population means, as well as determining sample sizes needed to estimate population means to the desired degree of accuracy. The following section deals with the same concepts, but the population parameter will be a proportion instead of a mean.

6-2 Exercises A: Basic Skills and Concepts

1. a. If $\alpha = 0.05$, find $z_{\alpha/2}$.
  b. If $\alpha = 0.02$, find $z_{\alpha/2}$.
  c. Find $z_{\alpha/2}$ for the value of α corresponding to a confidence level of 96%.
  d. If $\alpha = 0.05$, find $t_{\alpha/2}$ for a sample of 20 scores.
  e. If $\alpha = 0.01$, find $t_{\alpha/2}$ for a sample of 15 scores.

2. a.   If $\alpha = 0.10$, find $z_{\alpha/2}$.
  b.   Find $z_{\alpha/2}$ for the value of α corresponding to a confidence level of 95%.
  c.   Find $z_{\alpha/2}$ for the value of α corresponding to a confidence level of 80%.
  d.   If $\alpha = 0.10$, find $t_{\alpha/2}$ for a sample of 10 scores.
  e.   If $\alpha = 0.02$, find $t_{\alpha/2}$ for a sample of 25 scores.

In Exercises 3–6, use the given data to find the margin of error E. Be sure to use the correct formula for E, depending on whether the normal distribution or the Student t distribution applies.

3. $\alpha = 0.01$, $\sigma = 40$, $n = 25$
4. $\alpha = 0.05$, $s = 30$, $n = 25$
5. $\alpha = 0.01$, $s = 15$, $n = 100$
6. $\alpha = 0.01$, $s = 30$, $n = 64$
7. Find the 95% confidence interval for the population mean μ if the sample mean is $\bar{x} = 70.4$ and the sample size is $n = 36$. Assume that the population standard deviation is known from a previous study to be $\sigma = 5$.
8. Find the 99% confidence interval for the population mean μ if the sample mean is $\bar{x} = 84.2$ and the sample size is $n = 40$. Assume that the population standard deviation is known from a previous study to be $\sigma = 7.3$.
9. The National Center for Education Statistics surveyed 4400 college graduates about the lengths of time required to earn their bachelor's degrees. The mean is 5.15 years, and the standard deviation is 1.68 years. Based on these sample data, construct the 99% confidence interval for the mean time required by all college graduates.
10. The United States Marine Corps is reviewing its orders for uniforms because it has a surplus of uniforms for tall recruits and a shortage for shorter recruits. Its review involves data for 772 men between the ages of 18 and 24. That sample group has a mean height of 69.7 in. with a standard deviation of 2.8 in. (see USDHEW publication 79-1659). Use these sample data to find the 99% confidence interval for the mean height of all men between the ages of 18 and 24.
11. In a time-use study, 20 randomly selected managers were found to spend a mean of 2.40 h each day on paperwork. The standard deviation of the 20 scores is 1.30 h (based on data from Adia Personnel Services). Construct the 95% confidence interval for the mean time spent on paperwork by all managers.
12. Data Set 6 in Appendix B includes 17 weights (in grams) of red M&Ms. For this sample, the mean is 0.9043 g and the standard deviation is 0.0414 g. Construct a 90% confidence interval estimate of the mean weight of all red M&Ms.
13. The standard IQ test is designed so that the mean is 100 and the standard deviation is 15 for the population of normal adults. We want to determine the sample size necessary to estimate the mean IQ score of statistics instructors. We want to be 98% confident that our sample mean is within 1.5 IQ points of the true mean. The mean for this population is clearly greater than 100. The standard deviation for this population is probably less than 15 because it is a group with less variation than a group randomly selected from the general population; therefore, if we use $\sigma = 15$, we are being conservative by using a

value that will make the sample size at least as large as necessary. Assume then that $\sigma = 15$ and determine the required sample size.

14. The Washington Vending Machine Company must adjust its machines to accept only coins with specified weights. We will obtain a sample of quarters and weigh them to determine the mean. How many quarters must we randomly select and weigh if we want to be 99% confident that the sample mean is within 0.025 g of the true population mean for all quarters? Assume that the population standard deviation is 0.068 g (based on Data Set 8 in Appendix B).

15. If we refer to the weights (in grams) of quarters listed in Data Set 8 in Appendix B, we will find 50 weights with a mean of 5.622 g and a standard deviation of 0.068 g. Based on this random sample of quarters in circulation, construct a 99% confidence interval estimate of the population mean of all quarters in circulation. The U.S. Department of the Treasury claims that it mints quarters to yield a mean weight of 5.670 g. Is this claim consistent with the confidence interval? If not, what is a possible explanation for the discrepancy?

16. In a study of the amounts of time required for room service delivery at a newly opened Radisson Hotel, 20 deliveries had a mean time of 24.2 min and a standard deviation of 8.7 min. Construct the 90% confidence interval for the mean of all deliveries.

17. A psychologist has developed a new test of spatial perception, and she wants to estimate the mean score achieved by adult male pilots. How many people must she test if she wants the sample mean to be in error by no more than 2.0 points, with 95% confidence? An earlier study suggests that $\sigma = 21.2$.

18. To plan for the proper handling of household garbage, the City of Providence must estimate the mean weight of garbage discarded by households in one week. Find the sample size necessary to estimate that mean if you want to be 98% confident that the sample mean is within 2 lb of the true population mean. For the population standard deviation σ, use the value of 12.46 lb, which is the standard deviation of the sample of 62 households included in the Garbage Project study conducted at the University of Arizona.

19. Data Set 1 in Appendix B includes the total weights of garbage discarded by 62 households in one week (based on data collected as part of the Garbage Project at the University of Arizona). For that sample, the mean is 27.44 lb and the standard deviation is 12.46 lb. Construct a 99% confidence interval estimate of the mean weight of garbage discarded by all households. If the City of Providence can handle the garbage as long as the mean is less than 35 lb, is there any cause for concern that there might be too much garbage to handle?

20. The U.S. Department of Health, Education, and Welfare collected sample data for 1525 women, aged 18 to 24. That sample group has a mean serum cholesterol level (measured in mg/100 mL) of 191.7 with a standard deviation of 41.0 (see USDHEW publication 78-1652). Use these sample data to find the 97% confidence interval for the mean serum cholesterol level of all women in the 18–24 age bracket. If a doctor claims that the mean serum cholesterol for women in that age bracket is 200, does that claim seem to be consistent with the confidence interval?

21. In an insurance company study of New York State licensed drivers' ages, 570 randomly selected ages have a mean of 41.8 years and a standard deviation of 16.7 years (based on data from the New York State Department of Motor Vehicles).

   a.   Construct a 99% confidence interval for the mean age of all New York State licensed drivers.

   b.   Use the sample standard deviation as an estimate of the population standard deviation and determine the sample size necessary to estimate the mean age of New York State licensed drivers, assuming that you want 99% confidence and a margin of error of 0.5 year.

22. In a study of physical attractiveness and mental disorders, 231 subjects were rated for attractiveness, and the resulting sample mean and standard deviation are 3.94 and 0.75, respectively. (See "Physical Attractiveness and Self-Perception of Mental Disorder," by Burns and Farina, *Journal of Abnormal Psychology*, Vol. 96, No. 2.)

   a.   Use these sample data to construct the 95% confidence interval for the population mean.

   b.   Use the sample standard deviation as an estimate of the population standard deviation and determine the sample size necessary to estimate the population mean, assuming that you want 95% confidence and a margin of error of 0.05.

23. In a study of the use of hypnosis to relieve pain, sensory ratings were measured for 16 subjects, with the results given below. (See "An Analysis of Factors That Contribute to the Efficacy of Hypnotic Analgesia," by Price and Barber, *Journal of Abnormal Psychology*, Vol. 96, No. 1.) Use these sample data to construct the 95% confidence interval for the mean sensory rating for the population from which the sample was drawn.

8.8	6.6	8.4	6.5	8.4	7.0	9.0	10.3
8.7	11.3	8.1	5.2	6.3	11.6	6.2	10.9

24. The Internal Revenue Service conducted a study of estates valued at more than $300,000 and determined the value of bonds for a randomly selected sample, with the following results (in dollars). Find a 90% interval estimate of the mean value of bonds for all such estates. Assume that the sample standard deviation can be used as an estimate of the population standard deviation.

45,300	36,200	72,500	50,500	15,300	58,500
26,200	97,100	74,200	83,700	72,000	10,000
63,000	15,000	49,200	37,500	81,000	24,000
145,000	27,900	53,100	27,500	94,000	23,800
74,600	36,800	65,900	29,400	86,300	25,600
53,200	47,200	61,800	33,200	18,200	75,000

25. Refer to Data Set 1 in Appendix B for the 62 weights (in pounds) of *paper* discarded by households (based on data from the Garbage Project at the University of Arizona). Using that sample, construct a 95% confidence interval estimate of the mean weight of paper discarded by all households.

26. Refer to the ages of U.S. commercial aircraft listed in Data Set 10 of Appendix B. Use that sample to construct a 99% confidence interval estimate of the mean age of all U.S. commercial aircraft.

6-2 Exercises B: Beyond the Basics

27. Refer to Data Set 5 in Appendix B.
    a. Construct a 95% confidence interval for the mean length of movies rated R.
    b. Construct a 95% confidence interval for the mean length of movies rated PG or PG-13.
    c. Compare the methods and results of parts a and b.
28. Based on the confidence interval obtained in Exercise 20, should an *individual* woman be concerned if her serum cholesterol level is 200? Why or why not?
29. It is found that a sample size of 810 is necessary to estimate the mean weight (in mg) of Bufferin tablets. That sample size is based on a 95% degree of confidence and a population standard deviation that is estimated by the sample standard deviation for Data Set 9 in Appendix B. Find the margin of error E.
30. A 95% confidence interval for the lives (in minutes) of Kodak AA batteries is $430 < \mu < 470$. (See Program 1 of *Against All Odds: Inside Statistics*.) Assume that this result is based on a sample of size 100.
    a. Construct the 99% confidence interval.
    b. What is the value of the sample mean?
    c. What is the value of the sample standard deviation?
    d. If the confidence interval $432 < \mu < 468$ is obtained from the same sample data, what is the degree of confidence?
31. The development of Formula 6-1 assumes that either the population is infinite, we are sampling with replacement, or the population is very large. If we have a relatively small population and sample without replacement, we should modify E to include the finite population correction factor as follows:

$$E = z_{\alpha/2} \frac{\sigma}{\sqrt{n}} \sqrt{\frac{N - n}{N - 1}}$$

where N is the population size. Show that the preceding expression can be solved for n to yield

$$n = \frac{N\sigma^2(z_{\alpha/2})^2}{(N - 1)E^2 + \sigma^2(z_{\alpha/2})^2}$$

Repeat Exercise 13, assuming that the statistics instructors are randomly selected without replacement from a population of $N = 200$ statistics instructors.

32. The standard error of the mean is $\sigma/\sqrt{n}$, provided that the population size is infinite. If the population size is finite and is denoted by N, then the correction factor

$$\sqrt{\frac{N - n}{N - 1}}$$

should be used whenever $n > 0.05N$. This correction factor multiplies the standard error of the mean, as shown in Exercise 31. Find the 95% confidence interval for the mean of 100 IQ scores if a sample of 30 of those scores produces a mean and standard deviation of 132 and 10, respectively.

6-3   Estimating a Population Proportion

In this section we consider the same concepts of estimating and sample size determination that were discussed in Section 6-2, but we apply the concepts to *proportions* instead of means. For example, Nielsen Media Research might want to estimate the proportion of households with their television sets tuned to *60 Minutes*. The marketing department of Pepsi might want to estimate the proportion of consumers who recognize the Pepsi logo. Because we are dealing with proportions, we can use the binomial distribution introduced in Section 4-3.

We assume in this section that the conditions given in Section 4-3 for the binomial distribution are essentially satisfied. We also assume that the normal distribution can be used as an approximation to the distribution of sample proportions. Recall that in a binomial experiment we have all outcomes classified into one or the other of two different categories, typically referred to as *success* and *failure*. Also, we have n independent trials, and in each trial the probability of success is denoted by p while the probability of failure is denoted by q. The symbol x is used to denote the number of successes among the n trials. Recall from Section 5-5 that if the conditions $np \geq 5$ and $nq \geq 5$ are both satisfied, we can use the normal distribution as an approximation to the binomial distribution.

Although we make repeated references to proportions in this section, keep in mind that the theory and procedures also apply to probabilities and percents. Proportions and probabilities are both expressed in decimal or fraction form. If we intend to deal with percents, we can easily convert them to proportions by deleting the percent sign and dividing by 100. For example, the 48.7% rate of people who don't buy any books can be expressed in decimal form as 0.487. The symbol p may therefore represent a proportion, a probability, or the decimal equivalent of a percent. We continue to use p as the population proportion in the same way that we use μ to represent the population mean. We now introduce new notation.

Notation for Sample Proportion

$$\hat{p} = \frac{x}{n}$$

In this way, p represents the *population* proportion and $\hat{p}$ (called "p hat") represents the *sample* proportion. In previous chapters we stipulated that $q = 1 - p$, so it is natural to stipulate here that $\hat{q} = 1 - \hat{p}$.

The term $\hat{p}$ denotes the sample proportion that is analogous to the relative frequency definition of a probability. As an example, suppose that a pollster is hired to determine the proportion of adult Americans who favor a national health care plan. Let's assume that 2000 adult Americans are surveyed, with 1347 favorable reactions. The pollster seeks the value of p, the true proportion of all adult Americans favoring that plan. Sample results indicate that $x = 1347$ and $n = 2000$, so

$$\hat{p} = \frac{x}{n} = \frac{1347}{2000} = 0.6735$$

Just as $\overline{x}$ was selected as the point estimate of μ, we now select $\hat{p}$ as the best point estimate of p. Of the various estimators that could be used for p, $\hat{p}$ is deemed best because it is unbiased and the most consistent. It is unbiased in the sense that the distribution of sample proportions tends to center about the value of p. It is the most consistent in the sense that the variance of sample proportions tends to be smaller than the variance of any other unbiased estimators.

The sample proportion $\hat{p}$ is the best point estimate of the population proportion p.

Because we are assuming that the binomial conditions are essentially satisfied and that the normal distribution can be used as an approximation to the distribution of sample proportions, we can draw from results established in Section 5-5 and conclude that the mean number of successes μ and the standard deviation of the number of successes σ are given by

$$\mu = np$$
$$\sigma = \sqrt{npq}$$

where p is the probability of a success. Both of these parameters pertain to n trials, and we now convert them to a *per trial* basis simply by dividing by n as follows:

$$\text{Mean of sample proportions} = \frac{np}{n} = p$$

$$\text{Standard deviation of sample proportions} = \frac{\sqrt{npq}}{n} = \sqrt{\frac{npq}{n^2}} = \sqrt{\frac{pq}{n}}$$

The first result may seem trivial because we have already stipulated that the true population proportion is p. The second result is nontrivial and very useful. In the last section we saw that the sample mean $\bar{x}$ has a probability of $1 - \alpha$ of being within $z_{\alpha/2}\sigma/\sqrt{n}$ of μ. Similar reasoning leads us to conclude that $\hat{p}$ has a probability of $1 - \alpha$ of being within $z_{\alpha/2}\sqrt{pq/n}$ of p. Because we do not know the value of p (that's what we want to estimate) or q, we must replace p and q by their point estimates of $\hat{p}$ and $\hat{q}$ so that a margin of error can be computed in real situations. This leads to the following formula.

Margin of Error of the Estimate of p

Formula 6-4
$$E = z_{\alpha/2}\sqrt{\frac{\hat{p}\,\hat{q}}{n}}$$

Remember, p is the population proportion, and $\hat{p}$ is the sample proportion that typically differs from p. The probability that $\hat{p}$ differs from p by less than E is $1 - \alpha$. We can now describe the confidence interval used to estimate a population proportion.

Confidence Interval (or Interval Estimate) for the Population Proportion p

$$\hat{p} - E < p < \hat{p} + E$$

When constructing confidence intervals for proportions, round results to three significant digits—the same roundoff rule used for probabilities in Chapter 3.

Roundoff Rule for Confidence Interval Estimates of p

Round the confidence interval limits to three significant digits.

The following example illustrates the determination of point and interval estimates of a population proportion and uses the given roundoff rule.

TV Sample Sizes

A *Newsweek* article described the use of people meters as a way of determining how many people are watching different television programs. The article stated, "Statisticians have long argued that the household samples used by the rating services are simply too small to accurately determine what America is watching. In that light, it may be illuminating to note that the 4000 homes reached by the people meters constitute exactly 0.0045% of the wired nation." This implied claim that the sample size of 4000 homes is too small is not valid. Methods of this chapter can be used to show that a random sample size of 4000 can provide results that are quite good, even though the sample might be only 0.0045% of the population.

Capture-Recapture

Ecologists need to determine population sizes of endangered species. One method is to capture a sample of some species, mark all members of this sample, and then free them. Later, another sample is captured, and the ratio of marked subjects, coupled with the size of the first sample, can be used to estimate the population size. This capture-recapture method was used with other methods to estimate the blue whale population, and the result was alarming: The population was as small as 1000. That led to the International Whaling Commission to ban the killing of blue whales to prevent their extinction.

▶ **EXAMPLE**

When 500 college students are randomly selected and surveyed, it is found that 135 of them own personal computers (based on data from the America Passage Media Corporation).

a. Find the point estimate of the true proportion of all college students who own personal computers.

b. Find a 95% confidence interval for the true proportion of all college students who own personal computers.

● **SOLUTION**

a. The point estimate of p is

$$\hat{p} = \frac{x}{n} = \frac{135}{500} = 0.270$$

b. Construction of the confidence interval requires that we evaluate the margin of error E. The value of E can be found from $\hat{p} = 0.270$ (from part a), $\hat{q} = 0.730$ (from $\hat{q} = 1 - \hat{p}$), and $z_{\alpha/2} = 1.96$ (from Table A-2 in which 95% converts to $\alpha = 0.05$, which is then divided equally between the two tails so that $z = 1.96$ corresponds to an area of 0.4750).

$$E = z_{\alpha/2}\sqrt{\frac{\hat{p}\,\hat{q}}{n}} = 1.96\sqrt{\frac{(0.270)(0.730)}{500}} = 0.039$$

We can now find the confidence interval because we know that $\hat{p} = 0.270$ and $E = 0.039$.

$$\hat{p} - E < p < \hat{p} + E$$
$$0.270 - 0.039 < p < 0.270 + 0.039$$
$$0.231 < p < 0.309$$

If we wanted the 95% confidence interval for the true population *percentage*, we could express this result as 23.1% < p < 30.9%. This result is often reported in the following format: "Among college students, the percentage who own personal computers is estimated to be 27%, with a margin of error of plus or minus 3.9 percentage points." The level of confidence should also be reported, but it rarely is in the media. The media typically use a 95% degree of confidence but omit any reference to it.

Determining Sample Size

Having discussed point estimates and confidence intervals for p, we will now determine how large a sample should be when we want to find the approximate value of a population proportion. In the previous section we started with the expression for the error E and solved for n. Following that reasonable precedent, we begin with

$$E = z_{\alpha/2}\sqrt{\frac{\hat{p}\,\hat{q}}{n}}$$

and solve for n to get the sample size.

**Sample Size for Estimating Proportion p
(When an Estimate $\hat{p}$ Is Known)**

Formula 6-5
$$n = \frac{[z_{\alpha/2}]^2 \hat{p}\,\hat{q}}{E^2}$$

In finding the sample size necessary to estimate p, Formula 6-5 requires $\hat{p}$ as an estimate of the true population proportion p. But this is circular logic, because Formula 6-5 requires knowledge of the sample proportion $\hat{p}$ before we have determined the sample size and before the sample data have been collected. However, reasonable guesses of $\hat{p}$ can sometimes be made by using previous samples, a pilot study, or someone's expert knowledge. When no such guess can be made, mathematicians have cleverly devised an alternative based on the principle that, in the absence of $\hat{p}$ and $\hat{q}$, we can assign the value of 0.5 to each of those statistics and the resulting sample size will be at least sufficient. The underlying reason for the assignment of 0.5 is found in the conclusion that the product $\hat{p} \cdot \hat{q}$ achieves a maximum possible value of 0.25 when $\hat{p} = 0.5$ and $\hat{q} = 0.5$. (See the accompanying table, which lists some values of $\hat{p}$ and $\hat{q}$.) In practice, this means that lack of knowledge of the value of $\hat{p}$ or $\hat{q}$ requires that the preceding expression for n be replaced by the following one:

$\hat{p}$	$\hat{q}$	$\hat{p}\,\hat{q}$
0.1	0.9	0.09
0.2	0.8	0.16
0.3	0.7	0.21
0.4	0.6	0.24
0.5	0.5	0.25
0.6	0.4	0.24
0.7	0.3	0.21
0.8	0.2	0.16
0.9	0.1	0.09

**Sample Size for Estimating Proportion p
(When No Estimate $\hat{p}$ Is Known)**

Formula 6-6
$$n = \frac{(z_{\alpha/2})^2 \cdot 0.25}{E^2}$$

Here the occurrence of 0.25 reflects the substitutions of $\hat{p} = 0.5$ and $\hat{q} = 0.5$. If we have evidence supporting specific known values of $\hat{p}$ or $\hat{q}$, we can substitute those values and thereby reduce the sample size accordingly. The following example illustrates a case in which an estimate of the sample proportion $\hat{p}$ is known and another case in which no estimate of $\hat{p}$ is known.

▶ **EXAMPLE**

We want to estimate, with a margin of error of 0.03, the true proportion of all college students who own personal computers, and we want 95% confidence in our results. How many college students must we survey?

a. Assume that we have an estimate $\hat{p}$ derived from a prior study which revealed a percentage of 27%.

b. Assume that we have no prior information suggesting a possible value of $\hat{p}$.

● **SOLUTION**

a. The prior study suggests that $\hat{p} = 0.27$, so $\hat{q} = 1 - 0.27 = 0.73$. With a 95% degree of confidence, we have $\alpha = 0.05$, so $z_{\alpha/2} = 1.96$. Also, the margin of error is $E = 0.03$. Because we do have an estimated value of $\hat{p}$, we use Formula 6-5.

$$n = \frac{[z_{\alpha/2}]^2 \hat{p}\,\hat{q}}{E^2} = \frac{[1.96]^2(0.27)(0.73)}{0.03^2}$$
$$= 841.3104 = 842 \qquad \text{(rounded up)}$$

b. As in part a, we again use $z_{\alpha/2} = 1.96$ and $E = 0.03$, but with no prior knowledge of $\hat{p}$ (or $\hat{q}$), we use Formula 6-6 and get

$$n = \frac{[z_{\alpha/2}]^2 \cdot 0.25}{E^2} = \frac{[1.96]^2 \cdot 0.25}{0.03^2}$$
$$= 1067.1111 = 1068 \qquad \text{(rounded up)}$$

To be 95% confident that our sample proportion is within 0.03 of the true proportion of college students who own personal computers, we should randomly select 842 college students (if we can use the results from the prior study) or 1068 college students (if we have no prior information). Note that if we lack the knowledge of the prior study, a larger sample is required to achieve the same results as when the value of $\hat{p}$ can be estimated.

As in Section 6-2, we round the sample size n *up* to the next higher whole number so that the sample size is at least as large as it should be, instead of being slightly too small.

Roundoff Rule for Determining Sample Size

If the computed sample size is not a whole number, round it up to the next *higher* whole number.

Part b of the preceding example involved application of Formula 6-6, the same formula frequently used by Nielsen, Gallup, and other professional pollsters. Many people incorrectly believe that we should sample some percentage of the population, but Formula 6-6 shows that the population size is irrelevant. (In reality, the population size is sometimes used, but only in cases in which we sample without replacement from a relatively small population. See Exercise 35.) Most of the polls featured in newspapers, magazines, and broadcast media involve polls with sample sizes in the range of 1000 to 2000. Even though such polls may involve a very small percentage of the total population, they can provide results that are quite good. When Nielsen surveys 1068 TV households from the population of 93 million, only 0.001% of the households are surveyed; still, we can be 95% confident that the sample percentage will be within 3 percentage points of the population percentage.

Polling is an important and common practice in the United States. Polls can affect the television shows we watch, the leaders we elect, the legislation that governs us, and the products we consume. An understanding of the concepts of this section should remove much of the mystery and misunderstanding surrounding polls.

How One Telephone Survey Was Conducted

A *New York Times*/CBS News survey was based on 1417 interviews of adult men and women in the United States. It reported a 95% certainty that the sample results differed by no more than 3 percentage points from the percentage that would have been obtained if *every* adult American had been interviewed. A computer was used to select telephone exchanges in proportion to the population distribution. After selecting an exchange number, the computer generated random numbers to develop a complete phone number so that listed and unlisted numbers could be included.

6-3 Exercises A: Basic Skills and Concepts

In Exercises 1–4, a trial is repeated n times with x successes. In each case use a 95% degree of confidence and find the margin of error E.

1. $n = 500, x = 100$ **2.** $n = 2000, x = 300$

3. $n = 1068, x = 325$ **4.** $n = 1776, x = 50$

In Exercises 5–8, use the given data and degree of confidence to construct the confidence interval that is an estimate of the population proportion p.

5. $n = 400, x = 100,$ 95% confidence

6. $n = 900, x = 400,$ 95% confidence

7. $n = 512$, $x = 309$, 98% confidence

8. $n = 12,485$, $x = 3456$, 99% confidence

9. How many TV households must Nielsen survey to estimate the percentage that are tuned to *The Late Show with David Letterman*? Assume that you want 97% confidence that your sample percentage has a margin of error of 2 percentage points. Also assume that nothing is known about the proportion of households tuned in to television after 11 PM.

10. The South Carolina Sports Supply Company manufactures golfing equipment and plans to advertise its new line along with general information about golfing in America. The marketing manager finds that when 1180 randomly selected adults were surveyed, 79 indicated that they play golf (based on data from a Roper Organization poll). Construct the 95% confidence interval estimate of the true proportion of all adults who play golf.

11. Because radio station WMFT is considering a radical change in its format, the station manager is asked to investigate the preferences of listeners. He finds that in a survey of 1500 randomly selected persons, 63% indicated that they listen to their favorite radio station because of the music (based on data from a Strategic Radio Research poll). Construct the 98% interval estimate of the true population percentage of all people who listen to their favorite radio station because of the music.

12. The A. J. Collins Investment Company is planning a major telephone campaign in California. The campaign will begin with an automated dialing system that delivers a message and records responses to questions. However, answering machines could prove to be an obstacle to this approach. You've been hired to conduct a preliminary study for the purpose of estimating the percentage of households in California having answering machines. You want an error of no more than 4 percentage points and a 99% degree of confidence. A previous study by the U.S. Consumer Electronics Industry indicates that the percentage should be about 24%. How large should your sample be?

13. The president of the Jefferson Valley Bank wants to confirm her belief that the bank has a low rate of customers with regular savings accounts. She reads about a Roper Organization poll of 2000 randomly selected adults and learns that 1280 of them have money in regular savings accounts. Find the 95% confidence interval for the true proportion of adults who have money in regular savings accounts.

14. In considering the production of a new car accessory, the Michigan Manufacturing Company wants to do a marketing study to determine the proportion of cars with cellular phones. How many cars must be sampled to have 92% confidence that the sample proportion is in error by no more than 0.03?

15. The Telektronic Company's personnel manager is concerned that high employee turnover results in too many employees with job experience of less than one year. The company president wants you to conduct a survey to determine the nationwide rate of employees who have been with their current employer for one year or less. Data from the U.S. Department of Labor suggest that this

percentage is around 29%, but the president wants you to conduct a survey so that you are 90% confident that your randomly selected sample of employees leads to a sample percentage that is off by no more than 2 percentage points. How many employees must you survey?

16. You are writing a newspaper editorial on crime, and you read about a poll showing that among 24,350 randomly selected felons convicted in state courts, 67% were given jail sentences (based on data from the U.S. Department of Justice). Construct the 98% confidence interval for the proportion of all such felons who were given jail sentences.

17. A demographic study of Columbia County shows a high number of workers approaching retirement age. You are considering part-time work in real estate sales and want to determine if many houses will be sold as a result of workers' retiring and moving. According to Merrill Lynch, when 600 randomly selected full-time workers in the 45–64 age bracket were surveyed, 75% of them planned to remain in their current homes after they retired. Construct the 90% confidence interval for the true proportion of all such workers with that intent.

18. The tobacco industry closely monitors all surveys that involve smoking. One survey showed that among 785 randomly selected subjects who completed 4 years of college, 18.3% smoke (based on data from the American Medical Association). Construct the 98% confidence interval for the true percentage of smokers among all people who completed 4 years of college.

19. The Newton Car Park is a dealership considering newspaper advertising targeted at women buyers. A marketing study found that 312 of 650 randomly selected buyers of compact cars were women (based on data from the Ford Motor Company). Construct the 95% interval estimate for the true percentage of all compact car buyers who are women.

20. The West America Communications Company is considering a bid to provide long-distance phone service. You are asked to conduct a poll to estimate the percentage of consumers who are satisfied with their long-distance phone service. You want to be 90% confident that your sample percentage is within 2.5 percentage points of the true population value—a Roper poll suggests that this percentage should be about 85%. How large must your sample be?

21. You have been given a grant by a women's organization to estimate the percentage of small (10 or fewer employees) incorporated businesses that are owned by women. A previous study suggests that this percentage is 27%. How many of these small businesses must be surveyed if you want 94% confidence that your sample percentage is within 3 percentage points of the true population percentage?

22. The Life Trust Insurance Company is suspicious about a cluster of Orange County deaths that resulted from falls. You are asked to estimate the proportion of home accident deaths nationwide that are caused by falls. How many such deaths must you survey to be 95% confident that your sample proportion is within 0.04 of the true population proportion?

23. You randomly select 650 home accident deaths and find that 180 of them are caused by falls (based on data from the National Safety Council). Construct the 95% confidence interval for the true population proportion of all home accident deaths caused by falls.

24. The Airport Transit Association is considering a poll to determine the percentage of adults who have never flown in an airplane. How many adults must be surveyed to get a sample percentage with a margin of error of 1.5 percentage points? Use a 95% degree of confidence.

25. Air America is considering a promotional campaign to attract adults who will certify that they never have flown in an airplane. In an Airport Transit Association poll of 4664 randomly selected adults, 72% indicated that they have flown in an airplane. Find the 99% confidence interval for the percentage of all adults who have flown in an airplane.

26. The American Resorts hotel chain gives an aptitude test to job applicants and considers a multiple-choice test question to be easy if at least 80% of the responses are correct. A random sample of 6503 responses to one question indicates that 5463 of those responses were correct. Construct the 99% confidence interval for the true proportion of correct responses. Is it likely that this question is really easy?

27. Data Set 3 in Appendix B consists of survey data obtained from employed full-time workers. Find the sample proportion with some education beyond high school. Construct the 95% confidence interval estimate of the proportion of all such workers with some education beyond high school.

28. Refer to Data Set 6 in Appendix B and find the sample proportion of M&Ms that are red. Use that result to construct a 95% confidence interval estimate of the population proportion of M&Ms that are red. Is the result consistent with the proportion of 0.2 that is reported by the candy maker Mars?

29. Columbia Pictures chairman Mark Canton claims that 58% of the movies made are R-rated. Refer to the movie data in Data Set 5 of Appendix B and construct the 95% confidence interval for the percentage of movies with R ratings. Is the resulting confidence interval consistent with Canton's claim?

30. To encourage recycling, the City of Providence is considering a penalty for households discarding more than 10 lb of paper in a week. Refer to Data Set 1 in Appendix B and find the sample proportion of households discarding more than 10 lb of paper. Also construct the 95% confidence interval estimate of the percentage of all households that would be penalized under the new provision.

6-3   Exercises B: Beyond the Basics

31. A *New York Times* article about poll results states, "In theory, in 19 cases out of 20, the results from such a poll should differ by no more than one percentage point in either direction from what would have been obtained by interviewing

all voters in the United States." Find the sample size suggested by this statement.

32. Special tables are available for finding confidence intervals for proportions involving small numbers of cases where the normal distribution approximation cannot be used. For example, given 3 successes among 8 trials, the 95% confidence interval is $0.085 < p < 0.755$. Find the confidence interval that would result if you were to incorrectly use the normal distribution as an approximation to the binomial distribution. Are the results reasonably close?

33. A newspaper article indicates that an estimate of the unemployment rate involves a survey of 47,000 people (a typical sample size for Bureau of Labor Statistics surveys). If the reported unemployment rate must have an error no larger than 0.2 percentage point and the rate is known to be about 8%, find the corresponding confidence level.

34. a.   If IQ scores of adults are normally distributed with a mean of 100 and a standard deviation of 15, use the methods presented in Chapter 5 to find the percentage of IQ scores above 130.

    b.   Now assume that you plan to test a sample of adults with the intention of estimating the percentage of IQ scores above 130. How many adults must you test if you want to be 98% confident that your error is no more than 2.5 percentage points? (Use the result from part a.)

35. In this section we developed two formulas used for determining sample size, and in both cases we assumed that either the population is infinite, we are sampling with replacement, or the population is very large. If we have a relatively small population and sample without replacement, we should modify E to include the finite population correction factor as follows:

$$E = z_{\alpha/2}\sqrt{\frac{\hat{p}\,\hat{q}}{n}}\sqrt{\frac{N - n}{N - 1}}$$

Here N is the size of the population.

    a.   Show that the above expression can be solved for n to yield

$$n = \frac{N\hat{p}\,\hat{q}[z_{\alpha/2}]^2}{\hat{p}\,\hat{q}[z_{\alpha/2}]^2 + (N - 1)E^2}$$

    b.   Repeat Exercise 21, assuming that there is a finite population of size $N = 500$ businesses with 10 or fewer employees.

36. A *one-sided confidence interval* for p can be written as

$$p < \hat{p} + E \text{ or } p > \hat{p} - E$$

where z_α replaces $z_{\alpha/2}$ in the expression for E. If Air America wants to report an on-time performance of at least x percent with 95% confidence, construct the appropriate one-sided confidence interval and then find the percent in question. Assume that a random sample of 750 flights results in 630 that are on time.

6-4   Estimating a Population Variance

Many real situations, such as quality control in a manufacturing process, require that we estimate values of population variances or standard deviations. In addition to making products with measurements yielding a desired mean, the manufacturer must make products of *consistent* quality that do not run the gamut from extremely good to extremely poor. As this consistency can often be measured by the variance or standard deviation, these become vital statistics in maintaining the quality of products. Also, there are many other circumstances in which variance and standard deviation play an important role. For example, when banks changed their customer waiting lines from a separate line at each service window to one single main feeder line for all service windows, that change did not affect the mean waiting time but did reduce the variation among waiting times. On average, customers were not processed any faster, but the greater consistency among waiting times resulted in happier customers who were less frustrated by being caught in an exceptionally slow line. In this case, the system was improved through a reduced standard deviation, even though the mean wasn't affected; at no cost to the bank, customers became happier.

In this section we assume that the population has normally distributed values. This assumption was made in earlier sections, but it is a more critical assumption here. In using the Student t distribution in Section 6-2, for example, we required that the population of values be approximately normal, but we accepted departures from normality that were not too severe. However, when we deal with variances by the methods of this section, the use of populations with very nonnormal distributions can lead to gross errors. Consequently, the assumption of normality must be adhered to much more strictly, and we should check the distribution of the data by constructing a histogram to see if it is symmetrical and bell shaped. We describe this sensitivity to a normal distribution by saying that inferences about the population variance σ^2 (or population standard deviation σ) made on the basis of the chi-square distribution (defined below) are not *robust,* meaning that the inferences can be very misleading if the population does not have a normal distribution. In contrast, inferences made about the population mean μ based on the Student t distribution are reasonably robust, because departures from normality that are not too extreme will not lead to gross errors.

When we considered estimates of means and proportions in Sections 6-2 and 6-3, we used the normal and Student t distributions. When developing estimates of variances or standard deviations, we need another distribution referred to as the chi-square distribution.

Chi-Square Distribution

In a normally distributed population with variance σ^2, we randomly select independent samples of size n and compute the sample variance s^2 (see Formula 2-5) for each sample. The sample statistic $\chi^2 = (n - 1)s^2/\sigma^2$ has a distribution called the **chi-square distribution.**

> ### Chi-Square Distribution
>
> Formula 6-7
>
> $$\chi^2 = \frac{(n - 1)s^2}{\sigma^2}$$
>
> where
> n = sample size
> s^2 = sample variance
> σ^2 = population variance

We denote *chi-square* by χ^2, pronounced "kigh square." The specific mathematical equations used to define this distribution are not given here because they are beyond the scope of this text. Instead, you can refer to Table A-4 for critical values of the chi-square distribution. (*Note:* We use Formula 6-7 to calculate chi-square *statistics*, but we use Table A-4 to find chi-square *critical values*.) The chi-square distribution is determined by the number of degrees of freedom, and in this chapter we use $n - 1$ degrees of freedom.

degrees of freedom = $n - 1$

In later chapters we will encounter situations in which the degrees of freedom are not $n - 1$. For that reason, we should not universally equate degrees of freedom with $n - 1$.

Properties of the Distribution of the Chi-Square Statistic

1. The chi-square distribution is not symmetric, unlike the normal and Student t distributions (see Figure 6-6). (As the number of degrees of freedom increases, the distribution becomes more symmetric, as Figure 6-7 illustrates.)

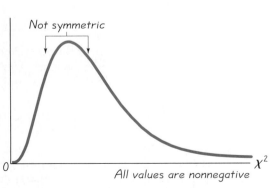

FIGURE 6-6 Chi-Square Distribution

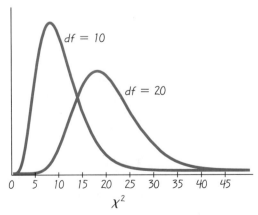

FIGURE 6-7 Chi-Square Distribution for df = 10 and df = 20

2. The values of chi-square can be zero or positive, but they cannot be negative (see Figure 6-6).

3. The chi-square distribution is different for each number of degrees of freedom (see Figure 6-7), which is df $= n - 1$ in this section. As the number of degrees of freedom increases, the chi-square distribution approaches a normal distribution.

In previous sections of this chapter we focused on the topics of estimating population parameters and determining sample size. In this section we again consider those same topics as they relate to variances. However, because of the nature of the chi-square distribution, the techniques discussed here will not closely parallel those in the earlier sections. One difference can be seen in the procedure for finding critical values, illustrated in the following example. Note that when Table A-4 is used, each critical value of χ^2 separates an area to the *right* that corresponds to the value given in the top row of that table.

▶ **EXAMPLE**

Find the critical values of χ^2 that determine critical regions containing an area of 0.025 in each tail. Assume that the relevant sample size is 10; therefore, the number of degrees of freedom is $10 - 1$, or 9.

● **SOLUTION**

See Figure 6-8 and refer to Table A-4. The critical value to the right ($\chi^2 = 19.023$) is obtained in a straightforward manner by locating 9 in the degrees-of-freedom column at the left and 0.025 across the top. The critical value of $\chi^2 = 2.700$ to the left once again corresponds to 9 in the degrees-of-freedom column, but we must locate 0.975 (found by subtracting 0.025 from 1) across the top because the values in the top row are always *areas to the right* of the critical value.

Figure 6-8 shows that, for a sample of 10 scores taken from a normally distributed population, the chi-square *statistic* $(n - 1)s^2/\sigma^2$ has a 0.95 probability of falling between the chi-square *critical values* of 2.700 and 19.023.

When obtaining critical values of chi-square from Table A-4, note that the numbers of degrees of freedom are consecutive integers from 1 to 30, followed by 40, 50, 60, 70, 80, 90, and 100. When a number of degrees of freedom (such as 52) is not found on the table, you can usually use the closest critical value. For example, if the number of degrees of freedom is 52, refer to Table A-4 and use 50 degrees of freedom. For numbers of degrees of freedom greater than

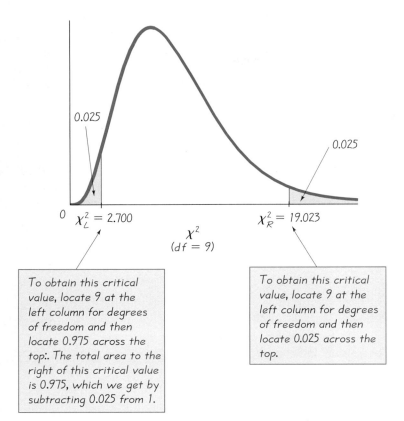

To obtain this critical value, locate 9 at the left column for degrees of freedom and then locate 0.975 across the top:. The total area to the right of this critical value is 0.975, which we get by subtracting 0.025 from 1.

To obtain this critical value, locate 9 at the left column for degrees of freedom and then locate 0.025 across the top.

FIGURE 6-8   Critical Values of the Chi-Square Distribution

100, use the equation given in Exercise 19 or a more detailed table or statistical software package.

Because sample variances s^2 (found by using Formula 2-5) tend to center on the value of the population variance σ^2, we say that s^2 is an *unbiased estimator* of σ^2. Also, the variance of s^2 values tends to be smaller than the variance of the other unbiased estimators. For these reasons s^2 is the best of the various possible statistics we could use to estimate σ^2.

> **The sample variance s^2 is the best point estimate of the population variance σ^2.**

Because s^2 is the best point estimate of σ^2, it would be natural to expect s to be the best point estimate of σ, but this is not the case; s is a biased estimator of σ. If the sample size is large, however, the bias is so small that we can use s as a reasonably good estimate of σ.

Although s^2 is the best point estimate of σ^2, there is no indication of how good it actually is. To compensate for that deficiency, we develop an interval estimate (or confidence interval) that is more revealing.

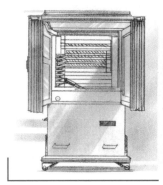

Polls and Psychologists

Poll results can be dramatically affected by the wording of questions. A phrase such as "over the last few years" is interpreted differently by different people. Over the last few years (actually, since 1980), survey researchers and psychologists have been working together to improve surveys by decreasing bias and increasing accuracy. In one case, psychologists studied the finding that 10 to 15 percent of those surveyed say they voted in the last election when they did not. They experimented with theories of faulty memory, a desire to be viewed as responsible, and a tendency of those who usually vote to say that they voted in the most recent election, even if they did not. Only the last theory was actually found to be part of the problem.

Confidence Interval (or Interval Estimate) for the Population Variance σ^2

$$\frac{(n-1)s^2}{\chi_R^2} < \sigma^2 < \frac{(n-1)s^2}{\chi_L^2}$$

The confidence interval (or interval estimate) for σ is found by taking the square root of each component of the preceding inequality:

$$\sqrt{\frac{(n-1)s^2}{\chi_R^2}} < \sigma < \sqrt{\frac{(n-1)s^2}{\chi_L^2}}$$

The notations χ_R^2 and χ_L^2 in the above expressions are described as follows. (Note that some books use $\chi_{\alpha/2}^2$ instead of χ_R^2 and $\chi_{1-\alpha/2}^2$ for χ_L^2.)

Notation

With a total area of α divided equally between the two tails of a chi-square distribution, χ_L^2 denotes the left-tailed critical value and χ_R^2 denotes the right-tailed critical value. (See Figure 6-9.)

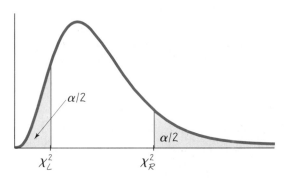

FIGURE 6-9 Chi-Square Distribution with Critical Values χ_L^2 and χ_R^2

The critical values χ_L^2 and χ_R^2 separate the extreme areas corresponding to sample variances that are unlikely (with probability α).

Let's now consider why the confidence intervals for σ^2 and σ have the forms given above. Figure 6-9 shows that there is a probability of $1 - \alpha$ that the statistic $(n-1)s^2/\sigma^2$ will fall between the critical values of χ_L^2 and χ_R^2. In other words (and symbols), there is a $1 - \alpha$ probability that both of the following are true:

$$\frac{(n-1)s^2}{\sigma^2} < \chi_R^2 \quad \text{and} \quad \frac{(n-1)s^2}{\sigma^2} > \chi_L^2$$

If we multiply both of the preceding inequalities by σ^2 and divide each inequality by the appropriate critical value of χ^2, we see that the two inequalities can be expressed in the equivalent forms

$$\frac{(n-1)s^2}{\chi_R^2} < \sigma^2 \quad \text{and} \quad \frac{(n-1)s^2}{\chi_L^2} > \sigma^2$$

These last two inequalities can be combined into one inequality:

$$\frac{(n-1)s^2}{\chi_R^2} < \sigma^2 < \frac{(n-1)s^2}{\chi_L^2}$$

There is a probability of $1 - \alpha$ that the population variance σ^2 is contained in the above interval, and this result corresponds to the definition of the confidence interval for σ^2. We can now construct confidence intervals for σ as illustrated in the next example.

Confidence interval limits for σ^2 and σ should be rounded by using the same basic roundoff rule given in Section 6-2.

Roundoff Rule for Confidence Interval Estimates of σ or σ^2

1. When using the original set of data to construct a confidence interval, round the confidence interval limits to one more decimal place than is used for the original set of data.

2. When the original set of data is unknown and only the summary statistics (n, s) are used, round the confidence interval limits to the same number of decimal places used for the sample standard deviation or variance.

▶ **EXAMPLE**

A container of car antifreeze is supposed to hold 3785 mL of the liquid. Realizing that fluctuations are inevitable, the quality control manager wants to be quite sure that the standard deviation is less than 30 mL. Otherwise, some

(continued)

★★★

The Wisdom of Hindsight

The Roper Organization erred by 10 percentage points when it predicted the outcome of the Mondale-Reagan presidential race. Roper noted that failure to update eight-year-old census data resulted in a sample with a disproportionate number of Democrats. Also, respondents were first asked to identify the candidate they favored and then asked if they planned to vote. This order of questions resulted in an inflated figure for the second question: Respondents who had identified a favorite candidate were more inclined to say that they would vote, even if they were not really inclined to do so. The Roper Organization is usually quite accurate in its results, and steps were taken to prevent such errors in the future.

How Valid Are Crime Statistics?

Police departments are often judged by their number of arrests because that statistic is readily available and clearly understood. However, this encourages police to concentrate on easily solved minor crimes (such as marijuana smoking) at the expense of more serious crimes, which are difficult to solve. A study of Washington, D.C., police records revealed that police can also manipulate statistics another way. The study showed that more than 1000 thefts in excess of $50 were intentionally valued at less than $50 so that they would be classified as petty larceny and not major crimes.

containers would overflow while others would not have enough of the coolant. She randomly selects a sample, with the results given below. Use these sample results to construct the 99% confidence interval for the true value of σ. Does this confidence interval suggest that the fluctuations are at an acceptable level? Assume that the distribution of fill amounts for the containers of antifreeze is a normal distribution.

3761	3861	3769	3772	3675	3861	$n = 18$
3888	3819	3788	3800	3720	3748	$\bar{x} = 3787.0$
3753	3821	3811	3740	3740	3839	$s = 55.4$

● SOLUTION

Based on the sample data, the mean of $\bar{x} = 3787.0$ appears to be acceptable because it is so close to 3785. However, note what happens as we construct the 99% confidence interval for σ^2 and for σ. With a sample size of $n = 18$, we have $n - 1 = 17$ degrees of freedom. Because we want 99% confidence, we let $\alpha = 0.01$ and divide it equally between the two tails to get an area of 0.005 in each tail. In the 17th row of Table A-4, we find that 0.005 in the left and right tails corresponds to $\chi_L^2 = 5.697$ and $\chi_R^2 = 35.718$. With these values and $n = 18$ and $s = 55.4$, we get

$$\frac{(18 - 1)(55.4)^2}{35.718} < \sigma^2 < \frac{(18 - 1)(55.4)^2}{5.697}$$

This becomes $1460.8 < \sigma^2 < 9158.5$. Taking the square root of each part yields $38.2 < \sigma < 95.7$. It appears that the standard deviation is too large, and corrective action must be taken to ensure more consistent container fillings. (Corrective action might involve checking line pressures, replacing worn parts, or retraining operators.)

 ## Using Computer Software for Confidence Intervals about σ or σ^2

If using statistical software, be aware that STATDISK can be easily used to construct confidence intervals for standard deviations or variances. The STATDISK display for the preceding example is shown on page 325.

Minitab does not have a command specifically designed to produce such confidence intervals. Although it does have commands that would allow you to indirectly obtain these confidence interval limits, we will not consider those commands here.

STATDISK DISPLAY OF CONFIDENCE INTERVAL FOR σ

```
About    File    Edit    Statistics    Data    View

                    Confidence Intervals

  Confidence Level ................................. =    .99
  Sample size ...................................n =   18
  Population standard deviation ................... =   55.38


              ┌────────────────────────────────────────┐
              │         The Confidence Interval is      │
              │    ==============================       │
              │                                         │
              │  1457.39 < Population variance < 9239.97 │
              │                    or                   │
              │  38.1757 < Population st. dev. < 96.1248 │
              └────────────────────────────────────────┘



              Use Menu Bar to change screen        Page 3 of 3
     F1: Help       F2: Menu                        ESC: Quit
```

Determining Sample Size

The task of determining the sample size necessary to estimate σ^2 to within given tolerances and confidence levels is much more complex than it was in similar problems that dealt with means and proportions. Instead of developing very complicated procedures, we will use Table 6-2, which lists approximate sample sizes. (Table 6-2 is on page 326.)

▶ **EXAMPLE**

With 99% confidence, you wish to estimate σ^2 to within 10%. How large should your sample be? Assume that the population is normally distributed.

● **SOLUTION**

From Table 6-2, we can see that 99% confidence and an error of 10% for σ^2 correspond to a sample of size 1400. You should randomly select 1400 values from the population.

TABLE 6-2 **Sample Size for σ^2 and σ**

To be 95% confident that s^2 is within	of the value of σ^2, the sample size n should be at least	To be 95% confident that s is within	of the value of σ, the sample size n should be at least
1%	77,210	1%	19,205
5%	3,150	5%	767
10%	806	10%	192
20%	210	20%	47
30%	97	30%	21
40%	57	40%	12
50%	38	50%	8
To be 99% confident that s^2 is within	of the value of σ^2, the sample size n should be at least	To be 99% confident that s is within	of the value of σ, the sample size n should be at least
1%	133,362	1%	33,196
5%	5,454	5%	1,335
10%	1,400	10%	336
20%	368	20%	85
30%	172	30%	38
40%	101	40%	22
50%	67	50%	14

6-4 Exercises A: Basic Skills and Concepts

In Exercises 1 through 16, assume that each sample is obtained by randomly selecting values from a population with a normal distribution.

1. a. Use Table 6-2 to find the approximate minimum sample size necessary to estimate σ^2 with a 30% maximum error and 95% confidence.
   b. If a sample is described by the statistics $n = 27$, $\overline{x} = 75.0$, and $s^2 = 144.0$, find the best point estimate of σ^2.
   c. Find the values of χ_L^2 and χ_R^2 for a sample of 27 scores and a confidence level of 95%.
   d. Use the sample data and results of parts b and c to construct a 95% confidence interval estimate of σ^2.
   e. Use the result of part d to construct a 95% confidence interval estimate of σ.

2. a. Use Table 6-2 to find the approximate minimum sample size necessary to estimate σ with a 10% maximum error and 99% confidence.
   b. If a sample is described by the statistics $n = 18$, $\overline{x} = 16.0$, and $s = 1.50$, find the best point estimate of σ^2. *(continued)*

c. Find the values of χ_L^2 and χ_R^2 for a sample of 18 scores and a confidence level of 99%.

d. Use the sample data and results of parts b and c to construct a 99% confidence interval estimate of σ^2.

e. Use the result of part d to construct a 99% confidence interval estimate of σ.

3. The following IQ scores are obtained from a randomly selected sample.

   85   91   93   99   103   107   111   115   122   92

a. Find the best point estimate of the population variance σ^2.

b. Construct a 95% confidence interval estimate of the population standard deviation σ.

c. Does the confidence interval contain the standard deviation value of 15? (IQ tests are typically designed to produce a mean of $\mu = 100$ and a standard deviation of $\sigma = 15$.)

4. The following are heights (in inches) of randomly selected women.

   63.7   61.2   66.0   62.5   65.7   64.4   63.0   61.9

a. Find the best point estimate of the population variance σ^2.

b. Construct a 99% confidence interval estimate of the population standard deviation σ.

c. Does the confidence interval contain the standard deviation value of 2.5 in.? (Women's heights are known to have a standard deviation of 2.5 in.)

5. The National Center for Education Statistics surveyed college graduates about the lengths of time required to earn their bachelor's degrees. The mean is 5.15 years, and the standard deviation is 1.68 years. Assume that the sample size is 101. Based on these sample data, construct the 99% confidence interval for the standard deviation of the times required by all college graduates.

6. The United States Marine Corps is reviewing its orders for uniforms because it has a surplus of uniforms for tall recruits and a shortage for shorter recruits. Its review involves sample data for men between the ages of 18 and 24. That sample group has a mean height of 69.7 in. with a standard deviation of 2.8 in. (see USDHEW publication 79-1659). Assume that the sample size is 81, and use these sample data to find the 99% confidence interval for the standard deviation of the heights of all men between the ages of 18 and 24.

7. In a time-use study, 20 randomly selected managers were found to spend a mean of 2.40 h each day on paperwork. The standard deviation of the 20 scores is 1.30 h (based on data from Adia Personnel Services). Construct the 95% confidence interval for the standard deviation of the times spent on paperwork by all managers.

8. Data Set 6 in Appendix B includes 17 weights (in grams) of red M&Ms. For this sample, the mean is 0.9043 g and the standard deviation is 0.0414 g. Construct a 90% confidence interval estimate of the standard deviation of the weights of all red M&Ms.

9. If we refer to the weights (in grams) of quarters listed in Data Set 8 in Appendix B, we will find 50 weights with a mean of 5.622 g and a standard deviation of 0.068 g. Based on this random sample of quarters in circulation, construct a 99% confidence interval estimate of the population variance for the weights of all quarters in circulation.

10. Data Set 1 in Appendix B includes the total weights of garbage discarded by 62 households in one week (based on data collected as part of the Garbage Project at the University of Arizona). For that sample, the mean is 27.44 lb and the standard deviation is 12.46 lb. Construct a 99% confidence interval estimate of the standard deviation for the weights of garbage discarded by all households.

11. The U.S. Department of Health, Education, and Welfare collected sample data for 101 women, aged 18 to 24. That sample group has a mean serum cholesterol level (measured in mg/100 mL) of 191.7 with a standard deviation of 41.0 (see USDHEW publication 78-1652). Use these sample data to find the 98% confidence interval for the standard deviation of serum cholesterol levels of all women in the 18–24 age bracket.

12. In a study of physical attractiveness and mental disorders, subjects were rated for attractiveness, and the resulting sample mean and standard deviation are 3.94 and 0.75, respectively. (See "Physical Attractiveness and Self-Perception of Mental Disorder," by Burns and Farina, *Journal of Abnormal Psychology*, Vol. 96, No. 2.) Assume that the sample size is 91, and construct the 95% confidence interval for the population variance.

13. In a study of the use of hypnosis to relieve pain, sensory ratings were measured for 16 subjects, with the results given below. (See "An Analysis of Factors That Contribute to the Efficacy of Hypnotic Analgesia," by Price and Barber, *Journal of Abnormal Psychology*, Vol. 96, No. 1.) Use these sample data to construct the 95% confidence interval for the standard deviation of the population of sensory ratings.

| 8.8 | 6.6 | 8.4 | 6.5 | 8.4 | 7.0 | 9.0 | 10.3 |
| 8.7 | 11.3 | 8.1 | 5.2 | 6.3 | 11.6 | 6.2 | 10.9 |

14. The Internal Revenue Service conducted a study of estates valued at more than $300,000 and determined the value of bonds for a randomly selected sample, with the following results (in dollars). Find a 90% interval estimate of the standard deviation of the bond values for all such estates.

45,300	36,200	72,500	50,500	15,300	58,500
26,200	97,100	74,200	83,700	72,000	10,000
63,000	15,000	49,200	37,500	81,000	24,000
145,000	27,900	53,100	27,500	94,000	23,800
74,600	36,800	65,900	29,400	86,300	25,600
53,200	47,200	61,800	33,200	18,200	75,000

15. Refer to Data Set 1 in Appendix B for the 62 weights (in pounds) of paper discarded by households (based on data from the Garbage Project at the University of Arizona). Using that sample, construct a 95% confidence interval estimate of the standard deviation of weights of paper discarded by all households.

16. Refer to the ages of U.S. commercial aircraft listed in Data Set 10 of Appendix B. Use that sample to construct a 99% confidence interval estimate of the standard deviation of ages of all U.S. commercial aircraft.

6-4   Exercises B: Beyond the Basics

17. A random sample is drawn from a normally distributed population, and it is found that $n = 20$, $\bar{x} = 45.2$, and $s = 3.8$. Based on this sample, the following confidence interval is constructed. Find the degree of confidence.

$$2.8 < \sigma < 6.0$$

18. A random sample of 12 scores is drawn from a normally distributed population, and the 95% confidence interval is as follows. Find the standard deviation of the sample.

$$19.1 < \sigma < 45.8$$

19. In constructing confidence intervals for σ or σ^2, we use Table A-4 to find the critical values χ_L^2 and χ_R^2, but that table applies only to cases in which $n \le 101$, so the number of degrees of freedom is 100 or fewer. For larger numbers of degrees of freedom, we can approximate χ_L^2 and χ_R^2 by using

$$\chi^2 = \frac{1}{2}\left[\pm z_{\alpha/2} + \sqrt{2k - 1} \right]^2$$

Here, k = number of degrees of freedom and $z_{\alpha/2}$ is as described in the preceding sections. Construct the 95% confidence interval about σ by using the following sample data: The measured heights of 772 men between the ages of 18 and 24 have a standard deviation of 2.8 in. (based on data from the National Health Survey, USDHEW publication 79-1659).

20. When 500 tree heights are randomly selected from a normally distributed population, the mean is 253.7 in. and the standard deviation is 4.8 in. Based on these data, the confidence interval shown below is obtained. What is the degree of confidence? (*Hint:* See Exercise 19.)

$$4.5459 < \sigma < 5.0788$$

VOCABULARY LIST

Define and give an example of each term.

estimator	critical value
estimate	margin of error
point estimate	confidence interval limits
confidence interval	Student t distribution
interval estimate	t distribution
degree of confidence	degrees of freedom
level of confidence	chi-square distribution
confidence coefficient	

Important Formulas

Parameter	Point Estimate	Confidence Interval	Sample Size
μ	$\bar{x}$	$\bar{x} - E < \mu < \bar{x} + E$ where $E = z_{\alpha/2}\dfrac{\sigma}{\sqrt{n}}$ if σ is known or if $n > 30$, in which case we use s for σ or $E = t_{\alpha/2}\dfrac{s}{\sqrt{n}}$ if σ is unknown and $n \leq 30$	$n = \left[\dfrac{z_{\alpha/2}\sigma}{E}\right]^2$
p	$\hat{p} = \dfrac{x}{n}$	$\hat{p} - E < p < \hat{p} + E$ where $E = z_{\alpha/2}\sqrt{\dfrac{\hat{p}\,\hat{q}}{n}}$	$n = \dfrac{[z_{\alpha/2}]^2 \hat{p}\,\hat{q}}{E^2}$ or $n = \dfrac{[z_{\alpha/2}]^2 \cdot 0.25}{E^2}$
σ^2	s^2	$\dfrac{(n-1)s^2}{\chi_R^2} < \sigma^2 < \dfrac{(n-1)s^2}{\chi_L^2}$	See Table 6-1.

REVIEW

This chapter introduced important and fundamental concepts of inferential statistics. The main objective of this chapter was to develop procedures for estimating values of these population parameters: means (Section 6-2), proportions (Section 6-3), and variances (Section 6-4). We saw that these parameters have best point estimates. The best point estimate (or single-valued estimate) of a population mean μ is the value of the sample mean $\bar{x}$. The best point estimate of a population proportion p is the value of the sample proportion $\hat{p}$. The best point estimate of a population variance σ^2 is the sample variance s^2. (However, the sample standard deviation s is a *biased* estimate of the population standard deviation σ.) As single values, the point estimates don't convey any real sense of how reliable they are, so we introduced confidence intervals (or interval estimates) as more informative estimates. We also considered ways of determining the sample sizes necessary to estimate parameters to within given tolerance factors. This chapter also introduced the Student t and chi-square distributions. We must be careful to use the correct distribution for each set of circumstances.

Fundamental and important measures used to estimate population parameters were introduced in this chapter, as well as some very important concepts that will be used in later chapters. The concept of a critical value is important in the hypothesis-testing procedures discussed in Chapters 7, 8, 9, 10, 11, and 13. The Student t and chi-square distributions will be used in Chapters 7, 8, 9, and 10. Understanding confidence intervals will help you grasp the rationale underlying the important methods of hypothesis testing, and we will make correspondences between confidence intervals and hypothesis tests.

REVIEW EXERCISES

1. When working high school students were randomly selected and surveyed about the amounts of time they spend at after-school jobs, the mean and standard deviation were found to be 17.6 h and 9.3 h, respectively (based on data from the National Federation of State High School Associations). Assume that the given statistics are based on a sample of size 50 drawn from a normally distributed population.
   a. Find the best point estimate of the population mean.
   b. Find a 95% confidence interval estimate of the population mean.
2. Refer to the data in Exercise 1.
   a. Find the best point estimate of the population variance.
   b. Find a 95% confidence interval for the population standard deviation.
3. The Newton Car Park is a large car dealership that wants information about how long car owners plan to keep their cars. A random sample of 25 car owners results in a sample mean and standard deviation of 7.01 years and 3.74 years, respectively (based on data from a Roper poll). Assume that the sample is drawn

from a normally distributed population. Find a 95% confidence interval estimate of the population standard deviation.

4. Refer to the data given in Exercise 3 and find a 95% confidence interval for the population mean.

5. In clinical trials of the allergy medication Seldane, 70 subjects experienced drowsiness, while 711 did not (based on data from Merrell Dow Pharmaceuticals). Construct the 95% confidence interval for the proportion of Seldane users who experience drowsiness.

6. A psychologist wants to determine the proportion of students in the Providence school district who have divorced parents. How many students must be surveyed if the psychologist wants 96% confidence that the sample proportion is in error by no more than 0.06?

7. A sociologist wants to determine the mean value of cars owned by retired people. If the sociologist wants to be 96% confident that the mean value of the sample group is off by no more than $250, how many retired people must be sampled? A pilot study suggests that the standard deviation is $3050.

8. In a Gallup poll of 1004 adults, 93% indicated that restaurants and bars should refuse service to patrons who have had too much to drink. If you plan to conduct a new poll to confirm that the percentage continues to be correct, how many randomly selected adults must you survey if you want 98% confidence that the margin of error is 4 percentage points?

9. Verbal PSAT scores of a random sample of 40 college-bound high school juniors have a mean of 40.7, a standard deviation of 10.2, and a distribution that is normal (based on data from Educational Testing Service). Find a 99% confidence interval estimate of the population mean.

10. Refer to the data in Exercise 9 and construct a 99% confidence interval for the population standard deviation.

11. You want to determine the percentage of individual tax returns that include capital gains deductions. How many such returns must be randomly selected and checked? You want to be 90% confident that your sample percent is in error by no more than 4 percentage points.

12. Of 1475 transportation workers randomly selected, 32.0% belong to unions (based on data from the U.S. Bureau of Labor Statistics). Construct the 95% confidence interval for the true proportion of all transportation workers who belong to unions.

13. A medical researcher wishes to estimate the serum cholesterol level (in mg/100 mL) of all women aged 18 to 24. There is strong evidence suggesting that $\sigma = 41.0$ mg/100 mL (based on data from a survey of 1524 women aged 18 to 24, as part of the National Health Survey, USDHEW publication 78-1652). If the researcher wants to be 95% confident of obtaining a sample mean that is off by no more than 4 units, how large must the sample be?

14. Based on recent data from the U.S. Bureau of the Census, 13.5% of Americans have incomes below the poverty level. A researcher wants to verify that figure by conducting an independent survey. Assuming that 13.5% is approximately correct, how many randomly selected Americans must be surveyed? The researcher wants to be 96% confident that the sample proportion is within 0.015 of the true population proportion.

15. In a Roper survey of 1,998 randomly selected adults, 24% included loud commercials among the annoying aspects of television. Construct the 99% confidence interval for the percentage of all adults who are annoyed by loud commercials.

16. When 28 seventeen-year-old women are randomly selected and tested with the Goodenough-Harris Drawing Test, the mean of their scores is 37.9 and their standard deviation is 7.3. The distribution of scores for all seventeen-year-old women is known to be normal (based on data from the National Health Survey, USDHEW publication 74-1620). Find a 90% confidence interval for the population mean.

17. The manager at the Dayton Machine Company measures arm lengths of a random sample of male machine operators, and the following values (in centimeters) are obtained. Construct the 95% confidence interval for the mean arm length of all such employees.

$$\begin{array}{cccccccc}
76.8 & 75.6 & 69.3 & 75.7 & 75.5 & 71.2 & 72.5 & 71.9 \\
70.9 & 69.4 & 71.7 & 72.5 & 72.2 & 68.5 & 75.9 & 73.0
\end{array}$$

18. Use the sample data given in Exercise 17 to find the 95% confidence interval for the standard deviation of the population from which the sample was drawn.

COMPUTER PROJECT

In 1977, Stanford University's Bradley Efron introduced the *bootstrap method*, which can be used to construct confidence intervals for situations in which traditional methods cannot (or should not) be used. For example, the following sample of 10 scores was randomly selected from a very nonnormal distribution, so the methods previously discussed cannot be used.

$$\begin{array}{cccccccccc}
2.9 & 564.2 & 1.4 & 4.7 & 67.6 & 4.8 & 51.3 & 3.6 & 18.0 & 3.6
\end{array}$$

The methods of this chapter require that the population have a distribution that is at least approximately normal. The bootstrap method, which makes no assumptions about the original population, typically requires a computer to build a bootstrap population by replicating (duplicating) a sample many times. We can draw from the sample with replacement, thereby creating an approximation of the original population. In this way, we pull the sample up "by its own bootstraps" to simulate the original population. This is a relatively new technique which has been gaining much attention.

Using the sample data given above, construct a 95% confidence interval estimate of the population mean μ by using the bootstrap method as described in the following steps.

a. Create 500 new samples, each of size 10, by selecting 10 scores *with replacement* from the 10 sample scores given above. (This can be accomplished with the first 8 lines in the Minitab commands that follow.)   *(continued)*

b. Find the means of the 500 bootstrap samples generated in part a. (This can be accomplished by using Minitab's RMEAN command, as shown on the 9th line in the Minitab commands.)

c. Rank the 500 means (arrange them in order). (This can be accomplished by using Minitab's SORT command, as shown on the 10th line in the Minitab commands.)

d. Find the percentiles $P_{2.5}$ and $P_{97.5}$ for the ranked means from Step 3. ($P_{2.5}$ is the mean of the 12th and 13th scores in the ranked list; $P_{97.5}$ is the mean of the 487th and 488th scores.) These two values are the limits of the desired confidence interval. Identify the resulting confidence interval by substituting the values for $P_{2.5}$ and $P_{97.5}$ in

$$P_{2.5} < \mu < P_{97.5}$$

The following Minitab commands accomplish Steps a, b, and c. The last command PRINT C21 results in a display of the 500 ranked means. For Step d, the values of $P_{2.5}$ and $P_{97.5}$ can be easily found by computing the mean of the 12th and 13th scores and the mean of the 487th and 488th scores.

```
MTB > SET C1
DATA> 2.9 564.2 1.4 4.7 67.6 4.8 51.3 3.6 18.0 3.6
DATA> ENDOFDATA
MTB > SET C2
DATA> 0.1 0.1 0.1 0.1 0.1 0.1 0.1 0.1 0.1 0.1
DATA> ENDOFDATA
MTB > RANDOM 500 C11-C20;
SUBC> DISCRETE C1 C2.
MTB > RMEAN C11-C20 C21
MTB > SORT C21 C21
MTB > PRINT C21
```

Find the confidence interval estimate of μ, then use the bootstrap method to find a 95% confidence interval about the population standard deviation σ. (Use the three steps previously listed, but in Step b find standard deviations instead of means, and in Step c rank the 500 standard deviations. If using Minitab, change RMEAN to RSTDEV.) Compare your result to the interval

$$318.4 < \sigma < 1079.6$$

which was obtained by incorrectly using the methods described in Section 6-4. (Their use is incorrect because the population distribution is very nonnormal.) This incorrect confidence interval for σ does not contain the true value of σ, which is 232.1. Does the bootstrap procedure yield a confidence interval for σ that contains 232.1, verifying that the bootstrap method is effective? (An alternative to using Minitab is to use special software designed specifically for bootstrap resampling methods. The author recommends *Resampling Stats*, available from Resampling Stats, Inc., 612 N. Jackson St., Arlington, VA 22201.)

FROM DATA TO DECISION: He's angry, but is he right?

The following excerpt is taken from a letter written by a corporation president and sent to the Associated Press.

> When you or anyone else attempts to tell me and my associates that 1223 persons account for our opinions and tastes here in America, I get mad as hell! How dare you! When you or anyone else tells me that 1223 people represent America, it is astounding and unfair and should be outlawed.

The writer then goes on to claim that because the sample size of 1223 people represents 120 million people, his letter represents 98,000 people (120 million divided by 1223) who share the same views.

a. Given that the sample size is 1223 and the degree of confidence is 95%, find the margin of error for the proportion. Assume that there is no prior knowledge about the value of that proportion.

b. The writer is taking the position that a sample size of 1223 taken from a population of 120 million people is too small to be meaningful. Do you agree or disagree? Write a response that either supports or refutes the writer's position that the sample is too small.

c. The writer also makes the claim that because the poll of 1223 people was projected to reflect the opinions of 120 million, any 1 person actually represents 98,000 other people. As the writer is 1 person, he claims to represent 98,000 other people. Is this claim correct? Explain why or why not.

7 Hypothesis Testing

7-1 Overview

Chapter objectives are defined. Hypothesis testing, a major topic of inferential statistics, is introduced.

7-2 Fundamentals of Hypothesis Testing

An informal example of a hypothesis test is presented, and its important components are described. The types of errors that can be made are also discussed.

7-3 Testing a Claim about a Mean: Large Samples

The traditional approach to hypothesis testing is presented, along with the P-value approach and an approach based on confidence intervals.

7-4 Testing a Claim about a Mean: Small Samples

The t test is presented—the method of choice for testing a claim about the mean of a population when the sample is small and the value of the population standard deviation is not known.

7-5 Testing a Claim about a Proportion

The method is described for testing a hypothesis made about a population proportion or percentage.

7-6 Testing a Claim about a Standard Deviation or Variance

The chi-square distribution is described—the distribution of choice for testing a hypothesis made about the standard deviation or variance of a population.

Chapter Problem:
Testing the claim that the mean body temperature of healthy adults is not 98.6°F

Methods of statistics are truly fascinating when they are used to show that a common belief is incorrect. For example, it is currently a common belief that the mean body temperature of healthy adults is 98.6°F. In Chapter 2 we considered a sample of 106 temperatures (see Table 2-1), obtained in a study conducted by University of Maryland researchers, and found that $\bar{x} = 98.2$, $s = 0.62$, and the distribution is approximately bell shaped. In Chapter 6 we used the same set of temperatures to estimate μ, the mean body temperature, and developed the 95% confidence interval $98.08°F < \mu < 98.32°F$. That is, if we were to select many different samples of size 106 and construct the confidence intervals, 95% of them would actually contain the value of the population mean μ. Note that the confidence interval $98.08°F < \mu < 98.32°F$ does *not* contain 98.6°F, the value generally believed to be the mean body temperature.

The confidence interval constitutes an *estimate* of the mean body temperature, but the researchers went further when they made the *claim* that 98.6°F "should be abandoned as a concept having any particular significance for the normal body temperature." That statement implies that we should reject the common belief that the mean body temperature of healthy adults is equal to 98.6°F. We now have a different issue to address. Instead of simply developing an estimate of the mean body temperature, we need to test the claim that $\mu \neq 98.6$. There is a standard procedure for testing such claims, and this chapter will describe that procedure.

7-1 Overview

In Chapter 6 we studied a major topic of inferential statistics—using sample statistics to *estimate* values of population parameters. In this chapter we study another major topic of inferential statistics as we use sample statistics to *test hypotheses* made about population parameters.

DEFINITION

In statistics, a **hypothesis** is a statement that something is true.

The following statements are examples of hypotheses that will be tested by the procedures we develop in this chapter.

- A medical researcher claims that the mean body temperature of healthy adults is not equal to 98.6°F.
- The president of CBS claims that the majority of all adults are not annoyed by violence on television.
- The president of the Jefferson Valley Bank claims that with a single line, customers have more consistent waiting times, with less than the 6.2-min standard deviation for multiple waiting lines.

Before you begin Section 7-2, it would be very helpful to have a general sense of the thinking used in tests of such statements, called **hypothesis tests** or **tests of significance.** Try to follow the reasoning behind the following example.

▶ **EXAMPLE**

Suppose that you take a dime from your pocket and claim that it favors heads when it is flipped. That claim is a hypothesis, and we can test it by flipping the dime 100 times. We would expect to get around 50 heads with a fair coin. If heads occur 94 times out of 100 tosses, most people would agree that the coin favors heads. If heads occur 51 times out of 100 tosses, we should not conclude that the dime favors heads because we could easily get 51 heads with a fair and unbiased coin. But what about 60 heads out of 100 tosses? Or 75? Here is the key point: We should conclude that the dime favors heads only if we get *significantly* more heads than we would expect with an unbiased dime. A result of 51 heads out of 100 tosses could easily happen by chance—it is not significant. But an outcome of 94 heads out of 100 tosses is not likely to happen by chance and is therefore significant. Later, we will establish exact criteria for identifying results that are significant.

This brief example illustrates the basic approach used in testing hypotheses. The formal method involves a variety of standard terms and conditions incorporated in an organized procedure. It is recommended that you begin the study of this chapter by first reading Sections 7-2 and 7-3 casually to obtain a general idea of their concepts. Then re-read Section 7-2 more carefully to become familiar with the terminology.

7-2   Fundamentals of Hypothesis Testing

This section introduces the concept of hypothesis testing by presenting an informal example. Then the important components of hypothesis testing are described. In Section 7-3 formal methods of a complete hypothesis test will be considered.

Because hypothesis testing involves a sequence of steps, there is a temptation to memorize the steps and apply them in a rote and mechanical way. It is far better to *understand* the procedure because (1) you will be more likely to remember it, (2) you will be prepared to apply it to different circumstances, and (3) you will be less likely to make gross errors. The following informal example of a hypothesis test excludes formal terminology and focuses instead on the type of reasoning we will use. Read it slowly, carefully, and repeatedly until you thoroughly understand it, and you will have captured a major concept of statistics.

▶ **EXAMPLE**

Consider the claim at the beginning of the chapter that the mean body temperature of healthy adults is not equal to 98.6°F ($\mu \neq 98.6$). The standard statistical procedure of hypothesis testing is designed to test such claims. The University of Maryland researchers collected sample data with the following characteristics: $n = 106$, $\bar{x} = 98.2$, $s = 0.62$, and the shape of the distribution is approximately normal. Here is the key question: *Do the sample data constitute sufficient evidence to warrant rejection of the common belief that $\mu = 98.6$?* Our answer will be based on the *probability* of getting a sample mean such as $\bar{x} = 98.2$, assuming that the population mean really is $\mu = 98.6$, as is commonly believed. With the assumption that $\mu = 98.6$, we will use the probability of getting a sample mean such as 98.2 to determine whether the sample mean of 98.2 is *likely* to occur, or whether it is so *unlikely* to occur that it constitutes significant evidence against the assumption that $\mu = 98.6$; that is, we want to determine whether 98.2 represents a *chance* difference or a *significant* difference. It's like using probabilities to

(continued)

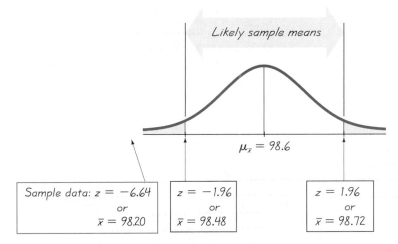

FIGURE 7-1 Central Limit Theorem: The Distribution of Sample Means

determine that the occurrence of 94 heads in 100 tosses of a dime is very unlikely, whereas the outcome of 51 heads in 100 tosses is likely.

Because we are concerned with the probability of a sample mean, we can use the central limit theorem (Section 5-4), which states that sample means tend to be normally distributed with mean $\mu_{\bar{x}} = \mu$ and standard deviation $\sigma_{\bar{x}} = \sigma/\sqrt{n}$. We assume that $\mu = 98.6$, and we know that the sample size is $n = 106$. We don't know the value of σ, but we can approximate it by using $s = 0.62$. If $\mu = 98.6$ (as we are assuming for the purpose of the test), the distribution of the sample means (for $n = 106$) is as shown in Figure 7-1. From that figure we see that if $\mu = 98.6$, then 95% of sample means should fall within 1.96 standard deviations of $\mu = 98.6$. The standard deviation is estimated to be

$$\sigma_{\bar{x}} = \frac{\sigma}{\sqrt{n}} \approx \frac{s}{\sqrt{n}} = \frac{0.62}{\sqrt{106}} = 0.06$$

and the values of 98.48 and 98.72 are values of $98.6 - (1.96)(0.06)$ and $98.6 + (1.96)(0.06)$. While 95% of sample means should fall between 98.48 and 98.72, our sample mean of 98.2 falls below 98.48, indicating that it is unlikely; that is, assuming that $\mu = 98.6$, there is less than a 0.05 probability of getting a sample with a mean of 98.2. Because this probability is so small, we *conclude that there is sufficient evidence to reject our assumption that* $\mu = 98.6$.

Again, it should be stressed that you need to thoroughly understand the preceding informal example. Read it several times until the reasoning makes sense. Focus on the key idea that there is an assumption that $\mu = 98.6$, but the sample mean is $\bar{x} = 98.2$; by using the central limit theorem, we determine that the probability of getting a sample with a mean of 98.2 is very small, suggesting that the assumption of $\mu = 98.6$ should be rejected. Section 7-3 will describe the specific steps used in hypothesis testing, but we will first describe the components of a formal hypothesis test.

Components of a Formal Hypothesis Test

- The **null hypothesis** (denoted by H_0) is a statement about the value of a population parameter (such as the mean μ), and it must contain the condition of equality (that is, it must be written with the symbol $=$, $\leq$, or $\geq$). For the mean, the null hypothesis will be stated in only one of three possible forms: H_0: $\mu = $ some value, H_0: $\mu \leq $ some value, or H_0: $\mu \geq $ some value. For example, the null hypothesis corresponding to the common belief that the mean body temperature is 98.6°F is expressed as H_0: $\mu = 98.6$. We test the null hypothesis *directly* in the sense that the conclusion will be either a rejection of H_0 or a failure to reject H_0.

- The **alternative hypothesis** (denoted by H_1) is the statement that must be true if the null hypothesis is false. For the mean, the alternative hypothesis will be stated in only one of three possible forms: H_1: $\mu \neq $ some value, H_1: $\mu < $ some value, or H_1: $\mu > $ some value. Note that H_1 is the opposite of H_0. For example, if H_0: $\mu = 98.6$, then it follows that the alternative hypothesis is given by H_1: $\mu \neq 98.6$.

Very Important Note 1: Depending on the original wording of the problem, the original claim will sometimes be the null hypothesis H_0, and at other times it will be the alternative hypothesis H_1. Regardless of whether the original claim corresponds to H_0 or H_1, the null hypothesis H_0 must always contain equality (with the symbolic form of $=$, $\leq$, or $\geq$). Table 7-1 illustrates some common cases. See Table 7-1 on the next page.

Very Important Note 2: Even though we sometimes express H_0 with the symbol $\leq$ or $\geq$ as in H_0: $\mu \leq 98.6$ or H_0: $\mu \geq 98.6$, we conduct the test by assuming that H_0: $\mu = 98.6$ is true. We must have a fixed and specific value for μ so that we can work with a single distribution having a specific mean.

Very Important Note 3: If we are making our own claims, we should arrange the null and alternative hypotheses so that the most serious error would be the rejection of a true null hypothesis (see the type I error described below and also Exercise 14). In this text we assume that we are testing a claim made by someone else. Ideally, all claims would be made so that they would all be null hypotheses. Unfortunately, our real world is not ideal. There is poverty, war, crime, and people who make claims that are actually alternative hypotheses. This text was written with the understanding that not all original claims are as they should be. As a result, some examples and exercises involve claims that are null hypotheses, whereas others involve claims that are alternative hypotheses.

TABLE 7-1   Examples of Claims and the Corresponding Null and Alternative Hypotheses

Original Claim: The mean temperature is . . .

	Equal to 98.6	Not 98.6	At least 98.6	Above 98.6	At most 98.6	Below 98.6
Symbolic Form of Original Claim	$\mu = 98.6$	$\mu \neq 98.6$	$\mu \geq 98.6$	$\mu > 98.6$	$\mu \leq 98.6$	$\mu < 98.6$
Symbolic Form That Is True When Original Claim Is False	$\mu \neq 98.6$	$\mu = 98.6$	$\mu < 98.6$	$\mu \leq 98.6$	$\mu > 98.6$	$\mu \geq 98.6$
Null Hypothesis H_0 (must contain equality)	$H_0:$ $\mu = 98.6$	$H_0:$ $\mu = 98.6$	$H_0:$ $\mu \geq 98.6$	$H_0:$ $\mu \leq 98.6$	$H_0:$ $\mu \leq 98.6$	$H_0:$ $\mu \geq 98.6$
Alternative Hypothesis H_1 (cannot contain equality)	$H_1:$ $\mu \neq 98.6$	$H_1:$ $\mu \neq 98.6$	$H_1:$ $\mu < 98.6$	$H_1:$ $\mu > 98.6$	$H_1:$ $\mu > 98.6$	$H_1:$ $\mu < 98.6$

For example, a Sears salesperson might claim that Diehard batteries last more than 5 years. The original claim of $\mu > 5$ does not contain equality, so it is an alternative hypothesis and the null hypothesis becomes $\mu \leq 5$. If the salesperson claimed that the mean was at least 5 years ($\mu \geq 5$), then the original claim would become the null hypothesis because it would contain equality.

When testing a null hypothesis, we arrive at a conclusion of rejecting it or failing to reject it. Such conclusions are sometimes correct and sometimes wrong, but there are two different types of errors that can be made. Table 7-2 summarizes the different possibilities and shows that we make a correct decision when we either reject a null hypothesis that is false or fail to reject a null hypothesis that is true. However, we make an error when we reject a true null hypothesis or fail to reject a false null hypothesis. The type I and type II errors are described below.

- **Type I error:** The mistake of rejecting the null hypothesis when it is true. For the preceding informal example, a type I error is the mistake of rejecting the null hypothesis that the mean body temperature is 98.6 ($\mu = 98.6$) when that mean really is 98.6. The type I error is not a miscalculation or procedural misstep; it is an actual error that can occur when a rare event happens by chance. The probability of rejecting the null hypothesis when it is true is called the **significance level;** that is, the significance level is the probability of a type I error. The symbol $\boldsymbol{\alpha}$ **(alpha)** is used to represent the significance level. The values of $\alpha = 0.05$ and $\alpha = 0.01$ are commonly used.

TABLE 7-2   Type I and Type II Errors

		True State of Nature	
		The null hypothesis is true.	The null hypothesis is false.
Decision	We decide to reject the null hypothesis.	**Type I error**	Correct decision
	We fail to reject the null hypothesis.	Correct decision	**Type II error**

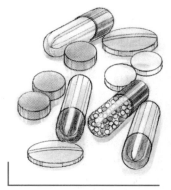

Drug Screening: False Positives

The American Management Association reports that more than half of all companies test for drugs. For a job applicant undergoing drug screening, a *false positive* is an indication of drug use when he or she does not use drugs. A *false negative* is an indication of no drug use when he or she is a user. The *test sensitivity* is the probability of a positive indication for a drug user; the *test specificity* is the probability of a negative indication for a nonuser. Suppose that 3% of job applicants use drugs, and a test has a sensitivity of 0.99 and a specificity of 0.98. The high probabilities make the test seem reliable, but 40% of the positive indications will be false positives.

- **Type II error:** The mistake of failing to reject the null hypothesis when it is false. For the preceding informal example, a type II error is the mistake of failing to reject the null hypothesis ($\mu = 98.6$) when it is actually false (that is, the mean is not 98.6). The symbol β (**beta**) is used to represent the probability of a type II error.

The following terms are associated with key components in the hypothesis-testing procedure.

- **Test statistic:** A sample statistic or a value based on the sample data. A test statistic is used in making the decision about the rejection of the null hypothesis. For the data in the preceding informal example, we can use the central limit theorem to get a test statistic of $z = -6.64$ as follows:

$$z = \frac{\overline{x} - \mu_{\overline{x}}}{\frac{\sigma}{\sqrt{n}}} = \frac{98.2 - 98.6}{\frac{0.62}{\sqrt{106}}} = -6.64$$

- **Critical region:** The set of all values of the test statistic that would cause us to reject the null hypothesis. For the preceding informal example, the critical region is represented by the shaded part of Figure 7-1 and consists of values of the test statistic less than $z = -1.96$ or greater than $z = 1.96$.

- **Critical value:** The value or values that separate the critical region from the values of the test statistic that would not lead to rejection of the null hypothesis. The critical values depend on the nature of the null hypothesis, the relevant sampling distribution, and the level of significance α. For the preceding informal example, the critical values of $z = -1.96$ and $z = 1.96$ separate the shaded critical regions. See Figure 7-1.

We will now discuss some important issues in hypothesis testing.

Controlling Type I and Type II Errors   We noted that α is the probability of a type I error (rejecting a true null hypothesis) whereas β is the probability of a type II error (failing to reject a false null hypothesis). One step in our procedure for testing hypotheses involves the selection of α [or P(type I error)]. However, we don't select β [P(type II error)]. It would be great if we could always have $\alpha = 0$ and $\beta = 0$, but in reality that is not possible and we must attempt to manage the α and β error probabilities. Mathematically, it can be shown that α, β, and the sample size n are all related, so when you choose or determine any two of them, the third is automatically determined. We could select both α and β (the required sample size n would then be determined), but the usual practice in research and industry is to determine in advance the values of α and n, so the value of β is determined. Depending on the seriousness of a type I error, try to use the largest α that you can tolerate. For type I errors with more serious consequences, select smaller values of α. Then choose a sample size n as large as is reasonable, based on considerations of time, cost, and other such relevant factors. (Sample size determinations were discussed in Section 6-2.) The following practical considerations may be relevant:

1. For any fixed α, an increase in the sample size n will cause a decrease in β. That is, a larger sample will lessen the chance that you fail to reject a false null hypothesis.

2. For any fixed sample size n, a decrease in α will cause an increase in β. Conversely, an increase in α will cause a decrease in β.

3. To decrease both α and β, increase the sample size.

As a practical example, let's consider M&Ms (produced by Mars, Inc.) and Bufferin brand aspirin tablets (produced by Bristol-Myers Products). The M&M package contains 496 candies. The mean weight of the individual candies is 0.916 g, and the M&M package is labeled as containing 453 g. The Bufferin package is labeled as holding 30 tablets, each of which contains 325 mg of aspirin. (See the sample data in Data Sets 6 and 9 of Appendix B.) Because M&Ms are candies used for enjoyment whereas Bufferin tablets are drugs used for treatment of health problems, we are dealing with two very different levels of seriousness. If the M&Ms don't have a population mean weight of 0.916 g, the consequences are not very serious, but if the Bufferin tablets don't have a mean weight of 325 mg of aspirin, the consequences could be very serious. If the M&Ms have a mean that is too large, Mars will lose some money but consumers will not complain. In contrast, if the Bufferin tablets have too much aspirin, Bristol-Myers could be faced with consumer lawsuits and Federal Drug Administration actions. Consequently, in testing the claim that $\mu = 0.916$ g for M&Ms, we might choose $\alpha = 0.05$ and a sample size of $n = 100$; in testing the claim that $\mu = 325$ mg for Bufferin tablets, we might choose $\alpha = 0.01$ and a sample size of $n = 500$. The smaller significance level α and larger sample size n are chosen because of the more serious consequences associated with a commercial drug.

Conclusions in Hypothesis Testing

We have already noted that the original claim sometimes becomes the null hypothesis and at other times becomes the alternative hypothesis. However, our procedure requires that we always test the null hypothesis. In Section 7-3 we will see that our initial conclusion will always be one of the following:

1. Fail to reject the null hypothesis H_0.
2. Reject the null hypothesis H_0.

The conclusion of failing to reject the null hypothesis or rejecting it is fine for those of us with the wisdom to take a statistics course, but then it's usually necessary to use simple, nontechnical terms in stating what the conclusion suggests. Students often have difficulty formulating this final nontechnical statement, which describes the practical consequence of the data and computations. It's important to be precise in the language used; the implications of words such as "support" and "fail to reject" are very different. Figure 7-2 shows how to formulate the correct wording of the final conclusion. Note that

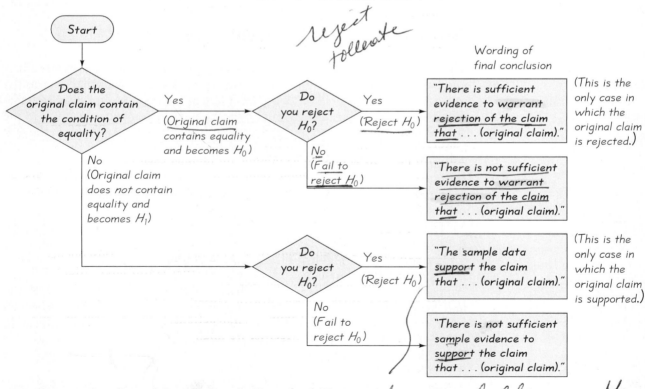

FIGURE 7-2   Wording of Conclusions in Hypothesis Tests

only one case leads to wording indicating that the sample data actually *support* the conclusion. If you want to justify some claim, state it in such a way that it becomes the *alternative* hypothesis and then hope that the null hypothesis gets rejected. For example, if you want to justify the claim that the mean body temperature is different from 98.6°F, then make the claim that $\mu \neq 98.6$. This claim will be an alternative hypothesis that will be supported if you reject the null hypothesis of $H_0: \mu = 98.6$. If, on the other hand, you claim that $\mu = 98.6$, you will either reject or fail to reject the claim; in either case you will not support the original claim.

Some texts say "accept the null hypothesis," instead of "fail to reject the null hypothesis." Whether we use the term *accept* or *fail to reject,* we should recognize that *we are not proving the null hypothesis;* we are merely saying that the sample evidence is not strong enough to warrant rejection of the null hypothesis. It's like a jury's saying that there is not enough evidence to convict a suspect. The term *accept* is somewhat misleading because it seems to incorrectly imply that the null hypothesis has been proved. The phrase *fail to reject* says more correctly that the available evidence isn't strong enough to warrant rejection of the null hypothesis. In this text we will use the conclusion *fail to reject the null hypothesis,* instead of *accept the null hypothesis.*

Two-Tailed, Left-Tailed, Right-Tailed

The *tails* in a distribution are the extreme regions bounded by critical values. Our informal example of hypothesis testing involved a **two-tailed** test in the sense that the critical region of Figure 7-1 is in the two extreme regions (tails) under the curve. We reject the null hypothesis H_0 if our test statistic is in the critical region because that indicates a significant discrepancy between the null hypothesis and the sample data. Some tests will be **left-tailed,** with the critical region located in the extreme left region under the curve. Other tests may be **right-tailed,** with the critical region in the extreme right region under the curve.

In two-tailed tests, the level of significance α is divided equally between the two tails that constitute the critical region. For example, in a two-tailed test with a significance level of $\alpha = 0.05$, there is an area of 0.025 in each of the two tails. In tests that are right- or left-tailed, the area of the critical region is α.

By examining the null hypothesis H_0, we should be able to deduce whether a test is right-tailed, left-tailed, or two-tailed. The tail will correspond to the critical region containing the values that would conflict significantly with the null hypothesis. A useful check is summarized in the margin figures, which show how the inequality sign in H_1 points in the direction of the critical region. The symbol $\neq$ is often expressed in programming languages as $<>$, and this reminds us that an alternative hypothesis such as $\mu \neq 98.6$ corresponds to a *two*-tailed test.

Section 7-3 will discuss formal procedures for testing hypotheses. For now, you should be able to deal with the components included in the following examples.

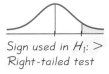

Sign used in H_1: $>$
Right-tailed test

Sign used in H_1: $<$
Left-tailed test

Sign used in H_1: $\neq$
Two-tailed test

► EXAMPLE

After analyzing 106 body temperatures of healthy adults, a medical researcher makes a claim that the mean body temperature is *less than* 98.6°F.

a. Express this claim in symbolic form.

b. Identify the null hypothesis H_0.

c. Identify the alternative hypothesis H_1.

d. Identify this test as being two-tailed, left-tailed, or right-tailed.

e. Identify the type I error for this test.

f. Identify the type II error for this test.

g. Assume that the conclusion is to reject the null hypothesis. State the conclusion in nontechnical terms; be sure to address the original claim.

h. Assume that the conclusion is failure to reject the null hypothesis. State the conclusion in nontechnical terms; be sure to address the original claim.

● SOLUTION

a. The claim that the mean body temperature is less than 98.6 is expressed as $\mu < 98.6$.

b. The original claim of $\mu < 98.6$ does not contain equality, as required by the null hypothesis. The original claim is therefore the alternative hypothesis; the null hypothesis is H_0: $\mu \geq 98.6$.

c. See part b. The alternative hypothesis is H_1: $\mu < 98.6$.

d. This test is left-tailed because the null hypothesis is rejected if the sample mean is significantly *less than* (or to the left of) 98.6. (As a double check, note that the alternative hypothesis $\mu < 98.6$ contains the sign $<$, which points to the left.)

e. The type I error (rejection of a true null hypothesis) is to reject the null hypothesis $\mu \geq 98.6$ when the population mean is really equal to or greater than 98.6.

f. The type II error (failure to reject a false null hypothesis) is to fail to reject the null hypothesis $\mu \geq 98.6$ when the population mean is really less than 98.6.

g. See Figure 7-2. If we reject the null hypothesis, we then conclude that there is sufficient evidence to support the claim that the mean is less than 98.6.

h. See Figure 7-2. If we fail to reject the null hypothesis, we conclude that there is not sufficient evidence to support the claim that the mean is less than 98.6.

▶ **EXAMPLE**

A claim about the population mean weight of all aspirin tablets is tested with a significance level of $\alpha = 0.05$. The conditions are such that the standard normal distribution can be used (because the central limit theorem applies). Find the critical value(s) of z if the test is (a) two-tailed, (b) left-tailed, and (c) right-tailed.

● **SOLUTION**

a. In a two-tailed test, the significance level of $\alpha = 0.05$ is divided equally between the two tails, so there is an area of 0.025 in each tail. We can find the critical values in Table A-2 as the values corresponding to areas of 0.4750 (found by subtracting 0.025 from 0.5) to the right and left of the mean. We get critical values of $z = -1.96$ and $z = 1.96$, as shown in Figure 7-3a.

b. In a left-tailed test, the significance level of $\alpha = 0.05$ is the area of the critical region at the left, so the critical value corresponds to an area of 0.4500 (found from $0.5 - 0.05$) to the left of the mean. Using Table A-2, we get a critical value of $z = -1.645$, as shown in Figure 7-3b.

c. In a right-tailed test, the significance level of $\alpha = 0.05$ is the area of the critical region to the right, so the critical value corresponds to an area of 0.4500 (found from $0.5 - 0.05$) to the right of the mean. Using Table A-2, we get a critical value of $z = 1.645$, as shown in Figure 7-3c.

(a)

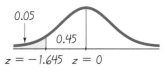

(b)

(c)

FIGURE 7-3
Finding Critical Values

7-2 Exercises A: Basic Skills and Concepts

In Exercises 1 through 6, assume that a hypothesis test of the given claim will be conducted.

    a. Express the claim in symbolic form.
    b. Identify the null hypothesis H_0.
    c. Identify the alternative hypothesis H_1.
    d. Identify the test as being two-tailed, left-tailed, or right-tailed.
    e. Identify the type I error for the test.
    f. Identify the type II error for the test.
    g. Assume that the conclusion is to reject the null hypothesis. State the conclusion in nontechnical terms; be sure to address the original claim.
    h. Assume that the conclusion is failure to reject the null hypothesis. State the conclusion in nontechnical terms; be sure to address the original claim.

1. The mean IQ of statistics instructors is equal to 120.

2. The mean weight of paper discarded by households each week is less than 10 lb.
3. The mean time required for undergraduates to earn a bachelor's degree is greater than 5 years.
4. The mean annual income of self-employed doctors is $191,000.
5. The mean age of U.S. commercial aircraft is at least 10 years.
6. The mean fuel consumption rate for Chevrolet Corvettes is no more than 17 mi/gal.

In Exercises 7 through 12, find the critical z values for the given conditions. In each case assume that the normal distribution applies, so Table A-2 can be used. Also, draw a graph showing the critical value and critical region.

7. Two-tailed test; $\alpha = 0.01$
8. Two-tailed test; $\alpha = 0.10$
9. Right-tailed test; $\alpha = 0.01$
10. Left-tailed test; $\alpha = 0.025$
11. Left-tailed test; $\alpha = 0.05$
12. Right-tailed test; $\alpha = 0.02$

7-2 Exercises B: Beyond the Basics

13. Identify the null hypotheses suggested by the given conclusions.
    a. There is sufficient evidence to support the claim that the mean age of college professors is greater than 30 years.
    b. There is not sufficient evidence to support the claim that the mean weight of quarters is less than 5 g.
    c. There is sufficient evidence to warrant rejection of the claim that adult men have a mean height of 70 in.
14. We stated that when making our own claim, we should arrange the null and alternative hypotheses so that the most serious error would be the rejection of a true null hypothesis. Because the probability of that (type I) error is the significance level α, we should make α very small if the result of a type I error would be very serious. Suppose that we must make a claim about the mean weight of all garbage discarded by households in a week. Using Data Set 1 in Appendix B as a pilot study, we see that the mean appears to be about 27.44 lb. However, underestimating the true mean would be a very serious error because garbage pickup and landfill provisions would be insufficient. On the other hand, overestimating the true mean would lead to waste in labor, equipment, and land costs.
    a. Which error is more serious?
    b. Based on the result of part a, which null and alternative hypotheses should be used?
    c. If the more serious error is *extremely* serious because it will cost you your job, which of these significance levels should you use: 0.05 or 0.01?
15. In a hypothesis test, can the significance level α be set equal to 0? How?

7-3   Testing a Claim about a Mean: Large Samples

The preceding section presented one informal example of a hypothesis test, but the emphasis was on introducing some of the basic components of hypothesis testing. This section makes use of those components in three different procedures for hypothesis testing. The first procedure is the traditional method, and it will be the procedure used most often throughout the remainder of this book. The second procedure, based on *P*-values, will be referred to often, whereas the third procedure, based on confidence intervals, will be used in only a few selected cases.

Before considering the three procedures of hypothesis testing, we will identify the two assumptions that apply to the methods of this section.

Assumptions for Testing the Claim about the Mean of a Single Population

1. The sample is large ($n > 30$), so the central limit theorem applies and we can use the normal distribution.

2. When applying the central limit theorem, we can use the sample standard deviation s as an estimate of the population standard deviation σ whenever σ is unknown and the sample size is large ($n > 30$).

The Traditional Method of Testing Hypotheses

Figure 7-4 summarizes the steps used in the **traditional** (or **classical**) **method** of testing hypotheses. This procedure uses the components described in Section 7-2 as part of a scheme to identify a sample result that is *significantly* different from the claimed value. The relevant sample statistic (such as $\bar{x}$) is converted to a test statistic, which we compare to a critical value. For the hypothesis tests of this section, the test statistic required in Step 6 can be calculated as follows:

Test Statistic for Claims about μ When $n > 30$

$$z = \frac{\bar{x} - \mu_{\bar{x}}}{\frac{\sigma}{\sqrt{n}}}$$

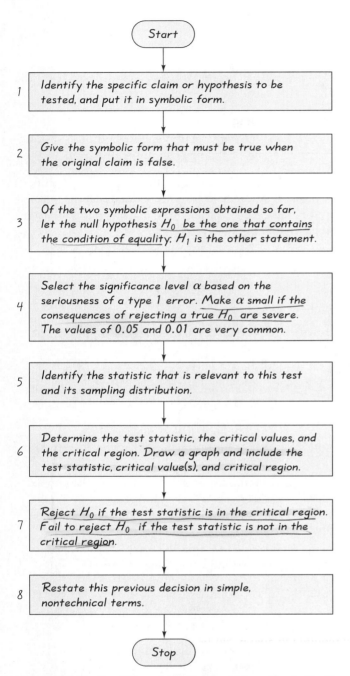

FIGURE 7-4   Traditional Method of Hypothesis Testing

Note that Step 7 in Figure 7-4 uses this decision criterion:

> **Reject the null hypothesis if the test statistic is in the critical region.**
> **Fail to reject the null hypothesis if the test statistic is not in the**
> **critical region.**

Section 7-2 presented an informal example of a hypothesis test; the following example formalizes that test.

▶ **EXAMPLE**

Using the sample data given at the beginning of the chapter ($n = 106$, $\bar{x} = 98.2$, $s = 0.62$) and a 0.05 significance level, test the claim that the mean body temperature of healthy adults is equal to 98.6°F. Use the traditional method by following the procedure outlined in Figure 7-4.

● **SOLUTION**

Refer to Figure 7-4 and follow these steps.

Step 1: The claim that the mean is equal to 98.6 is expressed in symbolic form as $\mu = 98.6$.

Step 2: The alternative (in symbolic form) to the original claim is $\mu \neq 98.6$.

Step 3: The statement $\mu = 98.6$ contains the condition of equality, and so it becomes the null hypothesis. We have

$$H_0: \mu = 98.6 \qquad H_1: \mu \neq 98.6$$

Step 4: As specified in the statement of the problem, the significance level is $\alpha = 0.05$.

Step 5: Because the claim is made about the population mean μ, the sample statistic most relevant to this test is $\bar{x} = 98.2$. Because $n > 30$, the central limit theorem indicates that the distribution of sample means can be approximated by a *normal* distribution.

Step 6: The test statistic is found by converting the sample mean of $\bar{x} = 98.2$ to $z = -6.64$ through the following computation.

$$z = \frac{\bar{x} - \mu_{\bar{x}}}{\frac{\sigma}{\sqrt{n}}} = \frac{98.2 - 98.6}{\frac{0.62}{\sqrt{106}}} = -6.64$$

The critical z values are found by first noting that the test is two-tailed because a sample mean significantly less than or greater than 98.6 is strong evidence against the null hypothesis that $\mu = 98.6$. We now divide $\alpha = 0.05$ equally between the two tails to get 0.025 in each tail. We then refer to Table A-2 (because the central limit theorem lets us assume a normal distribution) to find the z value corresponding to $0.5 - 0.025$, or 0.4750. After finding $z = 1.96$, we use the property of symmetry to conclude that the left critical value is $z = -1.96$. The test statistic, critical region, and critical values are shown in Figure 7-5.

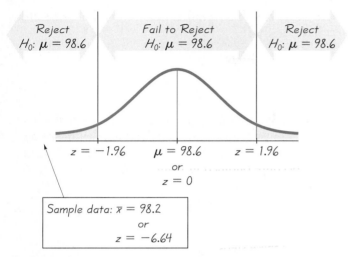

FIGURE 7-5   **Distribution of Means of Body Temperatures**

Step 7:   The sample mean of $\bar{x} = 98.2$ is converted to a test statistic of $z = -6.64$, which falls within the critical region, so we *reject the null hypothesis.*

Step 8:   To restate the Step 7 conclusion in nontechnical terms, we might refer to Figure 7-2 in the preceding section. We are rejecting the null hypothesis, which is the original claim. We conclude that there is sufficient evidence to warrant rejection of the claim that the mean body temperature of healthy adults is 98.6°F.

In hypothesis testing, it is easy to get lost in the mechanics of the test and lose sight of what is really happening. We should recognize that the test statistic is just a z score representing the *number of standard deviations* that our sample mean $\bar{x}$ is away from the claimed population mean μ. If the claimed value of μ is in fact the true value, then we would expect $\bar{x}$ to be reasonably close to μ. But if $\bar{x}$ is very far away from μ, then this unusual result leads us to believe that the claimed value of μ is incorrect, and so we reject the null hypothesis.

In the preceding example, the sample mean of 98.2 led to a test statistic of $z = -6.64$. That is, *the sample mean is 6.64 standard deviations away from the claimed mean of $\mu = 98.6$.* In Chapter 2 we saw that a z score of 6.64 is quite large and corresponds to an unusual result if in fact $\mu = 98.6$. This evidence is strong enough to make us believe that the mean is actually different from 98.6. The critical values and critical region clearly identify the range of unusual values that cause us to reject the claimed value of μ.

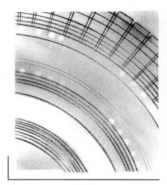

Commercials

Television networks have their own clearance departments for screening commercials and verifying claims. The National Advertising Division, a branch of the Council of Better Business Bureaus, investigates advertising claims. The Federal Trade Commission and local district attorneys also become involved. In the past, Firestone had to drop a claim that its tires resulted in 25% faster stops, and Warner Lambert had to spend $10 million informing customers that Listerine doesn't prevent or cure colds. Many deceptive ads are voluntarily dropped, and many others escape scrutiny simply because the regulatory mechanisms can't keep up with the flood of commercials.

The preceding example illustrated a two-tailed hypothesis test; the following example illustrates a left-tailed test.

▶ **EXAMPLE**

The Mill Valley Brewery distributes beer in bottles labeled 32 oz. The Bureau of Weights and Measures randomly selects 50 of these bottles, measures their contents, and obtains a sample mean of 31.8 oz and a sample standard deviation of 0.75 oz. Using a 0.01 significance level, test the Bureau's claim that the brewery is cheating consumers.

● **SOLUTION**

The brewery is cheating consumers if it distributes bottles containing significantly less than 32 oz of beer. Again refer to Figure 7-4 and follow these steps.

Step 1: The claim that the mean is less than 32 oz becomes $\mu < 32$ oz in symbolic form.

Step 2: The opposite (in symbolic form) of the original claim is $\mu \geq 32$ oz.

Step 3: The statement $\mu \geq 32$ oz contains the condition of equality and therefore becomes the null hypothesis H_0. (Original claim: $\mu < 32$.)

$$H_0: \mu \geq 32 \text{ oz} \qquad H_1: \mu < 32 \text{ oz}$$

Step 4: With a 0.01 significance level, we have $\alpha = 0.01$.

Step 5: The sample mean $\bar{x} = 31.8$ oz should be used to test a claim made about the population mean μ. Because $n > 30$, the central limit theorem indicates that the distribution of sample means can be approximated by a normal distribution.

Step 6: In calculating the test statistic, we can use $s = 0.75$ as a reasonable estimate of σ (because $n > 30$), so the test statistic of $z = -1.89$ is computed as follows:

$$z = \frac{\bar{x} - \mu_{\bar{x}}}{\frac{\sigma}{\sqrt{n}}} = \frac{31.8 - 32}{\frac{0.75}{\sqrt{50}}} = -1.89$$

The critical value of $z = -2.33$ is found in Table A-2 as the z score corresponding to an area of 0.4900. The test statistic, critical region, and critical value are shown in Figure 7-6.

Step 7: From Figure 7-6 we see that the sample mean of 31.8 oz does not fall within the critical region, so we fail to reject the null hypothesis H_0.

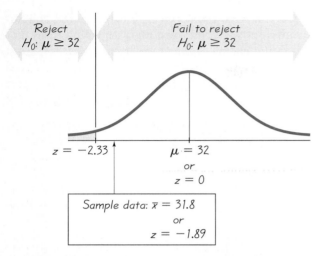

Reject
$H_0: \mu \geq 32$

Fail to reject
$H_0: \mu \geq 32$

$z = -2.33$

$\mu = 32$
or
$z = 0$

Sample data: $\bar{x} = 31.8$
or
$z = -1.89$

FIGURE 7-6   Distribution of Means of Bottle Contents

Step 8:   There is not sufficient evidence to support the claim that the mean is less than 32 oz. The brewery might be cheating consumers, but there isn't enough evidence yet to justify that conclusion.

In presenting the results of a hypothesis test, it is not always necessary to show all of the steps included in Figure 7-4. However, the results should include the null hypothesis, the alternative hypothesis, the calculation of the test statistic, a graph such as Figure 7-6, the initial conclusion (reject H_0 or fail to reject H_0), and the final conclusion stated in nontechnical terms. The graph should show the test statistic, critical value(s), critical region, and significance level.

The *P*-Value Method of Testing Hypotheses

Many professional articles and software packages use another approach to hypothesis testing that is based on the calculation of a probability value, or *P*-value.

DEFINITION

A *P*-value (or **probability value**) is the probability of getting a value of the sample test statistic that is at least as extreme as the one found from the sample data, assuming that the null hypothesis is true.

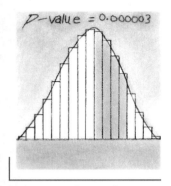

Beware of *P*-Value Misuse

John P. Campbell, editor of the *Journal of Applied Psychology*, wrote the following on the subject of *P*-values. "Books have been written to dissuade people from the notion that smaller *P*-values mean more important results or that statistical significance has anything to do with substantive significance. It is almost impossible to drag authors away from their *P*-values, and the more zeros after the decimal point, the harder people cling to them." Although it might be necessary to provide a statistical analysis of the results of a study, we should place strong emphasis on the significance of the results themselves.

Whereas the traditional approach results in a "reject/fail to reject" conclusion, *P*-values measure how confident we are in rejecting a null hypothesis. For example, a *P*-value of 0.0002 would lead us to reject the null hypothesis, but it would also suggest that the sample results are *extremely* unusual if the claimed value of μ is in fact correct. In contrast, given a *P*-value of 0.40, we fail to reject the null hypothesis because the sample results can *easily* occur if the claimed value of μ is correct.

The *P*-value approach uses most of the same basic procedures as the traditional approach, but Steps 6 and 7 are different:

Step 6: Find the *P*-value.

Step 7: Report the *P*-value. Some statisticians prefer to simply report the *P*-value and leave the conclusion to the reader. Others prefer to use the following decision criterion:

- *Reject* the null hypothesis if the *P*-value is less than or equal to the significance level α.
- *Fail to reject* the null hypothesis if the *P*-value is greater than the significance level α.

In Step 7 above, if the conclusion is based on the *P*-value alone, the following guide may be helpful.

P-value	Interpretation
Less than 0.01	Highly statistically significant Very strong evidence against the null hypothesis
0.01 to 0.05	Statistically significant Adequate evidence against the null hypothesis
Greater than 0.05	Insufficient evidence against the null hypothesis

Many statisticians consider it good practice to always select a significance level *before* doing a hypothesis test. This is a particularly good procedure when using *P*-values because we may be tempted to adjust the significance level based on the results. For example, with a 0.05 level of significance and a *P*-value of 0.06, we should fail to reject the null hypothesis, but it is sometimes tempting to say that a probability of 0.06 is small enough to warrant rejection of the null hypothesis. Other statisticians believe that prior selection of a significance level reduces the usefulness of *P*-values. They contend that no significance level should be specified and that the conclusion should be left to the reader. We will use the decision criterion that involves a comparison of a significance level and the *P*-value.

Figure 7-7 outlines key steps and decisions that lead to the *P*-value. The figure shows that in a right-tailed test, the *P*-value is the area to the right of the

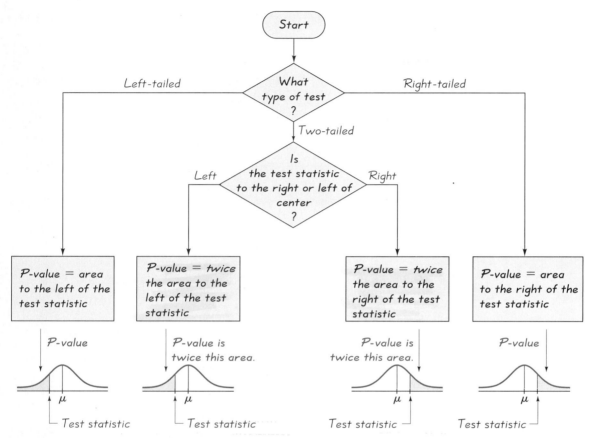

FIGURE 7-7 Finding *P*-Values

test statistic. In a left-tailed test, the *P*-value is the area to the left of the test statistic. However, we must be careful to note that *in a two-tailed test, the P-value is twice the area of the extreme region bounded by the test statistic.* This makes sense when we recognize that the *P*-value gives us the probability of getting a sample mean that is *at least as extreme* as the sample mean actually obtained, and the two-tailed case has critical, or extreme, regions in *both* tails.

Figure 7-7 shows how to find *P*-values, and Figure 7-8 summarizes the *P*-value method for testing hypotheses. A comparison of the traditional method (summarized in Figure 7-4) and the *P*-value method (Figure 7-8) shows that they are essentially the same, but they differ in the decision criterion. The traditional method compares the test statistic to the critical values, whereas the *P*-value method compares the *P*-value to the significance level. However, the traditional and *P*-value methods are equivalent in the sense that they will always result in the same conclusion.

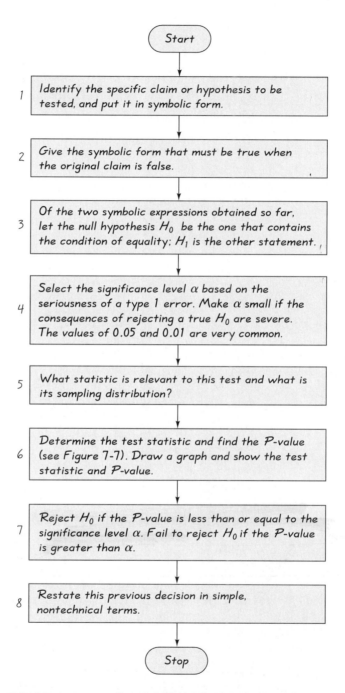

FIGURE 7-8 *P*-Value Method of Testing Hypotheses

▶ **EXAMPLE**

Use the *P*-value method to test the claim that the mean body temperature of healthy adults is equal to 98.6°F. As before, use a 0.05 significance level and the sample data summarized in the chapter opening problem ($n = 106$, $\bar{x} = 98.2$, $s = 0.62$, a bell-shaped distribution).

● **SOLUTION**

Except for Steps 6 and 7, this solution is the same as the one developed earlier in this section using the traditional method. Steps 1, 2, and 3 led to the following hypotheses:

$$H_0:\ \mu = 98.6 \qquad H_1:\ \mu \neq 98.6$$

In Steps 4 and 5 we noted that the significance level is $\alpha = 0.05$ and that the central limit theorem indicates use of the normal distribution. We now proceed to Steps 6 and 7.

Step 6:   The test statistic of $z = -6.64$ was already found in a preceding example. We can now find the *P*-value by referring to Figure 7-7. Because the test is two-tailed, the *P*-value is *twice* the area to the left of the test statistic $z = -6.64$. Using Table A-2, we find that the area to the left of $z = -6.64$ is 0.0001, so the *P*-value is $2 \cdot 0.0001 = 0.0002$ (see Figure 7-9). (Tables more precise than Table A-2 would

(continued)

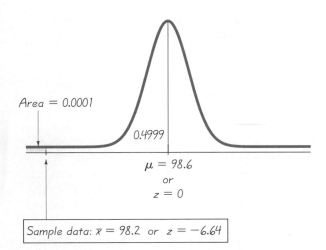

FIGURE 7-9   *P*-Value Method of Testing H_0: $\mu = 98.6$
Because the test is two-tailed, the *P*-value is *twice* the shaded area.

show that the area to the left of $z = -6.64$ is actually much closer to 0.)

Step 7:   Because the *P*-value of 0.0002 is less than the significance level of $\alpha = 0.05$, we reject the null hypothesis.

As with the traditional method, we conclude in Step 8 that there is sufficient evidence to warrant rejection of the claim that the mean body temperature is 98.6°F.

The following example applies the *P*-value approach to the Mill Valley Brewery example presented earlier in this section.

▶ **EXAMPLE**

Use the *P*-value approach and a 0.01 significance level to test the claim that the Mill Valley Brewery is cheating consumers by filling bottles with less than 32 oz of beer. The Bureau of Weights and Measures randomly selects 50 of these bottles, measures their contents, and obtains a sample mean of 31.8 oz and a sample standard deviation of 0.75 oz.

● **SOLUTION**

Again, refer to the steps used earlier to solve this problem by the traditional method. Steps 1, 2, and 3 resulted in the following hypotheses:

$$H_0: \mu \geq 32 \text{ oz} \qquad H_1: \mu < 32 \text{ oz}$$

Steps 4 and 5 led to a significance level of $\alpha = 0.01$ and the decision that the normal distribution is relevant to this test of a claim about a sample mean. We now proceed to Steps 6 and 7.

Step 6:   The test statistic of $z = -1.89$ was computed in the earlier solution to this same problem. To find the *P*-value, we refer to Figure 7-7, which indicates that for this left-tailed test, the *P*-value is the area to the left of the test statistic. Referring to Table A-2, we find that the area to the left of $z = -1.89$ is $0.5 - 0.4706$, or 0.0294. The *P*-value is therefore 0.0294.

Step 7:   Because the *P*-value of 0.0294 is not less than the significance level of $\alpha = 0.01$, we fail to reject the null hypothesis that $\mu \geq 32$.

As in the previous solution of this same problem, we conclude that there is not sufficient evidence to support the claim that the Mill Valley Brewery is cheating consumers. (If we had chosen a significance level of $\alpha = 0.05$, there would have been sufficient evidence to support the claim of cheating.) The P-value of 0.0294 shows us that it is unlikely that we would get a sample mean of 31.8 if the mean were really 32, but the probability isn't small enough (less than $\alpha = 0.01$) to be regarded as significant.

The next procedure for testing hypotheses is based on confidence intervals and therefore requires the concepts discussed in Section 6-2.

Testing Claims with Confidence Intervals

Let's again consider the hypothesis-testing problem described at the beginning of the chapter. We want to test the claim that the mean body temperature of healthy adults is equal to 98.6°F. Sample data consist of $n = 106$ temperatures with mean $\bar{x} = 98.2$ and standard deviation $s = 0.62$. Chapter 6 described methods of constructing confidence intervals. In particular, we used the body temperature sample data to construct the following 95% confidence interval:

$$98.08 < \mu < 98.32$$

We are 95% confident that the limits of 98.08 and 98.32 contain the population mean μ. This confidence interval suggests that it is very *unlikely* that the population mean is $\mu = 98.6$. That is, based on the above confidence interval, we reject the common belief that the mean body temperature of healthy adults is 98.6°F. We can generalize this procedure as follows: First use the sample data to construct a confidence interval, and then apply the following decision criterion.

> **A confidence interval estimate of a population parameter contains the likely values of that parameter. We should therefore reject a claim that the population parameter has a value that is not included in the confidence interval.**

Using this criterion, we note that the confidence interval given above does not contain the claimed value of 98.6, and we therefore reject the claim that the population mean equals 98.6. (*Note:* We can make a *direct* correspondence between a confidence interval and a hypothesis test only when the test is two-tailed. A one-tailed hypothesis test with significance level α corresponds to a confidence interval with degree of confidence $1 - 2\alpha$. For example, a right-tailed hypothesis test with a 0.05 significance level corresponds to a 90% confidence interval.)

This use of confidence intervals gives us one method for identifying results that are highly unlikely, so that we can distinguish between chance differences and significant differences.

💾 Using Computers to Test Hypotheses

STATDISK and Minitab are two of the many statistical software packages capable of conducting hypothesis tests for a wide variety of different circumstances. Shown below are the STATDISK and Minitab displays for the test of $\mu = 98.6$, using the 106 body temperatures listed in Table 2-1. For STATDISK, select the main menu bar item `Statistics`, then select `Hypothesis Testing`, then select `Mean - One Sample`, and proceed to enter the data as requested. In addition to the data shown, STATDISK displays a graph of the normal distribution, which shows the critical values and test statistic.

For Minitab, first use the `SET` command to enter the data in column C1 (as described in Section 1-5), and then enter the command `TTEST MU=98.6 C1`. The `TTEST` command uses the sample mean $\overline{x}$ and standard deviation s in the calculation of the test statistic. The result will be as shown in the Minitab display. (Minitab's test statistic of -6.61 is slightly different because Minitab uses a more precise value of the standard deviation.)

STATDISK DISPLAY

```
 About    File    Edit    Statistics    Data    View

         Hypothesis test for a claim about a SINGLE POPULATION MEAN

                  NULL HYPOTHESIS:  Population mean = 98.6

              ┌──────────────────────────────────────────────────┐
              │  Sample mean .............. =  98.2               │
              │  Standard deviation ...... =  .62                 │
              │  Sample size ........... n =  106                 │
              │  Degrees of freedom ... df =  105                 │
              │  Test statistic ........ z = -6.64234            │
              │  Critical value ........ z = -1.96039 , 1.96039   │
              │  P-value ................. =  .00000             │
              │  Significance level ...... =  .05                 │
              └──────────────────────────────────────────────────┘


                  CONCLUSION: REJECT the null hypothesis



                     Use Menu Bar to change screen        Page 3 of 3
    F1: Help       F2: Menu                                ESC: Quit
```

MINITAB DISPLAY

```
MTB > TTEST MU=98.6 C1

TEST OF MU = 98.6000 VS MU N.E. 98.6000

               N      MEAN     STDEV    SE MEAN         T    P VALUE
C1           106    98.2000    0.6229     0.0605     -6.61     0.0000
```

At the beginning of this section, we noted that we are testing a claim made about the mean of a single population and that the sample size is large ($n > 30$). There are many important cases involving claims about a mean in which the sample size is small ($n \leq 30$). Such cases will be considered in the following section.

In the following sections and chapters, we will apply methods of hypothesis testing to other circumstances, such as those involving claims about proportions or standard deviations or those involving more than one population. It is easy to become entangled in a complex web of steps without ever understanding the underlying rationale of hypothesis testing. The key to that understanding lies in recognizing the following concept:

> **When testing a claim, we make an assumption (null hypothesis) that contains equality. If the sample results can easily occur when that assumption is true, we attribute the relatively small discrepancy between the assumption and the sample results to chance. But if the sample results appear to be unusual, we explain that discrepancy by concluding that the assumption is not true.**

If, when we are testing the claim that a dime is fair, 100 tosses result in 51 heads, we conclude that the coin is fair. The small discrepancy between 51 heads obtained and 50 heads expected is attributed to chance. But if 100 tosses yield 94 heads, we explain the result by concluding that the assumption of a fair coin is not true. If you keep this idea in mind as you examine various examples, hypothesis testing will become meaningful instead of a rote mechanical process.

In interpreting the results of hypothesis tests, we should also keep in mind the distinction between *statistical* significance and *practical* significance. If a diet is tested on 1000 subjects with an average weight loss of 0.5 lb, a hypothesis test may well show that the diet is effective. But such a diet would not be considered practical because the average weight loss of 0.5 lb is too small, even though it is statistically significant. As always, we should use common sense to interpret results.

7-3   Exercises A: Basic Skills and Concepts

In Exercises 1–24, test the given claim using the *traditional* method of hypothesis testing. Also identify the *P*-value. Assume that all samples have been randomly selected.

1. Test the claim that the population mean $\mu = 100$, given a sample of $n = 81$ for which $\bar{x} = 100.8$ and $s = 5$. Test at the $\alpha = 0.01$ significance level.
2. Test the claim that $\mu \geq 20$, given a sample of $n = 100$ for which $\bar{x} = 18.7$ and $s = 3$. Use a significance level of $\alpha = 0.05$.
3. Test the claim that a population mean equals 500. You have a sample of 300 items for which the sample mean is 510 and the sample standard deviation is 50. Test at the $\alpha = 0.10$ significance level.

4. Test the claim that a population mean exceeds 40. You have a sample of 50 items for which the sample mean is 42 and the sample standard deviation is 8. Use a significance level of 0.05.

5. The Boston Bottling Company distributes cola in cans labeled 12 oz. The Bureau of Weights and Measures randomly selected 36 cans, measured their contents, and obtained a sample mean of 11.82 oz and a sample standard deviation of 0.38 oz. Use a 0.01 significance level to test the claim that the company is cheating consumers.

6. A study analyzed the use of different crime prevention strategies. A standardized scale was used to measure the frequency with which a random sample of 807 adults locked the doors of their cars when parking at their homes. The sample mean and standard deviation for locking cars at home are 2.61 and 1.60, respectively. (See "A Longitudinal Study of the Effects of Various Crime Prevention Strategies on Criminal Victimization, Fear of Crime, and Psychological Distress," by Norris and Kaniasty, *American Journal of Community Psychology,* Vol. 20, No. 5.) At the 0.05 level of significance, test the claim that the population mean score for locking cars parked at home is greater than 2.52, which is the mean score for the use of home deadbolt locks.

7. In a study of consumer habits, researchers designed a questionnaire to identify compulsive buyers. For a sample of consumers who identified themselves as compulsive buyers, questionnaire scores have a mean of 0.83 and a standard deviation of 0.24 (based on data from "A Clinical Screener for Compulsive Buying," by Faber and Guinn, *Journal of Consumer Research,* Vol. 19). Assume that the subjects were randomly selected and that the sample size was 32. At the 0.01 level of significance, test the claim that the self-identified compulsive-buyer population has a mean greater than 0.21, the mean for the general population.

8. In a study of job satisfaction, 244 college administrators were given a standardized survey to measure satisfaction with their work. Their mean is 39.213, and their standard deviation is 7.567 (based on data from "Job Satisfaction Among Academic Administrators," by Glick, *Research in Higher Education,* Vol. 33, No. 5). At the 0.05 level of significance, test the claim that the mean for all college administrators is 47.0, the national norm for people of comparable education levels.

9. A study included 123 children who were wearing seat belts when injured in motor vehicle crashes. The amounts of time spent in an intensive care unit have a mean of 0.83 day and a standard deviation of 0.16 day (based on data from "Morbidity Among Pediatric Motor Vehicle Crash Victims: The Effectiveness of Seat Belts," by Osberg and Di Scala, *American Journal of Public Health,* Vol. 82, No. 3). Using a 0.01 significance level, test the claim that the seat belt sample comes from a population with a mean of less than 1.39 days, which is the mean for the population who were not wearing seat belts when injured in motor vehicle crashes.

10. The mean amount of disposable per capita income in Colorado is $13,901 (based on data from the U.S. Bureau of Economic Analysis). Tom Phelps plans to open a Cadillac car dealership and wants to verify that amount for a particular region of Colorado. He finds results from a recent survey of 200

people, with a mean of $13,447 and a standard deviation of $4883. At the 0.05 level of significance, test the claim that the sample was drawn from a population with a mean of $13,901.

11. The effectiveness of a test preparation course was studied for a random sample of 75 subjects who took the SAT before and after coaching. The differences between the scores resulted in a mean increase of 0.6 and a standard deviation of 3.8. (See "An Analysis of the Impact of Commercial Test Preparation Courses on SAT Scores," by Sesnowitz, Bernhardt, and Kwain, *American Education Research Journal*, Vol. 19, No. 3.) At the 0.05 significance level, test the claim that the population mean increase is greater than 0, indicating that the course is effective in raising scores.

12. In a study of distances traveled by buses before the first major engine failure, a sampling of 191 buses resulted in a mean of 96,700 mi and a standard deviation of 37,500 mi (based on data in *Technometrics*, Vol. 22, No. 4). At the 0.05 level of significance, test the manufacturer's claim that mean distance traveled before a major engine failure is more than 90,000 mi.

13. For the past several years, when third-grade boys at Columbus Elementary School were given the reading portion of the Wide Range Achievement test, the mean score was 51.0. The third-grade teachers recently instituted a new reading program and then tested 150 randomly selected third-grade boys. Their mean score is 52.4, and the standard deviation is 13.14 (based on data from the National Health Survey, USDHEW publication 72-1011). At the 0.05 level of significance, test the claim that this sample is from a population with a mean greater than 51.0 (indicating that the new reading program is effective).

14. A poll of 100 randomly selected car owners revealed that the mean length of time that they plan to keep their cars is 7.01 years and the standard deviation is 3.74 years (based on data from a Roper poll). The president of the Newton Car Park is trying to plan a sales campaign targeted at car owners who are ready to buy a different car. Test the claim of the sales manager, who authoritatively states that the mean length of time for all car owners to keep their cars is less than 7.5 years. Use a 0.05 significance level.

15. When 200 convicted embezzlers were randomly selected, the mean length of prison sentence was found to be 22.1 months and the standard deviation was found to be 8.6 months (based on data from the U.S. Department of Justice). Kim Patterson is running for political office on a platform of tougher treatment of convicted criminals. Test her claim that prison terms for convicted embezzlers have a mean of less than 2 years. Use a 0.05 significance level.

16. The nighttime cold medicine Dozenol bears a label indicating the presence of 600 mg of acetaminophen in each fluid ounce of the drug. The Food and Drug Administration randomly selected 65 1-oz samples and found that the mean acetaminophen content is 589 mg, whereas the standard deviation is 21 mg. Using $\alpha = 0.01$, test the claim of the Medassist Pharmaceutical Company that the population mean is equal to 600 mg.

17. The New England Insurance Company is reviewing the driving habits of women aged 16–24 to determine whether they should continue to pay higher premiums than women in the 25–34 age bracket. In a study of 750 randomly selected women drivers aged 16–24, the mean driving distance for one year is

6047 mi and the standard deviation is 2944 mi (based on data from the Federal Highway Administration). Use a 0.05 significance level to test the claim that the population mean for women in the 16–24 age bracket is less than 7124 mi, which is the known mean for women in the 25–34 age bracket.

18. In the article "Multiple Spans in Transcription Typing" (*Journal of Applied Psychology*, Vol. 72, No. 2), data were given for a sample of 45 typists. Their mean normal typing score is 182, while the standard deviation is 52. Test the claim that the sample is from a population with a mean of 180. Use a 0.05 significance level.

19. *The Late Show with David Letterman* is seen by a relatively large percentage of household members who videotape the show for viewing at a more convenient time. The show's marketing manager claims that the mean income of households with VCRs is greater than $40,000. Test that claim using a 0.05 significance level. A sample of 1700 households with VCRs produces a sample mean of $41,182 and a sample standard deviation of $19,990 (based on data from Nielsen Media Research).

20. The mean time between failures (in hours) for a Telektronic Company radio used in light aircraft is 420 h. After 35 new radios were modified in an attempt to improve reliability, tests showed that the mean time between failures for this sample is 385 h and the standard deviation is 24 h. Use a 0.05 significance level to test the claim that the modifications improved reliability. (Note that improved reliability should result in a *longer* mean time between failures.)

21. Refer to Data Set 10 in Appendix B and use the aircraft age sample data ($\bar{x}$ = 13.41 years, s = 8.28 years) to test the *Time* magazine claim that the mean age of aircraft in the U.S. fleet is 14 years. Use a significance level of 0.05.

22. Refer to Data Set 1 in Appendix B for the total weights of garbage discarded by households in one week (based on data collected as part of the Garbage Project at the University of Arizona). For that data set, the mean is 27.44 lb and the standard deviation is 12.46 lb. At the 0.05 level of significance, test the claim of the City of Providence supervisor that the mean weight of all garbage discarded by households each week is less than 35 lb, the amount that can be handled by the town. Based on the result, is there any cause for concern that there might be too much garbage to handle?

23. If we refer to the weights (in grams) of quarters listed in Data Set 8 in Appendix B, we find 50 weights with a mean of 5.622 g and a standard deviation of 0.068 g. The U.S. Department of the Treasury claims that the procedure it uses to mint quarters yields a mean weight of 5.670 g. Use a 0.01 significance level to test the claim that the mean weight of quarters in circulation is 5.670 g. If the claim is rejected, what is a possible explanation for the discrepancy?

24. Mars, Inc. labels a package of M&Ms as containing 453 g. Assume that each package contains 496 M&Ms, as is the case for the package used for Data Set 6 in Appendix B. The implied mean weight of individual M&Ms is therefore 453 g ÷ 496. At the 0.01 level of significance, test the implied claim that the mean weight of individual M&Ms is equal to 453/496 g. Data Set 6 consists of a sample of 100 M&Ms with a mean of 0.916 g and a standard deviation of 0.043 g.

7-3   Exercises B: Beyond the Basics

25. A sample of 93 IQ scores was randomly selected from a certain population; the results are described by the accompanying boxplot. At the 0.03 level of significance, use the P-value approach to test the claim that the population has a mean IQ score equal to 105.

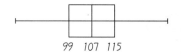

99 107 115

26. A journal article reported that a null hypothesis of $\mu = 100$ was rejected because the P-value was less than 0.01. The sample size was given as 62, and the sample mean was given as 103.6. Find the largest possible standard deviation.

27. Find the smallest mean above \$23,460 that leads to rejection of the claim that the mean annual household income is \$23,460. Assume a 0.02 significance level. The sample consists of 50 randomly selected households with a sample standard deviation of \$3750.

28. For a given hypothesis test, the probability α of a type I error is fixed, whereas the probability β of a type II error depends on the particular value of μ that is used as an alternative to the null hypothesis. For hypothesis tests of the type found in this section, we can find β as follows:

Step 1:   Find the value(s) of $\overline{x}$ that correspond to the critical value(s). In

$$z = \frac{\overline{x} - \mu_{\overline{x}}}{\sigma_{\overline{x}}}$$

substitute the critical value(s) for z, enter the values for $\mu_{\overline{x}}$ and $\sigma_{\overline{x}}$, and solve for $\overline{x}$.

Step 2:   Given a particular value of μ that is an alternative to the null hypothesis H_0, draw the normal curve with this new value of μ at the center. Also plot the value(s) of $\overline{x}$ found in Step 1.

Step 3:   Refer to the graph in Step 2 and find the area of the new critical region bounded by $\overline{x}$. This is the probability of rejecting the null hypothesis, given that the new value of μ is correct.

Step 4:   The value of β is 1 minus the area from Step 3. This is the probability of failing to reject the null hypothesis, given that the new value of μ is correct.

The preceding steps allow you to find the probability of failing to reject H_0 when it is false. You are determining the area under the curve that *excludes* the critical region in which you reject H_0; this area corresponds to a failure to reject a false H_0, because we use a particular value of μ that goes against H_0. Refer to the body temperature example discussed in this section and find the value of β corresponding to the following:
a. $\mu = 98.7$ b. $\mu = 98.4$

29. The **power** of a test is $1 - \beta$, the probability of rejecting a false null hypothesis. Refer to the Mill Valley Brewery example discussed in this section. If that test has a power of 0.8, find the mean μ (see Exercise 28).

7-4   Testing a Claim about a Mean: Small Samples

Sections 7-2 and 7-3 introduced the general method for testing a claim about the mean of a population. However, each example and exercise involved a large ($n > 30$) sample, so according to the central limit theorem, the sample standard deviation s could be used as an estimate of the population standard deviation σ. In such cases we can use the normal distribution as an approximation to the distribution of sample means. The estimation of σ by s is reasonable because large random samples tend to be representative of the population. But small random samples may exhibit unusual behavior and thus cannot be so trusted. In this section we consider tests of hypotheses made about a population mean when the samples are small. We begin by referring to Figure 7-10, which outlines the theory we are describing.

Starting at the top of Figure 7-10, we see that our immediate concerns lie only with the hypotheses made about one population mean. (In following sections we will consider hypotheses made about population parameters other than the mean.) Figure 7-10 summarizes the following observations:

1. In *any* population, the distribution of sample means can be approximated by the normal distribution as long as the random samples are large ($n > 30$). This is justified by the central limit theorem.

2. In populations with distributions that are essentially normal, samples of *any* size will yield means having a distribution that is approximately normal. The value of the population mean μ corresponds to the null hypothesis, and the value of the population standard deviation σ must be known. If σ is unknown and the samples are large, we can use the sample standard deviation s as a substitute for σ because large random samples tend to be representative of the populations from which they come.

3. In populations with distributions that are essentially normal, assume that we randomly select *small* samples and do not know the value of σ. For this case, we can use the Student t distribution, first introduced in Section 6-2. When employing the Student t distribution, we use s instead of σ (which is unknown).

4. If our random samples are small, σ is unknown, and the population distribution is grossly nonnormal, we can use nonparametric methods, some of which are discussed in Chapter 13.

Section 6-2 introduced the Student t distribution and noted several important features. The features particularly relevant to this section are given below.

Conditions for Using the Student *t* Distribution

1. The sample is small ($n \leq 30$); and
2. The value of σ is unknown; and
3. The parent population has a distribution that is essentially normal.

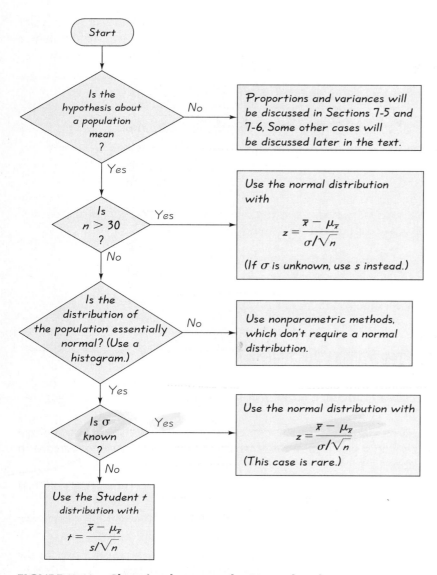

FIGURE 7-10 Choosing between the Normal and Student *t* Distributions

Examination of the above three conditions and Figure 7-10 shows that the Student *t* distribution is used only for small sample cases, but not all small sample cases automatically require the Student *t* distribution. If we have a small sample randomly selected from a normally distributed population and we know the value of σ, Figure 7-10 indicates that we should use the normal distribution, not the Student *t* distribution. However, it is very unusual to know the population standard deviation σ without also knowing the population

Small Sample

The Children's Defense Fund was organized to promote the welfare of children. The group published *Children Out of School in America,* which reported that in one area, 37.5% of the 16- and 17-year-old children were out of school. This statistic received much press coverage, but it was based on a sample of only 16 children. Another statistic was based on a sample size of only 3 students. (See "Firsthand Report: How Flawed Statistics Can Make an Ugly Picture Look Even Worse," *American School Board Journal,* Vol. 162.)

mean μ. Consequently, small sample cases for which we can use the normal distribution will be rare. Most small sample cases involve the Student t distribution with the test statistic and critical values described as follows.

Test Statistic for Claims about μ When $n \leq 30$ and σ Is Unknown

If a population is essentially normal, then the distribution of

$$t = \frac{\overline{x} - \mu_{\overline{x}}}{\dfrac{s}{\sqrt{n}}}$$

is essentially a *Student t distribution* for all samples of size n. (The Student t distribution is often referred to as the t *distribution*.)

Critical Values

1. Critical values are found in Table A-3.
2. Degrees of freedom $= n - 1$.

Important Properties of the Student t Distribution

1. The Student t distribution is different for different sample sizes (see Figure 6-4 in Section 6-2).

2. The Student t distribution has the same general bell shape as the normal distribution; its wider shape reflects the greater variability that is expected with small samples.

3. The Student t distribution has a mean of $t = 0$ (just as the standard normal distribution has a mean of $z = 0$).

4. The standard deviation of the Student t distribution varies with the sample size and is greater than 1 (unlike the standard normal distribution, which has $\sigma = 1$).

5. As the sample size n gets larger, the Student t distribution gets closer to the normal distribution. For values of $n > 30$, the differences are so small that we can use the critical z values instead of developing a much larger table of critical t values. (The values in the bottom row of Table A-3 are equal to the corresponding critical z values from the normal distribution.)

In the preceding section we used the body temperature data (Table 2-1) for two examples (traditional and P-value methods) of a hypothesis test of the claim that healthy adults have a mean body temperature equal to 98.6°F. Figure 7-10 indicates that we were correct in using the normal distribution because the sample size of $n = 106$ is large. The following example involves a small sample drawn from a normally distributed population for which the standard deviation σ is not known.

Statistics

▶ **EXAMPLE**

In one part of a test developed by a psychologist, the test subject is asked to form a word by unscrambling the letters "ciatttsss." (Try it yourself.) Given below are the times (in seconds) required by 15 randomly selected persons to unscramble the letters. At the $\alpha = 0.05$ level of significance, test the claim that the mean time is equal to 60 s.

$$
\begin{array}{cccccccc}
68.7 & 27.4 & 26.0 & 60.5 & 34.6 & 61.1 & 68.6 & 48.4 \\
43.6 & 39.5 & 85.3 & 26.3 & 43.4 & 83.7 & 68.9 &
\end{array}
$$

● **SOLUTION**

For the given sample data, we can use the procedures of Chapter 2 to find that $n = 15$, $\bar{x} = 52.40$, and $s = 20.12$. We now follow the steps of the traditional approach to hypothesis testing, as outlined in Figure 7-4 of the preceding section. As in the preceding section, Steps 1, 2, and 3 result in the following null and alternative hypotheses:

$$H_0: \mu = 60 \qquad H_1: \mu \neq 60$$

Step 4:   The significance level is $\alpha = 0.05$.

Step 5:   The sample mean should be used in testing a claim about a population mean. Using Figure 7-10, we see that the sample is small (because $n = 15$ does not exceed 30), it's reasonable to conclude that the population has a normal distribution (because we're dealing with test measurements), and σ is unknown. Figure 7-10 indicates that we should use the Student t distribution.

Step 6:   The test statistic is

$$t = \frac{\bar{x} - \mu_{\bar{x}}}{\frac{s}{\sqrt{n}}} = \frac{52.40 - 60}{\frac{20.12}{\sqrt{15}}} = -1.463$$

The critical values of $t = -2.145$ and $t = 2.145$ are found by referring to Table A-3. Use $n - 1 = 14$ degrees of freedom in the column at the left and refer to the column with the heading of 0.05 (two tails). The test statistic and critical value are shown in Figure 7-11, which is on the following page.

Step 7:   Because the test statistic of $t = -1.463$ does not fall in the critical region, we fail to reject H_0.

Step 8:   (Refer to Figure 7-2 for help in wording the final conclusion.) There is not sufficient evidence to warrant rejection of the claim that the mean time required to unscramble the letters is 60 s.

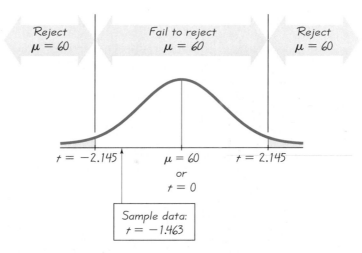

FIGURE 7-11 t-Test of $\mu = 60$ s

The critical values in the preceding example were $t = \pm 2.145$; if the preceding example had involved a large ($n > 30$) sample, however, the critical values would have been $z = \pm 1.96$. The critical values for the Student t distribution are farther away from the center, indicating that the sample evidence must be *more extreme* before we consider the difference to be a *significant* difference rather than a *chance* difference.

▶ **EXAMPLE**

The Carolina Tobacco Company advertises that its best-selling cigarettes contain at most 40 mg of nicotine, but *Consumer Advocate* magazine conducts tests of 10 randomly selected cigarettes and finds that $\overline{x} = 43.3$ mg and $s = 3.8$ mg. Other evidence suggests that the distribution of nicotine content is a normal distribution. The sample is small because the laboratory work required to extract the nicotine is time consuming and expensive. It's a serious matter to charge that the company advertising is wrong, so the magazine editor chooses a significance level of $\alpha = 0.01$ in testing the company's claim that the mean nicotine content is at most 40 mg. Using a 0.01 significance level, test the company's claim that the mean is at most 40 mg.

● **SOLUTION**

We use the traditional method and follow the procedure outlined in Figure 7-4.

Step 1:   The original claim expressed in symbolic form is $\mu \leq 40$.

Step 2:   The opposite of the original claim is $\mu > 40$.

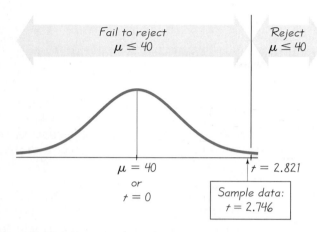

FIGURE 7-12   *t*-Test of $\mu > 40$

Step 3: H_0 must contain the condition of equality, so we get

$$H_0: \mu \leq 40 \qquad \text{(Null hypothesis)}$$
$$H_1: \mu > 40 \qquad \text{(Alternative hypothesis)}$$

Step 4:   The significance level is $\alpha = 0.01$.

Step 5:   The sample mean should be used in testing a claim about a population mean. Because the sample is small and σ is unknown, we use the Student t distribution.

Step 6:   The test statistic is

$$t = \frac{\bar{x} - \mu_{\bar{x}}}{\dfrac{s}{\sqrt{n}}} = \frac{43.3 - 40}{\dfrac{3.8}{\sqrt{10}}} = 2.746$$

The critical value of $t = 2.821$ is found in Table A-3 by using the row for $n - 1 = 9$ degrees of freedom and the column for $\alpha = 0.01$ (one tail). The test statistic of $t = 2.746$ and the critical value are shown in Figure 7-12.

Step 7:   Because the test statistic of $t = 2.746$ does not fall in the critical region, we fail to reject H_0.

Step 8:   There is not sufficient evidence to warrant rejection of the company's claim that the mean amount of nicotine is at most 40 mg.

In the preceding example, if the magazine had *incorrectly* used the normal distribution instead of the Student t distribution, the critical value would have been $z = 2.33$ and the magazine would have concluded (incorrectly) that there was sufficient evidence to reject the company's claim of a mean that is at most 40 mg.

P-Values

In the examples presented in this section, we used the traditional approach to hypothesis testing. However, much of the literature and many computer packages (including STATDISK and Minitab) will display *P*-values.

Because the *t* distribution table (Table A-3) includes only selected values of the significance level α, we cannot usually find the specific *P*-value from Table A-3. Instead, we can use that table to identify limits that contain the *P*-value. In the last example we found the test statistic to be $t = 2.746$, and we know that the test is one-tailed with 9 degrees of freedom. By examining the row of Table A-3 corresponding to 9 degrees of freedom, we see that the test statistic of 2.746 falls between the table values of 2.821 and 2.262, which in a one-tailed test correspond to $\alpha = 0.01$ and $\alpha = 0.025$. Although we cannot determine the exact *P*-value from Table A-3, we do know that it must fall between 0.01 and 0.025; we can therefore conclude that

$$0.01 < P\text{-value} < 0.025$$

(Some calculators and computer programs allow us to find exact *P*-values. For the preceding example, STATDISK displays a *P*-value of 0.01131 and Minitab displays a *P*-value of 0.011.) With a significance level of 0.01 and a *P*-value greater than 0.01, we fail to reject the null hypothesis, as we did with the traditional method in the preceding example.

So far, we have discussed tests of hypotheses made about population means only. In the next section we will test hypotheses made about population proportions or percentages.

7-4   Exercises A: Basic Skills and Concepts

In Exercises 1 and 2, find the critical *t* values suggested by the given information.

1. a. $H_0: \mu = 12$ b. $H_0: \mu \leq 50$ c. $H_1: \mu < 1.36$
      $n = 27$ $n = 17$ $n = 6$
      $\alpha = 0.05$ $\alpha = 0.10$ $\alpha = 0.01$
2. a. $H_0: \mu \leq 100$ b. $H_1: \mu \neq 500$ c. $H_1: \mu < 67.5$
      $n = 27$ $n = 16$ $n = 12$
      $\alpha = 0.10$ $\alpha = 0.05$ $\alpha = 0.05$

In Exercises 3–6, assume that the sample is randomly selected from a population with a normal distribution. Test the given claim by using the traditional method of testing hypotheses.

3. Test the claim that $\mu \leq 10$, given a sample of 9 for which $\bar{x} = 11$ and $s = 2$. Use a significance level of $\alpha = 0.05$.
4. Test the claim that $\mu \geq 100$, given a sample of 22 for which $\bar{x} = 95$ and $s = 18$. Use a significance level of $\alpha = 0.05$.

5. Test the claim that $\mu = 75$, given a sample of 15 for which $\bar{x} = 77.6$ and $s = 5$. Use a significance level of $\alpha = 0.05$.

6. Test the claim that $\mu = 500$, given a sample of 20 for which $\bar{x} = 541$ and $s = 115$. Use a significance level of $\alpha = 0.10$.

For hypothesis tests in Exercises 7–30, follow the traditional procedure summarized in Figure 7-4. Draw the appropriate graph. In each case, assume that the population has a distribution that is approximately normal and that the sample is randomly selected. *Caution:* Some of the exercises require the use of the normal distribution (as described in the preceding section) instead of the Student t distribution (as described in this section); be sure to check the conditions to determine which distribution is appropriate.

7. Using only the first 25 body temperatures listed in Table 2-1, test the claim that the mean body temperature of all healthy adults is equal to 98.6°F. For the level of significance, use $\alpha = 0.05$. (For the first 25 scores, $\bar{x} = 98.24$ and $s = 0.56$.)

8. Using only the first 35 body temperatures listed in Table 2-1, test the claim that the mean body temperature of all healthy adults is equal to 98.6°F. Use a 0.05 significance level. (For the first 35 scores, $\bar{x} = 98.27$ and $s = 0.65$.)

9. At the 0.10 level of significance, test the claim of the Mill Valley Brewery that it fills bottles with amounts having a mean greater than 32 oz. A sample of 27 bottles produces a mean of 32.2 oz and a standard deviation of 0.4 oz.

10. Using Table A-3, what can be concluded about the P-value in Exercise 9?

11. For each of 12 organizations, the cost of operation per client was found. The 12 scores have a mean of $2133 and a standard deviation of $345 (based on data from "Organizational Communication and Performance," by Snyder and Morris, *Journal of Applied Psychology,* Vol. 69, No. 3). At the 0.01 significance level, test the claim of a critic who complains that the mean for all such organizations exceeds $1800 per client.

12. The skid properties of a snow tire have been tested, and a mean skid distance of 154 ft has been established for standardized conditions. A new, more expensive tire is developed, but tests on a sample of 20 new tires yield a mean skid distance of 141 ft with a standard deviation of 12 ft. Because of the cost involved, the new tires will be purchased only if it can be shown at the $\alpha = 0.005$ significance level that they skid less than the current tires. Based on the sample, will the new tires be purchased?

13. Using Table A-3, what can you conclude about the P-value in Exercise 12?

14. The Wichita Aircraft Company is introducing a new line of light aircraft and must provide the Federal Aviation Administration with the required takeoff distance. The manufacturer randomly selected 12 of the planes and tested them to determine the distance (in meters) they require for takeoff. The sample mean and standard deviation were computed to be 524 m and 23 m, respectively. Can the manufacturer justifiably claim that the mean is equal to 500 m? Use a 0.05 significance level.

15. The Medassist Pharmaceutical Company makes a pill intended for children susceptible to seizures. The pill is supposed to contain 20.0 mg of phenobarbital. A random sample of 30 pills yielded a mean and standard deviation of 20.5 mg and 1.5 mg, respectively. Are these sample pills acceptable at the $\alpha = 0.02$ significance level?

16. In a study of factors affecting hypnotism, visual analogue scale (VAS) sensory ratings were obtained for 16 subjects. For these sample ratings, the mean is 8.33 while the standard deviation is 1.96 (based on data from "An Analysis of Factors That Contribute to the Efficacy of Hypnotic Analgesia," by Price and Barber, *Journal of Abnormal Psychology*, Vol. 96, No. 1). At the 0.01 level of significance, test the claim that this sample comes from a population with a mean rating of less than 10.00.

17. The Bank of New England is concerned about the amount of debt being accrued by customers using its credit cards. The Board of Directors voted to institute an expensive monitoring system if the mean for all of the bank's customers is greater than $2000. The bank randomly selected 100 credit card holders and determined the amounts they charged. For this sample group, the mean is $2177 and the standard deviation is $843. Use a 0.025 level of significance to test the claim that the mean amount charged is greater than $2000. Based on the result, will the monitoring system be implemented?

18. A long-range missile missed its target by an average of 0.88 mi. A new steering device is supposed to increase accuracy, and a random sample of 8 missiles were equipped with this new mechanism and tested. These 8 missiles missed by distances with a mean of 0.76 mi and a standard deviation of 0.04 mi. At $\alpha = 0.01$, does the new steering mechanism lower the miss distance?

19. Using Table A-3, what can you conclude about the *P*-value in Exercise 18?

20. As principal of the John F. Kennedy High School, Susan Giangrasso was concerned with the amount of time her students devote to working at after-school jobs. She randomly selected 25 students, obtained their working hours for one week, and computed $\bar{x} = 12.3$ and $s = 11.2$. She claims that her students work significantly more than the mean of 10.7 h obtained from a study conducted by the National Federation of State High School Associations. Test her claim by using a 0.05 level of significance.

21. A study was conducted to determine whether a standard clerical test would need revision for use on video display terminals (VDTs). The VDT scores of 22 subjects have a mean of 170.2 and a standard deviation of 35.3 (based on data from "Modification of the Minnesota Clerical Test to Predict Performance on Video Display Terminals," by Silver and Bennett, *Journal of Applied Psychology*, Vol. 72, No. 1). At the 0.05 level of significance, test the claim that the mean for all subjects taking the VDT test differs from the mean of 243.5 for the standard printed version of the test. Based on the result, should the VDT test be revised?

22. Using Table A-3, what can you conclude about the *P*-value in Exercise 21?

23. Gina Thompson is a high school senior who is concerned about attending college because she knows that many college students require more than 4 years to earn a bachelor's degree. At the 0.10 level of significance, test the claim of her guidance counselor, who states that the mean time is greater than 5 years. Sample data consist of a mean of 5.15 years and a standard deviation of 1.68 years for 80 randomly selected college graduates (based on data from the National Center for Education Statistics).

24. The following reading test results were obtained for a sample of 15 third-grade students: $\bar{x} = 31.0$, $s = 10.5$. (The data are based on "A Longitudinal Study of the Effects of Retention/Promotion on Academic Achievement," by Peterson and others, *American Educational Research Journal,* Vol. 24, No. 1.) Does this third-grade sample mean differ significantly from a first-grade population mean of 41.9? Assume a 0.01 level of significance.

25. Using the weights of the Bufferin tablets in Data Set 9 of Appendix B, test the claim that the mean weight is equal to 650 mg. Use a 0.05 significance level. (For the sample data, $\bar{x} = 665.41$ and $s = 7.26$.)

26. Using the weights of the *red* M&Ms listed in Data Set 6 of Appendix B, test the claim that the mean is at least 0.913 g, the mean value necessary for the 496 M&Ms to produce a total of 453 g as the package indicates. Use a 0.05 significance level. (For the red M&Ms, $\bar{x} = 0.9043$ and $s = 0.0414$.)

27. A standard final examination in an elementary statistics course produced a mean score of 75. At the 0.05 level of significance, test the claim that the following sample scores reflect an above-average class:

79	79	78	74	82	89	74	75	78	73
74	84	82	66	84	82	82	71	72	83

28. Listed below are the total electric energy consumption amounts (in kWh) for the author's home during 7 different years.

11,943	11,463	10,789	9907	9012	9942	11,153

The utility company claims that the mean annual consumption amount is 11,000 kWh and offers a budget payment plan based on that amount. At the 0.05 significance level, test the utility company's claim that the mean is equal to 11,000 kWh.

29. Using Table A-3, what can you conclude about the *P*-value in Exercise 28?

30. Given below are the birth weights (in kilograms) of male babies born to mothers on a special vitamin supplement (based on data from the New York State Department of Health).

3.73	4.37	3.73	4.33	3.39	3.68	4.68	3.52
3.02	4.09	2.47	4.13	4.47	3.22	3.43	2.54

At the 0.05 level of significance, test the claim that the mean for all male babies of mothers given vitamins is equal to 3.39 kg, which is the mean for the population of all males. Based on the result, does the vitamin supplement appear to have an effect on birth weight?

7-4   Exercises B: Beyond the Basics

31. Referring to Data Set 3 in Appendix B, test the claim that married males between the ages of 35 and 44 have a mean weight equal to 176 lb, which is the mean weight of males in that age bracket selected from the general population. Use a 0.05 level of significance.

32. Because of certain conditions, a hypothesis test requires the Student t distribution, as described in this section. Assume that the standard normal distribution was incorrectly used instead. Does using the standard normal distribution make you more or less likely to reject the null hypothesis, or does it not make a difference? Explain.

33. When finding critical values, we often need to use significance levels other than those available in Table A-3. Some computer programs approximate critical t values by

$$t = \sqrt{\mathrm{df} \cdot (e^{A^2/\mathrm{df}} - 1)}$$

where

$$\mathrm{df} = n - 1$$
$$e = 2.718$$
$$A = z\left(\frac{8 \ \mathrm{df} + 3}{8 \ \mathrm{df} + 1}\right)$$

and z is the critical z score. Use this approximation to find the critical t score corresponding to $n = 10$ and a significance level of 0.05 in a right-tailed case. Compare the results to the critical t value found in Table A-3.

34. Refer to Exercise 9 and assume that you're testing the null hypothesis of $\mu \leq$ 32.0 oz. Find β (the probability of a type II error), given that $\mu = 32.3$. (See Exercise 28 from Section 7-3.)

7-5   Testing a Claim about a Proportion

Sections 7-2 and 7-3 presented basic methods for testing hypotheses. Those procedures can be easily modified for many circumstances. In this section we consider methods for testing hypotheses made about a population proportion (rather than the population mean, as in Sections 7-2 through 7-4). The methods used in this section closely parallel those used in the earlier sections of this chapter. With the methods of this section, we can test claims such as these:

- The majority of Americans are in favor of legislation restricting the ownership of automatic guns.

- Among all televisions manufactured by Sony, fewer than 1% are defective.

- Among all late-night television viewers, 18% watch *The Late Show with David Letterman*.

The assumptions we make in this section are listed as follows.

Assumptions Used When Testing a Claim about a Population Proportion, Probability, or Percentage

1. The conditions for a *binomial experiment* are satisfied. That is, we have a fixed number of independent trials having constant probabilities, and each trial has two outcome categories, which we classify as "success" and "failure."

2. The conditions $np \geq 5$ and $nq \geq 5$ are both satisfied, so **the binomial distribution of sample proportions can be approximated by a normal distribution with $\mu = np$ and $\sigma = \sqrt{npq}$** (as described in Section 5-5).

If $np \geq 5$ and $nq \geq 5$ are not both true (as in the second assumption), we may be able to use Table A-1 or the binomial probability formula described in Section 4-4; this section, however, deals only with situations in which the normal distribution is a suitable approximation for the distribution of sample proportions.

If the above conditions are all satisfied, the value of the test statistic will be found by computing z, as follows.

Test Statistic for Testing a Claim about a Proportion

$$z = \frac{\hat{p} - p}{\sqrt{\dfrac{pq}{n}}}$$

where n = number of trials
  p = population proportion (given in the null hypothesis)
  $q = 1 - p$
  $\hat{p} = x/n$ (sample proportion)

The sample proportion $\hat{p}$ is sometimes given directly. For example, the statement "40% of those surveyed are men" translates to $\hat{p} = 0.40$. In other cases, we may have to calculate the sample proportion by using $\hat{p} = x/n$. For example, from the statement "20 of the 50 persons surveyed are men," we can calculate the value of the sample proportion as follows:

$$\hat{p} = \frac{x}{n} = \frac{20}{50} = 0.40$$

In Section 5-5 we included a correction for continuity. We do not include such a correction in the above test statistic because its effect is negligible with large samples.

The critical value is found from Table A-2 (standard normal distribution) by using the same procedures described in Section 7-2. (For example, in a two-tailed test with significance level $\alpha = 0.05$, divide α equally between the two tails, then refer to Table A-2 for the z score corresponding to an area of

Muzak Commercials

Muzak is the provider of the music typically heard by elevator passengers and patients waiting for root canal work, among many others. The company is now entering the in-store advertising market, so that audio commercials can be sandwiched between sessions of the light music we have all heard so often. Muzak claims that when combined with visual displays, its audio ads increase sales of cereal by 12%, laundry detergent by 27%, and toothpaste by 29%. Such claims can be tested by using the standard methods of hypothesis testing introduced in this text. Advertisers seek promotional methods whose success can be proved, and statistical methods can be used to provide such proof.

$0.5 - 0.025 = 0.475$; the result is $z = 1.96$, so the critical values are $z = -1.96$ and $z = 1.96$.)

The test statistic is justified by noting that when using the normal distribution to approximate a binomial distribution, we substitute $\mu = np$ and $\sigma = \sqrt{npq}$ to get

$$z = \frac{x - \mu}{\sigma} = \frac{x - np}{\sqrt{npq}}$$

Here x is the number of successes among n trials. Divide the numerator and denominator of this last expression by n, then replace x/n by the symbol $\hat{p}$, and you will get the test statistic given above. In other words, the above test statistic is simply $z = (x - \mu)/\sigma$ modified for the binomial notation by incorporating the fact that the distribution of sample proportions $\hat{p}$ is a normal distribution with mean p and standard deviation $\sqrt{pq/n}$.

We can now test hypotheses made about population proportions. Simply follow the same general steps listed in Figure 7-4 (for the traditional method) or Figure 7-8 (for the P-value method) and use the test statistic given above.

▶ **EXAMPLE**

Environmental concerns often conflict with modern technology, as is the case with birds that pose a hazard to aircraft during takeoff. An environmental group states that incidents of bird strikes are too rare to justify killing the birds. A pilots' group claims that among aborted takeoffs leading to an aircraft's going off the end of the runway, 10% are due to bird strikes. Use a 0.05 level of significance to test that claim. Sample data consist of 74 aborted takeoffs in which the aircraft overran the runway. Among those 74 cases, 5 were due to bird strikes (based on data from the Air Line Pilots Association and Boeing, as reported in *USA Today*).

● **SOLUTION**

We will use the traditional method of testing hypotheses as outlined in Figure 7-4.

Step 1:   The original claim is that among aborted takeoffs leading to overruns, 10% are due to bird strikes. We express this in symbolic form as $p = 0.10$.

Step 2:   The opposite of the original claim is $p \neq 0.10$.

Step 3:   Because $p = 0.10$ contains equality, we have

$$H_0: p = 0.10 \qquad \text{(Null hypothesis)}$$
$$H_1: p \neq 0.10 \qquad \text{(Alternative hypothesis)}$$

Step 4:   The significance level is $\alpha = 0.05$.

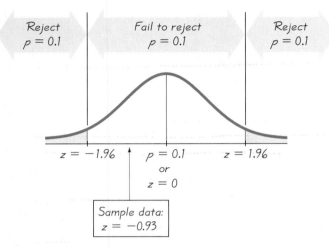

FIGURE 7-13   Hypothesis Test of $p = 0.10$

Step 5:   The statistic relevant to this test is $\hat{p} = 5/74 = 0.0676$. The sampling distribution of sample proportions is approximated by the normal distribution. (The requirements that $np \geq 5$ and $nq \geq 5$ are satisfied because $np = 74 \cdot 0.10 = 7.4$ and $nq = 74 \cdot 0.90 = 66.6$.)

Step 6:   The test statistic of $z = -0.93$ is found as follows:

$$z = \frac{\hat{p} - p}{\sqrt{\dfrac{pq}{n}}} = \frac{0.0676 - 0.1}{\sqrt{\dfrac{(0.1)(0.9)}{74}}} = -0.93$$

The critical values of $z = -1.96$ and $z = 1.96$ are found from Table A-2. The critical values correspond to a table entry of 0.4750, which is $0.5 - 0.05/2$. The test statistic and critical values are shown in Figure 7-13.

Step 7:   Because the test statistic does not fall within the critical region, we fail to reject the null hypothesis.

Step 8:   There is not sufficient evidence to warrant rejection of the claim that 10% of the aborted takeoff and overrun incidents are due to bird strikes. (Although the sample data suggest that the percentage of bird strikes is about 7% (from 5/74), there isn't sufficient evidence to warrant rejection of the claim that it's 10%.) .

The claim in the preceding example involved a percentage, but we must use the equivalent decimal form. The methods presented in this section can be used to test claims made about proportions, probabilities, or percentages. Whether we have a proportion, probability, or percentage, the value of p must be between 0 and 1, and the sum of p and q must be exactly 1.

The Conqueror

The Conqueror is a 1954 movie, filmed near St. George, Utah, a year after an atomic bomb was exploded nearby. Among the 220 people who worked on the movie, John Wayne and 97 others eventually died of cancer. Methods of statistics can be used to show that the 45% cancer death rate is significantly higher than the rate that would have been expected under normal circumstances. However, methods of statistics cannot be used to show that the *cause* of the cancer deaths was the explosion of the atomic bomb.

▶ EXAMPLE

A television executive claims that "fewer than half of all adults are annoyed by the violence shown on television." Test this claim by using sample data from a Roper poll in which 48% of 1,998 surveyed adults indicated their annoyance with television violence. Use a 0.05 significance level.

● SOLUTION

We summarize the key components of the hypothesis test.

H_0: $p \geq 0.5$
H_1: $p < 0.5$ (from the claim that "fewer than half are annoyed")

Test statistic:

$$z = \frac{\hat{p} - p}{\sqrt{\frac{pq}{n}}} = \frac{0.48 - 0.5}{\sqrt{\frac{(0.5)(0.5)}{1998}}} = -1.79$$

The test statistic, critical value, and critical region are shown in Figure 7-14. Because the test statistic is in the critical region, we reject the null hypothesis. There is sufficient evidence to support the claim that fewer than half of all adults are annoyed by the violence shown on television.

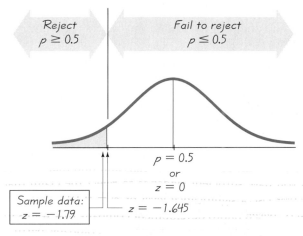

FIGURE 7-14   Hypothesis Test of $p < 0.5$

The *P*-Value Method

The examples in this section followed the traditional approach to hypothesis testing, but it would be just as easy to use the *P*-value approach because the test statistic is a *z* score. The *P*-value is obtained by using the procedure described in Section 7-3. In a right-tailed test, the *P*-value is the area to the right of the test statistic. In a left-tailed test, the *P*-value is the area to the left of the test statistic. In a two-tailed test, the *P*-value is twice the area of the extreme region bounded by the test statistic (see Figure 7-7). As in Section 7-3, we use the decision criterion of rejecting the null hypothesis if the *P*-value is less than or equal to the significance level α.

The last example was left-tailed, so the *P*-value is the area to the left of the test statistic $z = -1.79$. Table A-2 indicates that the area between $z = 0$ and $z = -1.79$ is 0.4633, so the *P*-value is $0.5 - 0.4633 = 0.0367$. Because the *P*-value of 0.0367 is less than the significance level of 0.05, we reject the null hypothesis and again conclude that there is sufficient sample evidence to support the claim that fewer than half of all adults are annoyed with television violence. Again, the *P*-value method is simply another way of arriving at the same conclusion reached using the traditional approach.

7-5   Exercises A: Basic Skills and Concepts

In Exercises 1–20, use the traditional method to test the given hypothesis. Include the steps listed in Figure 7-4, and draw the appropriate graph. Also identify the *P*-value. Assume that all samples have been randomly selected.

1. The quality control manager at the Telektronic Company considers production of telephone answering machines to be "out of control" when the overall rate of defects exceeds 4%. Testing of a random sample of 150 machines revealed 9 defects, so the sample percentage of defects is 6%. The production manager claims that this is only a chance difference, so production is not really out of control and no corrective action is necessary. Use a 0.05 significance level to test the production manager's claim. Does it appear that corrective action is necessary?

2. In a *Sports Illustrated for Kids* survey of 603 children, 43% preferred McDonald's for fast food. The next highest rating went to Taco Bell, preferred by 13% of the children. An advertising executive claims that McDonald's is preferred by half of all children. Test that claim at the 0.05 significance level.

3. According to a Harris poll, 71% of Americans believe that the cost of lawsuits is too high. If you survey 500 randomly selected Americans and find that 74% of them hold that belief, do you have sufficient evidence to reject the claim that the actual percentage is 71%? Use a significance level of $\alpha = 0.10$.

4. Bay Photo is a San Francisco–based photo development company that wants to use automated telephone messages to solicit customers. A company partner argues that potential customers with unlisted numbers will not be reached, but

the marketing manager claims that fewer than one-half of residential telephones in San Francisco have unlisted numbers. A random sample of 400 residential phones in San Francisco resulted in an unlisted rate of 39%. Use a 0.005 level of significance to test the claim that fewer than one-half have unlisted numbers.

5. In a survey of randomly selected households, 288 had computers and 962 did not (based on data from the Electronic Industries Association). At the 0.02 level of significance, test the claim that computers are in 20% of all households.

6. Recently TWA reported an on-time arrival rate of 78.4%. Assume that in a later random sample of 750 flights, 630 were on time. If TWA were to claim that its on-time arrival rate was now higher than 78.4%, would that claim be supported at the 0.01 level of significance?

7. Test the claim that more than one-fourth of all white collar criminals have attended college. U.S. Bureau of Justice statistics reveal that of a sample of 1400 randomly selected white collar criminals, 33% had attended college. Use a 0.02 level of significance.

8. In a survey by Media General and the Associated Press, 813 of the 1084 respondents indicated support for a ban on household aerosols. In lobbying Congress for action to correct the ozone depletion problem, the environmental group Save Planet Earth claims that more than 70% of the population supports the ban on household aerosols. Test that claim at the 0.01 significance level.

9. In a genetics experiment, the Mendelian law is followed as expected if one-eighth of the offspring exhibit a certain recessive trait. Analysis of 500 randomly selected offspring indicated that 83 exhibit the necessary recessive trait. Is the Mendelian law being followed as expected? Use a 2% level of significance.

10. In attempting to gain approval for increased advertising expenditures, the president of the Jefferson Valley Bank tells the Board of Directors that the percentage of adults with money in regular savings accounts is less than 65%. The board chairperson reports on a Roper Organization poll of 2000 adults, showing that 1280 have money in regular savings accounts. Using a 0.05 significance level, test the president's claim.

11. In clinical studies of the allergy drug Seldane, 70 of the 781 subjects experienced drowsiness (based on data from Merrell Dow Pharmaceuticals, Inc.). A competitor claims that more than 8% of Seldane users experience drowsiness. Use a 0.05 significance level to test that claim.

12. The Federal Aviation Administration will fund research of spatial disorientation of pilots if there is sufficient sample evidence (at the 0.01 significance level) to conclude that among aircraft accidents involving such disorientation, more than three-fourths result in fatalities. A study of 500 aircraft accidents involving spatial disorientation of the pilot found that 91% of those accidents resulted in fatalities (based on data from the Department of Transportation). Based on these sample results, will the funding be approved?

13. Dr. Kelley Roberts is dean of a medical school and must plan courses for incoming students. The college president is encouraging her to increase the emphasis on pediatrics, but Dr. Roberts argues that fewer than 10% of U.S. medical students prefer pediatrics. She refers to sample data indicating that 64 of 1068 randomly selected medical school students chose pediatrics (based on

data reported by the Association of American Medical Colleges). Is there sufficient sample evidence to support (at the 0.01 significance level) her argument?

14. A marketing strategist for the tobacco industry claims that the percentage of college graduates who smoke is less than the 32% rate for the general population of persons aged 21 and over (based on data from the U.S. National Institute on Drug Abuse). If that claim can be justified, the Carolina Tobacco Company is willing to underwrite an expensive ad campaign targeted at college graduates. Test that claim using a 0.04 significance level. Sample data reveal that of 785 subjects who completed four years of college, 18.3% smoke (based on data from the American Medical Association).

15. A reservations system for Air America suffered from a 7% rate of no-shows. A new procedure was instituted whereby reservations are confirmed on the day preceding the actual flight, and a study was then made of 5218 randomly selected reservations made under the new system. If 333 no-shows were recorded, test the claim that the no-show rate is lower with the new system. Does the new system appear to be effective in reducing no-shows?

16. In most states, the majority of car passengers use seat belts, but a survey of 850 car passengers in Vermont showed that only 400 of them used seat belts (based on data from the National Highway Traffic Safety Administration). A traffic safety specialist says that Vermont should launch a campaign of advertising and enforcement so that seat belt use rises above 50%. However, a state senator argues that the sample is too small to justify such a campaign. He claims that at least half of Vermont passengers use seat belts and that it's due to chance that the sample results don't reflect that. The senator states, "If the sample results were convincing in showing that the rate was below 50%, then I would support the campaign." At the 0.05 significance level, test the claim that the rate is below 50%. Will the results of that test be helpful in convincing the senator to support the campaign?

17. Ralph Carter is a high school history teacher who says that if students are not aware of the Holocaust, then the curriculum should be revised to correct that deficiency. A Roper Organization survey of 506 high school students showed that 268 of them did not know that the term "Holocaust" refers to the Nazi killing of about 6 million Jews during World War II. Using the sample data and a 0.05 significance level, test the claim that most (more than 50%) students don't know what "Holocaust" refers to. Based on these results, should Ralph Carter seek revisions to the curriculum?

18. The Kennedy-Nixon presidential race was extremely close. Kennedy won with 34,227,000 votes to Nixon's 34,108,000 votes. The closeness of the results caused some people to speculate that the event was like flipping a coin and Nixon might have won on another day. Although the votes aren't sample data randomly selected from a larger population, assume that they are. At the 0.01 level of significance, test the claim that the true population proportion for Kennedy exceeded 0.5. Does the difference in the votes appear to be a chance difference or a significant difference?

19. A Bruskin-Goldring Research poll of 1012 adults showed that among those who used fruitcakes, 28% ate them and 72% used them for other purposes,

such as doorstops, birdfeed, and landfill. This surprised fruitcake producers, who believe that a fruitcake is an appealing food. The president of the Kansas Food Products Company claims that the poll results are a fluke and, in reality, at <u>least half of all adults eat their fruitcakes</u>. Use a 0.005 level of significance to test that claim. Based on the result, does it appear that fruitcake producers should consider changes to make their product more appealing as a food or better suited for its uses as a doorstop, birdfeed, and so on?

20. In a nationally televised live taste test funded by the Joseph Schlitz Brewery Company, 100 regular drinkers of Budweiser beer were given blind samples of Budweiser and Schlitz; 48 of these subjects preferred Schlitz. At the 0.05 level of significance, test the claim that Schlitz is preferred by 50% of Budweiser beer drinkers who participate in such blind taste tests. (Critics of the test claimed that the subjects couldn't observe differences and guessed when making their choices, so the 50% rate is justified.)

7-5   Exercises B: Beyond the Basics

21. Referring to Data Set 6 in Appendix B, test the claim of Mars, Inc. that 20% of M&Ms are red. Use a 0.05 significance level.
    a. Use the traditional method
    b. Use the P-value method
    c. Test the claim by constructing and interpreting a 95% confidence interval.

22. Chemco, a supplier of chemical waste containers, finds that 3% of a sample of 500 units are defective. Being somewhat devious, the Chemco production manager wants to make a claim that the rate of defective units is no more than some specified percentage, and he doesn't want that claim rejected at the 0.05 level of significance if the sample data are used. What is the *lowest* defective rate he can claim under these conditions?

23. A reporter for the *Providence Journal* claims that 10% of the city's residents believe that the mayor is doing a good job. Test her claim if, in a random sample of 15 residents, there are none who believe that the mayor is doing a good job. Use a 0.05 level of significance. Because $np = 1.5$ and is not at least 5, the normal distribution is not a suitable approximation of the distribution of sample proportions.

24. A study of 500 aircraft accidents involving spatial disorientation of the pilot found that 91% of those accidents resulted in fatalities. Someone with a vested interest wants to claim that the percentage is at least some particular value, and this person doesn't want that claim rejected at the 0.01 level of significance if the sample data are used. What is the *highest* percentage that can be claimed under these conditions?

25. Refer to the example in this section that relates to television violence. If the true value of p is 0.45, find β, the probability of a type II error (see Exercise <u>28 from Section 7-3</u>). (*Hint:* In Step 3 use the values of $p = 0.45$ and $\sqrt{pq/n} = \sqrt{(0.45)(0.55)/1998}$.)

7-6 Testing a Claim about a Standard Deviation or Variance

Like Section 7-5, this section does not introduce a radically new procedure for testing hypotheses. Instead, we use the same general procedures introduced in Sections 7-2 and 7-3, but with a distribution, a test statistic, and critical values appropriate for testing a claim about a population standard deviation σ or variance σ^2.

Because σ is the square root of σ^2, if we know the value of one we also know the value of the other. We can therefore use the same procedure for testing claims about σ or σ^2. The preceding sections of this chapter used the normal distribution and the Student t distribution. Tests of claims about σ or σ^2 require that the population have normally distributed values, so the discussions of this section are based on the following assumption.

Assumption for Testing Claims about σ or σ^2

In testing a hypothesis made about a population standard deviation σ or variance σ^2, we assume that the population has values that are normally distributed.

Recall from Section 6-4 that when using the Student t distribution, we require that the population of values be approximately normal, but we can accept deviations away from normality that are not too severe. However, when standard deviations or variances are involved, populations with very non-normal distributions can lead to gross errors. Consequently, the assumption of normality must be adhered to much more strictly in such cases, and we should check the distribution of the data by constructing a histogram to see if it is symmetrical and bell shaped. We described this sensitivity to a normal distribution by saying that inferences about the population standard deviation σ (or variance σ^2) made on the basis of the chi-square distribution are not robust, meaning that the inferences can be very misleading if the population does not have a normal distribution. Given the assumption of a normal distribution, the following test statistic has a chi-square distribution with $n - 1$ degrees of freedom and critical values given in Table A-4.

Test Statistic for Testing Hypotheses about σ or σ^2

$$\chi^2 = \frac{(n - 1)s^2}{\sigma^2}$$

where n = sample size
s^2 = sample variance
σ^2 = population variance (given in the null hypothesis)

The chi-square distribution was introduced in Section 6-4, where we noted the following important properties.

Properties of the Chi-Square Distribution

1. All values of χ^2 are nonnegative, and the distribution is not symmetric (see Figure 7-15).

2. There is a different distribution for each number of degrees of freedom (see Figure 7-16).

3. The critical values are found in Table A-4 where

$$\text{degrees of freedom} = n - 1$$

In using Table A-4, it is essential to note that each critical value separates an *area to the right* that corresponds to the value given in the top row.

As noted in Section 6-4, later chapters will involve cases in which the degrees of freedom are not $n - 1$, so we should not universally equate degrees of freedom with $n - 1$. Once we have determined degrees of freedom, the significance level α, and the type of test (left-tailed, right-tailed, or two-tailed), we can use Table A-4 to find the critical chi-square values.

In a right-tailed test, the value of α will correspond exactly to the areas given in the top row of Table A-4. In a left-tailed test, the value of $1 - \alpha$ will correspond exactly to the areas given in the top row of Table A-4. In a two-tailed test, the values of $\alpha/2$ and $1 - \alpha/2$ will correspond exactly to the areas given in the top row of Table A-4. (See Figure 6-8 and the example on page 320.)

Many applications of statistics involve decisions or inferences about variances or standard deviations. In manufacturing, quality control engineers want

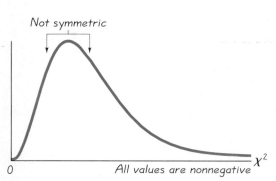

FIGURE 7-15 Properties of the Chi-Square Distribution

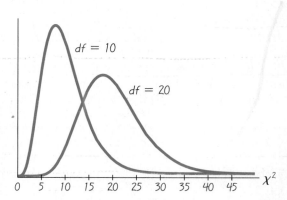

FIGURE 7-16 Chi-Square Distributions for 10 and 20 Degrees of Freedom

There is a different distribution for each number of degrees of freedom.

to ensure that a product is, on the average, acceptable. But they also want to produce items of *consistent* quality so that there will be few defective products. Consistency is often measured by variance or standard deviation. Let's consider aircraft altimeters. Because of mass production techniques and various other factors, these altimeters don't give readings that are exactly correct; some errors are to be expected. Federal Aviation Regulation 91.36 requires that aircraft altimeters be tested and calibrated to give a reading "within 125 feet (on a 95-percent probability basis)." Even if the mean altitude reading is exactly correct, an excessively large standard deviation will result in individual readings that are dangerously low or high. Such a large standard deviation would indicate that production was out of control and that unacceptable and dangerous altimeters were being manufactured. See the following example.

▶ **EXAMPLE**

The Stewart Aviation Products Company has been successfully manufacturing aircraft altimeters with errors that have a mean of 0 ft (achieved by calibration) and a standard deviation of 43.7 ft. After the installation of new production equipment, 30 altimeters were randomly selected from the new line. This sample group had errors with a standard deviation of $s = 54.7$ ft. Has the standard deviation changed with the new equipment, or has it remained the same, with $s = 54.7$ ft representing a chance sample fluctuation? At the $\alpha = 0.05$ level of significance, test the claim that the new population has errors with a standard deviation equal to 43.7 ft. Based on previous observations, we can assume that the errors are normally distributed.

● **SOLUTION**

We will use the traditional method of testing hypotheses as outlined in Figure 7-4.

Step 1: The claim is that the new production method has resulted in a population with a standard deviation equal to 43.7 ft. In symbolic form we have $\sigma = 43.7$ ft.

Step 2: If the original claim is false, then $\sigma \neq 43.7$ ft.

Step 3: Since the null hypothesis must contain equality, we have

$$H_0: \sigma = 43.7 \qquad H_1: \sigma \neq 43.7$$

Step 4: The significance level is $\alpha = 0.05$.

Step 5: Because this claim is about σ, we will use the chi-square distribution.

(continued)

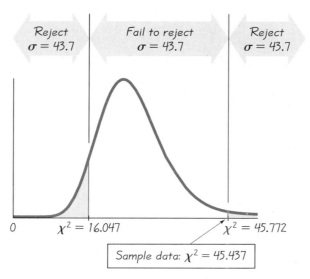

FIGURE 7-17   Hypothesis Test of $\sigma = 43.7$

Step 6: The test statistic is

$$\chi^2 = \frac{(n-1)s^2}{\sigma^2} = \frac{(30-1)(54.7)^2}{43.7^2} = 45.437$$

The critical values are 16.047 and 45.772. They are found in Table A-4, in the 29th row (degrees of freedom $= n - 1 = 29$) in the columns corresponding to 0.975 and 0.025. See the test statistic and critical values shown in Figure 7-17.

Step 7: Because the test statistic is not in the critical region, we fail to reject the null hypothesis.

Step 8: (For help in wording the final conclusion, refer to Figure 7-2.) There is not sufficient evidence to warrant rejection of the claim that the standard deviation is equal to 43.7 ft. However, it would be wise to continue monitoring and testing the new product line.

As discussed briefly in an earlier chapter, another application in which variability is a major concern involves the waiting lines of banks. In the past, customers traditionally entered a bank and selected one of several lines formed at various windows. A different system that has become popular involves one main waiting line, which feeds the various windows as vacancies occur. The mean waiting time isn't reduced, but the variability among waiting times is decreased and the irritation of being caught in a slow line is consequently diminished.

▶ **EXAMPLE**

With individual lines at its various windows, the Jefferson Valley Bank found that the standard deviation for normally distributed waiting times on Friday afternoons was 6.2 min. The bank experimented with a single main waiting line and found that for a random sample of 25 customers, the waiting times have a standard deviation of 3.8 min. Based on previous studies, we can assume that the waiting times are normally distributed. At the $\alpha = 0.05$ significance level, test the claim that a single line causes lower variation among the waiting times.

● **SOLUTION**

Again using the traditional method of testing hypotheses as outlined in Figure 7-4, we get the following results.

Step 1:   We wish to test the claim that the variation is lower than it was in the past. Using the standard deviation σ as the measure of variation, we can express the claim of lower variation as $\sigma < 6.2$.

Step 2:   If the original claim is false, then $\sigma \geq 6.2$.

Step 3:   Because the null hypothesis must contain equality, we have

$$H_0: \sigma \geq 6.2 \qquad H_1: \sigma < 6.2$$

Step 4:   The significance level is $\alpha = 0.05$.

Step 5:   Because this claim is about σ, we will use the chi-square distribution.

Step 6:   The test statistic is

$$\chi^2 = \frac{(n-1)s^2}{\sigma^2} = \frac{(25-1)(3.8)^2}{(6.2)^2} = 9.016$$

This test is left-tailed because H_0 will be rejected only for significantly small values of χ^2. With $\alpha = 0.05$ and $n = 25$, we go to Table A-4 and align 24 degrees of freedom with an area of 0.95 to obtain the critical χ^2 value of 13.848. The test statistic of $\chi^2 = 9.016$, the critical value of $\chi^2 = 13.848$, and the critical region are shown in Figure 7-18, which is on the next page.

Step 7:   Because the test statistic falls within the critical region, we reject H_0.

Step 8:   We conclude that the 3.8-min standard deviation is significantly less than the 6.2-min standard deviation that corresponds to multiple waiting lines. The sample data support the claim of lower variation; that is, the single main line does appear to lower the variation among waiting times. (Refer to Figure 7-2 for help in stating the final conclusion.)

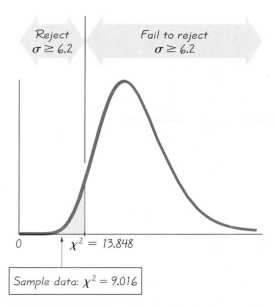

FIGURE 7-18   Hypothesis Test of $\sigma < 6.2$

One aspect of the preceding example that might cause some confusion is the reference to two different distributions. The population of actual waiting times represents a *normal* distribution, but in Step 6 of the preceding solution we use the *chi-square* distribution. When we randomly select samples of size n, the distribution of $(n - 1)s^2/\sigma^2$ is a chi-square distribution, given a normally distributed population. The methods of this section require that the population have a normal distribution, but we use the chi-square distribution in testing claims about standard deviations or variances.

The *P*-Value Method

The *P*-value approach to hypothesis testing (see Figures 7-7 and 7-8) can be used to test claims made about population standard deviations or variances. Because the chi-square distribution table (Table A-4) includes only selected values of α, however, we cannot usually determine the specific *P*-value from that table. Instead, we use the table to identify limits that contain the *P*-value. In the last example we found the test statistic to be $\chi^2 = 9.016$, and we know that the test is left-tailed with 24 degrees of freedom. By examining the 24th row of Table A-4, we see that the test statistic of 9.016 is less than the lowest table value of 9.886, so the *P*-value must be less than 0.005. With a significance level of $\alpha = 0.05$, we reject the null hypothesis as we did when using the traditional method in the preceding example. Again, the traditional method and *P*-value method are equivalent in the sense that they always lead to the same conclusion.

Using Computers to Find *P*-Values

Although Table A-4 can be used to find limits containing *P*-values, we can find good approximations of *P*-values using calculators or statistical software such as STATDISK or Minitab. The STATDISK display for the last example follows. Note the displayed *P*-value of 0.00257. A graph similar to that of Figure 7-18 can also be displayed by STATDISK. Minitab is not designed to test claims involving a standard deviation or variance.

STATDISK DISPLAY

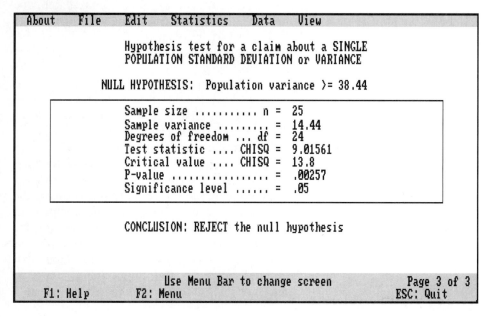

| About | File | Edit | Statistics | Data | View |

Hypothesis test for a claim about a SINGLE
POPULATION STANDARD DEVIATION or VARIANCE

NULL HYPOTHESIS: Population variance >= 38.44

Sample size .... n = 25
Sample variance .... = 14.44
Degrees of freedom ... df = 24
Test statistic .... CHISQ = 9.01561
Critical value .... CHISQ = 13.8
P-value .... = .00257
Significance level ...... = .05

CONCLUSION: REJECT the null hypothesis

Use Menu Bar to change screen   Page 3 of 3
F1: Help   F2: Menu   ESC: Quit

Product Testing

The United States Testing Company in Hoboken, New Jersey, is the world's largest independent product-testing laboratory. It's often hired to verify advertising claims. A vice president has said that the most difficult part of his job is "telling a client when his product stinks. But if we didn't do that, we'd have no credibility." He says that there have been a few clients who wanted positive results fabricated, but most want honest results. United States Testing Laboratory evaluates laundry detergents, cosmetics, insulation materials, zippers, pantyhose, football helmets, toothpastes, fertilizers, and a wide variety of other products.

7-6   Exercises A: Basic Skills and Concepts

In Exercises 1 and 2, use Table A-4 to find the critical values of χ^2 based on the given information.

1. a. $\alpha = 0.05$
   $n = 20$
   $H_0: \sigma = 16$

   b. $\alpha = 0.05$
   $n = 20$
   $H_0: \sigma^2 \geq 256$

   c. $\alpha = 0.05$
   $n = 75$
   $H_1: \sigma^2 \neq 31.5$

2. a. $\alpha = 0.10$
   $n = 6$
   $H_1: \sigma < 10$

   b. $\alpha = 0.05$
   $n = 41$
   $H_1: \sigma^2 > 500$

   c. $\alpha = 0.01$
   $n = 23$
   $H_0: \sigma^2 \leq 23.5$

In Exercises 3–18, use the traditional method to test the given hypotheses. Follow the steps outlined in Figure 7-4, and draw the appropriate graph. In all cases, assume that the population is normally distributed and that the sample has been randomly selected.

3. The Stewart Aviation Products Company uses a new production method to manufacture aircraft altimeters. A random sample of 81 altimeters resulted in errors with a standard deviation of $s = 52.3$ ft. At the 0.05 level of significance, test the claim that the new production line has errors with a standard deviation equal to 43.7 ft, which was the standard deviation for the old production method. If it appears that the standard deviation has changed, does the new production method appear to be better or worse than the old method?

4. In a study of the wide ranges in the academic success of college freshmen, one obvious factor is the amount of time spent studying. At the 0.05 significance level, test the claim that the standard deviation is more than 4.00 h. The sample consists of 70 randomly selected freshmen who have a standard deviation of 5.33 h (based on data reported by *USA Today*).

5. The Kansas Farm Products Company uses a machine that fills 50-lb corn seed bags. In the past, the machine has had a standard deviation of 0.75 lb. In an attempt to get more consistent weights, mechanics have replaced some worn machine parts. A random sample of 61 bags taken from the repaired machinery produced a sample mean of 50.13 lb and a sample standard deviation of 0.48 lb. At the $\alpha = 0.05$ significance level, test the claim that the standard deviation is lower with the repaired machinery than it was in the past.

6. The Collins Investment Company finds that if the standard deviation for the weekly downtimes of their computer is 2 h or less, then the computer is predictable and planning is facilitated. Twelve weekly downtimes for the computer were randomly selected, and the sample standard deviation was computed to be 2.85 h. The manager of computer operations claims that computer access times are unpredictable because the standard deviation exceeds 2 h. Test her claim, using a 0.025 significance level. Does the variation appear to be too high?

7. The Medassist Pharmaceutical Company uses a machine to pour cold medicine into bottles in such a way that the standard deviation of the weights is 0.15 oz. A new machine was tested on 71 bottles, and the standard deviation for this sample is 0.12 oz. The Dayton Machine Company, which manufactures the new machine, claims that it fills bottles with a lower variation. At the 0.05 significance level, test the claim made by the Dayton Machine Company. If Dayton's machine is being used on a trial basis, should its purchase be considered?

8. For randomly selected adults, IQ scores are normally distributed with a mean of 100 and a standard deviation of 15. A sample of 24 randomly selected college professors resulted in IQ scores having a standard deviation of 10. A psychologist is quite sure that college professors have IQ scores that have a mean greater than 100. He doesn't understand the concept of standard deviation very well and claims the standard deviation for IQ scores of college

professors is equal to 15, the same standard deviation as in the general population. Use a 0.05 level of significance to test that claim. Based on the result, what do you conclude about the standard deviation of IQ scores for college professors?

9. Systolic blood pressure results from contraction of the heart. In comparing systolic blood pressure levels of men and women, Dr. Jane Taylor obtained readings for a random sample of 50 women. The sample mean and standard deviation were found to be 130.7 and 23.4, respectively. If systolic blood pressure levels for men are known to have a mean and standard deviation of 133.4 and 19.7, respectively, test the claim that women have a larger standard deviation. Use a 0.05 level of significance. (All readings are in millimeters of mercury, and data are based on the National Health Survey, USDHHS publication 81-1671.)

10. The weights (in grams) of quarters listed in Data Set 8 of Appendix B have a mean of 5.622 g and a standard deviation of 0.068 g. Test the claim that the standard deviation exceeds 0.05 g. Use a 0.05 significance level.

11. In an attempt to design car seats that are comfortable for a wide variety of people, the engineering department of General Motors analyzed heights of men and women. For women in the 25–34 age bracket, heights have a mean of 64.1 in. and a standard deviation of 2.4 in. (based on data from the National Health Survey, USDHEW publication 79-1659). In order to verify those parameters, 50 women drivers aged 25–34 were randomly selected and measured; their heights have a mean of 64.3 in. and a standard deviation of 2.7 in. At the 0.05 level of significance, test the claim that the sample is drawn from a population with the same standard deviation of 2.4 in.

12. Data Set 6 in Appendix B includes weights (in grams) of red M&Ms. For this sample, the mean is 0.9043 g and the standard deviation is 0.0414 g. Use a 0.05 significance level to test the claim that the weights of red M&Ms have a standard deviation of less than 0.05 g.

13. Referring to Data Set 1 (from the Garbage Project at the University of Arizona) in Appendix B, use a 0.05 significance level to test the claim that the weights of food discarded by households have a standard deviation equal to 3 lb. The data set has a mean of 4.816 lb and a standard deviation of 3.297 lb.

14. Referring to Data Set 10 in Appendix B, use a 0.05 significance level to test the claim that the ages of U.S. commercial aircraft have a standard deviation of less than 10 years. The data set has a mean of 13.41 years and a standard deviation of 8.28 years.

15. The Gleason Supermarket chain has instituted a system for monitoring the quality of the products it sells. The Gleason specifications for its generic cola require that the caffeine contents have a standard deviation of less than 2.0 mg. Listed below are the caffeine contents (in mg) for a dozen randomly selected cans of the cola.

34.2	33.7	31.9	34.3	31.6	32.7
33.1	35.2	31.6	32.9	33.0	32.4

(continued)

At the 0.025 level of significance, test the claim that the cola meets the stated specification.

16. Based on data from the National Health Survey (USDHEW publication 79-1659), men aged 25 to 34 have heights with a standard deviation of 2.9 in. Test the claim that men aged 45 to 54 have heights with a standard deviation of less than 2.9 in. The heights of 25 randomly selected men in the 45 to 54 age bracket are listed below.

66.80	71.22	65.80	66.24	69.62	70.49	70.00
71.46	65.72	68.10	72.14	71.58	66.85	69.88
68.69	72.77	67.34	68.40	68.96	68.70	72.69
68.67	67.79	63.97	67.19			

17. Given below are birth weights (in kilograms) of male babies born to mothers on a special vitamin supplement (based on data from the New York State Department of Health).

3.73	4.37	3.73	4.33	3.39	3.68	4.68	3.52
3.02	4.09	2.47	4.13	4.47	3.22	3.43	2.54

Test the claim that this sample comes from a population with a standard deviation equal to 0.470 kg, which is the standard deviation for male birth weights in general. Does the vitamin supplement appear to affect the variation among birth weights?

18. If annual electric energy consumption amounts are consistent (with a low standard deviation), then the costs can be predicted reasonably well. Listed below are the total electric energy consumption amounts (in kWh) for the author's home during 7 different years.

11,943	11,463	10,789	9907	9012	9942	11,153

At the 0.10 significance level, test the claim that the standard deviation for all such years is equal to 1000 kWh.

7-6   Exercises B: Beyond the Basics

19. What do you know about the range of possible P-values in each of the following cases? (Assume use of Table A-4.)
    a. H_1: $\sigma > 15.0$; $n = 10$; test statistic is $\chi^2 = 19.735$.
    b. H_1: $\sigma < 45.0$; $n = 20$; test statistic is $\chi^2 = 7.337$.
    c. H_0: $\sigma = 1.52$; $n = 30$; test statistic is $\chi^2 = 54.603$.
20. Refer to Data Set 1 (from the Garbage Project at the University of Arizona) in Appendix B. First construct a histogram to confirm that the distribution is approximately normal, and then use a 0.05 significance level to test the claim

that the *total* weights of garbage discarded by households have a standard deviation equal to 15.5 lb. Test the claim using

a.   the traditional method

b.   the P-value method

c.   a 95% confidence interval estimate of σ.

21. For large numbers of degrees of freedom, we can approximate critical values of χ^2 as follows:

$$\chi^2 = \frac{1}{2}(z + \sqrt{2k - 1})^2$$

Here k is the number of degrees of freedom and z is the critical value, found in Table A-2.

For example, if we want to approximate the two critical values of χ^2 in a two-tailed hypothesis test with $\alpha = 0.05$ and a sample size of 150, we let $k = 149$ with $z = -1.96$, followed by $k = 149$ and $z = 1.96$.

a.   Use this approximation to estimate the critical values of χ^2 in a two-tailed hypothesis test when $n = 101$ and $\sigma = 0.05$. Compare the results to those found in Table A-4.

b.   Use this approximation to estimate the critical values of χ^2 in a two-tailed hypothesis test when $n = 150$ and $\alpha = 0.05$.

22. Repeat Exercise 21 using the approximation

$$\chi^2 = k\left(1 - \frac{2}{9k} + z\sqrt{\frac{2}{9k}}\right)^3$$

Here k and z are as described in that exercise.

23. Refer to the second example presented in this section (customer waiting times at the Jefferson Valley Bank). Assuming that σ is actually 4.0, find β (the probability of a type II error). See Exercise 28 from Section 7-3 and modify the procedure so that it applies to a hypothesis test involving σ instead of μ.

24. In Sections 7-2 and 7-3 we used the body temperature data in Table 2-1 and rejected the claim that the mean body temperature of healthy adults equals 98.6°F. Suppose critics argue that the claim should *not* be rejected because the population standard deviation is actually much larger than the value of $s = 0.62$°F that was used. They note that a larger standard deviation will result in a smaller test statistic.

a.   Find the smallest value of σ greater than 0.62 that results in a test statistic not falling in the critical region.

b.   Use the sample data from Table 2-1 to test the claim that σ is at least as large as the value from part *a*.

c.   Based on the results from parts a and b, is there any validity to the argument that the claim of $\mu = 98.6$ should not be rejected because the population standard deviation is much larger than 0.62?

VOCABULARY LIST

Define and give an example of each term.

hypothesis	test statistic
hypothesis test	critical region
test of significance	critical value
null hypothesis	two-tailed test
alternative hypothesis	left-tailed test
type I error	right-tailed test
type II error	traditional method
alpha (α)	classical method
beta (β)	P-value
significance level	probability value

Important Formulas

Parameter to Which Hypothesis Refers	Applicable Distribution	Assumption	Test Statistic	Table of Critical Values
μ (population mean)	Normal	σ is known and the population is normally distributed.	$z = \dfrac{\overline{x} - \mu}{\sigma/\sqrt{n}}$	Table A-2
	Normal	$n > 30$. (If σ is not known, use s for σ.)	$z = \dfrac{\overline{x} - \mu}{\sigma/\sqrt{n}}$	Table A-2
	Student t	σ is unknown, $n \le 30$, and the population is normally distributed.	$t = \dfrac{\overline{x} - \mu}{s/\sqrt{n}}$	Table A-3
p (population proportion)	Normal	$np \ge 5$ and $nq \ge 5$.	$z = \dfrac{\hat{p} - p}{\sqrt{pq/n}}$ where $\hat{p} = \dfrac{x}{n}$	Table A-2
σ^2 (population variance) σ (population standard deviation)	Chi-square	The population is normally distributed.	$\chi^2 = \dfrac{(n - 1)s^2}{\sigma^2}$	Table A-4

REVIEW

Chapters 6 and 7 introduced two of the most important concepts in using sample data to form inferences about population data. Whereas the major objective of Chapter 6 was estimating the values of population parameters, this chapter focused on methods for testing hypotheses (claims) made about population parameters. We considered claims made about population means, proportions, variances, and standard deviations. Hypothesis tests are also called tests of significance, reflecting the fact that we use them to decide whether sample differences are caused by chance fluctuations or whether the differences are so dramatic that they are not likely to have occurred by chance. We are able to select exact levels of significance; 0.05 and 0.01 are common values. Sample results are said to reflect significant differences when their occurrences have probabilities less than the chosen level of significance.

Section 7-2 presented the fundamental concepts of a hypothesis test: null hypothesis, alternative hypothesis, type I error, type II error, test statistic, critical region, critical value, and significance level. We also discussed two-tailed tests, left-tailed tests, right-tailed tests, and the statement of conclusions. Section 7-3 used those components in identifying three different methods for testing hypotheses. Figure 7-4 summarized the traditional method, and Figure 7-8 summarized the P-value method. The third method is based on the construction of confidence intervals, which was discussed in Chapter 6.

Sections 7-2 and 7-3 introduced the method of testing hypotheses, using examples in which only the normal distribution applied. Other distributions were introduced in subsequent sections. The "Important Formulas" section that follows summarizes the hypothesis tests covered in this chapter.

In Section 7-3 we saw that with the traditional approach, we make a decision about the null hypothesis by comparing the test statistic and critical value. With the P-value method, we base that decision on a comparison of the significance level and the P-value, which represents the probability of getting a sample that is at least as extreme as the one obtained.

REVIEW EXERCISES

In Exercises 1 and 2, find the appropriate critical values. In all cases, assume that the population standard deviation σ is unknown.

1. a. $\alpha = 0.05$
      $n = 160$
      $H_0: p \geq 0.5$

   b. $\alpha = 0.01$
      $n = 35$
      $H_0: \mu \leq 16.5$

   c. $\alpha = 0.01$
      $n = 12$
      $H_0: \mu = 38.4$

   d. $\alpha = 0.01$
      $n = 25$
      $H_0: \sigma^2 \geq 225$

   e. $\alpha = 0.05$
      $n = 30$
      $H_0: \sigma^2 = 84.3$

2. a. $\alpha = 0.10$ b. $\alpha = 0.10$ c. $\alpha = 0.06$
       $n = 15$ $n = 15$ $n = 100$
       $H_0: \mu = 1.23$ $H_0: \sigma^2 = 123$ $H_0: \mu = 72.3$
   d. $\alpha = 0.05$ e. $\alpha = 0.01$
       $n = 10$ $n = 30$
       $H_0: \sigma = 15$ $H_1: \sigma < 5.8$

In Exercises 3 and 4, respond to each of the following:

   a. Give the null hypothesis in symbolic form.
   b. Is this test left-tailed, right-tailed, or two-tailed?
   c. In simple terms devoid of symbolism and technical language, describe the type I error.
   d. In simple terms devoid of symbolism and technical language, describe the type II error.
   e. Identify the probability of making a type I error.

3. At the 0.01 level of significance, the claim is that the mean treatment time for a dentist is at least 20.0 min.

4. At the 0.05 level of significance, the claim is that the mean reading in a biofeedback experiment is 6.2 mV. $a\ H_0 : \mu = 6.2$ Two tail e .05

5. A *USA Today*/CNN/Gallup survey was the basis for a report that "57% of gun owners favor stricter gun laws." Assume that the survey group consisted of 504 randomly selected gun owners. Test the claim that the majority (more than 50%) of gun owners favor stricter gun laws. Use a 0.01 significance level.

6. In a study of birth weights (in grams), a random sample of 30 baby girls produced a mean of 3264 g and a standard deviation of 485 g. Assume that the mean and standard deviation for birth weights of boys are known to be 3393 g and 470 g, respectively (based on data from the New York State Department of Health, Monograph No. 11). At the 0.05 level of significance, test the claim that boys and girls have birth weights with the same standard deviation.

TS $\chi^2 = 30.581$
CV $\chi^2 = 16.047$
   45.72

7. Use the data in Exercise 6 to test the claim that boys and girls have birth weights with the same mean. Use a 0.05 significance level.

8. In a randomly selected group of people who bought compact cars, 312 were women and 338 were men (based on data from the Ford Motor Company). Test the claim that men constitute more than half of the buyers of compact cars. Use a 0.05 significance level.

TS $z = 1.02$
CV $z = 1.645$

9. A Telektronics dental X-ray machine bears a label stating that the machine gives radiation dosages with a mean of less than 5.00 milliroentgens. Sample data consist of 36 randomly selected observations with a mean of 4.13 milliroentgens and a standard deviation of 1.91 milliroentgens. Using a 0.01 level of significance, test the claim stated on the label.

10. Is Elvis alive? *USA Today* ran a report about a University of North Carolina poll of 1248 adults from the southern United States. It was reported that 8% of those surveyed believe that Elvis Presley still lives. The article began with the claim that "almost 1 out of 10" Southerners still thinks Elvis is alive. At the 0.01 significance level, test the claim that the true percentage is less than 10%. Based

on the result, determine whether the 8% sample result justifies the phrase "almost 1 out of 10." $TS \ Z = -2.36$, $C V Z = -2.33$

11. Use a 0.025 level of significance to test the claim that vehicle speeds on I-95 near Boston have a mean of above 55.0 mi/h. A random sample of 50 vehicles produces a mean of 61.3 mi/h and a standard deviation of 3.3 mi/h.

12. A standard test for braking reaction times (in seconds) has produced an average of 0.75 s for young females. A driving instructor claims that his class of young females exhibits an overall reaction time that is below the average. Test his claim if it is known that 13 of his students (randomly selected) produced a mean of 0.71 s and a standard deviation of 0.06 s. Test at the 0.01 level of significance.

$TS = t = -2.404$
$CV = t = -2.681$

13. One large high school has found that students taking a standard college aptitude test earn scores with a variance of 6410. A counselor claims that the current group of test subjects has more varied aptitudes. Test the claim that the variance is larger than 6410 if a random sample of 60 students produces a variance of 8464. Use a 0.10 level of significance and assume that the test scores are normally distributed.

14. A sociologist designed a test to measure prejudicial attitudes and claimed that the mean population score is 60. The test was then administered to 40 randomly selected subjects, and the results produced a mean and standard deviation of 69 and 12, respectively. At the 0.05 level of significance, test the sociologist's claim.

$TS: Z = 4.74$
$CV: Z = \pm 1.96$

15. Many companies rely heavily on market research for determining advertising strategies. In a study of brand recognition, 831 subjects recognized the Campbell's Soup brand, and 18 did not (based on data from Total Research Corporation). Use the sample data to test the claim that the recognition rate is equal to 98%. Use a 0.10 level of significance.

16. A student majoring in psychology has developed her own intelligence test. She claims that it will produce the same mean of 100 that is obtained with standard IQ tests. Given below are the results for randomly selected subjects.

$TS: t = -0.921$
$CV: t = \pm 2.977$

| 101 | 106 | 98 | 92 | 97 | 80 | 89 | 88 |
| 110 | 112 | 100 | 100 | 103 | 97 | 97 | |

At the 0.01 significance level, test the claim that the population has a mean of 100.

17. Using the sample data from Exercise 16, test the claim that the standard deviation is less than 10.0. Use a 0.05 level of significance.

18. Test the claim that the mean female reaction time to a highway signal is less than 0.700 s. When 18 females are randomly selected and tested, their mean is 0.668 s and their standard deviation is 0.100 s. Use a 0.05 level of significance.

$TS \ t = -1.358$
$CV: t = -1.740$

19. A national conference of governors issued the optimistic claim that the annual rate of household burglaries is now less than 5%. Assume that the claim was based on a survey of 1500 randomly selected households, 72 of which experienced a burglary last year (based on data from the U.S. Department of Justice). Using a 0.10 significance level, test the given claim. Also identify the P-value.

20. At the 0.05 significance level, test the claim that attorneys have starting salaries

$TS = Z = -.48$

$CV = Z = -1.645$

of at least $50,000 each year. Sample data consist of starting salaries of 35 randomly selected attorneys, which have a mean of $48,817 and a standard deviation of $14,558 (based on data from *The Jobs Rated Almanac*, by Krantz).

COMPUTER PROJECT

One advantage of computers is that they make possible some approaches that are otherwise impractical. This project will use a computer simulation to test the hypothesis that the mean body temperature of healthy adults is 98.6°F. Refer to Table 2-1 for the sample data ($n = 106$, $\bar{x} = 98.2$, $s = 0.62$) collected by University of Maryland researchers. Here's the key question: Given a normally distributed population of body temperatures with mean 98.6 and standard deviation 0.62, how unusual is it to get a sample of 106 temperatures with a mean such as 98.2? If such a result is very unusual (with a probability below 0.05), we can conclude that either the sample data represent very rare results or the claimed value of 98.6 is incorrect. Proceed as follows:

a. Use a software package such as STATDISK or Minitab to generate 20 samples, each consisting of 106 values randomly selected from a normally distributed population with a mean of 98.6 and a standard deviation of 0.62. Find and record the means of the 20 samples. If you are using STATDISK, select `Data` from the main menu bar, then select the option `New Generated Sample`, choose the normal distribution with a sample size of 106, choose data with 2 decimal places, and then proceed to enter the desired mean and standard deviation of 98.6 and 0.62. If you are using Minitab, the commands

```
MTB > RANDOM 106 C1-C20;
SUBC> NORMAL 98.6 0.62.
MTB > DESCRIBE C1-C20
```

will produce the desired 20 sample means. With STATDISK you generate one sample at a time, and it will take about 10 minutes to obtain the 20 sample means. With Minitab you can generate all 20 samples at once, and it will take less than 1 minute to obtain the 20 sample means.

b. How many of the sample means differ from the claimed value of 98.6 by 0.4 or more? Divide that number by 20 to express the result as a proportion. (This is an initial indication of how likely we are to get a sample mean such as 98.2 if the true mean is 98.6. Here's the rationale: The data from Table 2-1 have a sample mean of $\bar{x} = 98.2$ that differs from the claimed mean of $\mu = 98.6$ by the amount of 0.4. By examining the 20 sample means, we can decide whether a sample mean such as 98.2—a sample mean that is at least 0.4 away from 98.6—can easily occur by chance or whether it constitutes a significant difference from 98.6.)

c. The proportion from part b is an indication of how likely we are to get a sample mean such as 98.2 when in fact the population mean is 98.6. Is that proportion less than a significance level of 0.05? (If so, we conclude that the sample mean of 98.2 represents a significant difference from 98.6, and we therefore reject the claim that the mean body temperature of healthy adults is 98.6°F.) What do you conclude? How does your conclusion compare to those resulting from the traditional and *P*-value methods of hypothesis testing?

FROM DATA TO DECISION: Developing a Plan for Scheduling Movies

You are about to become owner of the County Cinema Complex, which contains four movie theaters. You are determined to develop a movie schedule that is easy for viewers to remember but that is also efficient in its use of theaters. You would be delighted if all movies had lengths of 110 min or less so that you could schedule movies to begin on even hours (12:00, 2:00, 4:00, and so on) with at least 10 min available for emptying the theater and allowing the next audience to enter. Unfortunately, some movies are longer. Refer to Data Set 5 in Appendix B.

a. Construct a histogram for the lengths of times of all movies listed in the sample.

b. Construct a histogram for the lengths of R-rated movies.

c. Construct a histogram for the lengths of movies with ratings other than R.

d. Based on sample size and/or shape of the distribution, which of the following claims can be tested using either the normal or the Student *t* distribution?

- The mean length of all movies is 110 min or less.

- The mean length of R-rated movies is 110 min or less.

- The mean length of movies with ratings other than R (that is, G, PG, or PG-13) is 110 min or less.

e. For the claims in part d that can be tested, conduct the tests using a 0.05 level of significance.

f. Find the sample proportion of all movies in Data Set 5 that have lengths of 110 min or less. At the 0.05 significance level, test the claim that most movies run 110 min or less.

g. Based on the preceding results, does it seem practical to implement a scheduling plan based on the assumption that most movies have lengths of 110 min or less?

h. Another factor to consider in scheduling movies is their rating. You would like to have R-rated movies available for adult audiences, as well as movies with other ratings for younger viewers. Columbia Pictures chairman Mark Canton has claimed that "58% of films made are R-rated." Test that claim using a 0.05 level of significance.

Interview

Jay Dean

Senior Vice President at the San Francisco Office of Young & Rubicam Advertising

Jay Dean is Director of the Consumer Insights Department at the San Francisco office of Young & Rubicam. He has worked with many well-known advertisers, including AT&T, Chevron, Clorox, Coors, Dr. Pepper, Ford Motor Co., General Foods, Gillette, Gulf Oil, H.J. Heinz, Kentucky Fried Chicken, and Warner-Lambert.

How extensive is your use of statistics at Young & Rubicam?

We use statistics every day. We do a lot of consumer research, and we conduct many consumer attitude and usage surveys. For some clients we do product tests, taste tests, basic strategy studies—anything they need. I'm now working on a typical design, an advertising test for Take Heart salad dressing, one of the brands we handle for The Clorox Company. Two commercials were shown to independent samples of consumers. We then asked questions about what was being communicated by each commercial, perceptions of the brand, likes and dislikes, and so on. Significance testing will be used to compare the results and to choose the best commercial.

On larger surveys we use multivariate techniques such as factor analysis and multiple regression. Marketing is becoming tougher and tougher today. There are more brands, increasing price competition, and more sophisticated consumers. A tougher marketing environment requires more marketing research, and that means we're using statistics more often as well.

Could you cite a case in which the use of statistics was instrumental in determining a successful strategy?

We recently conducted a major creative exploratory for Pine-Sol, another Clorox brand. About half a dozen commercials were produced in rough form and shown to independent samples of consumers. Statistics helped us to identify the best commercial for the new Pine-Sol advertising campaign. Early marketplace results are very encouraging. At Y&R we believe consumer research helps us to produce the most effective advertising. Statistics helps us to make informed decisions, based upon the results of the research.

How do you typically collect data for statistical analysis?

There are many ways to collect data, but typically it's either random telephone sampling for survey work, or if we want to show people something like a commercial, we use a central location interview. For example, respondents are intercepted in a shopping mall by an interviewer, screened for eligibility, asked to go into a testing facility to be shown a commercial, and so on.

Do you find it difficult to obtain representative and unbiased samples?

Yes. The marketing research industry now is very concerned about the growing refusal rate among the general public. That is driven in part by salespeople who use marketing research as a guise for selling. Mail surveys are subject to a huge self-selection bias and response rates are typically quite low. Shopping mall interviews have other problems. Not everyone goes to malls, and there is a certain degree of interviewer bias in approaching prospective respondents.

What is your typical sample size?

For a national survey the rule of thumb is about a thousand people, although you might go with as few as 600 in some cases. For an advertising test a sample on the order of 200 would be a pretty healthy sample size—sometimes as few as a hundred are used.

Do you feel that job applicants in your field are viewed more favorably if they have studied some statistics?

Yes, absolutely. Everyone has to understand and use statistics at some level. It's very important for entry-level people to know statistics well because they do much of our research project work. In the beginning they do a lot of number crunching and a lot of data analysis.

Do you have any advice for today's students?

If I could go back to school, I would certainly study more math, statistics, and computer science. I studied a lot of it, but I would like to know even more. There's an enormous data explosion in business these days. All businesses are becoming much more quantitative than they ever were in the past. Today there is more information than you know what to do with, and you've got to have the analytical tools and knowledge to deal with this information if you want to be successful.

"If I could go back to school, I would certainly study more math, statistics, and computer science."

8 Inferences from Two Samples

8-1 Overview

Objectives are identified for this chapter, in which the methods of inferential statistics are extended to two populations. Both hypothesis testing and the construction of confidence intervals are discussed.

8-2 Inferences about Two Means: Dependent Samples

Two dependent samples (consisting of paired data) are used to test hypotheses about the means of two populations. Methods are presented for constructing confidence intervals as estimates of the difference between two population means.

8-3 Inferences about Two Means: Independent and Large Samples

Independent and large ($n > 30$) samples are used to test hypotheses and construct confidence interval estimates of the difference between two population means.

8-4 Comparing Two Variances

A method is presented for testing hypotheses made about two population variances or standard deviations.

8-5 Inferences about Two Means: Independent and Small Samples

This section focuses on samples that are independent and small ($n \leq 30$). Methods of testing hypotheses and constructing confidence intervals of the difference between two population means are discussed.

8-6 Inferences about Two Proportions

Pooled estimates of p_1 and p_2 are used to test hypotheses made about the population proportions p_1 and p_2. Confidence intervals are constructed for estimating the difference between two population proportions.

Chapter Problem:
Do glycol ethers cause miscarriages?

When one study suggested that pregnant animals exposed to glycol ethers were inclined to develop reproductive problems, computer manufacturer IBM became concerned because those same chemicals were used to make computer chips. (Glycol ethers are also used in the printing and aerospace industries.) IBM requested that researchers from Johns Hopkins University investigate the potential problem. The researchers found that among 30 pregnant women who worked with glycol ethers, 10 (or 33.3%) had miscarriages; among 750 pregnant workers not exposed to glycol ethers, 120 (or 16.0%) had miscarriages (based on data from *Johns Hopkins Magazine*). The 33.3% miscarriage rate for the exposed workers is greater than the 16.0% rate for workers not exposed, but are the results really significant? Is the sample evidence sufficient to conclude that there is a problem with glycol ethers? Based on the sample results, should IBM take corrective action, or should the company simply continue to monitor the situation until more information is available? This is a major decision with important and serious consequences. If it does appear that there is a problem with glycol ethers, then IBM should warn its employees, notify the Environmental Protection Agency, and try to eliminate or reduce workers' exposure to glycol ethers.

This problem involves two different samples from two different populations. One population consists of all pregnant women who work closely with glycol ethers in the manufacture of computer chips. The second population consists of all pregnant women who are not exposed to glycol ethers in their work. In Section 7-5 we tested claims about the values of single population proportions, and in this chapter we extend those methods to include claims made about the means of two population proportions. In particular, we will use the sample data to test the claim that glycol ethers have no effect on pregnancies, so the two populations have the same rate of miscarriage. If there is sufficient sample evidence, we will reject that claim and conclude that the difference between 33.3% and 16.0% is significant.

8-1   Overview

We use inferential statistics when we use sample data to form conclusions about populations. Chapters 6 and 7 introduced two of the most important and practical topics in the field of inferential statistics: estimating population parameters and testing hypotheses made about population parameters. Both chapters shared this feature—all examples and exercises involved the use of *one* sample to form an inference about *one* population. In reality, there are many cases in which the main objective is the comparison of *two* sets of data. The following are typical cases.

- Determine whether the dropout rate for college students who smoke is significantly different from the dropout rate for college students who do not smoke.

- Determine whether there is a difference between the means of the lives of Sears Diehard car batteries and those of Montgomery Ward Quickstart car batteries.

- Determine whether there is a difference between the waiting times of the customers at the Jefferson Valley Bank (where customers enter a single waiting line) and those of the customers at the Bank of Providence (where customers enter different lines formed at the individual teller stations).

This chapter presents methods for using data from two samples so that inferences can be made about the populations from which they came. We begin with Section 8-2, which deals with dependent samples consisting of paired data. In Section 8-3 we consider samples that are independent and large. Section 8-4 presents a method for testing hypotheses about the equality of two population variances or standard deviations. Section 8-5 deals with independent and small samples. This section is the most difficult section of the chapter because we must consider issues such as the independence of samples, knowledge of σ_1 and σ_2, equality of σ_1 and σ_2, and sample sizes before we can select the best procedure. To help you through what might seem at times to be a confusing maze, Section 8-5 includes flowcharts that summarize the procedures. The chapter concludes with Section 8-6, which describes methods for dealing with two proportions.

8-2   Inferences about Two Means: Dependent Samples

Chapters 6 and 7 presented formal and standard methods for constructing confidence intervals and testing hypotheses made about the population parameters of means, proportions, variances, and standard deviations. Section 6-2 discussed the construction and interpretation of confidence interval estimates of a population mean. Sections 7-2 through 7-4 considered claims made about the mean of a single population. In this section we use dependent samples to consider claims made about the means of two populations. We first define *independent* and *dependent* and identify the assumptions that apply to this section.

DEFINITION

Two samples are **independent** if the sample selected from one population is not related to the sample selected from the other population. If one sample is related to the other, the samples are **dependent.** (With dependent samples we get two values from each person or item, so such samples are often referred to as **paired samples** or **matched samples.**)

Consider the paired sample data given below. The sample of pretraining weights and the sample of posttraining weights are *dependent samples*, because each pair is matched according to the person involved. "Before/after" data are usually matched and are usually dependent.

Subject	A	B	C	D	E	F
Pretraining weights (kg)	99	62	74	59	70	73
Posttraining weights (kg)	94	62	66	58	70	76

Based on data from the *Journal of Applied Psychology,*
Vol. 62, No. 1.

For the following data, however, the two samples are *independent* because the sample of females is not related to the sample of males. The data are not matched as they are in the table above.

Weights of females (lb)	115	107	110	128	130		
Weights of males (lb)	128	150	160	140	163	155	175

Assumptions

For the hypothesis tests and confidence intervals described in this section, we assume that

1. Two dependent samples are selected from two populations in a way that is random.

2. Each of the two populations is normally distributed.

The two involved populations must be normally distributed. If they depart radically from normal distributions, we should not use the methods given in this section. Instead, we may be able to use a nonparametric method such as the sign test or the Wilcoxon signed-ranks test (discussed in Chapter 13).

In dealing with two dependent samples, it would be inefficient to consider each sample separately (using $\bar{x}_1$, s_1, n_1, $\bar{x}_2$, s_2, and n_2), because the relationship

x	99	62	74	59	70	73
y	94	62	66	58	70	76
d	5	0	8	1	0	−3

between matched pairs of values would be completely lost. Instead of comparing the sample means $\bar{x}_1$ and $\bar{x}_2$, we compute the *differences* (d) between the pairs of data as shown to the left.

Notation for Two Dependent Samples

$\bar{d}$ = mean value of the differences d (or $x - y$) for the paired *sample* data

μ_d = mean value of the differences d (or $x - y$) for the *population* of paired data

s_d = standard deviation of the d (or $x - y$) values for the paired sample data

n = number of *pairs* of data

For example, using the d values of 5, 0, 8, 1, 0, −3 taken from the preceding table to calculate $\bar{d}$, we get

$$\bar{d} = \frac{\Sigma d}{n} = \frac{5 + 0 + 8 + 1 + 0 - 3}{6} = \frac{11}{6} = 1.8$$

Also, for the d values of 5, 0, 8, 1, 0, −3, we get $s_d = 4.0$ as follows:

$$s_d = \sqrt{\frac{\Sigma(d - \bar{d})^2}{n - 1}} = \sqrt{\frac{78.83}{5}} = 4.0$$

Finally, for the data of the last table, $n = 6$.

Hypothesis Tests

We now use the preceding notation to describe the test statistic to be used in hypothesis tests of claims made about the means of two populations, given that the two samples are dependent. When we randomly select two dependent samples from normally distributed populations in which the population mean of the paired differences is μ_d, the following test statistic possesses a Student t distribution.

Test Statistic for Two Dependent Samples

$$t = \frac{\bar{d} - \mu_d}{\frac{s_d}{\sqrt{n}}}$$

where   degrees of freedom = $n - 1$

This test statistic closely resembles the standard score introduced in Section 2-6 and the test statistic introduced in Section 7-4. It serves as a convenient way of expressing the number of standard errors by which $\bar{d}$ differs from μ_d. This test statistic follows the same format as many test statistics based on the normal and Student t distributions:

$$\frac{\text{(sample statistic)} - \text{(claimed population parameter)}}{\text{(standard deviation of sample statistics)}}$$

Note that if the number of pairs of data is large ($n > 30$), the number of degrees of freedom will be at least 30, so critical values will be z scores instead of t scores. As previously stated, differences between the t distribution and the normal distribution are negligible when the sample size is large.

If we claim that there is no difference between the two population means, then we are claiming that $\mu_d = 0$. This makes sense if we recognize that $\bar{d}$ should be around 0 if there is no difference between the two population means.

▶ **EXAMPLE**

Does it pay to take preparatory courses for standardized tests such as the SAT? Using the sample data in the following table, test the claim that the Allan Preparation Course has no effect on the SAT score. Use a 0.05 level of significance.

Subject	A	B	C	D	E	F	G	H	I	J
SAT score before course (x)	700	840	830	860	840	690	830	1180	930	1070
SAT score after course (y)	720	840	820	900	870	700	800	1200	950	1080
Difference of before $-$ after $(x - y)$	-20	0	10	-40	-30	-10	30	-20	-20	-10

Based on data from the College Board and "An Analysis of the Impact of Commercial Test Preparation Courses on SAT Scores," by Sesnowitz, Bernhardt, and Knain, *American Educational Research Journal*, Vol. 19, No. 3.

● **SOLUTION**

Inspection of these data shows that among the 10 subjects, 7 had higher scores after the Allan course, 2 had lower scores, and 1 score didn't change. (Note that we are finding the differences of before $-$ after, so an effective course should raise scores, producing negative differences.) Using the traditional method of hypothesis testing, we will test the claim that the course has no effect. We follow the steps summarized in Figure 7-4.

Step 1:   If the course has no effect, we expect the mean of the differences to be 0. This is expressed in symbolic form as $\mu_d = 0$.   *(continued)*

Crest and Dependent Samples

In the late 1950s, Procter and Gamble introduced Crest toothpaste as the first such product with fluoride. To test the effectiveness of Crest in reducing cavities, researchers conducted experiments with several sets of twins. One of the twins in each set was given Crest with fluoride, while the other twin continued to use ordinary toothpaste without fluoride. It was believed that each pair of twins would have similar eating, brushing, and genetic characteristics. Results showed that the twins who used Crest had significantly fewer cavities than those who did not. This use of twins as dependent samples allowed the researchers to control many of the different variables affecting cavities.

Step 2: If the original claim is not true, we have $\mu_d \neq 0$.

Step 3: The null hypothesis must contain equality, so we have

$$H_0: \mu_d = 0 \qquad H_1: \mu_d \neq 0$$

Step 4: The significance level is $\alpha = 0.05$.

Step 5: Because we are testing the claim that paired dependent data have a mean of 0, we use the Student t distribution.

Step 6: Before finding the value of the test statistic, we must first find the values of $\overline{d}$ and s_d. Using Formulas 2-1 and 2-6, we find those values as follows:

$$\overline{d} = \frac{\Sigma d}{n} = \frac{-110}{10} = -11.0$$

$$s_d = \sqrt{\frac{n(\Sigma d^2) - (\Sigma d)^2}{n(n-1)}} = \sqrt{\frac{10(4900) - (-110)^2}{10(10 - 1)}} = 20.2$$

Using these statistics and the claim of $\mu_d = 0$, we can now find the value of the test statistic.

$$t = \frac{\overline{d} - \mu_d}{\frac{s_d}{\sqrt{n}}} = \frac{-11.0 - 0}{\frac{20.2}{\sqrt{10}}} = -1.722$$

The critical values of $t = -2.262$ and $t = 2.262$ are found from Table A-3; use the column for 0.05 (two tails), and use the row with degrees of freedom of $n - 1 = 9$. Figure 8-1 shows the test statistic, critical values, and critical region.

Step 7: Because the test statistic does not fall in the critical region, we fail to reject the null hypothesis of $\mu_d = 0$.

Step 8: There is not sufficient evidence to warrant rejection of the claim that the mean difference is equal to zero; that is, there is not sufficient evidence to warrant rejection of the claim that the course has no effect. Based on the available evidence, it appears that the Allan course does not have an effect on SAT scores.

The preceding example is a two-tailed test, but left-tailed tests and right-tailed tests follow the same basic procedure. For example, if we want to test the claim that the Allan course *raises* SAT scores, we have a claim that $\mu_d < 0$. (If the Allan course is effective in raising scores, then the after scores should be higher. That is, the values of before − after should be negative.) This

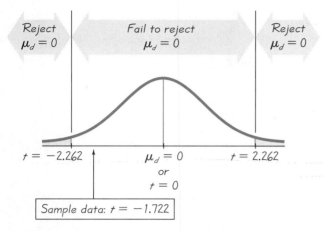

FIGURE 8-1   Distribution of SAT Score Differences

claim leads to a left-tailed test with a critical value of $t = -1.833$; the test statistic is again $t = -1.722$. With the given test statistic and critical value, we fail to reject the null hypothesis of $\mu_d \geq 0$. That is, this left-tailed test leads to the conclusion that there isn't sufficient evidence to support the claim that the Allan course raises SAT scores.

The preceding two-tailed example used the traditional method, but the P-value approach could be used by modifying Steps 6 and 7. In Step 6, use the test statistic of $t = -1.722$ and refer to the ninth row of Table A-3 to find that 1.722 is between the table values of 1.833 and 1.383. The P-value is therefore between the corresponding areas of 0.10 and 0.20; that is,

$$0.10 < P\text{-value} < 0.20$$

In Step 7, we again fail to reject the null hypothesis because the P-value is greater than the significance level of $\alpha = 0.05$.

We can develop a confidence interval estimate of the population mean difference μ_d by using the sample mean $\overline{d}$, the standard deviation of sample means $\overline{d}$ (which is $s_d/\sqrt{n}$), and the critical value $t_{\alpha/2}$.

Confidence Intervals

The confidence interval estimate of the mean difference μ_d is as follows:

$$\overline{d} - E < \mu_d < \overline{d} + E$$

where

$$E = t_{\alpha/2}\frac{s_d}{\sqrt{n}}$$

$$\text{degrees of freedom} = n - 1$$

Statistical Significance versus Practical Significance

The hypothesis tests described in this text address the issue of statistical significance. We try to determine if the observations are so unlikely that we are led to believe that differences are due to factors other than chance sample fluctuations. Experimental results can sometimes be statistically significant without being practically significant. A diet causing a weight loss of 1/2 lb might be statistically significant if the sample size is 10,000, but such a diet would not have practical significance. Nobody would bother with a diet that results in a loss of only 1/2 lb.

▶ **EXAMPLE**

Use the sample data from the preceding example to construct a 95% confidence interval estimate of μ_d.

● **SOLUTION**

Using the values of $\overline{d} = -11.0$, $s_d = 20.2$, $n = 10$, and $t_{\alpha/2} = 2.262$, we first find the value of the margin of error E.

$$E = t_{\alpha/2} \frac{s_d}{\sqrt{n}} = 2.262 \frac{20.2}{\sqrt{10}} = 14.4$$

The confidence interval can now be found.

$$\overline{d} - E < \mu_d < \overline{d} + E$$
$$-11.0 - 14.4 < \mu_d < -11.0 + 14.4$$
$$-25.4 < \mu_d < 3.4$$

In the long run, 95% of such samples will lead to confidence limits that actually do contain the true population mean of the differences. Note that the confidence interval limits contain 0, indicating that the true value of μ_d is not significantly different from 0. That is, the mean value of the before − after differences is 0 or close to 0, suggesting that there is *no difference* between SAT scores before and after the Allan Preparation Course. Based on the confidence interval, we would therefore fail to reject a claim that the Allan course is ineffective. This conclusion agrees with the conclusion in the preceding example.

Computer Usage for Two Dependent Samples

Statistical software packages such as STATDISK and Minitab can be used to test hypotheses and construct confidence intervals for situations involving two dependent samples. Shown below are the STATDISK and Minitab displays for the example of this section. If you are using STATDISK, select `Statistics` from the main menu bar, then select `Hypothesis Testing`, then select the menu item for `Mean - Two Samples; DEPENDENT Case`, and proceed to enter the data requested. The result will be the display shown here. (STATDISK automatically provides confidence interval limits for two-tailed hypothesis tests.) If you are using Minitab, enter the data as shown, then create column C3 consisting of the differences. Proceed to use the `TTEST` and `TINTERVAL` commands as shown. The hypothesis test results and the confidence interval limits will be displayed as shown.

STATDISK DISPLAY

```
About    File    Edit    Statistics    Data    View

        Hypothesis test for a claim about two DEPENDENT populations

            NULL HYPOTHESIS:  Mean 1 = Mean 2

            ┌──────────────────────────────────────────────────────────┐
            │   Mean of differences ..... = -11                         │
            │   St. Dev. of differences . =  20.2485                    │
            │   Degrees of freedom ... df =  9                          │
            │   Conf. interval limits ... = -25.4959 , 3.4959           │
            │   Test statistic ........ t = -1.71791                    │
            │   Critical value ........ t = -2.26388 , 2.26388          │
            │   P-value ................. =  .11994                     │
            │   Significance level ...... =  .05                        │
            └──────────────────────────────────────────────────────────┘


        CONCLUSION: FAIL TO REJECT the null hypothesis



                    Use Menu Bar to change screen
    F1: Help        F2: Menu                              ESC: Quit
```

MINITAB DISPLAY

```
MTB > READ C1 C2
DATA> 700  720
DATA> 840  840
DATA> 830  820
DATA> 860  900
DATA> 840  870
DATA> 690  700
DATA> 830  800
DATA> 1180 1200
DATA> 930  950
DATA> 1070 1080
DATA> ENDOFDATA
MTB > LET C3 = C1 - C2
MTB > TTEST MU=0 C3

TEST OF MU =  0.000 VS MU N.E.  0.000

            N      MEAN    STDEV   SE MEAN       T    P VALUE
C3         10   -11.000   20.248     6.403   -1.72       0.12

MTB > TINTERVAL 95 PERCENT C3

            N      MEAN    STDEV   SE MEAN   95.0 PERCENT C.I.
C3         10   -11.00    20.25      6.40 (  -25.49,    3.49)
```

By using Table A-3, we concluded that the P-value is between 0.10 and 0.20, but from the STATDISK and Minitab displays we can see that the P-value is actually 0.12.

8-2 Exercises A: Basic Skills and Concepts

In Exercises 1 and 2, assume that you want to test the claim that the paired sample data come from a population for which the mean difference is $\mu_d = 0$. Assuming a 0.05 level of significance, find (a) $\overline{d}$, (b) s_d, (c) the t test statistic, and (d) the critical values.

1.

x	3	4	4	6	8	10	12	11	14	17	20
y	2	5	4	5	9	8	9	8	12	15	17

2.

x	15	17	20	26	26	27	28	29	30	31	34	37
y	16	16	20	26	27	25	24	27	28	25	29	30

3. Using the sample paired data in Exercise 1, construct a 95% confidence interval for the mean of all $x - y$ values.

4. Using the sample paired data in Exercise 2, construct a 99% confidence interval for the mean of all $x - y$ values.

5. The Malloy Advertising Company has prepared two different television commercials for Taylor's women's jeans. One commercial is humorous, and the other is serious. A test screening involves 8 consumers who are asked to rate the commercials by using a standard scale with higher scores indicating more favorable responses. The results are listed below. At the 0.05 significance level, test the claim that the differences between the commercials have a mean of 0. Based on the result, does one commercial seem to be better?

Consumer	A	B	C	D	E	F	G	H
Humorous commercial	26.2	20.3	25.4	19.6	21.5	28.3	23.7	24.0
Serious commercial	24.1	21.3	23.7	18.0	20.1	25.8	22.4	21.4

6. Air America experimented with two different reservation systems and recorded the times (in seconds) required to process randomly selected passenger requests. The results are listed below. At the 0.05 significance level, test the claim that the differences in times have a mean of 0.

MicroAir Software	21	23	25	27	27	29	31	32	30	41	34
Flight Services Software	18	20	21	26	27	24	22	33	27	34	47

7. A dose of the drug Captopril, designed to lower systolic blood pressure, is administered to 10 randomly selected volunteers, with the following results. Construct the 95% confidence interval for μ_d, the mean of the differences between the before and after scores.

Subject	A	B	C	D	E	F	G	H	I	J
Before pill	120	136	160	98	115	110	180	190	138	128
After pill	118	122	143	105	98	98	180	175	105	112

8. Using the sample data from Exercise 7, test the claim that systolic blood

pressure is not affected by the pill. Use a 0.05 significance level. Does it appear that the drug has an effect?

9. A study was conducted to investigate the effectiveness of hypnotism in reducing pain. Results for randomly selected subjects are given below. At the 0.05 significance level, test the claim that the sensory measurements are lower after hypnotism. (The values are before and after hypnosis; the measurements are in centimeters on a pain scale.) Does hypnotism appear effective in reducing pain?

Subject	A	B	C	D	E	F	G	H
Before	6.6	6.5	9.0	10.3	11.3	8.1	6.3	11.6
After	6.8	2.4	7.4	8.5	8.1	6.1	3.4	2.0

Based on "An Analysis of Factors That Contribute to the Efficacy of Hypnotic Analgesia," by Price and Barber, *Journal of Abnormal Psychology*, Vol. 96, No. 1.

10. Using the data in Exercise 9, construct a 95% confidence interval for the mean difference of before − after.

11. A teacher proposes a course designed to increase reading speed and comprehension. To evaluate the effectiveness of this course, the teacher tests students before and after the course; the sample results follow. Construct a 95% confidence interval for the mean of the differences between the before and after scores.

Student	A	B	C	D	E	F	G	H	I	J
Before	100	170	135	167	200	118	127	95	112	136
After	136	160	120	169	200	140	163	101	138	129

12. Use the data from Exercise 11 to test the claim that the scores are higher after the course. Use a 0.05 level of significance. Does the course appear to be effective in increasing reading speed and comprehension? If you were responsible for the curriculum, would you support the course proposal?

13. A study was conducted to investigate some effects of physical training. Sample data are listed in the margin. (See "Effect of Endurance Training on Possible Determinants of VO_2 During Heavy Exercise," by Casaburi and others, *Journal of Applied Physiology*, Vol. 62, No. 1.) At the 0.05 level of significance, test the claim that the mean pretraining weight equals the mean posttraining weight. All weights are given in kilograms. What do you conclude about the effect of training on weight?

Pre-training	Post-training
99	94
57	57
62	62
69	69
74	66
77	76
59	58
92	88
70	70
85	84

14. Use the data from Exercise 13 to construct a 95% confidence interval for the mean of the differences between pretraining and posttraining weights.

15. Refer to Data Set 2 in Appendix B. Use the paired data consisting of body temperatures of women at 8:00 AM and at 12:00 AM on Day 2. Using a 0.05 level of significance, test the claim that for those temperatures, the mean difference is 0. Based on the result, do morning and night body temperatures appear to be about the same?

16. Using the sample data described in Exercise 15, construct a 95% confidence interval for the mean difference.

8-2 **Exercises B: Beyond the Basics**

17. Refer to the indicated exercise and find the *P*-value.
  a. Exercise 5   b. Exercise 9

18. You have collected 50 randomly selected pairs of data. Find the critical value(s) for testing the given claim at the specified significance level.
  a. $\mu_d = 0$; $\alpha = 0.10$
  b. $\mu_d \geq 0$; $\alpha = 0.025$
  c. $\mu_d > 0$; $\alpha = 0.02$

19. The 95% confidence interval for a collection of paired sample data is

$$0.0 < \mu_d < 1.2$$

Based on this confidence interval, the traditional method of hypothesis testing leads to the conclusion that the claim of $\mu_d > 0$ is *supported*. What is the smallest possible value of the significance level of the hypothesis test?

8-3 **Inferences about Two Means: Independent and Large Samples**

Section 8-2 described methods of using inferential statistics for dependent (paired) data, but in this section our samples are *independent*. We defined samples to be independent if the sample selected from one population is not related to the sample selected from the other population.

In addition to considering only independent samples, this section will be restricted to *large* ($n > 30$) samples. That is, the sample size n_1 for sample 1 will be greater than 30, and n_2 for sample 2 will also be greater than 30. In Section 8-5 we will deal with cases involving small samples. For the purposes of this section, the following assumptions apply.

Assumptions

In testing hypotheses about two population means and constructing confidence intervals for the difference between two population means, we make the following assumptions.

1. The two samples are *independent*.

2. The two sample sizes are large. That is, $n_1 > 30$ and $n_2 > 30$.

Hypothesis Tests

We know from the central limit theorem that sample means tend to be normally distributed. It follows that the differences between sample means tend to be normally distributed. Therefore, with the preceding assumptions, the following test statistic applies to the test of a hypothesis made about the means of two populations.

Test Statistic for Two Means: Independent and Large Samples

$$z = \frac{(\overline{x}_1 - \overline{x}_2) - (\mu_1 - \mu_2)}{\sqrt{\dfrac{\sigma_1^2}{n_1} + \dfrac{\sigma_2^2}{n_2}}}$$ *variance*

As in Chapter 7, if the values of σ_1 and σ_2 are not known, we can use s_1 and s_2 in their places, provided that both samples are large. If σ_1 and σ_2 are known, we use those values in calculating the test statistic, but realistic cases will usually require the use of s_1 and s_2. It is rare that we know the value of population standard deviations without knowing the value of the population means.

The above test statistic is very similar to several of the test statistics that we have already encountered. It has the same basic format of

$$\frac{\text{(sample statistic)} - \text{(claimed population parameter)}}{\text{(standard deviation of sample statistics)}}$$

Later in this section we will justify the particular form of the above test statistic, but we will first present an example illustrating the hypothesis-testing procedure.

▶ **EXAMPLE**

The Medassist Pharmaceutical Company wants to test Dozenol, a new cold medicine intended for night use. Tests for such products often include a "treatment group" of people who use the drug and a "control group" of people who don't use the drug. Fifty people with colds are given Dozenol, and 100 others are not. The systolic blood pressure is measured for each subject, and the sample statistics are summarized to the right. The head of research at Medassist claims that Dozenol does not affect blood pressure—that is, the treatment population mean μ_1 and the control population mean μ_2 are equal. Test that claim using the traditional method of testing hypotheses. Use a significance level of $\alpha = 0.05$.

Treatment Group	Control Group
$n_1 = 50$	$n_2 = 100$
$\overline{x}_1 = 203.4$	$\overline{x}_2 = 189.4$
$s_1 = 39.4$	$s_2 = 39.0$

● **SOLUTION**

Step 1:   The claim can be expressed symbolically as $\mu_1 = \mu_2$.

Step 2:   If the original claim is false, then $\mu_1 \neq \mu_2$.

Step 3:   The null hypothesis must contain equality, so we have

$$H_0\colon \mu_1 = \mu_2 \qquad H_1\colon \mu_1 \neq \mu_2 \qquad \text{\textit{(continued)}}$$

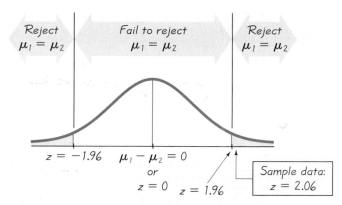

FIGURE 8-2 Distribution of Differences between Means of Treatment and Control Groups

Step 4: The significance level is $\alpha = 0.05$.

Step 5: The stated problem satisfies the assumptions of this section, and we use a normal distribution with the test statistic given above.

Step 6: The test statistic is

$$z = \frac{(\bar{x}_1 - \bar{x}_2) - (\mu_1 - \mu_2)}{\sqrt{\dfrac{\sigma_1^2}{n_1} + \dfrac{\sigma_2^2}{n_2}}} = \frac{(203.4 - 189.4)(-0)}{\sqrt{\dfrac{39.4^2}{50} + \dfrac{39.0^2}{100}}} = 2.06$$

Because we are using a normal distribution, the critical values of $z = -1.96$ and $z = 1.96$ are found from Table A-2. [Divide $\alpha = 0.05$ equally between the two tails, and find the z score corresponding to an area of $0.5 - (0.05/2)$, or 0.4750.] The test statistic, critical values, and critical region are shown in Figure 8-2.

Step 7: Because the test statistic does fall within the critical region, reject the null hypothesis of $\mu_1 = \mu_2$.

Step 8: There is sufficient evidence to warrant rejection of the claim that the treatment group and control group have the same mean. The bad news for the Medassist Company is that it appears that Dozenol does affect blood pressure. (We conclude that the difference between the two means is *statistically* significant, but medical experts need to determine whether that difference has *practical* importance.)

Determination of P-values is easy because the test statistics are from the standard normal distribution. Following the method outlined in Figures 7-7 and 7-8, we see that for a two-tailed test the P-value is twice the area to the right of the test statistic $z = 2.06$. We get

$$P\text{-value} = 2 \cdot (0.5 - 0.4803) = 0.0394$$

Because the P-value is less than the significance level of 0.05, we have an indication that the difference between the two population means is significant.

Confidence Intervals

The confidence interval estimate of the difference $\mu_1 - \mu_2$ is as follows:

$$(\overline{x}_1 - \overline{x}_2) - E < (\mu_1 - \mu_2) < (\overline{x}_1 - \overline{x}_2) + E$$

where

$$E = z_{\alpha/2} \sqrt{\frac{\sigma_1^2}{n_1} + \frac{\sigma_2^2}{n_2}}$$

▶ **EXAMPLE**

Using the sample data given in the preceding example, construct a 95% confidence interval estimate of the difference between the means of the treatment and control populations.

● **SOLUTION**

We first find the value of the margin of error E.

$$E = z_{\alpha/2} \sqrt{\frac{\sigma_1^2}{n_1} + \frac{\sigma_2^2}{n_2}} = 1.96 \sqrt{\frac{39.4^2}{50} + \frac{39.0^2}{100}} = 13.3$$

We then find the desired confidence interval as follows:

$$(\overline{x}_1 - \overline{x}_2) - E < (\mu_1 - \mu_2) < (\overline{x}_1 - \overline{x}_2) + E$$
$$(203.4 - 189.4) - 13.3 < (\mu_1 - \mu_2) < (203.4 - 189.4) + 13.3$$
$$0.7 < (\mu_1 - \mu_2) < 27.3$$

We are 95% confident that the limits of 0.7 and 27.3 actually do contain the difference between the two population means. Because those limits do not contain 0, it is very unlikely that the two population means are equal. It appears that the treatment group has a mean that is different from that of the control group.

Why do the test statistic and confidence interval have the particular forms we have presented? Both forms are based on a normal distribution of $\bar{x}_1 - \bar{x}_2$ values with mean $\mu_1 - \mu_2$ and standard deviation $\sqrt{\sigma_1^2/n_1 + \sigma_2^2/n_2}$. This follows from the central limit theorem introduced in Section 5-4. The central limit theorem tells us that sample means $\bar{x}$ are normally distributed with mean μ and standard deviation $\sigma/\sqrt{n}$. Also, when samples have a size of 31 or larger, the normal distribution serves as a reasonable approximation to the distribution of sample means. By similar reasoning, the values of $\bar{x}_1 - \bar{x}_2$ also tend to approach a normal distribution with mean $\mu_1 - \mu_2$. When both samples are large, the following property of variances leads us to conclude that the values of $\bar{x}_1 - \bar{x}_2$ will have a standard deviation of

$$\sqrt{\frac{\sigma_1^2}{n_1} + \frac{\sigma_2^2}{n_2}}$$

The variance of the *differences* between two independent random variables equals the variance of the first random variable *plus* the variance of the second random variable.

That is, the variance of sample values $\bar{x}_1 - \bar{x}_2$ will tend to equal

$$\sigma_{\bar{x}_1}^2 + \sigma_{\bar{x}_2}^2$$

provided that $\bar{x}_1$ and $\bar{x}_2$ are independent. This is not an intuitively obvious concept, and we therefore illustrate it by the specific data given in Table 8-1. (This is an illustration, not a formal proof.) The population x values of 12, 24, 36 have a variance of 96, the population y values of 0, 3, 6 have a variance of 6,

TABLE 8-1   Variance of Differences

x	y		$x - y$
12	0	Find every possible $x - y$ value:	$12 - 0 = 12$
24	3		$12 - 3 = 9$
36	6		$12 - 6 = 6$
			$24 - 0 = 24$
			$24 - 3 = 21$
			$24 - 6 = 18$
			$36 - 0 = 36$
			$36 - 3 = 33$
			$36 - 6 = 30$

| The x values have variance $\sigma_x^2 = 96$. | + | The y values have variance $\sigma_y^2 = 6$. | = | The $x - y$ values have variance $\sigma_{x-y}^2 = 102$. |

and the population $x - y$ values have a variance of $6 + 96 = 102$. That is,

$$\sigma^2_{x-y} = \sigma^2_x + \sigma^2_y = 96 + 6 = 102$$

For means of large random samples, the standard deviation of those sample means is $\sigma/\sqrt{n}$, so the variance is σ^2/n. We can now combine our additive property of variances with the central limit theorem's expression for variance of sample means to obtain the following:

$$\sigma^2_{\overline{x}_1 - \overline{x}_2} = \sigma^2_{\overline{x}_1} + \sigma^2_{\overline{x}_2} = \frac{\sigma^2_1}{n_1} + \frac{\sigma^2_2}{n_2}$$

In this expression, we assume that we have population 1 with variance σ^2_1 and population 2 with variance σ^2_2. Samples of size n_1 are randomly drawn from population 1, and the mean $\overline{x}$ is computed. The same is done for population 2. Here, $\sigma^2_{\overline{x}_1 - \overline{x}_2}$ denotes the variance of $\overline{x}_1 - \overline{x}_2$ values. This result shows that the standard deviation of $\overline{x}_1 - \overline{x}_2$ values is

$$\sqrt{\frac{\sigma^2_1}{n_1} + \frac{\sigma^2_2}{n_2}}$$

Since z is a standard score that corresponds in general to

$$z = \frac{(\text{sample statistic}) - (\text{population mean})}{(\text{standard deviation of sample statistics})}$$

we get

$$z = \frac{(\overline{x}_1 - \overline{x}_2) - (\mu_1 - \mu_2)}{\sqrt{\frac{\sigma^2_1}{n_1} + \frac{\sigma^2_2}{n_2}}}$$

by noting that the sample values of $\overline{x}_1 - \overline{x}_2$ will have a mean of $\mu_1 - \mu_2$ and the standard deviation just given.

💾 Using Computers for Hypothesis Tests and Confidence Intervals

The hypothesis tests and confidence interval construction of this section can be accomplished by using statistical software packages. To use STATDISK, select Statistics from the main menu bar, then select Hypothesis Testing, then select Mean - Two Samples: Independent Case, and proceed to enter the data requested. The result will be as shown below. You can see that the STATDISK results agree quite well with those obtained in this section.

We are faced with an obstacle in the use of Minitab: Minitab is designed for raw data and does not deal directly with summary statistics, such as those given in the examples of this section. There is a way to circumvent this problem. Minitab can be used to first generate a normally distributed data set from a

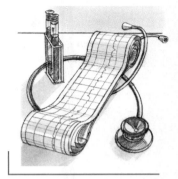

Gender Gap in Drug Testing

A study of the relationship between heart attacks and doses of aspirin involved 22,000 male physicians. This study, like many others, excluded women. The General Accounting Office recently criticized the National Institutes of Health for not including both sexes in many studies because results of medical tests on males do not necessarily apply to females. For example, women's hearts are different from men's in many important ways. When forming conclusions based on sample results, we should be wary of an inference that extends to a population larger than the one from which the sample was drawn.

STATDISK DISPLAY

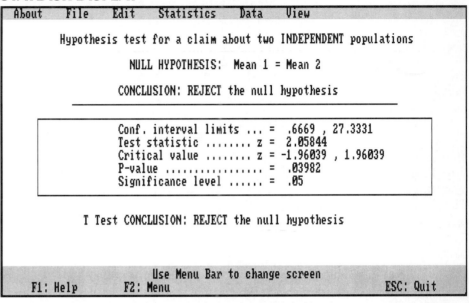

```
About    File    Edit    Statistics    Data    View

         Hypothesis test for a claim about two INDEPENDENT populations

                    NULL HYPOTHESIS:  Mean 1 = Mean 2

                 CONCLUSION: REJECT the null hypothesis
         _____

             Conf. interval limits ... =  .6669 , 27.3331
             Test statistic ........ z =  2.05844
             Critical value ........ z = -1.96039 , 1.96039
             P-value .................. =  .03982
             Significance level ...... =  .05


             T Test CONCLUSION: REJECT the null hypothesis



                    Use Menu Bar to change screen
     F1: Help        F2: Menu                            ESC: Quit
```

population with a mean of 0 and a standard deviation of 1. Minitab can then be used to transform that sample into another sample with the desired mean and standard deviation. For example, the following Minitab entries create a normally distributed sample with the treatment group statistics of $n_1 = 50$, $\bar{x}_1 = 203.4$, and $s_1 = 39.4$.

```
MTB > RANDOM 50 C1;
SUBC> NORMAL 0 1.
MTB > LET C3 = ((C1-MEAN(C1))/STDEV(C1))*39.4+203.4
```

A similar procedure can be used to create a column C4 of 100 sample values with the statistics for the control group. After creating columns C3 and C4, use the TWOSAMPLE Minitab command as shown in the display given below.

MINITAB DISPLAY

```
MTB > TWOSAMPLE Z FOR C3 VS C4

TWOSAMPLE T FOR C3 VS C4
      N      MEAN      STDEV    SE MEAN
C3    50     203.4     39.4       5.6
C4   100     189.4     39.0       3.9

95 PCT CI FOR MU C3 - MU C4: (0.5, 27.5)

TTEST MU C3 = MU C4 (VS NE): T= 2.06  P=0.042  DF= 97
```

8-3   Exercises A: Basic Skills and Concepts

In Exercises 1 and 2, use a 0.05 significance level to test the claim that the two samples come from populations with the <u>same mean</u>. In each case, the two samples are independent and have been randomly selected.

1.

Control Group	Experimental Group
$n_1 = 40$	$n_2 = 40$
$\bar{x}_1 = 79.6$	$\bar{x}_2 = 84.2$
$s_1 = 12.4$	$s_2 = 12.2$

2.

Treated	Untreated
$n_1 = 32$	$n_2 = 60$
$\bar{x}_1 = 8.49$	$\bar{x}_2 = 8.41$
$s_1 = 0.11$	$s_2 = 0.18$

3. Using the sample data given in Exercise 1, construct a 95% confidence interval estimate of the difference between the two population means.

4. Using the sample data given in Exercise 2, construct a 95% confidence interval estimate of the difference between the two population means.

5. For Data Set 3 of Appendix B, the 86 males have family incomes with a mean of $35,330 and a standard deviation of $23,260. The 39 females have family incomes with a mean of $29,610 and a standard deviation of $16,480. At the 0.01 level of significance, test the claim that the mean family income is greater for the population of males.

6. Using the sample data in Exercise 5, construct a 99% confidence interval for the difference between the mean family income of men and the mean family income of women.

7. As part of the National Health Survey, data were collected on the weights of men. For 804 men aged 25 to 34, the mean is 176 lb and the standard deviation is 35.0 lb. For 1657 men aged 65 to 74, the mean and standard deviation are 164 lb and 27.0 lb, respectively. (See the National Health Survey, USDHEW publication 79-1659.) Construct a 99% confidence interval for the difference between the means of the men in the two age brackets.

8. Use the data from Exercise 7 to test the claim that the older men come from a population with a mean that is less than the mean for men in the 25–34 age bracket. Use a 0.01 level of significance.

9. In a study of flight attendants, salaries paid by two different airlines were randomly selected. For 40 American Airlines flight attendants, the mean is $23,870 and the standard deviation is $2960. For 35 TWA flight attendants, the mean is $22,025 and the standard deviation is $3065 (based on data from the Association of Flight Attendants). At the 0.10 level of significance, test the claim that American and TWA have the same mean salary for flight attendants. Based on the result, should salary be an important factor for a prospective flight attendant in choosing between American and TWA?

10. Use the sample data from Exercise 9 to construct a 90% confidence interval for the difference between the population means ($\mu_1 - \mu_2$), where μ_1 is the mean salary for all American Airlines flight attendants and μ_2 is the mean salary for those at TWA. Do the confidence interval limits contain 0, suggesting that there is not a significant difference between the two population means?

11. Two different instructors teach a course in the use of the WordPerfect word processing system. Tom Bennet's class consists of 35 students who achieve test scores with a mean of 76.0 and a standard deviation of 14.1. Sue Stein's class consists of 40 students who achieve test scores with a mean of 73.4 and a standard deviation of 13.5. At the 0.05 level of significance, test the claim that the two teachers produce students with the same mean score.

12. Using the same sample data in Exercise 11, construct a 95% confidence interval for the difference between the means for the two teachers.

East	West
$n = 35$	$n = 50$
$\overline{x} = 421$	$\overline{x} = 347$
$s = 122$	$s = 85$

13. Investors are considering two possible locations for a new Locust Tree Restaurant. They commission a study of pedestrian traffic at both sites. At each location, pedestrians are observed in one-hour units, and an index of desirable characteristics is compiled for each hour. The sample results are shown in the margin. Construct a 95% confidence interval for the difference between the two mean indices. Do the confidence interval limits contain 0, suggesting that there is not a significant difference between the two means?

14. Use the sample data given in Exercise 13 to test the claim that both sites have the same mean index. Use a 0.05 level of significance. Based on the results, does either location seem better because it has a significantly higher mean index?

15. The Weston Rescue Service and the Mid-Valley Ambulance Service are tested for response times. A sample of 50 responses from the Weston firm produces a mean of 12.2 min and a standard deviation of 1.5 min. A sample of 50 responses from Mid-Valley produces a mean of 14.0 min with a standard deviation of 2.1 min. Test the claim that Weston and Mid-Valley have the same mean response time. Use a 0.05 level of significance. If you needed an ambulance, which service would you choose?

16. A study of seat belt use involved children who were hospitalized as a result of motor vehicle crashes. For a group of 290 children who were not wearing seat belts, the number of days spent in intensive care units (ICU) has a mean of 1.39 and a standard deviation of 3.06. For a group of 123 children who were wearing seat belts, the number of days in ICU has a mean of 0.83 and a standard deviation of 1.77 (based on data from "Morbidity Among Pediatric Motor Vehicle Crash Victims: The Effectiveness of Seat Belts," by Osberg and Di Scala, *American Journal of Public Health*, Vol. 82, No. 3). At the 0.01 significance level, test the claim that the population of children not wearing seat belts has a higher mean number of days spent in ICU. Based on the result, is there significant evidence in favor of seat belt use among children?

8-3   Exercises B: Beyond the Basics

17. Refer to Data Set 2 of Appendix B and use only the 12:00 AM body temperatures for Day 1. We want to test, at the 0.05 significance level, the claim that persons aged 18–24 inclusive have the same mean as those aged 25 and older.

   a.   Conduct the hypothesis test using the traditional method.

(continued)

   b.   Conduct the hypothesis test using the *P*-value method.
   c.   Conduct the hypothesis test by constructing a 95% confidence interval for
        the difference $\mu_1 - \mu_2$.
18. Refer to Exercise 8. If the actual difference between the population means is
    6.0, find β, the probability of a type II error. (See Exercise 28 in Section 7-3.)
    (*Hint:* In Step 1, replace $\bar{x}$ by $(\bar{x}_1 - \bar{x}_2)$, replace $\mu_{\bar{x}}$ by 0, and replace $\sigma_{\bar{x}}$ by

$$\sqrt{\frac{\sigma_1^2}{n_1} + \frac{\sigma_2^2}{n_2}}$$

19. a.   Find the variance for this *population* of *x* scores: 5, 10, 15.
    b.   Find the variance for this *population* of *y* scores: 1, 2, 3.
    c.   List the *population* of all possible $x - y$ scores, and find the variance of
         this population.
    d.   Use the results from parts a, b, and c to verify that

$$\sigma_{x-y}^2 = \sigma_x^2 + \sigma_y^2$$

    (This principle is used to derive the test statistic and confidence interval
    for several cases in this section.)
20. Consider 3 populations: a population of distinct *x* values, a population of
    distinct *y* values, and a population of all possible $x - y$ values. In this section
    we saw that the variance of the $x - y$ values is the *sum* of the variances of the *x*
    values and the *y* values. How is the *range* of the $x - y$ values related to the
    range of the *x* values and the range of the *y* values?

8-4   Comparing Two Variances

We have already noted that the characteristic of variation among data is
extremely important. This section presents a method for using two samples to
compare the variances of the populations from which the samples are drawn.
The method we use requires the following assumptions.

Assumptions

When testing a hypothesis about the *variances* of two populations, we assume
that

1.   The two populations are *independent* of each other. (Recall that two samples
     are independent if the sample selected from one population is not related to the
     sample selected from the other population.)
2.   The two populations are each *normally distributed*.

The second assumption—that both populations have normal distributions—
is critical for the test presented in this section. The relevant test statistic is very
sensitive to departures from normality, and this extreme sensitivity does not
diminish with large samples. As a result, if we use the hypothesis test of this

section with data sets having very nonnormal distributions, large errors could result. Other tests in this chapter are not so sensitive to departures from normality.

The computations of this section will be simplified if we stipulate that s_1^2 represents the *larger* of the two sample variances. This stipulation presents no logical difficulties because the identification of the samples through subscript notation is arbitrary. We choose the following notation.

Notation for Hypothesis Tests with Two Variances

$s_1^2 = $ *larger* of the two *sample variances*

$n_1 = $ size of the sample with the *larger* variance

$\sigma_1^2 = $ variance of the *population* from which the sample was drawn

The symbols s_2^2, n_2, and σ_2^2 correspond to the other sample and population.

Extensive analyses have shown that *for two normally distributed populations with equal variances (that is, $\sigma_1^2 = \sigma_2^2$), the sampling distribution of the following test statistic is the F distribution shown in Figure 8-3 with critical values listed in Table A-5*. If you continue to repeat an experiment of randomly selecting samples from two normally distributed populations with equal variances, the distribution of the ratio s_1^2/s_2^2 of the sample variances is the F distribution. In Figure 8-3, note that this distribution is not symmetric, it has values that cannot be negative, and its exact shape depends on two different degrees of freedom.

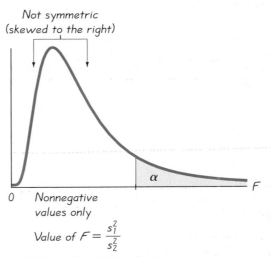

FIGURE 8-3   F Distribution
There is a different F distribution for each different pair of degrees of freedom for numerator and denominator.

Test Statistic for Hypothesis Tests with Two Variances

$$F = \frac{s_1^2}{s_2^2}$$

If the two populations really do have equal variances, then $F = s_1^2/s_2^2$ tends to be close to 1 because s_1^2 and s_2^2 tend to be close in value. But if the two populations have radically different variances, s_1^2 and s_2^2 tend to be very different numbers. Denoting the larger of the sample variances by s_1^2, we see that the ratio s_1^2/s_2^2 will be a large number whenever s_1^2 and s_2^2 are far apart in value. Consequently, a value of F near 1 will be evidence in favor of the conclusion that $\sigma_1^2 = \sigma_2^2$. A large value of F will be evidence against the conclusion of equality of the population variances.

We can use the same test statistic to test claims about two population standard deviations. Any claim about two population standard deviations can be restated in terms of the corresponding variances.

Using Table A-5, we obtain *critical F values* that are determined by the following three values:

1. The significance level α. (Table A-5 has six pages of critical values for $\alpha = 0.01$, 0.025, and 0.05.)

2. **Numerator degrees of freedom** $= n_1 - 1$

3. **Denominator degrees of freedom** $= n_2 - 1$

When using Table A-5, be sure that n_1 corresponds to the sample having variance s_1^2, and that n_2 is the size of the sample with variance s_2^2. Identify a level of significance, and determine whether the test is one-tailed or two-tailed. For a one-tailed test (H_0: $\sigma_1^2 \leq \sigma_2^2$), use the significance level found in Table A-5. (Because we stipulate that the larger sample variance is s_1^2, all one-tailed tests will be *right*-tailed.) For a two-tailed test (H_0: $\sigma_1^2 = \sigma_2^2$), first divide the area of the critical region (equal to the significance level) equally between the two tails, and then refer to the part of Table A-5 that represents *one-half* of the significance level. In that part of Table A-5, find the intersection of the column representing the degrees of freedom for s_1^2 with the row representing the degrees of freedom for s_2^2. (Unlike the normal and Student t distributions, the F distribution is not symmetric and does not have 0 at its center. Consequently, left-tailed critical values *cannot* be found by using the negative of the right-tailed critical values. Instead, the left-tailed critical value can be found by using the reciprocal of the right-tailed value with the numbers of degrees of freedom reversed. See Exercise 19.)

We sometimes have numbers of degrees of freedom that are not included in Table A-5. We can use linear interpolation to approximate the missing values, but in most cases that's not necessary because the F test statistic is either less than the lowest possible critical value or greater than the largest possible

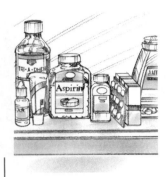

Drug Approval

The Pharmaceutical Manufacturing Association has reported that the development and approval of a new drug costs around $87 million and takes about eight years. Extensive laboratory testing is followed by FDA approval for human testing, which is done in three phases. Phase I human testing involves about 80 people, and phase II involves about 250 people. In phase III, between 1000 and 3000 volunteers are used. Overseeing such a complex, extensive, and time-consuming process would be enough to give anyone a headache, but the process does protect us from dangerous or worthless drugs.

critical value. For example, Table A-5 shows that for $\alpha = 0.025$ in the right tail, 20 degrees of freedom for the numerator, and 34 degrees of freedom for the denominator, the critical F value is between 2.0677 and 2.1952. Any F test statistic below 2.0677 will result in failure to reject the null hypothesis because the test statistic is less than the critical value. Similarly, any F test statistic above 2.1952 will result in rejection of the null hypothesis because the test statistic exceeds the critical value. Interpolation would be necessary only if the F test statistic were between 2.0677 and 2.1952. Here, we don't yet know whether the test statistic is above or below the critical value, and we must interpolate to find a specific critical value that can be compared to the test statistic.

▶ EXAMPLE

Let's consider the two sets of sample waiting times in Table 8-2. The Jefferson Valley Bank collects data by recording the waiting times of randomly selected customers on Tuesday afternoons. On one Tuesday afternoon, customers are allowed to select any one of several different waiting lines that have formed at the various teller stations. On another Tuesday, all customers enter a single main waiting line that feeds the individual teller stations as vacancies occur. Examination of the summary statistics shows that the multiple line system seems to have a higher standard deviation, but is the difference significant?

TABLE 8-2

Waiting times (in minutes) of customers at the Jefferson Valley Bank

Multiple Lines					One Line				
2.8	11.5	8.7	13.3	5.2	7.1	6.3	7.2	7.4	4.3
6.2	2.1	6.2	5.6	4.3	6.2	8.8	5.7	7.4	8.2
5.9	11.8	3.7	3.5	9.6	6.8	10.2	9.3	9.4	11.5
9.2	7.8	6.2	2.2	11.2	8.9	7.9	4.7	5.9	6.0
0.6	14.3	7.1	6.0	7.4					

$$n_1 = 25 \qquad n_2 = 20$$
$$\overline{x}_1 = 6.896 \qquad \overline{x}_2 = 7.460$$
$$s_1 = 3.619 \qquad s_2 = 1.841$$

Using the Jefferson Valley Bank sample data in Table 8-2, test the claim that the two samples come from populations with the same variance (or standard deviation). Use a 0.05 level of significance, and assume that the two samples are independent and are drawn from normally distributed populations. (Histograms of the samples will confirm that the distributions are normal.) Based on the results, does the line configuration (multiple or single) seem to affect the customer waiting times?

● **SOLUTION**

Because the larger variance is already denoted by s_1^2, we use the same subscript notation given previously. (If s_2^2 had been larger, we would have interchanged the subscripts to have s_1^2 represent the larger variance.) We now proceed to use the traditional method of testing hypotheses as outlined in Figure 7-4.

Step 1: The claim of equal variances is expressed symbolically as $\sigma_1^2 = \sigma_2^2$.

Step 2: If the original claim is false, then $\sigma_1^2 \neq \sigma_2^2$.

Step 3: Because the null hypothesis must contain equality, we have

$$H_0:\ \sigma_1^2 = \sigma_2^2 \qquad H_1:\ \sigma_1^2 \neq \sigma_2^2$$

Step 4: The significance level is $\alpha = 0.05$.

Step 5: Because this test involves two population variances, we use the F distribution.

Step 6: The test statistic is

$$F = \frac{s_1^2}{s_2^2} = \frac{3.619^2}{1.841^2} = 3.8643$$

For the critical values, we first note that this is a two-tailed test with 0.025 in each tail. As long as we are stipulating that the larger variance is placed in the numerator of the F test statistic, we need to find only the right-tailed critical value. From Table A-5 we get a critical value of $F = 2.4523$, which corresponds to 0.025 in the right tail, with 24 degrees of freedom for the numerator and 19 degrees of freedom for the denominator (see Figure 8-4). *(continued)*

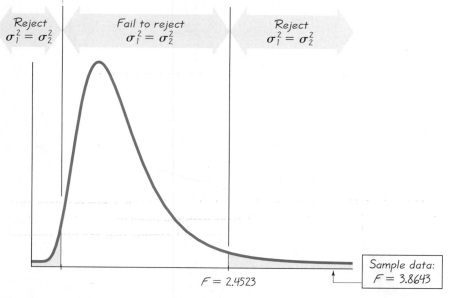

Reject $\sigma_1^2 = \sigma_2^2$ Fail to reject $\sigma_1^2 = \sigma_2^2$ Reject $\sigma_1^2 = \sigma_2^2$

Sample data: $F = 3.8643$

$F = 2.4523$

FIGURE 8-4 Distribution of s_1^2/s_2^2 for Multiple Line and Single Line Data

Step 7:   Figure 8-4 shows that the test statistic $F = 3.8643$ does fall within the critical region, so we reject the null hypothesis.

Step 8:   There is sufficient evidence to warrant rejection of the claim that the two variances are equal. It appears that whether customers are arranged in multiple lines or a single line does have an effect on the variability among their waiting times.

   In the preceding example we tested the claim of equality between the two variances. If we were to test the claim that the multiple line arrangement results in *greater* variation than the single line, then the preceding test would be right-tailed instead of two-tailed, and we would obtain a critical value of $F = 2.1141$ for a right-tailed area of 0.05. With a test statistic of $F = 3.8643$, we would have sufficient evidence to support the claim of a greater variance.

   We have described a method for using the traditional method of testing hypotheses made about two population variances. Exercise 18 deals with the *P*-value approach, and Exercise 20 deals with the construction of confidence intervals.

8-4   Exercises A: Basic Skills and Concepts

In Exercises 1–4, test the given claim. Use a significance level of $\alpha = 0.05$, and assume that all populations are normally distributed. Use the traditional method of testing hypotheses outlined in Figure 7-4, and draw the appropriate graphs.

1. Claim: Populations A and B have the same variance $(\sigma_1^2 = \sigma_2^2)$.

Sample A	Sample B
$n = 10$	$n = 10$
$\bar{x} = 200$	$\bar{x} = 185$
$s^2 = 50$	$s^2 = 25$

2. Claim: Populations A and B have the same variance $(\sigma_1^2 = \sigma_2^2)$.

Sample A	Sample B
$n = 10$	$n = 15$
$\bar{x} = 255$	$\bar{x} = 212$
$s^2 = 50$	$s^2 = 12$

3. Claim: Population A has a larger variance than population B.

Sample A	Sample B
$n = 16$	$n = 200$
$\bar{x} = 124.23$	$\bar{x} = 128.71$
$s = 15.00$	$s = 12.66$

4. Claim: Population A has a larger variance than population B.

Sample A	Sample B
$n = 35$	$n = 25$
$\bar{x} = 238$	$\bar{x} = 254$
$s^2 = 42.3$	$s^2 = 16.2$

5. The Telektronic Company manufactures car batteries using two different production methods. The lives (in years) of the batteries are found for a sample from each group, with the following results.

Traditional method	Experimental method
$n = 25$	$n = 30$
$\bar{x} = 4.31$	$\bar{x} = 4.07$
$s = 0.37$	$s = 0.31$

At the 0.05 significance level, test the claim that the two production methods yield batteries with the same variance. The experimental method is much more expensive and will be used only if it produces *significantly* less variation. Should the experimental method be used?

6. In attempting to improve service to students, a college investigates its registration process and finds that for 60 randomly selected students, the times required for registration have a mean of 73.2 min and a standard deviation of 14.2 min. Another college agrees to collect similar data and finds that for 30 randomly selected students, the times have a mean and standard deviation of 42.3 min and 9.8 min, respectively. At the 0.02 significance level, test the claim that the two samples come from populations with the same standard deviation.

7. In an insurance study of pedestrian deaths in New York State, monthly fatalities are totaled for two different time periods. Sample data for the first time period are summarized by these statistics: $n = 12$, $\bar{x} = 46.42$, $s = 11.07$. Sample data for the second time period are summarized by these statistics: $n = 12$, $\bar{x} = 51.00$, $s = 10.39$ (based on data from the New York State Department of Motor Vehicles). At the 0.05 significance level, test the claim that both time periods have the same variance.

8. Data Set 6 in Appendix B includes a sample of 9 green M&Ms with weights having a mean of 0.8901 g and a standard deviation of 0.0465 g. The data set also includes a sample of 30 brown M&Ms with weights having a mean of 0.9256 g and a standard deviation of 0.0521 g. Use a 0.05 level of significance to test for equality of the two population standard deviations. If you were responsible for controlling production so that green and brown M&Ms have weights with the same amount of variation, would you take corrective action?

9. In a study of flight attendants, salaries paid by two different airlines were randomly selected. For 40 American Airlines flight attendants, the mean is $23,870 and the standard deviation is $2960. For 35 TWA flight attendants, the mean is $22,025 and the standard deviation is $3065 (based on data from the Association of Flight Attendants). At the 0.10 level of significance, test the claim that salaries of American and TWA flight attendants have the same standard deviation.

10. As part of the National Health Survey, data were collected on the weights of men. For 804 men aged 25 to 34, the mean is 176 lb and the standard deviation is 35.0 lb. For 1657 men aged 65 to 74, the mean and standard deviation are 164 lb and 27.0 lb, respectively. (The data are based on the National Health Survey, USDHEW publication 79-1659.) At the 0.01 significance level, test the claim that the older men come from a population with a standard deviation less than that for men in the 25 to 34 age bracket.

11. The Gleason Supermarket's manager experiments with two methods of checking out customers. One method requires that the cashier manually key each price into the register, while the other method uses a scanner that automatically registers prices. The times (in minutes) are given below for samples of goods checked out. At the 0.02 significance level, test the claim that both processes have the same variance.

Manual Keying	Scanning
$n_1 = 16$	$n_2 = 10$
$\overline{x}_1 = 157.6$	$\overline{x}_2 = 112.4$
$s_1^2 = 225.0$	$s_2^2 = 56.0$

12. The Wisconsin Bottling Company employs two shifts of workers to fill bottles with iced tea. A random sample of 25 bottles from the day shift yields a mean and standard deviation of 12.02 oz and 0.14 oz, respectively. A random sample of 30 bottles from the night shift yields a mean and standard deviation of 11.85 oz and 0.22 oz, respectively. Use a 0.05 significance level to test equality of the two population standard deviations. Based on the result, should corrective action be taken to achieve a lower standard deviation with the night shift?

13. The effectiveness of a mental training program was tested in a military training program. In an antiaircraft artillery examination, scores for an experimental group and a control group were recorded. Use the given data to test the claim that both groups come from populations with the same variance. Use a 0.05 significance level.

Experimental				Control			
60.83	117.80	44.71	75.38	122.80	70.02	119.89	138.27
73.46	34.26	82.25	59.77	118.43	54.22	118.58	74.61
69.95	21.37	59.78	92.72	121.70	70.70	99.08	120.76
72.14	57.29	64.05	44.09	104.06	94.23	111.26	121.67
80.03	76.59	74.27	66.87				

Based on "Routinization of Mental Training in Organizations: Effects on Performance and Well-Being," by Larsson, *Journal of Applied Psychology,* Vol. 72, No. 1.

14. Many students have had the unpleasant experience of panicking on a test because the first question was exceptionally difficult. The arrangement of test items was studied for its effect on anxiety. The following scores are measures of "debilitating test anxiety." At the 0.05 significance level, test the claim that the two given samples come from populations with the same variance.

Questions arranged from easy to difficult				Questions arranged from difficult to easy			
24.64	39.29	16.32	32.83	33.62	34.02	26.63	30.26
28.02	33.31	20.60	21.13	35.91	26.68	29.49	35.32
26.69	28.90	26.43	24.23	27.24	32.34	29.34	33.53
7.10	32.86	21.06	28.89	27.62	42.91	30.20	32.54
28.71	31.73	30.02	21.96				
25.49	38.81	27.85	30.29				
30.72							

Based on data from "Item Arrangement, Cognitive Entry Characteristics, Sex and Test Anxiety as Predictors of Achievement in Examination Performance," by Klimko, *Journal of Experimental Education,* Vol. 52, No. 4.

15. Sample data were collected in a study of calcium supplements and their effects on blood pressure. A placebo group and a calcium group began the study with blood pressure measurements. At the 0.05 significance level, test the claim that the two sample groups come from populations with the same standard deviation.

Placebo				Calcium			
124.6	104.8	96.5	116.3	129.1	123.4	102.7	118.1
106.1	128.8	107.2	123.1	114.7	120.9	104.4	116.3
118.1	108.5	120.4	122.5	109.6	127.7	108.0	124.3
113.6				106.6	121.4	113.2	

Based on data from "Blood Pressure and Metabolic Effects of Calcium Supplementation in Normotensive White and Black Men," by Lyle and others, *Journal of the American Medical Association,* Vol. 257, No. 13.

16. A prospective home buyer is investigating differences in home-selling prices between Zone 1 (southern Dutchess County) and Zone 7 (northern Dutchess County). The given random sample data are in thousands of dollars. (This will not affect the value of the test statistic F, but it allows us to work with more manageable numbers.) At the 0.05 significance level, test the claim that both zones have the same variance. Assume that the sample data come from normally distributed populations. (In many cases, home-selling prices might not

be normally distributed. But histograms show that the assumption of normal distributions is reasonable here.)

Zone 7 (north)			
270.000	107.000	148.000	125.000
127.500	125.500	126.000	109.000
113.500	147.000	167.000	

$n = 11, \bar{x} = 142.32, s^2 = 2122$

Zone 1 (south)			
115.000	136.900	121.000	164.000
175.000	128.500	147.500	147.000
105.000	163.750	115.000	149.165
120.500	147.000		

$n = 14, \bar{x} = 138.24, s^2 = 455$

8-4 Exercises B: Beyond the Basics

17. An experiment is devised to study the variability of grading procedures among college professors of organic chemistry. Two different professors are asked to grade the same set of 25 exam solutions, and their grades have variances of 103.4 and 39.7, respectively. At the 0.05 significance level, test the claim that the first professor's grading exhibits greater variance. Given that a student is very weak in organic chemistry, and assuming that both professors give grades with the same mean, is there any advantage in the student's choosing one professor over the other? If so, which one? Why?

18. Refer to the Jefferson Valley Bank example in this section. Using Table A-5, what can you determine about the P-value for that test?

19. For hypothesis tests in this section that were two-tailed, we found only the upper critical value. Let's denote that value by F_R, where the subscript suggests the critical value for the *right* tail. The lower critical value F_L (for the *left* tail) can be found as follows: First interchange the degrees of freedom, and then take the reciprocal of the resulting F value found in Table A-5. (F_R is often denoted by $F_{\alpha/2}$, and F_L is often denoted by $F_{1-\alpha/2}$.) Find the critical values F_L and F_R for two-tailed hypothesis tests based on the following values.
    a. $n_1 = 10, n_2 = 10, \alpha = 0.05$ b. $n_1 = 10, n_2 = 7, \alpha = 0.05$
    c. $n_1 = 7, n_2 = 10, \alpha = 0.05$ d. $n_1 = 25, n_2 = 10, \alpha = 0.02$
    e. $n_1 = 10, n_2 = 25, \alpha = 0.02$

20. In addition to testing claims involving σ_1^2 and σ_2^2, we can also construct *interval estimates* of the ratio σ_1^2/σ_2^2 using the following expression.

$$\frac{s_1^2}{s_2^2} \cdot \frac{1}{F_R} < \frac{\sigma_1^2}{\sigma_2^2} < \frac{s_1^2}{s_2^2} \cdot \frac{1}{F_L}$$

Here F_L and F_R are as described in Exercise 19. Construct the 95% confidence interval estimate for the ratio of the experimental group variance to the control group variance for the data in Exercise 13.

21. Sample data consist of temperatures recorded for two different groups of items that were produced by two different production techniques. A quality control specialist plans to analyze the results. She begins by testing for equality of the two population standard deviations. *(continued)*

a. If she adds the same constant to every temperature from both groups, is the value of the test statistic F affected? Explain.

b. If she multiplies every score from both groups by the same constant, is the value of the test statistic F affected? Explain.

c. If she converts all temperatures from the Fahrenheit scale to the Celsius scale, is the value of the test statistic F affected? Explain.

22. a. Two samples of equal size produce variances of 37 and 57. At the 0.05 significance level, we test the claim that the variance of the second population exceeds that of the first, and this claim is upheld by the data. What is the approximate minimum size of each sample?

b. A sample of 21 scores produces a variance of 67.2, and another sample of 25 produces a variance that causes rejection of the claim that the two populations have equal variances. If this test is conducted at the 0.02 level of significance, find the maximum variance of the second sample if you know that it is smaller than that of the first sample.

8-5 Inferences about Two Means: Independent and Small Samples

This section deals with two independent samples, at least one of which is small ($n \leq 30$). We will discuss testing hypotheses and constructing confidence intervals for situations in which the following assumptions apply.

Assumptions

In testing hypotheses about the means of two populations or in constructing confidence interval estimates of the difference between those means, the methods of this section apply to cases in which

1. The two samples are *independent.*

2. The two samples are randomly selected from *normally distributed* populations.

3. At least one of the two samples is small ($n \leq 30$).

Unlike the preceding sections of this chapter, this section requires the use of different approaches, depending on circumstances. In Section 8-2 we discussed the case of dependent samples, and in Section 8-3 we discussed the case of independent and large samples. This section includes the following additional cases that apply to independent samples when at least one of the samples is small.

Case 1: Both population variances are known.

Case 2: Based on a hypothesis test of $\sigma_1^2 = \sigma_2^2$, we fail to reject equality of the two population variances.

Case 3: Based on a hypothesis test of $\sigma_1^2 = \sigma_2^2$, we reject equality of the two population variances.

This section also discusses procedures that are somewhat controversial in the sense that statisticians do not universally agree with any single approach. However, the methods we present are commonly used.

Case 1: Both Population Variances Are Known

In reality, Case 1 doesn't occur very often because it's rare that we know the population variances σ_1^2 and σ_2^2 but don't know the values of the population means μ_1 and μ_2. (Usually, the population variances are computed from the known population data, and if we can find σ_1^2 and σ_2^2, we should be able to find μ_1 and μ_2. If we can find μ_1 and μ_2, there is no need to test claims or construct confidence intervals.) If some strange set of circumstances leads to this case, hypothesis testing of claims about μ_1 and μ_2 can be done with the following test statistic.

Test Statistic: Known Population Variances

$$z = \frac{(\overline{x}_1 - \overline{x}_2) - (\mu_1 - \mu_2)}{\sqrt{\dfrac{\sigma_1^2}{n_1} + \dfrac{\sigma_2^2}{n_2}}}$$

Because the above test statistic refers to the standard normal distribution, it will be easy to find P-values.

Confidence Intervals

Confidence interval estimates of the difference $\mu_1 - \mu_2$ can be found by using

$$(\overline{x}_1 - \overline{x}_2) - E < (\mu_1 - \mu_2) < (\overline{x}_1 - \overline{x}_2) + E$$

where

$$E = z_{\alpha/2} \sqrt{\frac{\sigma_1^2}{n_1} + \frac{\sigma_2^2}{n_2}}$$

The above test statistic and confidence interval reflect the property of variances that was discussed in Section 8-3: The variance of the *differences* between two independent random variables equals the variance of the first random variable plus the variance of the second random variable. From this property, we get the variances used above.

Because a case of known population variances is so unlikely to occur in reality, we do not illustrate it with examples. However, the calculations closely parallel those of Section 8-3, which used the same test statistic and confidence interval format.

If the assumptions listed at the beginning of this section are satisfied and if we don't know the values of the population standard deviations, we will use the following procedure.

1. Use the F test described in Section 8-4 to test the null hypothesis that $\sigma_1^2 = \sigma_2^2$.

2. If we fail to reject $\sigma_1^2 = \sigma_2^2$, then treat the populations as if they have equal variances. (See Case 2.) If we reject $\sigma_1^2 = \sigma_2^2$, then treat the populations as if they have unequal variances. (See Case 3.)

The following two cases require that we conduct a preliminary F test of the claim that $\sigma_1^2 = \sigma_2^2$ before we can identify the case that applies. Not all statisticians agree that this preliminary F test should be conducted. Some argue that if we apply the F test with a certain level of significance and then do a t test at the same level, the overall result will not be at the same level of significance. Also, in addition to being sensitive to differences in population variances, the F statistic is sensitive to departures from normal distributions, so it's possible to reject a null hypothesis for the wrong reason. (For one argument against the preliminary F test, see "Homogeneity of Variance in the Two-Sample Means Test," by Moser and Stevens, *The American Statistician*, Vol. 46, No. 1.)

Case 2: Equal Variances (Fail to Reject $\sigma_1^2 = \sigma_2^2$)

If we don't have sufficient evidence to warrant rejection of equal variances (that is, if we fail to reject $\sigma_1^2 = \sigma_2^2$), we pool (combine) the sample variances to get a weighted average of s_1^2 and s_2^2, which is the best possible estimate of the variance σ^2 that is common to both populations. This pooled estimate, denoted by s_p^2, is used in the test statistic and confidence interval as shown below.

Test Statistic (Small Independent Samples and Equal Variances)

$$t = \frac{(\bar{x}_1 - \bar{x}_2) - (\mu_1 - \mu_2)}{\sqrt{\dfrac{s_p^2}{n_1} + \dfrac{s_p^2}{n_2}}}$$

where

$$s_p^2 = \frac{(n_1 - 1)s_1^2 + (n_2 - 1)s_2^2}{(n_1 - 1) + (n_2 - 1)}$$

In this case, the degree of freedom is given by $df = n_1 + n_2 - 2$.

Confidence Interval (Small Independent Samples and Equal Variances)

$$(\overline{x}_1 - \overline{x}_2) - E < (\mu_1 - \mu_2) < (\overline{x}_1 - \overline{x}_2) + E$$

where
$$E = t_{\alpha/2}\sqrt{\frac{s_p^2}{n_1} + \frac{s_p^2}{n_2}}$$

and s_p^2 is as given in the test statistic above.

▶ **EXAMPLE**

Age 7	Age 6
$n_1 = 16$	$n_2 = 12$
$\overline{x}_1 = 44.0$	$\overline{x}_2 = 27.5$
$s_1 = 13.2$	$s_2 = 10.2$

Samples of girls aged 6 and 7 are given the Wide Range Achievement Test, with the results shown to the left (based on data from the National Health Survey, USDHEW publication 72-1011). At the 0.05 level of significance, test the claim that the mean score for 7-year-old girls is greater than the mean for 6-year-old girls. (Given the Section 8-4 method for testing equality of variances, it will be convenient to identify the 7-year-old group as Group 1 because its variance is larger.)

● **SOLUTION**

We begin by analyzing the types of samples we are using. Because we are dealing with two separate groups of subjects, we know that we have independent samples. The sample sizes are small because both n_1 and n_2 are less than 31. Because of the nature of the data (test scores and physical measurements typically result in normally distributed populations), it is reasonable to assume that the samples come from normally distributed populations. We don't know the values of σ_1 and σ_2, so the Case 1 approach is ruled out, and we will use a preliminary F test to choose between Case 2 and Case 3. With H_0: $\sigma_1^2 = \sigma_2^2$, H_1: $\sigma_1^2 \neq \sigma_2^2$, and $\alpha = 0.05$, we find the F test statistic as follows:

$$F = \frac{s_1^2}{s_2^2} = \frac{13.2^2}{10.2^2} = 1.6747$$

With $\alpha = 0.05$ in a two-tailed F test and with 15 and 11 as the degrees of freedom for the numerator and denominator, respectively, we use Table A-5 to obtain the critical F value of 3.3299. Because the computed test statistic of $F = 1.6747$ does not fall within the critical region, we fail to reject the null hypothesis of equal variances and proceed by using the approach outlined in Case 2.

In using the Case 2 approach, we test the given claim that $\mu_1 > \mu_2$ with the following null and alternative hypotheses.

H_0: $\mu_1 \leq \mu_2$ (or $\mu_1 - \mu_2 \leq 0$) H_1: $\mu_1 > \mu_2$ (or $\mu_1 - \mu_2 > 0$)

The test statistic requires the value of the pooled variance s_p^2, so we find that value and then find the test statistic as shown below.

$$s_p^2 = \frac{(n_1 - 1)s_1^2 + (n_2 - 1)s_2^2}{(n_1 - 1) + (n_2 - 1)} = \frac{(16 - 1) \cdot 13.2^2 + (12 - 1) \cdot 10.2^2}{(16 - 1) + (12 - 1)}$$

$$= \frac{3758.04}{26} = 144.54$$

$$t = \frac{(\bar{x}_1 - \bar{x}_2) - (\mu_1 - \mu_2)}{\sqrt{\dfrac{s_p^2}{n_1} + \dfrac{s_p^2}{n_2}}} = \frac{(44.0 - 27.5) - 0}{\sqrt{\dfrac{144.54}{16} + \dfrac{144.54}{12}}} = 3.594$$

Noting that $\mu_1 > \mu_2$ is equivalent to $\mu_1 - \mu_2 > 0$, we see more clearly that the test is right-tailed. With $\alpha = 0.05$ in this right-tailed t test and $n_1 + n_2 - 2 = 16 + 12 - 2 = 26$ degrees of freedom, we obtain a critical t value of 1.706 by referring to Table A-3. The computed test statistic of $t = 3.594$ falls within the critical region, so we reject the null hypothesis. The given sample data support the claim that the mean for 7-year-old girls is greater than that for 6-year-old girls.

P-values can be found by using the procedures presented in Section 7-3 (see Figures 7-7 and 7-8). The preceding hypothesis test example is right-tailed with a test statistic of $t = 3.594$ and 26 degrees of freedom. Refer to the 26th row of Table A-3, where a test statistic of $t = 3.594$ corresponds to a P-value of less than 0.005. The P-value is further confirmation that we are highly unlikely to get sample means with values like those obtained if the two populations have the same mean.

▶ **EXAMPLE**

Using the data in the preceding example, construct a 99% confidence interval estimate of $\mu_1 - \mu_2$.

● **SOLUTION**

With $s_p^2 = 144.54$ (as found in the preceding example) and with $t_{\alpha/2} = 2.779$ (df = 26 and 0.01 in two tails), the margin of error is

$$E = t_{\alpha/2} \sqrt{\frac{s_p^2}{n_1} + \frac{s_p^2}{n_2}} = 2.779 \sqrt{\frac{144.54}{16} + \frac{144.54}{12}} = 12.8$$

(continued)

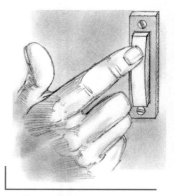

Poll Resistance

Surveys based on relatively small samples can be quite accurate, provided the sample is random or representative of the population. However, increasing survey refusal rates are now making it more difficult to obtain random samples. The Council of American Survey Research Organizations reported that in a recent year, 38% of consumers refused to respond to surveys. The head of one market research company said, "Everyone is fearful of self-selection and worried that generalizations you make are based on cooperators only." Results from the multibillion-dollar market research industry affect the products we buy, the television shows we watch, and many other facets of our lives.

With $\bar{x}_1 - \bar{x}_2 = 16.5$ and $E = 12.8$, the confidence interval is as shown below.

$$(\bar{x}_1 - \bar{x}_2) - E < (\mu_1 - \mu_2) < (\bar{x}_1 - \bar{x}_2) + E$$

$$16.5 - 12.8 < (\mu_1 - \mu_2) < 16.5 + 12.8$$

$$3.7 < (\mu_1 - \mu_2) < 29.3$$

That is, we are 99% confident that the difference between the mean score for 7-year-old girls and the mean score for 6-year-old girls is between 3.7 and 29.3. This confidence interval does not contain 0, suggesting that the difference between the two means is significant.

Case 3:
Unequal Variances (Reject $\sigma_1^2 = \sigma_2^2$)

If we do have sufficient evidence to warrant rejection of equal variances (that is, if we reject $\sigma_1^2 = \sigma_2^2$), there is no *exact* method for testing equality of means and constructing confidence intervals. An *approximate* method, however, uses the following test statistic and confidence interval.

Test Statistic (Small Independent Samples and Unequal Variances)

$$t = \frac{(\bar{x}_1 - \bar{x}_2) - (\mu_1 - \mu_2)}{\sqrt{\dfrac{s_1^2}{n_1} + \dfrac{s_2^2}{n_2}}}$$

In this case, the degree of freedom is given by

$$df = \text{smaller of } n_1 - 1 \text{ and } n_2 - 1$$

One or other
Whichever is smaller

Confidence Interval (Small Independent Samples and Unequal Variances)

$$(\bar{x}_1 - \bar{x}_2) - E < (\mu_1 - \mu_2) < (\bar{x}_1 - \bar{x}_2) + E$$

where
$$E = t_{\alpha/2}\sqrt{\frac{s_1^2}{n_1} + \frac{s_2^2}{n_2}}$$
$$df = \text{smaller of } n_1 - 1 \text{ and } n_2 - 1$$

As stated, this use of the Student t distribution has a number of degrees of freedom equal to the smaller of $n_1 - 1$ and $n_2 - 1$. It represents a more conservative and simplified alternative to computing the number of degrees of freedom by using Formula 8-1.

Formula 8-1

$$df = \frac{(A + B)^2}{\dfrac{A^2}{n_1 - 1} + \dfrac{B^2}{n_2 - 1}}$$

where

$$A = \frac{s_1^2}{n_1} \text{ and } B = \frac{s_2^2}{n_2}$$

More exact results are obtained by using Formula 8-1, but they continue to be only approximate.

▶ **EXAMPLE**

Refer to the Jefferson Valley Bank sample data in Table 8-2, which include the following statistics. At the 0.05 significance level, test the claim that the population of waiting times has a mean that is the same for multiple lines as for a single line. Assume that the two samples are independent and are randomly selected from normally distributed populations.

Multiple Lines	Single Line
$n_1 = 25$	$n_2 = 20$
$\bar{x}_1 = 6.896$	$\bar{x}_2 = 7.460$
$s_1 = 3.619$	$s_2 = 1.841$

● **SOLUTION**

We are dealing with two independent samples, each of which is small. We don't know the values of the population variances, but in Section 8-4 we already tested and rejected the claim that $\sigma_1^2 = \sigma_2^2$. *Because we rejected equality of variances, we assume that the variances are unequal and proceed to test $\mu_1 = \mu_2$ by using the test statistic given for the case of unequal variances (Case 3).* The test statistic is

$$t = \frac{(\bar{x}_1 - \bar{x}_2) - (\mu_1 - \mu_2)}{\sqrt{\dfrac{s_1^2}{n_1} + \dfrac{s_2^2}{n_2}}} = \frac{(6.896 - 7.460) - 0}{\sqrt{\dfrac{3.619^2}{25} + \dfrac{1.841^2}{20}}} = -0.677$$

The critical values of $t = -2.093$ and $t = 2.093$ are found from Table A-2 by referring to the column for $\alpha = 0.05$ (two tails) and to the row for df = 19 (the

(continued)

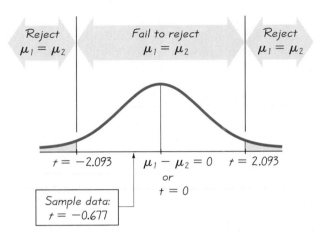

FIGURE 8-5   Distribution of Differences between Means of Multiple Line Times and Single Line Times

smaller of $25 - 1$ and $20 - 1$). Figure 8-5 shows the test statistic, critical values, and critical region. We can see that the test statistic does not fall in the critical region, so we fail to reject the null hypothesis H_0: $\mu_1 = \mu_2$. There is not sufficient evidence to warrant rejection of the claim that the mean waiting time is the same for the multiple line system and the single line system. That is, it appears that the two waiting lines have the same mean.

▶ **EXAMPLE**

Using the same Jefferson Valley sample data, construct a 95% confidence interval estimate of the difference $\mu_1 - \mu_2$, where μ_1 is the mean waiting time for customers using the multiple line system and μ_2 is the mean waiting time for customers using the single line system.

● **SOLUTION**

We can easily find that $\overline{x}_1 - \overline{x}_2 = 6.896 - 7.460 = -0.564$. Next, we need to find the value of the margin of error E.

$$E = t_{\alpha/2} \sqrt{\frac{s_1^2}{n_1} + \frac{s_2^2}{n_2}} = 2.093 \sqrt{\frac{3.619^2}{25} + \frac{1.841^2}{20}} = 1.743$$

We can now proceed to find the desired confidence interval.

$$(\overline{x}_1 - \overline{x}_2) - E < (\mu_1 - \mu_2) < (\overline{x}_1 - \overline{x}_2) + E$$
$$-0.564 - 1.743 < (\mu_1 - \mu_2) < -0.564 + 1.743$$
$$-2.307 < (\mu_1 - \mu_2) < 1.179$$

We are 95% confident that the difference between the mean waiting time with multiple lines and the mean waiting time with a single line is between -2.307 min and 1.179 min. Because the confidence interval contains 0, we can conclude that the difference between those means is not significant; that is, there is not a significant difference between the mean waiting time for customers using multiple lines and the mean waiting time for customers using the single waiting line. This is no surprise; the bank doesn't become more efficient and process customers faster by simply reconfiguring the way customers wait in line. However, we did see in Section 8-4 that the single line system has much less *variability* than the multiple line system. The lower variability is the important redeeming characteristic that justifies the use of the single waiting line instead of the multiple lines. On average, the customers don't get processed any faster, but they are happier because they can avoid the frustration of being caught in a slow line.

This section and Sections 8-2 and 8-3 all deal with inferences about the means of two populations. Determining the correct procedure can be difficult because we must consider issues such as the independence of the samples, the sizes of the samples, and whether we know σ_1 and σ_2. We can avoid confusion by using the organized and systematic procedures summarized in Figures 8-6 and 8-7 on the next pages.

Using Computer Software for Two Populations

The calculations in the examples of this section are complicated, but a calculator or computer can be used to simplify them. STATDISK, for example, can be used to conduct tests of the type included in this section. Simply select Statistics, then Hypothesis Testing, then the option indicating a hypothesis test involving the means of two independent samples, and you will be prompted for the necessary data. When relevant, the results of the prerequisite F test are included in the STATDISK display.

Minitab does not have a command to execute an F test of equality between two variances, but we can use the Minitab command STDEV to find the sample standard deviations used in computing the F test statistic. If we fail to reject equality of the two population variances, Minitab allows us to use a pooled estimate of the common variance by using the subcommand POOLED.

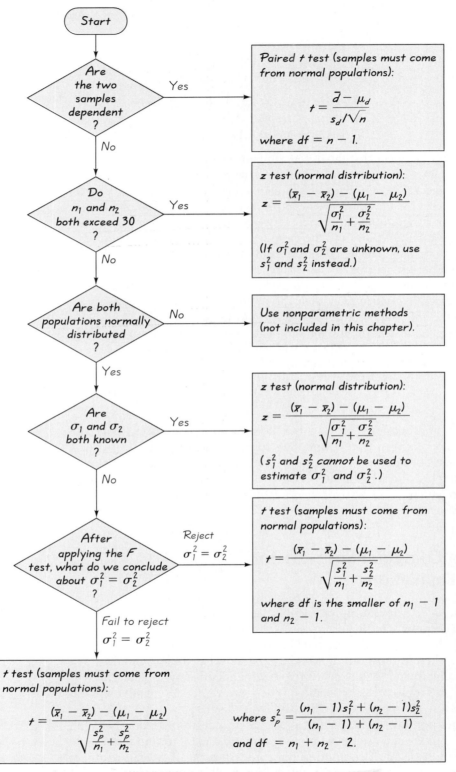

FIGURE 8-6   Testing Hypotheses Made about the Means of Two Populations

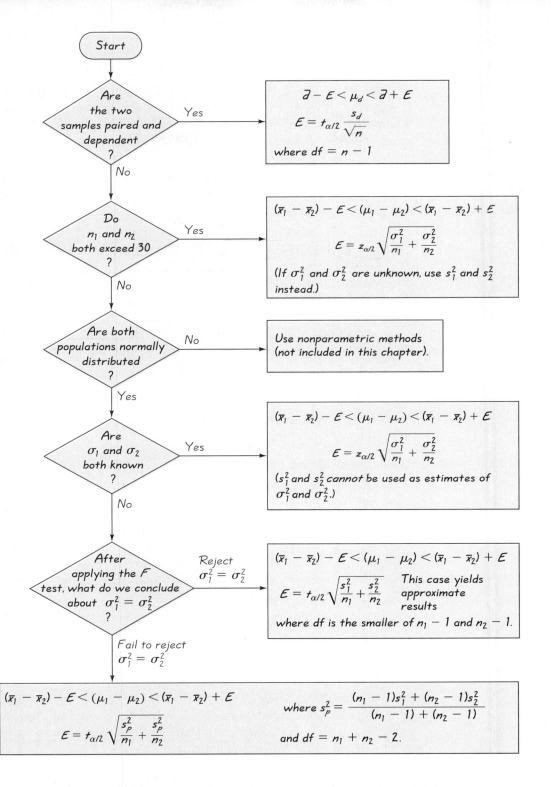

FIGURE 8-7 Confidence Intervals for the Difference between Two Population Means

After the sample data in columns C1 and C2 have been entered, the following Minitab commands will yield the test statistic, *P*-value, and 95% confidence interval limits.

```
MTB > TWOSAMPLE 95 PERCENT C1 C2;
SUBC> POOLED.
```

If we reject equality of the two population variances, we can simply enter the command TWOSAMPLE 95 PERCENT C1 C2 without the POOLED subcommand.

8-5   Exercises A: Basic Skills and Concepts

In Exercises 1 and 2, test the given claim. Use a significance level of $\alpha = 0.05$, and assume that all populations are normally distributed. Use the traditional method of testing hypotheses outlined in Figure 7-4, and draw the appropriate graphs.

1. Claim: Populations A and B have the same mean ($\mu_1 = \mu_2$).

Sample A	Sample B
$n = 10$	$n = 10$
$\bar{x} = 200$	$\bar{x} = 185$
$s^2 = 50$	$s^2 = 25$

2. Claim: Populations A and B have the same mean ($\mu_1 = \mu_2$).

Sample A	Sample B
$n = 10$	$n = 15$
$\bar{x} = 255$	$\bar{x} = 212$
$s^2 = 50$	$s^2 = 12$

3. Using the sample data in Exercise 1, construct a 95% confidence interval estimate of the difference between the means of the two populations.

4. Using the sample data in Exercise 2, construct a 95% confidence interval estimate of the difference between the means of the two populations.

5. The Telektronic Company manufactures car batteries with two different production methods. The lives (in years) of the batteries are found for a sample from each group, with the following results.

Traditional method	Experimental method
$n = 25$	$n = 30$
$\bar{x} = 4.31$	$\bar{x} = 4.07$
$s = 0.37$	$s = 0.31$

   At the 0.05 significance level, test the claim that the two production methods yield batteries with the same mean. Based on the results, if you were buying a battery for your car, would you prefer a battery manufactured by the traditional method or the experimental method?

6. Using the sample data given in Exercise 5, construct a 95% confidence interval estimate of the difference between the mean life of car batteries for the traditional method and the mean life for the experimental method.

7. In an insurance study of pedestrian deaths in New York State, monthly fatalities are analyzed for two different time periods. Sample data for the first time period are summarized by these statistics: $n = 12$, $\bar{x} = 46.42$, $s = 11.07$. Sample data for the second time period are summarized by these statistics: $n = 12$, $\bar{x} = 51.00$, $s = 10.39$ (based on data from the New York State Department of Motor Vehicles). At the 0.05 significance level, test the claim that both time periods have the same mean.

8. Data Set 6 in Appendix B includes a sample of 9 green M&Ms with weights having a mean of 0.8901 g and a standard deviation of 0.0465 g. The data set also includes a sample of 30 brown M&Ms with weights having a mean of 0.9256 g and a standard deviation of 0.0521 g. Use a 0.05 level of significance to test for equality of the two population means. If you were responsible for controlling production so that green and brown M&Ms have weights with the same mean, would you take corrective action?

9. The Gleason Supermarket's manager experiments with two methods of checking out customers. One method requires that the cashier manually key each price into the register, and the other method uses a scanner that automatically registers prices. The times (in minutes) are given below for samples of goods checked out. At the 0.02 significance level, test the claim that the two methods have the same mean.

Manual keying	Scanning
$n_1 = 16$	$n_2 = 10$
$\bar{x}_1 = 157.6$	$\bar{x}_2 = 112.4$
$s_1^2 = 225.0$	$s_2^2 = 56.0$

10. Using the sample data from Exercise 9, construct a 98% confidence interval estimate of the difference between the mean for manual keying times and the mean for scanning times.

11. The Wisconsin Bottling Company employs two shifts of workers to fill bottles with iced tea. A random sample of 25 bottles from the day shift yields a mean and standard deviation of 12.02 oz and 0.14 oz, respectively. A random sample of 30 bottles from the night shift yields a mean and standard deviation of 11.85 oz and 0.22 oz, respectively. Construct a 95% confidence interval estimate of the difference between the mean fill amount for the day shift and the mean fill amount for the night shift.

12. Refer to the sample data given in Exercise 11 and use a 0.05 significance level to test for equality of the two population means.

13. Are severe psychiatric disorders related to biological factors that can be physically observed? One study used X-ray computed tomography (CT) to collect data on brain volumes for a group of patients with obsessive-compulsive disorders and a control group of healthy persons. Sample results (in mL) follow for volumes of the right cordate (based on data from "Neuroanatomical Abnormalities in Obsessive-Compulsive Disorder Detected with Quantitative X-Ray Computed Tomography," by Luxenberg and others, *American Journal of Psychiatry*, Vol. 145, No. 9). At the 0.01 significance level, test the claim

that obsessive-compulsive patients and healthy persons have the same mean. Based on this result, does it seem that obsessive-compulsive disorders have a biological basis?

Obsessive-compulsive patients	Control group
$n = 10$	$n = 10$
$\overline{x} = 0.34$	$\overline{x} = 0.45$
$s = 0.08$	$s = 0.08$

14. Using the sample data from Exercise 13, construct a 99% confidence interval estimate of the difference between the mean brain volume for the patient group and the mean brain volume for the healthy control group.

15. The same study cited in Exercise 13 resulted in the values summarized below for total brain volumes (in mL). Using these sample statistics, construct a 95% confidence interval for the difference between the mean brain volume of obsessive-compulsive patients and the mean brain volume of healthy persons.

Obsessive-compulsive patients	Control group
$n = 10$	$n = 10$
$\overline{x} = 1390.03$	$\overline{x} = 1268.41$
$s = 156.84$	$s = 137.97$

16. Refer to the sample data in Exercise 15 and use a 0.05 significance level to test the claim that there is no difference between the mean for obsessive-compulsive patients and the mean for healthy persons. Based on this result, does it seem that obsessive-compulsive disorders have a biological basis?

17. The Lectrolyte Company collects sample data on the lengths of telephone calls (in minutes) made by employees in two different divisions, and the results are as shown below. At the 0.02 level of significance, test the claim that there is no difference between the mean times of all long distance calls made in the two divisions.

Sales division	Customer service division
$n_1 = 40$	$n_2 = 20$
$\overline{x}_1 = 10.26$	$\overline{x}_2 = 6.93$
$s_1 = 8.65$	$s_2 = 4.93$

18. Use the sample data in Exercise 17 to construct a 98% confidence interval for the difference between the two population means.

19. Refer to Data Set 6 in Appendix B and test the claim that red M&Ms and brown M&Ms have the same mean weight. If you were part of a quality control team with responsibility for ensuring that red M&Ms and brown M&Ms have the same mean weight, would you take corrective action?

20. Refer to Data Set 5 in Appendix B and test the claim that there is no difference in length between movies rated R and movies with G, PG, or PG-13 ratings.

8-5   Exercises B: Beyond the Basics

21. In constructing the confidence interval for Exercise 11, the value of $t_{\alpha/2}$ was found by setting the degrees of freedom equal to the smaller of $n_1 - 1$ and $n_2 - 1$. A better result is obtained by using Formula 8-1 to find the number of degrees of freedom. Use Formula 8-1 to determine the number of degrees of freedom, then use it to find the confidence interval requested in Exercise 11. Compare the results to the confidence interval originally obtained in Exercise 11.

22. Refer to the indicated exercise and find the P-value.

   a. Exercise 1   b. Exercise 5

   c. Exercise 9   d. Exercise 12

23. Assume that two samples have the same standard deviation and that both are independent, small, and randomly selected from normally distributed populations. Also assume that we want to test the claim that the samples come from populations with the same mean.

   a. Is it necessary to conduct a preliminary F test?

   b. If both samples have standard deviation s, what is the value of s_p^2 expressed in terms of s?

8-6   Inferences about Two Proportions

In this section we will consider methods of using two sample proportions to make inferences about two population proportions. The concepts and procedures we will develop can be used to answer questions such as the following, which relate to a very common type of statistical experiment involving a control group and a treatment group.

- When one group is given an experimental drug and another group is given a placebo, is there a difference between the cure rates for the two groups?

- When one group is given an experimental drug and another group is given a placebo, what is an estimate of the difference between the two cure rates?

When testing a hypothesis made about two population proportions or when constructing a confidence interval for the difference between two population proportions, we make the following assumptions and use the following notation.

Assumptions

1. We have two *independent* sets of randomly selected sample data.

2. For both samples, the conditions $np \geq 5$ and $nq \geq 5$ are satisfied. (In many cases, we will test the claim that two populations have the *same* proportion, but

no value will be specified. In such cases, the conditions $np \geq 5$ and $nq \geq 5$ can be checked by replacing p by the estimated pooled proportion $\bar{p}$, which will be described later in this section.)

Notation for Two Proportions

For population 1 we let

$$p_1 = \textit{population} \text{ proportion}$$
$$n_1 = \text{size of the sample}$$
$$x_1 = \text{number of successes}$$
$$\hat{p}_1 = \frac{x_1}{n_1} \quad \text{(or the } \textit{sample} \text{ proportion)}$$
$$\hat{q}_1 = 1 - \hat{p}_1$$

The corresponding meanings are attached to p_2, n_2, x_2, and $\hat{p}_2$, which come from population 2.

Hypothesis Tests

In Sections 6-3 and 7-5 we considered the construction of confidence intervals and hypothesis testing for cases involving a single population proportion p, and in both sections we used a sample proportion of successes $\hat{p}$ that resulted from a random sample of size n. In both sections we used a normal distribution as an approximation to the binomial distribution. This section extends those same methods to two population proportions. We know from Section 5-5 that for the first population, if $n_1 p_1 \geq 5$ and $n_1 q_1 \geq 5$, then we can approximate the binomial distribution by the normal distribution. This applies to the second population as well. As a result, our test statistic will be approximately normally distributed. We will be testing only claims including the assumption that $p_1 = p_2$, and we will use the following pooled (or combined) estimate of their common value. (For other cases, see Exercise 17 of this section.)

Pooled Estimate of p_1 and p_2

The **pooled estimate of p_1 and p_2** is denoted by $\bar{p}$ and is given by

$$\bar{p} = \frac{x_1 + x_2}{n_1 + n_2}$$

We denote the complement of $\bar{p}$ by $\bar{q}$, so $\bar{q} = 1 - \bar{p}$.

With a null hypothesis of $p_1 = p_2$ or $p_1 \geq p_2$ or $p_1 \leq p_2$, we can use the following test statistic.

Test Statistic for Two Proportions

For H_0: $p_1 = p_2$, H_0: $p_1 \geq p_2$, or H_0: $p_1 \leq p_2$:

$$z = \frac{(\hat{p}_1 - \hat{p}_2) - (p_1 - p_2)}{\sqrt{\dfrac{\overline{p}\,\overline{q}}{n_1} + \dfrac{\overline{p}\,\overline{q}}{n_2}}}$$

where
$$p_1 - p_2 = 0$$

$$\hat{p}_1 = \frac{x_1}{n_1} \quad \text{and} \quad \hat{p}_2 = \frac{x_2}{n_2}$$

$$\overline{p} = \frac{x_1 + x_2}{n_1 + n_2}$$

$$\overline{q} = 1 - \overline{p}$$

We will first illustrate the procedure for testing hypotheses and then justify the test statistic given above. The following example applies to the problem stated at the beginning of this chapter. As you read the example, the symbols x_1, x_2, n_1, n_2, $\hat{p}_1$, $\hat{p}_2$, $\overline{p}$, and $\overline{q}$ should become more meaningful. In particular, you should recognize that under the assumption of equal proportions, the best estimate of the common proportion is obtained by pooling both samples into one large sample, so that $\overline{p}$ becomes a more obvious estimate of the common population proportion.

▶ **EXAMPLE**

Johns Hopkins researchers conducted a study of pregnant IBM employees. Among 30 who worked with glycol ethers, 10 (or 33.3%) had miscarriages, but among 750 who were not exposed to glycol ethers, 120 (or 16.0%) had miscarriages. At the 0.05 significance level, test the claim that the miscarriage rate is the same for both groups.

The fundamental question is really this: Is there a significant difference between the 33.3% miscarriage rate and the 16.0% miscarriage rate?

● **SOLUTION**

For notation purposes, we stipulate that sample 1 is the group that worked

(continued)

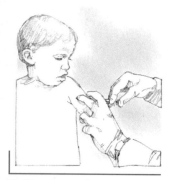

Polio Experiment

In 1954 an experiment was conducted to test the effectiveness of the Salk vaccine as protection against the devastating effects of polio. Approximately 200,000 children were injected with an ineffective salt solution, and 200,000 other children were injected with the vaccine. The experiment was "double blind" because the children being injected didn't know whether they were given the real vaccine or the placebo, and the doctors giving the injections and evaluating the results didn't know either. Only 33 of the 200,000 vaccinated children later developed paralytic polio, whereas 115 of the 200,000 injected with the salt solution later developed paralytic polio. Statistical analysis of these and other results led to the conclusion that the Salk vaccine was indeed effective against paralytic polio.

with glycol ethers and sample 2 is the group not exposed. We can therefore summarize the sample data as follows:

$$n_1 = 30 \qquad\qquad n_2 = 750$$
$$x_1 = 10 \qquad\qquad x_2 = 120$$
$$\hat{p}_1 = \frac{10}{30} = 0.333 \qquad \hat{p}_2 = \frac{120}{750} = 0.160$$

We will now proceed to follow the same basic hypothesis-testing procedure introduced in Chapter 7. We use the traditional method summarized in Figure 7-4.

Step 1: The claim of no difference in miscarriage rates between the two groups can be represented by $p_1 = p_2$.

Step 2: If $p_1 = p_2$ is false, then $p_1 \neq p_2$.

Step 3: Because our claim of $p_1 = p_2$ contains equality, it becomes the null hypothesis, and we have

$$H_0:\ p_1 = p_2 \qquad H_1:\ p_1 \neq p_2$$

Step 4: The significance level is $\alpha = 0.05$.

Step 5: We will use the normal distribution, with the test statistic given above, as an approximation to the binomial distribution. We have two independent samples, and the conditions $np \geq 5$ and $nq \geq 5$ are satisfied for each of the two samples. To check this, we note that we are assuming $p_1 = p_2$, where their common value is the pooled estimate $\bar{p}$, calculated as

$$\bar{p} = \frac{x_1 + x_2}{n_1 + n_2} = \frac{10 + 120}{30 + 750} = \frac{130}{780} = 0.1667$$

With $\bar{p} = 0.1667$, it follows that $\bar{q} = 1 - \bar{p} = 0.8333$. We can now verify that $np \geq 5$ and $nq \geq 5$ for both samples as follows:

Sample 1	Sample 2
$n_1 p = (30)(0.1667) = 5 \geq 5$	$n_2 p = (750)(0.1667) = 125 \geq 5$
$n_1 q = (30)(0.8333) = 25 \geq 5$	$n_2 p = (750)(0.8333) = 625 \geq 5$

Step 6: With $\hat{p}_1 = 0.333$, $\hat{p}_2 = 0.160$, $\bar{p} = 0.1667$, $\bar{q} = 0.8333$, $n_1 = 30$, and $n_2 = 750$ and with the claim of $p_1 = p_2$ (or $p_1 - p_2 = 0$), we can find the value of the test statistic as follows:

$$z = \frac{(\hat{p}_1 - \hat{p}_2) - (p_1 - p_2)}{\sqrt{\dfrac{\bar{p}\,\bar{q}}{n_1} + \dfrac{\bar{p}\,\bar{q}}{n_2}}}$$

$$= \frac{(0.333 - 0.160) - 0}{\sqrt{\dfrac{(0.1667)(0.8333)}{30} + \dfrac{(0.1667)(0.8333)}{750}}} = 2.49$$

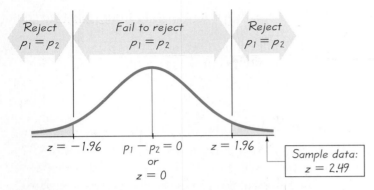

FIGURE 8-8   Distribution of Differences between Proportions for Group Exposed to Glycol Ethers and Group Not Exposed

The critical values of $z = -1.96$ and $z = 1.96$ are found from Table A-2 by observing that we have a two-tailed test with $\alpha/2 = 0.025$ in each of the two tails. The value of $z = 1.96$ is found from Table A-2 as the z score corresponding to an area of $0.5 - 0.025 = 0.4750$. [The test is two-tailed because rejection of $p_1 - p_2 = 0$ occurs if we have a value of $\hat{p}_1 - \hat{p}_2$ that is significantly large (or to the right) or significantly small (or to the left).] Figure 8-8 shows the test statistic of $z = 2.49$ and the critical values of $z = -1.96$ and $z = 1.96$.

Step 7:   In Figure 8-8 we see that the test statistic falls within the critical region, so we reject the null hypothesis of $p_1 = p_2$.

Step 8:   We conclude that there is sufficient evidence to warrant rejection of the claim that the miscarriage rate is the same for both groups of workers.

These steps use the traditional approach to hypothesis testing, but it would be quite easy to use the P-value approach. In Step 6, instead of finding the critical values of z, we would find the P-value by using the procedure summarized in Figures 7-7 and 7-8. With a test statistic of $z = 2.49$ and a two-tailed test, we get

$$P\text{-value} = 2 \cdot (\text{area beyond } z = 2.49)$$
$$= 2 \cdot (0.5 - 0.4936) = 0.0128$$

Again, we reject the null hypothesis because the P-value of 0.0128 is less than the significance level α.

With this evidence, the Johns Hopkins researchers concluded that women employees exposed to glycol ethers "have a significantly increased risk of miscarriage." Based on these results, IBM warned its employees of the danger, notified the Environmental Protection Agency, and greatly reduced its use of glycol ethers.

[handwritten margin note: to find x_1 & x_2 → $x_1 = \hat{p_1} \cdot n_1$]

In the preceding example we were given information from which we could easily see that $x_1 = 10$ and $x_2 = 120$. In other cases, we may have to calculate the values of x_1 and x_2, which can be found by noting that $\hat{p}_1 = x_1/n_1$ implies that $x_1 = \hat{p}_1 \cdot n_1$. The following example provides an illustration.

▶ **EXAMPLE**

In a study of people who stop to help drivers with disabled cars, researchers hypothesized that more people would stop to help someone if they first saw another driver with a disabled car getting help. In one experiment, 2000 drivers first saw a woman being helped with a flat tire and then saw a second woman who was alone, farther down the road, with a flat tire; 2.90% of those 2000 drivers stopped to help the second woman. Among 2000 other drivers who did not see the first woman being helped, only 1.75% stopped to help. (See "Help on the Highway," by McCarthy, *Psychology Today*, July 1987.) At the 0.05 significance level, test the claim that the percentage of people who stop after first seeing a driver with a disabled car being helped is greater than the percentage of people who stop without first seeing someone else being helped.

● **SOLUTION**

The sample data can be summarized as follows:

Saw earlier help	Didn't see earlier help
$n_1 = 2000$	$n_2 = 2000$
$x_1 = 2.90\%$ of $2000 = 58$	$x_2 = 1.75\%$ of $2000 = 35$

We now use the traditional method to test the given claim.

Step 1:   The claim that the first percentage is greater than the second is expressed in symbols as $p_1 > p_2$.

Step 2:   If the original claim is false, then $p_1 \leq p_2$.

Step 3:   The null hypothesis must contain equality, so we have

$$H_0: p_1 \leq p_2 \quad * \quad H_1: p_1 > p_2$$

Step 4:   The significance level is $\alpha = 0.05$ (a very common value in social science experiments).

Step 5:   The normal distribution will be used. The two samples are independent, and both satisfy $np \geq 5$ and $nq \geq 5$. (This can be checked by using the value of

$$\bar{p} = \frac{x_1 + x_2}{n_1 + n_2} = \frac{58 + 35}{2000 + 2000} = 0.02325$$

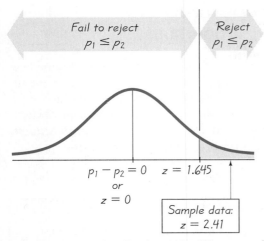

FIGURE 8-9 **Distribution of Differences between Proportions for Group Who Saw Earlier Help and Group Who Didn't See Earlier Help**

and the value of $\overline{q} = 1 - \overline{p} = 0.97675$; that is, $n_1\overline{p}$, $n_1\overline{q}$, $n_2\overline{p}$, and $n_2\overline{q}$ all have values of at least 5.)

Step 6: The test statistic is

$$z = \frac{(\hat{p}_1 - \hat{p}_2) - (p_1 - p_2)}{\sqrt{\dfrac{\overline{p}\,\overline{q}}{n_1} + \dfrac{\overline{p}\,\overline{q}}{n_2}}} = \frac{(0.0290 - 0.0175) - 0}{\sqrt{\dfrac{(0.02325)(0.97675)}{2000} + \dfrac{(0.02325)(0.97675)}{2000}}}$$

$$= 2.41$$

The critical value is $z = 1.645$, found in Table A-2 as the z score corresponding to an area of 0.45. (With $\alpha = 0.05$, we refer to the area of $0.5 - 0.05$ between the mean and the critical value.) The test statistic and critical value are shown in Figure 8-9.

Step 7: Figure 8-9 shows that the test statistic of $z = 2.41$ falls within the critical region, so we reject the null hypothesis of $p_1 \leq p_2$.

Step 8: There is sufficient evidence to support the claim that the percentage of people who stop after first seeing someone else being helped is greater than the percentage of people who stop without first seeing someone else being helped.

With the P-value approach, we first use the test statistic $z = 2.41$ to find that the P-value is 0.0080 (the area to the right of the test statistic). Because the P-value of 0.0080 is less than the significance level of 0.05, we reject the null hypothesis and arrive at the same conclusion given in Step 8.

Using Computer Software for Testing Hypotheses about Two Proportions

The STATDISK display for the preceding example can be easily obtained by selecting the Hypothesis Testing option from the Statistics menu. Then select Proportion - Two Samples, and proceed to select the case of p1 > p2 and answer the questions to obtain the display shown. The use of Minitab is a bit tricky here, and we recommend that you consult the *Minitab Student Laboratory Manual and Workbook* for a procedure that will work.

STATDISK DISPLAY

```
 About    File    Edit    Statistics    Data    View

        Hypothesis test for a claim about TWO POPULATION PROPORTIONS

                    NULL HYPOTHESIS:  p1 <= p2

        Sample 1:    x1 = 58       n1 = 2000     x1/n1 =  .02900
        Sample 2:    x2 = 35       n2 = 2000     x2/n2 =  .01750

              ┌─────────────────────────────────────────────┐
              │   Test statistic ........ z =  2.41321       │
              │   Critical value ........ z =  1.64522       │
              │   P-value ................... =  .00796      │
              │   Significance level ...... =  .05           │
              └─────────────────────────────────────────────┘


              CONCLUSION: REJECT the null hypothesis



                        Use Menu Bar to change screen
      F1: Help        F2: Menu                              ESC: Quit
```

The test statistic we are using in this section can be justified as follows:

1. With $n_1 p_1 \geq 5$ and $n_1 q_1 \geq 5$, the distribution of $\hat{p}_1$ can be approximated by a normal distribution with mean p_1, standard deviation $\sqrt{p_1 q_1/n_1}$, and variance $p_1 q_1/n_1$. These conclusions follow from Sections 5-5 and 6-3. They also can be applied to the second sample and population.

2. Because $\hat{p}$ and $\hat{q}$ are each approximated by a normal distribution, $\hat{p}_1 - \hat{p}_2$ will also be approximated by a normal distribution with mean $p_1 - p_2$ and variance

$$\sigma^2_{(\hat{p}_1 - \hat{p}_2)} = \sigma^2_{\hat{p}_1} + \sigma^2_{\hat{p}_2} = \frac{p_1 q_1}{n_1} + \frac{p_2 q_2}{n_2}$$

(In Section 8-2 we established that the variance of the differences between two independent random variables is the sum of their individual variances.)

3. Because the values of p_1, p_2, q_1, and q_2 are typically unknown and from the null

8-6   Inferences about Two Proportions   **459**

hypothesis we assume that $p_1 = p_2$, we can pool (or combine) the sample data. The pooled estimate of the common value of p_1 and p_2 is

$$\overline{p} = \frac{x_1 + x_2}{n_1 + n_2}$$

If we replace p_1 and p_2 by $\overline{p}$ and replace q_1 and q_2 by $\overline{q} = 1 - \overline{p}$, the variance from Step 2 leads to the following standard deviation.

$$\sigma_{(\hat{p}_1 - \hat{p}_2)} = \sqrt{\frac{\overline{p}\,\overline{q}}{n_1} + \frac{\overline{p}\,\overline{q}}{n_2}}$$

4. We now know that the distribution of $\hat{p}_1 - \hat{p}_2$ is approximately normal, with mean $p_1 - p_2$ and standard deviation as given above. This corresponds to the test statistic

$$z = \frac{\text{(sample statistic)} - \text{(population mean)}}{\text{(standard deviation of sample statistics)}}$$

or

$$z = \frac{(\hat{p}_1 - \hat{p}_2) - (p_1 - p_2)}{\sqrt{\dfrac{\overline{p}\,\overline{q}}{n_1} + \dfrac{\overline{p}\,\overline{q}}{n_2}}}$$

which is the test statistic given earlier.

Remember that to use this test statistic, we must assume that $p_1 = p_2$, $p_1 \le p_2$, or $p_1 \ge p_2$. With this assumption, $p_1 - p_2$ in the test statistic will always be 0. For testing claims that the difference $p_1 - p_2$ is equal to a nonzero constant, a different test statistic is used (see Exercise 17).

Confidence Intervals

In this section we have discussed hypothesis testing, but we sometimes need an estimate of the actual difference between two proportions. Suppose that an advertising specialist uses hypothesis testing and concludes that there is a difference between the proportions of men and women who react favorably to a television commercial. The next logical question might be "How large is the difference?" Is the difference large enough to address with different ad campaigns targeted at the different market segments? We can estimate the size of the difference by constructing a confidence interval.

With the same assumptions given at the beginning of this section, a confidence interval for the difference between population proportions $p_1 - p_2$ can be constructed by evaluating

$$(\hat{p}_1 - \hat{p}_2) - E < (p_1 - p_2) < (\hat{p}_1 - \hat{p}_2) + E$$

where

$$E = z_{\alpha/2}\sqrt{\frac{\hat{p}_1\hat{q}_1}{n_1} + \frac{\hat{p}_2\hat{q}_2}{n_2}}$$

The Lead Margin of Error

Authors Stephen Ansolabehere and Thomas Belin wrote in their article "Poll Faulting" (*Chance* magazine) that "our greatest criticism of the reporting of poll results is with the margin of error of a single proportion (usually ±3%) when media attention is clearly drawn to the *lead* of one candidate." They point out that the lead is really the *difference* between two proportions $(p_1 - p_2)$ and go on to explain how they developed the following rule of thumb: The lead is approximately $\sqrt{3}$ times larger than the margin of error for any one proportion. For a typical pre-election poll, a reported ±3% margin of error translates to about ±5% for the lead of one candidate over the other. They write that the margin of error for the lead should be reported.

First we will give an example illustrating the construction of a confidence interval, then we will justify the format given above.

▶ **EXAMPLE**

Use the sample data given in the preceding example (summarized in the margin) to construct the 95% confidence interval for the difference between the two population proportions.

● **SOLUTION**

With a 95% degree of confidence, $z_{\alpha/2} = 1.96$ (from Table A-2). With $\hat{p}_1 = 58/2000 = 0.0290$ and $\hat{p}_2 = 35/2000 = 0.0175$, we first evaluate the margin of error E.

$$E = z_{\alpha/2}\sqrt{\frac{\hat{p}_1\hat{q}_1}{n_1} + \frac{\hat{p}_2\hat{q}_2}{n_2}} = 1.96\sqrt{\frac{(0.0290)(0.9710)}{2000} + \frac{(0.0175)(0.9825)}{2000}}$$
$$= 0.0093$$

With $\hat{p}_1 = 0.0290$, $\hat{p}_2 = 0.0175$, and $E = 0.0093$, the confidence interval

$$(\hat{p}_1 - \hat{p}_2) - E < (p_1 - p_2) < (\hat{p}_1 - \hat{p}_2) + E$$

becomes

$$(0.0290 - 0.0175) - 0.0093 < (p_1 - p_2)$$
$$< (0.0290 - 0.0175) + 0.0093$$

or
$$0.0022 < (p_1 - p_2) < 0.0208$$

We can conclude, with 95% confidence, that the limits of 0.0022 and 0.0208 actually do contain the difference in proportions. This suggests that the first group has a higher proportion, and it appears that more people stop to give help if they have seen someone else getting help earlier.

As always, we should be careful when interpreting confidence intervals. Because p_1 and p_2 have fixed values and are not variables, it is wrong to state that there is a 95% chance that the value of $p_1 - p_2$ falls between 0.0022 and 0.0208. It is correct to state that if we repeat the same sampling process and construct 95% confidence intervals, in the long run 95% of the intervals will actually contain the value of $p_1 - p_2$.

If we had reversed the order of the samples in the preceding example, the result would have been

$$-0.0208 < (p_1 - p_2) < -0.0022$$

When there does appear to be a difference, be sure that you know which proportion is larger.

Saw
earlier help

$n_1 = 2000$
$x_1 = 58$

Didn't see
earlier help

$n_2 = 2000$
$x_2 = 35$

The form of the confidence interval comes directly from the test statistic if we use the variance

$$\sigma^2_{(\hat{p}_1 - \hat{p}_2)} = \sigma^2_{\hat{p}_1} + \sigma^2_{\hat{p}_2} \left(= \frac{p_1 q_1}{n_1} + \frac{p_2 q_2}{n_2} \right)$$

to estimate the standard deviation as

$$\sqrt{\frac{\hat{p}_1 \hat{q}_1}{n_1} + \frac{\hat{p}_2 \hat{q}_2}{n_2}}$$

(We don't use pooled estimates of proportions because we're not assuming from a null hypothesis that $p_1 = p_2$.) In the test statistic

$$z = \frac{(\hat{p}_1 - \hat{p}_2) - (p_1 - p_2)}{\sqrt{\frac{\hat{p}_1 \hat{q}_1}{n_1} + \frac{\hat{p}_2 \hat{q}_2}{n_2}}}$$

let z be positive and negative (for two tails) and solve for $p_1 - p_2$. The results are the limits of the confidence interval given earlier.

Using Computer Software for Confidence Intervals

Users of STATDISK can get confidence interval limits for the problems in this section. From the main menu bar item `Statistics`, select `Hypothesis Testing` and proceed as if you were testing the claim $p_1 = p_2$. If you choose that null hypothesis, the confidence interval limits will be included in the output.

8-6   Exercises A: Basic Skills and Concepts

In Exercises 1 and 2, assume that you want to test the claim that $p_1 = p_2$ at the $\alpha = 0.05$ significance level. Use the given sample data to find (a) the pooled estimate $\bar{p}$, (b) the z test statistic, (c) the critical z values, and (d) the P-value.

1. Sample 1   Sample 2

   $n_1 = 100$ $n_2 = 200$
   $x_1 = 45$ $x_2 = 115$

2. Sample 1   Sample 2

   $n_1 = 250$ $n_2 = 800$
   $x_1 = 30$ $x_2 = 44$

3. A public relations expert and consultant for the airline industry is planning a strategy to influence voter perception of government regulation of airfares. In a *New York Times*/CBS News survey, it is found that 35% of 552 Democrats believe that the government should regulate airline prices, compared to 41% of the 417 Republicans surveyed. At the 0.05 significance level, test the claim that there is no difference between the proportions of Democrats and Republicans who believe in government regulation of airfares. Based on the result, should the consultant consider different approaches for Democrats and Republicans?

4. Use the sample data from Exercise 3 to construct a 95% confidence interval for the difference between the proportions of Democrats and Republicans who believe that the government should regulate airline prices.

5. A study was made of 413 children who were hospitalized as a result of motor vehicle crashes. Among 290 children who were not using seat belts, 50 were injured severely. Among 123 children using seat belts, 16 were injured severely (based on data from "Morbidity Among Pediatric Motor Vehicle Crash Victims: The Effectiveness of Seat Belts," by Osberg and Di Scala, *American Journal of Public Health*, Vol. 82, No. 3). Is there sufficient sample evidence to conclude, at the 0.05 level, that the rate of severe injuries is lower for children wearing seat belts?

6. In initial tests of the Salk vaccine, 33 of 200,000 vaccinated children later developed polio. Of 200,000 children vaccinated with a placebo, 115 later developed polio. At the 0.01 level of significance, test the claim that the Salk vaccine is effective in lowering the polio rate. Does it appear that the vaccine is effective?

7. As part of a campaign for the presidency, one candidate plans to wage a voter registration drive. In considering whether to target specific age groups, this candidate finds survey results showing that among 200 randomly selected persons aged 18–24, 36.0% voted. Among 250 persons in the 25–44 age bracket, 54.0% voted (based on data from the U.S. Bureau of the Census). Construct a 95% confidence interval for the difference between the proportions of voters in the two age brackets. Are the percentages of voters in the two groups different enough that the groups should be targeted differently?

8. Using the sample data given in Exercise 7, test the claim that the proportions of voters in the two age brackets are the same. Use a 0.05 significance level.

9. The *New York Times* ran an article about a study in which Professor Denise Korniewicz and other Johns Hopkins researchers subjected laboratory gloves to stress. Among 240 vinyl gloves, 63% leaked viruses. Among 240 latex gloves, 7% leaked viruses. At the 0.005 significance level, test the claim that vinyl gloves have a larger virus leak rate than latex gloves. (It might seem obvious that 63% is larger than 7%, but we must justify our conclusion by considering the sample sizes, the significance level, and the nature of the distribution.)

10. Using the sample data from Exercise 9, construct the 99% confidence interval for the difference between the two proportions of gloves that leak viruses. Does that difference appear to warrant a decision to use latex gloves, even if they are more expensive?

11. Professional pollsters are becoming concerned about the growing rate of refusals among potential survey subjects. In analyzing the problem, there is a need to know if the refusal rate is universal or if there is a difference between the rates for central city residents and those not living in central cities. Specifically, it was found that when 294 central city residents were surveyed, 28.9% refused to respond. A survey of 1015 residents not in a central city resulted in a 17.1% refusal rate (based on data from "I Hear You Knocking But You Can't Come In," by Fitzgerald and Fuller, *Sociological Methods and*

Research, Vol. 11, No. 1). At the 0.01 significance level, test the claim that the central city refusal rate is the same as the refusal rate in other areas.

12. The newly appointed head of the state mental health agency claims that a greater proportion of the crimes committed by persons younger than 21 years of age are violent crimes (when compared to the crimes committed by persons 21 years of age or older). Of 2750 randomly selected arrests of criminals younger than 21 years of age, 4.25% involve violent crimes. Of 2200 randomly selected arrests of criminals 21 years of age or older, 4.55% involve violent crimes (based on data from the Uniform Crime Reports). Construct a 95% confidence interval for the difference between the two proportions of violent crimes. Do the confidence interval limits contain 0, indicating that there isn't a significant difference between the two rates of violent crimes?

13. Tom Knight is a high school senior who plans to attend college, but he is concerned about dropout rates among first-year college students. The American College Testing Program provides data showing that 30% of freshmen at four-year public colleges drop out, whereas the dropout rate at four-year private colleges is 26%. Assume that these results are based on observations of 1000 four-year public college freshmen and 500 four-year private college freshmen. At the 0.05 significance level, test the claim that four-year public and private colleges have the same freshman dropout rate.

14. Use the data from Exercise 13 to construct a 95% confidence interval for the difference between the dropout rates for four-year public and private colleges. Based on the result, does the difference appear to be a factor that might affect someone's choice between a four-year public and private college?

15. Refer to Data Set 3 in Appendix B for the survey results of employed full-time workers. Test the claim that the proportion of males with at least some college education is equal to the proportion of females with at least some college education. Use a 0.05 level of significance.

8-6   Exercises B: Beyond the Basics

16. Refer to the movie ratings in Data Set 5 of Appendix B. At the 0.05 level of significance, we want to test the claim that the proportion of R-rated movies with more than two stars is equal to the proportion of G/PG/PG-13 movies with more than two stars. Can we use the methods of this section to test that claim? Why or why not?

17. To test the null hypothesis that the difference between two population proportions is equal to a nonzero constant c, use

$$z = \frac{(\hat{p}_1 - \hat{p}_2) - c}{\sqrt{\dfrac{\hat{p}_1(1 - \hat{p}_1)}{n_1} + \dfrac{\hat{p}_2(1 - \hat{p}_2)}{n_2}}}$$

As long as n_1 and n_2 are both large, the sampling distribution of the above test statistic z will be approximately the standard normal distribution. Suppose a winery is conducting market research on a new wine in New York and California. A sample of 500 New Yorkers reveals that 120 like the wine, and a sample of 500 Californians shows that 210 like the wine. Use a 0.05 level of significance to test the claim that the percentage of Californians who like the wine is 25% more than the percentage of New Yorkers who like it.

18. Sample data are randomly drawn from three independent populations. The sample sizes and the numbers of successes follow:

Population 1	Population 2	Population 3
$n = 100$	$n = 100$	$n = 100$
$x = 40$	$x = 30$	$x = 20$

   a. At the 0.05 significance level, test the claim that $p_1 = p_2$.
   b. At the 0.05 significance level, test the claim that $p_2 = p_3$.
   c. At the 0.05 significance level, test the claim that $p_1 = p_3$.
   d. In general, if hypothesis tests lead to the conclusions that $p_1 = p_2$ and $p_2 = p_3$ are reasonable, does it follow that $p_1 = p_3$ is also reasonable?

19. The *sample size* needed to estimate the difference between two population proportions to within E with a confidence level of $1 - \alpha$ can be found as follows. In the expression

$$E = z_{\alpha/2} \sqrt{\frac{p_1 q_1}{n_1} + \frac{p_2 q_2}{n_2}}$$

replace n_1 and n_2 by n (assuming that both samples have the same size) and replace each of $p_1, q_1, p_2,$ and q_2 by 0.5 (since their values are not known). Then solve for n.

   Use this approach to find the size of each sample if you want to estimate the difference between the proportions of men and women who own cars. Assume that you want 95% confidence that your error is no more than 0.03.

20. Refer to Exercise 9 and assume that you want to test the given claim by constructing an appropriate confidence interval. Find that confidence interval, identify its degree of confidence, and state the conclusion.

VOCABULARY LIST

Define and give an example of each term.

independent samples	F distribution
dependent samples	numerator degrees of freedom
paired samples	denominator degrees of freedom
matched samples	pooled estimate of p_1 and p_2

REVIEW

Chapters 6 and 7 introduced two major concepts of inferential statistics: the estimation of population parameters and the methods of testing hypotheses made about population parameters. The examples and exercises in Chapters 6 and 7 were restricted to cases involving a single sample drawn from a single population. This chapter extended our coverage to include two samples drawn from two populations.

In Section 8-2 we used two dependent (paired) samples to make inferences about two population means. In Section 8-3 we considered independent samples, with both samples having (large) sizes of 31 or greater. Section 8-4 presented a method for testing a claim about the variances (or standard deviations) of two populations. The test for equality of variances was also used in Section 8-5, which described methods for dealing with two independent and small samples. We saw that when testing claims about population means with small independent samples, we must conduct a preliminary F test of equal variances to determine which of two cases should be used. (Another case in which σ_1^2 and σ_2^2 are both known is unrealistic.) Finally, Section 8-6 presented a method for testing claims made about two population proportions.

Sections 8-2, 8-3, and 8-5 were devoted to testing claims about two population means. There are several different sets of circumstances requiring several different approaches, so it is easy to experience some confusion. Figures 8-6 and 8-7 summarize those cases in flowcharts that are helpful in determining the correct form for the hypothesis test or confidence interval.

Important formulas for this chapter are listed on pages 466 and 467.

REVIEW EXERCISES

1. Orange County and Washington County use two different procedures for selecting jurors. The mean waiting time for the 40 randomly selected subjects from Orange County is 183.0 min with a standard deviation of 21.0 min. The mean waiting time for the 50 randomly selected subjects from Washington County is 253.1 min with a standard deviation of 29.2 min.
   a. At the 0.05 significance level, test the claim that the mean waiting time of prospective jurors is the same for both counties.
   b. Construct a 95% confidence interval estimate of the difference between the two mean waiting times.
2. The Delaney and Taylor advertising company has contracted to promote country music for a local radio station. It studies the proportions of radio listeners who prefer country music and finds that in market region A, 38% of the 250 listeners surveyed indicated a preference for country music. In market region B, country music was preferred by 14% of the 400 listeners surveyed.
   a. Construct a 98% confidence interval for the difference between the proportions of listeners who prefer country music.
   b. Using a 0.02 significance level, test the claim that region A has a greater proportion of listeners who prefer country music.

Important Formulas

Parameters	Applicable Distribution	Testing Hypotheses (Test statistic)
μ_1, μ_2 Dependent samples	Student t	$t = \dfrac{\overline{d} - \mu_d}{\dfrac{s_d}{\sqrt{n}}}$
μ_1, μ_2 Large independent samples	Normal	$z = \dfrac{(\overline{x}_1 - x_2) - (\mu_1 - \mu_2)}{\sqrt{\dfrac{\sigma_1^2}{n_1} + \dfrac{\sigma_2^2}{n_2}}}$
μ_1, μ_2 Equal variances	Student t	$t = \dfrac{(\overline{x}_1 - \overline{x}_2) - (\mu_1 - \mu_2)}{\sqrt{\dfrac{s_p^2}{n_1} + \dfrac{s_p^2}{n_2}}}$ where $s_p^2 = \dfrac{(n_1 - 1)s_1^2 + (n_2 - 1)s_2^2}{n_1 + n_2 - 2}$
μ_1, μ_2 Unequal variances	Student t	$t = \dfrac{(\overline{x}_1 - \overline{x}_2) - (\mu_1 - \mu_2)}{\sqrt{\dfrac{s_1^2}{n_1} + \dfrac{s_2^2}{n_2}}}$
σ_1, σ_2 Two standard deviations or σ_1^2, σ_2^2 Two variances	F	$F = \dfrac{s_1^2}{s_2^2}$ where $s_1^2 \geq s_2^2$
p_1, p_2 Two proportions	Normal	$z = \dfrac{(\hat{p}_1 - \hat{p}_2) - (p_1 - p_2)}{\sqrt{\dfrac{\overline{p}\,\overline{q}}{n_1} + \dfrac{\overline{p}\,\overline{q}}{n_2}}}$

Confidence Intervals	Table of Critical Values
$\bar{d} - E < \mu_d < \bar{d} + E$ where $E = t_{\alpha/2}\dfrac{s_d}{\sqrt{n}}$	Table A-3
$(\bar{x}_1 - \bar{x}_2) - E < (\mu_1 - \mu_2) < (\bar{x}_1 - \bar{x}_2) + E$ where $E = z_{\alpha/2}\sqrt{\dfrac{\sigma_1^2}{n_1} + \dfrac{\sigma_2^2}{n_2}}$	Table A-2
$(\bar{x}_1 - \bar{x}_2) - E < (\mu_1 - \mu_2) < (\bar{x}_1 - \bar{x}_2) + E$ where $E = t_{\alpha/2}\sqrt{\dfrac{s_p^2}{n_1} + \dfrac{s_p^2}{n_2}}$	Table A-3
$(\bar{x}_1 - \bar{x}_2) - E < (\mu_1 - \mu_2) < (\bar{x}_1 - \bar{x}_2) + E$ where $E = t_{\alpha/2}\sqrt{\dfrac{s_1^2}{n_1} + \dfrac{s_2^2}{n_2}}$	Table A-3
See Exercises 19 and 20 in Section 8-4.	Table A-5
$(\hat{p}_1 - \hat{p}_2) - E < (p_1 - p_2) < (\hat{p}_1 - \hat{p}_2) + E$ where $E = z_{\alpha/2}\sqrt{\dfrac{\hat{p}_1\hat{q}_1}{n_1} + \dfrac{\hat{p}_2\hat{q}_2}{n_2}}$	Table A-2

3. The Minton Motor Products Company produces a new type of spark plug and mails advertisements stating that the spark plug improves fuel consumption. The U.S. Postal Service charges mail fraud on the basis that these spark plugs are no better than ordinary spark plugs. Sample data are collected by testing the new plugs and ordinary plugs in a random sample of cars. The results are given below as distances (in miles) that the cars travel on 1 gallon of gas. At the 0.05 significance level, test Minton's claim that the new plugs have a higher mean. Based on the results, does the U.S. Postal Service have a case?

Car	A	B	C	D	E	F	G	H	I
Minton plugs	18.2	23.4	19.7	14.7	28.7	23.4	19.0	21.2	18.2
Ordinary plugs	16.1	21.3	19.2	14.8	29.3	20.2	18.6	19.7	16.4

4. Twelve different tablets from each of two competing cold medicines are randomly selected and tested for the amount of acetaminophen each contains, and the results (in milligrams) follow.

Dozenol	472	487	506	512	489	503	511	501	495	504	494	462
Niteze	562	512	523	528	554	513	516	510	524	510	524	508

   a. At the 0.05 significance level, test the claim that the mean amount of acetaminophen is the same in each brand.

   b. Construct a 95% confidence interval estimate of the difference between the mean for Dozenol and the mean for Niteze.

5. The New York State Department of Motor Vehicles provided the following motor vehicle conviction data for a recent year.

	Albany County	Queens County
DWI convictions	558	1,214
Total convictions	24,384	166,197

   At the 0.01 level of significance, test the claim that the proportion of DWI (driving while intoxicated) convictions is lower in Queens County. Assume that the given data represent random samples drawn from a larger population.

6. A study was conducted to investigate relationships between different types of standard test scores. On the Graduate Record Examination verbal test, 68 women had a mean of 538.82 and a standard deviation of 114.16, and 86 men had a mean of 525.23 and a standard deviation of 97.23. (See "Equivalencing MAT and GRE Scores Using Simple Linear Transformation and Regression Methods," by Kagan and Stock, *Journal of Experimental Education*, Vol. 49, No. 1.) At the 0.02 significance level, test the claim that the two groups come from populations with the same standard deviation.

7. Using the sample data given in Exercise 6, test the claim that the mean score for women is equal to the mean score for men. Use a 0.02 significance level.

8. Tests are being conducted on two separate and independent computer systems used for air traffic control. The numbers of operations for 30 different randomly selected hours are listed below.

System 1	63 62 67 66 53 72 62 57 49 57 60 68 58 64 61
System 2	62 63 67 61 68 64 66 62 67 66 62 62 65 65 66

a. At the 0.05 significance level, test the claim that the use of system 2 results in a mean number of operations per hour that exceeds the mean for system 1.

b. Construct a 95% confidence interval for the difference between the means of the two systems.

9. Air America is experimenting with its training program for flight attendants. With the traditional 6-week program, a random sample of 60 flight attendants achieve competency test scores with a mean of 83.5 and a standard deviation of 16.3. With a new 10-day program, a random sample of 35 flight attendants achieve competency test scores with a mean of 79.8 and a standard deviation of 19.2. At the 0.01 significance level, test the claim that the 10-day program results in scores with a lower mean.

10. In a study of techniques used to measure lung volumes, physiological data were collected for 10 subjects. The values given in the table below are in liters, representing the measured forced vital capacities of the 10 subjects in a sitting position and in a supine (lying) position. At the 0.05 significance level, test the claim that both positions have the same mean.

Subject	A	B	C	D	E	F	G	H	I	J
Sitting	4.66	5.70	5.37	3.34	3.77	7.43	4.15	6.21	5.90	5.77
Supine	4.63	6.34	5.72	3.23	3.60	6.96	3.66	5.81	5.61	5.33

Based on data from "Validation of Esophageal Balloon Technique at Different Lung Volumes and Postures," by Baydur, Cha, and Sassoon, *Journal of Applied Physiology*, Vol. 62, No. 1.

11. A test question is considered good if it discriminates between prepared and unprepared students. The first question on a test was answered correctly by 62 of 80 prepared students and by 23 of 50 unprepared students. At the 0.05 level of significance, test the claim that this question was answered correctly by a greater proportion of prepared students.

12. The Dayton Machine Company and the Pit Metal Manufacturing Company both manufacture garage door springs that are designed to produce a tension of

68 kg. Random samples are selected from each of these two suppliers, and tension test results are as follows:

	Dayton Machine	Pit Metal
	$n = 20$	$n = 32$
	$\bar{x} = 66.0$ kg	$\bar{x} = 68.3$ kg
	$s = 2.1$ kg	$s = 0.4$ kg

a. At the 0.05 level of significance, test the claim that both firms produce the same standard deviation.

b. Test the claim that both firms' springs produce the same mean tension. Use a 0.05 level of significance.

c. Construct a 95% confidence interval for the difference between the mean for Dayton Machine and the mean for Pit Metal.

13. To test the effectiveness of a lesson, a teacher gives randomly selected students a pretest and a follow-up test. The results are given below.

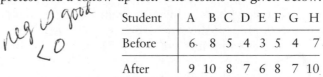

Student	A	B	C	D	E	F	G	H
Before	6	8	5	4	3	5	4	7
After	9	10	8	7	6	8	7	10

a. At the 0.025 level of significance, test the claim that the lesson was effective.

b. Construct a 95% confidence interval for the mean of the differences between the before and after scores.

Dover	Baltimore
$n = 40$	$n = 60$
$\bar{x} = 43.7$	$\bar{x} = 48.2$
$s = 16.2$	$s = 16.5$

14. The Bank of Delmarva has branches in Dover and Baltimore. The bank uses a standard credit-rating system for all loan applicants. Randomly selected applicants are chosen from each branch, and the results are summarized in the margin.

a. At the 0.05 level of significance, test the claim that both populations have the same mean.

b. Construct a 95% confidence interval for the difference between the mean for Dover and the mean for Baltimore.

Winston	Barrington
$n = 18$	$n = 24$
$\bar{x} = 85.7$	$\bar{x} = 80.6$
$s = 5.8$	$s = 6.1$

15. Researchers are testing commercial air filtering systems made by the Winston Industrial Supply Company and the Barrington Filter Company. Random samples are tested for each company, and the filtering efficiency is scored on a standard scale, with the results shown at left. (Higher scores correspond to better filtering.) At the 0.05 level of significance, test the claim that both systems have the same mean.

16. In a survey, 500 men and 500 women were randomly selected. Among the men, 52 had been ticketed for speeding within the last year, whereas 27 of the women had been ticketed for speeding during the same period (based on data from R. H. Bruskin Associates).

a. At the 0.01 significance level, test the claim that women have a lower rate of speeding tickets.

b. Construct a 99% confidence interval for the difference between the two proportions.

COMPUTER PROJECT

Among the data sets listed in Appendix B, Data Set 2 (consisting of body temperatures obtained by University of Maryland researchers) is the only one not included on the disk that is available with this book. Using the 8:00 AM times for Day 1, test for equality between the mean body temperature of men and the mean body temperature of women. Use STATDISK or Minitab. Obtain a printed copy of the computer displays, and write a summary that interprets the results.

FROM DATA TO DECISION: Does aspirin help prevent heart attacks?

In a recent study of 22,000 male physicians, half were given regular doses of aspirin while the other half were given placebos. The study ran for six years at a cost of $4.4 million. Among those who took the aspirin, 104 suffered heart attacks. Among those who took the placebos, 189 suffered heart attacks. (The figures are based on data from *Time* and the *New England Journal of Medicine*, Vol. 318, No. 4.) Do these results show a statistically significant decrease in heart attacks among the sample group who took

aspirin? The issue is clearly quite important because it can affect many lives.

Use the methods of this chapter to determine whether the use of aspirin seems to help prevent heart attacks. Write a report that summarizes your findings. Include any relevant factors you can think of that might affect the validity of the study. For example, is it noteworthy that the study involved only male physicians? Is it noteworthy that aspirin sometimes causes stomach problems?

9 Correlation and Regression

9-1 Overview

Objectives are identified for this chapter, which presents methods for analyzing the relationship between two variables.

9-2 Correlation

Scatter diagrams are presented to graphically depict relationships between two variables. The linear correlation coefficient is used to determine whether a linear relationship exists between two variables.

9-3 Regression

Linear relationships between two variables are described using the equation and graph of the regression line. The determination of the equation for the regression line is based on the least-squares property, which is described in this section. A method for determining predicted values of a variable is also presented.

9-4 Variation and Prediction Intervals

Building on Section 9-3, which illustrates how to determine predicted values of a variable, this section analyzes the differences between predicted values and actual observed values. In addition, prediction intervals, which are confidence interval estimates of predicted values, are constructed.

9-5 Multiple Regression

Methods are presented for finding a linear equation that relates three or more variables. Because of the calculations involved, this section emphasizes the use of computer software and the interpretation of computer displays. The multiple coefficient of determination is also discussed.

Chapter Problem:
Can we predict population size by analyzing discarded household garbage?

Researchers involved in the Garbage Project at the University of Arizona use archaeological methods to analyze and study garbage. They have obtained some interesting and important results, including the realization that surprisingly little of our garbage biodegrades in landfills. By studying garbage, archaeologists can determine what pets people have, how many babies they have, how much they drink, and the level of their economic status. Authors William Rathje and Cullen Murphy write in *Rubbish! The Archeology of Garbage* that "demographic data derived unobtrusively from garbage have a variety of real-world uses—in marketing and consumer research; in the rational governance of communities; in any endeavor that demands detailed knowledge of the behavior (including the relatively private behavior) of large groups of people."

The data in Table 9-1 are taken from Data Set 1 in Appendix B. (For the purposes of this chapter, it will be easier to work with Table 9-1, but better results can be obtained form the more complete data set.) Based on the data in Table 9-1, we can formulate important questions such as these: Is there a relationship between household size and the weight of discarded plastic? If so, what is it? Using the additional data of Data Set 1 in Appendix B, we might question whether there is a relationship between household size and the weight of discarded paper and, if so, try to find that relationship. Such questions are important to the Census Bureau because its population counts as the basis for distributing billions of dollars to individual states. In fact, the Census Bureau funded garbage research in the hope that the date derived could be used to verify its survey data in specific neighborhoods. We will address the issue of whether garbage data can be used to accurately determine the size of a population. This issue represents a fascinating and powerful application of statistics—the use of data from certain variables (such as plastic and paper) to determine the value of some other related variable (such as population size).

Table 9-1 Data from the Garbage Project

x: Plastic (lb)	0.27	1.41	2.19	2.83	2.19	1.81	0.85	3.05
y: Household size	2	3	3	6	4	2	1	5

Data provided by Masakazu Tani and the Garbage Project at the University of Arizona.

9-1   Overview

Chapters 6 and 7 introduced the important concepts of estimating parameters and testing hypotheses, as applied to a sample from a single population. In Chapter 8 we again considered those concepts, but dealt with two samples selected from two populations. This chapter also involves estimating parameters and testing hypotheses, but the methods we will use are very different because of the very different issue we will be considering: Given paired data, we want to investigate the *relationship* between the two variables. Specifically, we want to determine whether there is a relationship between the two variables and, if so, identify what the relationship is. For example, using the data in Table 9-1, we want to determine whether there is a relationship between plastic and household size. If such a relationship exists, we want to identify it with an equation. With the plastic/household size data, the goal is to develop a process for predicting household size based on the weight of plastic discarded.

We considered paired data in Section 8-2, but the objective there was very different: We wanted to estimate μ_d, the *mean difference* between two variables, or test a claim about the two means; in this chapter we want to determine whether there is a *relationship* between the two variables and describe it when it exists.

In this chapter we investigate ways of analyzing the relationship between two variables. Sample data will be paired, as in Table 9-1. We begin Section 9-2 by considering the concept of correlation, which is used to decide whether there is a statistically significant relationship between two variables. We describe the scatter diagram, which serves as a graph of the sample data. We also describe the linear correlation coefficient, which is a measure of the strength of association between two variables. Section 9-3 investigates regression analysis, as we attempt to identify the exact nature of the relationship between two variables. Specifically, we show how to determine an equation that relates the two variables and then proceed to use that equation to predict values of a variable. In Section 9-4 we analyze the differences between predicted values and actual observed values of a variable. Sections 9-2 through 9-4 deal with relationships between *two* variables, but Section 9-5 uses concepts of multiple regression to describe the relationship among *three or more* variables. Note that throughout this text we deal only with *linear* (or straight-line) relationships between two or more variables.

9-2   Correlation

This section introduces the concept of correlation and presents methods for determining whether there is a correlation between two variables. Table 9-1 contains paired data consisting of weights of discarded plastic matched with household sizes. (Such paired data are sometimes referred to as **bivariate data.**)

Is there a relationship between those two variables? If so, we can proceed to determine a way to predict household size based on the amount of plastic discarded. Discarded plastic can be weighed at resource recovery centers and the weights used to establish population size—a statistic of vital interest to organizations such as the Census Bureau. When we question the presence or absence of a relationship between two variables, we are really trying to determine whether there is a correlation between them. For example, there is a known correlation between smoking and health problems, but there is no correlation between IQ scores and hat sizes.

DEFINITION

A **correlation** exists between two variables when one of them is related to the other in some way.

As we work with sample data and develop methods of forming inferences about populations, we make the following assumptions.

Assumptions

1. The sample of paired (x, y) data is a *random* sample.
2. The pairs of (x, y) data have a **bivariate normal distribution**. The key feature of such a distribution is that for any fixed value of x, the values of y have a normal distribution, and for any fixed value of y, the values of x have a normal distribution.

The second assumption is usually difficult to check, but a partial check can be made by determining whether the values of both x and y are normally distributed. For example, if one of the variables consists of the ranks 1, 2, 3, . . . , 10, then the second assumption is clearly violated and you shouldn't use the techniques discussed in this section. Instead, you should consider *rank correlation*, which will be discussed in Section 13-6. In general, if either variable has a distribution that is very nonnormal, then you know that the second assumption is not satisfied.

We can often form intuitive and qualitative conclusions about paired data by constructing a graph similar to the one shown in the accompanying Minitab display, which represents the data in Table 9-1. (To obtain this display, enter the x values in column C1, enter the corresponding y values in column C2, and then enter the Minitab command PLOT C2 VS C1.) The Minitab display is an example of a **scatter diagram**, which is a plot of paired (x, y) data with a horizontal x axis and a vertical y axis. The points in the figure seem to follow a pattern, so we might conclude that there is a relationship between discarded plastic and household size. This conclusion is largely subjective because it is based on our perception of whether a pattern is present.

Manatees Saved

Manatees are large mammals that like to float just below the water's surface, where they are in danger from powerboat propellers. A Florida study of the number of powerboat registrations and the numbers of accidental manatee deaths confirmed that there was a significant positive correlation. As a result, Florida created coastal sanctuaries where powerboats are prohibited so that manatees could thrive. (See Program 1 from the series *Against All Odds: Inside Statistics* for a discussion of this case.) This is one of many examples of the beneficial use of statistics.

MINITAB DISPLAY

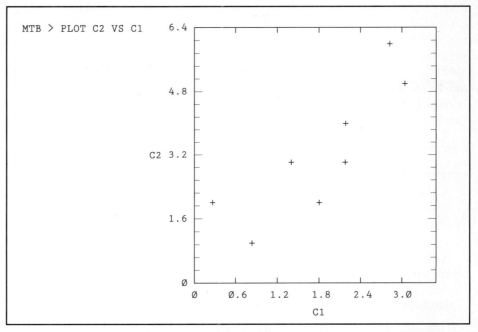

Other examples of scatter diagrams are shown in Figure 9-1. The graphs in Figure 9-1(a), (b), and (c) depict a pattern of increasing values of y that correspond to increasing values of x. As you proceed from (a) to (c), the pattern becomes closer to a straight line, suggesting that the relationship between x and y becomes stronger. The scatter diagrams in (d), (e), and (f) depict patterns in which the y values decrease as the x values increase. Again, as you proceed from (d) to (f), the relationship becomes stronger. In contrast to the first six graphs, the scatter diagram of (g) shows no pattern and suggests that there is no correlation (or relationship) between x and y. Finally, the scatter diagram of (h) shows a pattern, but it is not a straight-line pattern. In general, scatter diagrams often reveal patterns and are easy to plot, but conclusions drawn from them tend to be subjective. More precise and objective methods use the linear correlation coefficient, which is useful for detecting straight-line patterns, but not nonlinear patterns such as the one in Figure 9-1(h).

DEFINITION

The **linear correlation coefficient** r measures the strength of the linear relationship between the paired x and y values in a *sample*. Its value is computed by using Formula 9-1, which follows. [The linear correlation coefficient is sometimes referred to as the **Pearson product moment correlation coefficient** in honor of Karl Pearson (1857–1936), who originally developed it.]

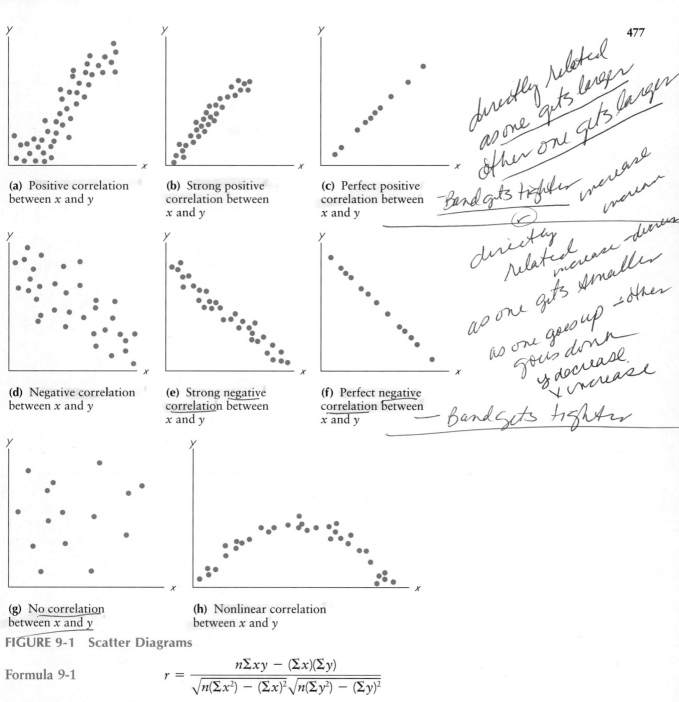

(a) Positive correlation between x and y

(b) Strong positive correlation between x and y

(c) Perfect positive correlation between x and y

(d) Negative correlation between x and y

(e) Strong negative correlation between x and y

(f) Perfect negative correlation between x and y

(g) No correlation between x and y

(h) Nonlinear correlation between x and y

FIGURE 9-1   Scatter Diagrams

Formula 9-1
$$r = \frac{n\Sigma xy - (\Sigma x)(\Sigma y)}{\sqrt{n(\Sigma x^2) - (\Sigma x)^2}\sqrt{n(\Sigma y^2) - (\Sigma y)^2}}$$

Because r is calculated using sample data, it is a sample statistic. We might think of r as a point estimate of the population parameter ρ (rho), which is the linear correlation coefficient for all pairs of data in a population.

We will describe how to compute and interpret the linear correlation coefficient r given a list of paired data, but we will first identify the notation relevant to Formula 9-1. Later in this section we will present the underlying theory that led to the development of Formula 9-1.

Notation for the Linear Correlation Coefficient

n	=	*number of pairs* of data present
Σ		denotes the addition of the items indicated
Σx		denotes the sum of all x scores.
Σx^2		indicates that each x score should be squared and then those squares added.
$(\Sigma x)^2$		indicates that the x scores should be added and the total then squared. It is extremely important to avoid confusing Σx^2 and $(\Sigma x)^2$.
Σxy		indicates that each x score should be first multiplied by its corresponding y score. After obtaining all such products, find their sum.
r	=	linear correlation coefficient for a *sample*
ρ	=	linear correlation coefficient for a *population*

Rounding the Linear Correlation Coefficient *r*

Round the linear correlation coefficient r to three decimal places. When you are calculating r and other statistics in this chapter, rounding errors can sometimes wreak havoc with your results. Try using your calculator's memory to store intermediate results, and round off only at the end. Don't round off intermediate results. Many inexpensive calculators have Formula 9-1 built in so that you can automatically evaluate r after entering the sample data.

▶ **EXAMPLE**

Using the data in Table 9-1, find the value of the linear correlation coefficient r. (A later example will use this value to determine whether there is a relationship between discarded plastic and household size.)

● **SOLUTION**

For the sample paired data in Table 9-1, we get $n = 8$ because there are 8 pairs of data. The other components required in Formula 9-1 are found from the calculations in Table 9-2. Note how this vertical format makes the calculations easier.

TABLE 9-2 Finding Statistics Used to Calculate *r*

Weight of Plastic (lb): x	Household Size: y	x · y	x²	y²
0.27	2	0.54	0.0729	4
1.41	3	4.23	1.9881	9
2.19	3	6.57	4.7961	9
2.83	6	16.98	8.0089	36
2.19	4	8.76	4.7961	16
1.81	2	3.62	3.2761	4
0.85	1	0.85	0.7225	1
3.05	5	15.25	9.3025	25
Total 14.60	26	56.80	32.9632	104
↑	↑	↑	↑	↑
Σx	Σy	Σxy	Σx^2	Σy^2

Using the calculated values, we can now evaluate *r* as follows:

$$r = \frac{n(\Sigma xy) - (\Sigma x)(\Sigma y)}{\sqrt{n(\Sigma x^2) - (\Sigma x)^2}\sqrt{n(\Sigma y^2) - (\Sigma y)^2}}$$

$$= \frac{8(56.80) - (14.60)(26)}{\sqrt{8(32.9632) - (14.60)^2}\sqrt{8(104) - (26)^2}}$$

$$= \frac{74.8}{\sqrt{50.5456}\sqrt{156}}$$

$$= 0.842$$

After calculating *r*, how do we interpret the result? Given the way in which Formula 9-1 was derived, it can be shown that the computed value of *r* must always fall between −1 and +1 inclusive. If *r* is close to 0, we conclude that there is no significant linear correlation between *x* and *y*. If *r* is close to −1 or +1, we conclude that there is significant linear correlation between *x* and *y*. To objectively determine whether there is a significant linear correlation, we use a formal hypothesis test that uses the same basic procedure first described in Chapter 7. We will consider the formal hypothesis test after briefly discussing computer software and properties of *r*.

Using Computer Software for the Linear Correlation Coefficient

The linear correlation coefficient is so common that it is included as a function on many calculators, as well as being a feature of most statistical software packages. STATDISK and Minitab both accept paired data as input and provide the linear correlation coefficient as output. With STATDISK, simply select `Correlation and Regression` from the `Statistics` menu, and then enter the data in pairs. In addition to the linear correlation coefficient, a scatter diagram will also be displayed. With Minitab, enter the paired data in columns C1 and C2, enter the command `CORRELATION C1 AND C2` to get the linear correlation coefficient, and then enter `PLOT C2 VS C1` to get a scatter diagram.

The format of Formula 9-1 leads to the following properties of the linear correlation coefficient r.

Properties of the Linear Correlation Coefficient r

1. *The value of r is always between -1 and 1.* That is,

$$-1 \le r \le 1$$

2. *The value of r does not change if all values of either variable are converted to a different scale.* For example, if the units of x are so large that they cause calculator errors, you can divide them all by a constant number such as 1000.

3. *The value of r is not affected by the choice of x or y.* Interchange all x and y values and the value of r will not change.

4. *r measures the <u>strength</u> of a linear relationship.* It is not designed to measure the strength of a relationship that is not linear.

Hypothesis Test of the Significance of r

We present two procedures for using a formal hypothesis test to determine whether there is a significant linear correlation between two variables. These methods are summarized in Figure 9-2. Some instructors prefer Method 1 because it reinforces concepts introduced in earlier chapters. Others prefer Method 2 because it eliminates the step of calculating the test statistic in terms of t.

Figure 9-2 shows that for the null and alternative hypotheses, we use the following:

$$H_0: \rho = 0 \qquad H_1: \rho \ne 0$$

For the test statistic, we use one of the following methods.

Method 1: Test Statistic Is t   This method follows the format presented in earlier chapters. It uses the Student t distribution with a test statistic having the

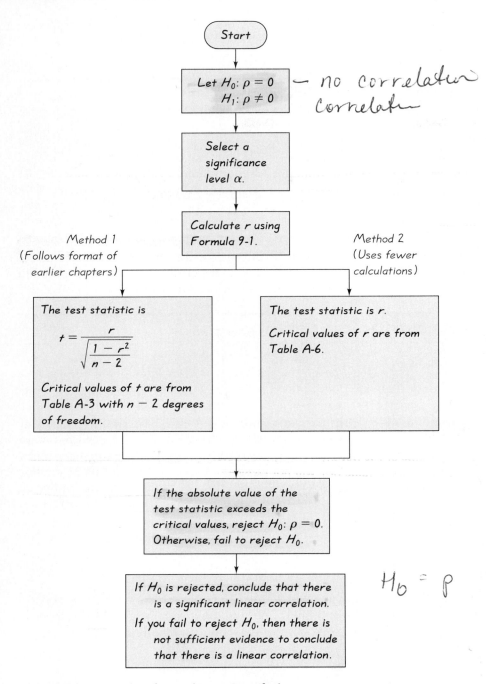

no correlation
correlation

$H_0 = \rho$

FIGURE 9-2   Testing for a Linear Correlation

form $t = (r - \mu_r)/s_r$, where μ_r and s_r denote the claimed value of the mean and the sample standard deviation of r values. Because we assume that $\rho = 0$, it follows that $\mu_r = 0$. Also, it can be shown that s_r, the standard deviation of linear correlation coefficients, can be expressed as $\sqrt{(1 - r^2)/(n - 2)}$. We can therefore use the following test statistic.

Test Statistic *t* for Linear Correlation

$$t = \frac{r}{\sqrt{\dfrac{1 - r^2}{n - 2}}}$$

Critical values: Use Table A-3 with degrees of freedom = $n - 2$.

Method 2: Test Statistic Is *r*   This method requires fewer calculations. Instead of calculating the test statistic given above, we use the computed value of *r* as the test statistic. If we use *r* as the test statistic, the critical values can be found in Table A-6, which lists critical values for various sample sizes *n*.

Test Statistic *r* for Linear Correlation

Test statistic: *r*
Critical values: Refer to Table A-6.

Figure 9-2 shows that the decision criterion is to reject the null hypothesis of $\rho = 0$ if the absolute value of the test statistic exceeds the critical values; rejection of $\rho = 0$ means that there is sufficient evidence to support a claim of a linear correlation between the two variables. If the absolute value of the test statistic does not exceed the critical values, then we fail to reject $\rho = 0$; that is, there is not sufficient evidence to conclude that there is a linear correlation between the two variables.

▶ **EXAMPLE**

Using the sample data in Table 9-1, test the claim of University of Arizona researchers that there is a linear correlation between weights of discarded plastic and household sizes. For the test statistic, use both (a) Method 1 and (b) Method 2.

● **SOLUTION**

Refer to either the general method of testing hypotheses outlined in Figure 7-4 or the more specific method outlined in Figure 9-2. To claim that there is a correlation is to claim that the population linear correlation ρ is different from 0. We therefore get the following hypotheses.

H_0: $\rho = 0$ (No correlation) H_1: $\rho \neq 0$ (Correlation)

reject Ho if
r > cv

fail to reject Ho
if cv > r

No significance level α was specified, so we use the common value of $\alpha = 0.05$.

We need the value of the linear correlation coefficient r, but in the preceding example we found that $r = 0.842$. With that value, we now find the test statistic and critical value, using each of the two methods just described.

a. *Method 1:* The test statistic is

$$t = \frac{r}{\sqrt{\frac{1 - r^2}{n - 2}}} = \frac{0.842}{\sqrt{\frac{1 - 0.842^2}{8 - 2}}} = 3.823$$

The critical values of $t = -2.447$ and $t = 2.447$ are found in Table A-3, where 2.447 corresponds to 0.05 in two tails and the number of degrees of freedom is $n - 2 = 6$. See Figure 9-3 for the graph that includes the test statistic and critical values.

b. *Method 2:* The test statistic is $r = 0.842$. The critical values of $r = -0.707$ and $r = 0.707$ are found in Table A-6 with $n = 8$ and $\alpha = 0.05$. See Figure 9-4 on the following page for a graph that includes this test statistic and critical values.

Using either of the two methods, we find that the absolute value of the test statistic does exceed the critical value (Method 1: 3.823 > 2.447; Method 2: 0.842 > 0.707); that is, the test statistic does fall within the critical region. We therefore reject $H_0: \rho = 0$. There is sufficient evidence to support the claim of a linear correlation between weights of discarded plastic and household size. The size of a household does seem to correspond to the weight of plastic the household discards.

Power Lines Correlate with Cancer

Interesting, surprising, and useful results sometimes occur when correlations are found between variables. Several scientific studies suggest that there is a correlation between exposure to electromagnetic fields and the occurrence of cancer. Epidemiologists from Sweden's Karolinska Institute researched 500,000 Swedes who lived within 300 meters of a high-tension power line during a 25-year period. They found that the children had a higher incidence of leukemia. Their findings led Sweden's government to consider regulations that would reduce housing in close proximity to high-tension power lines. In an article on this study, *Time* magazine reported that "Although the research does not prove cause and effect, it shows an unmistakable correlation between the degree of exposure and the risk of childhood leukemia."

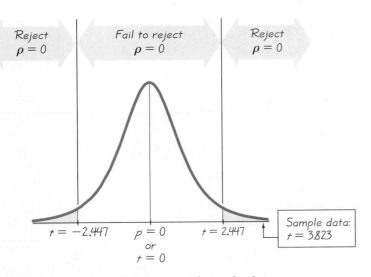

FIGURE 9-3 Testing $H_0: \rho = 0$ with Method 1

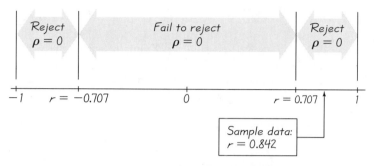

FIGURE 9-4   Testing H_0: $\rho = 0$ with Method 2

The preceding example and Figures 9-3 and 9-4 correspond to a two-tailed hypothesis test. The examples and exercises in this section will generally involve only two-tailed tests. One-tailed tests can occur with a claim of a positive correlation or a claim of a negative correlation. In such cases, the hypotheses will be as shown below.

Claim of Negative Correlation (Left-tailed test)	Claim of Positive Correlation (Right-tailed test)
H_0: $\rho \geq 0$	H_0: $\rho \leq 0$
H_1: $\rho < 0$	H_1: $\rho > 0$

For these one-tailed tests, Method 1 can be handled as in earlier chapters. For Method 2, either calculate the critical value as described in Exercise 26 or modify Table A-6 by replacing the column headings of $\alpha = 0.05$ and $\alpha = 0.01$ by the one-sided critical values of $\alpha = 0.025$ and $\alpha = 0.005$, respectively.

Common Errors Involving Correlation

We now identify three of the most common errors made in interpreting results involving correlation.

1. *We must be careful to avoid concluding that a significant linear correlation between two variables is proof that there is a cause-effect relationship between them.* The statistical correlation between smoking and cancer is not proof that smoking *causes* cancer. The techniques in this chapter can be used only to establish a linear relationship. We cannot establish the existence or absence of any inherent cause-effect relationship between the variables. Such a relationship can be established only by using concrete evidence supplied by various professionals such as medical researchers, psychologists, sociologists, and biologists.

2. *Another source of potential error arises with data based on rates or averages.* When we use rates or averages for data, we suppress the variation

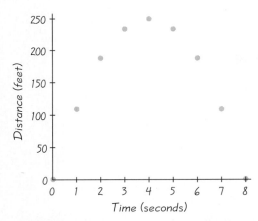

FIGURE 9-5   Scatter Diagram of Distance above Ground and Time for Object Thrown Upward

among the individuals or items, and this may lead to an inflated correlation coefficient. One study produced a 0.4 linear correlation coefficient for paired data relating income and education among *individuals*, but the correlation coefficient became 0.7 when regional *averages* were used.

3. A *third error involves the property of linearity*. The conclusion that there is no significant linear correlation does not mean that x and y are not related in any way. The data depicted in Figure 9-5 result in a value of $r = 0$, which is an indication of no *linear* relationship between the two variables. However, we can easily see from Figure 9-5 that there is a pattern reflecting a very strong (nonlinear) relationship. (Figure 9-5 is a scatter diagram that depicts the relationship between distance above ground and time elapsed for an object thrown upward.)

could be other relationships other than linear relationship

We have presented Formula 9-1 for calculating r and illustrated its use; we will now give a justification for it. Formula 9-1 simplifies the calculations used in this equivalent formula:

$$r = \frac{\Sigma(x - \overline{x})(y - \overline{y})}{(n - 1)s_x s_y}$$

We will temporarily use this latter version of Formula 9-1 because its form relates more directly to the underlying theory. We will consider the following paired data, which are depicted in the scatter diagram shown on the following page in Figure 9-6.

x	1	1	2	4	7
y	4	5	8	15	23

Figure 9-6 includes the point $(\overline{x}, \overline{y}) = (3, 11)$, which is called the centroid of the sample points.

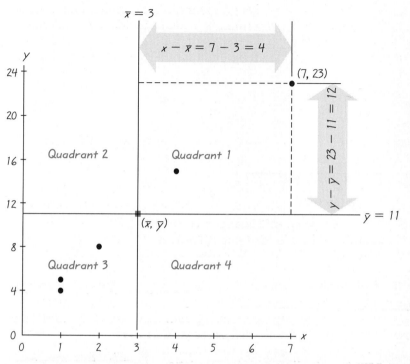

FIGURE 9-6   Scatter Diagram Partitioned into Quadrants

DEFINITION

Given a collection of paired (x, y) data, the point $(\overline{x}, \overline{y})$ is called the **centroid**.

We have noted that sometimes r is called the Pearson product moment. This title reflects both that it was first developed by Karl Pearson and that it is based on the product of the moments $(x - \overline{x})$ and $(y - \overline{y})$; that is, Pearson based his measure of scattering on the statistic $\Sigma(x - \overline{x})(y - \overline{y})$. In any scatter diagram, vertical and horizontal lines through the *centroid* $(\overline{x}, \overline{y})$ divide the diagram into four quadrants (see Figure 9-6). If the points of the scatter diagram tend to approximate an uphill line (as in the figure), individual values of the product $(x - \overline{x})(y - \overline{y})$ tend to be positive because most of the points are found in the first and third quadrants, where the products of $(x - \overline{x})$ and $(y - \overline{y})$ are positive. If the points of the scatter diagram approximate a downhill line, most of the points are in the second and fourth quadrants, where $(x - \overline{x})$ and $(y - \overline{y})$ are opposite in sign, so $\Sigma(x - \overline{x})(y - \overline{y})$ tends to be negative. Points that follow no linear pattern tend to be scattered among the four quadrants, so the value of $\Sigma(x - \overline{x})(y - \overline{y})$ tends to be close to 0.

The sum $\Sigma(x - \bar{x})(y - \bar{y})$ depends on the magnitude of the numbers used. For example, if you change x from pounds to ounces, that sum will change. However, r should not be affected by the particular scale used. We can make r independent of the scale used by incorporating the sample standard deviations as follows:

$$r = \frac{\Sigma(x - \bar{x})(y - \bar{y})}{(n - 1)s_x s_y}$$

This expression can be algebraically manipulated into the equivalent form of Formula 9-1 (see Exercise 30, part a).

Preceding chapters discussed methods of inferential statistics by addressing methods of hypothesis testing, as well as methods for constructing confidence interval estimates. A similar procedure may be used to find confidence intervals for ρ. However, because the construction of such confidence intervals involves somewhat complicated transformations, that process is presented in Exercise 32 (Beyond the Basics).

We can use the linear correlation coefficient to determine whether there is a linear relationship between two variables. Using the data of Table 9-1, we have concluded that there is a correlation between weights of discarded plastic and sizes of households. Now that we have concluded that a relationship exists, we would like to determine what that relationship is. The Census Bureau would like to know how to actually estimate population size based on weights of garbage. This next stage of analysis is addressed in the following section.

9-2   Exercises A: Basic Skills and Concepts

For each part of Exercises 1 and 2, a sample of paired data produces a linear correlation coefficient r with the given value. Find the corresponding critical values. Assuming a significance level of $\alpha = 0.05$, do you conclude that there is no significant linear correlation or do you conclude that there is a significant linear correlation?

1. a. $n = 20, r = 0.502$
    b. $n = 20, r = 0.203$
    c. $n = 50, r = -0.281$

2. a. $n = 77, r = -0.351$
    b. $n = 22, r = -0.370$
    c. $n = 47, r = 0.312$

In Exercises 3 and 4, use the given list of paired data.

    a. Construct the scatter diagram.
    b. Find the values of $n, \Sigma x, \Sigma x^2, (\Sigma x)^2,$ and Σxy.
    c. Find the value of the linear correlation coefficient r.

3. | x | 1 1 2 3 |
    |-----|------------|
    | y | 1 5 4 2 |

4. | x | 0 1 1 2 5 |
    |-----|---------------|
    | y | 3 3 4 5 6 |

In Exercises 5–20,

 a. Construct the scatter diagram.

 b. Find the value of the linear correlation coefficient r.

 c. Using a significance level of $\alpha = 0.05$, test the claim that there is no linear correlation between the two variables.

 d. Save your work because the same data will be used in the next section.

5. The paired data below consist of weights (in pounds) of discarded paper and sizes of households.

x: Paper	2.41	7.57	9.55	8.82	8.72	6.96	6.83	11.42
y: Household size	2	3	3	6	4	2	1	5

Sample data obtained from the Garbage Project at the University of Arizona.

6. The paired data below consist of weights (in pounds) of discarded food and sizes of households.

x: Food	1.04	3.68	4.43	2.98	6.30	1.46	8.82	9.62
y: Household size	2	3	3	6	4	2	1	5

Sample data obtained from the Garbage Project at the University of Arizona.

7. The paired data below consist of the total weights (in pounds) of discarded garbage and sizes of households.

Total weight	10.76	19.96	27.60	38.11	27.90	21.90	21.83	49.27	33.27	35.54
Household size	2	3	3	6	4	2	1	5	6	4

Sample data obtained from the Garbage Project at the University of Arizona.

8. The paired data below consist of the total weights (in pounds) of paper and plastic discarded by households in a week.

Paper	2.41	7.57	9.55	8.82	8.72	6.96	6.83	11.42	16.08	6.83	13.05	11.36
Plastic	0.27	1.41	2.19	2.83	2.19	1.81	0.85	3.05	3.42	2.10	2.93	2.44

Sample data obtained from the Garbage Project at the University of Arizona.

9. A study was conducted to investigate any relationship between age (in years) and BAC (blood alcohol concentration) measured when convicted DWI jail inmates were first arrested. Sample data are given below for randomly selected subjects.

Age	17.2	43.5	30.7	53.1	37.2	21.0	27.6	46.3
BAC	0.19	0.20	0.26	0.16	0.24	0.20	0.18	0.23

Based on data from the Dutchess County STOP-DWI Program.

10. Randomly selected subjects ride a bicycle at 5.5 mi/h for one minute. Their weights (in pounds) are given with the numbers of calories used.

Weight	167	191	112	129	140	173	119
Calories used	4.23	4.69	3.21	3.47	3.72	4.45	3.36

Based on data from *Diet Free*, by Kuntzlemann.

11. In a study of the factors that affect success in a calculus course, data were collected for 10 different persons. Scores on an algebra placement test are given, along with calculus achievement scores.

Algebra score	17	21	11	16	15	11	24	27	19	8
Calculus score	73	66	64	61	70	71	90	68	84	52

Based on data from "Factors Affecting Achievement in the First Course in Calculus," by Edge and Friedberg, *Journal of Experimental Education*, Vol. 52, No. 3.

12. The accompanying table lists the number of registered automatic weapons (in thousands), along with the murder rate (in murders per 100,000), for randomly selected states.

Automatic weapons	11.6	8.3	3.6	0.6	6.9	2.5	2.4	2.6
Murder rate	13.1	10.6	10.1	4.4	11.5	6.6	3.6	5.3

Data provided by the FBI and the Bureau of Alcohol, Tobacco, and Firearms.

13. Emissions data for a sample of different vehicles are given for HC (hydrocarbon) and CO (carbon monoxide), with both measured in grams per meter.

HC	0.65	0.55	0.72	0.83	0.57	0.51	0.43	0.37
CO	14.7	12.3	14.6	15.1	5.0	4.1	3.8	4.1

Based on data from "Determining Statistical Characteristics of a Vehicle Emissions Audit Procedure," by Lorenzen, *Technometrics*, Vol. 22, No. 4.

14. For randomly selected homes recently sold in Dutchess County, New York, the living areas (in hundreds of square feet) are listed, along with the annual tax amounts (in thousands of dollars).

Living area	15	38	23	16	16	13	20	24
Taxes	1.9	3.0	1.4	1.4	1.5	1.8	2.4	4.0

15. There are many regions where the winter accumulation of snowfall is a primary source of water. Several investigations of snowpack characteristics have used satellite observations from NASA's Landsat series, along with measurements taken on Earth. Given here are ground measurements of snow depth (in centimeters), along with the corresponding temperatures (in degrees Celsius).

Temperature	−62	−41	−36	−26	−33	−56	−50	−66
Snow depth	21	13	12	3	6	22	14	19

Data based on information in Kastner's *Space Mathematics,* published by NASA.

16. The following table lists per capita cigarette consumption in the United States for various years, along with the percentage (in percentage points) of the population admitted to mental institutions as psychiatric cases.

Cigarette consumption	3522	3597	4171	4258	3993	3971	4042	4053
Psychiatric admissions	0.20	0.22	0.23	0.29	0.31	0.33	0.33	0.32

17. Refer to Data Set 3 in Appendix B. For persons who never married, use the paired data consisting of heights and incomes.

18. Refer to Data Set 3 in Appendix B. For men who never married, use the paired data consisting of weights and income.

19. Refer to Data Set 4 in Appendix B. Use the paired data consisting of weights of cars and their lengths.

20. Refer to Data Set 4 in Appendix B. Use the paired data consisting of the weights of cars and their highway fuel consumption amounts.

In Exercises 21–24, identify the error in the stated conclusion. (See the list of common errors included in this section.)

21. *Given:* The paired sample data of age and score result in a linear correlation coefficient very close to 0.

Conclusion: Older people tend to get lower scores.

22. *Given:* There is a significant positive linear correlation between per capita income and per capita spending.

Conclusion: Increased spending is caused by increased income.

23. *Given:* The paired sample data result in a linear correlation coefficient very close to 0.

Conclusion: The two variables are not related in any way.

24. *Given:* There is a significant linear correlation between state average tax burdens and state average incomes.

Conclusion: There is a significant linear correlation between individual tax burdens and individual incomes.

9-2   Exercises B: Beyond the Basics

25. In addition to testing for a linear correlation between x and y, we can often use *transformations* of data to explore for other relationships. For example, we might replace each x value by x^2 and use the methods of this section to determine whether there is a linear correlation between y and x^2. Given the paired data below, construct the scatter diagram and then test for a linear correlation between y and each of the following.

a. x b. x^2 c. $\log x$
d. $\sqrt{x}$ e. $1/x$

Which case results in the largest value of r?

x	1.3	2.4	2.6	2.8	2.4	3.0	4.1
y	0.11	0.38	0.41	0.45	0.39	0.48	0.61

26. The critical values of r in Table A-6 are found by solving

$$t = \frac{r}{\sqrt{\dfrac{1 - r^2}{n - 2}}}$$

for r to get

$$r = \frac{t}{\sqrt{t^2 + n - 2}}$$

Here the t value is found from Table A-3 by assuming a two-tailed case with $n - 2$ degrees of freedom. Table A-6 lists the results for selected values of n and α. Use the above formula for r and Table A-3 (with $n - 2$ degrees of freedom) to find the critical values of r for the given cases.

a. $H_0: \rho = 0$, $n = 50$, $\alpha = 0.05$
b. $H_1: \rho \neq 0$, $n = 75$, $\alpha = 0.10$
c. $H_0: \rho \geq 0$, $n = 20$, $\alpha = 0.05$
d. $H_0: \rho \leq 0$, $n = 10$, $\alpha = 0.05$
e. $H_1: \rho > 0$, $n = 12$, $\alpha = 0.01$

27. First plot the scatter diagram for the data in the accompanying table, then try to compute the linear correlation coefficient r. Comment on the results.

x	0	3	5	5	6
y	2	2	2	2	2

28. Repeat Exercise 15 after interchanging each x value with the corresponding y value and then changing all the depths from centimeters to millimeters by multiplying each depth by 10. How is the value of the linear correlation coefficient affected? How is it affected if we change from degrees Celsius to degrees Fahrenheit?

29. Repeat Exercise 15 after changing the depth of 19 cm to 190 cm. How much effect does an extreme value have on the value of the linear correlation coefficient?

30. a. Show that $\dfrac{\Sigma(x - \bar{x})(y - \bar{y})}{(n - 1)s_x s_y} = \dfrac{n\Sigma xy - (\Sigma x)(\Sigma y)}{\sqrt{n(\Sigma x^2) - (\Sigma x)^2}\sqrt{n(\Sigma y^2) - (\Sigma y)^2}}$.

   b. Show that Formula 9-1 is equivalent to

$$r = \frac{(\overline{xy}) - \bar{x} \cdot \bar{y}}{\sqrt{[(\overline{x^2}) - (\bar{x})^2][(\overline{y^2}) - (\bar{y})^2]}}$$

   where $\overline{xy} = \Sigma xy/n$ and $(\overline{x^2}) = \Sigma x^2/n$.

31. The graph of $y = x^2$ is a parabola, not a straight line, so we might expect that the value of r would not reflect a linear correlation between x and y. Using $y = x^2$, make a table of x and y values for $x = 0, 1, 2, \ldots, 10$ and calculate the value of r. What do you conclude? How do you explain the result?

32. Given n pairs of data from which the linear correlation coefficient r can be found, use the following procedure to construct a confidence interval about the population parameter ρ.

   Step a.   Use Table A-2 to find $z_{\alpha/2}$ that corresponds to the desired degree of confidence.

   Step b.   Evaluate the interval limits w_L and w_R:

$$w_L = \frac{1}{2}\ln\left(\frac{1 + r}{1 - r}\right) - z_{\alpha/2} \cdot \frac{1}{\sqrt{n - 3}}$$

$$w_R = \frac{1}{2}\ln\left(\frac{1 + r}{1 - r}\right) + z_{\alpha/2} \cdot \frac{1}{\sqrt{n - 3}}$$

   Step c.   Now evaluate the confidence interval limits in the expression below.

$$\frac{e^{2w_L} - 1}{e^{2w_L} + 1} < \rho < \frac{e^{2w_R} - 1}{e^{2w_R} + 1}$$

   Use this procedure to construct a 95% confidence interval for ρ, given 50 pairs of data for which $r = 0.600$.

9-3 Regression

In Section 9-2 we analyzed paired data with the goal of determining whether there is a correlation between the two variables. Using the data of Table 9-1, we found that there does appear to be a correlation between the weights of discarded plastic and the sizes of households. Many organizations, such as the Census Bureau, expressed interest in the study that resulted in the Table 9-1 data, but their interest goes beyond simply determining that a correlation exists. Specifically, such organizations would like to determine actual population sizes by using the weights of discarded garbage. Whereas our goal in Section 9-2 was determining whether there is a correlation, our goal in this section is to *identify* the relationship between variables so that we can *predict* the value of one variable (such as population size), given the value of the other variable (such as weight of discarded plastic). Our identification of the relationship between correlated variables will be in two forms: We will identify the *graph* of the straight line that best fits the points in the scatter diagram, and we will identify the *equation* of that same straight line. Such equations are very useful for predictions.

Why do we use the term *regression?* Sir Francis Galton (1822–1911), a cousin of Charles Darwin, studied the phenomenon of heredity in which certain characteristics *regress,* or revert to more typical values. Galton noted, for example, that children of tall parents tend to grow up to be shorter than their parents. Similarly, children of short parents tend to be taller than their parents. That is, when children are born to parents who are exceptionally tall or short, those children tend to have heights that go back (or "regress") to a mean height. The original studies of regressing characteristics explain why we now use the term *regression,* even though current applications involve much more than principles of heredity.

DEFINITIONS

Given a collection of paired sample data, the **regression equation**

$$\hat{y} = \hat{\beta}_0 + \hat{\beta}_1 x$$

describes the relationship between the two variables. (Recall that the symbol β is the Greek letter beta.) The graph of the regression equation is called the **regression line** (or *line of best fit,* or *least-squares line*).

This definition expresses a relationship between x (called the **independent variable**) and $\hat{y}$ (called the **dependent variable**). In the preceding definition, different symbols are used to represent the typical equation of a straight line: $y = mx + b$, where m is the slope and b is the y-intercept. See the notation box below, where the symbol ^ indicates a sample statistic that is used as an estimate of a population parameter. (We used this notation in Chapters 6, 7, and 8 when we stipulated that $\hat{p}$ represents a sample estimate of the population proportion p.) The notation box shows that instead of using m for the *slope* of a line, we use either $\hat{\beta}_1$ for the estimated slope or β_1 for the true slope. Instead of using b for the *y-intercept*, we use either $\hat{\beta}_0$ for the estimated y-intercept or β_0 for the true y-intercept. Don't be intimidated by the equation $\hat{y} = \hat{\beta}_0 + \hat{\beta}_1 x$, because it's really $y = mx + b$ expressed with different symbols.

Notation for Regression Equation

	Population Parameter	Point Estimate
y-intercept of regression equation	β_0	$\hat{\beta}_0$
Slope of regression equation	β_1	$\hat{\beta}_1$
Equation of the regression line	$y = \beta_0 + \beta_1 x$	$\hat{y} = \hat{\beta}_0 + \hat{\beta}_1 x$

Assumptions

For the regression methods given in this section, we assume that

1. We are investigating only *linear* relationships.
2. For each x value, y is a random variable having a normal distribution. All of these y distributions have the same variance. Also, for a given value of x, the distribution of y values has a mean that lies on the regression line. (Results are not seriously affected if departures from normal distributions and equal variances are not too extreme.)

An important goal of this section is to use sample paired data to estimate the regression equation. Using only sample data, we can't find the exact values of the population parameters β_0 and β_1, but we can find their point estimates $\hat{\beta}_0$ and $\hat{\beta}_1$ by using Formulas 9-2 and 9-3.

Formula 9-2
$$\hat{\beta}_0 = \frac{(\Sigma y)(\Sigma x^2) - (\Sigma x)(\Sigma xy)}{n(\Sigma x^2) - (\Sigma x)^2} \quad \text{y-intercept}$$

Formula 9-3
$$\hat{\beta}_1 = \frac{n(\Sigma xy) - (\Sigma x)(\Sigma y)}{n(\Sigma x^2) - (\Sigma x)^2} \quad \text{slope}$$

These formulas appear formidable, but many calculators accept entries of paired data and provide the $\hat{\beta}_0$ and $\hat{\beta}_1$ values directly. Statistical software packages such as STATDISK and Minitab accept paired sample data and provide these values. When we must use Formulas 9-2 and 9-3 instead of a calculator or computer, there are three observations that make the required computations easier. First, if the correlation coefficient r has been computed using Formula 9-1, the values of Σx, Σy, Σx^2, and Σxy have already been determined. These values can now be used again in Formulas 9-2 and 9-3. (Note that the numerator for r in Formula 9-1 is identical to the numerator for $\hat{\beta}_1$ in Formula 9-3.) Second, examine the denominators of the formulas for $\hat{\beta}_0$ and $\hat{\beta}_1$ and note that they are identical. This means that the computation of $n(\Sigma x^2) - (\Sigma x)^2$ need be done only once; the result can be used in both formulas. A third observation that helps simplify the calculations is that the regression line always passes through the centroid. We could use algebra (see Exercise 29) to prove that when the sample mean $\overline{x}$ is substituted for x and the sample mean $\bar{y}$ is substituted for $\hat{y}$, the regression equation $\hat{y} = \hat{\beta}_0 + \hat{\beta}_1 x$ becomes

$$\bar{y} = \hat{\beta}_0 + \hat{\beta}_1 \bar{x}$$

and this second equation is an identity (that is, it is always true). This would prove algebraically that the regression line always passes through the centroid. Because the equation $\bar{y} = \hat{\beta}_0 + \hat{\beta}_1 \bar{x}$ must always be true, it follows that $\hat{\beta}_0 = \bar{y} - \hat{\beta}_1 \bar{x}$, and we now have another expression for the y-intercept $\hat{\beta}_0$. It is usually easier to evaluate the y-intercept $\hat{\beta}_0$ by using Formula 9-4 than by using Formula 9-2.

Formula 9-4 $$\hat{\beta}_0 = \bar{y} - \hat{\beta}_1 \bar{x}$$

Formulas 9-2, 9-3, and 9-4 describe the computations necessary to obtain the estimated regression equation. By using these formulas, we get the equation with the following special property: *The regression line fits the sample points best.* (The specific criterion used to determine which line fits "best" is the least-squares property, which will be described later.) We will now briefly discuss rounding and then illustrate the procedure for finding and applying the regression equation.

Rounding the y-Intercept $\hat{\beta}_0$ and the Slope $\hat{\beta}_1$

It's difficult to provide a simple universal rule for rounding values of $\hat{\beta}_0$ and $\hat{\beta}_1$, but we usually try to round each of these values to three significant digits or use the values provided by a statistical software package such as STATDISK or Minitab. Because these values are very sensitive to rounding at intermediate steps of calculations, you should try to carry at least six significant digits (or use exact values) in the intermediate steps. Depending on how you round, answers to examples and exercises may be slightly different from your answers.

▶ **EXAMPLE**

In Section 9-2 we used the Table 9-1 data (x = weights of discarded plastic, y = household sizes) to find that the linear correlation coefficient of $r = 0.842$ indicates that at the 0.05 significance level, there is a significant linear correlation. Now find the regression equation of the straight line that relates x and y.

● **SOLUTION**

We will find the regression equation by using Formulas 9-3 and 9-4 to determine the estimated y-intercept $\hat{\beta}_0$ and slope $\hat{\beta}_1$, but first recall from Section 9-2 that the following statistics were found. (See Table 9-2.)

$$n = 8 \qquad \Sigma x = 14.60 \qquad \Sigma y = 26$$
$$\Sigma x^2 = 32.9632 \qquad \Sigma y^2 = 104 \qquad \Sigma xy = 56.80 \qquad \textit{(continued)}$$

Having determined the values of the individual components, we can now find the slope $\hat{\beta}_1$ by using Formula 9-3.

$$\hat{\beta}_1 = \frac{n(\Sigma xy) - (\Sigma x)(\Sigma y)}{n(\Sigma x^2) - (\Sigma x)^2}$$

$$= \frac{8(56.80) - (14.60)(26)}{8(32.9632) - (14.60)^2} = \frac{74.8}{50.5456} = 1.47985$$

$$= 1.48 \quad \text{(rounded)}$$

We now use Formula 9-4 to find the y-intercept $\hat{\beta}_0$.

$$\hat{\beta}_0 = \bar{y} - \hat{\beta}_1\bar{x}$$

$$= \frac{26}{8} - (1.47985)\frac{14.60}{8} = 0.549 \quad \text{(rounded)}$$

Having found the slope and y-intercept, we can now express the estimated equation of the regression line as follows:

$$\hat{y} = 0.549 + 1.48x$$

The accompanying STATDISK display shows the graph of the regression equation found in the preceding example. Later we will describe the use of STATDISK and Minitab for the procedures outlined in this section.

STATDISK DISPLAY

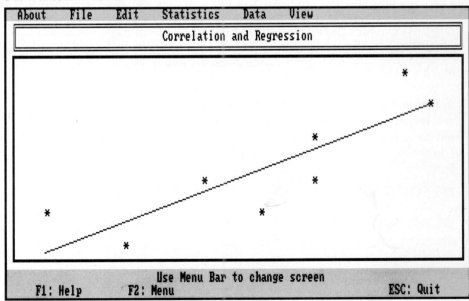

We should realize that $\hat{y} = 0.549 + 1.48x$ is an *estimate* of the true regression equation $y = \beta_0 + \beta_1 x$. This estimate is based on one particular set of sample data listed in Table 9-1. Another sample drawn from the same population would probably lead to a slightly different equation.

Marginal Change

We can use the regression equation to see the effect on one variable when the other variable changes by some specific amount.

DEFINITION

In working with two variables related by a regression equation, the **marginal change** in a variable is the amount that it changes when the other variable changes by one unit.

The slope $\hat{\beta}_1$ in the regression equation represents the *marginal change* resulting when x changes by one unit. For the garbage data of Table 9-1, we can see that an increase in x of one unit will cause $\hat{y}$ to change by 1.48 units (or persons). That is, if a household discards 1 lb more plastic than another household, the predicted size of the first household is 1.48 persons more than that of the second.

Predictions

Regression equations can be helpful when used in *predicting* the value of one variable, given some particular value for the other variable. If the regression line fits the data quite well, then it makes sense to use its equation for predictions, provided that we don't go beyond the scope of the available scores. However, *we should use the equation of the regression line only if r indicates that there is a significant linear correlation. In the absence of a significant linear correlation, we should not use the regression equation for projecting or predicting; instead, our best estimate of the second variable is simply its sample mean.*

In predicting a value of y based on some given value of x . . .

1. If there is *not* a significant linear correlation, the best predicted y value is $\bar{y}$.
2. If there *is* a significant linear correlation, the best predicted y value is found by substituting the x value into the regression equation.

Figure 9-7 (see the following page) summarizes this process, which is easier to understand if we think of r as a measure of how well the regression line fits the sample data. If r is near -1 or $+1$, then the regression line fits the data well, but if r is near 0, then the regression line fits poorly.

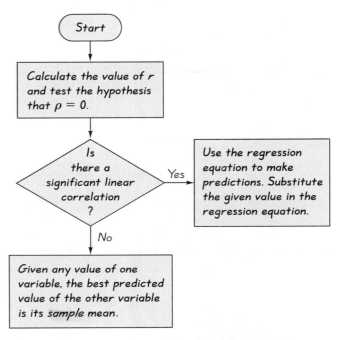

FIGURE 9-7 Predicting the Value of a Variable

▶ **EXAMPLE**

Using the sample data in Table 9-1, we found that there is a significant linear correlation between weights of discarded plastic and sizes of households. We also used the same sample data to find that the regression equation relating weight of plastic and household size is $\hat{y} = 0.549 + 1.48x$. Use these results to *predict* the size of a household that discards 2.50 lb of plastic in a week.

● **SOLUTION**

There's a strong temptation to jump in and substitute 2.50 for x in the regression equation, but we should first consider whether there is a significant correlation that justifies the use of that equation. In this example, we do have a significant linear correlation, so our predicted value is found as follows:

$$\hat{y} = \hat{\beta}_0 + \hat{\beta}_1 x$$
$$= 0.549 + 1.48(2.50) = 4.25$$

That is, we predict that a household that discards 2.50 lb of plastic in a week will have 4.25 persons. (If there had *not* been a significant linear correlation, our best predicted household size would have been $\bar{y} = 3.25$.)

► **EXAMPLE**

Data Set 3 in Appendix B is based on data from the Second National Health and Examination Survey, U.S. Center for Health Statistics. Make the following predictions, using only the data for the 86 men.

a. Find the best predicted *family income* for a man who is 70 in. tall.
b. Find the best predicted *weight* for a man who is 70 in. tall.

● **SOLUTION**

a. The method we use to predict the family income y depends on whether there is a significant correlation between that variable and the x variable of height. We therefore need to determine first whether such a correlation exists. Using the Data Set 3 paired data of *family incomes* and *heights* for the 86 males, we let the variable x represent height (in inches) and the variable y represent family income (in thousands of dollars). The following statistics can then be computed by using $\bar{x} = \Sigma x/n$ and Formulas 9-1, 9-3, and 9-4. (For data sets this large, we strongly recommend using statistical software.)

$$r = -0.024 \qquad \hat{\beta}_0 = 51.7 \qquad \hat{\beta}_1 = -0.236$$
$$\text{Regression Equation: } \hat{y} = 51.7 - 0.236x$$
$$n = 86 \qquad \bar{x} = 69.353 \qquad \bar{y} = 35.33$$

If we use a significance level of $\alpha = 0.05$ and test the null hypothesis of $H_0: \rho = 0$, we conclude that there is *not* a significant linear correlation between family incomes and heights. (Using Method 2 in Section 9-2, we find that the test statistic is $r = -0.024$ and the critical values are $r = \pm 0.207$.) *Because there is not a significant linear correlation between the two variables, we should not use the regression equation for predictions; instead, for any value of one variable we should use the mean of the other variable.* (See Figure 9-7.) The best predicted value for family income is $\bar{y} = 35.33$, or $35,330.

b. Again, the method we use for predictions depends on whether there is a significant correlation between weight and height. We therefore need to determine whether such a correlation exists. Using the Data Set 3 paired data of weights and heights for the 86 males, we let the variable x represent height (in inches) and the variable y represent weight (in pounds). The following statistics can then be computed by using $\bar{x} = \Sigma x/n$ and Formulas 9-1, 9-3, and 9-4. (Again, computers should be used for large data sets.)

$$r = 0.388 \qquad \hat{\beta}_0 = -131 \qquad \hat{\beta}_1 = 4.37$$
$$\text{Regression Equation: } \hat{y} = -131 + 4.37x$$
$$n = 86 \qquad \bar{x} = 69.353 \qquad \bar{y} = 171.67 \qquad \textit{(continued)}$$

Los Angeles Ozone

The South Coast Air Quality Management District monitors the ozone levels for the Los Angeles basin region. The ozone levels are affected by weather, as well as by pollutants from the infamous stream of dense traffic in the Los Angeles area. One useful indicator of an ozone problem is its level in parts per million for the worst hour of the year. Regression analysis shows a downward trend in that indicator, suggesting that despite a large increase in people and cars, ozone levels are currently at about half the level of 40 years ago. The downward trend can be seen in the "worst hour" levels for six recent consecutive years: 0.32, 0.32, 0.25, 0.21, 0.22, 0.22. Such statistical analyses verify the impact of clean air legislation, and they also raise issues of the cost effectiveness of new legislation.

If we use a significance level of $\alpha = 0.05$ and test the null hypothesis of H_0: $\rho = 0$, we conclude that there *is* a significant linear correlation between weights and heights. (Using Method 2 in Section 9-2, we find that the test statistic is $r = 0.388$ and the critical values are $r = \pm 0.207$.) *Because there is a significant linear correlation between the two variables, we should use the regression equation for predictions.* (See Figure 9-7.) The best predicted value for weight is

$$\hat{y} = -131 + 4.37x = -131 + 4.37(70) = 175 \text{ lb}$$

For a 70-in. tall male, the best predicted weight is 175 lb.

Carefully compare the solutions to parts a and b in the preceding example and note that we used the regression equation when there was a significant linear correlation, but in the absence of such a correlation, the best predicted value $\hat{y}$ is simply the value of the sample mean $\bar{y}$. If we think about it, this makes sense. We know there isn't a correlation between IQ and shoe size; for someone with a shoe size of 9.5, the best predicted IQ score is 100, which is the mean IQ score. But we know there is a correlation between the cost of filling a gas tank and the amount of gas required to fill that tank; if a fill-up requires 10 gal and the gas costs $1.50 per gal, then the total cost of $15 is calculated by multiplying 10 by $1.50 (or substituting $x = 10$ in the regression equation $\hat{y} = 0 + 1.50x$). In determining the cost of a fill-up, we normally wouldn't think of developing an estimate based on the mean amount of gas for a sample of cars. In general, we should use the regression equation if there is significant correlation between the two variables.

In part b of the preceding example, we were able to use the regression equation for predictions because there is a significant linear correlation; a common error is to use the regression equation when there is no significant linear correlation. That error violates the first of the following guidelines.

Guidelines for Using the Regression Equation

1. *If there is no significant linear correlation, don't use the regression equation to make predictions.*

2. *When using the regression equation for predictions, stay within the scope of the available sample data.* If you find a regression equation that relates women's heights and shoe sizes, it's absurd to predict the shoe size of a woman who is 10 ft tall.

3. *A regression equation based on old data is not necessarily valid now.* The regression equation relating used car prices and ages of cars is no longer usable if it's based on data from the 1950s.

4. *Don't make predictions about a population that is different from the population from which the sample data were drawn.* If we collect sample

data from men and develop a regression equation relating SAT math scores and SAT verbal scores, the results don't necessarily apply to women. <u>If we use *state averages* to develop a regression equation relating SAT math scores and SAT verbal scores, the results don't necessarily apply to *individuals*.</u>

Least-Squares Property

Formulas 9-2, 9-3, and 9-4 describe the computations necessary to obtain the regression equation. The criterion used to arrive at these particular formulas is the least-squares property.

DEFINITION

The **least-squares property** is the property that the sum of the squares of the vertical deviations of the sample points from the regression line is the smallest sum possible.

The *least-squares property* can be understood (believe it or not!) by examining Figure 9-8 and the following text. The figure shows the paired data contained in the table in the margin. Figure 9-8 also shows the regression line ($\hat{y} = 5 + 4x$) found by using Formulas 9-3 and 9-4 for the y-intercept $\hat{\beta}_0$ and the slope $\hat{\beta}_1$. The distances identified as errors in Figure 9-8 are the differences between the actual observed y values and the predicted $\hat{y}$ values on the

x	1	2	4	5
y	4	24	8	32

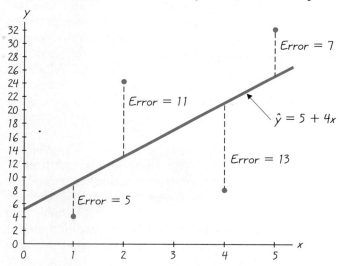

FIGURE 9-8 **Scatter Diagram with Regression Line and Errors**

regression line. For example, if $x = 1$, the *predicted* value of y is $\hat{y} = 9$, but the *actual* value of y is $y = 4$. The distance identified as the error is therefore 5. Likewise, if $x = 2$, then the predicted y value is 13, whereas the actual y value is 24, so the error is 11. The least-squares property dictates that the sum of the squares of those vertical errors

$$5^2 + 11^2 + 13^2 + 7^2 = 364$$

is the *minimum* sum that is possible with the given data. (If you retain signs in your error values, any negative signs become positive when the error is squared.) Any other line will yield a sum of squares larger than 364. It is in this sense that the regression line $\hat{y} = 5 + 4x$ fits the data best. (For example, $\hat{y} = 8 + 3x$ will have vertical errors of 7, 10, 12, and 9; the sum of their squares is 374, which exceeds the minimum of 364, indicating that this line doesn't fit the points as well as $\hat{y} = 5 + 4x$.)

Fortunately, we need not deal directly with the least-squares property when we want to find the equation of the regression line. Calculus has been used to build the least-squares property into the formulas for the y-intercept $\hat{\beta}_0$ and the slope $\hat{\beta}_1$. Because the derivations of these formulas require calculus, we don't include them in this text.

Using Computer Software for Correlation and Regression

For larger collections of paired data, a calculator or computer is required to find the values of the linear correlation coefficient r, the y-intercept $\hat{\beta}_0$, and the slope $\hat{\beta}_1$. We have already mentioned that some calculators will provide those values, as they allow the entry of paired data. Many computer software packages take paired data as input and produce more complete results. Section 9-2 provided a Minitab scatter diagram for the data of Table 9-1, and this section provided the STATDISK scatter diagram with a graph of the regression line included. Following are other STATDISK and Minitab displays obtained by using the same data. With STATDISK, select `Correlation and Regression` from the `Statistics` menu, then proceed to enter the information requested. With Minitab, enter the paired data in columns C1 (for x) and C2 (for y) by using the `SET` or `READ` command as described in Section 1-5. The Minitab command `PLOT C2 VS C1` generates a scatter diagram, the command `CORRE-LATION C1 C2` generates the value of the linear correlation coefficient, and the command `REGRESSION C2 1 C1` generates the regression equation. (The 1 in the command `REGRESSION C2 1 C1` indicates that the dependent variable is to be expressed in terms of only 1 independent variable.) In addition to the correlation coefficient and regression equation, STATDISK and Minitab automatically display some additional statistics, such as the coefficient of determination and the standard error of estimate, which will be discussed in the following section.

```
About    File    Edit    Statistics    Data    View
                    Correlation and Regression

        Linear correlation coefficient ....... r =   .84236
        Coefficient of determination ........... =   .70957
        Standard error of estimate ............. =   .971544

        Explained variation .................... =   13.8366
        Unexplained variation .................. =   5.66339
        Total variation ........................ =   19.5

        Equation of regression line .. y =  .54927 +  1.47985 x

        ------------ Test for linear correlation --------------
        Level of significance .................. =   .05
        Test statistic is .................... r =   .84236
        Critical value is .................... r =   .707365

        REJECT the claim of no significant linear correlation

                    Use Menu Bar to change screen
     F1: Help          F2: Menu                        ESC: Quit
```

MINITAB DISPLAY

```
MTB > READ  C1  C2
DATA> 0.27  2
DATA> 1.41  3
DATA> 2.19  3
DATA> 2.83  6
DATA> 2.19  4
DATA> 1.81  2
DATA> 0.85  1
DATA> 3.05  5
DATA> ENDOFDATA
MTB > CORRELATION  C1  C2

Correlation of C1 and C2  =  0.842

MTB > REGRESSION  C2  1  C1

The regression equation is
C2  =  0.549 +  1.48  C1
```

Confidence Intervals

It is possible to construct confidence interval estimates of the population parameters β_0 and β_1, which represent the y-intercept and slope of the regression line for the entire population of data. The difficult computations can be simplified somewhat by using the standard error of estimate s_e, which will be introduced in the next section. (See Exercise 17 in Section 9-4 for a procedure that allows you to construct confidence interval estimates of β_0 and β_1.)

9-3 Exercises A: Basic Skills and Concepts

In Exercises 1–4, use the given data to find the equation of the regression line.

1.

x	1	3	3	4	5	5
y	5	3	2	2	0	1

$(n = 6, \Sigma x = 21, \Sigma y = 13, \Sigma x^2 = 85, \Sigma y^2 = 43, \Sigma xy = 33)$

2.

x	0	3	4	5	9	12	15	15	17	20
y	1	7	9	11	19	24	31	31	35	41

$(n = 10, \Sigma x = 100, \Sigma y = 209, \Sigma x^2 = 1414, \Sigma y^2 = 6017, \Sigma xy = 2916)$

3.

x	1	1	2	3
y	1	5	4	2

4.

x	0	1	1	2	5
y	3	3	4	5	6

Exercises 5–20 use the same data sets as the exercises in Section 9-2. In each case, find the regression equation, letting the first variable be the independent (x) variable. Find predicted values where they are requested.

5.

x: Paper	2.41	7.57	9.55	8.82	8.72	6.96	6.83	11.42
y: Household size	2	3	3	6	4	2	1	5

6.

x: Food	1.04	3.68	4.43	2.98	6.30	1.46	8.82	9.62
y: Household size	2	3	3	6	4	2	1	5

7.

Total weight	10.76	19.96	27.60	38.11	27.90	21.90	21.83	49.27	33.27	35.54
Household size	2	3	3	6	4	2	1	5	6	4

8.

Paper	2.41	7.57	9.55	8.82	8.72	6.96	6.83	11.42	16.08	6.83	13.05	11.36
Plastic	0.27	1.41	2.19	2.83	2.19	1.81	0.85	3.05	3.42	2.10	2.93	2.44

9.

Age	17.2	43.5	30.7	53.1	37.2	21.0	27.6	46.3
BAC	0.19	0.20	0.26	0.16	0.24	0.20	0.18	0.23

10.

Weight	167	191	112	129	140	173	119
Calories used	4.23	4.69	3.21	3.47	3.72	4.45	3.36

11.

Algebra	17	21	11	16	15	11	24	27	19	8
Calculus	73	66	64	61	70	71	90	68	84	52

12.

Automatic weapons	11.6	8.3	3.6	0.6	6.9	2.5	2.4	2.6
Murder rate	13.1	10.6	10.1	4.4	11.5	6.6	3.6	5.3

13.

HC	0.65	0.55	0.72	0.83	0.57	0.51	0.43	0.37
CO	14.7	12.3	14.6	15.1	5.0	4.1	3.8	4.1

Find the best predicted value of CO (carbon monoxide, in grams per meter) given that the HC (hydrocarbon) amount is 0.75 g/m.

14.

Living area (hundreds of sq ft)	15	38	23	16	16	13	20	24
Taxes (thousands of dollars)	1.9	3.0	1.4	1.4	1.5	1.8	2.4	4.0

Find the best predicted value of taxes (in dollars) for a home with a living area of 1800 sq ft.

15.

Temperature (°C)	−62	−41	−36	−26	−33	−56	−50	−66
Snow depth (cm)	21	13	12	3	6	22	14	19

Find the best predicted snow depth given a temperature of −60°C.

16.

Cigarette consumption	3522	3597	4171	4258	3993	3971	4042	4053
Psychiatric admissions (in percentage points)	0.20	0.22	0.23	0.29	0.31	0.33	0.33	0.32

Find the best predicted percentage of psychiatric admissions given per capita cigarette consumption of 3650 (equivalent to 10 cigarettes per day).

17. Refer to Data Set 3 in Appendix B. For persons who never married, use the paired data consisting of heights (x) and incomes (y).

18. Refer to Data Set 3 in Appendix B. For men who never married, use the paired data consisting of weights (x) and incomes (y).

19. Refer to Data Set 4 in Appendix B and use the sample data consisting of car lengths (x) and weights (y).

20. Refer to Data Set 4 in Appendix B and use the sample data consisting of weights of cars (x) and amounts of highway fuel consumption (y).

21. For a collection of paired sample data, the regression equation is found to be $\hat{y} = 10.0 + 50.0x$ and the sample means are $\bar{x} = 0.30$ and $\bar{y} = 25.0$. In each of the following cases, use the additional information to find the best predicted estimate of y when $x = 2.0$. Assume a significance level of $\alpha = 0.05$.
   a. $n = 100$; $r = 0.999$
   b. $n = 10$; $r = 0.005$
   c. $n = 15$; $r = 0.519$
   d. $n = 25$; $r = 0.393$
   e. $n = 22$; $r = 0.567$

22. For a collection of paired sample data, the regression equation is found to be $\hat{y} = 50.0 - 20.0x$ and the sample means are $\bar{x} = 0.50$ and $\bar{y} = 40.0$. In each of the following cases, use the additional information to find the best predicted estimate of y when $x = 1.00$. Assume a significance level of $\alpha = 0.05$.
   a. $n = 5$; $r = -0.102$
   b. $n = 50$; $r = -0.997$
   c. $n = 20$; $r = -0.403$
   d. $n = 20$; $r = -0.449$
   e. $n = 65$; $r = -0.229$

23. In each of the following cases, find the best predicted estimate of y when $x = 5$. The given statistics are summarized from paired sample data. Assume a significance level of $\alpha = 0.01$.

    a. $n = 40$, $\bar{y} = 6$, $r = 0.01$, and the equation of the regression line is $\hat{y} = 2 + 3x$.

    b. $n = 40$, $\bar{y} = 6$, $r = 0.93$, and the equation of the regression line is $\hat{y} = 2 + 3x$.

    c. $n = 20$, $\bar{y} = 6$, $r = -0.654$, and the equation of the regression line is $\hat{y} = 2 - 3x$.

    d. $n = 20$, $\bar{y} = 6$, $r = 0.432$, and the equation of the regression line is $\hat{y} = 3.7 + 1.2x$.

    e. $n = 100$, $\bar{y} = 6$, $r = -0.175$, and the equation of the regression line is $\hat{y} = 16.7 - 2.4x$.

24. In each of the following cases, find the best predicted point estimate of y when $x = 8.0$. The given statistics are summarized from paired sample data. Assume a significance level of $\alpha = 0.05$.

    a. $n = 10$, $\bar{y} = 8.40$, $r = -0.236$, and the equation of the regression line is $\hat{y} = 3.5 - 2.0x$.

    b. $n = 10$, $\bar{y} = 8.40$, $r = -0.654$, and the equation of the regression line is $\hat{y} = 3.5 - 2.0x$.

    c. $n = 10$, $\bar{y} = 8.40$, $r = 0.602$, and the equation of the regression line is $\hat{y} = 3.5 + 2.0x$.

    d. $n = 32$, $\bar{y} = 8.40$, $r = -0.304$, and the equation of the regression line is $\hat{y} = 3.5 - 2.0x$.

    e. $n = 75$, $\bar{y} = 8.40$, $r = 0.257$, and the equation of the regression line is $\hat{y} = 3.5 + 2.0x$.

9-3   Exercises B: Beyond the Basics

25. Large numbers, such as those in the table below, often cause computational problems. Find the equation of the regression line. Now divide each x entry by 1000, and find the equation of the regression line. How are the results affected by the change in x? How would the results be affected if each y entry were divided by 1000?

x	924,736	832,985	825,664	793,427	857,366
y	142	111	109	95	119

x	1	2	2	3
y	5	4	3	1

26. According to the least-squares property, the regression line minimizes the sum of the squares of the errors between the actual y values and the predicted y values. Using the data in the margin and the equation of the regression line, find the sum of the squares of the vertical deviations for the given points. Show that this sum is less than the corresponding sum obtained by replacing the regression equation with $\hat{y} = 6 - x$.

27. If the scatter diagram reveals a nonlinear (not a straight line) pattern that you recognize as another type of curve, you may be able to apply the methods of this section. For the data given in the margin, find an equation of the form $\hat{y} = \hat{\beta}_0 + \hat{\beta}_1 x^2$ by using the values of x^2 and y instead of the values of x and y. Use Formulas 9-3 and 9-2 or 9-4, but enter the values of x^2 wherever x occurs.

28. Prove that the linear correlation coefficient r and the slope $\hat{\beta}_1$ have the same sign.

29. Prove that the centroid $(\overline{x}, \overline{y})$ must always lie on the regression line.

30. We have noted that ρ is the linear correlation coefficient for a population, and the regression equation for the population is $y = \beta_0 + \beta_1 x$. Explain why a test of the null hypothesis $H_0: \rho = 0$ is *equivalent* to a test of the null hypothesis $H_0: \beta_1 = 0$.

x	4	5	6	7
y	3	7	11	20

9-4   Variation and Prediction Intervals

The preceding two sections introduced the fundamental concepts of linear correlation and regression. We saw that the linear correlation coefficient r could be used to determine whether there is a significant statistical relationship between two variables. In Section 9-2 we interpreted r in a very limited way when we concluded that there is or is not a significant linear correlation between the two variables. The actual values of r can provide us with additional information about the variation of sample points about the regression line. We begin with a sample case, which leads to an important definition.

Let's assume that we have a large collection of paired data, which yields the following results:

- There is a significant linear correlation.
- The equation of the regression line is $\hat{y} = 3 + 2x$.
- $\overline{y} = 9$
- The scatter diagram contains the point $(5, 19)$, which comes from the original set of paired sample data.

When $x = 5$, we can find the predicted value $\hat{y}$ as follows:

$$\hat{y} = 3 + 2x = 3 + 2(5) = 13$$

Because the point $(5, 13)$ was determined by assuming that $x = 5$ and by calculating the predicted value of $\hat{y} = 13$ using the regression equation, the point $(5, 13)$ lies on the regression line. Whereas the point $(5, 13)$ is on the regression line, the point $(5, 19)$ is from the original data set and does not lie on the regression line because it does not satisfy the regression equation. Take time to carefully examine Figure 9-9 and note the differences defined as follows.

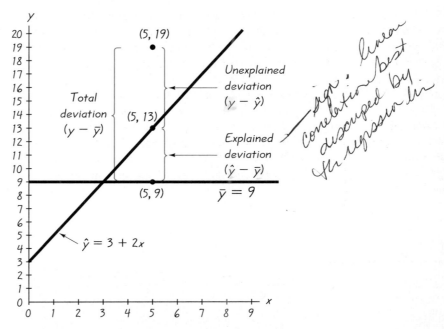

FIGURE 9-9   Unexplained, Explained, and Total Deviation

DEFINITIONS

Assume that we have a collection of paired data containing the particular point (x, y), that $\hat{y}$ is the predicted value of y (obtained by using the regression equation), and that the mean of the sample y values is $\bar{y}$.

The **total deviation** (<u>from the mean</u>) of the particular point (x, y) is the vertical distance $y - \bar{y}$, which is the distance between the point (x, y) and the horizontal line passing through the sample mean $\bar{y}$.

The **explained deviation** is the vertical distance $\hat{y} - \bar{y}$, which is the distance between the predicted y value and the horizontal line passing through the sample mean $\bar{y}$.

The **unexplained deviation** is the vertical distance $y - \hat{y}$, which is the vertical distance between the point (x, y) and the regression line.

For the specific data under consideration, we get

$$\text{\textit{total deviation}} \text{ of } (5, 19) = y - \bar{y} = 19 - 9 = 10$$
$$\text{\textit{explained deviation}} \text{ of } (5, 19) = \hat{y} - \bar{y} = 13 - 9 = 4$$
$$\text{\textit{unexplained deviation}} \text{ of } (5, 19) = y - \hat{y} = 19 - 13 = 6$$

If we were totally ignorant of correlation and regression concepts and wanted to predict a value of y given a value of x and a collection of paired (x, y) data, our best guess would be $\bar{y}$. But we are *not* totally ignorant of correlation and regression concepts: We know that in this case the way to predict the value of y when $x = 5$ is to use the regression equation, which yields $\hat{y} = 13$, as calculated above. We can explain the discrepancy between $\bar{y} = 9$ and $\hat{y} = 13$ by simply noting that there is a significant linear correlation best described by the regression line. Consequently, when $x = 5$, y *should be* 13 and not 9. But whereas y *should be* 13, it *is* 19. The discrepancy between 13 and 19 cannot be explained by the regression line and is called an unexplained deviation or a residual. The specific case illustrated in Figure 9-9 can be generalized as follows:

(total deviation) = (explained deviation) + (unexplained deviation)

or $\qquad (y - \bar{y}) = (\hat{y} - \bar{y}) \qquad\qquad + (y - \hat{y})$

This last expression applies to a particular point (x, y), but it can be further generalized and modified to include all of the pairs of sample data, as shown below in Formula 9-6. In that formula, the **total variation** is expressed as the sum of the squares of the total deviation values, the **explained variation** is the sum of the squares of the explained deviation values, and the **unexplained variation** is the sum of the squares of the unexplained deviation values.

Formula 9-6   (total variation) = (explained variation) + (unexplained variation)

or $\qquad \Sigma(y - \bar{y})^2 \qquad = \Sigma(\hat{y} - \bar{y})^2 \qquad + \Sigma(y - \hat{y})^2$

The components of this last expression are used in the next important definition.

DEFINITION

The **coefficient of determination** is the amount of the variation in y that is explained by the regression line. It is computed as

$$r^2 = \frac{\text{explained variation}}{\text{total variation}} \qquad \text{or just} \quad \boxed{r^2}$$

We can compute r^2 by using the above definition with Formula 9-6, or we can simply square the linear correlation coefficient r, which is found by using the methods given in Section 9-2. As an example, if $r = 0.8$, then the coefficient of determination is $r^2 = 0.8^2 = 0.64$, which means that 64% of the total variation in y can be explained by the regression line. It follows that 36% of the total variation in y remains unexplained.

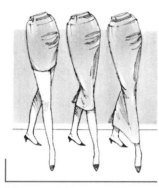

Unusual Economic Indicators

Forecasting and predicting are important goals of statistics. Investors seek indicators that can be used to forecast stock market behavior. Some of them are quite colorful. The hemline index is based on heights of women's skirts; rising hemlines supposedly precede a rise in the Dow Jones Industrial Average. According to the Super Bowl omen, a Super Bowl victory by a team with NFL origins is followed by a year in which the New York Stock Exchange index rises; otherwise, it falls. (In 1970 the NFL and AFL merged into the current NFL.) This indicator has been correct in 24 of the past 27 years. Other indicators: aspirin sales, limousines on Wall Street, and elevator traffic at the New York Stock Exchange.

▶ **EXAMPLE**

Referring to the sample garbage data in Table 9-1, find the percentage of the variation in y that can be explained by the regression line.

● **SOLUTION**

Recall that Table 9-1 contains 8 pairs of sample data, with each pair consisting of the weight (in pounds) of plastic discarded by a household and the corresponding household size. In Section 9-2 we found that the linear correlation coefficient is $r = 0.842$. The coefficient of determination is $r^2 = 0.842^2 = 0.709$, indicating that the ratio of explained variation in y to total variation in y is 0.709. We can therefore state that 70.9% of the total variation in y can be explained by the regression line. That is, 70.9% of the total variation in household size can be explained by the variation in discarded plastic; the other 29.1% is attributable to other factors.

Prediction Intervals

In Section 9-3 we used the Table 9-1 sample data to find the regression equation $\hat{y} = 0.549 + 1.48x$, where $\hat{y}$ represents predicted household size and x represents the weight of discarded plastic. We then used that equation to predict the y value, given that $x = 2.50$ lb. We found that the best predicted size for a household that discards 2.50 lb of plastic is 4.25 persons. Because 4.25 is a single value, it is referred to as a *point estimate*. In Chapter 6 we saw that point estimates have the disadvantage of not conveying any sense of how accurate they might be. Here, we know that 4.25 is the best predicted value, but we don't know how accurate that value is. In Chapter 6 we developed confidence interval estimates to overcome that disadvantage, and in this section we follow that precedent. We will use a **prediction interval,** which is a confidence interval estimate of a predicted value of y.

The development of a prediction interval requires a measure of the spread of sample points about the regression line. Recall that the unexplained deviation is the vertical distance between a sample point and the regression line, as illustrated in Figure 9-9. The standard deviation of those vertical distances is a collective measure of the spread of the sample points about the regression line; it is formally defined as follows.

DEFINITION

The **standard error of estimate** is a measure of the errors that are the differences (or distances) between the observed sample y values and the predicted values of $\hat{y}$ that are obtained from the regression equation. It is denoted by s_e and is given as

$$s_e = \sqrt{\frac{\Sigma(y - \hat{y})^2}{n - 2}}$$

where $\hat{y}$ is the predicted y value.

The development of the standard error of estimate s_e closely parallels that of the ordinary standard deviation introduced in Chapter 2. Just as the standard deviation is a measure of how scores deviate from their mean, the standard error of estimate s_e is a measure of how sample data points deviate from their regression line. The reasoning behind dividing by $n - 2$ is similar to the reasoning that led to division by $n - 1$ for the ordinary standard deviation, and we will not pursue the complex details. It is important to note that smaller values of s_e reflect points that stay close to the regression line, and larger values indicate greater dispersion of points away from the regression line.

Formula 9-7 can also be used to compute the standard error of estimate. It is algebraically equivalent to the expression in the definition, but this form is generally easier to work with because it doesn't require that we compute each of the predicted values $\hat{y}$ by substitution in the regression equation. However, Formula 9-7 does require that we find the y-intercept $\hat{\beta}_0$ and the slope $\hat{\beta}_1$ of the estimated regression line.

Formula 9-7
$$s_e = \sqrt{\frac{\Sigma y^2 - \hat{\beta}_0 \Sigma y - \hat{\beta}_1 \Sigma xy}{n - 2}}$$

▶ **EXAMPLE**

Use Formula 9-7 to find the standard error of estimate s_e for the garbage data listed in Table 9-1.

● **SOLUTION**

In Section 9-2 we used the Table 9-1 data to find the value of the linear correlation coefficient. We determined the following values:

$$n = 8 \qquad \Sigma y^2 = 104 \qquad \Sigma y = 26 \qquad \Sigma xy = 56.80 \qquad \textit{(continued)}$$

In Section 9-3 we used the Table 9-1 data to find the y-intercept and slope of the regression line. Those values are given here with extra decimal places for greater precision.

$$\hat{\beta}_0 = 0.549270 \qquad \hat{\beta}_1 = 1.47985$$

We can now use these values to find the standard error of estimate s_e.

$$s_e = \sqrt{\frac{\Sigma y^2 - \hat{\beta}_0 \Sigma y - \hat{\beta}_1 \Sigma xy}{n-2}}$$

$$= \sqrt{\frac{104 - (0.549270)(26) - (1.47985)(56.80)}{8-2}}$$

$$= 0.971554$$

$$= 0.972 \quad \text{(rounded)}$$

We can measure the spread of the sample points about the regression line with the standard error of estimate $s_e = 0.972$.

We can use the standard error of estimate s_e to construct interval estimates that will help us see how dependable our point estimates of y really are. Assume that for each fixed value of x, the corresponding sample values of y are normally distributed about the regression line, and those normal distributions have the same variance. The following interval estimate applies to an *individual y*. (For a confidence interval used to predict the *mean* of all y values for some given x value, see Exercise 20.)

Prediction Interval for *y*

Given the fixed value x_0, the *prediction interval for an individual y* is

$$\hat{y} - E < y < \hat{y} + E$$

where the margin of error E is

$$E = t_{\alpha/2} s_e \sqrt{1 + \frac{1}{n} + \frac{n(x_0 - \overline{x})^2}{n(\Sigma x^2) - (\Sigma x)^2}}$$

and

x_0 represents the given value of x, $t_{\alpha/2}$ has $n - 2$ degrees of freedom, and s_e is found from Formula 9-7.

▶ **EXAMPLE**

Refer to the Table 9-1 sample data listing weights (x) of discarded plastic along with household sizes (y). In previous sections we have shown that

- There is a significant linear correlation (at the 0.05 significance level).
- The regression equation is $\hat{y} = 0.549 + 1.48x$.
- When $x = 2.50$, the predicted y value is 4.25 persons.

Construct a 95% prediction interval for the size of an individual household that discards 2.50 lb of plastic. This will give you a sense of how reliable the estimate of 4.25 persons really is.

● **SOLUTION**

We have already used the Table 9-1 sample data to find the following values:

$$n = 8 \qquad \bar{x} = 1.825 \qquad \Sigma x = 14.60 \qquad \Sigma x^2 = 32.9632$$
$$s_e = 0.971554$$

From Table A-3 we find $t_{\alpha/2} = 2.447$. (We used $8 - 2 = 6$ degrees of freedom with $\alpha = 0.05$ in two tails.) We can now calculate the margin of error E by letting $x_0 = 2.50$, because we want the prediction interval of y for $x = 2.50$.

$$E = t_{\alpha/2} s_e \sqrt{1 + \frac{1}{n} + \frac{n(x_0 - \bar{x})^2}{n(\Sigma x^2) - (\Sigma x)^2}}$$

$$= (2.447)(0.971554) \sqrt{1 + \frac{1}{8} + \frac{8(2.50 - 1.825)^2}{8(32.9632) - (14.60)^2}}$$

$$= (2.447)(0.971554)(1.09413) = 2.60$$

With $\hat{y} = 4.25$ and $E = 2.60$, we get the prediction interval

$$\hat{y} - E < y < \hat{y} + E$$
$$4.25 - 2.60 < y < 4.25 + 2.60$$
$$1.65 < y < 6.85$$

That is, for a household that discards 2.50 lb of plastic in a week, the predicted household size is between 1.65 persons and 6.85 persons. That's a relatively large range.

In addition to knowing that the predicted household size is 4.25 persons, we now have a sense of how reliable that estimate really is. The 95% prediction interval found above shows that the 4.25 estimate can vary substantially.

9-4 Exercises A: Basic Skills and Concepts

In Exercises 1–4, use the value of the linear correlation coefficient r to find the coefficient of determination and the percentage of the total variation that can be explained by the regression line.

1. $r = 0.333$
2. $r = -0.444$
3. $r = 0.800$
4. $r = -0.700$

In Exercises 5–8, find the (a) explained variation, (b) unexplained variation, (c) total variation, (d) coefficient of determination, and (e) standard error of estimate s_e.

5. The following table lists numbers x of patio tiles and costs y (in dollars) of having them manually cut to fit.

x	1	2	3	5	6
y	5	8	11	17	20

   (The equation of the regression line is $\hat{y} = 2 + 3x$.)

6. The paired data below consist of the total weights (in pounds) of discarded garbage and sizes of households.

Total weight	10.76	19.96	27.60	38.11	27.90	21.90	21.83	49.27	33.27	35.54
Household size	2	3	3	6	4	2	1	5	6	4

   (The equation of the regression line is $\hat{y} = 0.182817 + 0.119423x$.)

7. The table lists emissions data (in grams per meter) of hydrocarbon (HC) and carbon monoxide (CO) for a sample of different vehicles.

HC	0.65	0.55	0.72	0.83	0.57	0.51	0.43	0.37
CO	14.7	12.3	14.6	15.1	5.0	4.1	3.8	4.1

   (The equation of the regression line is $\hat{y} = -8.21713 + 30.1160x$.)

8. The table lists NASA's Landsat ground measurements of snow depth (in centimeters) along with the corresponding temperatures (in degrees Celsius).

Temperature	−62	−41	−36	−26	−33	−56	−50	−66
Snow depth	21	13	12	3	6	22	14	19

   (The equation of the regression line is $\hat{y} = -6.49522 - 0.437735x$.)

9. Refer to the data given in Exercise 5 and assume that the necessary conditions of normality and variance are met.
   a. For $x = 4$, find $\hat{y}$, the predicted value of y.
   b. How does the value of s_e affect the construction of the 95% prediction interval of y for $x = 4$?

10. Refer to Exercise 6 and assume that the necessary conditions of normality and variance are met.
   a. For a household discarding $x = 20.0$ lb of garbage, find the predicted household size.
   b. Find the 99% prediction interval of y for $x = 20.0$.

11. Refer to the data given in Exercise 7 and assume that the necessary conditions of normality and variance are met.
    a. For $x = 0.75$, find $\hat{y}$, the predicted value of y.
    b. Find the 95% prediction interval of y for $x = 0.75$.

12. Refer to the data given in Exercise 8 and assume that the necessary conditions of normality and variance are met.
    a. For $x = -50°C$, find $\hat{y}$, the predicted value of y.
    b. Find the 99% prediction interval of y for $x = -50°C$.

In Exercises 13–16, refer to the Table 9-1 sample data. Let x represent the weight of discarded plastic (in pounds), and let y represent the size of the household. Construct a prediction interval estimate of an individual household size that corresponds to the given weight of discarded plastic. Use the given degree of confidence. (See the example in this section.)

13. 1.00 lb; 95% confidence
14. 3.00 lb; 90% confidence
15. 1.50 lb; 90% confidence
16. 0.75 lb; 99% confidence

9-4 Exercises B: Beyond the Basics

17. Confidence intervals for the y-intercept β_0 and slope β_1 for a regression line $(y = \beta_0 + \beta_1 x)$ can be found by evaluating the limits in the intervals below.

$$\hat{\beta}_0 - E < \beta_0 < \hat{\beta}_0 + E$$

where

$$E = t_{\alpha/2} \cdot s_e \cdot \sqrt{\frac{1}{n} + \frac{\overline{x}^2}{\Sigma x^2 - \dfrac{(\Sigma x)^2}{n}}}$$

$$\hat{\beta}_1 - E < \beta_1 < \hat{\beta}_1 + E$$

where

$$E = t_{\alpha/2} \cdot \frac{s_e}{\sqrt{\Sigma x^2 - \dfrac{(\Sigma x)^2}{n}}}$$

In these expressions, the y-intercept $\hat{\beta}_0$ and the slope $\hat{\beta}_1$ are found from the sample data and $t_{\alpha/2}$ is found from Table A-3 by using $n - 2$ degrees of freedom. Using the garbage data in Table 9-1, find the 95% confidence interval estimates of β_0 and β_1.

18. a. Assuming that a collection of paired data includes at least 3 pairs of values, what do you know about the linear correlation coefficient if $s_e = 0$?
    b. If a collection of paired data is such that the total explained variation is 0, what can be deduced about the slope of the regression line?

19. a. Find an expression for the unexplained variation in terms of the sample size n and the standard error of estimate s_e.

b. Find an expression for the explained variation in terms of the coefficient of determination r^2 and the unexplained variation.

c. Suppose we have a collection of paired data for which $r^2 = 0.900$ and the regression equation is $\hat{y} = 3 - 2x$. Find the linear correlation coefficient.

20. The formula

$$s_{\hat{y}} = s_e \sqrt{1 + \frac{1}{n} + \frac{n(x_0 - \bar{x})^2}{n(\Sigma x^2) - (\Sigma x)^2}}$$

gives the **standard error of the prediction** when predicting for a *single y*, given that $x = x_0$. When predicting for the *mean* of *all* values of y for which $x = x_0$, the point estimate $\hat{y}$ is the same, but $s_{\hat{y}}$ is described as follows:

$$s_{\hat{y}} = s_e \sqrt{\frac{1}{n} + \frac{n(x_0 - \bar{x})^2}{n(\Sigma x^2) - (\Sigma x)^2}}$$

Extend the last example of this section to find a point estimate and a 95% confidence interval estimate of the *mean* size of *all* households that discard 2.50 lb of plastic. (Use the same data from Table 9-1.)

9-5 Multiple Regression

So far, this chapter has used examples and exercises involving *two* variables, but there are cases in which better results can be obtained by using three or more variables. Consider, for example, the garbage data in Table 9-1, which consists of weights of discarded plastic and the corresponding household sizes. We have analyzed this data set in an attempt to develop a method for predicting population size based on the amount of garbage that is discarded. As we have mentioned, the ability to make such population predictions is important to the Census Bureau and those interested in market research. We have found that there is a significant correlation between weights of discarded plastic and household sizes; we have also found that the regression equation $\hat{y} = 0.549 + 1.48x$ is useful for predicting household size based on the discarded plastic. Can better results be obtained if, instead of expressing household size in terms of only plastic, we express household size in terms of plastic and some other variables, such as paper, metal, and glass? To address this question, we must develop an equation with the dependent variable y (household size) expressed in terms of independent variables x_1, x_2, and so on. As in the previous sections of this chapter, we will work with linear relationships only. The following are examples of linear equations:

$$y = mx + b \qquad \hat{y} = \hat{\beta}_0 + \hat{\beta}_1 x \qquad \hat{y} = \hat{\beta}_0 + \hat{\beta}_1 x_1 + \hat{\beta}_2 x_2$$

DEFINITION

A **multiple regression equation** expresses a linear relationship between a dependent variable y and two or more independent variables $(x_1, x_2, \ldots, x_k)$.

We will use the following notation, which follows naturally from the notation used in Section 9-3.

Notation

$$\hat{y} = \hat{\beta}_0 + \hat{\beta}_1 x_1 + \hat{\beta}_2 x_2 + \cdots + \hat{\beta}_k x_k \qquad \text{General form of the estimated multiple regression equation}$$

n = sample size

k = number of *independent* variables. (The independent variables are also called **predictor variables** or x variables.)

$\hat{y}$ = predicted value of the dependent variable y ~ *response*

$x_1, x_2, \ldots, x_k$ are the independent variables.

β_0 = constant

$\hat{\beta}_0$ = sample estimate of the constant β_0

$\beta_1, \beta_2, \ldots, \beta_k$ are the coefficients of the independent variables $x_1, x_2, \ldots, x_k$.

$\hat{\beta}_1, \hat{\beta}_2, \ldots, \hat{\beta}_k$ are the sample estimates of the coefficients $\beta_1, \beta_2, \ldots, \beta_k$.

█ Using Computer Software for Multiple Regression

In previous sections of this chapter we were able to develop procedures that could be used with a calculator, but the computations required for multiple regression are so complex that *we will assume that you have access to a statistical software package.* We will describe procedures for entering the data and interpreting the computer results.

STATDISK and Minitab both produce multiple regression equations. With STATDISK, use the `Data` option from the main menu bar to create different data sets for the different variables. Then select the main menu item `Statistics`, then `Multiple Regression`, and proceed to select the independent and dependent variables as requested. STATDISK will provide the multiple regression equation, the multiple coefficient of determination R^2, and the adjusted R^2. Minitab provides much more.

We will describe how to obtain and interpret the Minitab results. Let's consider the data of Table 9-3, which lists only the first eight values for each

Constant

TABLE 9-3   Data Obtained from the Garbage Project

Variable	Minitab Column	Name								
y *dependent*	C1	HH Size	2	3	3	6	4	2	1	5
x_2	C2	Metal	1.09	1.04	2.57	3.02	1.50	2.10	1.93	3.57
x_3	C3	Paper	2.41	7.57	9.55	8.82	8.72	6.96	6.83	11.42
x_4	C4	Plastic	0.27	1.41	2.19	2.83	2.19	1.81	0.85	3.05
x_5	C5	Glass	0.86	3.46	4.52	4.92	6.31	2.49	0.51	5.81
x_6	C6	Food	1.04	3.68	4.43	2.98	6.30	1.46	8.82	9.62
x_7	C7	Yard	0.38	0.00	0.24	0.63	0.15	4.58	0.07	4.76
x_8	C8	Textile	0.05	0.46	0.50	2.26	0.55	0.36	0.60	0.21
x_9	C9	Other	4.66	2.34	3.60	12.65	2.18	2.14	2.22	10.83
x_{10}	C10	Total	10.76	19.96	27.60	38.11	27.90	21.90	21.83	49.27

Independent Predictor Variables

variable included in Data Set 1 of Appendix B (data obtained from the Garbage Project at the University of Arizona). The values of y are the sizes of the households surveyed, and the values for x_2 through x_{10} are weights (in pounds) of different categories of garbage discarded by a household in one week. (By not using a variable x_1, we have a nice relationship with x_2 corresponding to Minitab's column C2, x_3 corresponding to C3, and so on.) Let's assume that we want to obtain and analyze the regression equation that expresses the household size y in terms of the independent variables x_4 (for plastic) and x_5 (for glass). With Minitab, enter the sample data using the READ command, as shown below.

```
MTB > READ C1 C4 C5
DATA> 2   0.27   0.86
DATA> 3   1.41   3.46
         .
         .
         .
DATA> 5   3.05   5.81
DATA> ENDOFDATA     .
```

The columns can be given meaningful names by using Minitab's NAME command.

```
MTB > NAME C1 'HHSIZE'
MTB > NAME C2 'METAL'
MTB > NAME C3 'PAPER'
         .
         .
MTB > NAME C10 'TOTAL'
```

[handwritten margin note: measures how sample data points deviate from regression line]

[handwritten note: smaller = closer   larger = further away]

Now enter the Minitab command REGRESSION C1 2 C4 C5, which requests multiple regression data for the variable C1 (household size) expressed in terms of the two variables C4 (plastic) and C5 (glass). The result will be the Minitab display shown.

MINITAB DISPLAY

[handwritten note: Coeff of variable x_1]   *[handwritten note: Coeff of y & z]*

```
The regression equation is
HHSIZE = 0.568 + 0.870 PLASTIC + 0.303 GLASS  ←   ①  Multiple regression
                                                       equation

Predictor      Coef      Stdev     t-ratio        p
Constant       0.5676    0.8052    0.70        0.512
PLASTIC        0.8698    0.8296    1.05        0.342
GLASS          0.3033    0.3623    0.84        0.441

s = 0.9967        R-sq = 74.5%      R-sq(adj) = 64.3%

Analysis of Variance                            ②  Adjusted R²= 0.643

SOURCE         DF        SS        MS          F         p
Regression     2      14.5329    7.2665     7.31      0.033
Error          5       4.9671    0.9934
Total          7      19.5000               ③  Overall significance
                                               of multiple regression
SOURCE         DF      SEQ SS                   equation
PLASTIC        1      13.8366
GLASS          1       0.6963
```

[handwritten annotations: Constant; B_4; B_5; adjusts for the # of variables & sample size; good = 1 poor = 0; $R^2 = 0.745$ amount of variation explained by regression line; Small is good #; measures how well the equation fits the data / sample]

The annotated Minitab display has these three key components, identified with circled numbers:

1. Multiple regression equation:

$$HHSIZE = 0.568 + 0.870 \ PLAS + 0.303 \ GLASS$$

2. Adjusted $R^2 = 0.643$

3. Overall significance of multiple regression equation: $p = 0.033$

We will now briefly address each of these three important components. Because we are focusing on interpretation of computer displays and because the underlying theory is quite complicated, we will include very little theory. For a more thorough discussion, refer to a book with more thorough coverage, such as *Elementary Business Statistics* by Triola and Franklin.

Multiple Regression Equation   The multiple regression equation of

$$HHSIZE = 0.568 + 0.870 \ PLAS + 0.303 \ GLASS$$

can be expressed in our standard notation as

$$\hat{y} = 0.568 + 0.870x_4 + 0.303x_5$$

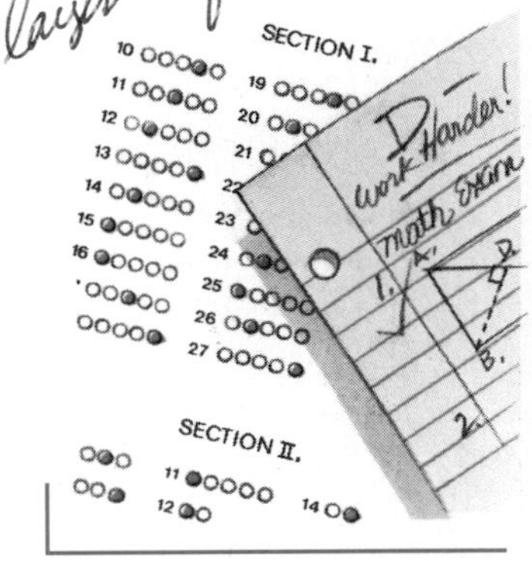

Predictors for Success

When a college accepts a new student, it would like to have some positive indication that the student will be successful in his or her studies. College admissions deans consider SAT scores, standard achievement tests, rank in class, difficulty of high school courses, high school grades, and extracurricular activities. In a study of characteristics that make good predictors of success in college, it was found that class rank and scores on standard achievement tests are better predictors than SAT scores. A multiple regression equation with college grade-point average predicted by class rank and achievement test score was not improved by including another variable for SAT score. This particular study suggests that SAT scores should not be included among the admissions criteria, but supporters argue that SAT scores are useful for comparing students from different geographic locations and high school backgrounds.

Making Music with Multiple Regression

Sony manufactures millions of compact discs in Terre Haute, Indiana. At one step in the manufacturing process, a laser exposes a photographic plate so that a musical signal is transferred into a digital signal coded with 0s and 1s. This process was statistically analyzed to identify the effects of different variables, such as the length of exposure and the thickness of the photographic emulsion. Methods of multiple regression showed that among all of the variables considered, four were most significant. The photographic process was adjusted for optimal results based on the four critical variables. As a result, the percentage of defective discs dropped and the tone quality was maintained. The use of multiple regression methods led to lower production costs and better control of the manufacturing process.

This equation best fits the given data according to the same *least-squares* criterion described in Section 9-3. That is, when we use the Table 9-3 data for household size, weight of plastic, and weight of glass, the above equation fits the data best; if we used *other* data, we might find another equation that fits the data better. (Later, we will discuss determination of the regression equation that fits the data best.) If the equation fits the data well, it can be used for predictions. For example, if we determine that the above equation is suitable for predictions, we can predict the size of a household that discards 3.10 lb of plastic and 5.25 lb of glass by substituting $x_4 = 3.10$ and $x_5 = 5.25$ in the multiple regression equation to get a predicted household size of $\hat{y} = 4.86$ persons. Also, the coefficients of $\hat{\beta}_4 = 0.870$ and $\hat{\beta}_5 = 0.303$ can be used to determine *marginal change*, as described in Section 9-3. For example, the coefficient of $\hat{\beta}_4 = 0.870$ shows that when the weight of glass is held constant, the predicted household size increases by 0.870 person for each 1-lb increase in the weight of discarded plastic.

Adjusted R^2 R^2 denotes the **multiple coefficient of determination,** which is a measure of how well the multiple regression equation fits the sample data. A perfect fit would result in $R^2 = 1$. A very good fit results in a value near 1. A very poor fit results in a value of R^2 close to 0. The value of $R^2 = 0.745$ in the Minitab display indicates that 74.5% of the variation in household size can be explained by the weight of plastic x_4 and the weight of glass x_5. The multiple coefficient of determination R^2 is a measure of how well the regression equation fits the sample data, but it has a serious flaw: As more variables are included, R^2 increases. (Actually, R^2 could remain the same, but it usually increases.) Although the largest R^2 is thus achieved by simply including all of the available variables, the best multiple regression equation does not necessarily use all of the available variables. Therefore it is better to use the adjusted coefficient of determination when comparing different multiple regression equations, because it adjusts the value based on the number of variables and the sample size.

DEFINITION

The **adjusted coefficient of determination** is the multiple coefficient of determination R^2 modified to account for the number of variables and the sample size. It is calculated by using Formula 9-8.

Formula 9-8
$$\text{Adjusted } R^2 = 1 - \frac{(n-1)}{[n-(k+1)]}(1 - R^2)$$

The Minitab display shows the adjusted coefficient of determination as R-sq(adj) = 64.3%. If we use Formula 9-8 with the R^2 value of 0.745, $n = 8$, and $k = 2$, we find that the adjusted R^2 value is 0.643, confirming Minitab's displayed value of 64.3%. For the data in Table 9-3 for household size, plastic, and glass, the R^2 value of 74.5% indicates that 74.5% of the variation in household size can be explained by the weights of plastic x_4 and glass x_5; however, when we compare this multiple regression equation to others, it is better to use the adjusted R^2 of 64.3% (or 0.643).

Overall Significance: _P_-Value   The annotated Minitab display shows a P-value of 0.033. This value is a measure of the overall significance of the multiple regression equation. In this case, the _small_ value of 0.033 suggests that the multiple regression equation has _good_ overall significance and is usable for predictions. That is, it makes sense to predict household size based on the weights of discarded plastic and glass. Like the adjusted R^2, this P-value is a good measure of how well the equation fits the sample data. The value of 0.033 results from a test of the null hypothesis that $\beta_4 = \beta_5 = 0$. Rejection of $\beta_4 = \beta_5 = 0$ implies that at least one of β_4 and β_5 is not 0, suggesting that it is effective in determining the value of household size.

A complete analysis of the Minitab results might include other important elements, such as the significance of the individual coefficients, but we will limit our discussion to the three key components previously identified.

Finding the Best Multiple Regression Equation

So far, we have considered only the multiple regression equation relating household size to the independent variables of plastic and glass, but there are several other variables that could be included. Perhaps some other combination of variables would result in an equation that is better in the sense that it fits the sample data better and is more likely to provide better predictions of household size. Table 9-4 lists a few of the key components for various combinations of variables.

Although _determination of the best multiple regression equation is often quite difficult and beyond the scope of this book,_ we can provide some help with the following guidelines.

1.   Use common sense and practical considerations to include or exclude variables.

TABLE 9-4   Searching for the Best Multiple Regression Equation

	Plastic	Plastic/Glass	Metal/Plastic	Food/Yard	Paper/ Plastic/Glass	Metal/Paper/Plastic/ Glass/Text
R^2	0.710	0.745	0.723	0.036	0.844	0.915
Adjusted R^2	0.661	0.643	0.613	0.000	0.727	0.701
Overall significance	0.009	0.033	0.040	0.912	0.043	0.200

Predicting Wine Before Its Time

Princeton University economist Orley Ashenfelter applies regression analysis to use weather as a predictor of the quality and price of vintage wines. He includes these variables: rainfall preceding the growing season, average growing season temperature, and rainfall during harvest. Ashenfelter states, "Predicting the quality and price of a wine could be like predicting any other market item. All you need is the right equation and the right values for your variables." He successfully tested his multiple regression equation on past results, finding that wine auction prices confirmed his prediction of quality.

2. Instead of including almost every available variable, include relatively few independent (x) variables. In weeding out independent variables that don't have an effect on the dependent variable, it might be helpful to find the linear correlation coefficient r using the dependent variable and each individual independent variable. (This is discussed in more detail later in the section.)

3. Select an equation having a value of adjusted R^2 with this property: If an additional independent variable is included, the value of adjusted R^2 does not increase by a substantial amount.

4. For a given number of independent (x) variables, select the equation with the largest value of adjusted R^2. That is, choose those variables with the property that no other combination of the same number of independent variables will yield a larger value of adjusted R^2.

5. Select an equation having overall significance, as determined by the P-value in the computer display.

Using the above guidelines in an attempt to find the best equation for predicting population size based on discarded garbage, we might follow a reasoning process such as the following.

1. As we consider the sample data in Table 9-3, common sense suggests that we eliminate the category "Other" because it is difficult to find weights for this miscellaneous category. It is much easier to find weights for well-identified categories such as metal, paper, plastic, and glass, especially if recycling regulations require that they be separated. Also, we might question whether the category "Total" should be included because we are including the separate components that make up the total weight.

2. Instead of using all nine independent variables, we will try to omit those that have little or no effect on the household size. The Minitab command `CORRELATE C1 C2-C10` provides a display that includes the following linear correlation coefficients.

```
MTB > CORRELATE C1 C2-C10
              HHSIZE
    METAL      0.608
    PAPER      0.630
    PLAS       0.842
    GLASS      0.830
    FOOD       0.127
    YARD       0.156
    TEXT       0.610
    OTHER      0.817
    TOTAL      0.794
```

The above results suggest that we might exclude the variables for food and yard waste because their correlation coefficients are so small. The results also suggest that plastic is the best single predictor, with glass not too far behind.

3. Examination of Table 9-4 shows that when we use plastic alone, the adjusted R^2 is 0.661, and the inclusion of a second variable does not increase the adjusted

R^2. The adjusted R^2 increases from 0.661 to 0.727 when we use the *three* independent variables of paper, plastic, and glass, but we must weigh that gain against the cost of going from one independent variable to three. We subjectively judge that the increase in R^2 isn't worth the cost of using three independent variables instead of one.

4. If we use only one independent variable, it should be the variable representing the weight of plastic. The adjusted R^2 for plastic is 0.661, and that's higher than the adjusted R^2 for any other single variable.

5. With only plastic as the independent variable, the regression equation $\hat{y} = 0.549 + 1.48x$ is significant, as established in Section 9-3.

For cases involving a large number of independent variables, many statistical software packages include a program for performing **stepwise regression**, whereby different combinations are tried until the best model is obtained. In Minitab, for example, the command STEPWISE C1 C2-C8 (which excludes the variables for OTHER and TOTAL) will provide a display suggesting that the best regression equation is the one in which plastic is the only independent variable. It appears that we can estimate the population of a region based on the weight of discarded plastic. If the weight of plastic discarded in a week is 322 lb, then the population is estimated as $\hat{y} = 0.549 + 1.48(322) = 477$ persons.

Authors William Rathje and Cullen Murphy write in *Rubbish! The Archeology of Garbage* that "For a neighborhood of some 100 households, the projected total population estimate derived from the Garbage Project equations applied to five weeks' worth of garbage will be accurate to within plus or minus 2.5 percent. That is considerably better than the Census Bureau can do in many places—and indeed, better than the Census Bureau actually did for some groups."

When we discussed regression in Section 9-3, we listed four common errors that should be avoided when using regression equations to make predictions. These same errors should be avoided when using multiple regression equations. Be especially careful about concluding that a cause-effect relationship exists. For example, in this section we expressed the household size in terms of the weight of discarded plastic. It's reasonable to believe that a larger household produces more plastic garbage, but it would be foolish to think that population size can be reduced by getting people to discard less plastic.

9-5   Exercises A: Basic Skills and Concepts

In Exercises 1–4, use the following regression equation:

$$\hat{y} = 34.8 + 1.21x_1 + 0.23x_2$$

Here $\hat{y}$ is the predicted calculus grade, x_1 is the score on an algebra placement test, and x_2 is the high school rank expressed as a percentile. (The equation is based on data from "Factors Affecting Achievement in the First Course in Calculus," by Edge and Friedberg, *Journal of Experimental Education*, Vol. 52, No. 3.)

1. Find the predicted calculus grade if the score on the algebra pretest (x_1) is 24 and the high school rank is the 92nd percentile.
2. Find the predicted calculus grade if the score on the algebra pretest (x_1) is 12 and the high school rank is the 71st percentile.
3. Find the predicted calculus grade if the score on the algebra pretest (x_1) is 18 and the high school rank is the 81st percentile.
4. Find the predicted calculus grade if the score on the algebra pretest (x_1) is 31 and the high school rank is the 99th percentile.

In Exercises 5–8, refer to the accompanying Minitab display to answer the given questions. This display can be obtained from the data in Table 9-3 by entering the Minitab command REGRESSION C1 3 C2 C4 C6.

```
MTB > REGRESSION C1 3 C2 C4 C6

The regression equation is
HHSIZE = 0.92 - 0.244 METAL + 1.75 PLASTIC - 0.073 FOOD

Predictor        Coef       Stdev     t-ratio        p
Constant        0.924       1.093        0.85      0.445      —no corral
METAL          -0.2440      0.8264      -0.30      0.783      —no corral
PLASTIC         1.7476      0.7554       2.31      0.082      ← correlate
FOOD           -0.0732      0.1472      -0.50      0.645      no corral

s = 1.127    R-sq = 73.9%    R-sq(adj) = 54.4%

Analysis of Variance

SOURCE         DF         SS         MS        F         p
Regression      3      14.419      4.806     3.78     0.116
Error           4       5.081      1.270
Total           7      19.500

SOURCE         DF      SEQ SS
METAL           1       7.211
PLASTIC         1       6.894
FOOD            1       0.314
```

5. Identify the multiple regression equation that expresses the household size in terms of the weights of discarded metal, plastic, and food.
6. Identify
   a. The *P*-value corresponding to the overall significance of the multiple regression equation   .116
   b. The value of the multiple coefficient of determination R^2
   c. The adjusted value of R^2

7. Is the multiple regression equation usable for predicting household size based on weights of discarded metal, plastic, and food? Why or why not?

8. An individual household discards 2.25 lb of metal, 2.95 lb of plastic, and 4.27 lb of food.

   a. Find the predicted household size by substituting the given values in the multiple regression equation.

   b. Find the predicted household size by using the regression equation $\hat{y} = 0.549 + 1.48x$, with plastic as the only independent variable.

   c. Which of the two predicted household sizes is likely to be the better prediction? Why?

In Exercises 9 and 10, use the data in Table 9-3 and software such as STATDISK or Minitab.

9. a. Find the multiple regression equation that expresses household size in terms of metal and plastic.

   b. Identify the P-value corresponding to the overall significance of the multiple regression equation.

   c. Identify the value of the multiple coefficient of determination R^2.

   d. Identify the adjusted value of R^2.

   e. Does the multiple regression equation seem suitable for making predictions of household sizes based on weights of discarded metal and plastic?

10. a. Find the multiple regression equation that expresses household size in terms of paper and plastic.

   b. Identify the P-value corresponding to the overall significance of the multiple regression equation.

   c. Identify the value of the multiple coefficient of determination R^2.

   d. Identify the adjusted value of R^2.

   e. Does the multiple regression equation seem suitable for making predictions of household sizes based on weights of discarded paper and plastic?

9-5   Exercises B: Beyond the Basics

11. In some cases, the best-fitting multiple regression equation is of the form $\hat{y} = \hat{\beta}_0 + \hat{\beta}_1 x_1 + \hat{\beta}_2 x_1^2$. The graph of such an equation is a parabola. Let $x_1 = x$, and let $x_2 = x^2$. Use the given values of y, x_1, and x_2 to find the multiple regression equation for the parabola that best fits the given data. Based on the value of the multiple coefficient of determination, how well does this equation fit the given data?

x	1	3	4	7	5
y	5	14	19	42	26

12. Use the following data to answer the given questions.

x	−2.0	−1.0	0.0	1.0	2.0	3.0
y	13.0	4.0	5.0	4.0	13.0	68.0

(continued)

a. Find the linear correlation coefficient and the equation of the regression line.

b. Find the multiple regression equation

$$\hat{y} = \hat{\beta}_0 + \hat{\beta}_1 x + \hat{\beta}_2 x^2$$

(*Hint:* Let $x_1 = x$, and let $x_2 = x^2$.) Also find the value of the multiple coefficient of determination R^2 and the adjusted R^2.

c. Find the multiple regression equation

$$\hat{y} = \hat{\beta}_0 + \hat{\beta}_1 x + \hat{\beta}_2 x^2 + \hat{\beta}_3 x^3$$

(Let $x_1 = x$, let $x_2 = x^2$, and let $x_3 = x^3$.) Also find the value of the multiple coefficient of determination R^2 and the adjusted R^2.

d. Find the multiple regression equation

$$\hat{y} = \hat{\beta}_0 + \hat{\beta}_1 x + \hat{\beta}_2 x^2 + \hat{\beta}_3 x^3 + \hat{\beta}_4 x^4$$

Also find the value of the multiple coefficient of determination R^2 and the adjusted R^2.

e. Based on the preceding results, which regression equation seems to be best?

VOCABULARY LIST

Define and give an example of each term.

bivariate data
correlation
bivariate normal distribution
scatter diagram
linear correlation coefficient
Pearson product moment
   correlation coefficient
centroid
regression equation
regression line
independent variable
dependent variable
marginal change
least-squares property
total deviation

explained deviation
unexplained deviation
total variation
explained variation
unexplained variation
coefficient of determination
prediction interval
standard error of estimate
multiple regression equation
predictor variables
multiple coefficient of
   determination
adjusted coefficient of
   determination
stepwise regression

Important Formulas

$$r = \frac{n\Sigma xy - (\Sigma x)(\Sigma y)}{\sqrt{n(\Sigma x^2) - (\Sigma x)^2}\sqrt{n(\Sigma y^2) - (\Sigma y)^2}}$$ Linear correlation coefficient

$$\hat{\beta}_1 = \frac{n\Sigma xy - (\Sigma x)(\Sigma y)}{n(\Sigma x^2) - (\Sigma x)^2}$$ Slope of regression line

$$\hat{\beta}_0 = \bar{y} - \hat{\beta}_1 \bar{x} \text{ or}$$ y-intercept of regression line

$$\hat{\beta}_0 = \frac{(\Sigma y)(\Sigma x^2) - (\Sigma x)(\Sigma xy)}{n(\Sigma x^2) - (\Sigma x)^2}$$

$$r^2 = \frac{\text{explained variation}}{\text{total variation}}$$ Coefficient of determination

$$s_e = \sqrt{\frac{\Sigma y^2 - \hat{\beta}_0 \Sigma y - \hat{\beta}_1 \Sigma xy}{n - 2}}$$ Standard error of estimate

$$\hat{y} - E < y < \hat{y} + E$$ Prediction interval

where

$$E = t_{\alpha/2}s_e\sqrt{1 + \frac{1}{n} + \frac{n(x_0 - \bar{x})^2}{n(\Sigma x^2) - (\Sigma x)^2}}$$

REVIEW

In this chapter we studied the concepts of linear correlation and regression so that we could analyze paired sample data. We limited our discussion to linear relationships because consideration of nonlinear relationships requires more advanced mathematics. With correlation, we attempted to decide whether there is a significant linear relationship between the two variables. With regression, we attempted to identify the relationship with an equation. Although a scatter diagram provides a graphic display of the paired data, the linear correlation coefficient r and the equation of the regression line serve as more precise and objective tools for analysis.

Given a list of paired data, we can compute the linear correlation coefficient r by using Formula 9-1. We can use one of two methods of testing for a significant linear relationship. The presence of a significant linear correlation does not necessarily mean that there is a direct cause-and-effect relationship between the two variables.

In Section 9-3 we developed procedures for obtaining the equation of the regression line, which, by the least-squares criterion, is the straight line that best fits the paired data. When there is a significant linear correlation, the regression line can be used to predict the value of one variable, given some value of the other variable. The regression equation has the form $\hat{y} = \hat{\beta}_0 + \hat{\beta}_1 x$, where the y-intercept $\hat{\beta}_0$ and the slope $\hat{\beta}_1$ can be found by using the formulas given in this section.

Section 9-4 introduced the concept of total variation, with components of explained and unexplained variation. We defined the coefficient of determination r^2 to be the quotient obtained by dividing explained variation by total variation. We saw that we could measure the amount of spread of sample points about the regression line by the standard error of estimate, s_e. For estimated values of y, we developed prediction intervals, which are helpful in judging the accuracy of predicted values.

In Section 9-5 we considered multiple regression, which allows us to investigate relationships among several variables. We discussed procedures for obtaining a multiple regression equation, as well as the values of the multiple coefficient of determination R^2, the adjusted R^2, and a P-value for the overall significance of the equation. These values give us an indication of how well the multiple regression equation actually fits the available sample data. Because of the nature of the calculations involved, our treatment of multiple regression was based on the use and interpretation of computer software such as STATDISK or Minitab.

REVIEW EXERCISES

In Exercises 1–4, use the given paired data.
- a. Find the value of the linear correlation coefficient r.
- b. Assuming a 0.05 level of significance, find the critical value(s) of r from Table A-6.
- c. Use the results from parts a and b to decide whether there is a significant linear correlation.
- d. Find the equation of the regression line.
- e. Plot the regression line on the scatter diagram.

1. The costs (in thousands of dollars) of used Corvettes are listed along with their odometer readings (in thousands of miles).

Cost	18.9	48.0	26.9	18.0	21.5	17.0	26.0	14.0
Miles	37.1	40.0	65.0	17.8	14.0	34.0	39.0	35.0

Data randomly selected from ads in the *New York Times*.

2. Randomly selected girls are given the Wide Range Achievement Test. Their ages are listed along with their scores on the reading part of that test.

Age	6.1	7.2	5.9	6.3	10.5	11.0
Score	17.8	47.4	25.8	24.3	66.6	91.4

Based on data from the National Health Survey, USDHEW publication 72-1011.

3. At one point during a recent season of the National Basketball Association, *USA Today* reported the current statistics. Given below are the total minutes played and the total points scored by 9 randomly selected players.

Minutes	1364	53	457	717	384	1432	365	1626	840
Points	652	20	163	210	175	821	143	1098	459

4. Fifteen assembly-line workers are randomly selected from the Telektronic Company, a manufacturer of answering machines. Each is tested for dexterity and level of productivity. The results of standard tests are listed below.

Productivity	63	67	88	44	52	106	99	110	75	58	77	91	101	51	86
Dexterity	2	9	4	5	8	6	9	8	9	7	4	10	7	4	6

In Exercises 5–8, use the sample data in the accompanying table. The table includes "minutes before midnight on the doomsday clock," a measure of the perceived threat of nuclear war as established by the *Bulletin of the Atomic Scientists*. The "periodicals" values are indices of print media coverage of issues related to nuclear war. People's "savings" values are expressed in terms of a standard measure of the rate of savings. The data are matched for randomly selected years.

Minutes	2.0	7.0	12.0	12.0	9.25	11.17	9.0	9.0	4.0
Periodicals	0.16	0.21	0.13	0.12	0.12	0.05	0.05	0.06	0.14
Savings	7.2	10.2	13.5	13.5	12.0	13.1	11.6	11.6	8.5

Based on data from "Saving and the Fear of Nuclear War," by Slemrod, *Journal of Conflict Resolution*, Vol. 30, No. 3.

5. a. Use a 0.05 significance level to test for a linear correlation between minutes before midnight on the doomsday clock and the index of savings.
   b. Find the equation of the regression line that expresses savings (y) in terms of minutes before midnight (x).
   c. What is the best predicted index of savings for a year in which the doomsday clock is at 5.0 min before midnight?

6. a. Use a 0.05 significance level to test for a linear correlation between the periodical index and the index of savings.
    b. Find the equation of the regression line that expresses the savings index (y) in terms of the periodical index (x).
    c. What is the best predicted index of savings for a year in which the periodical index is 0.20?

7. Use only the paired minutes/savings data. For a year in which the number of minutes before midnight on the doomsday clock is 5.0, find a 95% prediction interval estimate of the savings index.

8. Let y = savings index, x_1 = minutes before midnight, and x_2 = index of periodicals. Use software such as STATDISK or Minitab to find the multiple regression equation of the form $\hat{y} = \hat{\beta}_0 + \hat{\beta}_1 x_1 + \hat{\beta}_2 x_2$. Also identify the value of the multiple coefficient of determination R^2, the adjusted R^2, and the P-value representing the overall significance of the multiple regression equation.

 COMPUTER PROJECT

Can a person's weight be predicted from his or her height? It's easier to measure the true height of a person, because it's necessary to remove only the shoes and hat and you can use a ruler instead of a scale. Data Set 3 in Appendix B includes data for 125 survey subjects. (The data are from the U.S. National Center for Health Statistics.)

a. Using the 125 pairs of sample data for weights and heights, employ a statistical software package such as STATDISK or Minitab to find the linear correlation coefficient, the coefficient of determination, the equation of the regression line, and a scatter diagram.

b. Using the same software package, test the claim that there is no correlation between weight and height. (That is, test H_0: $\rho = 0$.)

c. Find the predicted weight of someone who is 70 in. tall.

d. Find a 95% prediction interval for the predicted weight of someone who is 70 in. tall.

FROM DATA TO DECISION:

Is IQ the Result of Heredity or Environment?

In studying the effects of heredity and environment on intelligence, it has been helpful to analyze IQs of identical twins who were separated soon after birth. Identical twins share identical genes inherited from the same fertilized egg. By studying identical twins raised apart, we can eliminate the variable of heredity and better isolate the effects of environment. Given below are the IQs of identical twins (older twins are x) raised apart. Find the linear correlation coefficient r, the coefficient of determination r^2, and the equation of the regression line. Then construct a scatter diagram. Based on the results, write a summary statement about the effect of heredity and environment on intelligence. Note that your conclusions will be based on this relatively small sample of 12 pairs of identical twins.

x	107	96	103	90	96	113	86	99	109	105	96	89
y	111	97	116	107	99	111	85	108	102	105	100	93

Based on data from "IQs of Identical Twins Reared Apart," by Arthur Jensen, *Behavioral Genetics*.

Interview

Barry Cook
Senior Vice President at Nielsen Media Research

Barry Cook is a Senior Vice President and Chief Research Officer at Nielsen Media Research. He has taught at Yale and Hunter College and has worked for NBC and the USA Cable Network. He is now in charge of Nielsen's rating system, doing research to better understand how the measurements work, as well as developing new measurement systems.

How did you become involved in your current work with media research?

I used to teach statistics myself. Before getting into the commercial aspect of the research business, I was at Yale in the psychology department, then at Hunter College in the sociology department. I taught an introductory statistics course just about every semester. After teaching for a number of years, I was at Hunter College in an applied social research program where I was training people to go into market research. I looked at where my students were going and I felt it looked good, so I followed them. I got into media research at NBC doing market research on their news programs. I later did ratings work at NBC. After that, I became involved with a cable network and then Nielsen.

What major trends do you see in the way Americans watch TV?

In 1985 the average home received 18.8 channels. In 1990 the average home received 33.2 channels. That obviously has an effect on what people choose to view.

What is your sample size?

For the national survey we use "people meters" in 4000 homes with about 11,000 people. We increased the sample size because the use of television has changed. Instead of only three major sources of TV (ABC, NBC, CBS), there are now dozens of sources of programming, many of which get only a small piece of the audience. In order to measure those smaller pieces with enough precision, a larger sample is needed. In addition we also have meter services in 25 markets; the television sets are metered (but not the people) in 250 to 500 homes per market. Nielsen is still very big in the diary business—not for the national audience, but for measuring audiences in the 200 or so separate markets across the country. Those diaries amount to a combined sample size of 100,000, four times a year.

Have you been experiencing greater resistance to polls and surveys?

There's no question that we have seen a decline in cooperation in both telephone and in-person contacts with people.

It's across the entire survey industry. There are concerns about privacy. The data-gathering efforts are being mixed up with sales efforts and that probably is contributing to a decline in the cooperation rate. Also, answering machines make it harder to get through to people.

Do you weight sample results to better reflect population parameters?

We have a policy against that. We try to represent population parameters by doing the sampling correctly in the first place. We sample in a way that gives an equal probability of selection to all housing units in the 50 states. As a result there is a known amount of sampling error and there's also an unknown amount of nonsampling error, but we've done validation research to estimate how close the samples are to measures of the population.

What are some of the specific statistical methods you use?

We use a lot of statistics for our own understanding and our clients' understanding of sampling and trends. We get into hypothesis testing when we try to understand why things change. Confidence intervals are very important in interpreting the estimates of the population.

Could you cite a television programming strategy that is based on survey results?

The most important strategy is called "prime time." The biggest usage of television occurs in the evening hours when most people are home. The most general programming strategy is to put on your best shows when there's the greatest number of people there. With a miniseries what you get on the first episode serves as almost a cap on what you can get after that. You want to get the maximum possible potential audience for the first installment, so Sunday night does that.

"We try to represent population parameters by doing the sampling correctly in the first place."

10 Multinomial Experiments and Contingency Tables

10-1 Overview

Objectives are identified for this chapter, which introduces methods for dealing with data consisting of frequencies from three or more categories.

10-2 Multinomial Experiments

The goodness-of-fit procedure of hypothesis testing is presented. That procedure is used to test claims that observed frequencies fit, or conform to, a particular distribution. Multinomial experiments are defined and examined.

10-3 Contingency Tables

Contingency tables are defined and described. A standard method is presented for testing claims about the independence between the two variables that categorize rows and columns in contingency tables. Also presented is a method for a test of homogeneity, in which we test a claim that different populations have the same proportions of some characteristics.

Chapter Problem:
Which prosecuting strategy is best?

Kate Malloy is a district attorney assigned to prosecute a case involving "property offenses," more commonly referred to as white-collar crime. The defendant is a bank employee who stole money by altering the bank's computer records and mailing forged withdrawal slips. Because of the nature of the crime, Malloy can charge the defendant with embezzlement, fraud, and forgery. But in return for some cooperation, she has negotiated an arrangement whereby only one of those crimes will be charged. Malloy wants to obtain the maximum possible sentence because the defendant stole large sums of money from people who were hospitalized for long periods of time. Which crime should she choose? Does it make any difference? In attempting to answer these questions, Malloy finds that a researcher randomly selected 400 convictions for white-collar crimes and obtained the results summarized in Table 10-1.

Table 10-1 Crime and Punishment

	Embezzlement	Fraud	Forgery	Total
Sent to jail	22	130	20	172
Not sent to jail	57	146	25	228
Total	79	276	45	400

Based on data from the U.S. Department of Justice, Bureau of Justice Statistics.

She examines the data in the table and finds that for the 400 cases surveyed, the different categories of white-collar crimes are not punished equally. Table 10-1 shows that the rates of jail sentences are as follows: 28% (or 22 out of 79) for convicted embezzlers; 47% (or 130 out of 276) for persons convicted of fraud; and 44% (or 20 out of 45) for convicted forgers. These results seem to suggest that the choice of crime does make a difference in the sentence, because 28%, 47%, and 44% appear to be different. But are those differences *significant*? Is the likelihood of a jail sentence related to the type of white-collar crime committed? We will address these questions in this chapter. The answers will be helpful to Kate Malloy as she plans her strategy for prosecuting the defendant.

10-1 Overview

Chapter 1 defined *categorical* (or *qualitative*, or *attribute*) *data* as data that can be separated into different categories (often called **cells**) that are distinguished by some nonnumeric characteristic. For example, we might separate a sample of M&Ms into the color categories of red, orange, yellow, brown, tan, and green. After finding the frequency count for each category, we might proceed to test the claim that the frequencies fit the color distribution claimed by the manufacturer (Mars, Inc.). In general, this chapter focuses on the analysis of categorical data consisting of frequency counts for the different categories. In Section 10-2 we will consider multinomial experiments, which consist of observed frequency counts arranged in a single row or column. In Section 10-3 we will consider contingency tables, which consist of frequency counts arranged in a table such as Table 10-1.

We will determine that the two major topics of this chapter (multinomial experiments and contingency tables) share some common elements: Unlike many of the preceding methods of inferential statistics, the methods for analyzing multinomial experiments and contingency tables do not require that the data have a particular distribution, such as the normal distribution. Also, the test statistics for both multinomial experiments and contingency tables are approximated by the χ^2 (chi-square) distribution used in Chapters 6 and 7. Recall the following important properties of the chi-square distribution:

1. Unlike the normal and Student t distributions, the chi-square distribution is not symmetric. (See Figure 10-1.)

2. The values of the chi-square distribution can be 0 or positive, but they cannot be negative. (See Figure 10-1.)

3. The chi-square distribution is different for each number of degrees of freedom. (See Figure 10-2.)

Critical values of the chi-square distribution are found in Table A-4.

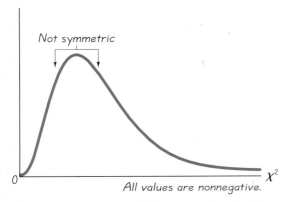

Not symmetric

All values are nonnegative.

χ^2

FIGURE 10-1   The Chi-Square Distribution

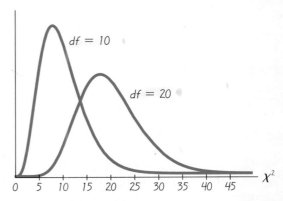

df = 10

df = 20

χ^2

FIGURE 10-2   Chi-Square Distributions for 10 and 20 Degrees of Freedom

10-2   Multinomial Experiments

Chapter 4 defined a binomial experiment to be an experiment with a fixed number of independent trials, probabilities of outcomes that remain constant from trial to trial, and outcomes that belong to one of two categories. The requirement of *two* categories is reflected in the prefix *bi,* which begins the term *binomial*. In this section we consider multinomial experiments, in which each trial yields outcomes belonging to one of several (more than two) categories. Except for this difference, binomial and multinomial experiments are essentially the same.

DEFINITION

A **multinomial experiment** is an experiment that meets the following conditions.

1.   The number of trials is fixed.
2.   The trials are independent.
3.   Each trial must have all outcomes classified into exactly one of several different categories.
4.   The probabilities for the different categories remain constant for each trial.

In this section we present a method for testing a claim that in a multinomial experiment, the frequencies observed in the different categories fit a particular theoretical distribution. Because we test for how well an observed frequency distribution conforms to some theoretical distribution, this method is often called a goodness-of-fit test.

DEFINITION

A **goodness-of-fit test** is used to test the hypothesis that an observed frequency fits (or conforms to) some claimed theoretical distribution.

For example, the Casino Supply Company manufactures dice and claims that they are "fair," meaning that the outcomes are all equally likely. An individual die can be tested by rolling it 60 times to determine whether the frequency of each outcome is reasonably close to the frequency expected with a fair and unbiased die. Here we can use the goodness-of-fit test to test the hypothesis that the die fits a uniform distribution, which is characterized by categories with the same frequencies. Following are other uses of the goodness-of-fit test.

- Test the claim that the proportion of defective cars manufactured at the Saturn assembly plant is the same each day of the week.
- Test Mars, Inc.'s claim that its M&Ms are made so that 30% are brown, 20% are yellow, 20% are red, 10% are orange, 10% are green, and 10% are tan.
- Test the claim that a sample of weights of quarters is drawn from a population having a normal distribution.

Our goodness-of-fit tests will incorporate the following notation.

Notation

O   represents the *observed frequency* of an outcome.

E   represents the *expected frequency,* or theoretical frequency, of an outcome. np (p = probability

k   represents the *number of different categories* or outcomes.

n   represents the total *number of trials.*

In the typical situation requiring a goodness-of-fit test, we have observed frequencies (denoted by O) and must use the claimed distribution to determine the expected frequencies (denoted by E). In many cases, the expected frequency can be found by multiplying the probability p for a category by the number of different trials n, so

$$E = np$$

For example, if we test the claim that a die is fair by rolling it 60 times, we have $n = 60$ (because there are 60 trials) and $p = 1/6$ (because a die is fair if the six possible outcomes are equally likely with the same probability of $1/6$). The expected frequency for each category or cell is therefore

$$E = np$$
$$= (60)(1/6) = 10$$

Assumptions

The following assumptions apply when we test a hypothesis that for the k categories of outcomes in a multinomial experiment, the population proportion for each of the k categories is as claimed.

1. The sample data consist of frequency counts for the k different categories, and the data constitute a random sample.

2. For each of the k categories, the *expected* frequency is at least 5. (There is no requirement that every *observed* frequency must be at least 5.)

We know that samples deviate from what we theoretically expect, so we now present the key question: Are the differences between the actual *observed* values O and the theoretically *expected* values E differences that occur just by chance, or are the differences statistically significant? To answer this question we need some way of measuring the significance of the differences between the observed values and the expected values. We use the following test statistic, which measures the difference between the observed frequencies and the expected frequencies.

Test Statistic for Goodness-of-Fit Tests in Multinomial Experiments

$$\chi^2 = \sum \frac{(O - E)^2}{E}$$

Critical Values

1. Critical values are found in Table A-4 by using $k - 1$ degrees of freedom.

2. Degrees of freedom $= k - 1$, where $k =$ number of categories

The number of degrees of freedom reflects the fact that we can freely assign frequencies to $k - 1$ categories before the frequency for every category is determined. (Although we say that we can "freely" assign frequencies to $k - 1$ categories, we cannot have negative frequencies nor can we have frequencies so large that their sum exceeds the total of the frequencies for all categories combined.)

Simply summing the differences between observed and expected values does not result in an effective measure because that sum is always 0, as shown below.

$$\Sigma(O - E) = \Sigma O - \Sigma E = n - n = 0$$

Squaring the $O - E$ values provides a better statistic, which reflects the differences between observed and expected frequencies, but $\Sigma(O - E)^2$ measures only the magnitude of the differences; we need to find the magnitude of the differences relative to what was expected. This relative magnitude is found through division by the expected frequencies, as in the above test statistic.

The theoretical distribution of $\Sigma(O - E)^2/E$ is a discrete distribution because the number of possible values is limited. The distribution can be approximated by a chi-square distribution, which is continuous. This approximation is generally considered acceptable, provided that all values of E are at least 5. We included this requirement with the assumptions that apply to this section. In Section 5-5 we saw that the continuous normal probability distribution can reasonably approximate the discrete binomial probability distribution, provided that np and nq are both at least 5. We now see that the continuous chi-square distribution can reasonably approximate the discrete distribution of $\Sigma(O - E)^2/E$, provided that all values of E are at least 5. (There are ways of circumventing the problem of an expected frequency that is less

Crime Prevention Strategies

Authors Fran Norris and Krzysztof Kaniasty wrote an article called "A Longitudinal Study of the Effects of Various Crime Prevention Strategies on Criminal Victimization, Fear of Crime, and Psychological Distress" (*American Journal of Community Psychology*). They reported on a study of crime prevention strategies and the effects on crime, fear of crime, and psychological distress. The chi-square distribution was used to verify that there was a good fit between the crime prevention scale and data already collected in another study. The Norris and Kaniasty study found that campaigns to increase personal and household protection have little or no effect on either crime or the fear of crime. The use of locks also seemed to have little effect. Although not highly effective, the best strategy seemed to be informal cooperation among neighbors.

than 5, such as combining categories so that all expected frequencies are at least 5.)

When we use the chi-square distribution as an approximation, we obtain the critical value from Table A-4 after determining the level of significance α and the number of degrees of freedom. In a multinomial experiment with k possible outcomes, the number of degrees of freedom is $k - 1$.

Note that *close agreement* between observed and expected values will lead to a *small* value of χ^2. A large value of χ^2 will indicate strong disagreement between observed and expected values. A significantly large value of χ^2 will thus cause rejection of the null hypothesis of no difference between observed and expected frequencies. Our test is therefore right-tailed because the critical value and critical region are located at the extreme right of the distribution. Unlike previous hypothesis tests in which we had to determine whether the test was left-tailed, right-tailed, or two-tailed, **these goodness-of-fit tests are all right-tailed.**

Once we know how to evaluate the test statistic and critical value, the method of testing hypotheses can be applied by using the same procedure summarized in Figure 7-4.

▶ **EXAMPLE**

A study was made of 147 industrial accidents that required medical attention. The sample data are summarized in Table 10-2. Test the claim that accidents occur with equal proportions on the 5 workdays.

TABLE 10-2   Frequency of Accidents

Day	Mon	Tues	Wed	Thurs	Fri
Observed accidents	31	42	18	25	31

Based on results from "Counted Data CUSUM's," by Lucas, *Technometrics*, Vol. 27, No. 2.

● **SOLUTION**

We will use the same method of testing hypotheses outlined in Figure 7-4, but let's first examine the claim that accidents occur with equal proportions on the 5 different days. This is really a claim about the distribution of accidents. If the claim is true, the probability of an accident on any given workday is

TABLE 10-3   Observed and Expected Frequencies

Day	Mon	Tues	Wed	Thurs	Fri
Observed accidents	31	42	18	25	31
Expected accidents	29.4	29.4	29.4	29.4	29.4

$p = 1/5$. Table 10-2 includes data for 147 trials (industrial accidents), so $n = 147$. With $n = 147$ and $p = 1/5$, we can find the expected number of accidents for each day as follows:

$$E = np = (147)(1/5) = 29.4$$

Table 10-3 lists the observed frequencies along with the frequencies we expect, assuming that the claim of equal proportions is true. (The claim is made about proportions, but we use frequencies instead.)

We want to determine whether the differences between observed and expected frequencies are significant. Having identified the observed frequencies O and having found the expected frequencies E, we will now follow our standard procedure for testing hypotheses.

Step 1: The original claim is that the accidents occur with equal proportions on the 5 days. This can be represented as $p_1 = p_2 = p_3 = p_4 = p_5$.

Step 2: If the original claim is false, then at least 1 of the 5 proportions is different.

Step 3: The null hypothesis must contain the condition of equality, so we have

H_0: $p_1 = p_2 = p_3 = p_4 = p_5$
H_1: At least 1 of the 5 proportions is different from the others.

Step 4: No significance level was specified, so we select $\alpha = 0.05$, a very common choice.

Step 5: Because we are dealing with categorical data that are frequencies and because we are testing a claim about the distribution of the data, we use the goodness-of-fit test described in this section. The χ^2 distribution is used with the test statistic given earlier.

Step 6: Using the observed frequencies O and the expected frequencies E as listed in Table 10-3, we find the value of the test statistic. It is often helpful to organize the calculations in a table such as Table 10-4. (More often, it is *really* helpful to use computer software.)

(continued)

TABLE 10-4   Calculating the χ^2 Test Statistic for Accident Data

Category	Observed Frequency O	Expected Frequency E	$O - E$	$(O - E)^2$	$\dfrac{(O - E)^2}{E}$
Mon	31	29.4	1.6	2.56	0.0871
Tues	42	29.4	12.6	158.76	5.4000
Wed	18	29.4	−11.4	129.96	4.4204
Thurs	25	29.4	−4.4	19.36	0.6585
Fri	31	29.4	1.6	2.56	0.0871

$$\chi^2 = \sum \frac{(O - E)^2}{E} = 10.6531$$

The test statistic is $\chi^2 = 10.653$ (rounded). The critical value is $\chi^2 = 9.488$ (found in Table A-4 with $\alpha = 0.05$ in the right tail and degrees of freedom equal to $k - 1 = 5 - 1 = 4$). The test statistic and critical value are shown in Figure 10-3.

Step 7:   Because the test statistic falls within the critical region, there is sufficient evidence to reject the null hypothesis.

Step 8:   There is sufficient evidence to warrant rejection of the claim that the accidents occur with equal proportions on the 5 workdays. (Although it appears that Wednesday has a lower accident rate, arriving at such a conclusion would require other methods of analysis.)

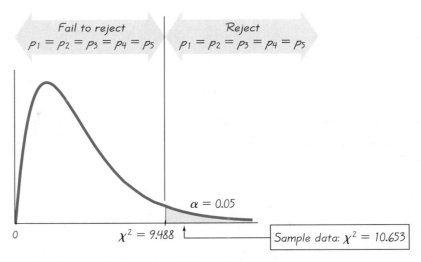

FIGURE 10-3   Goodness-of-Fit Test of $p_1 = p_2 = p_3 = p_4 = p_5$

The preceding example dealt with the null hypothesis that the frequencies of accidents on the 5 workdays are all equal. The theory and methods presented can also be used when the claimed frequencies are different, as in the next example.

► **EXAMPLE**

Mars, Inc. claims that its M&M candies are distributed with the color percentages of 30% brown, 20% yellow, 20% red, 10% orange, 10% green, and 10% tan. A classroom exercise resulted in the observed frequencies listed in Table 10-5. At the 0.05 significance level, test the claim that the color distribution is as claimed by Mars, Inc.

TABLE 10-5   Frequencies of M&Ms

	Brown	Yellow	Red	Orange	Green	Tan
Observed frequency	84	79	75	49	36	47
Expected frequency	111	74	74	37	37	37

Based on data from "Testing Colour Proportions of M&M's," by Roger W. Johnson, *Teaching Statistics*, Vol. 15, No. 1.

● **SOLUTION**

We extended Table 10-5 so that it includes the expected frequencies, which are calculated as follows. For n, we use the total number of trials (370), which is the total number of M&Ms observed. For the probabilities, we use the decimal equivalents of the claimed percentages (30%, 20%, . . . , 10%).

$$\text{Brown: } E = np = (370)(0.30) = 111$$
$$\text{Yellow: } E = np = (370)(0.20) = 74$$
$$\vdots$$
$$\text{Tan: } \quad E = np = (370)(0.10) = 37$$

In testing the given claim, Steps 1, 2, and 3 result in the following hypotheses:

H_0: $p_b = 0.3$ and $p_y = 0.2$ and $p_r = 0.2$ and $p_o = 0.1$ and $p_g = 0.1$ and $p_t = 0.1$

H_1: At least one of the above proportions is different from the claimed value.

Steps 4, 5, and 6 lead us to use the goodness-of-fit test with a 0.05 significance level and a test statistic calculated from Table 10-6.   *(continued)*

TABLE 10-6   Calculating the χ^2 Test Statistic for M&M data

Category	Observed Frequency O	Expected Frequency $E = np$	$O - E$	$(O - E)^2$	$\dfrac{(O - E)^2}{E}$
Brown	84	$370 \cdot 0.30 = 111$	-27	729	6.5676
Yellow	79	$370 \cdot 0.20 = 74$	5	25	0.3378
Red	75	$370 \cdot 0.20 = 74$	1	1	0.0135
Orange	49	$370 \cdot 0.10 = 37$	12	144	3.8919
Green	36	$370 \cdot 0.10 = 37$	-1	1	0.0270
Tan	47	$370 \cdot 0.10 = 37$	10	100	2.7027

$$\chi^2 = \sum \frac{(O - E)^2}{E} = 13.5405$$

The χ^2 test statistic is 13.541 (rounded). The critical value of χ^2 is 11.071, and it is found in Table A-4 (using $\alpha = 0.05$ in the right tail with $k - 1 = 5$ degrees of freedom). The test statistic and critical value are shown in Figure 10-4. Because the test statistic falls within the critical region, there is sufficient evidence to warrant rejection of the null hypothesis. There is sufficient evidence to warrant rejection of the claim that the colors are distributed with the percentages given.

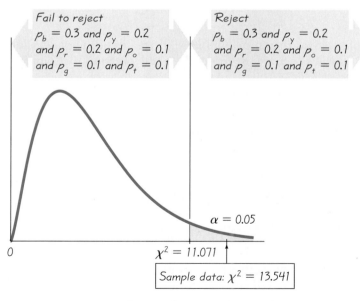

FIGURE 10-4   Goodness-of-Fit Test of $p_b = 0.3$ and $p_y = 0.2$ and $p_r = 0.2$ and $p_o = 0.1$ and $p_g = 0.1$ and $p_t = 0.1$

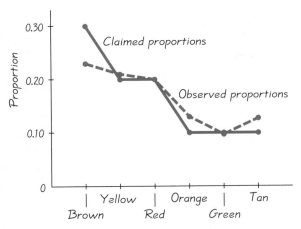

FIGURE 10-5   **Comparison of Claimed and Observed Proportions**

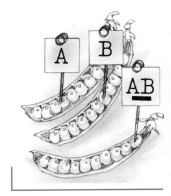

Did Mendel Fudge His Data?

R. A. Fisher analyzed the results of Mendel's experiments in hybridization. Fisher noted that the data were unusually close to theoretically expected outcomes. He says, "The data have evidently been sophisticated systematically, and after examining various possibilities, I have no doubt that Mendel was deceived by a gardening assistant, who knew only too well what his principal expected from each trial made." Fisher used chi-square tests and concluded that only about a 0.00004 probability exists of such close agreement between expected and reported observations.

In Figure 10-5 we graph the claimed proportions of 0.30, 0.20, ... , 0.10, along with the observed proportions of 84/370, 79/370, ... , 47/370, so that we can visualize the discrepancy between the distribution that was claimed and the frequencies that were observed. The points along the solid line represent the claimed proportions, and the points along the broken line represent the observed proportions. The corresponding pairs of points are all fairly close, except for the first pair, which represents brown M&Ms. If we refer to the calculation of the test statistic, we see that the largest contribution to the χ^2 test statistic was the value of 6.5676, which resulted from the category of brown M&Ms. In general, graphs such as Figure 10-5 are helpful in visually comparing expected frequencies and observed frequencies, as well as suggesting which categories result in the major discrepancies.

P-Values

The examples of this section used the traditional approach to hypothesis testing, but the P-value approach can also be used. P-values can be obtained by using the same methods described in Section 7-3. For instance, the preceding example resulted in a test statistic of $\chi^2 = 13.541$. That example had $k = 6$ categories, so there were $k - 1 = 5$ degrees of freedom. Referring to Table A-4, we see that for the row with 5 degrees of freedom, the test statistic of 13.541 falls between 12.833 and 15.086, so the P-value is between the corresponding areas of 0.025 and 0.01. That is,

$$0.01 < P\text{-value} < 0.025$$

Because the P-value is less than the significance level $\alpha = 0.05$, we again reject the null hypothesis.

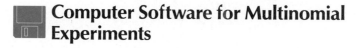

Computer Software for Multinomial Experiments

Multinomial experiments can be conducted with many statistical software packages such as STATDISK. (Minitab does not have a specific command that can be used with multinomial experiments.) To use STATDISK, select `Multinomial Experiments` from the `Statistics` menu and proceed to enter the data as requested. The resulting display will be as shown below. Note that more exact P-values can be obtained with STATDISK than can be found by using Table A-4; the STATDISK display includes the P-value of 0.0194703.

STATDISK DISPLAY

```
 About    File    Edit    Statistics    Data    View

                    Multinomial Experiments

                                          Observed   Expected
   Test statistic .... Chi Square = 13.5405    84        111
                                              79         74
   Critical Value .... Chi Square = 11        75         74
                                              49         37
   P-value .................... p = .0194703  36         37
                                              47         37
   Degrees of freedom ........ df = 5

   Level of significance ........ = .05


                      REJECT
        the null hypothesis of no difference
        between observed and expected frequencies


                 Use Menu Bar to change screen
   F1: Help        F2: Menu                       ESC: Quit
```

The techniques in this section can be used to test how well an observed frequency distribution conforms to some theoretical frequency distribution. For the employee accident data considered in this section, we used a goodness-of-fit test to decide whether the observed accidents conformed to a uniform distribution, and we found that the differences were significant. It appears that the observed frequencies do not make a good fit with a uniform distribution. Because many statistical analyses require a normally distributed population, we can use the chi-square test in this section to determine whether given samples are drawn from normally distributed populations (see Exercise 22).

10-2   Exercises A: Basic Skills and Concepts

1. One common way to test for authenticity of data is to analyze the frequencies of digits. When people are weighed and their weights are rounded to the nearest pound, we expect the last digits 0, 1, 2, . . . , 9 to occur with about the same frequency. In contrast, if people are *asked* how much they weigh, the digits 0 and 5 tend to occur at higher rates. When the author randomly selected 80 students, asked them for their weights, and recorded only the last digits, the following results were obtained. At the 0.01 significance level, test the claim that the last digits occur with the same frequency.

Last digit	0	1	2	3	4	5	6	7	8	9
Frequency	35	0	2	1	4	24	1	4	7	2

2. It is a common belief that more fatal car crashes occur on certain days of the week, such as Friday or Saturday. A sample of motor vehicle deaths in Montana is randomly selected for a recent year. The numbers of fatalities for the different days of the week are listed below. At the 0.05 significance level, test the claim that accidents occur with equal frequency on the different days.

Day	Sun	Mon	Tues	Wed	Thurs	Fri	Sat
Number of fatalities	31	20	20	22	22	29	36

Based on data from the Insurance Institute for Highway Safety.

3. Many people believe that fatal DWI crashes occur because of casual drinkers who tend to binge on Friday and Saturday nights, whereas others believe that fatal DWI crashes are caused by alcoholics who drink every day of the week. In a study of fatal car crashes, 216 cases are randomly selected from the pool in which the driver was found to have a blood alcohol content over 0.10. These cases are broken down according to the day of the week, with the results listed in the accompanying table. At the 0.05 significance level, test the claim that such fatal crashes occur on the different days of the week with equal frequency. Does the evidence support the theory that fatal DWI car crashes are due to casual drinkers or those who drink daily?

Day	Sun	Mon	Tues	Wed	Thurs	Fri	Sat
Number	40	24	25	28	29	32	38

Based on data from the Dutchess County STOP-DWI Program.

4. Statistical software packages commonly include random number generators that can be used to simulate the random selection of data. The Minitab command RANDOM and subcommand INTEGERS can be used to generate data from a population with a uniform distribution. Shown on the next page is a Minitab display summarizing the random selection of 60 digits, between 1 and 6 inclusive. This simulates the rolling of a die 60 times. Use the frequency

counts to test the claim that the random number generator in the Minitab software produces digits that are equally likely, as they should be with a fair die.

```
MTB > RANDOM 60 C1;
SUBC> INTEGER 1 6.
MTB > HISTOGRAM C1

Histogram of C1    N = 60

Midpoint    Count
      1         6   ******
      2        16   ****************
      3        11   ***********
      4        12   ************
      5         8   ********
      6         7   *******
```

5. Use a 0.05 significance level and the industrial accident data in Table 10-2 to test the claim of a safety expert that accidents are distributed on workdays as follows: 30% on Monday, 15% on Tuesday, 15% on Wednesday, 20% on Thursday, and 20% on Friday. Does rejection of that claim provide any help in correcting the industrial accident problem?

6. The Gleason Supermarket's manager must decide how much of each ice cream flavor he should stock so that customer demands are satisfied but unwanted flavors don't result in waste. The ice cream supplier claims that among the 4 most popular flavors, customers have these preference rates: 62% prefer vanilla, 18% prefer chocolate, 12% prefer neapolitan, and 8% prefer vanilla fudge. (The data are based on results from the International Association of Ice Cream Manufacturers.) A random sample of 200 customers produces the results below. At the $\alpha = 0.05$ significance level, test the claim that the supplier has correctly identified customer preferences.

Flavor	Vanilla	Chocolate	Neapolitan	Vanilla Fudge
Customers	120	40	18	22

7. The Gleason Supermarket's manager must make work assignments one week in advance, and he makes them according to the numbers of shoppers expected on different days. A marketing specialist claims that among supermarket shoppers who prefer a particular day, these rates apply: 7% prefer Sunday, 5% prefer Monday, 9% prefer Tuesday, 11% prefer Wednesday, 19% prefer Thursday, 24% prefer Friday, and 25% prefer Saturday. (The data are based on results from the Food Marketing Institute.) Sample results are given here. At the 0.05 significance level, test the claim that the marketing specialist has correctly identified customer preferences for shopping days.

Day	Sun	Mon	Tues	Wed	Thurs	Fri	Sat
Customers	9	6	10	8	19	23	28

8. Maureen Jenkins has taught a statistics course for several years and has found that the following grade distribution has been very consistent: A, 10%; B, 25%;

C, 35%; D, 20%; F, 10%. This semester, the grades of A, B, C, D, and F occurred with the frequencies of 15, 22, 25, 8, and 5, respectively. At the 0.05 significance level, test the claim that this semester's grades have the same distribution as past grades.

9. The number π is an *irrational* number with the property that when we try to express it in decimal form, it requires an infinite number of decimal places and there is no pattern of repetition. In the decimal representation of π, the first 100 digits occur with the frequencies described in the table below. At the 0.05 significance level, test the claim that the digits are uniformly distributed.

Digit	0	1	2	3	4	5	6	7	8	9
Frequency	8	8	12	11	10	8	9	8	12	14

10. The number 22/7 is similar to π in the sense that they both require an infinite number of decimal places. However, 22/7 is a *rational number* because it can be expressed as the ratio of two integers, whereas π cannot. When rational numbers such as 22/7 are expressed in decimal form, there is a pattern of repetition. In the decimal representation of 22/7, the first 100 digits occur with the frequencies described in the table below. At the 0.05 significance level, test the claim that the digits are uniformly distributed. Does the result differ from that found in Exercise 9?

Digit	0	1	2	3	4	5	6	7	8	9
Frequency	0	17	17	1	17	16	0	16	16	0

11. A genetics experiment involving 320 peas is designed to determine whether Mendelian principles hold for a certain list of characteristics. The following table summarizes the actual experimental results and the expected Mendelian results for the 5 characteristics being considered. At the $\alpha = 0.01$ significance level, test the claim that the characteristics actually occur with rates that agree with the rates expected from the Mendelian principles.

Characteristic	A	B	C	D	E
Observed frequency	30	15	58	83	134
Expected (Mendelian) frequency	20	20	40	120	120

12. Among drivers who have had a car crash in the last year, 88 are randomly selected and categorized by age, with the results listed below. If all ages have the same crash rate, we would expect (because of the age distribution of licensed drivers) the given categories to have 16%, 44%, 27%, and 13% of the subjects, respectively. At the 0.05 significance level, test the claim that the actual frequencies agree with the expected rates. Does any age group appear to have a disproportionate number of crashes?

Age	Under 25	25–44	45–64	Over 64
Drivers	36	21	12	19

Based on data from the Insurance Information Institute.

13. The second example of this section was based on data from a journal article. Instead of using the sample data given in the article, use Data Set 6 in Appendix

B to test the claim that 30% of M&Ms are brown, 20% are yellow, 20% are red, 10% are orange, 10% are green, and 10% are tan. Use a 0.05 significance level. Is there sufficient evidence to warrant rejection of the Mars, Inc. claim that the colors are distributed as stated?

14. Refer to Data Set 3 in Appendix B and use the education levels of all 125 subjects. Find the frequency for each of these three categories: (1) no education beyond elementary school; (2) some high school education, but no college; (3) at least some college education. Test the claim that the sample group has this distribution: 28% for the elementary school category, 51% for the high school category, and 21% for the college category (according to results from the Census Bureau, these percentages apply to the general population). Use a 0.05 level of significance. Based on this result, does it appear that the sample is representative of the U.S. population?

15. Refer to Data Set 5 in Appendix B and categorize the listed movies as poor (0.0 to 1.5 stars), fair (2.0 or 2.5 stars), good (3.0 or 3.5 stars), or excellent (4.0 stars). Movie critic James Harrington claims that movies are distributed evenly among the four categories. Test his claim at the 0.05 significance level.

16. Does Maryland select its lottery numbers in a way that is fair? Refer to Data Set 7 in Appendix B and use the 150 digits listed for the Pick Three lottery. Using 10 categories corresponding to the 10 possible digits, find the frequency for each category. At the 0.05 level of significance, test the claim that the lottery is fair in the sense that the 10 digits occur with the same frequency.

10-2   Exercises B: Beyond the Basics

17. What do you know about the P-value for the hypothesis test in Exercise 16?

18. In doing a test for goodness-of-fit as described in this section, suppose that we multiply each observed frequency by the same positive integer greater than 1. How is the critical value affected? How is the test statistic affected?

19. In this exercise we will show that a hypothesis test involving a multinomial experiment with only two categories is equivalent to a hypothesis test for a proportion (Section 7-5). Assume that a particular multinomial experiment has only two possible outcomes A and B with observed frequencies of f_1 and f_2, respectively.

   a.   Find an expression for the χ^2 test statistic, and find the critical value for a 0.05 significance level. Assume that we are testing the claim that both categories have the same frequency $(f_1 + f_2)/2$.

   b.   The test statistic

$$z = \frac{\hat{p} - p}{\sqrt{\dfrac{pq}{n}}}$$

is used to test the claim that a population proportion is equal to some value p. With the claim that $p = 0.5$, $\alpha = 0.05$, and

$$\hat{p} = \frac{f_1}{f_1 + f_2}$$

show that z^2 is equivalent to χ^2 (from part a). Also show that the square of the critical z score is equal to the critical χ^2 value from part a.

20. An observed frequency distribution is as follows:

Number of successes	0	1	2	3
Frequency	89	133	52	26

   a. Assuming a binomial distribution with $n = 3$ and $p = 1/3$, use the binomial probability formula to find the probability corresponding to each category of the table.

   b. Using the probabilities found in part a, find the expected frequency for each category.

   c. Use a 0.05 level of significance to test the claim that the observed frequencies fit a binomial distribution for which $n = 3$ and $p = 1/3$.

21. In a survey of radio listeners, data are collected for different time slots, beginning at 1:00 PM. The results are listed, along with expected frequencies based on past surveys. We cannot use the chi-square distribution because all expected values are not at least 5. However, we can combine some columns so that all expected values do equal or exceed 5. Use this suggestion to test the claim that the observed and expected frequencies are compatible. Try to combine categories in a meaningful way.

Time slot	1	2	3	4	5	6	7	8	9	10
Observed frequency	2	8	8	9	3	5	3	0	12	3
Expected frequency	4	5	8	7	4	6	5	2	9	3

22. An observed frequency distribution of sample IQ scores is as follows:

IQ score	Less than 80	80–95	96–110	111–120	More than 120
Frequency	20	20	80	40	40

   a. Assuming a normal distribution with $\mu = 100$ and $\sigma = 15$, use the methods given in Chapter 5 to find the probability of a randomly selected subject belonging to each class. (Use class boundaries of 79.5, 95.5, 110.5, 120.5.)

   b. Using the probabilities found in part a, find the expected frequency for each category.

   c. Use a 0.01 level of significance to test the claim that the IQ scores were randomly selected from a normally distributed population with $\mu = 100$ and $\sigma = 15$.

10-3   Contingency Tables

Section 10-2 involved frequencies listed in a single row (or column) according to category. This section also involves frequencies listed according to category, but here we consider cases with at least two rows and at least two columns. For example, see Table 10-1, which has two rows (jail, no jail) and three columns (embezzlement, fraud, forgery). Tables similar to Table 10-1 are generally called contingency tables, or two-way tables.

DEFINITIONS

A **contingency table** (or **two-way table**) is a table in which frequencies correspond to two variables. (One variable is used to categorize rows, and a second variable is used to categorize columns.) A **test of independence** tests the null hypothesis that the row variable and column variable in a contingency table are not related.

Section 10-2 and this section both involve the analysis of categorical data consisting of frequencies, so they are similar in many ways. However, the frequencies in Section 10-2 were all arranged in a single row or column, and we tested for goodness-of-fit with a claimed distribution. In this section our frequencies are arranged in a table with at least two rows and two columns, and we test for independence between the row and column variables. Using the methods of this section, we can address questions such as the following.

- Are sentences (jail, no jail) independent of the type of white-collar crime (embezzlement, fraud, forgery)? (See Table 10-1.)
- Is the quality of Ford transmissions (defective, acceptable) independent of the location of the plant in which they are manufactured (America, Japan)?
- Are the sexes (female, male) of survey respondents independent of their choice of car (subcompact, compact, intermediate, large)?

As we proceed to use contingency tables in tests of independence, we should recognize that in this context, the word *contingency* refers to dependence. But this is only a statistical dependence; it cannot be used to establish a direct cause-and-effect link between the two variables in question. For example, after analyzing the data of Table 10-1, we might conclude that there is a relationship between the sentence and the category of crime, but that doesn't mean that the crime category is a direct *cause* of the sentence.

Assumptions

When working with data in the form of a contingency table, we test the null hypothesis that the row variable and the column variable are *independent,* and the following assumptions apply. (Note that these assumptions do *not* require that the parent population have a normal distribution or any other particular distribution.)

1. The sample data are randomly selected.
2. For every cell in the contingency table, the *expected* frequency E is at least 5. (There is no requirement that every *observed* frequency must be at least 5.)

Our test of independence between the row and column variables uses the following test statistic.

Test Statistic for a Test of Independence

$$\chi^2 = \sum \frac{(O - E)^2}{E}$$

Critical Values

1. The critical values are found in Table A-4.
2. *Tests of independence with contingency tables involve only right-tailed critical regions.*
3. In a contingency table with r rows and c columns, the number of degrees of freedom is given by

$$\text{degrees of freedom} = (r - 1)(c - 1)$$

The test statistic allows us to measure the degree of disagreement between the frequencies actually observed and those that we would theoretically expect when the two variables are independent. The reasons underlying the development of the χ^2 statistic in the previous section also apply here. In repeated large samplings, *the distribution of the test statistic χ^2 can be approximated by the chi-square distribution, provided that all expected frequencies are at least 5.* We included this requirement with the assumptions that apply to this section.

Small values of the χ^2 test statistic support the claimed independence of the two variables. That is, when observed and expected frequencies are close, the χ^2 test statistic is small. Large values of χ^2 are to the right of the chi-square distribution, and they reflect significant differences between observed and expected frequencies.

The number of degrees of freedom $(r - 1)(c - 1)$ reflects the fact that because we know the total of all frequencies in a contingency table, we can freely assign frequencies to only $r - 1$ rows and $c - 1$ columns before the frequency for every cell is determined. (However, we cannot have negative frequencies or frequencies so large that their sum exceeds the total of the frequencies for all cells combined.)

In the preceding section we knew the corresponding probabilities and could easily determine the expected values, but the typical contingency table does not come with the relevant probabilities. Consequently, we need to devise a

Home Field Advantage

In the *Chance* magazine article "Predicting Professional Sports Game Outcomes from Intermediate Game Scores," authors Harris Cooper, Kristina DeNeve, and Frederick Mosteller used statistics to analyze two common beliefs: Teams have an advantage when they play at home, and only the last quarter of professional basketball games really counts. Using a random sample of hundreds of games, they found that for the four top sports, the home team wins about 58.6% of games. Also, basketball teams ahead after 3 quarters go on to win about 4 out of 5 times, but baseball teams ahead after 7 innings go on to win about 19 out of 20 times. The statistical methods of analysis included the chi-square distribution applied to a contingency table.

method for obtaining the corresponding expected values. We will first describe the procedure for finding the values of the expected frequencies, and then we will proceed to justify that procedure. For each cell in the frequency table, the expected frequency E can be calculated by using the following equation.

Expected Frequency for a Contingency Table

$$\text{Expected frequency } E = \frac{(\text{row total})(\text{column total})}{(\text{grand total})}$$

Here *grand total* refers to the total number of observations in the table. For example, in the upper left cell of Table 10-7 (a duplicate of Table 10-1 with expected frequencies inserted), we see the observed frequency of 22. The total of all frequencies for that row is 172, the total of the column frequencies is 79, and the total of all frequencies in the table is 400, so we get an expected frequency of

$$E = \frac{(\text{row total})(\text{column total})}{(\text{grand total})} = \frac{(172)(79)}{400} = 33.97$$

▶ **EXAMPLE**

The expected frequency for the upper left cell of Table 10-7 is 33.97. Using the row and column totals of that table and assuming that the sentence (jail, no jail) and the crime (embezzlement, fraud, forgery) are independent, find the expected frequency for the cell corresponding to those who were convicted of embezzlement but were not sent to jail. That is, calculate the expected frequency for the lower left cell of Table 10-7.

TABLE 10-7   Observed and Expected Frequencies

	Embezzlement	Fraud	Forgery	Row totals
Sent to jail	22 (33.97)	130 (118.68)	20 (19.35)	172
Not sent to jail	57 (45.03)	146 (157.32)	25 (25.65)	228
Column totals	79	276	45	(Grand total: 400)

● **SOLUTION**

The lower left cell lies in the second row and the first column. The second row total is 228 and the first column total is 79, so the expected frequency is

$$E = \frac{\text{(row total)(column total)}}{\text{(grand total)}} = \frac{(228)(79)}{400} = 45.03$$

So far, we have found the expected values of 33.97 and 45.03. It would be helpful to verify that the other expected values are 118.68, 19.35, 157.32, and 25.65.

To better understand the rationale for finding expected frequencies with this procedure, let's pretend that we know only the row and column totals and that we must fill in the cell frequencies by assuming independence (or no relationship) between the two variables involved—that is, pretend that we know only the totals shown in Table 10-8.

We begin with the cell in the upper left corner, which corresponds to embezzlers who were jailed. Because 172 of the 400 convicts were jailed, we have $P(\text{jail}) = 172/400$. Similarly, 79 of the 400 convicts are embezzlers, so $P(\text{embezzler}) = 79/400$. Because we assume that sentence and category of crime are independent, we conclude that

$$P(\text{jail and embezzler}) = \frac{172}{400} \cdot \frac{79}{400}$$

This equation follows from the multiplication rule of probability whereby $P(A \text{ and } B) = P(A) \cdot P(B)$ if A and B are independent events. To obtain the expected value for the upper left cell, we simply multiply the probability for that cell by the total number of subjects available to get

$$\frac{172}{400} \cdot \frac{79}{400} \cdot 400 = 33.97$$

TABLE 10-8   Row and Column Totals

	Embezzlement	Fraud	Forgery	Row totals
Sent to jail				172
Not sent to jail				228
Column totals	79	276	45 (Grand total: 400)	

Better Results with Smaller Class Size

An experiment at the State University of New York at Stony Brook found that students did significantly better in classes limited to 35 students than in large classes with 150 to 200 students. For a calculus course, failure rates were 19% for the small classes compared to 50% for the large classes. The percentages of As were 24% for the small classes and 3% for the large classes. These results suggest that students benefit from smaller classes, which allow for more direct interaction between students and teachers.

The form of this product suggests a general way to obtain the expected frequency of a cell:

$$\text{expected frequency } E = \frac{(\text{row total})}{(\text{grand total})}\ \frac{(\text{column total})}{(\text{grand total})} \cdot (\text{grand total})$$

This expression can be simplified to

$$E = \frac{(\text{row total}) \cdot (\text{column total})}{(\text{grand total})}$$

We can now proceed to use contingency table data for testing hypotheses, as in the following example, which uses the data given in the Chapter Problem.

▶ **EXAMPLE**

At the 0.05 significance level, use the data in Table 10-1 to test the claim that the sentence (jail, no jail) is independent of the category of crime (embezzlement, fraud, forgery).

● **SOLUTION**

The null hypothesis and alternative hypothesis are as follows:

H_0: The sentence and crime category are independent.
H_1: The sentence and crime category are dependent.

The significance level is $\alpha = 0.05$.

Because the data are in the form of a contingency table, we use the χ^2 distribution with this test statistic:

$$\chi^2 = \sum \frac{(O - E)^2}{E}$$
$$= \frac{(22 - 33.97)^2}{33.97} + \frac{(130 - 118.68)^2}{118.68} + \cdots + \frac{(25 - 25.65)^2}{25.65}$$
$$= 4.2179 + 1.0797 + \cdots + 0.0165$$
$$= 9.332$$

The critical value is $\chi^2 = 5.991$, and it is found from Table A-4 by noting that $\alpha = 0.05$ in the right tail and the number of degrees of freedom is given by $(r - 1)(c - 1) = (2 - 1)(3 - 1) = 2$. The test statistic and critical value are shown in Figure 10-6. Because the test statistic falls within the critical region, we reject the null hypothesis that the sentence and crime category are independent. It appears that the sentence and category of crime are depen-

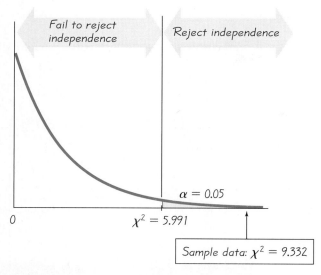

**FIGURE 10-6   Test of Independence between
                Sentence and Crime**

If the chi-square distribution has only 1 or 2 degrees of
freedom, the shape of the distribution is as shown here.

dent. Thus, the chance of going to jail appears to vary depending on the category of crime that has been committed.

Recall from the problem at the beginning of the chapter that Kate Malloy plans to prosecute a defendant charged with a white-collar crime and has agreed to select one charge of embezzlement, fraud, or forgery. The results of this example suggest that her choice *will* make a difference in determining whether the defendant is sent to jail if convicted. Kate Malloy might be wise to charge fraud because that crime has the highest jail rate of 47%, compared to 44% for forgery and 28% for embezzlement. However, justification of this conclusion involves methods that are beyond the scope of this chapter.

Computer Software and Contingency Tables

Because of the large number of calculations required, it is often helpful to use a statistical software package for tests of the type discussed in this section. The accompanying STATDISK and Minitab displays show the results obtained for the data in Table 10-1. If you are using STATDISK, select `Contingency Table` from the STATISTICS menu and proceed to enter the data as requested. If you are using Minitab, use the READ command to first enter the data in columns

C1, C2, and C3, and then use the command CHISQUARE C1 C2 C3 as shown in the display. Note that STATDISK provides the test statistic, critical value, *P*-value, and conclusion, whereas Minitab provides the test statistic, but not the critical value, *P*-value, or conclusion.

STATDISK DISPLAY

```
About    File    Edit    Statistics    Data    View
                      Contingency Tables

               22         130         20
             (33.97)    (118.7)    (19.35)

               57         146         25
             (45.03)    (157.3)    (25.65)


         Test statistic .... Chi Square =  9.33233
         Critical Value .... Chi Square =  5.9
         P-value ...................... p =  9.61406E-03
         Degrees of freedom ........ df =  2
         Level of significance ........ =  .05

                          REJECT
       the null hypothesis that row and column variables are independent


                   Use Menu Bar to change screen
       F1: Help         F2: Menu                              ESC: Quit
```

MINITAB DISPLAY

```
MTB > READ C1 C2 C3
DATA> 22 130 20
DATA> 57 146 25
DATA> ENDOFDATA
MTB > CHISQUARE C1 C2 C3

Expected counts are printed below observed counts

            C1       C2       C3     Total
    1       22      130       20      172
          33.97   118.68    19.35

    2       57      146       25      228
          45.03   157.32    25.65

Total       79      276       45      400

ChiSq =  4.218 +  1.080 +  0.022 +
         3.182 +  0.815 +  0.016 =  9.332
df = 2
```

P-Values

The preceding example used the traditional approach to hypothesis testing, but we can find *P*-values by using the same methods introduced earlier. The preceding example resulted in a test statistic of $\chi^2 = 9.332$, and the critical value involved two degrees of freedom. Refer to Table A-4 and note that for the row with 2 degrees of freedom, the test statistic of $\chi^2 = 9.332$ falls between the critical values of 9.210 and 10.597, so the *P*-value must fall between the corresponding areas of 0.01 and 0.005. We can express this as follows:

$$0.005 < P\text{-value} < 0.01$$

Based on this relatively small *P*-value, we again reject the null hypothesis and conclude that there is sufficient sample evidence to warrant rejection of the claim that the sentence and crime category are independent. It appears that they are dependent. If the *P*-value had been greater than the significance level of 0.05, we would have failed to reject the null hypothesis of independence.

Test of Homogeneity

In the preceding example, we illustrated a test of independence by using a sample of 400 convictions drawn from a *single* population. In some cases samples are drawn from *different* populations, and we want to determine whether those populations have the same proportions of the characteristics being considered. For example, we might want to test the claim that the proportion of voters who are Republicans is the same in New York, California, Texas, and Iowa.

DEFINITION

In a **test of homogeneity,** we test the claim that different populations have the same proportions of some characteristics.

When we sample from different populations, we have predetermined totals for either the rows or the columns in the contingency table. For example, we might choose to find the political party registrations for 200 New Yorkers, 250 Californians, 100 Texans, and 80 Iowans. When we construct the contingency table summarizing the results, either the row totals or the column totals (whichever represent the different states) will already be determined (200, 250, 100, 80). In general, *a test of homogeneity* involves random selections made in such a way that either the row totals are predetermined or the column totals are predetermined. In contrast, the preceding example was a test of independence because the researcher randomly selected 400 white-collar crimes and then constructed the contingency table, with no predetermined totals for rows or

Zip Codes Reveal Much

The Claritas Corporation has developed a way of obtaining considerable information about people from their zip codes. Zip code data are extracted from a variety of mailing lists, purchase orders, warranty cards, census data, and market research surveys. With people of the same social and economic levels tending to live in the same areas, it is possible to match zip codes with such factors as purchase patterns, types of cars driven, foods preferred, leisure time activities, and television viewing choices. The company helps clients target their advertising efforts to regions that are most likely to accept particular products.

columns. In trying to distinguish between a test for homogeneity and a test for independence, we can therefore pose the following question: Were predetermined sample sizes used for different populations (test of homogeneity), or was one big sample drawn so both row and column totals were determined randomly (test of independence)?

In conducting a test of homogeneity, we can use the same procedures already presented in this section, as illustrated in the following example.

▶ EXAMPLE

A study of seat belt use in taxi cabs involved 77 New York taxis, 129 Chicago taxis, and 72 Pittsburgh taxis. Table 10-9 summarizes the sample data. A spokesperson for the taxi cab industry argues that although it appears that very few taxis have usable seat belts, the total sample size from the 3 cities was only 278, so the results aren't significant. At the 0.05 level of significance, test the claim that the 3 cities have the same proportion of taxis with usable seat belts. That is, test for *homogeneity* of proportions of usable taxi seat belts in the 3 cities.

TABLE 10-9 Seat Belt Use in Taxi Cabs

		New York	Chicago	Pittsburgh
Taxi has usable seat belt?	Yes	3	42	2
	No	74	87	70

Based on "The Phantom Taxi Seat Belt," by Welkon and Reisinger, *American Journal of Public Health*, Vol. 67, No. 11.

● SOLUTION

The null and alternative hypotheses are as follows:

H_0: The 3 cities have the same proportion of taxis with usable seat belts.
H_1: The proportion of taxis with usable seat belts is not the same in all 3 cities.

The significance level is $\alpha = 0.05$, and we use the same χ^2 test described earlier.

Before finding the value of the χ^2 test statistic, we must find the expected frequency for each of the 6 cells. The expected frequency of the first (upper left) cell is 13.012, which is calculated as follows:

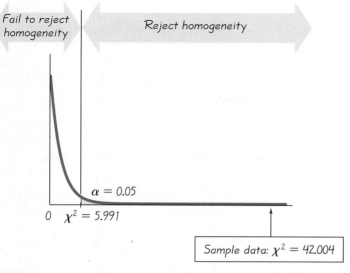

FIGURE 10-7   Test of Homogeneity for Three Cities

$$E = \frac{\text{(row total)(column total)}}{\text{(grand total)}} = \frac{(47)(77)}{278} = 13.0180$$

The other expected frequencies (listed in order by row) are 21.8094, 12.1727, 63.9820, 107.1906, and 59.8273.

We can now find the value of the test statistic.

$$\chi^2 = \sum \frac{(O - E)^2}{E}$$

$$= \frac{(3 - 13.0180)^2}{13.0180} + \frac{(42 - 21.8094)^2}{21.8094} + \cdots + \frac{(70 - 59.8273)^2}{59.8273}$$

$$= 7.7094 + 18.6920 + 8.5013 + 1.5686 + 3.8031 + 1.7297$$

$$= 42.004$$

The critical χ^2 value of 5.991 is found in Table A-4. This value corresponds to $(r - 1)(c - 1) = 2$ degrees of freedom with $\alpha = 0.05$ in the right tail. Figure 10-7 shows the test statistic and critical value. Because the test statistic of 42.004 falls well within the critical region, we reject the null hypothesis of equal proportions. There is sufficient evidence to warrant rejection of the claim that the 3 cities have the same proportion of usable seat belts in taxis. Although we have not established any further conclusions by formal methods of statistics, an informal examination of Table 10-9 shows that Chicago appears to have a much higher proportion of taxis with usable seat belts than do New York and Pittsburgh.

This example shows that the procedures for the test of independence and the test of homogeneity are essentially the same. The only real difference is the nature of the samples. If our contingency table results from a single sample, we test for independence between the row and column variables, but if our contingency table results from different samples drawn from different populations, we test for homogeneity of population proportions.

10-3   Exercises A: Basic Skills and Concepts

1. The following table lists survey results obtained from a random sample of different crime victims. At the 0.05 level of significance, test the claim that the type of crime is independent of whether the criminal is a stranger.

Relationship of Criminal to Victim	Homicide	Robbery	Assault
Stranger	12	379	727
Acquaintance or relative	39	106	642

Based on data from the U.S. Department of Justice.

2. Companies such as Campbell Soup and Procter & Gamble want to target their advertising to the consumers who actually do the shopping. The marketing manager for *Independent Gentleman* magazine is trying to attract more advertising by claiming that more men shop now than in the past and the type of item purchased isn't related to the gender of the buyer. Following are randomly selected sample survey results. At the 0.05 significance level, test the claim that the category of the item purchased is independent of the sex of the shopper.

	Frozen Foods	Detergent	Soup
Female	203	73	142
Male	97	27	58

Based on data from Nielsen Homescan.

3. Nicorette is a chewing gum designed to help people stop smoking cigarettes. Tests for adverse reactions yielded the results given in the table. At the 0.05 significance level, test the claim that the treatment (drug or placebo) is independent of the reaction (whether or not mouth or throat soreness was experienced). If you are thinking about using Nicorette as an aid to stop smoking, should you be concerned about mouth or throat soreness?

	Drug	Placebo
Mouth or throat soreness	43	35
No mouth or throat soreness	109	118

Based on data from Merrell Dow Pharmaceuticals, Inc.

4. Many people believe that criminals who plead guilty tend to get lighter sentences than those who are convicted in trials. The following table summarizes randomly selected sample data for San Francisco defendants in burglary cases. All of the subjects had prior prison sentences. At the 0.05 significance level, test the claim that the sentence (sent to prison or not sent to prison) is independent of the plea. If you were an attorney defending a guilty defendant, would these results suggest that you should encourage a guilty plea? (Table results are based on data from "Does It Pay to Plead Guilty? Differential Sentencing and the Functioning of the Criminal Courts," by Brereton and Casper, *Law and Society Review*, Vol. 16, No. 1.)

	Guilty Plea	Not Guilty Plea
Sent to prison	392	58
Not sent to prison	564	14

5. In the judicial case *United States v. City of Chicago*, fair employment practices were challenged. A minority group (group A) and a majority group (group B) took the Fire Captain Examination. At the 0.05 significance level, use the given results to test the claim that success on the test is independent of the group.

	Pass	Fail
Group A	10	14
Group B	417	145

6. A study of jail inmates charged with DWI provided the sample data in the table. At the 0.05 significance level, test the claim that the unconvicted/convicted status is independent of the prior DWI sentences. Do convictions seem to be affected by prior sentences?

	Unconvicted	Convicted
No prior DWI sentence	100	672
1 prior DWI sentence	56	394
2 prior DWI sentences	16	173
3 or more prior DWI sentences	11	72

Based on data from the U.S. Department of Justice.

7. A study was conducted of 531 persons injured in bicycle crashes, and randomly selected sample results are summarized in the table. At the 0.05 significance level, test the claim that wearing a helmet has no effect on whether facial injuries are received. Based on these results, does a helmet seem to be effective in helping to prevent facial injuries in a crash?

	Helmet Worn	No Helmet
Facial injuries received	30	182
All injuries nonfacial	83	236

Based on data from "A Case-Control Study of the Effectiveness of Bicycle Safety Helmets in Preventing Facial Injury," by Thompson, Thompson, Rivara, and Wolf, *American Journal of Public Health*, Vol. 80. No. 12.

8. The following table provides randomly selected sample data from a study of dropout rates among college freshmen. Is there sufficient evidence to warrant rejection of the claim that the type of college is independent of the dropout rate? Use a 0.05 significance level. Based on these results, should the type of college be considered as a factor by someone concerned about dropout rates?

	4-Year Public	4-Year Private	2-Year Public	2-Year Private
Freshmen dropouts	10	9	15	9
Freshmen who stay	26	28	18	27

Based on results from the American College Testing Program.

9. A study of randomly selected car accidents and drivers who use cellular phones provided the following sample data. At the 0.05 level of significance, test the claim that the occurrence of accidents is independent of the use of cellular phones. Based on these results, does it appear that the use of cellular phones affects driving safety?

	Had Accident in Last Year	Had No Accident in Last Year
Cellular phone user	23	282
Not cellular phone user	46	407

Data based on results from AT&T and the Automobile Association of America.

10. A study of seat belt users and non-users yielded the randomly selected sample data summarized in the following table. Test the claim that the amount of smoking is independent of seat belt use. A plausible theory is that people who smoke more are less concerned about their health and safety and are therefore less inclined to wear seat belts. Is this theory supported by the sample data?

	Number of Cigarettes Smoked per Day			
	0	1–14	15–34	35 and over
Wear seat belts	175	20	42	6
Don't wear seat belts	149	17	41	9

Based on data from "What Kinds of People Do Not Use Seat Belts?" by Helsing and Comstock, *American Journal of Public Health*, Vol. 67, No. 11.

11. Researchers were concerned about the causes of deaths being accurately reported for U.S. Army veterans. For two random samples of deaths of U.S. Army veterans, a medical panel established the underlying cause of death for one sample, and the original death certificates were used for the other sample. See the results summarized in the following table. At the 0.05 significance level, test to determine whether there is agreement between the original death certificates and the medical panel.

	Natural	Motor Vehicle	Suicide	Homicide or Other Causes
Original death certificate	102	130	54	140
Medical panel	101	132	60	133

Based on data from "Underreporting of Alcohol-Related Mortality on Death Certificates of Young U.S. Army Veterans," by Pollock and others, *Journal of the American Medical Association,* Vol. 258, No. 3.

12. Colleges often compile data describing the distribution of grades. The table below lists sample data randomly selected from the author's college. At the 0.10 significance level, test the claim that the two departments have the same grade distribution.

	A	B	C	D	F
Math	49	69	64	42	61
Physical Sciences	52	73	78	23	18

13. A study conducted to determine the rate of smoking among people from different age groups provided the randomly selected sample data summarized in the accompanying table. At the 0.05 significance level, test the claim that smoking is independent of the 4 age groups listed. Based on the data, does it make sense to target cigarette advertising to particular age groups?

	Age (years)			
	20–24	25–34	35–44	45–64
Smoke	18	15	17	15
Don't smoke	32	35	33	35

Based on data from the National Center for Health Statistics.

14. A study of people who refused to answer survey questions provided the randomly selected sample data in the table. At the 0.01 significance level, test the claim that the cooperation of the subject (response, refuse) is independent

of the age category. Does any particular age group appear to be particularly uncooperative?

	Age					
	18–21	22–29	30–39	40–49	50–59	60 and over
Responded	73	255	245	136	138	202
Refused	11	20	33	16	27	49

Based on data from "I Hear You Knocking But You Can't Come In," by Fitzgerald and Fuller, *Sociological Methods and Research*, Vol. 11, No. 1.

15. Randomly selected sample data collected by the New York State Department of Motor Vehicles provided the results summarized in the table. At the 0.10 significance level, test the claim that the time of day of fatal accidents is the same in New York City as in all other New York State locations.

	Time of Day of Fatal Accidents							
	AM				PM			
	1–4	4–7	7–10	10–1	1–4	4–7	7–10	10–1
New York City	73	60	53	68	80	67	87	81
Other New York locations	187	115	136	161	196	257	237	235

16. Maine Seafood Products market tested a new clam chowder to determine whether its appeal is the same in different geographic regions. Given the randomly selected sample data in the following table, use a 0.01 significance level to test the claim that consumers' preferences are the same in the different regions. Based on the results, would it be wise to use the same marketing strategies in the different regions?

	Like	Dislike	Uncertain
Northeast	30	15	15
Southeast	10	30	20
West	40	60	15

10-3 Exercises B: Beyond the Basics

17. What do you know about the *P*-value for the hypothesis test in Exercise 16?

18. The chi-square distribution is continuous, whereas the test statistic used in this

section is discrete. Some statisticians use **Yates' correction for continuity** in cells with an expected frequency of less than 10 or in all cells of a contingency table with two rows and two columns. With Yates' correction, we replace

$$\sum \frac{(O - E)^2}{E} \quad \text{with} \quad \sum \frac{(|O - E| - 0.5)^2}{E}$$

Given the contingency table in Exercise 9, find the value of the χ^2 test statistic with and without Yates' correction. In general, what effect does Yates' correction have on the value of the test statistic?

19. If each observed frequency in a contingency table is multiplied by a positive integer K (where $K \geq 2$), how is the value of the test statistic affected? How is the critical value affected?

20. a. For the contingency table with frequencies of a and b in the first row and frequencies of c and d in the second row, verify that the test statistic becomes

$$\chi^2 = \frac{(a + b + c + d)(ad - bc)^2}{(a + b)(c + d)(b + d)(a + c)}$$

b. Let $\hat{p}_1 = a/(a + c)$ and let $\hat{p}_2 = b/(b + d)$. Show that the test statistic

$$z = \frac{(\hat{p}_1 - \hat{p}_2) - 0}{\sqrt{\dfrac{\bar{p}\bar{q}}{n_1} + \dfrac{\bar{p}\bar{q}}{n_2}}}$$

where

$$\bar{p} = \frac{a + b}{a + b + c + d}$$

and

$$\bar{q} = 1 - \bar{p}$$

is such that $z^2 = \chi^2$ (the same result as in part a). (This result shows that the chi-square test involving a 2×2 table is equivalent to the test for the difference between two proportions, as described in Section 8-6.)

VOCABULARY LIST

Define and give an example of each term.

cells

multinomial experiment

goodness-of-fit test

contingency table

two-way table

test of independence

test of homogeneity

Important Formulas

Application	Applicable Distribution	Test Statistic	Degrees of Freedom	Table of Critical Values
Multinomial	chi-square	$\chi^2 = \sum \dfrac{(O - E)^2}{E}$	$k - 1$	Table A-4
Contingency table	chi-square	$\chi^2 = \sum \dfrac{(O - E)^2}{E}$ where $E = \dfrac{\text{(row total) (column total)}}{\text{(grand total)}}$	$(r - 1)(c - 1)$	Table A-4

REVIEW

Section 10-2 described the goodness-of-fit test for determining whether a single row (or column) of frequencies has some claimed distribution. For multinomial experiments we tested for goodness-of-fit, or agreement between observed and expected frequencies, by using the chi-square test statistic. In repeated large samplings, the distribution of the χ^2 test statistic can be approximated by the chi-square distribution. This approximation is generally considered acceptable as long as all expected frequencies are at least 5. In a multinomial experiment with k cells or categories, the number of degrees of freedom is $k - 1$.

In Section 10-3 we used the sample χ^2 test statistic to measure disagreement between observed and expected frequencies in contingency tables. A contingency table contains frequencies; the rows correspond to categories of one variable, and the columns correspond to categories of another variable. With contingency tables, we test the hypothesis that the two variables of classification are independent. In this test of independence, we can again approximate the sampling distribution of the statistic by the chi-square distribution, as long as all expected frequencies are at least 5. In a contingency table with r rows and c columns, the number of degrees of freedom is $(r - 1)(c - 1)$.

When the contingency table results from a sample drawn from a *single* population, we use the test of independence. If the contingency table results from different samples drawn from different populations, however, we use the test for homogeneity; that is, we test the null hypothesis that the different populations have the same proportion of the characteristic in question. The procedures for a test of independence and a test of homogeneity are the same.

REVIEW EXERCISES

1. When *Time* magazine tracked U.S. deaths by gunfire during a one-week period, the results in the table were obtained. At the 0.05 significance level, test the claim that gunfire death rates are the same for the different days of the week. Is

there any support for the theory that more gunfire deaths occur on weekends when more people are at home?

Weekday	Mon	Tues	Wed	Thurs	Fri	Sat	Sun
Number of deaths by gunfire	74	60	66	71	51	66	76

2. A car dealer plans to order new compact cars and wants to determine which colors to select. A study of color choices of buyers of compact cars claims that among the 5 most frequent choices, these preference rates apply: 22% prefer light red/brown, 22% prefer white, 20% prefer light blue, 18% prefer dark blue, and 18% prefer red. (The data are based on results from the Automotive Information Center.) When 270 compact cars were randomly selected from the population of compact cars recently sold, the following results were found. At the 0.05 level of significance, test the claim that the given percentages accurately reflect buyer preferences.

Color	Lt. red/brown	White	Lt. blue	Dk. blue	Red
Selling frequency	60	61	43	41	65

3. In an experiment on extrasensory perception, subjects were asked to identify the month showing on a calendar in the next room. If the results were as shown, test the claim that months were selected with equal frequencies. Assume a significance level of 0.05. If it appears that the months were not selected with equal frequencies, is the claim that the subjects have extrasensory perception supported?

Month	Jan	Feb	Mar	Apr	May	June	July	Aug	Sept	Oct	Nov	Dec
Number selected	8	12	9	15	6	12	4	7	11	11	5	20

4. Clinical tests of the allergy drug Seldane yielded results summarized in the table. At the 0.05 significance level, test the claim that the occurrence of headaches is independent of the group (Seldane, placebo, control). Based on these results, should Seldane users be concerned about getting headaches?

	Seldane Users	Placebo Users	Control
Headache	49	49	24
No headache	732	616	602

Based on data from Merrell Dow Pharmaceuticals, Inc.

5. The given table summarizes results for randomly selected drivers convicted of motor vehicle violations in New York State. At the 0.05 significance level, test the claim that the 4 counties have the same proportions of convictions for the different violations. (DWI is driving while intoxicated, and DTD is driving with a disabled traffic device.)

	Albany	Monroe	Orange	Westchester
DWI	6	19	6	10
Speeding	103	261	160	226
DTD	60	152	25	174

Data based on results from the New York State Department of Motor Vehicles.

6. Advertisements for golf balls often include claims that they "feel soft." In an experiment designed to test golfers' "feel" for different ball types, professional golfers hit equal numbers of 3 different types. They hit normally, with only sight suppressed (blind), with only hearing suppressed (deaf), and with both sight and hearing suppressed (blind and deaf). The number of correct identifications for each case is listed in the table. Test the claim that when golfers identify the type of golf ball used, the sensory state of the golfer is independent of the ball type. Based on these results, are the advertising claims of a different "feel" justified?

| | Ball Type | | |
	Three-Piece Balata	Three-Piece Surlyn	Two-Piece Surlyn
Normal	18	19	16
Deaf	19	18	9
Blind	19	12	19
Blind and deaf	16	11	9

Based on data from "The Fallacy of Feel," by Pelz, *Golf* magazine.

7. In Orange County, 80% of the drivers have no accidents in a given year, 16% have one accident, and 4% have more than one accident. A survey of 200 randomly selected teachers from the county produced 172 with no accidents, 23 with one accident, and 5 with more than one accident. At the 0.05 level of significance, test the claim that the teachers exhibit the same accident rate as the countywide population.

8. In random sampling, each member of the population has the same chance of being selected. When the author asked 60 students to "randomly" select 3 digits each, the results listed below were obtained. Use this sample of 180 digits to test the claim that students select digits randomly. Were the students successful in choosing their own random numbers?

213	169	812	125	749	137	202	344	496	348	714	765
831	491	169	312	263	192	584	968	377	403	372	123
493	894	016	682	390	123	325	734	316	357	945	208
115	776	143	628	479	316	229	781	628	356	195	199
223	114	264	308	105	357	333	421	107	311	458	007

 COMPUTER PROJECT

Because of the nature of the calculations required, statistical software packages have become very useful in testing hypotheses involving multinomial experiments and contingency tables.

a. Use existing software such as STATDISK or Minitab to do Review Exercise 6.

b. For the sample data used in part a above, multiply each frequency by 10 and note the effect on the test statistic, critical value, and conclusion. In general, what is the effect of multiplying each frequency in a contingency table by a positive integer greater than 1?

c. For the same sample data used in part a, add 100 to each frequency and note the effects on the test statistic, critical value, and conclusion. In general, what is the effect of adding a positive integer to each entry in a contingency table?

d. For the same sample data used in part a, interchange the first row (normal conditions) frequencies with those of the second row (deaf) and note the effect on the test statistic, critical value, and conclusion. In general, what is the effect of interchanging any two rows in a contingency table? What is the effect of interchanging any two columns in a contingency table?

FROM DATA TO DECISION: Does smoking affect the cause of death?

Many people (especially nonsmokers) believe that smoking is unhealthy. In a study of 1000 randomly selected deaths of men aged 45 to 64, the causes of death are listed as shown in the table below, along with the victims' smoking habits.

| | Cause of Death | | |
	Cancer	Heart Disease	Other
Smoker	135	310	205
Nonsmoker	55	155	140

Based on data from "Chartbook on Smoking, Tobacco, and Health," USDHEW publication CDC75-7511.

What can we conclude from these data? We cannot conclude that smoking *causes* deaths, because *every one* of the 1000 subjects died. There are two issues we might explore using the sample data.

1. Smokers have a disproportionately large representation among the 1000 deaths. The table shows that among 1000 randomly selected men aged 45–64 who died, a total of 650 (or 65%) were smokers. Other data from the U.S. Nation-

al Center for Health Statistics show that only 45% of males in the 45–64 age bracket are smokers. If tobacco lobbyists are correct when they claim that smoking doesn't affect health, we would expect 45% of the deceased to be smokers, not the 65% that the table shows. Using the sample proportion of $\hat{p} = 0.65$, test the claim that the proportion of deaths among smokers is $p = 0.45$. (*Hint:* See Section 7-5.) Does the result suggest that among the 1000 males in the 45–64 age bracket who died, smokers are disproportionately represented? Does this suggest that smoking is unhealthy?

2. Test the claim that smoking is independent of the cause of death. Based on the result, does it seem that the cause of death is related to whether the men smoked?

Note that we can use one method from Chapter 7 to investigate one issue (whether smoking is unhealthy), and we can use another method from this chapter to investigate a different issue (whether smoking is related to the cause of death). Given sample data, there are often different questions that can be considered using different statistical methods.

11 Analysis of Variance

11-1 Overview

Objectives are identified for this chapter, which introduces methods for testing the hypothesis that three or more populations have the same mean. In Chapter 8 we tested for equality between the means of two populations and used test statistics having normal or Student t distributions, but the methods of this chapter are very different.

11-2 One-Way ANOVA with Equal Sample Sizes

This section introduces the basic method of analysis of variance (ANOVA), which is used to test the claim that three or more populations have the same mean. This section is restricted to cases in which a sample is randomly selected from each population so that all samples have the same number of scores.

11-3 One-Way ANOVA with Unequal Sample Sizes

The methods presented in Section 11-2 are extended to apply to samples with different numbers of scores.

11-4 Two-Way ANOVA

Sections 11-2 and 11-3 deal with a single factor (or characteristic used to differentiate between populations), but this section considers two factors. Because the calculations are very difficult, the emphasis is on interpreting computer displays rather than performing manual calculations.

Chapter Problem:

Are bad movies as long as good movies, or does it just seem that way?

When we sit through a movie we don't like, it often seems to take forever to end. In contrast, a good movie seems to be over before we know it. Is it our imaginations, or are bad movies really longer than good movies? Are there really any differences in length between good and bad movies? Consider the data in Table 11-1, based on the randomly selected movies in Data Set 5 of Appendix B. In Table 11-1 we separate the movie lengths into four categories, depending on the number of rating stars each movie received from the critics. Examination of the summary statistics seems to suggest that there are differences in the mean lengths of movies, with movies rated as excellent tending to be longer. But are those differences significant? Specifically, we will test the claim that the four categories of movies have the same length; that is, we will test the claim that $\mu_1 = \mu_2 = \mu_3 = \mu_4$.

Table 11-1 Lengths (in min) of Movies Categorized by Star Ratings

Poor 0.0–1.5 Stars	Fair 2.0–2.5 Stars	Good 3.0–3.5 Stars		Excellent 4.0 Stars
105	110	93	123	72
108	114	115	97	120
96	98	133	104	106
91	100	94	82	104
	96	94	98	159
	123	106	107	125
	101	93	95	103
	92	129	94	160
	155	102	117	193
	92	90	104	168
	99	104	119	88
	100	105	96	121
	108	139	134	144
		111	100	90
		111		

$n_1 = 4$	$n_2 = 13$	$n_3 = 29$	$n_4 = 14$
$\bar{x}_1 = 100.00$	$\bar{x}_2 = 106.77$	$\bar{x}_3 = 106.52$	$\bar{x}_4 = 125.21$
$s_1 = 7.87$	$s_2 = 17.00$	$s_3 = 14.48$	$s_4 = 35.02$

11-1 Overview

In Sections 8-2, 8-3, and 8-5 we developed procedures for testing the hypothesis that two population means are equal (H_0: $\mu_1 = \mu_2$). In this chapter we will develop a procedure for testing the hypothesis that three or more population means are equal. (The methods of this chapter can be used to test for equality between two population means, but the methods of Chapter 8 are more efficient.) A typical null hypothesis in this chapter will be H_0: $\mu_1 = \mu_2 = \mu_3 = \mu_4$; the alternative hypothesis is H_1: At least one mean is different. The method we use is based on an analysis of sample variances.

DEFINITION

Analysis of variance (ANOVA) is a method of testing the equality of three or more population means by analyzing sample variances.

ANOVA is used in applications such as the following:

- When 6 different fields of corn are each divided into 6 plots, and each of the 6 plots is given a different fertilizer treatment, we can test to determine whether the crop yields are equal for the 6 populations corresponding to the 6 fertilizer treatments.

- When Kaplan, Princeton Review, and Collins Test Prep give students different SAT test preparation courses, we can test to determine whether there are differences in mean SAT scores for the 3 populations of students from the 3 different courses.

- When 3 different machines are used to make car keys, we can test to determine whether they have the same mean output of keys produced each day.

The method of analysis of variance derives its name from the fact that we analyze different types of variation among data. By comparing different types of sample variances, we can form conclusions about whether the population *means* are equal.

You might wonder why we should bother with a new procedure when we can test for equality of *two* means by using the methods presented in Chapter 8, where we developed tests of H_0: $\mu_1 = \mu_2$. For example, if we want to use the sample data from Table 11-1 to test the claim (at the $\alpha = 0.05$ level) that the four populations have the same mean, why not simply pair them off and do two at a time by testing H_0: $\mu_1 = \mu_2$, then H_0: $\mu_2 = \mu_3$, and so on? This approach (doing two at a time) requires six different hypothesis tests, so the degree of confidence could be as low as 0.95^6 (or 0.735). In general, as we increase the number of individual tests of significance, we increase the likelihood of finding a difference by chance alone. The risk of a type I error—finding a difference in one of the pairs when no such difference actually exists—is excessively high. The method of analysis of variance lets us avoid that particular pitfall (rejecting a true null hypothesis) by using *one* test for equality of several means.

F Distribution

The ANOVA methods use the *F* distribution, which was introduced in Section 8-4. Recall that the *F* distribution has the following important properties (see Figure 11-1):

1. The *F* distribution is not symmetric; it is skewed to the right.
2. The values of *F* can be 0 or positive, but they cannot be negative.
3. There is a different *F* distribution for each pair of degrees of freedom for the numerator and denominator.

Critical values of *F* are given in Table A-5.

Assumptions

In this chapter, the following assumptions apply when testing the hypothesis that three or more samples come from populations with the same mean:

1. The populations have normal distributions.
2. The populations have the same variance (or standard deviation).
3. The samples are random and independent of each other.

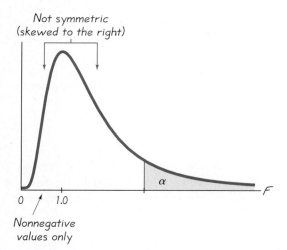

FIGURE 11-1   The *F* Distribution

There is a different *F* distribution for each different pair of degrees of freedom for numerator and denominator.

The requirements of normality and equal variances are somewhat relaxed, as the methods in this chapter work reasonably well unless a population has a distribution that is very nonnormal or the population variances differ by large amounts. University of Wisconsin statistician George E. P. Box showed that as long as the sample sizes are equal (or nearly equal), the variances can be up to nine times as large and the results of ANOVA will continue to be essentially reliable. (If the samples are independent but the distributions are very nonnormal, we can use the Kruskal-Wallis test presented in Section 13-5.)

Analysis of variance (or ANOVA) is based on a comparison of two different estimates of the variance common to the different populations. Those estimates (the *variance between samples* and the *variance within samples*) will be described in Section 11-2, which begins with a description of one-way analysis of variance. The term *one-way* is used because the sample data are separated into groups according to one characteristic, or factor. For example, the movie length problem described at the beginning of the chapter involves sample data separated into four different groups according to the one characteristic (or factor) of star ratings. The method of Section 11-2 can be applied to cases in which the sample data are obtained from independent and random samples that are all of the *same size*. In Section 11-3 we will consider cases in which the sample sizes are not all equal, as in Table 11-1. Section 11-4 will introduce *two-way analysis of variance*, which allows us to compare populations separated into categories using *two* characteristics (or factors). For example, we might separate the lengths of movies according to (1) their star ratings and (2) their viewer discretion ratings (G, PG, PG-13, R).

11-2 One-Way ANOVA with Equal Sample Sizes

In this section we consider tests of hypotheses that three or more population means are all equal, as in H_0: $\mu_1 = \mu_2 = \mu_3$.

Assumptions

As stated in Section 11-1, we assume that the populations have normal distributions, the populations have the same variance (or standard deviation), and the samples are random and independent of each other. In this section we make the following additional assumptions:

1. The different samples are all the same size.

2. The different samples are from populations that are categorized in only one way.

The method we use is called **one-way analysis of variance** (or **single-factor analysis of variance**) because we use a single property, or characteristic, for categorizing the populations. This characteristic is sometimes referred to as a treatment or factor.

DEFINITION

A **treatment** (or **factor**) is a property, or characteristic, that allows us to distinguish the different populations from one another.

For example, the movie lengths in Table 11-1 are sample data drawn from four different populations that are distinguished according to the treatment (or factor) of star rating. The term *treatment* is used because early applications of analysis of variance involved agricultural experiments in which different plots of farmland were treated with different fertilizers, seed types, insecticides, and so on.

We will introduce the basic concepts underlying the method of analysis of variance, and then we will use these general concepts as we consider cases with unequal sample sizes in Section 11-3 and cases with more than one treatment (that is, more than one way to classify the populations) in Section 11-4.

In using analysis of variance to test claims such as $H_0: \mu_1 = \mu_2 = \mu_3$, we employ the same general method of testing hypotheses that was introduced in Chapter 7. We adapt that method by developing a test statistic suitable for the circumstances we are now considering; the test statistic uses the following notation.

Notation for One-Way ANOVA with Equal Sample Sizes

n = size of each sample

k = number of samples

$s_{\bar{x}}^2$ = variance of the sample means

s_{p}^2 = pooled variance obtained by calculating the mean of the sample variances

---

The test statistic is the ratio of two different estimates of the variance σ^2 that is assumed to be common to the populations involved—the variance between samples and the variance within samples, terms that we now define.

More Police, Fewer Crimes?

Does an increase in the number of police officers result in lower crime rates? The question was studied in a New York City experiment that involved a 40% increase in police officers in one precinct while adjacent precincts maintained a constant level of officers. Statistical analysis of the crime records showed that crimes visible from the street (such as auto thefts) did decrease, but crimes not visible from the street (such as burglaries) were not significantly affected.

DEFINITIONS

The **variance between samples** (also called **variation due to treatment**) is an estimate of σ^2 based on the sample *means*. This estimate measures the variability caused by differences among the sample means that correspond to the different treatments, or categories of classification. With all samples of the same size n,

$$\text{variance between samples} = ns_{\bar{x}}^2$$

where $s_{\bar{x}}^2$ = variance of the sample means

The **variance within samples** (also called **variation due to error**) is an estimate of σ^2 based on the sample *variances*. With all samples of the same size n,

$$\text{variance within samples} = s_p^2$$

where s_p^2 = *pooled variance* obtained by finding the mean of the sample variances

Keep in mind that among the assumptions listed in Section 11-1 is this one: The populations being considered have the same variance (or standard deviation). The variance *between* samples constitutes one estimation of the variance σ^2 that is common to the populations, whereas the variance *within* samples is a second estimation of that same variance. The expression given above for variance between samples is justified as an estimate of σ^2 because $\sigma_{\bar{x}} = \sigma/\sqrt{n}$ (from the central limit theorem); squaring both sides and solving for σ^2 results in the equation $\sigma^2 = n\sigma_{\bar{x}}^2$, which indicates that σ^2 can be estimated by $ns_{\bar{x}}^2$. The expression given above for variance within samples is the mean of the sample variances; it indicates that we estimate σ^2 with the mean of the individual sample variances. We use the variance between samples and the variance within samples in the test statistic.

Test Statistic for One-Way ANOVA with Equal Sample Sizes

$$F = \frac{\text{variance between samples}}{\text{variance within samples}} = \frac{ns_{\bar{x}}^2}{s_p^2}$$

Interpreting the Test Statistic *F*

As the preceding test statistic indicates, we use both estimates of σ^2 to find the value of the *F* test statistic. The numerator measures variation between sample means. The estimate of variance in the denominator (s_p^2) depends only on the sample variances and is not affected by differences among the sample means.

Consequently sample means that are close in value result in an F test statistic that is close to 1, and we conclude that there are no significant differences among the sample means. But if the value of F is excessively large, then we reject the claim of equal means. (The vague terms *close to 1* and *excessively large* are made objective by the use of a specific critical value, which clearly differentiates between an F test statistic that is in the critical region and one that is not.) **Because excessively large values of F reflect unequal means, the test is right-tailed.**

To see how the F test statistic works, consider the two collections of sample data in Table 11-2. Note that the three samples in part A are identical to the three samples in part B, except that in part B we have added 10 to each value of sample 1 from part A. Comparing the sample statistics, we see that the three sample means in part A are very close, whereas there are substantial differences in part B. Comparing the sample variances, we see that the three values in part A are identical to those in part B.

TABLE 11-2   Testing Equality of Means: Two Cases

	A add 10			B		
	Sample 1	Sample 2	Sample 3	Sample 1	Sample 2	Sample 3
	7	6	4	17	6	4
	3	5	7	13	5	7
	6	5	6	16	5	6
	6	8	7	16	8	7
	$n_1 = 4$	$n_2 = 4$	$n_3 = 4$	$n_1 = 4$	$n_2 = 4$	$n_3 = 4$
	$\bar{x}_1 = 5.5$	$\bar{x}_2 = 6.0$	$\bar{x}_3 = 6.0$	$\bar{x}_1 = 15.5$	$\bar{x}_2 = 6.0$	$\bar{x}_3 = 6.0$
	$s_1^2 = 3.0$	$s_2^2 = 2.0$	$s_3^2 = 2.0$	$s_1^2 = 3.0$	$s_2^2 = 2.0$	$s_3^2 = 2.0$

Variance between samples	$ns_{\bar{x}}^2 = 4(0.0833) = 0.3332$	$ns_{\bar{x}}^2 = 4(30.0833) = 120.3332$
Variance within samples	$s_p^2 = \dfrac{3.0 + 2.0 + 2.0}{3} = 2.3333$	$s_p^2 = \dfrac{3.0 + 2.0 + 2.0}{3} = 2.3333$
F test statistic	$F = \dfrac{ns_{\bar{x}}^2}{s_p^2} = \dfrac{0.3332}{2.3333} = 0.1428$	$F = \dfrac{ns_{\bar{x}}^2}{s_p^2} + \dfrac{120.3332}{2.3333} = 51.5721$

Adding 10 to each score in the first sample changed our test statistic from $F = 0.1428$ to $F = 51.5721$. Note that the variance between samples for part A is 0.3332, but for part B it is 120.3332 (indicating that the sample means in part B are farther apart). Note also that the variance within samples is 2.3333 in both parts; the variance within a sample isn't affected when we add a constant to every score. The change in F is attributable only to the change in $\bar{x}_1$. This illustrates that the F test statistic is very sensitive to sample *means*, even though it is obtained through two different estimates of the common population *variance*.

The critical value of F that separates excessive values from acceptable values is found in Table A-5, where α is again the level of significance and the numbers of degrees of freedom are as follows (assuming that there are k sets of separate samples with n scores in each set). For cases with equal sample sizes:

$$\text{numerator degrees of freedom} = k - 1$$
$$\text{denominator degrees of freedom} = k(n - 1)$$

We have noted that our ANOVA tests are right-tailed. Because the individual components of n, $s_{\bar{x}}^2$, and s_p^2 are all positive, F is always positive. With the same traditional method for testing hypotheses (see Chapter 7), a value of the F test statistic indicates a significant difference among the sample means when the test statistic exceeds the critical value obtained from Table A-5.

▶ **EXAMPLE**

Do different age groups have different mean body temperatures? Table 11-3 lists the body temperatures of 5 randomly selected subjects from each of 3 different age groups. Informal examination of the 3 sample means (97.940, 98.580, 97.800) seems to suggest that the 3 samples come from populations with means that are not significantly different. In addition to the values of the 3 sample means, however, we should consider their standard deviations and sample sizes. We need to conduct a formal hypothesis test to determine whether the sample means are significantly different. Using a significance level of $\alpha = 0.05$, we will test the claim that the 3 age-group populations have the same mean body temperature.

● **SOLUTION**

Step 1:   The claim of equal mean body temperatures can be expressed in symbolic form as $\mu_1 = \mu_2 = \mu_3$.

Step 2:   If the original claim is false, then the 3 means are not all equal. (This is difficult to express in symbols, so we will express it verbally.)

TABLE 11-3	**Body Temperatures (°F) Categorized by Age**	
18–20	21–29	30 and older
98.0	99.6	98.6
98.4	98.2	98.6
97.7	99.0	97.0
98.5	98.2	97.5
97.1	97.9	97.3

$n_1 = 5$ $n_2 = 5$ $n_3 = 5$ (Statistics are rounded
$\overline{x}_1 = 97.940$ $\overline{x}_2 = 98.580$ $\overline{x}_3 = 97.800$ with extra digits for
$s_1 = 0.568$ $s_2 = 0.701$ $s_3 = 0.752$ greater precision)

Data obtained by Dr. Philip Mackowiak, Dr. Steven Wasserman, and Dr. Myron Levine of the University of Maryland.

Delaying Death

University of California sociologist David Phillips has studied the ability of people to postpone their death until after some important event. Analyzing death rates of Jewish men who died near Passover, he found that the death rate dropped dramatically in the week before Passover, but rose the week after. He found a similar phenomenon occurring among Chinese-American women; their death rate dropped the week before their important Harvest Moon Festival, then rose the week after.

Step 3: Because the null hypothesis must contain the condition of equality, we have

$$H_0: \mu_1 = \mu_2 = \mu_3$$
H_1: The 3 means are not all equal.

Step 4: The significance level is $\alpha = 0.05$.

Step 5: Because we are testing the claim that 3 or more means are equal, we use analysis of variance with an F test statistic.

Step 6: For one-way analysis of variance with equal sample sizes, the test statistic ($F = 1.8803$) is calculated after first finding the values of n, $s_{\overline{x}}^2$, and s_p^2.

$s_{\overline{x}}^2 = 0.172933$ [found by computing the variance of the sample means (97.940, 98.580, 97.800)]

$$s_p^2 = \frac{0.568^2 + 0.701^2 + 0.752^2}{3}$$

$= 0.459843$ (found by calculating the mean of the sample variances)

$$F = \frac{\text{variance between samples}}{\text{variance within samples}} = \frac{ns_{\overline{x}}^2}{s_p^2}$$

$$= \frac{5(0.172933)}{0.459843} = \frac{0.864665}{0.459843} = 1.8803$$

(continued)

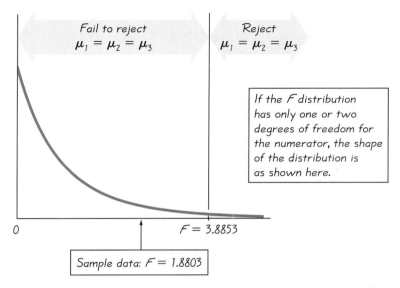

FIGURE 11-2 *F* **Distribution for Body Temperature Example**

The critical value of $F = 3.8853$ is found by referring to the portion of Table A-5 for which $\alpha = 0.05$. The degrees of freedom are as follows:

$$\text{numerator degrees of freedom} = k - 1 = 3 - 1 = 2$$
$$\text{denominator degrees of freedom} = k(n - 1) = 3(5 - 1) = 12$$

The test statistic and critical value are shown in Figure 11-2.

Step 7: Because the test statistic of $F = 1.8803$ does not fall in the critical region bounded by $F = 3.8853$, we fail to reject the null hypothesis of $\mu_1 = \mu_2 = \mu_3$.

Step 8: There is not sufficient evidence to warrant rejection of the claim that the 3 populations of different age groups have the same mean body temperature. Perhaps there really is a difference, but the sample size is too small and/or the sample differences are not large enough to justify that conclusion. (We used small sample sizes of $n = 5$ to keep the calculations manageable.)

In the preceding example, we assumed that the three populations have the same variance σ^2, which we estimated by two different methods. One method used the variance *between* samples: $ns_{\bar{x}}^2 = 0.864665$. The other method used the variance *within* samples: $s_p^2 = 0.459843$, which is simply the mean of the three sample variances obtained by averaging the three sample variances of 0.568^2, 0.701^2, and 0.752^2. Because the F ratio of these two estimates of σ^2 does not exceed 1 by a significant amount (is not in the critical region), we failed to reject the null hypothesis of equal means. If it could talk, the test statistic of

$F = 1.8803$ would say to us something like: "The two estimates of the population variance σ^2 are roughly the same because their ratio doesn't exceed 1 by a significant amount; therefore, the sample means are not significantly different."

Using Computer Software for One-Way ANOVA

The calculations of this section did not appear to be too formidable because we used relatively small sample sizes. Larger data sets could easily lead to messy calculations, but, fortunately, statistical software packages are available. STATDISK has an option of One-Way Analysis of Variance in its Statistics menu, and Minitab has a command AOVONEWAY for one-way analysis of variance. With Minitab, enter the different categories of sample data in the different columns C1, C2, C3 (and so on), then enter the command AOVONEWAY C1 C2 C3 (and so on). The result will include the F test statistic and P-value. If the P-value is less than the significance level, reject the null hypothesis; otherwise, fail to reject the null hypothesis. We will illustrate the use of Minitab in the next section.

We now summarize the key components of the hypothesis test for equality of means in three or more populations. Note that the following test statistic and numbers of degrees of freedom apply only to cases involving samples of the same size; that is, we have k samples, each with n scores.

Testing $\mu_1 = \mu_2 = \cdots = \mu_k$ with k Samples, Each Having n Scores

H_0: $\mu_1 = \mu_2 = \cdots = \mu_k$ H_1: The means are not all equal.

Test statistic:

$$F = \frac{\text{variance between samples}}{\text{variance within samples}} = \frac{ns_{\bar{x}}^2}{s_p^2}$$

where $s_{\bar{x}}^2$ = variance of sample means

and s_p^2 = mean of sample variances

Critical value: Test is always right-tailed. Refer to Table A-5, with

numerator degrees of freedom = $k - 1$
denominator degrees of freedom = $k(n - 1)$

We must also keep in mind that we should satisfy the assumptions described in Section 11-1. That is, the populations should have nearly normal distributions with approximately the same variance, and the samples must be independent of each other.

11-2   Exercises A: Basic Skills and Concepts

1. Refer to the data in Table 11-3 and add 1° to each temperature listed in the 18–20 age group. The summary statistics are as shown below. At the 0.05 significance level, test the claim that the 3 different age populations have the same mean body temperature.

18–20	21–29	30 and older
$n_1 = 5$	$n_2 = 5$	$n_3 = 5$
$\bar{x}_1 = 98.940$	$\bar{x}_2 = 98.580$	$\bar{x}_3 = 97.800$
$s_1 = 0.568$	$s_2 = 0.701$	$s_3 = 0.752$

2. The Wendt Brewing Company plans to launch a major media campaign. Three advertising companies prepared trial commercials in an attempt to win a $2 million contract. The commercials were tested on randomly selected consumers, whose reactions were measured; the results are summarized below. (Higher scores indicate more positive reactions to the commercial.) At the 0.05 significance level, test the claim that the 3 populations have the same mean reaction score. If you were responsible for advertising at Wendt Brewing, is there a company you would select on the basis of these results?

Delaney and Taylor Advertising Company	Madison Advertising	Kelly, Grey, and Landry Advertising
$n = 25$	$n = 25$	$n = 25$
$\bar{x} = 77.2$	$\bar{x} = 81.1$	$\bar{x} = 78.4$
$s = 15.4$	$s = 18.2$	$s = 12.8$

3. The City Resource Recovery Company (CRRC) collects the waste discarded by households in a region. Discarded waste must be separated into categories of metal, paper, plastic, and glass. In planning for the equipment needed to collect and process the garbage, CRRC refers to the data we have summarized in Data Set 1 in Appendix B (provided by the Garbage Project at the University of Arizona). The results (weights in pounds) are summarized below. At the 0.05 level of significance, test the claim that the 4 specific populations of garbage have the same mean. Based on the results, does it appear that these 4 categories require the same collection and processing resources? Does any single category seem to be a particularly large part of the waste management problem?

Metal	Paper	Plastic	Glass
$n = 62$	$n = 62$	$n = 62$	$n = 62$
$\bar{x} = 2.218$	$\bar{x} = 9.428$	$\bar{x} = 1.911$	$\bar{x} = 3.752$
$s = 1.091$	$s = 4.168$	$s = 1.065$	$s = 3.108$

4. Jane Corrigan is a pilot who does extensive flying in a variety of weather conditions. She has decided to buy a battery-powered radio as an independent

backup for her regular radios, which depend on her plane's electrical system. Jane randomly selects 5 batteries from each of 3 different battery suppliers, whose costs vary, and then tests the batteries for their operating time (in hours) before recharging is necessary. Do the 3 suppliers provide batteries that have the same mean usable time before recharging is necessary? Based on these results, does one of the suppliers seem to be superior to the others?

Telektronic	Lectrolyte	Electrocell
26.0	29.0	30.0
28.5	28.8	26.3
27.3	27.6	29.2
25.9	28.1	27.1
28.2	27.0	29.8
↑	↑	↑
$\bar{x} = 27.18$	$\bar{x} = 28.10$	$\bar{x} = 28.48$
$s^2 = 1.46$	$s^2 = 0.69$	$s^2 = 2.81$

5. Five socioeconomic groups were studied by a sociologist, and randomly selected members from each group were rated on their adjustment to society. A standard scale was designed in which higher scores indicated better adjusted persons. The sociologist used the sample data summarized in the accompanying table. At the $\alpha = 0.05$ level of significance, test the claim that the 5 populations have equal means. The sociologist's next stage of research is designed for a group that is not as well adjusted to society as other groups. Based on these results, can the sociologist proceed with the next stage, using 1 or more of the 5 groups?

Low	Low Middle	Middle	Upper Middle	Upper
$n = 10$	$n = 10$	$n = 10$	$n = 10$	$n = 10$
$\bar{x} = 103$	$\bar{x} = 97$	$\bar{x} = 102$	$\bar{x} = 100$	$\bar{x} = 110$
$s^2 = 230$	$s^2 = 75$	$s^2 = 200$	$s^2 = 150$	$s^2 = 100$

6. A unit on elementary algebra was taught to 5 different classes of randomly selected students with the same academic backgrounds. A different method of teaching was used in each class, and the final averages of the 20 students in each class were compiled. The results yielded the following data. At the $\alpha = 0.05$ level of significance, test the claim that the 5 population means are equal. Based on these results, does it make any difference which teaching method is used?

Traditional	Programmed	Audio	Audiovisual	Visual
$n = 20$	$n = 20$	$n = 20$	$n = 20$	$n = 20$
$\bar{x} = 76$	$\bar{x} = 74$	$\bar{x} = 70$	$\bar{x} = 75$	$\bar{x} = 74$
$s^2 = 60$	$s^2 = 50$	$s^2 = 100$	$s^2 = 36$	$s^2 = 40$

Zone	SP	LA	Acres	Taxes
1	147	20	0.50	1.9
1	160	18	1.00	2.4
1	128	27	1.05	1.5
1	162	17	0.42	1.6
1	135	18	0.84	1.6
1	132	13	0.33	1.5
1	181	24	0.90	1.7
1	138	15	0.83	2.2
1	145	17	2.00	1.6
1	165	16	0.78	1.4
4	160	18	0.55	2.8
4	140	20	0.46	1.8
4	173	19	0.94	3.2
4	113	12	0.29	2.1
4	85	9	0.26	1.4
4	120	18	0.33	2.1
4	285	28	1.70	4.2
4	117	10	0.50	1.7
4	133	15	0.43	1.8
4	119	12	0.25	1.6
7	215	21	3.04	2.7
7	127	16	1.09	1.9
7	98	14	0.23	1.3
7	147	23	1.00	1.7
7	184	17	6.20	2.2
7	109	17	0.46	2.0
7	169	20	3.20	2.2
7	110	14	0.77	1.6
7	68	12	1.40	2.5
7	160	18	4.00	1.8

7. Five car models were studied in a test that involved 4 cars of each model. For each of the 4 cars in each of the 5 samples, exactly one gallon of gas was placed in the tank and the car was driven until the gas was used up. The distances (in miles) traveled are given in the table. At the $\alpha = 0.05$ significance level, test the claim that the 5 population means are all equal. If you were responsible for purchasing a fleet of cars for your company's sales team, would it make any difference (in terms of gas mileage) which car model you selected?

A	B	C	D	E
16	18	18	19	15
22	23	18	21	16
17	15	20	22	20
17	20	20	22	17

8. A sociologist randomly selected subjects from 3 types of family structures: two-parent, single-parent, and families in transition. The selected subjects were interviewed and rated for their social adjustment. The sample results are given in the table. At the $\alpha = 0.05$ level of significance, test the claim that the 3 population means are equal.

Two-Parent	Single-Parent	Transition
140	115	90
135	105	120
130	110	125
125	130	100
150	105	105

For Exercises 9–12, use the data given in the table in the margin. The randomly selected data represent 30 different homes sold in Dutchess County, New York. The zones (1, 4, 7) correspond to different geographical regions of the county. The values of SP are the selling prices in thousands of dollars. The values of LA are the living areas in hundreds of square feet. The values of Acres are the lot sizes in acres, and the Taxes values are the annual tax bills in thousands of dollars. For example, the first home is in zone 1; it sold for $147,000; it has a living area of 2000 square feet; it is on a 0.50-acre lot; and the annual taxes are $1900.

9. At the 0.05 significance level, test the claim that the means of the selling prices are the same in zone 1, zone 4, and zone 7. Does any zone seem to have homes with higher selling prices?

10. At the 0.05 significance level, test the claim that the means of the living areas are the same in zone 1, zone 4, and zone 7. Does any zone seem to have larger homes?

11. At the 0.05 significance level, test the claim that the means of the lot sizes (in acres) are the same in zone 1, zone 4, and zone 7. Does any zone seem to have larger lots?

12. At the 0.05 significance level, test the claim that the means of the tax amounts are the same in zone 1, zone 4, and zone 7. Does any zone seem to have higher taxes?

11-2   Exercises B: Beyond the Basics

13. A study was made of 3 police precincts to determine the time (in seconds) required for a police car to be dispatched after a crime is reported. Sample results are given below.

Precinct 1	Precinct 2	Precinct 3
$n = 50$	$n = 50$	$n = 50$
$\bar{x} = 197$	$\bar{x} = 202$	$\bar{x} = 208$
$s = 18$	$s = 20$	$s = 23$

   a. At the 0.05 level of significance, test the claim that $\mu_1 = \mu_2$. Use the methods discussed in Chapter 8.
   b. At the 0.05 level of significance, test the claim that $\mu_2 = \mu_3$. Use the methods discussed in Chapter 8.
   c. At the 0.05 level of significance, test the claim that $\mu_1 = \mu_3$. Use the methods discussed in Chapter 8.
   d. At the 0.05 level of significance, test the claim that $\mu_1 = \mu_2 = \mu_3$. Use analysis of variance.
   e. Compare the methods and results of parts a, b, and c to those of part d. Is it better to use the approach of part d or the approach of combining parts a, b, and c?

14. Five independent samples of 50 scores are randomly drawn from populations that are normally distributed with equal variances. We wish to test the claim that $\mu_1 = \mu_2 = \mu_3 = \mu_4 = \mu_5$.
   a. If we used only the methods given in Chapter 8, we would test the individual claims $\mu_1 = \mu_2, \mu_1 = \mu_3, \ldots, \mu_4 = \mu_5$. How many ways could we pair off 5 means?
   b. Assume that for each test of equality between 2 means, there is a 0.95 probability of not making a type I error. If all possible pairs of means are tested for equality, what is the probability of making no type I errors? (Although the tests are not actually independent, assume that they are.)
   c. If we use analysis of variance to test the claim that $\mu_1 = \mu_2 = \mu_3 = \mu_4 = \mu_5$ at the 0.05 level of significance, what is the probability of not making a type I error?
   d. Compare the results of parts b and c. Which approach is better in the sense of giving us a greater chance of not making a type I error?

15. Five independent samples of 50 scores were randomly drawn from populations that are normally distributed with equal variances, and the values of n, $\bar{x}$, and s were obtained in each case. Analysis of variance was then used to test the claim that $\mu_1 = \mu_2 = \mu_3 = \mu_4 = \mu_5$.
   a. If a constant is added to each of the 5 sample means, how is the value of the test statistic affected?   *(continued)*

   b. If each of the 5 means is multiplied by a constant, how is the value of the test statistic affected?

   c. If the 5 samples have the same mean, what is the value of the F test statistic?

16. In this exercise you will verify that when you have 2 sets of sample data, the t test for independent samples and the ANOVA method of this section are equivalent. Refer to the real estate data used in Exercises 9–12, but exclude the zone 7 data.

   a. Use a 0.05 level and the method of Section 8-5 to test the claim that zones 1 and 4 have the same mean selling prices. (Assume that both zones have the same variance.)

   b. Use a 0.05 level and the ANOVA method of this section to test the claim made in part a.

   c. Verify that the squares of the t test statistic and critical value from part a are equal to the F test statistic and critical value from part b.

11-3 One-Way ANOVA with Unequal Sample Sizes

In Section 11-2 our discussion of analysis of variance involved only examples with the same number of scores in each sample. In this section we will consider the analysis of variance (ANOVA) method for working with samples of unequal sizes, but we will continue to deal with *one-way* ANOVA because a single characteristic is used for categorizing the data. Two-way ANOVA, which uses two different characteristics for categorizing data, is discussed in Section 11-4.

As in Section 11-2, the assumptions listed in Section 11-1 must apply: The populations should have data with nearly normal distributions with approximately the same variance, and the samples must be independent of each other. We again proceed to test claims of equal means, such as H_0: $\mu_1 = \mu_2 = \mu_3$, by analyzing the significance of the ratio of *variance between samples* to *variance within samples*. As in Section 11-2, we use the test statistic

$$F = \frac{\text{variance between samples}}{\text{variance within samples}}$$

For unequal sample sizes, however, we must "weight" each of the two estimates of variance to account for the different sample sizes. In expressing the test statistic that compensates for unequal sample sizes, we use the following notation.

Notation for One-Way ANOVA with Unequal Sample Sizes

$\overline{\overline{x}}$ = overall mean (sum of all sample scores divided by the total number of scores)

k = number of population means being compared

n_i = number of values in the ith sample

N = total number of values in all samples combined
$(N = n_1 + n_2 + \cdots + n_k)$

$\overline{x}_i$ = mean of values in the ith sample

s_i^2 = variance of values in the ith sample

Using the preceding notation, we can now express the test statistic as follows:

$$F = \frac{\text{variance between samples}}{\text{variance within samples}} = \frac{\left[\dfrac{\Sigma n_i(\overline{x}_i - \overline{\overline{x}})^2}{k - 1}\right]}{\left[\dfrac{\Sigma(n_i - 1)s_i^2}{\Sigma(n_i - 1)}\right]}$$

Note that the numerator in the above test statistic is really a form of the formula

$$s^2 = \frac{\Sigma(x - \overline{x})^2}{n - 1}$$

for variance that was given in Chapter 2. The factor of n_i is included so that larger samples carry more weight. The denominator of the test statistic is simply the mean of the sample variances, but it is a weighted mean with the weights based on the sample sizes. If all samples have the same number of scores, the test statistic simplifies to $F = ns_{\overline{x}}^2/s_p^2$, as given in Section 11-2.

Because the calculations required for ANOVA with unequal sample sizes are typically quite tedious, we strongly recommend using a statistical software package. Because calculating the above test statistic can lead to large rounding errors, the various software packages typically use a different (but equivalent) expression that involves SS (for sum of squares) and MS (for mean square) notation. Whereas Section 11-2 emphasized manual calculations that directly reveal the nature of the test statistic, this section emphasizes the interpretation of computer displays. Instead of using the above test statistic, we therefore introduce the test statistic used most often by statistical software packages. Although the following notation and components are complicated and involved, the basic idea is the same as that introduced in Section 11-2: The test statistic F is a ratio with a numerator reflecting variation *between* the means of the samples and a denominator reflecting variation *within* the samples. If the populations have equal means, the F ratio tends to be close to 1, but if the population means are not equal, the F ratio tends to be significantly larger than 1. Key components in our ANOVA method are identified below.

> **SS(total),** or total sum of squares, is a measure of the total variation (around $\overline{\overline{x}}$) in all of the sample data combined.
>
> Formula 11-1 $$SS(total) = \Sigma(x - \overline{\overline{x}})^2$$

SS(total) can be broken down into the components of SS(treatment) and SS(error), described as follows:

> **SS(treatment)** is a measure of the variation between the sample means. [In one-way ANOVA, SS(treatment) is sometimes referred to as SS(factor). Because it is a measure of variability *between* the sample means, it is also referred to as SS(*between* groups) or SS(*between* samples)].
>
> Formula 11-2
> $$SS(treatment) = n_1(\overline{x}_1 - \overline{\overline{x}})^2 + n_2(\overline{x}_2 - \overline{\overline{x}})^2 + \cdots + n_k(\overline{x}_k - \overline{\overline{x}})^2$$
> $$= \Sigma n_i(\overline{x}_i - \overline{\overline{x}})^2$$

If the population means ($\mu_1, \mu_2, \ldots, \mu_k$) are equal, then the sample means $\overline{x}_1, \overline{x}_2, \ldots, \overline{x}_k$ will all tend to be close together and also close to $\overline{\overline{x}}$. The result will be a relatively small value of SS(treatment). If the population means are not all equal, however, then at least one of $\overline{x}_1, \overline{x}_2, \ldots, \overline{x}_k$ will tend to be far apart from the others and also far apart from $\overline{\overline{x}}$. The result will be a relatively large value of SS(treatment).

> **SS(error)** is a sum of squares representing the variability that is assumed to be common to all the populations being considered.
>
> Formula 11-3
> $$SS(error) = (n_1 - 1)s_1^2 + (n_2 - 1)s_2^2 + (n_3 - 1)s_3^2 + \cdots + (n_k - 1)s_k^2$$
> $$= \Sigma(n_i - 1)s_i^2$$

SS(error) is the numerator of the expression for the *pooled variance* s_p^2 from Section 8-5. Because SS(error) is a measure of the variance within groups, it is sometimes denoted as SS(*within* groups) or SS(*within* samples).

Given the above expressions for SS(total), SS(treatment), and SS(error), the following relationship will always hold.

Formula 11-4 $$SS(total) = SS(treatment) + SS(error)$$

▶ **EXAMPLE**

Table 11-1 (based on Data Set 5 in Appendix B) includes sample data with movie lengths arranged according to the numbers of stars the movies were given. Use the data in Table 11-1 to find the values of SS(treatment), SS(error), and SS(total).

● **SOLUTION**

In the following calculations, extra decimal places are carried for greater precision of results.

$$k = 4 \qquad \text{number of samples}$$

$$\overline{\overline{x}} = \text{mean of all 60 sample scores} = \frac{6630}{60} = 110.5000$$

We now evaluate SS(treatment), SS(error), and SS(total) by using Formulas 11-2, 11-3, and 11-4 as follows:

$$\text{SS(treatment)} = n_1(\overline{x}_1 - \overline{\overline{x}})^2 + n_2(\overline{x}_2 - \overline{\overline{x}})^2 + n_3(\overline{x}_3 - \overline{\overline{x}})^2 + n_4(\overline{x}_4 - \overline{\overline{x}})^2$$

$$= 4(100.0000 - 110.5000)^2 + 13(106.7692 - 110.5000)^2$$
$$+ 29(106.5172 - 110.5000)^2 + 14(125.2143 - 110.5000)^2$$

$$= 441.0000 + 180.9453 + 460.0182 + 3031.1487$$

$$= 4113.1122$$

$$\text{SS(error)} = (n_1 - 1)s_1^2 + (n_2 - 1)s_2^2 + (n_3 - 1)s_3^2 + (n_4 - 1)s_4^2$$

$$= (4 - 1)(7.8740)^2 + (13 - 1)(17.001)^2$$
$$+ (29 - 1)(14.476)^2 + (14 - 1)(35.021)^2$$

$$= 185.9996 + 3468.4080 + 5867.5281 + 15944.1157$$

$$= 25466.0514$$

$$\text{SS(total)} = \text{SS(treatment)} + \text{SS(error)}$$

$$= 4113.1122 + 25466.0514 = 29579.1636$$

SS(treatment) and SS(error) are both sums of squares, and if we divide each by its corresponding number of degrees of freedom, we get *mean squares*, as defined on the next page.

Quality at Ford

In the early 1980s, the Ford Motor Company invested over $3 billion in development of the Taurus model. Subsequent sales of the Taurus proved the gamble to be successful, as Ford losses were turned into profits. The success of the Taurus is largely attributable to the priority placed on quality. One approach to the quality issue was developing a measurable standard referred to as "Best in Class." Fifty foreign and domestic car models were analyzed for their best features. After identifying 400 such features, the Taurus team set out to incorporate as many of them as possible. For example, seats from a dozen competing cars were tested with women and men drivers of all ages. Analysis of the results enabled the team to develop a seat design that was as good as or better than the best in its class. Statistical analysis of results of this type is often accomplished with analysis of variance techniques.

MS(treatment) is a mean square for treatment, obtained as follows:

Formula 11-5
$$MS(treatment) = \frac{SS(treatment)}{k - 1}$$

MS(error) is a mean square for error, obtained as follows:

Formula 11-6
$$MS(error) = \frac{SS(error)}{N - k}$$

MS(total) is a mean square for the total variation, obtained as follows:

Formula 11-7
$$MS(total) = \frac{SS(total)}{N - 1}$$

▶ **EXAMPLE**

Use the sample data in Table 11-1 to find the values of MS(treatment), MS(error), and MS(total).

● **SOLUTION**

We find the required values by using Formulas 11-5, 11-6, and 11-7, along with the results from the preceding example.

$$MS(treatment) = \frac{SS(treatment)}{k - 1} = \frac{4113.1122}{4 - 1} = 1371.0374$$

$$MS(error) = \frac{SS(error)}{N - k} = \frac{25466.0514}{60 - 4} = 454.7509$$

$$MS(total) = \frac{SS(total)}{N - 1} = \frac{29579.1636}{60 - 1} = 501.3418$$

We can now use the SS and MS values to identify the relevant test statistic.

Test Statistic for ANOVA with Unequal Sample Sizes

In testing the null hypothesis H_0: $\mu_1 = \mu_2 = \cdots = \mu_k$ against the alternative hypothesis that these means are not all equal, the test statistic

Formula 11-8
$$F = \frac{MS(\text{treatment})}{MS(\text{error})}$$

has an F distribution (when the null hypothesis H_0 is true) with degrees of freedom given by

numerator degrees of freedom = $k - 1$
denominator degrees of freedom = $N - k$

Interpreting the *F* Test Statistic

The interpretation of the F test statistic is essentially the same as in Section 11-2. For the test statistic F given above, the denominator depends only on the sample variances that measure variation within the treatments and is not affected by the differences among the sample means. In contrast, the numerator does depend on differences among the sample means. If the differences among the sample means are extreme, they will cause the numerator to be excessively large, so F will also be excessively large. Consequently **very large values of F suggest unequal means, and the ANOVA test is therefore right-tailed.**

▶ **EXAMPLE**

Are bad movies as long as good movies, or does it just seem that way? Refer to the sample data given in Table 11-1. Examination of the summary statistics seems to suggest that there are differences in the mean lengths of movies, with movies rated as excellent tending to be longer. But are those differences significant? Test the claim that the 4 categories of movies have the same mean length. That is, test the claim that $\mu_1 = \mu_2 = \mu_3 = \mu_4$.

● **SOLUTION**

In Chapter 7 we identified specific steps for testing hypotheses, and we will follow those same steps in this example. The first three steps are intended to identify the null and alternative hypotheses. In this case we have

H_0: $\mu_1 = \mu_2 = \mu_3 = \mu_4$
H_1: The preceding means are not all equal. *(continued)*

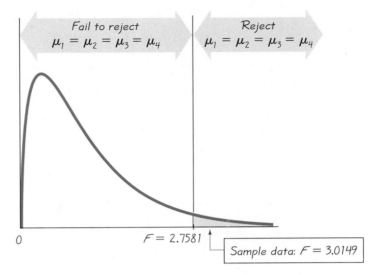

FIGURE 11-3 *F* **Distribution for Movie Example**

Step 4: No significance level was specified, so in the absence of any overriding considerations, we will use the common choice of $\alpha = 0.05$.

Step 5: Because of the nature of the data and the claim being tested, we use the *F* distribution in analysis of variance.

Step 6: In calculating the test statistic, we use the values of MS(treatment) = 1371.0374 and MS(error) = 454.7509 that were found in the preceding example.

$$F = \frac{\text{MS(treatment)}}{\text{MS(error)}} = \frac{1371.0374}{454.7509} = 3.0149$$

The critical value of $F = 2.7581$ is obtained from Table A-5 by noting that $\alpha = 0.05$, and the degrees of freedom are as follows:

numerator degrees of freedom = $k - 1 = 4 - 1 = 3$
denominator degrees of freedom = $N - k = 60 - 4 = 56$

(The actual critical *F* value is between 2.7581 and 2.8387, but we select the closest table value of 2.7581. Interpolation would have been necessary only if the test statistic had been between 2.7581 and 2.8387.)

Step 7: Because the test statistic of $F = 3.0149$ does exceed the critical value of $F = 2.7581$ (see Figure 11-3), we reject the null hypothesis that the means are equal.

Step 8: There is sufficient sample evidence to warrant rejection of the claim that the 4 population means are equal. It appears the mean movie length is not the same for poor, fair, good, and excellent movies. It seems that movies rated with 4 stars are longer than other movies, but we need other methods to formally justify this conclusion.

TABLE 11-4   ANOVA Table

Source of Variation	Sum of Squares SS	Degrees of Freedom	Mean Square MS	F Test Statistic
Treatments	4113.1122	3	1371.0374	3.0149
Error	25466.0514	56	454.7509	
Total	29579.1636	59		

Tables are a standard and convenient format for summarizing key results in ANOVA calculations. Table 11-4 includes the data found in the preceding example.

In Step 8 of the preceding example, we use our ANOVA methods to conclude that there is sufficient evidence to reject the claim of equal population means, but we cannot conclude that movies rated with four stars are longer than other movies. In general, when we use analysis of variance and conclude that three or more treatment groups have sample means that are not all equal, we do not identify the particular means that are different. There are several other tests that can be used to make such identifications. Procedures for specifically identifying the means that are different are called **multiple comparison procedures.** Comparison of confidence intervals, the Scheffé test, the extended Tukey test, and the Bonferroni test are four common multiple comparison procedures that are usually included in more advanced texts.

 ## Using Computer Software for One-Way ANOVA

We will now describe the use of Minitab in ANOVA calculations and present a Minitab display resulting from the data we are considering. Note that the display includes a format very similar to that of Table 11-4.

Refer to the Minitab printout that results from the sample data in Table 11-1. Note that Minitab uses the term *factor* instead of the term *treatment*. Under the column heading of SS, we therefore find the value of SS(factor) = 4113, indicating that SS(treatment) = 4113. Also, SS(error) = 25466, and SS(total) = 29579. Under the column heading of MS, we find the value of MS(factor) = 1371, indicating that MS(treatment) = 1371. Also, MS(error) = 455. The display includes the test statistic $F = 3.01$, as well as the P-value of 0.037. The display does not include the critical value, but we can reject the null hypothesis of $\mu_1 = \mu_2 = \mu_3 = \mu_4$ because the P-value of 0.037 is less than the significance level of $\alpha = 0.05$.

Death Penalty as Deterrent

A common argument supporting the death penalty is that it discourages others from committing murder. Jeffrey Grogger of the University of California analyzed daily homicide data in California for a four-year period during which executions were frequent. Among his conclusions published in the *Journal of the American Statistical Association* (Vol. 85, No. 410): "The analyses conducted consistently indicate that these data provide no support for the hypothesis that executions deter murder in the short term." This is a major social policy issue, and the efforts of people such as Professor Grogger help to dispel misconceptions so that we have accurate information with which to address such issues.

MINITAB DISPLAY

```
MTB > SET C1                                    Data entered here
DATA> 105 108  96  91
DATA> SET C2
DATA> 110 114  98 100   96 123 101   92 155   92   99 100 108
DATA> SET C3
DATA>  93 115 133  94   94 106   93 129 102   90 104 105 139 111 111
DATA> 123  97 104  82   98 107   95   94 117 104 119   96 134 100

DATA> SET C4
DATA>  72 120 106 104 159 125 103 160 193 168   88 121 144   90
DATA> ENDOFDATA


MTB > AOVONEWAY  C1 C2 C3 C4     Test statistic     P-value for test of
                                                    μ₁ = μ₂ = μ₃ = μ₄
ANALYSIS OF VARIANCE
SOURCE        DF        SS        MS        F        p
FACTOR         3      4113      1371     3.01    0.037
ERROR         56     25466       455
TOTAL         59     29579
```

$MTB > AOVONEWAY \ C1 \ C2 \ C3 \ C4$ Test statistic   P-value for test of $\mu_1 = \mu_2 = \mu_3 = \mu_4$

Minitab uses *factor* instead of *treatment*.

```
                                        INDIVIDUAL 95 PCT CI'S FOR
                                        MEAN BASED ON POOLED STDEV

LEVEL   N    MEAN   STDEV   -+-------+-------+-------+----
C1      4  100.00   7.87   (-------------×-------------)
C2     13  106.77  17.00           (-------×-------)
C3     29  106.52  14.48            (----×----)
C4     14  125.21  35.02                        (----×----)
                           -+-------+-------+-------+----
POOLED STDEV = 21.32       80      96     112     128
```

Individual sample means   Individual standard deviations

To use Minitab, begin by entering the lengths of poor movies (0.0 – 1.5 stars) in column C1, the lengths of fair movies (2.0 – 2.5 stars) in column C2, the lengths of good movies (3.0 – 3.5 stars) in column C3, and the lengths of excellent movies (4.0 stars) in column C4; then enter the command AOVONEWAY C1 C2 C3 C4. Following that command will be a display that includes the standard format of the ANOVA table given above, except for the order of the columns. The lower portion of the Minitab display will also include the individual sample means, standard deviations, and graphs of the 95% confidence interval estimates of each of the four population means. These confidence intervals are computed by the same methods used in Chapter 6, except that the pooled variance s_p^2 or MS(error) is used in place of the individual sample variances. (The last entry in the Minitab display shows that the pooled standard deviation is $s_p = 21.32$.) Examination of the confidence intervals shows that four-star movies appear to be longer than the other movies, because their graph is located farther to the right than the other confidence intervals.

As efficient and reliable as computer programs may be, they are totally worthless if you don't understand the relevant concepts. You should recognize that the methods in this section are used to test the claim that several samples come from populations with the same mean. *These methods require normally distributed populations with the same variance, and the samples must be independent.* We reject or fail to reject the null hypothesis of equal means by analyzing these two estimates of variance: the variance *between* samples and the variance *within* samples. MS(treatment) is an estimate of the variation between samples, and MS(error) is an estimate of the variation within samples. If MS(treatment) is significantly greater than MS(error), we reject the claim of equal means; otherwise, we fail to reject that claim.

11-3   Exercises A: Basic Skills and Concepts

High School Record		
Good	Fair	Poor
90	80	60
86	70	60
88	61	55
93	52	62
80	73	50
	65	70
	83	

1. Professor Jane Newman teaches an introductory calculus course. She wanted to test the belief that success in her course is affected by high school performance. Professor Newman collected the randomly selected sample data listed in the margin, which consist of final numerical averages in her course, separated into 3 groups according to high school records. At the $\alpha = 0.05$ level of significance, test the claim that the mean scores are equal in the 3 groups. Based on the results, does it appear that high school performance has an effect on success in Professor Newman's calculus course?

2. Three car models were studied in a test of fuel consumption. For each car, exactly one gallon of gas was placed in the tank and the car was driven until the gas was used up. The distances (in miles) traveled are given in the table. At the $\alpha = 0.05$ significance level, test the claim that the 5 population means are all equal. If you were responsible for purchasing a fleet of cars for your company's sales team, would it make any difference which car model you selected?

A	B	C
16	14	20
20	16	21
18	16	19
18	17	22
	15	18
	18	24
		18
		20

3. The numbers of program errors were recorded for 4 different programmers on randomly selected days. Test the claim that the programmers produced the same mean number of errors. Use a 0.01 level of significance. If you were supervising these programmers and needed to assign one to a special project, would it make any difference which programmer you selected? If so, which one would you select?

A	B	C	D
14	3	17	16
16	5	20	18
18	12	22	20
14	8	24	17
22	7	26	21
9	6	18	
	6	9	
	4	11	
	7		
	16		

4. The Wendt Brewing Company plans to launch a major media campaign. Three advertising companies prepared trial commercials in an attempt to win a $2 million contract. The commercials were tested on randomly selected consumers, whose reactions were measured; the results are summarized below. (Higher scores indicate more positive reactions to the commercial.) At the 0.05 significance level, test the claim that the 3 populations have the same mean reaction score. If you were responsible for advertising at Wendt Brewing, is there a company you would select on the basis of these results?

Delaney and Taylor Advertising Company	Madison Advertising	Kelly, Grey, and Landry Advertising
77 76 75 77	80 80 78 81	72 55 81 92
70 82 90 81	72 85 96 84	80 73 77 75
65 70	61 75	77 77 71 72

5. Flammability tests were conducted on children's sleepwear. The Vertical Semirestrained Test was used, in which pieces of fabric were burned under controlled conditions. After the burning stopped, the length of the charred portion was measured and recorded. Results are given in the margin for the same fabric tested at different laboratories. Because the same fabric was used, the different laboratories should have obtained the same results. Did they? Use a 0.05 significance level to test the claim that the different laboratories have the same population mean. (The data were provided by Minitab, Inc.)

6. Three different course preparation programs claim that they raise the SAT scores of their students. Scores for randomly selected test subjects were recorded before and after the preparatory course; the increases are listed below. (A negative sign indicates that the subject received a lower score after the preparatory course.) At the 0.05 significance level, test the claim that the mean increase is the same for subjects from all three courses. Does it appear that the course results are basically the same? Can analysis of variance be used to test the *effectiveness* of the courses?

Global Test Prep	Boston School for Tests	The Glenn School
10 0 10 40	90 60 20 10	80 70 30 50
−20 30 70 90	30 70 0 −10	130 0 10 80
40 70	120 40	40 20 60 90

Laboratory

1	2	3	4	5
2.9	2.7	3.3	3.3	4.1
3.1	3.4	3.3	3.2	4.1
3.1	3.6	3.5	3.4	3.7
3.7	3.2	3.5	2.7	4.2
3.1	4.0	2.8	2.7	3.1
4.2	4.1	2.8	3.3	3.5
3.7	3.8	3.2	2.9	2.8
3.9	3.8	2.8	3.2	
3.1	4.3	3.8	2.9	
3.0	3.4	3.5		
2.9	3.3			

7. Refer to Data Set 6 in Appendix B. At the 0.05 significance level, test the claim that the mean weight of M&Ms is the same for each of the 6 different color populations. If it is the intent of Mars, Inc. to make the candies so that the different color populations have the same mean, do these results suggest that the company has a problem that requires corrective action?

8. Refer to Data Set 4 in Appendix B. Create the separate weight classes of "under 2750 lb," "2750–3000 lb," and "over 3000 lb." Using the highway fuel consumption rate (in mi/gal), test the claim that the 3 weight categories have the same mean fuel consumption. Use a 0.05 significance level.

11-3   Exercises B: Beyond the Basics

9. Complete the following ANOVA table if it is known that there are 3 samples with sizes of 5, 7, and 7, respectively.

Source of Variation	Sum of Squares SS	Degrees of Freedom	Mean Square MS	Test Statistic
Treatments	?	?	?	
Error	112.57	?	?	$F = ?$
Total	114.74	?		

10. Complete the following ANOVA table, which resulted from samples of sizes 12, 25, 50, and 32.

Source of Variation	Sum of Squares SS	Degrees of Freedom	Mean Square MS	Test Statistic
Treatments	21.34	?	?	
Error	144.45	?	?	$F = ?$
Total	?	?		

11. Refer to the sample data given in Exercise 3. Assume that the first value of 14 is incorrectly entered as 1400. How are the results affected by this outlier?

12. A researcher plans to conduct an analysis of variance on 3 samples of temperatures. Does it make any difference whether the temperatures are entered in the Fahrenheit scale or the Celsius scale? In general, is the test statistic affected by the scale used?

11-4   Two-Way ANOVA

Sections 11-2 and 11-3 illustrated the use of analysis of variance in deciding whether three or more populations have the same mean. Those sections used procedures referred to as one-way analysis of variance or single-factor analysis of variance. These terms remind us that the data are categorized into groups according to a single factor (or treatment). (Recall that a factor, or treatment, is a property that is the basis for categorizing the different groups of data.) For example, in Table 11-5 (on the next page) the lengths of 30 movies are partitioned into three categories according to the single factor of star rating—a fair movie is rated with 2.0 or 2.5 stars, a good movie is rated with 3.0 or 3.5 stars, and an excellent movie is rated with 4.0 stars. (Table 11-5 uses some, but not all, of the data listed in Table 11-1.)

Because the three samples in Table 11-5 are all of the same size ($n = 10$), we could use the methods from Section 11-2 or 11-3 to test the claim that $\mu_1 = \mu_2 = \mu_3$. Following is the Minitab display for the data in Table 11-5.

TABLE 11-5	Lengths (in min) of Movies Categorized by Star Ratings	
Fair 2.0–2.5 Stars	Good 3.0–3.5 Stars	Excellent 4.0 Stars
98	93	103
100	94	193
123	94	168
92	105	88
99	111	121
110	115	72
114	133	120
96	106	106
101	93	104
155	129	159

MINITAB DISPLAY FOR ONE FACTOR OF STAR RATING

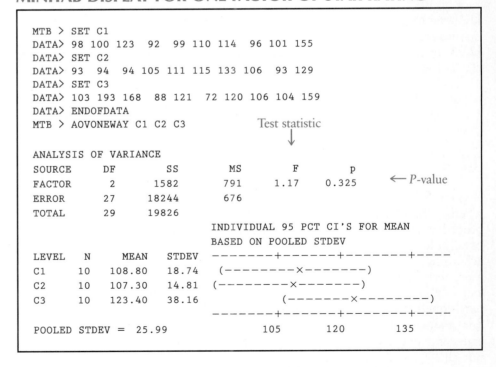

```
MTB > SET C1
DATA> 98 100 123  92  99 110 114  96 101 155
DATA> SET C2
DATA> 93  94  94 105 111 115 133 106  93 129
DATA> SET C3
DATA> 103 193 168  88 121  72 120 106 104 159
DATA> ENDOFDATA
MTB > AOVONEWAY C1 C2 C3              Test statistic
                                         ↓

ANALYSIS OF VARIANCE
SOURCE      DF       SS        MS         F        p
FACTOR       2      1582      791       1.17    0.325    ← P-value
ERROR       27     18244      676
TOTAL       29     19826
                              INDIVIDUAL 95 PCT CI'S FOR MEAN
                              BASED ON POOLED STDEV
LEVEL    N     MEAN    STDEV   ------+------+-------+----
C1      10   108.80   18.74   (--------×-------)
C2      10   107.30   14.81  (--------×-------)
C3      10   123.40   38.16         (-------×--------)
                              ------+------+-------+----
POOLED STDEV =   25.99                105      120      135
```

From the Minitab display we can see that the test statistic is

$$F = \frac{MS(\text{treatment})}{MS(\text{error})} = \frac{791}{676} = 1.17$$

(Again note that Minitab uses FACTOR, instead of TREATMENT.) The *P*-value of 0.325 indicates that at the 0.05 significance level, we fail to reject the claim of equal means. Based on our sample data, it appears that the three categories of fair, good, and excellent movies have the same mean length. In Section 11-3 we used the larger sample of 60 scores listed in Table 11-1 with 4 categories and rejected the claim of equal means, but when we use the 30 scores in Table 11-5, there is not sufficient evidence to warrant rejection of equal means.

The one-way analysis of variance for Table 11-5 involves the single factor of star rating. If we were investigating lengths of movies, a more complex analysis might include additional factors such as the nature of the movie (adventure, drama, thriller, comedy, and so on), the studio that produced the movie, the budget for the movie, the actors and actresses who are featured, the director, and the musical score. To do this kind of analysis we would use **two-way analysis of variance,** which involves two factors, as shown in Table 11-6. Table 11-6 uses (1) the star rating, by which we categorize a movie as fair, good, or excellent, and (2) the Motion Picture Association of America (MPAA) rating, by which movies are separated into a category of R or another category of G, PG, or PG-13. Using the two factors of star rating and MPAA rating, we partition the data into six categories, as shown in Table 11-6. Such subcategories are often called **cells,** so Table 11-6 consists of 6 cells containing 5 scores each. Tables 11-5 and 11-6 include the same numbers, but Table 11-5 classifies the data with the single factor of star rating, whereas Table 11-6 classifies the data with the two factors of star rating and MPAA rating.

Because we've already discussed the one-way analysis of variance for the single factor of star rating, it might seem reasonable to simply proceed with

TABLE 11-6 Lengths (in min) of Movies Categorized by Star Ratings and MPAA Ratings (G/PG/PG-13, R)

	Fair 2.0–2.5 Stars	Good 3.0–3.5 Stars	Excellent 4.0 Stars
MPAA Rating: G/PG/PG-13	98	93	103
	100	94	193
	123	94	168
	92	105	88
	99	111	121
MPAA Rating: R	110	115	72
	114	133	120
	96	106	106
	101	93	104
	155	129	159

another one-way ANOVA for the factor of MPAA rating. Unfortunately, that approach wastes information and totally ignores any effect from an interaction between the two factors.

DEFINITION

There is an **interaction** between two factors if the effect of one of the factors changes for different categories of the other factor.

In using ANOVA for the data of Table 11-6, we will consider the effect of an interaction between star ratings and MPAA ratings, as well as the effects of star ratings and the effects of MPAA ratings on movie length. The calculations are quite involved, so *we will assume that a software package is being used.* Shown below is the Minitab display for the data in Table 11-6.

MINITAB DISPLAY

```
MTB > READ C1 C2 C3
DATA>  98  1  1
DATA> 100  1  1
DATA> 123  1  1
DATA>  92  1  1
DATA>  99  1  1        93 is in the first row
DATA>  93  1  2 ←      and second column
     . . .
DATA> 104  2  3
DATA> 159  2  3
DATA> ENDOFDATA
MTB > TWOWAY C1 C2 C3


ANALYSIS OF VARIANCE C1
```

	SOURCE	DF	SS	MS
MPAA rating →	C2	1	32	32
	C3	2	1582	791
Star rating	INTERACTION	2	2256	1128
	ERROR	24	15956	665
	TOTAL	29	19826	

The Minitab display includes SS (sum of square) components similar to those introduced in Section 11-3. Because the circumstances of Section 11-3 involved only a single factor, we used SS(treatment) as a measure of the variation due to the different treatment categories and we used SS(error) as a measure of the variation due to sampling error. We now use SS(star rating), indicated by Minitab as the SS value for C3, as a measure of variation among the star rating means. We use SS(MPAA rating), indicated by Minitab as the SS

> **TABLE 11-7   Comparison of Components Used in One-Way ANOVA and Two-Way ANOVA**

One-Way ANOVA	Two-Way ANOVA
	Sample data: Table 11-6
Sample data: Table 11-5	Two factors: star rating
One factor: star rating	and MPAA rating
SS(star rating) = 1582	Same
MS(star rating) = 791	Same
df(star rating) = 2	Same
SS(total) = 19,826	Same
df(total) = 29	Same
	┌→SS(MPAA rating) = 32
SS(error) = 18,244 ————	→SS(interaction) = 2256
	└→SS(error) = 15,956
	┌→df(MPAA rating) = 1
df(error) = 27 ————	→df(interaction) = 2
	└→df(error) = 24
	MS(MPAA rating) = 32
MS(error) = 676	MS(interaction) = 1128
	MS(error) = 665

value for C2, as a measure of variation among the MPAA means. We continue to use SS(error) as a measure of variation due to sampling error. Similarly, we use MS(star rating) and MS(MPAA rating) for the two different mean squares and continue to use MS(error) as before. Also, we use df(star rating) and df(MPAA rating) for the two different degrees of freedom.

Table 11-7 compares the analyses of the data in Tables 11-5 and 11-6. Because Tables 11-5 and 11-6 have the same data in the star rating categories, the calculated values of SS(star rating), MS(star rating), and df(star rating) are also the same in both cases. But note that as we go from the one-factor case (Table 11-5) to the two-factor case (Table 11-6) by partitioning the data according to the second factor of MPAA rating, the value of SS(error) is partitioned into SS(MPAA rating), SS(interaction), and SS(error). Also, df(error) is partitioned into df(MPAA rating), df(interaction), and df(error). Similar partitioning does not apply to MS(error).

In executing a two-way analysis of variance, we consider three effects:

1. The effect due to the *interaction* between the two factors of star rating and MPAA rating

2. The effect due to the *row* factor (MPAA rating)

3. The effect due to the *column* factor (star rating)

Discrimination Case Uses Statistics

Statistics often play a key role in discrimination cases. One such case involved Matt Perez and more than 300 other FBI agents who won a class action suit charging that Hispanics in the FBI were discriminated against in the areas of promotions, assignments, and disciplinary actions. The plaintiff employed statistician Gary Lafree, who showed that the FBI's upper management positions had significantly low proportions of Hispanic employees. Statistics were instrumental in the plaintiff's victory in this case. (See Program 20 from the series *Against All Odds: Inside Statistics* for a discussion of this case.)

The following comments summarize the basic procedure for two-way analysis of variance. The procedure described below is actually quite similar to the procedures presented in Sections 11-2 and 11-3. We form conclusions about equal means by analyzing two estimates of variance, and the test statistic F is the ratio of the two estimates. A significantly large value for F indicates that there is a statistically significant difference in means. The following procedure for two-way ANOVA is summarized in Figure 11-4.

Procedure for Two-Way ANOVA

Step 1: In two-way analysis of variance, begin by testing the null hypothesis that there is no interaction between the two factors. Using Minitab for the data in Table 11-6, we get the following test statistic:

$$F = \frac{\text{MS(interaction)}}{\text{MS(error)}} = \frac{1128}{665} = 1.6962$$

With df(interaction) = 2 and df(error) = 24, we get a critical value of $F =$ 3.4028 (assuming a 0.05 significance level). Because the test statistic does not exceed the critical value, we fail to reject the null hypothesis of no interaction between the two factors. There is not sufficient evidence to conclude that movie length is affected by an interaction between star rating and MPAA rating.

Step 2: If we fail to reject the null hypothesis of no interaction between factors, then we should proceed to test the following two hypotheses:

H_0: There are no effects from the row factor
(that is, the row means are equal).
H_0: There are no effects from the column factor
(that is, the column means are equal).

If we do reject the null hypothesis of no interaction between factors, then we should stop now; we should not proceed with the two additional tests. (If there is an interaction between factors, we shouldn't consider the effects of either factor without considering those of the other.)

In Step 1, we failed to reject the null hypothesis of no interaction between factors, so we proceed with the next two hypothesis tests identified in Step 2. For the row factor,

$$F = \frac{\text{MS(MPAA rating)}}{\text{MS(error)}} = \frac{32}{665} = 0.0481$$

This value is not significant because the critical value, based on df(MPAA rating) = 1, df(error) = 24, and a 0.05 significance level, is $F = 4.2597$. We fail to reject the null hypothesis of no effects from MPAA rating. The MPAA rating does not appear to have an effect on movie length.

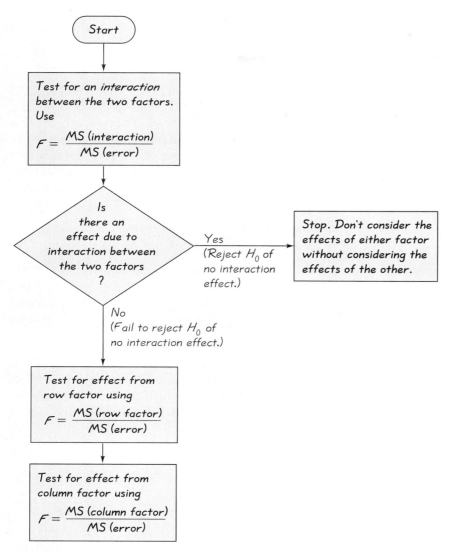

FIGURE 11-4   Procedure for Two-Way ANOVA

For the column factor,

$$F = \frac{MS(\text{star rating})}{MS(\text{error})} = \frac{791}{665} = 1.1895$$

This value is not significant because the critical value, based on df(star rating) = 2, df(error) = 24, and a 0.05 significance level, is $F = 3.4028$. We fail to reject the null hypothesis of no effects from star rating. The star rating of a movie does not appear to have an effect on its length.

Special Case: One Observation per Cell and No Interaction

Table 11-6 contains five observations per cell. If our sample data consist of only one observation per cell, we lose MS(interaction), SS(interaction), and df(interaction) because those values are based on sample variances computed for each individual cell. If there is only one observation per cell, there is no variation within individual cells and those sample variances cannot be calculated. Here's how we proceed when there is one observation per cell: *If it seems reasonable to assume (based on knowledge about the circumstances) that there is no interaction between the two factors, make that assumption and then proceed as before to test the following two hypotheses separately:*

H_0: There are no effects from the row factor.

H_0: There are no effects from the column factor.

As an example, suppose that we have only the first score in each cell of Table 11-6. Note that the two row means are 98.0 and 99.0. Is that difference significant, suggesting that there is an effect due to MPAA rating? The column means are 104.0, 104.0, and 87.5. Are those differences significant, suggesting that there is an effect due to star rating? If we believe the number of stars a movie receives seems to be totally unrelated to whether it is rated G, PG, PG-13, or R, we can then assume that there is no interaction between star rating and MPAA rating. (If we believe there is an interaction, the method described here does not apply.) To illustrate the method, we will make the assumption of no interaction. Following is the Minitab display for the data in Table 11-6, with only the first score from each cell.

MINITAB DISPLAY

```
        MTB > READ C1 C2 C3
        DATA>  98  1  1
        DATA>  93  1  2
        DATA> 103  1  3
        DATA> 110  2  1
        DATA> 115  2  2
        DATA>  72  2  3
        DATA> ENDOFDATA
        MTB > TWOWAY C1 C2 C3

        ANALYSIS OF VARIANCE  C1
MPAA
rating  SOURCE       DF       SS       MS
        C2            1        2        2
Star    C3            2      363      181
rating  ERROR         2      793      397
        TOTAL         5     1157
```

We first use the results from the Minitab display to test the null hypothesis of no effects from the row factor of MPAA rating. We get the test statistic

$$F = \frac{MS(MPAA\ rating)}{MS(error)} = \frac{2}{397} = 0.0050$$

Assuming a 0.05 significance level, we find that the critical value is $F = 18.513$. (Refer to Table A-5 and note that the Minitab display provides degrees of freedom of 1 and 2 for numerator and denominator.) Because the test statistic of $F = 0.0050$ does not exceed the critical value of $F = 18.513$, we fail to reject the null hypothesis; it seems that the MPAA rating does not affect movie length.

We now use the Minitab display to test the null hypothesis of no effect from the column factor of star rating. The test statistic is

$$F = \frac{MS(star\ rating)}{MS(error)} = \frac{181}{397} = 0.4559$$

which does not exceed the critical value of $F = 19.000$. [Refer to Table A-5 with df(star rating) = 2 and df(error) = 2.] We fail to reject the null hypothesis; it seems that the star rating does not affect movie length. Again, these conclusions are based on very limited data and might not be valid when more sample data are acquired.

Randomized Block Design

When we use the one-way (or single factor) analysis of variance technique and conclude that the differences among the means are significant, we cannot necessarily conclude that the given factor is responsible for the differences. It is possible that the variation of some other unknown factor is responsible for the differences. One way to reduce the effect of the extraneous factors is to design the experiment so that it has a **completely randomized design,** in which each element is given the same chance of belonging to the different categories, or treatments. Another way to reduce the effect of extraneous factors is to use a **rigorously controlled design,** in which elements are carefully chosen so that all other factors have no variability. That is, select elements that are the same in every characteristic except for the single factor being considered.

Yet another way to control for extraneous variation is to use a **randomized block design,** in which a measurement is obtained for each treatment on each of several individuals that are matched according to similar characteristics. A **block** is a group of similar individuals. For example, suppose that we want to use four different cars to test the mileage produced by three different grades (regular, extra, premium) of a gasoline. We want to control for the differences among the cars. Using a randomized block design, we burn each grade of gas in each of the four cars, randomly selecting the order in which this is done. We have three treatments corresponding to the three grades of gas. We have four blocks corresponding to the four cars used. Suppose the results are as given in Table 11-8.

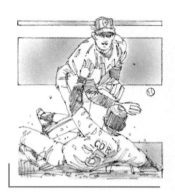

Statistics and Baseball Strategy

Statisticians are using computers to develop very sophisticated measures of baseball performance and strategy. They have found, for example, that sacrifice bunts, sacrifice fly balls, and stolen bases rarely help win games and that it is seldom wise to have a pitcher intentionally walk the batter. They can identify the ballparks that favor pitchers and those that favor batters. Instead of simply comparing the batting averages of two different players, they can develop better measures of offensive strength by taking into account such factors as pitchers faced, position in the lineup, ballparks played, and weather conditions.

TABLE 11-8	Mileage (mi/gal) for Different Grades of Gasoline

		Block			
		Car 1	Car 2	Car 3	Car 4
	Regular	19	33	23	27
Treatment	Extra	19	34	26	29
	Premium	22	39	26	34

With this configuration, we can use the two-way analysis of variance software to test the claim that the three row-factor treatments (regular, extra, premium) produce the same mean. The method we use is the same as the method used for the special case of one observation per cell. However, we need not use special knowledge of the circumstances to assume that there is no interaction; we can assume that there is no interaction because of the way we obtained the data.

Shown below is the Minitab display for the data in Table 11-8. Note that the command TWOWAY C1 C2 C3 is used; the table entries are entered in column C1, and the entries for columns C2 and C3 indicate the location of each score in the table.

MINITAB DISPLAY

```
         MTB > READ C1 C2 C3
         DATA> 19  1  1
         DATA> 33  1  2
         DATA> 23  1  3
         DATA> 27  1  4
         DATA> 19  2  1
         DATA> 34  2  2
         DATA> 26  2  3
         DATA> 29  2  4
         DATA> 22  3  1
         DATA> 39  3  2
         DATA> 26  3  3
         DATA> 34  3  4
         DATA> ENDOFDATA
         MTB > TWOWAY C1 C2 C3

         ANALYSIS OF VARIANCE   C1

         SOURCE        DF        SS         MS
         C2            2       47.17      23.58
         C3            3      390.25     130.08
         ERROR         6       11.50       1.92
         TOTAL        11      448.92
```

Grade of gasoline → C2

Car → C3

► **EXAMPLE**

Refer to the sample data in Table 11-8 and the corresponding Minitab display. Test the claim that the grade of gasoline does not affect mileage, and test the claim that the different cars do not affect mileage.

● **SOLUTION**

Considering the treatment (row) effects, we calculate the test statistic

$$F = \frac{MS(\text{grade})}{MS(\text{error})}$$

$$= \frac{23.58}{1.92}$$

$$= 12.2813$$

From the display we get df(grade) = 2 and df(error) = 6, and we can now refer to Table A-5. With a 0.05 significance level, the critical value is $F = 5.1433$. Because the test statistic of $F = 12.2813$ exceeds the critical value of $F = 5.1433$, we reject the null hypothesis. The grade of gas does seem to have an effect on the mileage.

For the block (column) effects, the test statistic is

$$F = \frac{MS(\text{car})}{MS(\text{error})}$$

$$= \frac{130.08}{1.92}$$

$$= 67.7500$$

and the critical value is $F = 4.7571$. We reject the null hypothesis of equal block means and conclude that the different cars have different mileage values. (If we interchange the roles played by blocks and treatments by transposing Table 11-8, we obtain the same results.)

In this section we have briefly discussed an important branch of statistics. We have emphasized the interpretation of computer displays while omitting the manual calculations and formulas, which are quite formidable. More advanced texts typically discuss this topic in much greater detail, but our intent here is to give some general insight into the nature of two-way analysis of variance.

11-4   Exercises A: Basic Skills and Concepts

In Exercises 1–4, use the following Minitab display, which corresponds to the data in the following table. (C2 represents MPAA rating, and C3 represents star rating.)

```
MTB > TWOWAY C1 C2 C3

ANALYSIS OF VARIANCE C1

SOURCE          DF        SS        MS
C2              1         14        14
C3              3       1049       350
INTERACTION     3       3790      1263
ERROR           8       5560       695
TOTAL          15      10413
```

Lengths (in min) of Movies Categorized by Star Ratings and MPAA Ratings (G/PG/PG-13, R)

	Poor 0.0–1.5 Stars	Fair 2.0–2.5 Stars	Good 3.0–3.5 Stars	Excellent 4.0 Stars
MPAA Rating: G/PG/PG-13	108	98	93	103
	91	100	94	193
MPAA Rating: R	105	110	115	72
	96	114	133	120

1. Identify the indicated values.
   a. MS(interaction)
   b. MS(error)
   c. MS(star rating)
   d. MS(MPAA rating)

2. Find the test statistic and critical value for the null hypothesis of no *interaction* between star rating and MPAA rating. What do you conclude?

3. Assume that the length of a movie is not affected by an *interaction* between its star rating and MPAA rating. Find the test statistic and critical value for the null hypothesis that star rating has no effect on movie length. What do you conclude?

4. Assume that the length of a movie is not affected by an *interaction* between its star rating and MPAA rating. Find the test statistic and critical value for the null hypothesis that MPAA rating has no effect on movie length. What do you conclude?

In Exercises 5 and 6, use only the first value from each of the 8 cells in the table used for Exercises 1–4. When we use only these first values, the Minitab display is as follows.

```
MTB > TWOWAY C1 C2 C3

ANALYSIS OF VARIANCE   C1

SOURCE       DF        SS        MS
C2            1         0         0
C3            3       459       153
ERROR         3       799       266
TOTAL         7      1258
```

5. Assuming that there is no effect on movie length from the interaction between star rating and MPAA rating, test the null hypothesis that MPAA rating has no effect on movie length. Identify the test statistic and critical value, and state the conclusion. Use a 0.05 significance level.

6. Assuming that there is no effect on movie length from the interaction between star rating and MPAA rating, test the null hypothesis that star rating has no effect on movie length. Identify the test statistic and critical value, and state the conclusion. Use a 0.05 significance level.

Exercises 7 and 8 refer to the sample data in the following table and the corresponding Minitab display. The table entries are the times (in minutes) required to complete a document. The same document was entered by 4 different typists using each of 3 different word processors. (In the Minitab results, typists are represented by C2 and word processors are represented by C3.)

```
MTB > TWOWAY C1 C2 C3

ANALYSIS OF VARIANCE   C1

SOURCE       DF         SS         MS
C2            3     22.000      7.333
C3            2    120.167     60.083
ERROR         6      2.500      0.417
TOTAL        11    144.667
```

		Word Processor		
		I	II	III
Typist	A	16	21	13
	B	20	25	16
	C	18	21	14
	D	19	22	15

7. At the 0.05 significance level, test the claim that the choice of typist has no effect on the time. Identify the test statistic and critical value, and state the conclusion.

8. At the 0.05 significance level, test the claim that the choice of word processor has no effect on the time. Identify the test statistic and critical value, and state the conclusion.

11-4 Exercises B: Beyond the Basics

9. Use a statistics software package, such as Minitab or SPSS/PC, that can produce results for two-way analysis of variance. First enter the data in the table used for Exercises 1–4 and verify that the results are as given in this section. Then transpose the table by making the MPAA rating the column factor and making the star rating the row factor. Obtain the computer display for the transposed table, and compare the results to those given above.

10. Refer to the data in the table used for Exercises 1–4 and subtract 10 from each table entry. Using a statistics software package with a two-way analysis of variance capability, determine the effects of subtracting 10 from each entry.

11. Refer to the data in the table used for Exercises 1–4 and multiply each table entry by 10. Using a statistics software package with a two-way analysis of variance capability, determine the effects of multiplying each entry by 10.

12. In analyzing Table 11-6, we concluded that movie length is not affected by an interaction between star rating and MPAA rating; it is not affected by star rating; and it is not affected by MPAA rating.

   a. Change the table entries so that there is an effect from the interaction between star rating and MPAA rating.

   b. Change the table entries so that there is no effect from the interaction between star rating and MPAA rating and there is no effect from star rating, but there is an effect from MPAA rating.

   c. Change the table entries so that there is no effect from the interaction between star rating and MPAA rating and there is no effect from MPAA rating, but there is an effect from star rating.

VOCABULARY LIST

Define and give an example of each term.

analysis of variance (ANOVA)	variation due to error
one-way analysis of variance	multiple comparison procedures
single-factor analysis of variance	two-way analysis of variance
treatment	interaction
factor	completely randomized design
variance between samples	rigorously controlled design
variation due to treatment	randomized block design
variance within samples	block

Important Formulas

Application	Distribution	Test Statistic	Degrees of Freedom	Critical Values
One-way ANOVA with equal sample sizes	F	$F = \dfrac{ns_{\bar{x}}^2}{s_p^2}$	numerator: $k-1$ denominator: $k(n-1)$	Table A-5
One-way ANOVA for all cases	F	$F = \dfrac{MS(\text{treatment})}{MS(\text{error})}$	numerator: $k-1$ denominator: $N-k$	Table A-5
Two-way ANOVA	F	(1) Interaction: $F = \dfrac{MS(\text{interaction})}{MS(\text{error})}$ (2) Row factor: $F = \dfrac{MS(\text{row factor})}{MS(\text{error})}$ (3) Column factor: $F = \dfrac{MS(\text{column factor})}{MS(\text{error})}$	See computer display	Table A-5 Table A-5 Table A-5

REVIEW

In this chapter we used analysis of variance (or ANOVA) to test for equality of three or more population means. This method requires (1) normally distributed populations, (2) populations with the same standard deviation (or variance), and (3) random samples that are independent of each other.

Our test statistics are based on the ratio of two different estimates of the common population variance. In repeated samplings, the distribution of the F test statistic can be approximated by the F distribution, which has critical values given in Table A-5.

In Section 11-2 we considered one-way analysis of variance for samples with the same number of scores. In Section 11-3 we extended that method to include samples of unequal sizes.

In Section 11-4 we considered two-way analysis of variance. The data were categorized by two factors instead of only one. We also considered two-way analysis of variance with one observation per cell and two-way analysis of variance with randomized block design. These cases use the same ANOVA methods, but in a randomized block experiment we try to control for extraneous variation by making blocks one of the factors. Because of the nature of the calculations required, this last section emphasized the interpretation of computer displays.

REVIEW EXERCISES

1. The Scholastic Science Publishing Company used readability studies to determine the clarity of 4 different high school biology textbooks; the sample readability scores are given in the table. At the $\alpha = 0.05$ level of significance, test the claim that the 4 textbooks produce the same mean readability score.

Text A	Text B	Text C	Text D
50	59	48	60
51	60	51	65
53	58	47	62
58	57	49	68
53	61	50	70

2. Medical researchers at the New England Institute used 3 different treatments in an experiment involving volunteer patients suffering from severe skin infections. The recovery times (in days) are given in the table. Using a 0.01 significance level, test the claim that the different treatments result in the same mean recovery time.

A	B	C
6	9	11
8	8	9
12	7	10
9	6	8
7	9	11

3. The Associated Insurance Institute sponsors studies of the effects of drinking and driving. In one such study, 3 groups of adult men were randomly selected for an experiment designed to measure their blood alcohol levels after consuming 5 drinks. Members of group A were tested after 1 hour, members of group B were tested after 2 hours, and members of group C were tested after 4 hours. The results are given in the accompanying table. At the 0.05 level of significance, test the claim that the 3 groups have the same mean level.

A	B	C
0.11	0.08	0.04
0.10	0.09	0.04
0.09	0.07	0.05
0.09	0.07	0.05
0.10	0.06	0.06
		0.04
		0.05

4. Three different transmission types were installed on different cars, all having 6-cylinder engines and an engine size of 3.0 liters. The fuel consumption (in

mi/gal) was found for highway conditions; the results are given below (based on data from the Environmental Protection Agency). The letters A, M, and L represent automatic, manual, and lockup torque converter, respectively. At the 0.05 significance level, test the claim that the mean fuel consumption values are the same for all 3 transmission types.

A	M	L
23	27	24
23	29	26
20	25	24
21	23	
	24	
$\bar{x} = 21.8$	$\bar{x} = 25.6$	$\bar{x} = 24.7$
$s^2 = 2.3$	$s^2 = 5.8$	$s^2 = 1.3$

5. Twelve different 4-cylinder cars were tested for fuel consumption (in mi/gal) after being driven under identical highway conditions; the results are listed in the table and accompanying Minitab display. At the 0.05 significance level, test the claim that fuel consumption is not affected by an *interaction* between engine size and transmission type.

```
MTB > READ    MPG IN C1    TRANS IN C2    SIZE IN C3
DATA> 31 1 1
DATA> 32 1 1
DATA> 28 1 2
        .
        .
        .
DATA> 34 2 3
DATA> ENDOFDATA
MTB > NAME C1 'MPG'    C2 'TRANS'    C3 'SIZE'
MTB > TWOWAY C1 C2 C3

ANALYSIS OF VARIANCE    MPG

SOURCE          DF        SS        MS
TRANS           1        40.3      40.3
SIZE            2        43.2      21.6
INTERACTION     2         1.2       0.6
ERROR           6        68.0      11.3
TOTAL          11       152.7
```

Highway Fuel Consumption (mi/gal) of Different 4-Cylinder Compact Cars

	Engine Size (liters)		
	1.5	2.2	2.5
Automatic Transmission	31, 32	28, 26	31, 23
Manual Transmission	33, 36	33, 30	27, 34

6. Refer to the same data used in Exercise 5 and assume that fuel consumption is not affected by an interaction between engine size and type of transmission. Use a 0.05 level of significance to test the claim that fuel consumption is not affected by engine size.

7. Refer to the same data used in Exercise 5 and assume that fuel consumption is not affected by an interaction between engine size and type of transmission. Use a 0.05 level of significance to test the claim that fuel consumption is not affected by type of transmission.

 COMPUTER PROJECT

a. Minitab-generated sample data and the ANOVA display for 3 different diets, with 5 subjects using each diet, are included in the following printout. Use a 0.05 level of significance to test the claim that the 3 diets have the same mean. (Note that we generated the sample data with the condition that the 3 diets really do have the same mean, so $\mu_1 = \mu_2 = \mu_3$.)

```
MTB > RANDOM K=5 C1;
SUBC> NORMAL MU=11 SIGMA=3.          ⎫
MTB > RANDOM K=5 C2;                 ⎪
SUBC> NORMAL MU=11 SIGMA=3.          ⎬  Commands
MTB > RANDOM K=5 C3;                 ⎪  for part b
SUBC> NORMAL MU=11 SIGMA=3.          ⎭
MTB > PRINT C1 C2 C3


  ROW       MPG       TRANS        SIZE

    1    9.6184     11.1828     14.2407
    2   11.1063     11.2295     11.4653
    3    9.5664     11.5676     11.6366
    4   10.2376      8.1651     15.2357
    5   11.6201      7.7958     10.3612
```

b. The Minitab printout shown on the next page begins with a set of Minitab commands that simulates weight losses (in pounds) for 3 different diets, with 5 subjects using each diet. We assume that the 3 diets all have the same effect, which consists of weight losses (over a 3-month period) that are normally distributed with a mean of 11 lb and a standard deviation of 3 lb. Repeat these Minitab commands to generate your own set of sample results, and conduct your own analysis of variance. Then repeat the test described in part a. Rejecting $H_0: \mu_1 = \mu_2 = \mu_3$ would be a type I error, because we know that the given null hypothesis is true. In a class of 40 students, how many of these type I errors can be expected if each student runs this simulation?

```
MTB > AOVONEWAY C1 C2 C3

ANALYSIS OF VARIANCE
SOURCE      DF        SS        MS        F        p     ⎫ Results
FACTOR       2      19.35      9.68     3.44    0.066   ⎬ for part a
ERROR       12      33.77      2.81                     ⎭
TOTAL       14      53.13
                             INDIVIDUAL 95 PCT CI'S FOR MEAN
                             BASED ON POOLED STDEV
LEVEL    N     MEAN    STDEV  ------+-------+-------+----
C1       5    10.430   0.910      (--------X--------)
C2       5     9.988   1.843  (--------X--------)
C3       5    12.588   2.053        (--------X--------)
                             ------+-------+-------+----
POOLED STDEV = 1.678               9.6     11.2    12.8
```

| *FROM DATA TO DECISION:* | **Does your education pay?** |

The media periodically present reports claiming that the level of a person's education has an effect on the amount of his or her income. Refer to Data Set 3 in Appendix B (based on data from the U.S. National Center for Health Statistics) and list the family incomes for persons in each of three education categories: elementary, high school, and college. Using one-way analysis of variance, test the claim that the mean income is the same for the three populations (elementary, high school, college). Does the result support the common belief that "education pays"? If not, survey the data and identify any unusually high scores. In particular, identify the two exceptional family incomes in the high school category. After deleting those two incomes, repeat the analysis of variance test of equal means. Do the results now support the common belief that "education pays"? Did the removal of only 2 scores from a collection of 125 scores have much of an impact on the ANOVA results?

12 Statistical Process Control

12-1 Overview

The importance of quality control and monitoring processes is discussed.

12-2 Control Charts for Variation and Mean

The construction of run charts and control charts is described. Illustrations are provided of the construction and interpretation of control charts for R (range), which are used to monitor the *variation* in a process, and control charts for $\overline{x}$, which are used to monitor the *mean* in a process.

12-3 Control Charts for Attributes

This section describes the construction of p charts, which are control charts that can be used to monitor the *proportion* of items having some attribute, such as being defective.

Chapter Problem:
Is the U.S. Mint doing its job?

The United States Mint manufactures all of our coins. The minting of quarters is especially important because quarters are used in many coin-operated devices, including telephones, washing machines, parking meters, and vending machines. The physical properties of quarters must be consistent so that the quarters will be accepted in these coin-operated devices. For example, if a quarter is too heavy, it might be rejected when we insert it into a vending machine and we might not get the item we want. Wouldn't that be a novel experience? Because it's impossible to make every quarter weigh precisely the targeted weight of 5.670 g, the Mint considers as acceptable any weight between 5.443 g and 5.897 g.

A new minting machine was recently placed into service. The weights were recorded when a quarter was randomly selected every 12 minutes for 20 consecutive hours of operations. The results are listed in Table 12-1. Based on these results, is the minting process proceeding as desired?

Table 12-1 Weights (in grams) of Minted Quarters

Hour	Weight (grams)					$\bar{x}$	s	Range
1	5.639	5.636	5.679	5.637	5.691	5.6564	0.0265	0.055
2	5.655	5.641	5.626	5.668	5.679	5.6538	0.0211	0.053
3	5.682	5.704	5.725	5.661	5.721	5.6986	0.0270	0.064
4	5.675	5.648	5.622	5.669	5.585	5.6398	0.0370	0.090
5	5.690	5.636	5.715	5.694	5.709	5.6888	0.0313	0.079
6	5.641	5.571	5.600	5.665	5.676	5.6306	0.0443	0.105
7	5.503	5.601	5.706	5.624	5.620	5.6108	0.0725	0.203
8	5.669	5.589	5.606	5.685	5.556	5.6210	0.0545	0.129
9	5.668	5.749	5.762	5.778	5.672	5.7258	0.0520	0.110
10	5.693	5.690	5.666	5.563	5.668	5.6560	0.0534	0.130
11	5.449	5.464	5.732	5.619	5.673	5.5874	0.1261	0.283
12	5.763	5.704	5.656	5.778	5.703	5.7208	0.0496	0.122
13	5.679	5.810	5.608	5.635	5.577	5.6618	0.0909	0.233
14	5.389	5.916	5.985	5.580	5.935	5.7610	0.2625	0.596
15	5.747	6.188	5.615	5.622	5.510	5.7364	0.2661	0.678
16	5.768	5.153	5.528	5.700	6.131	5.6560	0.3569	0.978
17	5.688	5.481	6.058	5.940	5.059	5.6452	0.3968	0.999
18	6.065	6.282	6.097	5.948	5.624	6.0032	0.2435	0.658
19	5.463	5.876	5.905	5.801	5.847	5.7784	0.1804	0.442
20	5.682	5.475	6.144	6.260	6.760	6.0642	0.5055	1.285

12-1   Overview

The early chapters of this book stressed that data have three important characteristics: (1) the nature or shape of the *distribution,* such as bell shaped; (2) a *representative value,* such as the mean; and (3) a measure of the scattering or *variation,* such as the standard deviation. In this chapter we consider a fourth characteristic of some data that is often critically important: its *pattern over time.* By monitoring this characteristic, we are better able to control the production of goods and services, thereby ensuring better quality.

There is currently a strong trend to try to improve the quality of American goods and services, and the methods presented in this chapter are being used by growing numbers of businesses. Evidence of the increasing importance of quality is found in its greater role in advertising and the growing number of books and articles that focus on the issue of quality. In many cases, job applicants enjoy a distinct advantage when they can tell employers that they have studied statistics and methods of quality control.

This chapter will present some of the basic tools commonly used to monitor quality. Section 12-2 introduces run charts for monitoring patterns over time and then discusses two types of control charts: R charts used to monitor variation and $\bar{x}$ charts used to monitor means. Finally, in Section 12-3, we will consider p charts, which are used to monitor the proportions of defective items.

12-2   Control Charts for Variation and Mean

For many data sets, the characteristic of a pattern over time is so important that serious consequences can occur if it is ignored. Consider, for example, the filling of 12-oz cola containers at a now-defunct bottling plant of a major soft drink company. Plant managers would run production for a full day, but they would wait until the end of the day to collect and analyze samples. Early one day, a main pressure line developed a leak that went unnoticed. Almost all of the day's production consisted of bottles that were only two-thirds full. The entire evening was spent opening thousands of bottles and pouring out the contents. This costly breakdown could have been quickly corrected if the filling process had been monitored throughout the day.

Now consider the weights of newly minted quarters, as shown in Table 12-1. The U.S. Mint requires weights between 5.443 g and 5.897 g; the mean of the 100 weights in Table 12-1 is 5.7098 g, suggesting that all is well. However, the U.S. Mint is concerned about the *pattern over time* (or the process data) for the quarters being produced by its new machine, because some patterns suggest serious problems that must be corrected.

DEFINITION

Process data are data arranged according to some time sequence. They are measurements of a characteristic of goods or services that results from some combination of equipment, people, materials, methods, and conditions.

For example, we collect *process data* when we measure the contents of a sample of cola bottles every 30 min. We collect process data when, during each hour of operation, we sample telephone customers and record the amounts of time it takes for calls to be completed. Because the data in Table 12-1 are arranged according to the time at which they were selected, they are another example of process data. Note this important point:

Important characteristics of process data can change over time.

In manufacturing quarters, the U.S. Mint might use a machine that begins with the desired mean of 5.670 g, but if the machine wears with use, the mean might gradually increase to the unacceptable level of 6.000 g. There are various methods that can be used to monitor a process to ensure that the important desired characteristics don't change—the run chart is one such method.

DEFINITION

A **run chart** is a sequential plot of *individual* data values over time. One axis (usually the vertical axis) is used for the data values, and the other axis (usually the horizontal axis) is used for the time sequence.

▶ **EXAMPLE**

Using the weights of the quarters in Table 12-1, construct a run chart by using a vertical axis for the weights of the quarters and a horizontal axis for the time sequence.

● **SOLUTION**

Figure 12-1 (on the next page) is the run chart for the data of Table 12-1. We have constructed a vertical scale that is suitable for the range of weights of the quarters. These weights fluctuate between a low of 5.059 g and a high of 6.760 g, so a scale ranging from 5.0 to 7.0 g is appropriate. Because there are

(continued)

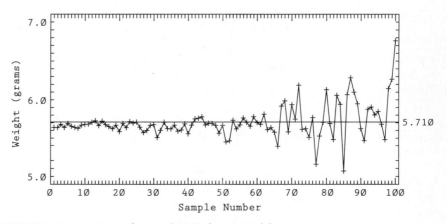

FIGURE 12-1   Run Chart of Weights in Table 12-1

100 values arranged in sequence, the horizontal scale is designed to include the numbers between 0 and 100. The first point corresponds to an x value of 1, and it is plotted to have a height of 5.639, which is the weight of the first quarter in grams. The other points are plotted in the same way.

Examine the run chart and note the pattern of increasing variation among scores. Observe in Figure 12-1 that the weights of the quarters manufactured at the beginning of the process are relatively stable, but as you proceed farther to the right, the values tend to fluctuate more. The *center* seems to be stable, indicating that the mean is not changing; however, the pattern of increasing fluctuations suggests that the variation of the weights is increasing. This is a serious problem that should be corrected as soon as possible because it indicates that some individual quarters are becoming unacceptably heavy or light.

 ## Using Computer Software to Generate Run Charts

Minitab can be used to construct run charts, such as the one shown in Figure 12-1. Begin by entering all of the sample data in column C1 by using the SET command as described in Section 1-5. Next find the location of the center line by entering the command MEAN C1. When we use the data in Table 12-1, the command MEAN C1 will result in a value of 5.7098. Next, enter the commands

```
MTB > GICHART C1;
SUBC> HLINES 5.7098.
```

The prefix G in GICHART specifies high resolution graphics, and ICHART is

Minitab's command for a chart of "individual" values. The subcommand HLINES specifies the location of the "horizontal line" in the run chart.

Whether we construct the run chart manually or by using computer software, it is important to interpret the chart carefully so that we can learn about how the process is behaving. The pattern of increasing variation in Figure 12-1 suggests that the process of manufacturing quarters is not statistically stable; characteristics of the weights seem to be changing over time.

DEFINITION

A process is **statistically stable** (or **within statistical control**) if it has only natural variation, with no patterns, cycles, or any unusual points.

A run chart is one of several tools used to determine whether a process is statistically stable. *Only when a process is statistically stable can its data be treated as if they came from a population; we can then construct a histogram and find the values of important statistics* such as the mean and standard deviation.

Figure 12-2 (on the next two pages) illustrates some common patterns indicating that the process of filling 16-oz cola cans is not statistically stable. In Figure 12-2(a) there is an obvious upward trend that corresponds to values that are increasing over time. If the filling process were to follow this type of pattern, the cans would be filled with more and more cola until they began to overflow, eventually leaving the employees swimming in cola. Figure 12-2(b) shows an obvious downward trend that corresponds to steadily decreasing values. The cans would be filled with less and less cola until they were extremely underfilled. Such a process would require a complete reworking of the cans in order to get them full enough for distribution to consumers. In Figure 12-2(c) we see an upward shift. A run chart such as this one might result from an adjustment to the filling process, making all subsequent values higher. In Figure 12-2(d) we note a downward shift—the first few values are relatively stable, and then something happened so that the last several values are relatively stable, but at a much lower level. Figure 12-2(e) shows a process that is stable except for one exceptionally high value. The cause of that unusual value should be investigated. Perhaps the cans became temporarily stuck and one particular can was filled twice instead of once. Figure 12-2(f) shows another unusual value, but in this case it's exceptionally low. The run chart of Figure 12-2(g) indicates a pattern that is cyclical—there's a repeating cycle. This pattern is clearly nonrandom and therefore reveals a statistically unstable process. Perhaps periodic overadjustments are being made to the machinery, with the effect that some desired value is continually being chased but never quite captured. Finally, in Figure 12-2(h) we see a common problem in quality

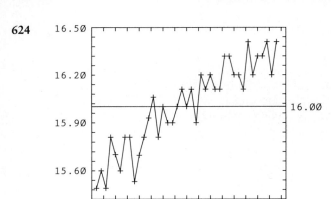

(a) Trend up

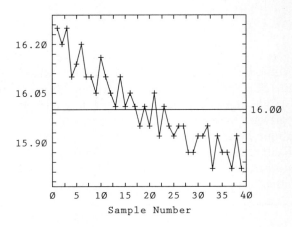

(b) Trend down

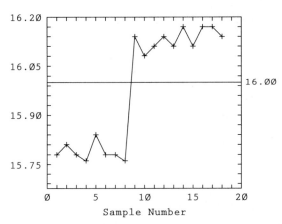

(c) Shift up

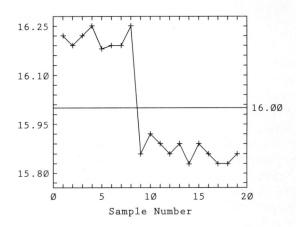

(d) Shift down

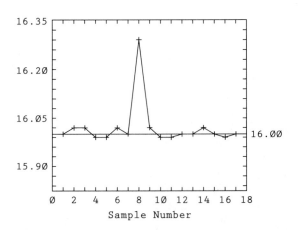

(e) Unusually high value

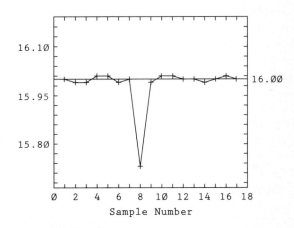

(f) Unusually low value

FIGURE 12-2   Processes with Patterns That Aren't Statistically Stable

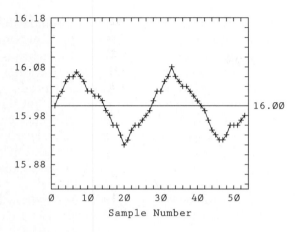

(g) Cyclical pattern

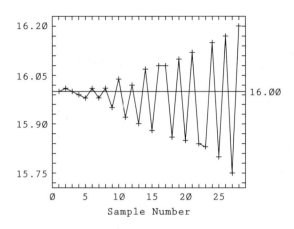

(h) Increasing variation

FIGURE 12-2 (continued)

control. The overall average value is remaining about the same, but variation is increasing over time. Some values are too high, whereas others are too small, and the gap is increasing. The net effect will be products that vary more and more until almost all of them are worthless. For example, some cola cans will be overflowing with wasted cola and some will be underfilled and unsuitable for distribution to consumers. Such trends have driven companies out of business. Figure 12-2(h) illustrates the same type of statistical instability depicted in Figure 12-1.

One common goal of many different methods of quality control is to reduce variation in a product or service. For example, Ford became concerned with variability when it found that its transmissions required significantly more warranty repairs than the same type of transmissions made by Mazda in Japan. A study showed that the Mazda transmissions had substantially less variation in the gearboxes; that is, crucial gearbox measurements varied much less in the Mazda transmissions. Although the Ford transmissions were built within the allowable limits, the Mazda transmissions were more reliable because of their lower variability.

Variation in a process can be described in terms of two types of causes.

DEFINITIONS

Random variation is due to chance; it is the type of variation inherent in any process that is not capable of producing every good or service exactly the same way every time.

Assignable variation results from causes that can be identified.

As an example, let's again consider the U.S. Mint. When we exclude the quarters minted by the new machine (the sample data in Table 12-1) we can establish that the U.S. Mint has a stable process that produces new quarters. The physical characteristics of the quarters (weight, diameter, thickness, roundness) stay within specifications, and the result is a high-quality product. The quarters do vary in weight, diameter, thickness, and roundness, but the variations are small and the result of *random variation*. However, the quarters minted by the new machine have random variation along with *assignable variation*—variation caused specifically by the defective new machine. Later in the chapter we will look at objective ways to distinguish between assignable variation and random variation.

The run chart is one tool for monitoring the stability of a process. We will now consider control charts, which are also extremely useful for that same purpose.

DEFINITIONS

A **control chart** of a process characteristic consists of values plotted sequentially over time, and it includes a **center line** as well as a **lower control limit** (LCL) and an **upper control limit** (UCL). The center line represents a central value of the characteristic measurements, whereas the control limits are boundaries used for identifying points considered to be unusual.

We will assume that the population standard deviation σ is not known as we consider only two of several different types of *control charts:* **R charts** (or **range charts**) used to monitor variation and **$\overline{x}$ charts** used to monitor means. When using control charts to monitor a process, we would be wise to examine process variation *along with* the process center, because a statistically unstable process may be the result of increasing variation or changing means or both.

Control Chart for Monitoring Variation: The *R* Chart

In developing a control chart for monitoring variation, it makes sense to use some measure of variation, such as standard deviation or range. Although the use of standard deviations has some theoretical advantage, we will use ranges because they are currently used more often. Before the age of calculators and computers, control charts for ranges had the distinct advantage of requiring much simpler calculations than control charts for standard deviations. Consequently, control charts based on ranges were used much more often, and this preference has continued to current applications, even though the calculations are no longer a serious obstacle.

To construct a control chart for monitoring *variation,* we plot sample *ranges,* instead of individual values. Using the data of Table 12-1, for example, we could plot the 20 sample ranges corresponding to the 20 different hours in which 5 quarters were sampled. The center line would be located at $\overline{R}$, which denotes the mean of all sample ranges. Following is a summary of the new notation, along with other notation we will be using.

Notation

Given: Process data consist of a sequence of samples all of the same size, and the distribution of the process data is essentially normal.

n = size of each sample, or *subgroup*
$\overline{\overline{x}}$ = mean of the sample means (that is, the sum of the sample means, divided by the number of samples), which is equivalent to the mean of all sample scores combined
$\overline{R}$ = mean of the sample ranges (that is, the sum of the sample ranges, divided by the number of samples)

The basic format of the R chart will include the sample ranges plotted in sequence; there will be a center line and upper and lower control limits. The center line and control limits can be described as follows.

Monitoring Process Variation: Control Chart for R

Points plotted: Sample ranges
Center line: $\overline{R}$
Upper control limit: $D_4\overline{R}$
Lower control limit: $D_3\overline{R}$
where the values of D_4 and D_3 are found in Table 12-2. (See the next page.)

The values of D_4 and D_3 were computed by quality control experts, and they are intended to simplify calculations. The upper and lower control limits of $D_4\overline{R}$ and $D_3\overline{R}$ are values that are roughly equivalent to 99.7% confidence interval limits. It is therefore highly unlikely that values would fall beyond those limits. If a value does fall beyond the control limits, it's very likely that something is wrong with the process.

As we proceed to consider control charts in more detail, it is important to note that the upper and lower control limits are based on the *actual* variation in the process, not the *desired* variation. A particular control chart by itself might suggest that a process is statistically stable, but failure to meet manufacturing specifications and customer requirements could make that same process totally worthless.

TABLE 12-2   Control Chart Constants

Observations in Subgroup, n	$\bar{x}$		$\bar{s}$		$\bar{R}$	
	A_2	A_3	B_3	B_4	D_3	D_4
2	1.880	2.659	0.000	3.267	0.000	3.267
3	1.023	1.954	0.000	2.568	0.000	2.574
4	0.729	1.628	0.000	2.266	0.000	2.282
5	0.577	1.427	0.000	2.089	0.000	2.114
6	0.483	1.287	0.030	1.970	0.000	2.004
7	0.419	1.182	0.118	1.882	0.076	1.924
8	0.373	1.099	0.185	1.815	0.136	1.864
9	0.337	1.032	0.239	1.761	0.184	1.816
10	0.308	0.975	0.284	1.716	0.223	1.777
11	0.285	0.927	0.321	1.679	0.256	1.744
12	0.266	0.886	0.354	1.646	0.283	1.717
13	0.249	0.850	0.382	1.618	0.307	1.693
14	0.235	0.817	0.406	1.594	0.328	1.672
15	0.223	0.789	0.428	1.572	0.347	1.653
16	0.212	0.763	0.448	1.552	0.363	1.637
17	0.203	0.739	0.466	1.534	0.378	1.622
18	0.194	0.718	0.482	1.518	0.391	1.608
19	0.187	0.698	0.497	1.503	0.403	1.597
20	0.180	0.680	0.510	1.490	0.415	1.585
21	0.173	0.663	0.523	1.477	0.425	1.575
22	0.167	0.647	0.534	1.466	0.434	1.566
23	0.162	0.633	0.545	1.455	0.443	1.557
24	0.157	0.619	0.555	1.445	0.451	1.548
25	0.153	0.606	0.565	1.435	0.459	1.541

Source: Adapted from *ASTM Manual on the Presentation of Data and Control Chart Analysis,* © 1976 ASTM, pp. 134–136. Reprinted with permission of American Society for Testing and Materials.

▶ **EXAMPLE**

Using the process data from Table 12-1, construct a control chart for R.

● **SOLUTION**

We begin by finding the value of $\bar{R}$, the mean of the sample ranges.

$$\bar{R} = \frac{0.055 + 0.053 + \cdots + 1.285}{20} = 0.3646$$

The center line for our R chart is therefore located at $\overline{R} = 0.3646$. To find the upper and lower control limits, we must first find the values of D_3 and D_4. Referring to Table 12-2 for $n = 5$, we get $D_3 = 0$ and $D_4 = 2.114$. The control limits are as follows:

Upper control limit: $D_4\overline{R} = (2.114)\ (0.3646) = 0.7708$
Lower control limit: $D_3\overline{R} = (0)\ (0.3646) = 0$

Using a center line value of $\overline{R} = 0.3646$ and control limits of 0 and 0.7708, we now proceed to plot the sample ranges. The result is shown in the Minitab display that follows.

Don't Tamper!

Nashua Corp. had trouble with its paper-coating machine and considered spending a million dollars to replace it. The machine was working well with a stable process, but samples were taken every so often and, based on the results, adjustments were made. These overadjustments, called *tampering,* caused shifts away from the distribution that had been good. The effect was an increase in defects. When statistician and quality expert W. Edwards Deming studied the process, he recommended that no adjustments be made unless warranted by a signal that the process had shifted or had become unstable. The company was better off with no adjustments than with the tampering that took place.

MINITAB DISPLAY

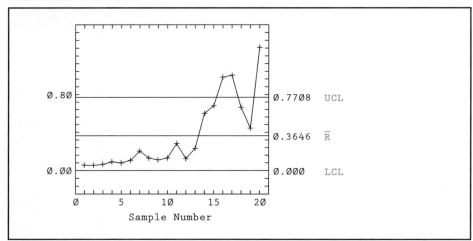

Using Computer Software to Generate *R* Charts

If Minitab is used with Table 12-1 to construct the control chart for R, we first enter the 100 individual weights in column C1 by using the procedure described in Section 1-5 and then enter the following Minitab commands:

```
MTB > GRCHART C1 5;
SUBC> HLINES 0 0.3646 0.7708.
```

The prefix G in GRCHART specifies high resolution graphics, the 5 indicates the

Quality Control at Perstorp

Perstorp Components, Inc. uses a computer that automatically generates control charts to monitor the thicknesses of the floor insulation the company makes for Ford Rangers and Jeep Grand Cherokees. The $20,000 cost of the computer was offset by a first-year savings of $40,000 in labor, which had been used to manually generate control charts to ensure that insulation thicknesses were between the specifications of 2.912 mm and 2.988 mm. Through the use of control charts and other quality control methods, Perstorp reduced its waste by more than two-thirds.

sample (or subgroup) size, and the subcommand HLINES specifies the locations of the horizontal lines in the control chart. If we enter the GRCHART C1 5 command followed by the subcommand RBAR (as shown with no values), we obtain a control chart similar to the one shown here. (If we don't use any subcommand, the control limits and center line will be different because Minitab's default is based on a pooled standard deviation instead of sample ranges.)

Interpreting Control Charts

We should clearly understand the criteria for determining whether a process is in statistical control (that is, whether it is statistically stable). So far, we have noted that a process is not statistically stable if its pattern resembles any of those shown in Figure 12-2. This criterion is included with some others in the following list.

Criteria for Using Control Charts to Determine That a Process Is Not Statistically Stable (out of statistical control)

1. There is a pattern, trend, or cycle that is obviously not random (such as those depicted in Figure 12-2).

2. There is a point lying beyond the upper or lower control limits.

3. Run of 8 Rule: There are 8 consecutive points all above or all below the center line. (With a statistically stable process, there is a 0.5 probability that a point will be above or below the center line, so it is very unlikely that 8 consecutive points will all be above the center line or all be below it. See Exercise 14.)

We will use only the three out-of-control criteria listed above, but some businesses use additional criteria such as these:

- There are 6 consecutive points all increasing or all decreasing.

- There are 14 consecutive points all alternating between up and down (such as up, down, up, down, and so on).

- Two out of 3 consecutive points are beyond control limits that are 1 standard deviation away from the center line.

- Four out of 5 consecutive points are beyond control limits that are 2 standard deviations away from the center line.

► **EXAMPLE**

Examine the R chart shown in the Minitab display and determine whether the process variation is within statistical control.

● **SOLUTION**

We can interpret control charts for R by applying the three out-of-control criteria given above. Applying the three listed criteria to the Minitab display, we conclude that variation in this process is out of control because there is a pattern of increasing ranges, there are points beyond the upper control limit, and there are 8 consecutive points lying below the center line. We therefore conclude that the *variation* (not necessarily the mean) of the process is out of statistical control; that is, the weights of the quarters are varying by increasing amounts. Although all three out-of-control criteria are satisfied, we can conclude that the process is out of control if any one of the three criteria is satisfied. With variation out of statistical control, the immediate priority should be to improve the process until variation is within statistical control. The U.S. Mint should not use this new machine until it has been repaired or adjusted so that the variation of the weights of the quarters is within statistical control.

Control Chart for Monitoring Means: The $\bar{x}$ Chart

In constructing a control chart for monitoring means, we can use various approaches to determine the locations of the control limits. One approach applies the central limit theorem by locating the control limits at

$$\bar{\bar{x}} + 3\bar{s}/\sqrt{n}$$

and

$$\bar{\bar{x}} - 3\bar{s}/\sqrt{n}$$

(The use of "3" reflects the very low probability of having a sample mean differ from a population mean by more than 3 standard deviations; the symbol $\bar{s}$ denotes the mean of the sample standard deviations.) We will use a similar approach, common in business and industry, in which the center line and control limits are based on ranges.

Monitoring Process Mean: Control Chart for $\bar{x}$

Points plotted: Sample means
Center line: $\bar{\bar{x}}$
Upper control limit (UCL): $\bar{\bar{x}} + A_2\bar{R}$
Lower control limit (LCL): $\bar{\bar{x}} - A_2\bar{R}$
where the values of A_2 are found in Table 12-2.

We have already concluded from the data in Table 12-1 that the process variation is out of control and the process should be improved as soon as possible. Although the process already appears to be out of control, we present the following example to illustrate the construction of $\bar{x}$ charts.

▶ **EXAMPLE**

Refer to the weights of quarters in Table 12-1. Using samples of size $n = 5$ collected each hour, construct a control chart for $\bar{x}$. Based on the control chart for $\bar{x}$ only, determine whether the process is within statistical control.

● **SOLUTION**

Before plotting the 20 points corresponding to the 20 values of $\bar{x}$, we must first find the value for the center line and the values for the control limits. We get

$$\bar{\bar{x}} = \frac{5.6564 + 5.6538 + \cdots + 6.0642}{20} = 5.7098 \qquad \text{Center line}$$

$$\bar{R} = \frac{0.055 + 0.053 + \cdots + 1.285}{20} = 0.3646$$

Referring to Table 12-2, we find that for $n = 5$, $A_2 = 0.577$. Knowing the values of $\bar{\bar{x}}$, A_2, and $\bar{R}$, we are now able to find the control limits.

Upper control limit: $\bar{\bar{x}} + A_2\bar{R} = 5.7098 + (0.577)(0.3646) = 5.920$
Lower control limit: $\bar{\bar{x}} - A_2\bar{R} = 5.7098 - (0.577)(0.3646) = 5.499$

The resulting control chart for $\bar{x}$ will be as shown in the accompanying Minitab display. Examination of the control chart shows that the process is not within statistical control because there appears to be a pattern of increasing variation [as in Figure 12-2(h)], there is a point beyond the upper control limit, and there are 8 consecutive points all below the center line.

MINITAB DISPLAY

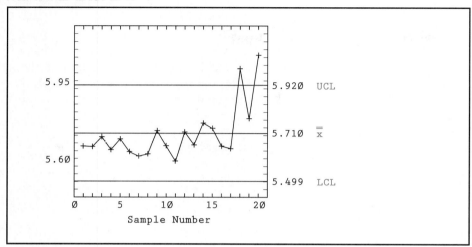

Potato Chip Control Charts

In the past, quality control was simply a matter of inspecting products after they were completed. Now, statistical process control (SPC) is used to analyze samples at different points in the production process so that we know when to make adjustments and when to leave things alone. Frito Lay's bags of potato chips now have much less variability because of the use of SPC and control charts. Frito Lay aims for 1.6% salt content and continually monitors that level by using control charts.

 ## Using Computer Software for $\overline{x}$ Charts

To use Minitab with the data of Table 12-1 for generating a control chart for $\overline{x}$, first enter the 100 individual weights in column C1 by using the procedure described in Section 1-5, and then enter the following Minitab commands:

```
MTB > GXBARCHART C1 5;
SUBC> HLINES 5.499 5.7098 5.920.
```

You can simply enter GXBARCHART C1 5, and Minitab will automatically generate a control chart for $\overline{x}$ with an upper control limit of 5.982 and a lower control limit of 5.438. These values are slightly different from the control limits we have calculated. If we don't specify values for the control limits and the mean (by using the HLINES subcommand), Minitab will automatically calculate the mean and control limits based on a slightly different approach; the locations of the control limits will be slightly different, but the points will be plotted in the same pattern.

It is important to recognize the fundamental issue presented through the use of the Chapter Problem in this section. If we treat the sample data in Table 12-1 as data coming from a fixed population, we are misled because the sample mean seems acceptable. By using a run chart and control charts for R and $\overline{x}$, we can see that the data come from a process that is not statistically stable, so we do not have a population with fixed parameters; rather, the variation is changing over time. Analysis of the pattern over time reveals that the process of minting quarters with the new machine is seriously flawed and requires immediate attention.

12-2   Exercises A: Basic Skills and Concepts

In Exercises 1–3, use the following information:

The York Chemical Company supplies epoxy cement hardener in containers that are designed to hold 5 oz of the liquid. Too much or too little of the hardener results in a mixture with little strength. Each hour, a sample of 5 containers of hardener are carefully measured; the results are given below for 20 consecutive hours of operation.

Amounts of Epoxy Cement Hardener

Sample	Amount (oz)					Mean	Range
1	4.95	5.01	5.00	4.95	4.87	4.956	0.14
2	4.90	4.96	4.89	5.00	5.00	4.950	0.11
3	5.04	5.01	5.05	5.04	5.08	5.044	0.07
4	4.89	4.96	5.00	5.03	5.03	4.982	0.14
5	5.04	4.90	5.02	5.04	5.04	5.008	0.14
6	5.00	5.03	4.97	5.04	4.87	4.982	0.17
7	4.99	5.01	5.10	5.06	5.00	5.032	0.11
8	4.95	5.02	4.92	5.01	4.99	4.978	0.10
9	5.01	5.05	5.07	5.06	4.95	5.028	0.12
10	4.96	5.01	5.03	4.94	4.98	4.984	0.09
11	5.06	5.07	5.05	5.01	5.04	5.046	0.06
12	5.13	4.94	4.97	5.05	4.98	5.014	0.19
13	5.01	5.02	5.12	4.97	5.03	5.030	0.15
14	4.96	4.95	4.98	4.96	4.86	4.942	0.12
15	5.04	5.02	5.05	5.01	5.05	5.034	0.04
16	5.04	4.98	5.06	4.94	4.98	5.000	0.12
17	5.00	5.03	4.97	4.96	5.00	4.992	0.07
18	5.02	4.97	4.90	4.98	5.07	4.988	0.17
19	4.95	5.06	5.13	4.88	5.03	5.010	0.25
20	4.81	4.86	4.85	4.85	4.93	4.860	0.12

1. Construct a run chart for the 100 values. Does there appear to be a pattern suggesting that the process is not within statistical control?

2. Construct an R chart and determine whether the process variation is within statistical control. If it is not, identify which of the three out-of-control criteria lead to rejection of statistically stable variation.

3. Construct an $\bar{x}$ chart and determine whether the process mean is within statistical control. Does this process appear to be working well, or does it require correction?

In Exercises 4–6, use the following information:

At the beginning of each semester, students purchase textbooks from the University Bookstore. During each hour of operation of the store, 4 students are randomly selected and the amounts of time (in minutes) they require to complete their purchases are recorded. The results are shown in the accompanying table.

Amounts of Time Required to Purchase Textbooks

Sample	Purchase Time (min)				Mean	Range
1	13	18	15	14	15.00	5.0
2	16	12	13	14	13.75	4.0
3	18	16	13	18	16.25	5.0
4	15	15	13	17	15.00	4.0
5	12	13	12	16	13.25	4.0
6	16	17	16	17	16.50	1.0
7	15	17	15	12	14.75	5.0
8	18	15	16	14	15.75	4.0
9	15	15	15	13	14.50	2.0
10	13	14	16	14	14.25	3.0
11	14	13	13	16	14.00	3.0
12	17	15	16	14	15.50	3.0
13	17	13	13	16	14.75	4.0
14	16	13	16	17	15.50	4.0
15	5	19	10	7	10.25	14.0
16	21	18	17	5	15.25	16.0
17	8	6	15	22	12.75	16.0
18	23	13	22	14	18.00	10.0
19	20	9	25	21	18.75	16.0
20	6	23	8	22	14.75	17.0
21	7	15	12	6	10.00	9.0
22	24	13	14	19	17.50	11.0
23	13	9	14	18	13.50	9.0
24	14	15	16	6	12.75	10.0

4. Construct a run chart for the 96 values. Is there a pattern suggesting that the process is not statistically stable?

5. Construct an R chart and determine whether the process variation is within statistical control. If it is not, identify which of the three out-of-control criteria lead to rejection of statistically stable variation.

6. Construct an $\bar{x}$ chart and determine whether the process mean is within statistical control. Does this process appear to be working well, or does it require correction?

In Exercises 7–9, use the following information:

Pisa Pizza provides home delivery and advertises that the order is free if it is not delivered in 30 minutes or less. To monitor the quality of delivery, the shop records delivery times for 6 orders randomly selected each day for 24 consecutive days. The delivery times (in minutes) are listed in the table.

Pizza Delivery Times

Sample	Time (min)						Mean	Range
1	26	12	25	12	17	23	19.17	14.0
2	14	25	24	28	24	13	21.33	15.0
3	22	27	12	29	25	27	23.67	17.0
4	24	25	10	17	14	21	18.50	15.0
5	24	27	27	30	10	29	24.50	20.0
6	18	10	11	30	28	11	18.00	20.0
7	12	27	19	15	10	29	18.67	19.0
8	17	24	16	14	14	19	17.33	10.0
9	16	17	20	14	14	30	18.50	16.0
10	23	18	21	13	20	24	19.83	11.0
11	15	18	20	19	11	11	15.67	9.0
12	27	12	23	23	29	18	22.00	17.0
13	26	15	12	23	26	16	19.67	14.0
14	26	30	25	25	23	21	25.00	9.0
15	27	22	24	27	16	19	22.50	11.0
16	28	20	25	28	22	30	25.50	10.0
17	27	11	13	17	18	10	16.00	17.0
18	11	13	17	23	23	10	16.17	13.0
19	30	29	13	10	23	24	21.50	20.0
20	10	24	19	30	29	25	22.83	20.0
21	16	17	14	21	17	30	19.17	16.0
22	17	20	29	23	25	27	23.50	12.0
23	12	17	11	12	11	28	15.17	17.0
24	13	29	30	26	13	28	23.17	17.0
							$\bar{\bar{x}}$ = 20.306	$\bar{R}$ = 14.958

7. Construct a run chart for the 144 values. Does there appear to be a pattern suggesting that the process is not within statistical control?

8. Construct an R chart and determine whether the process variation is within statistical control. If it is not, identify which of the three out-of-control criteria lead to rejection of statistically stable variation.

9. Construct an $\bar{x}$ chart and determine whether the process mean is within statistical control. Does this process appear to be working well, or does it require correction? If the $\bar{x}$ chart were to have a downward trend, as in Figure 12-2(b), should the process be modified to make the means stable?

In Exercises 10–12, use the following information:

The Baltimore Burger Palace is a fast-food restaurant that includes french fries on its limited menu. The process of cooking a batch of french fries is supposed to take 5 min. To monitor the quality of that process, the restaurant times 5 randomly selected batches on each of 20 consecutive days; the results (in minutes) are listed in the table.

Cooking Times of French Fries

Sample	Time (min)					Mean	Range
1	4.98	5.02	4.98	4.88	4.99	4.970	0.14
2	5.04	5.03	5.03	5.02	5.03	5.030	0.02
3	4.96	4.89	5.05	5.02	4.98	4.980	0.16
4	5.05	4.95	4.96	4.96	4.95	4.974	0.10
5	5.04	4.96	4.96	5.01	4.98	4.990	0.08
6	5.00	4.93	4.96	4.95	5.02	4.972	0.09
7	4.99	5.01	5.08	5.01	5.03	5.024	0.09
8	4.96	4.93	5.00	5.02	5.04	4.990	0.11
9	4.91	4.93	4.97	4.92	4.99	4.944	0.08
10	4.96	4.97	5.03	4.98	5.08	5.004	0.12
11	4.97	5.07	5.05	5.01	4.94	5.008	0.13
12	4.83	4.44	4.94	5.22	5.73	5.032	1.29
13	4.92	5.19	4.70	5.02	4.91	4.948	0.49
14	5.38	4.71	5.10	4.85	5.73	5.154	1.02
15	5.18	4.71	5.95	5.03	5.55	5.284	1.24
16	4.68	4.84	6.15	4.85	5.30	5.164	1.47
17	4.56	5.59	5.03	4.81	4.81	4.960	1.03
18	5.53	5.20	5.26	4.91	5.40	5.260	0.62
19	5.96	5.50	5.25	5.23	4.54	5.296	1.42
20	5.39	5.28	5.90	5.31	5.10	5.396	0.80

10. Construct a run chart for the 100 values. Is there a pattern indicating that the process is not statistically stable?

11. Construct an R chart and determine whether the process variation is within statistical control. If it is not, identify which of the three out-of-control criteria lead to rejection of statistically stable variation.

12. Construct an $\bar{x}$ chart and determine whether the process mean is within statistical control. If it is not, identify a consequence of not correcting the process to make it statistically stable.

12-2   Exercises B: Beyond the Basics

13. In this section we described control charts for R and $\bar{x}$ based on ranges. Control charts for monitoring variation and center (mean) can also be based on standard deviations.

    a. An *s chart* for monitoring variation is made by plotting sample standard deviations with a center line at $\bar{s}$ (the mean of the sample standard deviations) and control limits at $B_4\bar{s}$ and $B_3\bar{s}$, where B_4 and B_3 are found in Table 12-2. Construct an s chart for the data of Table 12-1. Compare the result to the R chart given in this section.

    b. An $\bar{x}$ chart (based on standard deviations) is made by plotting sample means with a center line at $\bar{\bar{x}}$ and control limits at $\bar{\bar{x}} + A_3\bar{s}$ and $\bar{\bar{x}} - A_3\bar{s}$, where A_3 is found in Table 12-2 and $\bar{s}$ is the mean of the sample standard deviations. Use the data in Table 12-1 to construct an $\bar{x}$ chart based on standard deviations. Compare the result to the $\bar{x}$ chart based on sample ranges.

14. One of the three out-of-control criteria is based on 8 consecutive points that are all above or all below the center line. For a statistically stable process, there is a 0.5 probability that a point will be above the center line and there is a 0.5 probability that a point will be below the center line. If the values are independent and the process is statistically stable, find the probability that 8 consecutive points will be all above or all below the center line.

12-3   Control Charts for Attributes

In this section we construct a control chart for an attribute, such as the proportion p of defective items. The objective is to monitor a process so as to maintain quality. Instead of tracking the *quantitative* characteristics of variation and mean, however, we now consider the *qualitative* attribute of whether an item has some particular characteristic, such as being defective, being acceptable, weighing under 16 oz, or being nonconforming. A good or a service is nonconforming if it doesn't meet specifications or requirements.

(Nonconforming goods are sometimes called "seconds" and sold at reduced prices.) We will describe control charts for the attribute of being defective, but any other attribute could be represented. As before, we select samples of size n at regular time intervals and plot points in a sequential graph with a center line and control limits. The **control chart for p** (or **p chart**) is a control chart used to monitor the proportion p; the relevant notation and control chart values are as follows.

Notation

$\overline{p}$ = pooled estimate of the proportion of defective items in the process

$\quad = \dfrac{\text{total number of defects found among all items sampled}}{\text{total number of items sampled}}$

$\overline{q}$ = pooled estimate of the proportion of process items that are without defects

$\quad = 1 - \overline{p}$

n = size of each sample (*not* the number of samples)

Control Chart for p

Center line: $\overline{p}$

Upper control limit: $\overline{p} + 3\sqrt{\dfrac{\overline{p}\,\overline{q}}{n}}$

Lower control limit: $\overline{p} - 3\sqrt{\dfrac{\overline{p}\,\overline{q}}{n}}$

We use $\overline{p}$ for the center line because it is the best estimate of the proportion of defects from the process. The expressions for the control limits correspond to 3 standard deviations away from the center line. In Section 6-3 we noted that the standard deviation of sample proportions is $\sqrt{pq/n}$, and this is expressed in the above control limits.

Motorola and the Six-Sigma (σ) Quality Standard

From the 1950s to the 1970s, U.S. manufacturers sought to make 99.8% of their components good, with only 0.2% defective. This rate corresponds to a "three-sigma" level of tolerance. This would seem to be a high level of quality, but Motorola recognized that when many such components are *combined,* the chance of getting a defective product is quite high. The company sought to reduce the variation in all processes by one-half. This standard became known as the "six-sigma" program. A six-sigma level corresponds to about two defects per *billion* items. Motorola found that it could achieve this level with many components and processes. The extra cost was more than offset by reduced costs of repair, scrap, and warranty work. In 1988 Motorola won the first Malcolm Baldridge National Quality Award.

▶ **EXAMPLE**

The Paris Cosmetics Company manufactures lipstick tubes that dispense (you guessed it) lipstick. Each day, 200 tubes are randomly selected and tested for defects. The results are given below for 20 consecutive days. Construct and interpret a control chart for p, where p represents the proportion of defective units.

Day	1	2	3	4	5	6	7	8	9	10	11	12	13	14	15	16	17	18	19	20
Number of defects	4	2	3	6	1	4	3	0	5	5	2	1	9	3	4	1	0	2	6	3
Proportion of defects																				

● **SOLUTION**

The center line for our control chart is located by the value

$$\overline{p} = \frac{\text{total number of defective items from all samples combined}}{\text{total number of items sampled}}$$

$$= \frac{4 + 2 + 3 + \cdots + 3}{20 \cdot 200} = \frac{64}{4000}$$

$$= 0.016$$

Because $\overline{p} = 0.016$, it follows that $\overline{q} = 1 - 0.016 = 0.984$. Using $\overline{p} = 0.016$, $\overline{q} = 0.984$, and $n = 200$, we find the control limits as follows:

Upper control limit: $\overline{p} + 3\sqrt{\dfrac{\overline{p}\,\overline{q}}{n}} = 0.016 + 3\sqrt{\dfrac{(0.016)(0.984)}{200}}$

$$= 0.016 + 0.027 = 0.043$$

Lower control limit: $\overline{p} - 3\sqrt{\dfrac{\overline{p}\,\overline{q}}{n}} = 0.016 - 0.027 = -0.011$

Because it's impossible to have a negative proportion, we will use 0 for the lower control limit. In this case, therefore, we can never have an out-of-control signal from a point that is below the lower control limit. Having found the values for the center line and control limits, we can proceed to plot the daily proportion of defects. The Minitab command `GPCHART C1 C2` results in the accompanying display; the column C1 consists of the numbers of defects (4, 2, 3, 6, . . .), and the column C2 consists of the sample sizes

(200, 200, 200, 200, . . .). The display plots the sample proportions, consisting of the C1 values divided by the C2 values.

We can interpret the control chart for p by considering the same three out-of-control criteria listed in Section 12-2. Using these criteria, we conclude that this process is out of statistical control because a point falls above the upper control limit. Recall that these limits represent values that are 3 standard deviations away from the center, so about 99.74% of all points should fall between them. Because we have a point beyond these limits, it's either an extremely rare exception or an indication that the process is not behaving as we expect.

AMP, Inc.

AMP, Inc. is a Pennsylvania company that makes electrical connectors. In 1983 it instituted a quality improvement plan that so far has resulted in savings of about $75 million. While some companies continue to use a system of management decree by higher-level supervisors, AMP and many other successful companies are committed to the total involvement of employees. One aspect of this total involvement is that each of the more than 10,000 AMP employees receives training in statistical procedures related to control of processes.

MINITAB DISPLAY

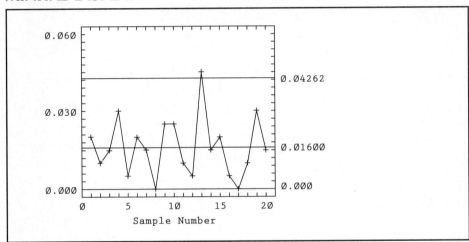

A variation of the control chart for p is the **np chart** in which the actual *numbers* of defects are plotted instead of the *proportions* of defects. The np chart will have a center line value of $n\bar{p}$, and the control limits will have values of $n\bar{p} + 3\sqrt{n\bar{p}\bar{q}}$ and $n\bar{p} - 3\sqrt{n\bar{p}\bar{q}}$. The p chart and the np chart differ only in the scale of values used for the vertical axis.

12-3   Exercises A: Basic Skills and Concepts

In Exercises 1–4, examine the given control chart for p and determine whether the process is within statistical control. If it is not, identify which of the three out-of-control criteria apply.

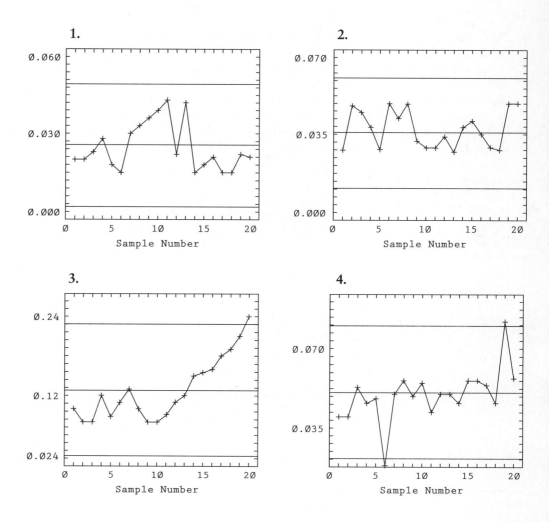

In Exercises 5–8, use the given process data to construct a control chart for p, the proportion of defects. In each case, use the three out-of-control criteria listed in Section 12-2 and determine whether the process is within statistical control. If it is not, identify which of the three out-of-control criteria apply.

5. Columbia House runs a membership club from Terre Haute, Indiana and uses control charts for p to monitor proportions of CD and tape orders that are incorrectly filled. Suppose that 300 orders are randomly selected and monitored each day; the results for 21 consecutive days are in order by row.

$$
\begin{array}{ccccccccccc}
3 & 2 & 4 & 7 & 3 & 15 & 18 & 2 & 6 & 4 & 3 \\
5 & 4 & 6 & 5 & 2 & 4 & 3 & 6 & 1 & 5 &
\end{array}
$$

(continued)

If further investigation revealed that on the 6th and 7th days temporary employees were hired to fill in for vacationing employees, what would you recommend?

6. The Telektronic Company produces 20-amp fuses used to protect car radios from too much electrical power. Each day 400 fuses are randomly selected and tested; the results (numbers of defects per 400 fuses tested) for 20 consecutive days are given below. The data are listed in order by row.

$$
\begin{array}{cccccccccc}
10 & 8 & 7 & 6 & 6 & 9 & 12 & 5 & 4 & 7 \\
9 & 6 & 11 & 4 & 6 & 5 & 10 & 5 & 9 & 11 \\
\end{array}
$$

7. The Riverside Building Supply Company manufactures concrete blocks to be used for home foundations. During each production run, which lasts for 6 hours, a random sample of 120 blocks is tested. The numbers of defective blocks for 18 consecutive production runs are given below.

$$
\begin{array}{cccccccccccccccccc}
4 & 3 & 3 & 2 & 5 & 4 & 3 & 16 & 6 & 3 & 3 & 2 & 5 & 7 & 8 & 4 & 6 & 5 \\
\end{array}
$$

8. A manager for Gleason Supermarket monitors customer waiting times at the checkout counter and considers a wait of more than 5 min to be a defect. Each week, 100 randomly selected customers are timed at the checkout counter. Given below are the numbers of defects found (per 100 customers timed) for 24 consecutive weeks. The data are listed in order by row.

$$
\begin{array}{cccccccccccc}
8 & 7 & 11 & 13 & 10 & 8 & 9 & 8 & 7 & 12 & 14 & 6 \\
5 & 9 & 11 & 8 & 7 & 10 & 9 & 8 & 10 & 7 & 11 & 10 \\
\end{array}
$$

12-3   Exercises B: Beyond the Basics

9. Refer to the data in Exercise 6. The control chart should indicate statistical stability so that the data can be treated as if they came from a population.
   a.   Using all of the data combined, construct a 95% confidence interval for the proportion of defects.
   b.   Using a 0.05 significance level, test the claim that the rate of defects is 1% or less.

10. Construct the np chart for the lipstick tube example in this section. Compare the result with the control chart for p given in this section.

11. a.   Identify the locations of the center line and control limits for a p chart representing a process that has been having a 5% rate of nonconforming items, based on samples of size 100.
    b.   Repeat part a after changing the sample size to 300.
    c.   Compare the two sets of results. What is an advantage and a disadvantage of using the larger sample size? Which chart would be better in detecting a shift from 5% to 10%?

VOCABULARY LIST

Define and give an example of each term.

process data
run chart
statistically stable
within statistical control
random variation
assignable variation
control chart
center line

lower control limit
upper control limit
R chart
range chart
$\bar{x}$ chart
control chart for p
p chart
np chart

Important Formulas

	Control Chart Values		
Characteristic	Center Line	Upper Control Limit	Lower Control Limit
Variation: R chart	$\overline{R}$	$D_4\overline{R}$	$D_3\overline{R}$
Mean: $\bar{x}$ chart	$\overline{\overline{x}}$	$\overline{\overline{x}} + A_2\overline{R}$	$\overline{\overline{x}} - A_2\overline{R}$
Proportion: p chart	$\overline{p}$	$\overline{p} + 3\sqrt{\dfrac{\overline{p}\,\overline{q}}{n}}$	$\overline{p} - 3\sqrt{\dfrac{\overline{p}\,\overline{q}}{n}}$

REVIEW

This chapter focused on process data and the important characteristic of pattern over time. *Run charts* and *control charts* were introduced to monitor processes. By sampling from a process at regular periods and constructing such charts, we can identify when a process begins to go out of control so that corrective action can be taken.

Section 12-2 defined and illustrated run charts and control charts. Random variation and assignable variation were also defined. We considered the basic format of a control chart, which consists of points plotted in sequence, a center line, and lines representing upper and lower control limits. We also identified specific criteria for determining whether a process is out of statistical control. We described the construction of R *charts* used to monitor process variation, as well as $\bar{x}$ charts used to monitor a process mean. We noted that the control charts presented in this chapter describe how a process *is behaving,* not how we might *like it to behave* because of such factors as manufacturer specifications.

In Section 12-3 we plotted sample proportions in control charts for p (or p charts) as we attempted to monitor an attribute such as defects or nonconforming goods or services in a process.

REVIEW EXERCISES

1. The Riverside Building Supply Company applies a coating designed to reduce the amount of light that passes through a window. During each day, a sample of 5 windows is randomly selected and the thickness of the coating (in mils) is measured. Given below are the results from 20 consecutive workdays. Construct a run chart and identify any patterns suggesting that the coating process is not within statistical control.

Sample	Thickness (mils)					Mean	Range
1	2.7	2.3	2.6	2.4	2.7	2.54	0.4
2	2.6	2.4	2.6	2.3	2.8	2.54	0.5
3	2.3	2.3	2.4	2.5	2.4	2.38	0.2
4	2.8	2.3	2.4	2.6	2.7	2.56	0.5
5	2.6	2.5	2.6	2.1	2.8	2.52	0.7
6	2.2	2.3	2.7	2.2	2.6	2.40	0.5
7	2.2	2.6	2.4	2.0	2.3	2.30	0.6
8	2.8	2.6	2.6	2.7	2.5	2.64	0.3
9	2.4	2.8	2.4	2.2	2.3	2.42	0.6
10	2.6	2.3	2.0	2.5	2.4	2.36	0.6
11	3.1	3.0	3.5	2.8	3.0	3.08	0.7
12	2.4	2.8	2.2	2.9	2.5	2.56	0.7
13	2.1	3.2	2.5	2.6	2.8	2.64	1.1
14	2.2	2.8	2.1	2.2	2.4	2.34	0.7
15	2.4	3.0	2.5	2.5	2.0	2.48	1.0
16	3.1	2.6	2.6	2.8	2.1	2.64	1.0
17	2.9	2.4	2.9	1.3	1.8	2.26	1.6
18	1.9	1.6	2.6	3.3	3.3	2.54	1.7
19	2.3	2.6	2.7	2.8	3.2	2.72	0.9
20	1.8	2.8	2.3	2.0	2.9	2.36	1.1

2. Using the same data from Exercise 1, construct an R chart to monitor variation. If the process does not appear to be within control, identify which of the three out-of-control criteria lead to rejection of statistically stable variation.

3. Using the same data from Exercise 1, construct an $\bar{x}$ chart to monitor the process mean. Does this process appear to be working well, or does it require correction?

4. The Telektronic Company manufactures electronic automatic shutoff devices for irons. During each work shift, 100 of these devices are randomly selected and tested. Listed below are the numbers of defects for each of 15 consecutive work shifts.

$$9 \quad 4 \quad 11 \quad 12 \quad 8 \quad 8 \quad 8 \quad 7 \quad 0 \quad 16 \quad 6 \quad 13 \quad 5 \quad 12 \quad 7$$

Construct a control chart for p, and then determine whether the process is within statistical control. If not, identify which criteria apply.

5. The Medassist Pharmaceutical Company provides discounted prescription medicine to retirees of several corporations. Orders received by mail each day are filled by 3 groups of workers in consecutive 8-hour shifts. In an attempt to monitor the quality of service, 50 orders are randomly selected every 2 hours and examined for correctness (correct drug, correct amount, etc.). The numbers of nonconforming orders were recorded for each shift over 2 consecutive days; the results are given below.

Shift	A	A	A	B	B	B	B	C	C	C	C	A	A	A	A	B	B	B	B	C	C	C	C
Number	2	6	3	4	11	10	9	12	2	3	4	2	3	5	4	6	12	10	11	12	5	2	3	4

Use a quality control chart to determine whether the process is statistically stable. Should any corrective action be taken to improve the process?

 COMPUTER PROJECT

a. Simulate 20 days of manufacturing lipstick tubes with a 1% nonconforming rate. If you are using Minitab, sample 200 tubes each day by entering the following Minitab commands:

```
MTB > RANDOM K = 20 C1;
SUBC> BINOMIAL N = 200 P = 0.01.
MTB > PRINT C1
```

If you are using STATDISK, use the New Generated Sample option from the File menu, and then choose Uniform 1 and generate 200 scores with a minimum of 1 and a maximum of 100. Display the ranked data by using View/Raw Data and Data/Rank; consider an outcome beginning with 1 to be defective and all other outcomes to be acceptable. Repeat this procedure until results for 20 days have been simulated.

b. Construct a p chart for the proportion of nonconforming tubes, and determine whether the process is within statistical control. Since we know the process is actually stable with $p = 0.01$, the conclusion that it is not stable would be a type I error; that is, we would have a false positive signal, causing us to believe that the process needed to be adjusted when in fact it should be left alone.

c. Simulate an additional 10 days of manufacturing lipstick tubes with a nonconforming rate of 3%. (Modify the above Minitab commands for the change from 1% to 3%.)

d. Combine the data generated from parts a and c to represent a total of 30 days of sample results. Construct a p chart for this combined data set. Is the process out of control? If we concluded that the process was not out of control, we would be making a type II error; that is, we would believe that the process was okay when in fact it should be repaired or adjusted to correct the shift to the 3% nonconforming rate.

FROM DATA TO DECISION: Are all Jeeps created equal?

In manufacturing crankshafts for Jeeps, specifications require that crankshafts have endplay between 35 ten-thousandths of an inch and 80 ten-thousands of an inch. Each hour, a random sample of four crankshafts are measured; the results are given below. Analyze the stability of this process. Include a plot of a histogram of all sample values, a plot of the R chart, and a plot of the $\bar{x}$ chart. Write a report that summarizes your results, and include any recommendations about whether the process needs to be modified.

Sample	Endplay			
1	51	65	52	47
2	62	49	66	60
3	79	73	79	61
4	119	96	104	110
5	79	64	69	73
6	72	60	32	41
7	68	55	64	72
8	66	49	55	62
9	54	48	55	71
10	60	66	43	49
11	48	58	66	55
12	65	51	74	49
13	53	58	61	62
14	68	70	59	80
15	61	68	62	59
16	51	61	52	70

Interview

David Hall
Division Statistical Manager at the Boeing Commercial Airplane Group

David Hall is Division Statistical Manager, Renton Division, Boeing Commercial Airplane Group. He manages the Statistical Methods Organization, which focuses on applying statistical and other quality technology techniques to continuous quality improvement. Before joining Boeing, he worked at Battelle Pacific Northwest Laboratories, where he was manager of a statistical applications group.

How extensive is the use of statistics at Boeing, and is it increasing, decreasing, or remaining about the same?

The use of statistics is extensive and definitely increasing. I'm sure that this is true of the aircraft industry in general. People are becoming very aware of the need to improve, and to improve you must have data. Statistics is riding the wave of quality improvement. The use of control charts and, more generally, statistical process control is increasing. Designed experiments are very common. In the beginning, 99% of the useful statistical tools are very, very simple. As the processes get refined and understanding increases, more sophisticated tools are required. Regression and correlation analysis, analysis of variance, contingency tables, hypothesis testing, confidence intervals, and time series analysis—virtually all techniques are used at some time. But given the state most American manufacturing is in now, incredible gains can be made with the simplest tools, such as Pareto diagrams and run charts.

How do you sample at Boeing?

Our Quality Assurance Department uses many of the well-known sampling schemes for inspection. Currently we check the daylights out of everything. However sampling is also involved in statistical process control applications. The sampling can be quite complex and is handled on a case-by-case basis.

Could you cite a specific example of how the use of statistics was helpful at Boeing?

We were working with our Fabrication Division to produce more consistent hydropress-formed parts. Through the use of designed experiments, we found that the type of rubber placed over the blanks during forming could drastically reduce the part-to-part variation. That's one simple example of what goes on all the time. An example of a more sophisticated application was the use of bootstrapping to estimate the variability in wind tunnel tests.

In maintaining quality, do you aim for zero defects or is it more efficient for you to allow for defects that must be rejected or reworked?

It's never more efficient to allow defects that get inspected out. That is far too costly. We constantly preach that we should produce to target. Things have to be right, and it's been our big push to reduce variation. Also, in dealing with outside suppliers, we have a program to get suppliers to continually reduce variation about the target value. Sometimes in electronics manufacturing they allow a high defect rate, but for us that is totally unacceptable.

Do you feel job applicants are viewed more favorably if they have studied some statistics?

We naturally have a great many engineers at Boeing and almost all of them have studied some statistics. We are now expecting our managers to have more familiarity with statistics than ever before. They are expected to understand variation and how to effectively use data.

Do you have any advice for today's students?

When I get new statisticians just out of school, whether they have a B.S., M.S., or Ph.D., they still have a tremendous amount to learn before they are effective in our environment. Most of what they need involves people skills, team building, planning, and communication. Right now, American industry is crying out for people with an understanding of statistics and the ability to communicate its use.

"Statistics is riding the wave of quality improvement."

13 Nonparametric Statistics

13-1 Overview

Chapter objectives are identified. The general nature and advantages and disadvantages of nonparametric methods are presented, and the concept of ranks is introduced.

13-2 Sign Test

The sign test is used to test the claim that two sets of dependent data have the same median.

13-3 Wilcoxon Signed-Ranks Test for Two Dependent Samples

The Wilcoxon signed-ranks test is used to test the claim that two sets of dependent data come from identical populations.

13-4 Wilcoxon Rank-Sum Test for Two Independent Samples

The Wilcoxon rank-sum test is used to test the claim that two independent samples come from identical populations.

13-5 Kruskal-Wallis Test

The Kruskal-Wallis test is used to test the claim that several independent samples come from identical populations.

13-6 Rank Correlation

The rank correlation coefficient is used to test for an association between two sets of paired data.

13-7 Runs Test for Randomness

The runs test is used to test for randomness in the way data are selected.

Chapter Problem:

Will our radio message to outer space be mistaken for "random noise"?

Dr. Frank Drake of Cornell University developed a radio message to send to intelligent life beyond our solar system. The message, transmitted as a series of pulses and gaps, is a sequence of 1271 0s and 1s, as listed in Figure 13-1. If we factor 1271 into the prime numbers of 41 and 31 and then make a 41 × 31 grid and put a dot at those positions corresponding to a 1 in the message, we get the pattern shown in Figure 13-1. This pattern contains information including the location of earth in the solar system; the symbols for hydrogen, carbon, and oxygen; and drawings of a man, woman, child, fish, and water. Suppose we send this message and it is intercepted by extraterrestrial intelligent life. Will they think that the pattern is random? Throughout this text we have made reference to "randomly" selected data, and in Section 13-7 we will consider a test for determining whether a sequence is random. Later in this chapter we will apply that test to the sequence of 0s and 1s in the radio message to determine whether it appears to be random.

```
10000000000000000000000000000000000000001
00001110000000000100000100000010000010000
000100C100000000000000000000001000000000
000100010000100000000000000000001000100010
00010001000001000000100000000010001001000
00001110000000000000010000000000001000000
000000000000000000000000000000001000000000
000000000000000100000100000010000010000
11000100000000000000000000001100001100001
00000000001100001100001100001100001100001
00000C0001001001001001001001001001001010
10100100100011000011000011000011000011000
00000001000000000001111101000000000000000
00000010000000000001111101000000000010110
11100100000000000001111101000000000000000
00000000000010000000000000001000100111
00000000000010100000000000000010100100001
10010101110010100000000000000010100100001
00000000001001000000000000000001001000001
0000C00000111110000000000000001111100000001
11010100000010101000000000010100010000000
00000000000100010100000000101000010000000
00000000000100010010001000100110110011101
10110100000100010001010101000100010000000
00000000000100010001001001001001000000000
00000000000111000001111100001110000001
11111010000101010000010100001000100001
000C00000100000100001110001000001000001
1000000000100000100010001001000001000001
00001100001000001000100010001000001000001
10000000011000001101100011011000001100111
```

Figure 13-1 Radio Message

Message shows water, fish, man, child, and woman; symbols for hydrogen, carbon, and oxygen; and the location of the earth.

13-1 Overview

Most of the methods of inferential statistics covered thus far in this text can be called *parametric methods* because their validity is based on sampling from a population with specific parameters, such as the mean, standard deviation, or proportion. Applications of parametric methods are usually limited to circumstances in which some fairly strict requirements are met. One typical requirement is that the sample data must come from a normally distributed population. What do we do when the necessary requirements are not satisfied? This chapter will address this question by introducing hypothesis tests that are classified as nonparametric.

DEFINITIONS

Parametric tests require assumptions about the nature or shape of the populations involved; **nonparametric tests** do not require such assumptions. Consequently nonparametric tests of hypotheses are often called **distribution-free tests.**

Although the term *nonparametric* strongly suggests that the test is not based on a parameter, there are some nonparametric tests that do depend on a parameter such as the median, but don't require a particular distribution. Although *distribution-free* is a more accurate description, the term *nonparametric* is commonly used. Sorry about that.

In this chapter we consider six of the more popular nonparametric methods currently used. As well as providing an alternative to parametric methods, nonparametric techniques are frequently valuable in their own right. These methods have important advantages and disadvantages.

Advantages of Nonparametric Methods

1. Nonparametric methods can be applied to a wide variety of situations because they do not have the more rigid requirements of their parametric counterparts. In particular, nonparametric methods do not require normally distributed populations.

2. Unlike the parametric methods, nonparametric methods can often be applied to nonnumerical data, such as the genders of survey respondents.

3. Nonparametric methods usually involve simpler computations than the corresponding parametric methods and are therefore easier to understand.

If all of these terrific advantages could be gained without any significant disadvantages, we could ignore the parametric methods and enjoy the much simpler nonparametric procedures. Unfortunately, the use of nonparametric methods does have some disadvantages.

Disadvantages of Nonparametric Methods

1. Nonparametric methods tend to waste information because exact numerical data are often reduced to a qualitative form. For example, in one nonparametric test (the sign test) weight losses by dieters are recorded simply as negative signs. With this particular method a weight loss of only 1 lb receives the same representation as a weight loss of 50 lb. This would not thrill dieters.

2. Nonparametric tests do not have the **efficiency** of parametric tests, so with a nonparametric test we generally need stronger evidence (such as a larger sample or greater differences) before we reject a null hypothesis.

When the requirements of population distributions are satisfied, nonparametric tests are generally less efficient than their parametric counterparts, but the reduced efficiency can be compensated for by an increased sample size. Section 13-6 will deal with a concept called the *rank correlation coefficient,* which has an efficiency rating of 0.91 when compared to the linear correlation coefficient presented in Chapter 9. This means that with all other things being equal, this nonparametric approach requires 100 sample observations to achieve the same results as 91 sample observations analyzed through the parametric approach, assuming the stricter requirements for using the parametric method are met. Not bad! The point, though, is that an increased sample size can overcome lower efficiency. Table 13-1 lists the nonparametric methods covered in this chapter, along with the corresponding parametric approach and efficiency rating. You can see from the table that several of these nonparametric tests have efficiency ratings above 0.90, in which cases the lower efficiency might not be a critical factor. However, because parametric tests have higher efficiency ratings than their nonparametric counterparts, we recommend the parametric tests when their required assumptions are satisfied.

TABLE 13-1   Comparison of Parametric and Nonparametric Tests

Application	Parametric Test	Nonparametric Test	Efficiency of Nonparametric Test with Normal Population
Two dependent samples (paired data)	t test or z test	Sign test Wilcoxon signed-ranks test	0.63 0.95
Two independent samples	t test or z test	Wilcoxon rank-sum test	0.95
Several independent samples	Analysis of variance (F test)	Kruskal-Wallis test	0.95
Correlation	Linear correlation	Rank correlation test	0.91
Randomness	No parametric test	Runs test	No basis for comparison

The decision to use either a parametric method or a nonparametric method should be governed by these major factors: cost, time, efficiency, amount of data available, type of data available, method of sampling, nature and distribution of the population, and probabilities (α and β) of making type I and type II errors. For an experiment in which we had abundant data and strong assurances that all of the requirements of a parametric test were satisfied, we would probably be wise to choose that parametric test. But given another experiment with a relatively small amount of data drawn from some mysterious population (that is, one for which we didn't know the distribution), we would probably fare better with a nonparametric test. Sometimes we don't really have a choice. For example, only nonparametric methods can be used on data consisting of observations that can only be ranked. In Section 13-6 we will use a nonparametric test to determine whether there is a correlation between job stress and job salaries; we will use a nonparametric method because the data consist of rankings.

Ranks

The procedure for finding the median of a set of values begins with ranking the data. In Section 2-6 we considered a procedure for finding a score that corresponds to a percentile, and that procedure also began with ranking the data. These earlier uses of ranked data simply involved arranging the data in numerical order, but in this chapter the ranks themselves often are the data. Ranks will now be discussed, so that you will be prepared to use them wherever they are required.

DEFINITION

Data are *ranked* when they are arranged according to some criterion, such as smallest to largest or best to worst. A **rank** is a number assigned to an individual sample item according to its order in the ranked list. The first item is assigned a rank of 1, the second item is assigned a rank of 2, and so on.

For example, the numbers 5, 3, 40, 10, and 12 can be arranged from lowest to highest as 3, 5, 10, 12, and 40, and these numbers have *ranks* of 1, 2, 3, 4, and 5, respectively:

5	3	40	10	12	Scores
3	5	10	12	40	Scores in order
↑	↑	↑	↑	↑	
1	2	3	4	5	Ranks

If a tie in ranks occurs, the usual procedure is to find the mean of the ranks involved and then assign this mean rank to each of the tied items. The numbers 3, 5, 5, 10, and 12 would be given ranks of 1, 2.5, 2.5, 4, and 5, respectively. In this case, ranks 2 and 3 were tied, so we found the mean of 2 and 3 (which is 2.5) and assigned it to the scores that created the tie:

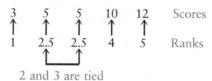

As another example, the scores 3, 5, 5, 7, 10, 10, 10, and 15 are ranked 1, 2.5, 2.5, 4, 6, 6, 6, and 8, respectively.

From these examples we can see how to convert numbers to ranks. In many situations, however, the original data consist of ranks. If a judge ranks five piano contestants, we get ranks of 1, 2, 3, 4, and 5 corresponding to five names; it's this type of data that precludes the use of parametric methods and demonstrates the importance of nonparametric methods.

13-2 Sign Test

Because it is based on plus and minus signs, which are easily determined, the **sign test** is one of the easiest nonparametric tests to use. It is applicable to several different situations, including the following:

1. *We can use a sign test to test claims about the median of the paired differences for two dependent samples.* For example, we will use the sign test with sample paired data consisting of SAT scores of subjects before and after they took the Allan Preparation Course. We will test the claim that the Allan Preparation Course has no effect on SAT scores; that is, we will test the claim that the differences between the before and after scores have a median of 0. (In Section 8-2 we used the parametric Student t test with the same data.)

2. *We can use the sign test to test claims about certain types of nominal data.* For example, later in this section we will test the claim of the Malloy and Turner Advertising Company that its hiring practices are fair. The company maintains that it does not discriminate on the basis of sex and the fact that 40 of the last 50 new employees are men is just a fluke.

3. *We can use the sign test to test a claim made about the median of a single population.* In Chapter 7 we used sample data to test the claim that the *mean* body temperature of healthy adults is not 98.6°F. In this section we will use the sign test with the same data to test the claim that the *median* is less than 98.6°F.

Claims Involving Two Dependent Samples

When using the sign test with two dependent populations, we begin by working with each pair of sample values: Subtract each value of the second variable from the corresponding value of the first variable, but record only the *sign* of the difference. In our sign test procedure, we exclude any ties

Biased Scholarship Test?

FairTest is an organization that opposes standardized tests. Its public education director, Bob Schaeffer, says that his company's recent review of 16,000 National Merit Scholarship semifinalists showed that 61% are males and 35% are females (the genders of the other 4% couldn't be determined). He says that the Preliminary Scholastic Aptitude Test, the first step in the National Merit Scholarship process, is to blame because it "puts a premium on guessing and is a type of game in which boys excel." A representative of the National Merit Scholarship Corporation argues that the gender difference is attributable to other factors, such as the greater proportion of males who take science and math courses.

(represented by 0s). (For other ways to handle ties, see Exercise 22.) The next step consists of recording the positive and negative signs. The key concept underlying the sign test is this: **If the two sets of data have equal medians, the number of positive signs should be approximately equal to the number of negative signs.** For consistency and ease, we will stipulate the following notation.

Notation for the Sign Test

---

x = the test statistic *representing the number of times the less frequent sign occurs*

n = the total number of positive and negative signs combined

Because the results fall into two categories (positive sign, negative sign) and we have a fixed number of independent cases (or pairs of values), we could use the binomial probability distribution (Section 4-3) to determine the likelihood of getting the test statistic x described above. However, we have already used the binomial probability formula to construct a separate table (Table A-7) that lists critical values for the sign test.

▶ **EXAMPLE**

Does it pay to take preparatory courses for standardized tests such as the SAT? Using the sample data in Table 13-2, test the claim that the Allan Preparation Course has no effect on SAT scores. Use a 0.05 level of significance.

● **SOLUTION**

Here's the basic idea: If the Allan Preparation Course has no effect, we expect the numbers of positive and negative signs to be approximately equal—but in Table 13-2 we have 7 negative signs and 2 positive signs. Are the numbers of positive and negative signs approximately equal, or are they *significantly* different? We follow the same basic steps for testing hypotheses as outlined in Figure 7-4.

Steps 1, 2, 3:   The claim that the course has no effect is the null hypothesis, and the alternative hypothesis is the claim that the course does have an effect on SAT scores.

        H_0: The course has no effect.
           (The median of the differences is equal to 0.)

        H_1: The course has an effect.
           (The median of the differences is not equal to 0.)

TABLE 13-2   SAT Scores Before and After the Allan Preparation Course

	Subject									
	A	B	C	D	E	F	G	H	I	J
SAT score before course	700	840	830	860	840	690	830	1180	930	1070
SAT score after course	720	840	820	900	870	700	800	1200	950	1080
Sign of difference (before − after)	−	0	+	−	−	−	+	−	−	−

Based on data from the College Board and "An Analysis of the Impact of Commercial Test Preparation Courses on SAT Scores," by Sesnowitz, Bernhardt, and Knain, *American Educational Research Journal,* Vol. 19, No. 3.

Step 4: The significance level is $\alpha = 0.05$.

Step 5: We are using the nonparametric sign test.

Step 6: The test statistic x is the number of times the less frequent sign occurs. The table includes differences with 7 negative signs, 2 positive signs, and 1 zero. We exclude the zero and let x equal the smaller of 7 and 2, so $x = 2$. Also, $n = 9$ (the total number of positive and negative signs combined). Our test is two-tailed with $\alpha = 0.05$, and reference to Table A-7 shows that the critical value is 1.

Step 7: With a test statistic of $x = 2$ and a critical value of 1, we fail to reject the null hypothesis of no effect. Under these circumstances, a total of 2 positive signs is not significantly less than the number expected; the critical value of 1 indicates that the number of the less frequent sign must be less than or equal to 1 in order to be significant. [See Note 2 included with Table A-7: "The null hypothesis is rejected if the number of the less frequent sign (x) is less than or equal to the value in the table." Because $x = 2$ is *not* less than or equal to 1, we fail to reject the null hypothesis.]

Step 8: There is not sufficient evidence to warrant rejection of the claim that the median of the difference is equal to 0; that is, there is not sufficient evidence to warrant rejection of the claim that the course has no effect. Based on the available evidence, it appears that the Allan course does not have an effect on SAT scores.

Using Computer Software for the Sign Test

Computer software packages such as STATDISK and Minitab can be used for the sign test. With STATDISK, select `Sign Test` from the `Statistics` menu and proceed to enter the data as requested. The displayed result will include the critical value for both a one-tailed test and a two-tailed test, as well as conclusions for one-and two-tailed tests. Shown below is a Minitab display for the preceding example. Note that the data are entered in columns designated as C1 and C2; column C3 is assigned the differences C1 − C2. The Minitab command STEST MEDIAN = 0 C3 is then entered, so we can proceed to test the claim that the differences in column C3 have a median equal to 0.

MINITAB DISPLAY

```
MTB > READ C1 C2
DATA> 700 720
DATA> 840 840
DATA> 830 820
DATA> 860 900
DATA> 840 870            Data entered here
DATA> 690 700
DATA> 830 800
DATA> 1180 1200
DATA> 930 950
DATA> 1070 1080
DATA> ENDOFDATA
MTB > LET C3 = C1 − C2  ←— C3 represents differences C1 − C2
MTB > STEST MEDIAN = 0 C3 ←
                             Minitab command for the sign test
SIGN TEST OF MEDIAN = 0.00000 VERSUS  N.E.  0.00000


            N  BELOW  EQUAL  ABOVE   P-VALUE MEDIAN
SIZE       10     7      1      2    0.1797 −15.00
```
Because the *P*-value is greater than 0.05, fail to reject null hypothesis

In the preceding example we arrived at the same conclusion obtained in Section 8-2, using the same data. We will now illustrate our previous assertion that nonparametric tests lack the sensitivity of parametric tests, so stronger evidence is required before a null hypothesis is rejected. Consider the new data set given in Table 13-3.

TABLE 13-3 SAT Scores Before and After the Allan Preparation Course

	Subject									
	A	B	C	D	E	F	G	H	I	J
SAT score before course	700	840	830	860	840	690	830	1180	930	1070
SAT score after course	800	840	820	980	980	800	800	1270	1080	1220
Sign of difference (before − after)	−	0	+	−	−	−	+	−	−	−

This data set again results in 7 negative signs, 2 positive signs, and 1 zero, so the sign test conclusion will be the same as in the previous example: Fail to reject the null hypothesis that the Allan Preparation Course has no effect on SAT scores. However, if we use the Student t test from Section 8-2, we get

$$t = \frac{\overline{d} - 0}{s_d/\sqrt{n}}$$

$$= \frac{-82 - 0}{69.089/\sqrt{10}}$$

$$= -3.753$$

This test causes *rejection* of the null hypothesis that the course has no effect, because the test statistic of $t = -3.753$ is in the critical region bounded by the critical values of $t = -2.262$ and $t = 2.262$. An intuitive analysis of Table 13-3 suggests that 7 of the 10 subjects did experience substantial increases in SAT scores, but because the sign test is blind to the *magnitude* of those increases, it is not as sensitive as the corresponding parametric t test.

When applying the sign test, be careful to avoid making the wrong conclusion when one sign occurs significantly more often than the other, but the sample data seem to *support* the null hypothesis. For example, assume that we have nine subjects who take the Allan Preparation Course, all of whom achieve a *lower* score after that course. We want to test the claim that the course *increases* SAT scores. The claim of increased scores becomes the alternative hypothesis, and the null hypothesis is the claim of no increase (that is, the scores are the same or lower). The sense of the data *supports* (instead of conflicts with) the null hypothesis because the nine sample scores are lower. We can now terminate the test and conclude that we fail to reject the null hypothesis of equal or lower scores. However, if we ignore common sense and proceed to blindly apply the sign test, we will get a test statistic of $x = 0$ (the smaller of 0 and 9) and a critical value of 1, and we will make the mistake of rejecting the null hypothesis of equal or lower scores. We will then make the mistake of supporting the claim of increased scores. With nine subjects achieving *lower* scores, we should never support a claim of *increased* scores. In all hypothesis tests, we should always check to be sure that the sample data are in conflict with the null hypothesis; we proceed with the test to determine whether that conflict is significant and thus the null hypothesis should actually be rejected. Figure 13-2 summarizes the procedure for the sign test and includes this check: Do the sample data support H_0? If the answer is yes, we fail to reject the null hypothesis.

In the preceding example we found the critical value in Table A-7, but examination of that table shows that it can be used only for values of n up to 25. When $n > 25$, we use a normal approximation to obtain critical values. For $n > 25$, the test statistic x is converted to a z score as shown below. (After the next example, we will justify the format of the test statistic for $n > 25$.)

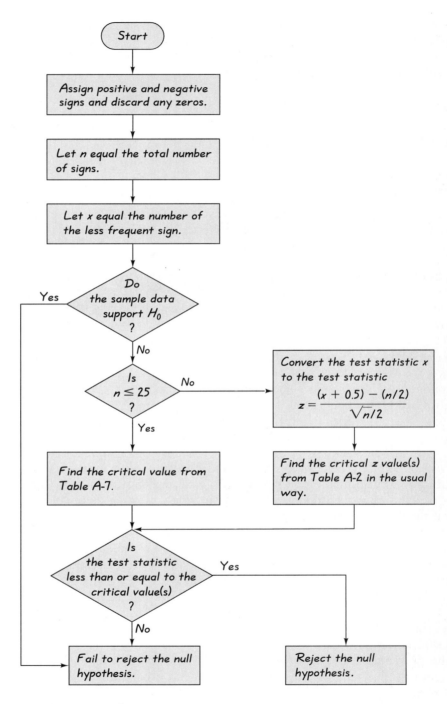

FIGURE 13-2   Sign Test

Test Statistic for the Sign Test

For $n \leq 25$: x (the number of times the less frequent sign occurs)

For $n > 25$: $z = \dfrac{(x + 0.5) - (n/2)}{\sqrt{n}/2}$

Critical values: The critical value corresponding to the x test statistic is found in Table A-7. The critical value corresponding to the z test statistic is found in Table A-2.

The next example involves a sample size greater than 25.

Claims Involving Nominal Data

The following example illustrates the fact that nonparametric methods can be used with nominal data. Also note that the sample size is greater than 25.

▶ **EXAMPLE**

The Malloy and Turner Advertising Company claims that its hiring practices are fair—it does not discriminate on the basis of sex, and the fact that 40 of the last 50 new employees are men is just a fluke. The company acknowledges that applicants are about half men and half women, and all have met the basic job qualification standards. Test the null hypothesis that men and women are hired equally by this company. Use a significance level of 0.05.

● **SOLUTION**

The null and alternative hypotheses are

$$H_0: p_1 = p_2 \quad \text{(the proportions of men and women are equal)}$$
$$H_1: p_1 \neq p_2$$

If we denote hired women by $+$ and hired men by $-$, we have 10 positive signs and 40 negative signs. Refer now to the flowchart in Figure 13-2. The test statistic x is the smaller of 10 and 40, so $x = 10$. This test involves two tails because a disproportionately low number of either sex will cause us to reject the claim of equality. The sample data do not support the null hypothesis

(continued)

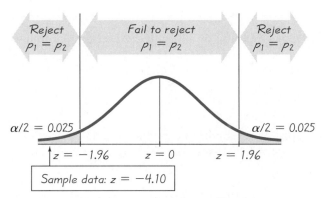

FIGURE 13-3   Testing the Claim That Hiring Practices Are Fair

because 10 and 40 are not precisely equal. Continuing with the procedure in Figure 13-2, we note that the value of $n = 50$ is above 25, so the test statistic x is converted (using a correction for continuity) to the test statistic z as follows:

$$z = \frac{(x + 0.5) - (n/2)}{\sqrt{n}/2}$$

$$= \frac{(10 + 0.5) - (50/2)}{\sqrt{50}/2} = -4.10$$

With $\alpha = 0.05$ in a two-tailed test, the critical values are $z = \pm 1.96$. The test statistic $z = -4.10$ is less than these critical values (see Figure 13-3), so we reject the null hypothesis of equality. There is sufficient sample evidence to warrant rejection of the claim that the hiring practices are fair. This company is in trouble.

When finding critical values for the sign test, we use Table A-7 only for n up to 25, so we need another procedure for finding critical values when $n > 25$. When $n > 25$, the test statistic z is based on a normal approximation to the binomial probability distribution with $p = q = 1/2$. Recall that in Section 5-5 we saw that the normal approximation to the binomial distribution is acceptable when both $np \geq 5$ and $nq \geq 5$. Recall also that in Section 4-4 we saw that $\mu = np$ and $\sigma = \sqrt{npq}$ for binomial experiments. Because this sign test assumes that $p = q = 1/2$, we meet the $np \geq 5$ and $nq \geq 5$ prerequisites whenever $n \geq 10$. Also, with the assumption that $p = q = 1/2$, we get $\mu = np = n/2$ and $\sigma = \sqrt{npq} = \sqrt{n/4} = \sqrt{n}/2$, so

$$z = \frac{x - \mu}{\sigma} \quad \text{becomes} \quad z = \frac{x - (n/2)}{\sqrt{n}/2}$$

Finally, we replace x by $x + 0.5$ as a correction for continuity. That is, the values of x are discrete, but because we are using a continuous probability distribution, a discrete value such as 10 is actually represented by the interval from 9.5 to 10.5. Because x represents the less frequent sign, we need to concern ourselves only with $x + 0.5$; we thus get the test statistic z, as given above and in Figure 13-2.

Claims About the Median of a Single Population

The previous examples involved applications of the sign test to comparisons of *two* sets of data, but we can also use the sign test to investigate a claim made about the *median of one* population, as the next example shows.

▶ **EXAMPLE**

In Chapter 7 we used sample data to test the claim that the *mean* body temperature of healthy adults is not 98.6°F. With the same data, use the sign test to test the claim that the *median* is less than 98.6°F. The data set used in Chapter 7 has 106 subjects—68 subjects with temperatures below 98.6° and 38 subjects with temperatures of at least 98.6°.

● **SOLUTION**

The claim that the median is less than 98.6° is the alternative hypothesis, while the null hypothesis is the claim that the median is at least 98.6°.

H_0: Median is at least 98.6°. (Median $\geq$ 98.6°)
H_1: Median is less than 98.6°. (Median $<$ 98.6°)

We select a significance level of 0.05, and we use the negative sign $-$ to denote each temperature that is below 98.6° and the positive sign $+$ to denote each temperature that is at least 98.6°. We therefore have 68 negative signs and 38 positive signs. We can now determine the significance of getting 38 positive signs out of a possible 106. Referring to Figure 13-2, we note that $n = 106$ and $x = 38$ (the smaller of 68 and 38). The sample data do conflict with the null hypothesis, because a median of at least 98.6° would require a minimum of 53 (half of 106) temperatures of 98.6° or higher, yet our sample data result in only 38 positive signs. We must now proceed to determine whether this conflict is significant. (If the sample data did not conflict with the null hypothesis, we could immediately terminate the test by concluding that we fail to reject the

(continued)

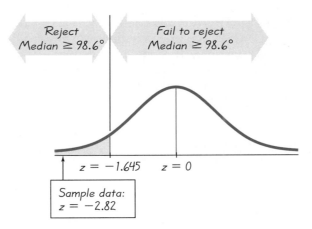

FIGURE 13-4   **Testing the Claim That the Median Is Less Than 98.6°F**

null hypothesis.) The value of n exceeds 25, so we convert the test statistic x to the test statistic z.

$$z = \frac{(x + 0.5) - (n/2)}{\sqrt{n}/2}$$

$$= \frac{(38 + 0.5) - (106/2)}{\sqrt{106}/2} = -2.82$$

In this one-tailed test with $\alpha = 0.05$, we use Table A-2 to get the critical z value of -1.645. From Figure 13-4 we can see that the test statistic of $z = -2.82$ does fall within the critical region. We therefore reject the null hypothesis. Based on the available sample evidence, we support the claim that the median body temperature of healthy adults is less than 98.6°.

We have shown that the sign test wastes information because it uses only information about the direction of the differences between pairs of data, ignoring the magnitudes of those differences. The next section introduces the Wilcoxon signed-ranks test, which largely overcomes that disadvantage.

13-2   Exercises A: Basic Skills and Concepts

In Exercises 1–20, use the sign test.

1. The Malloy and Turner Advertising Company prepared two different television commercials for Taylor women's jeans. One commercial is humorous, and the other is serious. A test screening of both commercials involved 8 randomly

selected consumers who were tested for their reactions; the results are listed below.

Consumer	A	B	C	D	E	F	G	H
Humorous commercial	26.2	20.3	25.4	19.6	21.5	28.3	23.7	24.0
Serious commercial	24.1	21.3	23.7	18.0	20.1	25.8	22.4	21.4

At the 0.05 significance level, test the claim that there is no difference between the reactions to the commercials. Based on the result, does either commercial seem to be better?

2. Air America experimented with two different reservation systems and recorded the times (in seconds) required to process randomly selected passenger requests. The results are listed below.

Passenger	A	B	C	D	E	F	G	H	I	J	K
MicroAir Software	21	23	25	27	27	29	31	32	30	41	47
Flight Services Software	18	20	21	26	27	24	22	33	27	34	34

At the 0.05 significance level, test the claim that there is no difference between the two systems. If you had to select the reservation system to be used by Air America, how would you decide which system to adopt?

3. The Medassist Pharmaceutical Company has developed a compound of the drug Captopril, which is designed to lower systolic blood pressure. It was administered to 10 randomly selected volunteers, with the following results.

Subject	A	B	C	D	E	F	G	H	I	J
Before drug	120	136	160	98	115	110	180	190	138	128
After drug	118	122	143	105	98	98	180	175	105	112

Test the claim that systolic blood pressure is not affected by the drug. Use a 0.05 significance level. Does it appear that the drug has an effect?

4. A study was conducted to investigate the effectiveness of hypnotism in reducing pain. Results for randomly selected subjects are given below. (The values are before and after hypnosis. The measurements are in centimeters on a pain scale.)

Subject	A	B	C	D	E	F	G	H
Before hypnosis	6.6	6.5	9.0	10.3	11.3	8.1	6.3	11.6
After hypnosis	6.8	2.4	7.4	8.5	8.1	6.1	3.4	2.0

Data based on "An Analysis of Factors That Contribute to the Efficacy of Hypnotic Analgesia," by Price and Barber, *Journal of Abnormal Psychology*, Vol. 96, No. 1.

At the 0.05 significance level, test the claim that the sensory measurements are lower after hypnotism. Does hypnotism appear to be effective in reducing pain?

5. A teacher proposed a course designed to increase reading speed and comprehension. To evaluate the effectiveness of this course, she gave randomly selected students a test both before and after this course; the sample results follow.

Before course	100 110 135 167 200 118 127 95 112 116
After course	136 160 120 169 200 140 163 101 138 129

Test the claim that scores are higher after the course. Use a 0.05 level of significance. Does the course appear to be effective in increasing reading speed and comprehension? If you were Dean of Curriculum, would you support the course proposal?

6. A study was conducted to investigate some effects of physical training. Sample data are listed in the margin. (See "Effect of Endurance Training on Possible Determinants of VO_2 During Heavy Exercise," by Casaburi and others, *Journal of Applied Physiology*, Vol. 62, No. 1.) At the 0.05 level of significance, test the claim that mean pretraining weight equals mean posttraining weight. All weights are given in kilograms. What do you conclude about the effect of training on weight?

Pre-training	Post-training
99	94
57	57
62	62
69	69
74	66
77	76
59	58
92	88
70	70
85	84

7. Two different firms designed their own IQ tests, and a psychologist administered both tests to randomly selected subjects. Their scores are given below.

Test I scores	98 94 111 102 108 105 92 88 100 99
Test II scores	105 103 113 98 112 109 97 95 107 103

At the 0.05 level of significance, test the claim that there is no significant difference between the two tests. If the tests appear to be different, further research must be funded to determine which test is better. Must that research be funded?

8. A test of race car driving ability was given to a random sample of 10 student drivers before and after they completed a formal driver education course at the Daytona School of Racing; the results follow.

Before course	100 121 93 146 101 109 149 130 127 120
After course	136 129 125 150 110 138 136 130 125 129

At the $\alpha = 0.05$ significance level, test the school's claim that the course affects scores. If you want to develop the skills of a race car driver, does this course seem helpful?

9. In a study of techniques used to measure lung volumes, physiological data were collected for 10 subjects. The values given in the table are in liters and represent the measured functional residual capacities of the 10 subjects both in a sitting position and in a supine (lying) position.

Sitting	2.96 4.65 3.27 2.50 2.59 5.97 1.74 3.51 4.37 4.02
Supine	1.97 3.05 2.29 1.68 1.58 4.43 1.53 2.81 2.70 2.70

Data based on "Validation of Esophageal Balloon Technique at Different Lung Volumes and Postures," by Baydur, Cha, and Sassoon, *Journal of Applied Physiology*, Vol. 62, No. 1.

(continued)

At the 0.05 significance level, test the claim that there is no significant difference between the measurements taken in the two positions.

10. The Acton Paper Company gives new employees a test for mathematics anxiety. The test is given twice: once before and once after an intensive program in which required mathematical skills are identified and the company's hiring and firing policies are discussed. Sample scores from the math anxiety test are given below.

Before test	104	97	60	88	39	82	91	87	41	43	82	58	67
After test	83	64	58	72	40	70	87	79	25	43	64	40	52

Test the claim that the program has no effect on the anxiety level.

11. A television commercial advertises that 7 out of 10 dentists surveyed prefer Covariant toothpaste over the leading competitor. Assume that 10 randomly selected dentists are surveyed; 7 do prefer Covariant, and 3 favor the other brand. Is this a reasonable basis for making the claim that most (more than half) dentists favor Covariant toothpaste? Use a significance level of 0.05.

12. Test the claim that the median life of a Telektronic Company battery is at least 40 hours, if a random sample of 75 batteries includes exactly 32 that lasted 40 hours or more. Assume a significance level of 0.05.

13. A college aptitude test was given to 100 randomly selected high school seniors. After a period of intensive training, a similar test was given to the same students; 59 students received higher grades, 36 received lower grades, and 5 received the same grades. At the 0.05 level of significance, test the claim that the training is effective.

14. The Grange Fertilizer Company experimented with a new fertilizer at 50 different locations. At 32 of the locations production increased, and at 18 locations it decreased. At the 0.05 level of significance, test the claim that the new fertilizer increases production.

15. A new diet is designed to lower cholesterol levels. In six months, 36 of the 60 randomly selected subjects on the diet had lower cholesterol levels, 22 had slightly higher levels, and 2 registered no change. At the 0.01 level of significance, test the claim that the diet produces no change in cholesterol levels. Do these results indicate that the diet is effective for those who seek lower cholesterol levels?

16. The Life Trust Insurance Company funded a university study of drinking and driving. After 30 randomly selected drivers were tested for reaction times, they were given two drinks and tested again, with the result that 22 had slower reaction times, 6 had faster reaction times, and 2 received the same scores as before the drinks. At the 0.01 significance level, test the claim that the drinks had no effect on the reaction times. Based on these very limited results, does it appear that the insurance company is justified in charging higher rates for those who drink and drive?

17. In target practice, 40 randomly selected police academy students used two different pistols. Analysis of the scores showed that 24 students got higher scores with the more expensive pistol and 16 students got better scores with the less expensive pistol. At the 0.05 level of significance, test the claim that both pistols are equally effective. Should the less expensive pistol be purchased?

18. The Chemco Company makes liquid industrial solvents that are poured into containers labeled 5 gal (or 640 oz). A sample was obtained from the filling process by selecting one container every 10 minutes; results are listed below.

    642   646   640   642   638   634   632   635
    631   638   636   628   625   630   623

    At the 0.05 level of significance, test the claim that the median fill amount is at least 640 oz.

19. Refer to Data Set 8 in Appendix B. If the weights of the sampled quarters are consistent with current manufacturing specifications of the U.S. Mint, the median should be 5.670 g. Use the data to test the claim that the sampled quarters have weights with that median.

20. Refer to Data Set 2 in Appendix B. Use the paired data consisting of body temperatures of women at 8:00 AM and at 12:00 AM on Day 2. Using a 0.05 level of significance, test the claim that for those temperatures, the median difference is 0. Based on the result, do morning and night body temperatures appear to be about the same?

13-2   Exercises B: Beyond the Basics

21. Given n sample scores sorted in ascending order $(x_1, x_2, \ldots, x_n)$, if we wish to find the approximate $1 - \alpha$ confidence interval for the population median M, we get

    $$x_{k+1} < M < x_{n-k}$$

    Here k is the critical value (Table A-7) for the number of signs in a two-tailed hypothesis test conducted at the significance level of α. Find the approximate 95% confidence interval for the sample scores listed below.

    3   8   6   2   1   7   9   11   17   23   25   10   14   8   30

22. In the sign test procedure described in this section, we excluded ties (represented by 0 instead of a sign of $+$ or $-$). A second approach is to treat half of the 0s as positive signs and half as negative signs. (If the number of 0s is odd, exclude one so that they can be divided equally.) With a third approach, in two-tailed tests make half the 0s positive and half negative; in one-tailed tests make all 0s either positive or negative, whichever supports the null hypothesis. Assume that in using the sign test on a claim that the median score is at least 100, we get 60 scores below 100, 40 scores above 100, and 21 scores equal to 100. Identify the test statistic and conclusion for the three different ways of handling differences of 0. Assume a 0.05 significance level in all three cases.

23. Of n subjects tested for high blood pressure, a majority of exactly 50 provided negative results. (That is, their blood pressure was not high.) This is sufficient for us to apply the sign test and reject (at the 0.01 level of significance) the claim that the median blood pressure level is high. Find the largest value n can assume.

24. Table A-7 lists critical values for limited choices of α. Use Table A-1 to add a new column in Table A-7 (down to $n = 15$) that represents a significance level of 0.03 in one tail or 0.06 in two tails. For any particular n we use $p = 0.5$, as the sign test requires the assumption that

$$P(\text{positive sign}) = P(\text{negative sign}) = 0.5$$

The probability of x or fewer like signs is the sum of the probabilities for values up to and including x.

13-3   Wilcoxon Signed-Ranks Test for Two Dependent Samples

In the preceding section we used the sign test to analyze the differences between sample paired data. The sign test used only the signs of the differences and ignored the actual magnitudes of the differences. This section introduces the **Wilcoxon signed-ranks test** for use with sample paired data; by using ranks, this test takes the magnitudes of the differences into account. Because the Wilcoxon signed-ranks test incorporates and uses more information than the sign test, it tends to yield conclusions that better reflect the true nature of the data. The Wilcoxon signed-ranks test requires the following assumption.

Assumption

In using the Wilcoxon signed-ranks test for two dependent (paired) samples, we assume that the population of differences (found from the pairs of data) has a distribution that is approximately symmetric.

Recall that a distribution is symmetric if its graph is such that the left half is approximately a mirror image of the right half. Unlike the Student t test for paired data (see Section 8-2), the Wilcoxon signed-ranks test does *not* require a normal distribution.

In general, the null hypothesis is the claim that both samples come from populations that have the same distribution. The null and alternative hypotheses can therefore be generalized as follows:

H_0: Both samples come from populations with the same distribution.

H_1: The two samples come from populations with different distributions.

The test statistic is based on ranks of the differences between the pairs of data. The rationale for the test statistic will be discussed later; for now the following steps can be used to determine its value.

Procedure for Finding the Value of the Test Statistic

Step 1: For each pair of data, find the difference d by subtracting the second score from the first. Retain signs, but discard any pairs for which $d = 0$.

Step 2: Ignoring the signs of the differences, rank the differences from lowest to highest. When differences have the same numerical value, assign to them the mean of the ranks involved in the tie. (See Section 13-1 for the method of ranking data.)

Step 3: Attach to each rank the sign of the difference from which it came.

Step 4: Find the sum of the absolute values of the negative ranks. Also find the sum of the positive ranks.

Step 5: Let T be the smaller of the two sums found in Step 4. Either sum could be used, but for a simplified procedure we arbitrarily select the *smaller* of the two sums. We therefore use the following notation.

Notation

T = the smaller of the sum of the absolute values of the negative ranks

or

the sum of the positive ranks

Step 6: Let n be the number of pairs of data for which the difference d is not 0.

Step 7: Choose the test statistic based on the sample size, as shown below.

Test Statistic for the Wilcoxon Signed-Ranks Test for Two Dependent Samples

---

For $n \leq 30$: T

For $n > 30$: $z = \dfrac{T - \dfrac{n(n + 1)}{4}}{\sqrt{\dfrac{n(n + 1)(2n + 1)}{24}}}$

Critical values:

If $n \leq 30$, the critical T value is found in Table A-8.

If $n > 30$, the critical z values are found in Table A-2.

When forming the conclusion, reject the null hypothesis if the sample data lead to a test statistic that is in the critical region—that is, the test statistic is less than or equal to the critical value(s). Otherwise, fail to reject the null hypothesis. We now present an example illustrating the Wilcoxon signed-ranks hypothesis testing procedure.

▶ **EXAMPLE**

Refer to the data in Table 13-4 (an extension of the Table 13-3 data used in Section 13-2) listing SAT scores for students before and after they took the Allan Preparation Course. Use the Wilcoxon signed-ranks test to test the claim that the course has no effect on SAT scores. Use a significance level of $\alpha = 0.05$.

● **SOLUTION**

We again follow the same basic procedure for hypothesis testing. Based on the stated claim, the null and alternative hypotheses are as follows:

H_0: The Allan Preparation Course has no effect on SAT scores.
H_1: The Allan Preparation Course has an effect on SAT scores.

The significance level is $\alpha = 0.05$. We are using the Wilcoxon signed-ranks test procedure, so the test statistic is calculated by using the seven-step procedure just presented.

Step 1:   In Table 13-4, the column of differences is obtained by subtracting each after score from the corresponding before score. We discarded the difference of 0.

Step 2:   Ignoring their signs, we rank the absolute differences from lowest to highest in the fifth column of Table 13-4; ties are treated in the manner described in Section 13-1.   *(continued)*

TABLE 13-4   SAT Scores Before and After the Allan Preparation Course

Subject	Before	After	Difference	Ranks of Differences	Signed Ranks
A	700	800	−100	4	−4
B	840	840	0	—	—
C	830	820	10	1	+1
D	860	980	−120	6	−6
E	840	980	−140	7	−7
F	690	800	−110	5	−5
G	830	800	30	2	+2
H	1180	1270	−90	3	−3
I	930	1080	−150	8.5	−8.5
J	1070	1220	−150	8.5	−8.5

Step 3:   The last column of Table 13-4 is created by attaching to each rank the sign of the corresponding difference. If the Allan course has no effect, we expect the number of positive ranks to be approximately equal to the number of negative ranks. If the course raises scores, the ranks with negative signs should outnumber those with positive signs. If the course lowers scores, we expect the number of positive signs to be greater than the number of negative signs. We can detect a dominance by either sign through analysis of the rank sums.

Step 4:   We find the sum of the absolute values of the negative ranks, and we also find the sum of the positive ranks. For the data in Table 13-4 we get

sum of absolute values of negative ranks
$$= 4 + 6 + 7 + 5 + 3 + 8.5 + 8.5 = 42$$

sum of positive ranks $= 1 + 2 = 3$

Step 5:   Letting T be the smaller of the two sums found in Step 4, we find that $T = 3$.

Step 6:   Letting n be the number of pairs of data for which the difference d is not 0, we have $n = 9$.

Step 7:   The number of pairs of data is $n = 9$. Because $n \leq 30$, we use a test statistic of $T = 3$.

We use Table A-8 to find the critical value because $n \leq 30$; the critical value is 6. The test statistic $T = 3$ is less than or equal to the critical value of 6, so we reject the null hypothesis. It appears that the Allan course has an effect on SAT scores. By examining the sample data, we see that there are 7 negative signs and only 2 positive signs. Because negative signs result from higher scores on the test taken after the course, it appears that this course is effective in raising SAT scores.

Using Computer Software for the Wilcoxon Signed-Ranks Test

Computer software such as STATDISK and Minitab can be used for the Wilcoxon signed-ranks test.

If you are using STATDISK, select the Statistics menu item of Wilcoxon Tests, then select the dependent case and proceed to enter the data as requested. The STATDISK display will include the sum of the negative ranks, sum of the positive ranks, sample size, test statistic, critical value, and conclusion.

If you are using Minitab, first enter the paired data in columns C1 and C2, as described in Section 1-5. Then create a column of differences by entering the command LET C3 = C1 − C2. Finally, enter the command WTEST C3, and the Minitab display will include the value of *n*, *T*, and a *P*-value. Shown below is the Minitab display for the data of the preceding example. The displayed *P*-value of 0.024 is less than the significance level of $\alpha = 0.05$, indicating that the null hypothesis of no effect should be rejected.

MINITAB DISPLAY

```
TEST OF MEDIAN = 0.000000 VERSUS MEDIAN N.E. 0.000000

                 N FOR   WILCOXON            ESTIMATED
            N    TEST   STATISTIC  P-VALUE     MEDIAN
   C3       10     9        3.0     0.024      −90.00
```

In this example the unsigned ranks of 1 through 9 have a total of 45. If the two sets of data have no significant differences, each of the two signed-rank totals should be in the neighborhood of 45 ÷ 2, or 22.5. However, for the given sample data one total is 42 and the other is 3; this 42-3 split is a significant departure from the 22.5-22.5 split expected with a true null hypothesis. The table of critical values shows that at the 0.05 level of significance with 9 pairs of data, a 6-39 split represents a significant departure from the null hypothesis, and any split that is farther apart (such as 5-40 or 4-41) will also represent a significant departure from the null hypothesis. Conversely, splits like 7-38, 8-37, or 25-20 do not represent significant departures away from a 22.5-22.5 split, and they would not be a basis for rejecting the null hypothesis. The Wilcoxon signed-ranks test is based on the lower rank total, so instead of analyzing both numbers constituting the split, it is only necessary to analyze the lower number.

The sum $1 + 2 + 3 + \cdots + n$ of all the ranks is equal to $n(n + 1)/2$; if this is a rank sum to be divided equally between two categories (positive and negative), each of the two totals should be near $n(n + 1)/4$, which is half of $n(n + 1)/2$. Recognition of this principle helps us understand the test statistic used when $n > 30$. The denominator in that expression represents a standard deviation of *T* and is based on the principle that

$$1^2 + 2^2 + 3^2 + \cdots + n^2 = \frac{n(n + 1)(2n + 1)}{6}$$

If we were to apply the sign test (Section 13-2) to the example given in this section, we would fail to reject the null hypothesis of no change in before and after scores. This is not the conclusion reached through the Wilcoxon signed-ranks test, which is more sensitive to the magnitudes of the differences and is therefore more likely to be correct.

The Wilcoxon signed-ranks test can only be used for dependent (matched) data. The next section will describe a rank-sum test that can be applied to two sets of independent data that are not paired.

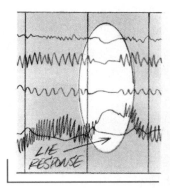

Lie Detectors

Why not require all criminal suspects to take lie detector tests and dispense with trials by jury? The Council of Scientific Affairs of the American Medical Association states, "It is established that classification of guilty can be made with 75% to 97% accuracy, but the rate of false positives is often sufficiently high to preclude use of this (polygraph) test as the sole arbiter of guilt or innocence." A "false positive" is an indication of guilt when the subject is actually innocent. Even with accuracy as high as 97%, the percentage of false positive results can be 50%, so half of the innocent subjects incorrectly appear to be guilty.

13-3   Exercises A: Basic Skills and Concepts

In Exercises 1–10, refer to the sample data for the given exercises in Section 13-2. Instead of the sign test, use the Wilcoxon signed-ranks test to test the claim that both samples come from populations having the same distribution. Use the significance level α that is given below.

1.   Exercise 1; $\alpha = 0.01$
2.   Exercise 2; $\alpha = 0.05$
3.   Exercise 3; $\alpha = 0.05$
4.   Exercise 4; $\alpha = 0.01$
5.   Exercise 5; $\alpha = 0.01$
6.   Exercise 6; $\alpha = 0.05$
7.   Exercise 7; $\alpha = 0.01$
8.   Exercise 8; $\alpha = 0.05$
9.   Exercise 9; $\alpha = 0.05$
10.   Exercise 10; $\alpha = 0.05$

13-3   Exercises B: Beyond the Basics

11. a.   Two checkout systems were tested at the Winthrop Department Store. One system used an optical scanner to record prices, and the other system had prices manually entered by the clerk. Randomly selected customers were paid to use both checkout systems, and their processing times were recorded. Listed here are the differences (in seconds) obtained when the times for the scanner system were subtracted from the corresponding times for the manual system. At the 0.01 significance level, use the Wilcoxon signed-ranks test to test the claim that both systems require the same times.

30	33	27	0	−5	−3	18	10	16	12	3	52	14
−8	−27	0	42	26	19	35	72	14	5	1	12	−6
23	52	47	33	19	16	0	−12	44	40	29	59	38

b.   Part a is a two-tailed test. Now test the claim that the scanner times are lower—that is, the distribution of scanner times is shifted to the left of the distribution of times for the manual system. Use the same 0.01 significance level.

12. a.   With $n = 8$ pairs of data, find the lowest and highest possible values of T.

b.   With $n = 10$ pairs of data, find the lowest and highest possible values of T.

c.   With $n = 50$ pairs of data, find the lowest and highest possible values of T.

13. Use the test statistic given in this section to find the critical value of T for a two-tailed hypothesis test with a significance level of 0.05. Assume that there are $n = 100$ pairs of data with no differences of 0 and no tied ranks.

14. The Wilcoxon signed-ranks test can be used to test the claim that a sample comes from a population with a specified median. This use of the Wilcoxon signed-ranks test requires that the population be approximately symmetrical; that is, when the population distribution is separated in the middle, the left half approximates a mirror image of the right half. The procedure used is the same as the one described in this section, except that the differences (Step 1) are obtained by subtracting the value of the hypothesized median from each score. At the 0.05 level of significance, test the claim that the following values are drawn from a population with a median of 10,000 lb. The scores are the weights (in pounds) of 50 different loads handled by a moving company in Dutchess County, New York.

8,090	9,110	17,810	12,350	3,670
14,800	10,100	26,580	17,330	15,970
8,800	11,860	7,770	8,450	12,430
10,780	13,260	5,030	10,220	11,430
13,490	11,600	13,520	7,470	4,510
14,310	14,760	13,410	4,480	7,450
7,540	3,250	10,630	6,400	10,330
8,160	10,510	9,310	12,700	9,900
7,200	6,170	12,010	16,200	11,450
8,770	9,140	6,820	7,280	6,390

13-4   Wilcoxon Rank-Sum Test for Two Independent Samples

Whereas Section 13-3 used ranks to analyze dependent, or paired, data, this section introduces the **Wilcoxon rank-sum test,** which is a nonparametric test that can be applied to data from two samples that are *independent* and not paired. This test is equivalent to the **Mann-Whitney *U* test,** which is included in some other textbooks (see Exercise 14). The basis for the procedure used in the Wilcoxon rank-sum test is the principle that if two samples are drawn from identical populations and the individual scores are all *ranked* as one combined collection of values, then the high and low ranks should be dispersed evenly between the two samples. If the low ranks are found predominantly in one sample and the high ranks are found predominantly in the other sample, we suspect that the two populations are not identical.

The procedure for using the Wilcoxon rank-sum test begins with the ranking of all the sample data combined. We then find the sum of the ranks for one of the two samples. In the following notation, either sample can be used as sample 1.

Notation for the Wilcoxon Rank-Sum Test

n_1 = size of sample 1
n_2 = size of sample 2
R_1 = sum of ranks for sample 1
R_2 = sum of ranks for sample 2
R = same as R_1 (sum of ranks for sample 1)
μ_R = mean of the sample R values that is expected when the two populations are identical
σ_R = standard deviation of the sample R values that is expected when the two populations are identical

If testing the null hypothesis of identical populations and if both sample sizes are greater than 10, then the sampling distribution of R is approximately normal with mean μ_R and standard deviation σ_R, and the test statistic is as follows.

Test Statistic for the Wilcoxon Rank-Sum Test for Two Independent Samples

$$z = \frac{R - \mu_R}{\sigma_R}$$

where $\mu_R = \dfrac{n_1(n_1 + n_2 + 1)}{2}$

$\sigma_R = \sqrt{\dfrac{n_1 n_2(n_1 + n_2 + 1)}{12}}$

n_1 = size of the sample from which the rank sum R is found
n_2 = size of the other sample
R = sum of ranks of the sample with size n_1

The expression for μ_R is a variation of the following result of mathematical induction: The sum of the first n positive integers is given by $1 + 2 + 3 + \cdots + n = n(n + 1)/2$. The expression for σ_R is a variation of a result that states that the integers $1, 2, 3, \ldots, n$ have standard deviation $\sqrt{(n^2 - 1)/12}$.

Because the test statistic is based on the normal distribution, critical values can be found in Table A-2. Note that our procedure requires that each of the two samples have more than 10 scores. This requirement is included among the other assumptions necessary for the Wilcoxon rank-sum test.

Assumptions

1. We have two independent samples.

2. We are testing the null hypothesis that the two independent samples come from the same distribution; the alternative hypothesis is the claim that the two distributions are different in some way.

3. Each of the two samples has more than 10 scores. (For samples with 10 or fewer values, special tables are available in reference books.)

Note that unlike the corresponding hypothesis tests in Section 8-3 and 8-5, the Wilcoxon rank-sum test does *not* require normally distributed populations. In addition, the Wilcoxon rank-sum test *can* be used with data at the ordinal level of measurement, such as data consisting of ranks.

In Table 13-1 we noted that the Wilcoxon rank-sum test has a 0.95 efficiency rating when compared with the parametric t test or z test. Because this test has such a high efficiency rating and involves easier calculations, it is often preferred over the parametric tests presented in Sections 8-3 and 8-5, even when the condition of normality is satisfied. The following example illustrates the procedure used for the Wilcoxon rank-sum test.

Massachusetts	Virginia	
$32,500 (10)	$30,200 (1)	
36,900 (24)	35,900 (21)	
37,100 (26)	30,600 (3)	
31,800 (7)	31,500 (4.5)	←
35,700 (20)	32,800 (11)	Tie Between
34,500 (17)	33,000 (12)	4 and 5
37,000 (25)	31,500 (4.5)	←
33,700 (15)	32,400 (9)	
43,000 (29)	35,600 (19)	
36,800 (23)	34,700 (18)	
38,600 (28)	36,700 (22)	
	37,200 (27)	
	30,500 (2)	
	33,400 (13)	
	31,600 (6)	
	34,000 (16)	
	32,200 (8)	
	33,500 (14)	
$n_1 = 11$	$n_2 = 18$	
$R_1 = 224$	$R_2 = 211$	

► **EXAMPLE**

Random samples of teachers' salaries from Massachusetts and Virginia are shown in the accompanying table. At the 0.05 level of significance, test the claim that the salaries of teachers are the same in both states.

● **SOLUTION**

The null and alternative hypotheses are as follows:

H_0: The populations of salaries are identical.
H_1: The populations are not identical.

There is a temptation to use the Student t test to compare the means of two independent samples (as in Sections 8-3 and 8-5), but there may be a question about the normality of the distribution of teachers' salaries. We therefore use the Wilcoxon rank-sum test, which does not require normal distributions.

We rank all 29 salaries, beginning with a rank of 1 (assigned to the lowest salary of $30,200). The ranks corresponding to the various salaries are shown in parentheses in the table. Note that the tie between the 4th and *(continued)*

Data based on a survey by the National Education Association.

The Placebo Effect

It has been a common belief that when patients are given a placebo (a treatment with no medicinal value), about one-third of them show some improvement. However, a more recent study of 6000 patients showed that for those with mild medical problems, the placebos seemed to result in improvement in about two-thirds of the cases. The placebo effect seems to be strongest when patients are very anxious and they like their physicians. Because it could cloud studies of new treatments, the placebo effect is minimized by using a *double-blind* experiment in which neither the patient nor the physician knows whether the treatment is a placebo or a real medicine.

5th salaries results in assigning the rank of 4.5 to each of these two salaries. R denotes the sum of the ranks for the sample we choose as sample 1. If we choose the Massachusetts salaries, we get

$$R = 10 + 24 + 26 + 7 + 20 + 17 + 25 + 15 + 29 + 23 + 28 = 224$$

Because there are 11 Massachusetts salaries, we have $n_1 = 11$. Then $n_2 = 18$, because there are 18 Virginia salaries. We can now determine the values of μ_R, σ_R, and the test statistic z.

$$\mu_R = \frac{n_1(n_1 + n_2 + 1)}{2} = \frac{11(11 + 18 + 1)}{2} = 165.00$$

$$\sigma_R = \sqrt{\frac{n_1 n_2(n_1 + n_2 + 1)}{12}} = \sqrt{\frac{(11)(18)(11 + 18 + 1)}{12}} = 22.25$$

$$z = \frac{R - \mu_R}{\sigma_R} = \frac{224 - 165.00}{22.25} = 2.65$$

A large positive value of z would indicate that the higher ranks were found disproportionately in the Massachusetts salaries, and a large negative value of z would indicate that Massachusetts had a disproportionate share of lower ranks. In either case, we would have strong evidence against the claim that the Massachusetts and Virginia salaries are identical. The test is therefore two-tailed.

The significance of the test statistic z can be treated in the same manner as in previous chapters. We are now testing (with $\alpha = 0.05$) the hypothesis that the two populations are the same, so we have a two-tailed test with critical z values of 1.96 and -1.96. The test statistic of $z = 2.65$ falls within the critical region, and we therefore reject the null hypothesis that the salaries are the same in both states. Massachusetts appears to have significantly higher teacher salaries than Virginia.

We can verify that if we interchange the two sets of salaries and consider Virginia to be first, $R = 211$, $\mu_R = 270.0$, $\sigma_R = 22.25$, and $z = -2.65$, so the same conclusion will be reached.

Using Computer Software for the Wilcoxon Rank-Sum Test

Computer software such as STATDISK and Minitab can be used for the Wilcoxon rank-sum test.

If you are using STATDISK, select the `Statistics` menu item `Wilcoxon Tests`, then select the independent case, and proceed to enter the data as requested. The STATDISK display will include the sum of the negative ranks, sum of the positive ranks, sample size, test statistic, critical value, and conclusion.

If you are using Minitab, first enter the two sets of sample data in columns C1 and C2, as described in Section 1-5. Then enter the Minitab command

MANN-WHITNEY C1 C2 for this Wilcoxon rank-sum test for two independent samples. (The Mann-Whitney U test is equivalent to this Wilcoxon test.) The Minitab display will be as shown. Note the inclusion of the P-value of 0.0086 and the comment that "the test is significant."

MINITAB DISPLAY

```
MTB > MANN-WHITNEY C1 C2

Mann-Whitney Confidence Interval and Test

C1         N =  11     Median =        36800
C2         N =  18     Median =        32900
Point estimate for ETA1-ETA2 is        2950
95.5 pct c.i. for ETA1-ETA2 is (900,5200)
W = 224.0
Test of ETA1 = ETA2  vs.  ETA1 n.e. ETA2 is significant at 0.0086
The test is significant at 0.0085 (adjusted for ties)
```

Like the Wilcoxon signed-ranks test, the rank-sum test considers the relative magnitudes of the sample data, whereas the sign test does not. In the sign test, a weight loss of 1 lb or 50 lb receives the same sign, so the actual magnitude of the loss is ignored. Although rank-sum tests do not directly involve quantitative differences between data from two samples, changes in magnitude do cause changes in rank, and these in turn affect the value of the test statistic. For example, if we change the Virginia salary of $30,200 to $40,200, then the value of the rank-sum R will change, and the value of the z test statistic will also change.

13-4 Exercises A: Basic Skills and Concepts

In Exercises 1–12, use the Wilcoxon rank-sum test.

1. Random samples of annual teachers' salaries (in hundreds of dollars) from California and Maryland are as follows. (The data are based on a survey by the National Education Association.)

California					Maryland				
452	446	424	343	438	317	354	345	427	332
411	459	330	329	420	400	378	379	409	354
548	397	331	441	370	325	351	410	307	361

At the 0.05 significance level, test the claim that salaries of teachers are the same in both states. If you had just earned a teaching degree and had friends in California and Maryland, would you prefer either state because of its teaching salaries?

2. The Jefferson Valley Bank collects data by recording waiting times of randomly selected customers on Tuesday afternoons. On one Tuesday, customers were allowed to select any one of several different waiting lines that had formed at the various teller stations. On another Tuesday, all customers entered a single main waiting line that fed the individual teller stations as vacancies occurred.

Multiple lines					One line				
2.8	11.5	8.7	13.3	5.2	7.1	6.3	7.2	7.4	4.3
6.2	2.1	6.2	5.6	4.3	6.2	8.8	5.7	7.4	8.2
5.9	11.8	3.7	3.5	9.6	6.8	10.2	9.3	9.4	11.5
9.2	7.8	6.2	2.2	11.2	8.9	7.9	4.7	5.9	6.0
0.6	14.3	7.1	6.0						

At the 0.05 significance level, test the claim that the waiting times are the same for both arrangements. Based on the results of this test, does it make any difference how the bank arranges its waiting lines?

3. Are severe psychiatric disorders related to biological factors that can be physically observed? One study used X-ray computed tomography (CT) to collect data on brain volumes for a group of patients with obsessive-compulsive disorders and a control group of healthy persons. Sample results (in milliliters) are given below for volumes of the right cordate (based on data from "Neuroanatomical Abnormalities in Obsessive-Compulsive Disorder Detected with Quantitative X-Ray Computed Tomography," by Luxenberg and others, *American Journal of Psychiatry*, Vol. 145, No. 9).

Obsessive-compulsive patients				Control group			
0.308	0.210	0.304	0.344	0.519	0.476	0.413	0.429
0.407	0.455	0.287	0.288	0.501	0.402	0.349	0.594
0.463	0.334	0.340	0.305	0.334	0.483	0.460	0.445

At the 0.01 significance level, test the claim that obsessive-compulsive patients and healthy persons have the same brain volumes. Based on this result, can we conclude that obsessive-compulsive disorders have a biological basis?

4. Michigan Auto Parts Supplies must send many shipments from its central warehouse to stores in the city, and the shipping manager wants to determine the faster of two railroad routes. A search of past records provided the following data. (The shipment times are in hours.)

Route A					Route B				
98	102	83	117	128	96	132	121	87	106
92	112	108	108	100	102	116	95	99	76
93	72	95	91		97	104	115	114	

At the 0.05 level of significance, determine whether there is a significant difference between the routes. Does it appear that shipping costs can be kept lower by always using one of these routes?

5. Given below are sample data consisting of BAC (blood alcohol concentration) levels at arrest of randomly selected jail inmates who were convicted of DWI or DUI offenses. The data are categorized by the type of drink consumed (based on data from the U.S. Department of Justice).

Beer					Liquor			
0.129	0.146	0.148	0.152		0.220	0.225	0.185	0.182
0.154	0.155	0.187	0.212		0.253	0.241	0.227	0.205
0.203	0.190	0.164	0.165		0.247	0.224	0.226	0.234
					0.190	0.257		

At the 0.05 significance level, test the claim that beer drinkers and liquor drinkers have the same BAC levels. Based on these results, do both groups seem equally dangerous, or is one group more dangerous than the other?

6. The Portland Police Department offers a refresher course on arrest procedures. The effectiveness of this course was examined by testing 15 randomly selected officers who had recently completed the course. The same test was given to 15 randomly selected officers who had not had the refresher course. The results are as follows:

Group completing the course						Group without the course				
173	141	219	157	163		159	124	170	148	135
165	178	200	154	189		133	137	189	181	111
192	201	157	168	181		144	127	138	151	162

At the 0.05 level of significance, test the claim that the course has no effect on the test grades. The course will be offered in the future only if it appears to be effective. If it were your decision, would you continue to offer the course in the future?

7. Military training in the use of antiaircraft artillery involves both physical training and education that is referred to as mental training. One study tested the effectiveness of mental training by recording scores earned by an experimental group and a control group on an antiaircraft artillery examination. Test the claim that both groups come from populations with the same scores. Use a 0.05 significance level. (See "Routinization of Mental Training in Organizations: Effects on Performance and Well-Being," by Larsson, *Journal of Applied Psychology*, Vol. 72, No. 1.)

Experimental group					Control group			
60.83	117.80	44.71	75.38		122.80	70.02	119.89	138.27
73.46	34.26	82.25	59.77		118.43	54.22	118.58	74.61
69.95	21.37	59.78	92.72		121.70	70.70	99.08	120.76
72.14	57.29	64.05	44.09		104.06	94.23	111.26	121.67
80.03	76.59	74.27	66.87					

8. Sample data were collected in a study of calcium supplements and their effects on blood pressure. A placebo group and a calcium group began the study with measures of blood pressures. At the 0.05 significance level, test the claim that the two sample groups come from populations with the same blood pressure levels. (The data are based on "Blood Pressure and Metabolic Effects of Calcium Supplementation in Normotensive White and Black Men," by Lyle and others, *Journal of the American Medical Association*, Vol. 257, No. 13.)

Placebo group				Calcium group			
124.6	104.8	96.5	116.3	129.1	123.4	102.7	118.1
106.1	128.8	107.2	123.1	114.7	120.9	104.4	116.3
118.1	108.5	120.4	122.5	109.6	127.7	108.0	124.3
113.6				106.6	121.4	113.2	

9. In a study involving motivation and test scores, data were obtained for men and women. Use the following data to test the claim that both samples come from populations with the same scores. Use a 0.05 significance level. (See "Relationships Between Achievement-Related Motives, Extrinsic Conditions, and Task Performance," by Schroth, *Journal of Social Psychology*, Vol. 127, No. 1.)

Men				Women			
12.27	39.53	32.56	23.93	31.13	18.71	14.34	23.90
19.54	25.73	32.20	19.84	13.96	13.88	29.85	20.15
20.20	23.01	25.63	17.98	6.66	19.20	15.89	
22.99	22.12	12.63	18.06				

10. The arrangement of test items was studied for its effect on anxiety. Sample results are as follows:

Easy to difficult				Difficult to easy			
24.64	39.29	16.32	32.83	33.62	34.02	26.63	30.26
28.02	33.31	20.60	21.13	35.91	26.68	29.49	35.32
26.69	28.90	26.43	24.23	27.24	32.34	29.34	33.53
7.10	32.86	21.06	28.89	27.62	42.91	30.20	32.54
28.71	31.73	30.02	21.96				
25.49	38.81	27.85	30.29				
30.72							

At the 0.05 level of significance, test the claim that the two samples come from populations with the same scores. (The data are based on "Item Arrangement, Cognitive Entry Characteristics, Sex and Test Anxiety as Predictors of Achievement Examination Performance," by Klimko, *Journal of Experimental Education*, Vol. 52, No. 4.)

11. Refer to Data Set 6 in Appendix B and test the claim that the weights of red M&Ms and the weights of yellow M&Ms are the same.

12. Refer to Data Set 2 in Appendix B and use only the body temperatures for 8:00 AM on Day 2. Test the claim of no difference between body temperatures of men and women.

13-4   Exercises B: Beyond the Basics

13. a.   The ranks for group A are 1, 2, . . . , 15, and the ranks for group B are 16, 17, . . . , 30. At the 0.05 level of significance, use the Wilcoxon rank-sum test to test the claim that both groups come from the same population.

   b.   The ranks for group A are 1, 3, 5, 7, . . . , 29, and the ranks for group B are 2, 4, 6, . . . , 30. At the 0.05 level of significance, use the Wilcoxon rank-sum test to test the claim that both groups come from the same population.

   c.   Compare parts a and b.

   d.   Interchange the rankings of the two groups in part a. What changes occur?

   e.   Using the two groups in part a, interchange the ranks of 1 and 30 and then note the changes that occur.

14. The Mann-Whitney U test is equivalent to the Wilcoxon rank-sum test for independent samples in the sense that they both apply to the same situations and always lead to the same conclusions. In the Mann-Whitney U test we calculate

$$z = \frac{U - \frac{n_1 n_2}{2}}{\sqrt{\frac{n_1 n_2 (n_1 + n_2 + 1)}{12}}}$$

where

$$U = n_1 n_2 + \frac{n_1(n_1 + 1)}{2} - R$$

Show that if the expression for U is substituted into the preceding expression for z, we get the same test statistic (with opposite sign) used in the Wilcoxon rank-sum test for two independent samples.

15. Assume that we have two treatments (A and B) that produce measurable results, and we have only two observations for treatment A and two observations for treatment B. We cannot use the test statistic given in this section because both sample sizes do not exceed 10.

Rank	Rank sum for
1 2 3 4	treatment A
A A B B	3

   a.   Complete the accompanying table by listing the 5 rows corresponding to the other 5 cases, and enter the corresponding rank sums for treatment A.

   b.   List the possible values of R, along with their corresponding probabilities. (Assume that the rows of the table from part a are equally likely.)

   c.   Is it possible, at the 0.10 significance level, to reject the null hypothesis that there is no difference between treatments A and B? Explain.

16. Repeat Exercise 15 for the case involving a sample of size 3 for treatment A and a sample of size 3 for treatment B.

13-5   Kruskal-Wallis Test

In Chapter 11 we used one-way analysis of variance (ANOVA) to test hypotheses that differences in means among several samples were significant. That particular F test requires that all the involved populations have normal

Class Attendance *Does* Help

In a study of 424 undergraduates at the University of Michigan, it was found that students with the worst attendance records tended to get the lowest grades. (Is anybody surprised?) Those who were absent less than 10% of the time tended to receive grades of B or above. The study also showed that students who sit in the front of the class tend to get significantly better grades.

distributions with variances that are approximately equal. In this section we introduce the **Kruskal-Wallis test** (also called the *H* **test**) as a nonparametric alternative that does not require a normal distribution. Assumptions for the Kruskal-Wallis test are as follows.

Assumptions

1. We have at least three samples, all of which are random.
2. We want to test the null hypothesis that the samples come from the same or identical populations.
3. Each sample has at least five observations. (If samples have fewer than five observations, refer to more advanced books for special tables of critical values.)
4. The Kruskal-Wallis test requires equal variances, so this test shouldn't be used if the different samples have variances that are very far apart.

Note that unlike the corresponding one-way analysis of variance method used in Chapter 11, the Kruskal-Wallis test does *not* require normally distributed populations. In addition, the Kruskal-Wallis test *can* be used with data at the ordinal level of measurement, such as data consisting of ranks.

In applying the Kruskal-Wallis test, we compute the test statistic *H,* **which has a distribution that can be approximated by the chi-square distribution as long as each sample has at least five observations.** When we use the chi-square distribution in this context, the number of degrees of freedom is $k - 1$, where k is the number of samples. (For a quick review of the key features of the chi-square distribution, see Section 6-4.) The relevant notation and test statistic are as follows.

Notation for the Kruskal-Wallis Test

N = total number of observations in all samples combined
R_1 = sum of ranks for sample 1
R_2 = sum of ranks for sample 2
R_k = sum of ranks for the kth sample
k = number of samples
n_k = number of observations in the kth sample

Test Statistic for the Kruskal-Wallis Test

$$H = \frac{12}{N(N + 1)}\left(\frac{R_1^2}{n_1} + \frac{R_2^2}{n_2} + \cdots + \frac{R_k^2}{n_k}\right) - 3(N + 1)$$

where   degrees of freedom = $k - 1$

When using the Kruskal-Wallis test, we replace the original scores by their corresponding ranks. We then proceed to calculate the test statistic H, which is basically a measure of the variance of the rank sums $R_1, R_2, \ldots, R_k$. If the ranks are distributed evenly among the sample groups, then H should be a relatively small number. If the samples are very different, then the ranks will be excessively low in some groups and high in others, with the net effect that H will be large. Consequently, only large values of H lead to rejection of the null hypothesis that the samples come from identical populations. **The Kruskal-Wallis test is therefore a right-tailed test.**

We begin by considering all observations together and assigning a rank to each one. We rank from lowest to highest, treating ties as we did in the previous sections of this chapter—the mean value of the ranks is assigned to each of the tied observations. We then take each individual sample and find the sum of the ranks and the corresponding sample size.

In order to apply the Kruskal-Wallis test to the motion picture lengths shown in Table 11-1, we must combine the first two categories so that all samples have at least five observations. Thus we get the data in Table 13-5. Because both tests are based on rank sums, we will see similarities between the Kruskal-Wallis test and the Wilcoxon rank-sum test.

TABLE 13-5   Lengths (in min) of Movies Categorized by Star Ratings

Poor/Fair 0.0–2.5 Stars		Good 3.0–3.5 Stars				Excellent 4.0 Stars	
Time	Rank	Time	Rank	Time	Rank	Time	Rank
105	(32.5)	93	(9.5)	123	(48.5)	72	(1)
108	(37.5)	115	(43)	97	(18)	120	(46)
96	(16)	133	(52)	104	(29.5)	106	(34.5)
91	(6)	94	(12)	82	(2)	104	(29.5)
110	(39)	94	(12)	98	(19.5)	159	(57)
114	(42)	106	(34.5)	107	(36)	125	(50)
98	(19.5)	93	(9.5)	95	(14)	103	(27)
100	(23)	129	(51)	94	(12)	160	(58)
96	(16)	102	(26)	117	(44)	193	(60)
123	(48.5)	90	(4.5)	104	(29.5)	168	(59)
101	(25)	104	(29.5)	119	(45)	88	(3)
92	(7.5)	105	(32.5)	96	(16)	121	(47)
155	(56)	139	(54)	134	(53)	144	(55)
92	(7.5)	111	(40.5)	100	(23)	90	(4.5)
99	(21)	111	(40.5)				
100	(23)						
108	(37.5)						

$n_1 = 17$	$n_2 = 29$	$n_3 = 14$
$R_1 = 457.5$	$R_2 = 841$	$R_3 = 531.5$

▶ **EXAMPLE**

As we asked in Chapter 11, are bad movies as long as good movies, or does it just seem that way? Are there really any differences in length between good and bad movies? In Table 13-5 we used Data Set 5 in Appendix B and partitioned the movie lengths into 3 categories, according to the number of stars they were given. Test the claim that the lengths of movies are the same in the 3 categories.

● **SOLUTION**

We use the same hypothesis testing procedure summarized in Figure 7-4.

Steps 1, 2, and 3:   The null and alternative hypotheses are expressed as

H_0: The 3 populations of movie star ratings are identical.
H_1: The populations are not identical.

Step 4:   No significance level was specified. In the absence of any overriding circumstances, we use $\alpha = 0.05$.

Step 5:   In Chapter 11 we used analysis of variance, but here we illustrate the method of the Kruskal-Wallis test.

Step 6:   In determining the value of the test statistic H, we must first rank all of the data. We begin with the lowest value of 72 min, which is assigned a rank of 1. Ranks are shown in parentheses with the original data in Table 13-5. Next we find the sample size, n, and sum of ranks, R, for each sample; these values are shown at the bottom of Table 13-5. Because the total number of observations is 60, we have $N = 60$. We can now evaluate the test statistic as follows:

$$H = \frac{12}{N(N+1)} \left(\frac{R_1^2}{n_1} + \frac{R_2^2}{n_2} + \frac{R_3^2}{n_3} \right) - 3(N+1)$$

$$= \frac{12}{60(60+1)} \left(\frac{457.5^2}{17} + \frac{841^2}{29} + \frac{531.5^2}{14} \right) - 3(60+1)$$

$$= \frac{12}{3660} (56{,}879.15) - 183 = 3.489$$

Because each sample has at least 5 observations, the distribution of H is approximately a chi-square distribution with $k - 1$ degrees of freedom. The number of samples is $k = 3$, so we have $3 - 1 = 2$ degrees of freedom. Refer to Table A-4 to find the critical value of 5.991, which corresponds to 2 degrees of freedom and a 0.05 significance level. (This use of the chi-square distribution is always right-tailed because only large values of H reflect disparity in the distribution of ranks among the samples.)

Step 7:   The test statistic $H = 3.489$ is not in the critical region bounded by 5.991, so we fail to reject the null hypothesis of identical populations.

Step 8:   The 3 categories of movies appear to have the same lengths. (When we considered the same data in Section 11-3, we *rejected* equality of population means, but here we fail to reject the hypothesis of identical populations. In Section 11-3 we used 4 categories instead of only 3, and we used a test that is more sensitive to differences among samples.)

Using Computer Software for the Kruskal-Wallis Test

Following is the Minitab display for the preceding example. When the KRUSKAL-WALLIS C4 C5 command is used, all the sample data must be stored in column C4; column C5 consists of subscripts or numbers that identify the different samples. After entering the three sets of sample scores in columns C1, C2, and C3, we proceed to STACK the 60 scores into one big column C4. The first 17 scores are from column C1, the second 29 are from column C2, and the last 14 are from column C3. Entry of the values in column C5 is automatic, and column C5 consists of these sample identifiers: 17 ones, 29 twos, and 14 threes. The Minitab display includes the test statistic of $H = 3.49$ and the P-value of 0.175. Because the P-value is not less than the significance level of $\alpha = 0.05$, we fail to reject the null hypothesis of identical populations.

MINITAB DISPLAY

```
MTB > SET C1
DATA> 105 108  96  91 110 114  98 100  96
DATA> 123 101  92 155  92  99 100 108
DATA> SET C2
DATA>  93 115 133  94  94 106  93 129 102  90
DATA> 104 105 139 111 111 123  97 104  82  98
DATA> 107  95  94 117 104 119  96 134 100
DATA> SET C3
DATA>  72 120 106 104 159 125 103
DATA> 160 193 168  88 121 144  90
DATA> ENDOFDATA
MTB > STACK C1 C2 C3 C4;
SUBC> SUBSCRIPTS C5.
MTB > KRUSKAL-WALLIS C4 C5


LEVEL     NOBS     MEDIAN   AVE. RANK   Z VALUE
    1       17      100.0       26.9     -1.00
    2       29      104.0       29.0     -0.64
    3       14      120.5       38.0      1.83
OVERALL     60                  30.5
                    Test statistic
H = 3.49   d.f. = 2   p = 0.175       P-value
H = 3.49   d.f. = 2   p = 0.175 (adj. for ties)
```

The test statistic H, as presented earlier, is the rank version of the test statistic F used in the analysis of variance discussed in Chapter 11. When we deal with ranks R instead of raw scores x, many components are predetermined. For example, the sum of all ranks can be expressed as $N(N + 1)/2$, where N is the total number of scores in all samples combined. The expression

$$H = \frac{12}{N(N + 1)}\Sigma n_i(\overline{R}_i - \overline{\overline{R}})^2$$

where
$$\overline{R}_i = \frac{R_i}{n_i} \qquad \overline{\overline{R}} = \frac{\Sigma R_i}{\Sigma n_i}$$

combines weighted variances of ranks to produce the test statistic H given here. This expression for H is algebraically equivalent to the expression for H given earlier as the test statistic. The earlier form of H (not the one given here) is easier to work with.

In comparing the procedures of the parametric F test for analysis of variance and the nonparametric Kruskal-Wallis test, we see that in the absence of computer software, the Kruskal-Wallis test is much simpler to apply. We need not compute the sample variances and sample means. We do not require normal population distributions. Life becomes so much easier. However, the Kruskal-Wallis test is not as efficient as the F test, and it may require more dramatic differences for the null hypothesis to be rejected.

13-5   Exercises A: Basic Skills and Concepts

In Exercises 1–12, use the Kruskal-Wallis test.

1. The accompanying table lists body temperatures of 5 randomly selected subjects from each of 3 different age groups.

Body Temperatures (°F) Categorized by Age

18–20	21–29	30 and older
98.0	99.6	98.6
98.4	98.2	98.6
97.7	99.0	97.0
98.5	98.2	97.5
97.1	97.9	97.5

Data obtained by Dr. Philip Mackowiak, Dr. Steven Wasserman, and Dr. Myron Levine of the University of Maryland.

Using a significance level of $\alpha = 0.05$, test the claim that the 3 age-group populations of body temperatures are identical.

2. The Wendt Brewing Company plans to conduct a major media campaign. Three advertising companies prepared trial commercials in an attempt to win a $2 million contract. The commercials were tested on randomly selected consumers whose reactions were measured. Their reactions are listed below.

Delaney and Taylor Advertising Co.				Madison Advertising				Kelly, Grey, and Landry Advertising			
77	76	75	77	80	80	78	81	72	55	81	92
70	82	90	81	72	85	96	84	80	73	77	75
65	70			61	75			77	77	71	72

At the 0.05 significance level, test the claim that reactions are the same for the 3 different commercials (that is, that the 3 populations are the same). If you were responsible for advertising at Wendt Brewing, is there an advertising company you would select on the basis of these results?

3. Jane Corrigan is a pilot who does extensive flying in a variety of weather conditions. She decided to buy a battery-powered radio as an independent backup for her regular radios, which depend on her plane's electrical system. Jane randomly selected 5 batteries from each of 3 different battery suppliers, whose costs vary, and then tested them for their operating time (in hours) before recharging was necessary. Do the 3 populations appear to be the same? Based on these results, does one of the battery suppliers seem to be superior to the others?

Telektronic	Lectrolyte	Electrocell
26.0	29.0	30.0
28.5	28.8	26.3
27.3	27.6	29.2
25.9	28.1	27.1
28.2	27.0	29.8

4. A sociologist randomly selected subjects from 3 types of family structures: two-parent, single-parent, and families in transition. The selected subjects were interviewed and rated for their social adjustment. The sample results are given in the accompanying table. At the $\alpha = 0.05$ level of significance, test the claim that the 3 populations are identical.

Two-Parent	Single-Parent	Transition
140	115	90
135	105	120
130	110	125
125	130	100
150	105	105

Laboratory				
1	2	3	4	5
2.9	2.7	3.3	3.3	4.1
3.1	3.4	3.3	3.2	4.1
3.1	3.6	3.5	3.4	3.7
3.7	3.2	3.5	2.7	4.2
3.1	4.0	2.8	2.7	3.1
4.2	4.1	2.8	3.3	3.5
3.7	3.8	3.2	2.9	2.8
3.9	3.8	2.8	3.2	
3.1	4.3	3.8	2.9	
3.0	3.4	3.5		
2.9	3.3			

Data provided by Minitab, Inc.

Sample Data (Exercises 7–10)

Zone	LA	Acres	Taxes
1	20	0.50	1.9
1	18	1.00	2.4
1	27	1.05	1.5
1	17	0.42	1.6
1	18	0.84	1.6
1	13	0.33	1.5
1	24	0.90	1.7
1	15	0.83	2.2
4	18	0.55	2.8
4	20	0.46	1.8
4	19	0.94	3.2
4	12	0.29	2.1
4	9	0.26	1.4
4	18	0.33	2.1
4	28	1.70	4.2
4	10	0.50	1.7
4	15	0.43	1.8
4	12	0.25	1.6
7	21	3.04	2.7
7	16	1.09	1.9
7	14	0.23	1.3
7	23	1.00	1.7
7	17	6.20	2.2
7	17	0.46	2.0
7	20	3.20	2.2

5. Flammability tests were conducted on children's sleepwear with the Vertical Semirestrained Test, in which a piece of fabric was burned under very controlled conditions. After the burning stopped, the length of the charred portion was measured and recorded. Results are given for the same fabric tested at different laboratories. Because the same fabric was used, the different laboratories should provide the same results. Do they? Use a 0.10 significance level to test the claim that the different laboratories produce identical results.

6. Randomly selected teachers' salaries from 4 different states are listed in the accompanying table. At the 0.10 significance level, test the claim that the salaries of teachers are the same in all 4 states.

Georgia	Vermont	Texas	Oregon
$29,050	$32,716	$27,493	$34,347
26,369	26,737	25,901	31,389
34,102	29,898	29,336	34,139
26,803	30,426	31,809	35,023
32,063	30,314	20,753	27,949
26,558	35,824	23,123	33,146
36,412		24,158	30,108
		32,466	

Data based on a survey by the National Education Association.

7. The accompanying data represent 25 different homes sold in Dutchess County, New York. The zones (1, 4, 7) correspond to different geographic regions of the county. LA values are living areas in hundreds of square feet, Acres values are lot sizes in acres, and Taxes values are the annual tax bills in thousands of dollars. At the 0.05 significance level, test the claim that living areas are the same in all 3 zones.

8. Use the data from Exercise 7. At the 0.05 level of significance, test the claim that lot sizes (as measured in acres) are the same in all 3 zones.

9. Use the data from Exercise 7. At the 0.05 level of significance, test the claim that taxes are the same in all 3 zones.

10. Refer to the data from Exercise 7 and change the zone 7 Tax values to 3.0, 1.8, 1.1, 1.4, 2.3, 2.0, 2.4. Note that the mean Tax amount does not change, but these values vary more. Now repeat Exercise 9 with this modified data set and note the effect of increasing the spread of the values in one of the samples.

11. Refer to Data Set 4 in Appendix B. Create the separate weight classes of "less than 2750 lb," "2750–2999," "3000–3499," and "more than 3500 lb." Using the highway fuel consumption rate (in mi/gal), test the claim that the 4 weight categories have the same fuel consumption amounts. Use a 0.05 significance level.

12. Refer to Data Set 6 in Appendix B. At the 0.05 significance level, test the claim that the weights of M&Ms are the same for each of the 6 different color populations. If it is the intent of Mars, Inc. to make the candies so that the different color populations are the same, do your results suggest that the company has a problem that requires corrective action?

13-5   Exercises B: Beyond the Basics

13. a.   Simplify the expression of the test statistic H for the special case of 8 samples, each consisting of exactly 6 observations.
    b.   In general, how is the value of the test statistic H affected if a constant is added to (or subtracted from) each score?
    c.   In general, how is the value of the test statistic H affected if each score is multiplied (or divided) by a positive constant?
14. For 3 samples, each of size 5, what are the largest and smallest possible values of H?
15. In using the Kruskal-Wallis test, there is a correction factor that should be applied whenever there are many ties: Divide H by

$$1 - \frac{\Sigma T}{N^3 - N}$$

Here $T = t^3 - t$. For each group of tied scores, find the number of observations that are tied and represent this number by t. Then compute $t^3 - t$ to find the value of T. Repeat this procedure for all cases of ties, and find the total of the T values, which is ΣT. For the example presented in this section, use this procedure to find the corrected value of H.

16. Show that for the case of two samples, the Kruskal-Wallis test is equivalent to the Wilcoxon rank-sum test. This can be done by showing that for the case of two samples, the test statistic H equals the square of the test statistic z used in the Wilcoxon rank-sum test. Also note that with 1 degree of freedom, the critical values of χ^2 correspond to the square of the critical z score.

13-6   Rank Correlation

Chapter 9 considered the concept of correlation and introduced the *linear correlation coefficient* as a measure of the strength of the association between two variables. In this section we study rank correlation, the nonparametric counterpart of that parametric measure. In Chapter 9 we used paired sample data to compute values for the linear correlation coefficient r; in this section we consider the **rank correlation coefficient,** which uses ranks as the basis for measuring the strength of the association between two variables. Rank

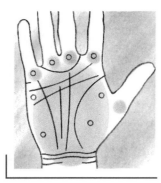

Palm Reading

Some people believe that the length of their palm's lifeline can be used to predict longevity. In a letter published in the *Journal of the American Medical Association,* authors M. E. Wilson and L. E. Mather refuted that belief with a study of cadavers. Ages at death were recorded, along with the lengths of palm lifelines. The authors concluded that there is no significant correlation between age at death and length of lifeline. Palmistry lost, hands down.

correlation has some distinct advantages over the parametric methods discussed in Chapter 9:

1. With rank correlation, we can analyze some types of data that can be ranked but not measured; such data could not be considered with the parametric linear correlation coefficient r of Chapter 9. For example, if two judges rank 30 different gymnasts, we can use those ranks to test for a correlation between the two judges; because the sample data consist of ranks, the methods of Chapter 9 do not apply.

2. Rank correlation can be used to detect some relationships that are not linear; an example will be given later in this section.

3. The computations for rank correlation are much simpler than those for the linear correlation coefficient r, as can be readily seen by comparing the formulas used to compute these statistics. Using a calculator or computer, we can get the value of r easily, but when neither is available, the rank correlation coefficient is easier to compute.

4. Rank correlation can be used when some of the more restrictive requirements of the linear correlation approach are not met; that is, the nonparametric approach can be used in a wider variety of circumstances than the parametric method can. For example, the parametric approach requires that the involved populations have normal distributions; the nonparametric approach does not require normality. We do assume that we have a random sample. If a sample is not random, it may be totally worthless.

A disadvantage of rank correlation is its efficiency rating of 0.91, as described in Section 13-1. This efficiency rating indicates that with all other circumstances being equal, the nonparametric approach of rank correlation requires 100 pairs of sample data to achieve the same results as only 91 pairs of sample observations analyzed through the parametric approach, assuming the stricter requirements of the parametric approach are met.

We will use the following notation, which closely parallels the notation used in Chapter 9 for linear correlation. (Recall from Chapter 9 that r denotes the linear correlation coefficient for sample paired data, ρ denotes the linear correlation coefficient for all paired data in a population, and n denotes the number of pairs of data.)

Notation

r_s = rank correlation coefficient for sample paired data (r_s is a sample statistic)

ρ_s = rank correlation coefficient for all the population data (ρ_s is a population parameter)

n = number of *pairs* of data

d = difference between ranks for the two observations within a pair

We use the notation r_s for the *rank* correlation coefficient so that we don't confuse it with the *linear* correlation coefficient r. The subscript s has nothing to do with standard deviation; it is used in honor of Charles Spearman (1863–1945), who originated the rank correlation approach. In fact, r_s is often called **Spearman's rank correlation coefficient.**

Given a collection of sample paired data, if we want to test for a relationship between the two variables, we can use the null hypothesis H_0: $\rho_s = 0$, which is the claim of no correlation. The test statistic is as follows:

Test Statistic for the Rank Correlation Coefficient

$$r_s = 1 - \frac{6\Sigma d^2}{n(n^2 - 1)}$$

Critical values: Found in Table A-9.

In addition to the form of test statistic given here, there is another way to calculate r_s: Replace each value by its corresponding rank, and then calculate the value of the linear correlation coefficient r as described in Chapter 9. The test statistic given here requires calculations that are much easier than those required for the linear correlation coefficient. However, both approaches are equivalent in the sense that they yield the same results.

Ties in the ranks of the original sample values can be handled as in the preceding sections of this chapter: Find the mean of the ranks involved in the tie, and then assign the mean rank to each of the tied items. The above test statistic yields the exact value of r_s only if there are no ties. With a relatively small number of ties, the above test statistic is a good approximation of r_s. (When ties occur, we can get an exact value of r_s by ranking the data and using Formula 9-1 for the linear correlation coefficient; after finding the value of r_s, we can proceed with the methods of this section.)

▶ **EXAMPLE**

Consider the data shown on the next page in Table 13-6, which is based on results in *The Jobs Rated Almanac*. Randomly selected jobs are ranked according to salary and stress. Does a relationship exist between salary and stress? You can't use the linear correlation coefficient r because the data consist of ranks and therefore do not satisfy the normal distribution requirement described in Section 9-2. Instead, use the rank correlation coefficient to test the claim that there is no relationship between salary and stress (that is, $\rho_s = 0$). Use a significance level of $\alpha = 0.05$.

(continued)

TABLE 13-6　Jobs Ranked by Salary and Stress

Job	Salary Rank	Stress Rank	Difference d	d^2
Stockbroker	2	2	0	0
Zoologist	6	7	1	1
Electrical engineer	3	6	3	9
School principal	5	4	1	1
Hotel manager	7	5	2	4
Bank officer	10	8	2	4
Occupational safety inspector	9	9	0	0
Home economist	8	10	2	4
Psychologist	4	3	1	1
Commercial airline pilot	1	1	0	0

Total: $\Sigma d^2 = \overline{24}$

● **SOLUTION**

We follow the same basic steps for testing hypotheses as outlined in Figure 7-4.

Step 1:　The claim of no correlation is expressed symbolically as $\rho_s = 0$.

Step 2:　The negation of the claim in Step 1 is $\rho_s \neq 0$.

Step 3:　Because the null hypothesis must contain the condition of equality, we have

$$H_0: \rho_s = 0 \qquad H_1: \rho_s \neq 0$$

Step 4:　The significance level is $\alpha = 0.05$.

Step 5:　As just stated, we cannot use the linear correlation approach of Section 9-2 because ranks do not satisfy the requirement of a normal distribution. We will use the rank correlation approach instead.

Step 6:　We now find the value of the test statistic. Table 13-6 shows the calculation of the differences d and their squares d^2, which results in a value of $\Sigma d^2 = 24$. With $n = 10$ (for 10 pairs of data) and $\Sigma d^2 = 24$, we can find the value of the test statistic r_s as follows:

$$r_s = 1 - \frac{6\Sigma d^2}{n(n^2 - 1)} = 1 - \frac{6(24)}{10(10^2 - 1)}$$

$$= 1 - 0.145 = 0.855$$

If we now refer to Table A-9 for the critical value, we see that with $n = 10$ and $\alpha = 0.05$, the critical value of r_s is 0.648. Because the test statistic $r_s = 0.855$ exceeds the critical value of 0.648, we reject the null hypothesis of no correlation. There appears to be a correlation between salary and stress. It seems that persons holding jobs with greater stress levels are compensated with higher salaries.

The hypothesis testing procedure used here is similar to the one used in Section 9-2, except for the calculation of the test statistic and the table used for the critical value. See Figure 13-5, which summarizes this procedure.

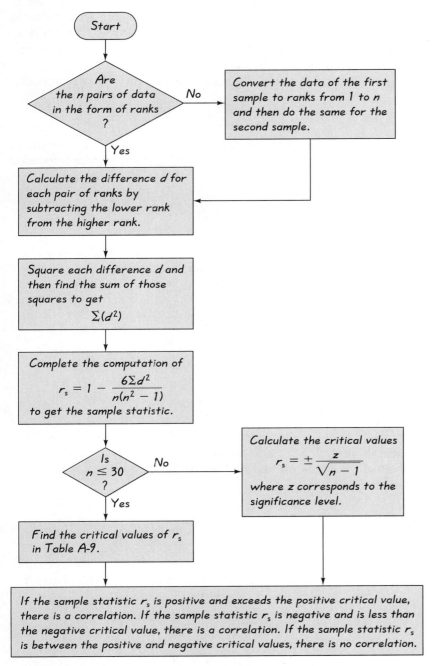

FIGURE 13-5   Rank Correlation H_0: $\rho = 0$, H_1: $\rho \neq 0$

For practical reasons, we are omitting the theoretical derivation of the test statistic r_s, but we can gain some insight by considering the three cases illustrated in Figure 13-6. If we examine the expression for r_s, we see that strong agreement between the two sets of ranks will lead to values of r_s close to 1 (see Case I). Conversely, when the ranks of one set are in an opposite direction from the ranks of the second set, then the values of r_s are near -1 (see Case II). If there is no relationship between the two sets of ranks, then the values of r_s tend to be near 0 (see Case III).

Cases I and II illustrate the most extreme cases, so the following property applies:

$$-1 \le r_s \le 1$$

When the number of pairs of ranks, n, exceeds 30, the sampling distribution of r_s is approximately a normal distribution with mean 0 and standard deviation $1/\sqrt{n-1}$. We therefore get

$$z = \frac{r_s - 0}{\dfrac{1}{\sqrt{n-1}}} = r_s \sqrt{n-1}$$

In a two-tailed case we use the positive and negative z values. Solving for r_s, we get the critical values by evaluating

Formula 13-1 $r_s = \dfrac{\pm z}{\sqrt{n-1}}$ (critical values when $n > 30$)

The value of z corresponds to the significance level.

▶ **EXAMPLE**

Find the critical values of Spearman's rank correlation coefficient r_s when the data consist of 40 pairs of ranks. Assume a two-tailed case with a significance level of $\alpha = 0.05$.

● **SOLUTION**

As there are 40 pairs of data, $n = 40$. Because n exceeds 30, we use Formula 13-1, instead of Table A-9. With $\alpha = 0.05$ in two tails, we let $z = 1.96$ to get

$$r_s = \frac{\pm 1.96}{\sqrt{40-1}} = \pm 0.314$$

Case I: Perfect Positive Correlation

Rank x	Rank y	Difference d	d^2
1	1	0	0
3	3	0	0
5	5	0	0
4	4	0	0
2	2	0	0
		$\Sigma d^2 = 0$	

$$r_s = 1 - \frac{6(0)}{5(5^2 - 1)} = 1$$

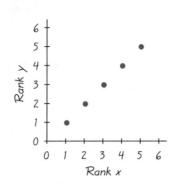

Case II: Perfect Negative Correlation

Rank x	Rank y	Difference d	d^2
1	5	4	16
2	4	2	4
3	3	0	0
4	2	2	4
5	1	4	16
		$\Sigma d^2 = 40$	

$$r_s = 1 - \frac{6(40)}{5(5^2 - 1)} = -1$$

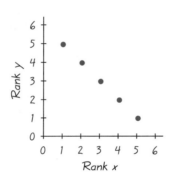

Case III: No Correlation

Rank x	Rank y	Difference d	d^2
1	2	1	1
2	5	3	9
3	3	0	0
4	1	3	9
5	4	1	1
		$\Sigma d^2 = 20$	

$$r_s = 1 - \frac{6(20)}{5(5^2 - 1)} = 0$$

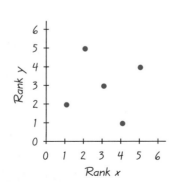

FIGURE 13-6   Positive, Negative, and No Correlation

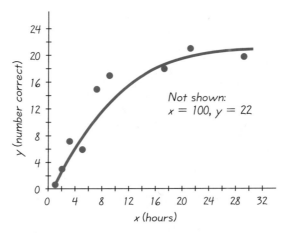

**FIGURE 13-7   Scatter Diagram for Hours Studied
and Correct Answers**

In the next example the data are graphed in the scatter diagram shown in Figure 13-7. The pattern is not linear, but watch what happens.

▶ **EXAMPLE**

Ten students study for a test; the following table lists the number of hours studied (x) and the corresponding number of correct answers (y).

Hours studied (x)	5	9	17	1	2	21	3	29	7	100
Correct answers (y)	6	16	18	1	3	21	7	20	15	22

At the 0.05 level of significance, use Spearman's rank correlation approach to determine whether there is a relationship between hours studied and the number of correct answers.

● **SOLUTION**

We will test the null hypothesis of no rank correlation ($\rho_s = 0$).

$$H_0: \rho_s = 0$$
$$H_1: \rho_s \neq 0$$

Refer to Figure 13-5, which we will follow in this solution. The given data are not ranks, so we convert them into ranks as shown in the table below. (Section 13-1 describes the procedure for converting scores into ranks.)

Ranks for hours studied	4	6	7	1	2	8	3	9	5	10
Ranks for correct answers	3	6	7	1	2	9	4	8	5	10
d	1	0	0	0	0	1	1	1	0	0
d^2	1	0	0	0	0	1	1	1	0	0

After expressing all data as ranks, we calculate the differences, d, and then square them. The sum of the d^2 values is 4. We now calculate

$$r_s = 1 - \frac{6\Sigma d^2}{n(n^2 - 1)}$$

$$= 1 - \frac{6(4)}{10(10^2 - 1)}$$

$$= 1 - 0.024$$

$$= 0.976$$

Proceeding with Figure 13-5, we have $n = 10$, so we answer yes when asked if $n \leq 30$. We use Table A-9 to get the critical values of -0.648 and 0.648. Finally, the sample statistic of 0.976 exceeds 0.648, so we conclude that there is significant correlation. More hours of study appear to be associated with higher grades. (You didn't really think we would suggest otherwise, did you?)

If we compute the linear correlation coefficient r (using Formula 9-1) for the original data in this last example, we get $r = 0.629$, which leads to the conclusion that there is no significant *linear* correlation at the 0.05 level of significance. If we examine the scatter diagram in Figure 13-7, we can see that there does seem to be a relationship, but it's not linear. This last example is intended to illustrate two advantages of the nonparametric approach over the parametric approach. We have already noted the advantage of detecting relationships that are not linear. The example also illustrates this additional advantage: *Spearman's rank correlation coefficient r_s is less sensitive to a value that is very far out of line,* such as the 100 hours in the preceding data.

 ## Using Computer Software for Rank Correlation

STATDISK and Minitab can be used to obtain the rank correlation coefficient.
If you are using STATDISK, select `Correlation and Regression` from the `Statistics` menu and indicate that you want rank correlation results.

Student Ratings of Teachers

Many colleges equate high student ratings with good teaching—an equation often fostered by the fact that student evaluations are easy to administer and measure.

However, one study that compared student evaluations of teachers with the amount of material learned found a strong *negative* correlation between the two factors. Teachers rated highly by students seemed to induce less learning.

In a related study, an audience gave a high rating to a lecturer who conveyed very little information but was interesting and entertaining.

If you are using Minitab, use the command CORRELATION, as described in Section 9-3. If the original data are not in the form of ranks, you can enter the data in columns C1 and C2, but the following two lines must be entered before the CORRELATION command:

```
MTB > RANK C1 AND PUT INTO C3
MTB > RANK C2 AND PUT INTO C4
```

Now enter the command CORRELATION C3 C4 to obtain the rank correlation coefficient. If the data are already in the form of ranks, you need not enter the above two lines. Shown below are the data entry and Minitab display for the preceding example.

MINITAB DISPLAY

```
MTB > READ C1 C2
DATA>   5   6
DATA>   9  16
DATA>  17  18
DATA>   1   1
DATA>   2   3
DATA>  21  21
DATA>   3   7
DATA>  29  20
DATA>   7  15
DATA> 100  22
DATA> ENDOFDATA
MTB > RANK C1 AND PUT INTO C3
MTB > RANK C2 AND PUT INTO C4
MTB > CORRELATION C3 C4

Correlation of C3 and C4 = 0.976
```

13-6 Exercises A: Basic Skills and Concepts

In Exercises 1 and 2, find the critical value(s) for r_s by using either Table A-9 or Formula 13-1, as appropriate. Assume two-tailed cases, where α represents the level of significance and n represents the number of pairs of data.

1. a. $n = 20$, $\alpha = 0.05$
   b. $n = 50$, $\alpha = 0.05$
   c. $n = 40$, $\alpha = 0.02$

2. a. $n = 15$, $\alpha = 0.01$
   b. $n = 82$, $\alpha = 0.04$
   c. $n = 50$, $\alpha = 0.01$

In Exercises 3 and 4, find the value of the rank correlation coefficient r_s.

3.

x	63	68	71	55	70	75
y	43	44	39	30	28	20

4.

x	28	28	35	37	40
y	16	17	12	19	20

In Exercises 5–24, use the rank correlation coefficient to test the claim of no correlation between the two variables. Use a significance level of $\alpha = 0.05$.

5. Table 13-6 in this section includes paired salary and stress level ranks for 10 randomly selected jobs. The physical demands of the jobs were also ranked; the salary and physical demand ranks are given below (based on data from *The Jobs Rated Almanac*).

Salary	2	6	3	5	7	10	9	8	4	1
Physical demand	5	2	3	8	10	9	1	7	6	4

6. Ten jobs were randomly selected and ranked according to stress level and physical demand, with the results given below (based on data from *The Jobs Rated Almanac*).

Stress level	2	7	6	4	5	8	9	10	3	1
Physical demand	5	2	3	8	10	9	1	7	6	4

7. The following table ranks 8 states according to teachers' salaries and students' SAT scores (based on data from the U.S. Department of Education).

	N.Y.	Cal.	Fla.	N.J.	Tex.	N.C.	Md.	Ore.
Teacher's salary	1	2	7	4	6	8	3	5
SAT score	4	3	5	6	7	8	2	1

8. The following table ranks 8 states according to SAT scores and cost per student (based on data from the U.S. Department of Education).

	N.Y.	Cal.	Fla.	N.J.	Tex.	N.C.	Md.	Ore.
SAT score	4	3	5	6	7	8	2	1
Cost per student	1	5	6	2	7	8	3	4

9. In studying the effects of heredity and environment on intelligence, scientists have learned much by analyzing the IQ scores of identical twins who were separated soon after birth. Identical twins have identical genes, which they inherited from the same fertilized egg. By studying identical twins raised apart, we can eliminate the variable of heredity and better isolate the effects of environment. Following are the IQ scores of identical twins (first-born twins are x) raised apart.

x	107	96	103	90	96	113	86	99	109	105	96	89
y	111	97	116	107	99	111	85	108	102	105	100	93

Based on data from "IQ's of Identical Twins Reared Apart," by Arthur Jensen, *Behavioral Genetics*.

10. *Consumer Reports* tested VHS tapes used in VCRs. Following are performance scores and prices (in dollars) of randomly selected tapes.

Performance	91	92	82	85	87	80	94	97
Price	4.56	6.48	5.99	7.92	5.36	3.32	7.32	5.27

11. In Chapter 9 we analyzed the data in the following table.

Plastic	0.27	1.41	2.19	2.83	2.19	1.81	0.85	3.05
Household size	2	3	3	6	4	2	1	5

Data provided by Masakazu Tani and the Garbage Project at the University of Arizona.

12. The following paired data consist of weights (in pounds) of discarded paper and sizes of households.

Paper	2.41	7.57	9.55	8.82	8.72	6.96	6.83	11.42
Household size	2	3	3	6	4	2	1	5

Sample data obtained from the Garbage Project at the University of Arizona.

13. A study was conducted to investigate any relationship between age (in years) and BAC (blood alcohol concentration) measured when convicted DWI jail inmates were first arrested. Sample data for randomly selected subjects are as follows.

Age	17.2	43.5	30.7	53.1	37.2	21.0	27.6	46.3
BAC	0.19	0.20	0.26	0.16	0.24	0.20	0.18	0.23

Based on data from the Dutchess County STOP-DWI Program.

14. Randomly selected subjects ride a bicycle at 5.5 mi/h for one minute. Their weights (in pounds) are given with the numbers of calories used during their bike rides.

Weight	167	191	112	129	140	173	119
Calories used	4.23	4.69	3.21	3.47	3.72	4.45	3.36

Based on data from *Diet Free*, by Kuntzlemann.

15. In a study of the factors that affect success in a calculus course, data were collected for 10 different persons. Scores on an algebra placement test are given below, along with calculus achievement scores.

Algebra score	17	21	11	16	15	11	24	27	19	8
Calculus score	73	66	64	61	70	71	90	68	84	52

Based on data from "Factors Affecting Achievement in the First Course in Calculus," by Edge and Friedberg, *Journal of Experimental Education*, Vol. 52, No. 3.

16. The accompanying table lists the number of registered automatic weapons (in thousands), along with the murder rate (in murders per 100,000), for randomly selected states.

Automatic weapons	11.6	8.3	3.6	0.6	6.9	2.5	2.4	2.6
Murder rate	13.1	10.6	10.1	4.4	11.5	6.6	3.6	5.3

Data provided by the FBI and the Bureau of Alcohol, Tobacco, and Firearms.

17. Emissions data for a sample of different vehicles are given for HC (hydrocarbon) and CO (carbon monoxide), with both measured in grams per meter.

HC	0.65	0.55	0.72	0.83	0.57	0.51	0.43	0.37
CO	14.7	12.3	14.6	15.1	5.0	4.1	3.8	4.1

Based on data from "Determining Statistical Characteristics of a Vehicle Emissions Audit Procedure," by Lorenzen, *Technometrics*, Vol. 22, No. 4.

18. There are many regions where the winter accumulation of snowfall is a primary source of water. Several investigations of snowpack characteristics have used satellite observations from the Landsat series, along with measurements taken on earth. Given here are ground measurements of snow depth (in centimeters), along with the corresponding temperatures (in degrees Celsius).

Temperature	−62	−41	−36	−26	−33	−56	−50	−66
Snow depth	21	13	12	3	6	22	14	19

Data based on information in Kastner's *Space Mathematics*, published by NASA.

19. The following table lists per capita cigarette consumption in the United States for various years, along with the percentage (in percentage points) of the population admitted to mental institutions as psychiatric patients.

Cigarette consumption	3522	3597	4171	4258	3993	3971	4042	4053
Psychiatric patients	0.20	0.22	0.23	0.29	0.31	0.33	0.33	0.32

20. At one point in a recent season of the National Basketball Association, *USA Today* reported the current statistics. Following are the total minutes played and the total points scored for 9 randomly selected players.

Minutes	1364	53	457	717	384	1432	365	1626	840
Points	652	20	163	210	175	821	143	1098	459

21. Refer to Data Set 3 in Appendix B. For men who never married, use the paired data consisting of heights and incomes.

22. Refer to Data Set 3 in Appendix B. For women who never married, use the paired data consisting of length of education and income.

23. Refer to Data Set 4 in Appendix B. Does there appear to be correlation between the weights of cars and their lengths?

24. Refer to Data Set 4 in Appendix B. Does there appear to be correlation between the weights of cars and their highway fuel consumption amounts?

13-6   Exercises B: Beyond the Basics

25. Two judges each rank 3 contestants, and the ranks for the first judge are 1, 2, and 3, respectively.
    a. List all possible ways that the second judge can rank the same 3 contestants. (No ties allowed.)
    b. Compute r_s for each of the cases found in part a.
    c. Assuming that all of the cases from part a are equally likely, find the probability that the sample statistic, r_s, is greater than 0.9.

26. One alternative to using Table A-9 involves an approximation of critical values for r_s given as

$$r_s = \pm \sqrt{\frac{t^2}{t^2 + n - 2}}$$

Here t is the t score from Table A-3 corresponding to the significance level and $n - 2$ degrees of freedom. Apply this approximation to find critical values of r_s for the following cases.

    a. $n = 8$, $\alpha = 0.05$ b. $n = 15$, $\alpha = 0.05$
    c. $n = 30$, $\alpha = 0.05$ d. $n = 30$, $\alpha = 0.01$
    e. $n = 8$, $\alpha = 0.01$

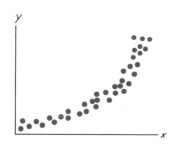

27. a. Given the bivariate data depicted in the scatter diagram, which would be more likely to detect the relationship between x and y: the linear correlation coefficient r or the rank correlation coefficient r_s? Explain.
    b. How is r_s affected if one variable is ranked from low to high while the other variable is ranked from high to low?
    c. One researcher ranks both variables from low to high, while another researcher ranks both variables from high to low. How will their values of r_s compare?
    d. Using the job salary/stress level data given in Table 13-6, test the claim that there is a *positive* correlation; that is, test the claim that $\rho_s > 0$. Use a 0.05 significance level.

13-7   Runs Test for Randomness

A classic example of the misuse of statistics involves a company president who was convinced that employees were stealing some of the pantyhose being produced. Further investigation of the production figures on which the president had based that assumption showed that the figures were derived from samples obtained with newly serviced machinery, which produced a finer mesh and more pantyhose. The production level dropped when the machinery became worn; there was no employee theft. The initial sampling with only newly serviced machinery was not random, and it led to misleading results and embarrassment for the president, who had proclaimed that pantyhose were being pilfered.

In many of the examples and exercises in this book we have assumed that data were randomly selected. In this section we describe the **runs test,** which is a systematic and standard procedure for testing the randomness of data. (Note that the runs test of this section is different from the run chart described in Section 12-2.)

DEFINITION

A **run** is a sequence of data that exhibit the same characteristic; the sequence is preceded and followed by different data or no data at all.

As an example, consider the cola choice of consumers in a market research project. D denotes a consumer who prefers *diet* cola, whereas R indicates a consumer who prefers *regular* cola. The following sequence contains exactly four runs.

$$\underbrace{D\,D\,D\,D}_{\text{1st run}} \quad \underbrace{R\,R}_{\text{2nd run}} \quad \underbrace{D\,D\,D}_{\text{3rd run}} \quad \underbrace{R}_{\text{4th run}}$$

We use the runs test in this situation to test for the randomness with which consumer preferences for diet and regular cola occur.

Let's use common sense to see how runs relate to randomness. Examine the following sequence, and then stop to consider how randomly diet and regular occur. Also count the number of runs.

D D D D D D D D D R R R R R R R R R R

It is reasonable to conclude here that diet and regular occur in a sequence that is *not* random. Note that in this sequence of 20 data, there are only 2 runs. This pattern suggests that if the number of runs is very low, randomness may be lacking.

Now consider the following sequence of 20 data. Try again to form your own conclusion about randomness before you continue reading.

D R D R D R D R D R D R D R D R D R D R

It should be apparent that the sequence of diet and regular is again *not* random because there is a distinct, predictable pattern. In this sample, the number of runs is 20. This pattern suggests that randomness is lacking when the number of runs is high.

Note that **this test for randomness is based on the *order* in which the data occur. This runs test is *not* based on the *frequency* of the data.** For example, a

Sports Hot Streaks

It is a common belief that athletes often have "hot streaks"—that is, brief periods of extraordinary success. Stanford University psychologist Amos Tversky and other researchers used statistics to analyze the thousands of shots taken by the Philadelphia 76ers for one full season and half of another. They found that the number of "hot streaks" was no different than you would expect from random trials with the outcome of each trial independent of any preceding results. That is, the probability of a hit doesn't depend on the preceding hit or miss.

particular sequence containing 3 men and 20 women might lead to the conclusion that the sequence is random. The issue of whether 3 men and 20 women constitute a *biased* sample is not addressed by the runs test.

The two sequences given above are obvious in their lack of randomness, but most sequences are not so obvious; we therefore need more sophisticated techniques for analysis. We begin by introducing some notation.

Notation

n_1 = number of elements in the sequence
that have the same characteristic

n_2 = number of elements in the sequence
that have the other characteristic

G = number of runs

We use G to represent the number of runs because n and r have already been used for other statistics, and G is a relatively innocuous letter that deserves more attention. For example, the sequence

$$D \quad D \quad D \quad D \quad R \quad R \quad D \quad D \quad D \quad R \quad R \quad R \quad R \quad D$$

results in the following values for n_1, n_2, and G.

$n_1 = 8$ because there are 8 diet cola consumers
$n_2 = 6$ because there are 6 regular cola consumers
$G = 5$ because there are 5 runs

We continue to use our standard procedure for hypothesis testing. **The null hypothesis H_0 is the claim that the sequence is random;** the alternative hypothesis H_1 will be the claim that the sequence is *not* random. The flowchart in Figure 13-8 summarizes the mechanics of the procedure. This flowchart directs us to a table (Table A-10) of critical G values when the following three conditions are all met:

1. $\alpha = 0.05$, and

2. $n_1 \leq 20$, and

3. $n_2 \leq 20$.

If one or more of these conditions is not satisfied, we use the property that G has a distribution that is approximately normal with mean and standard deviation as follows:

Formula 13-2
$$\mu_G = \frac{2n_1n_2}{n_1 + n_2} + 1$$

Formula 13-3
$$\sigma_G = \sqrt{\frac{(2n_1n_2)(2n_1n_2 - n_1 - n_2)}{(n_1 + n_2)^2(n_1 + n_2 - 1)}}$$

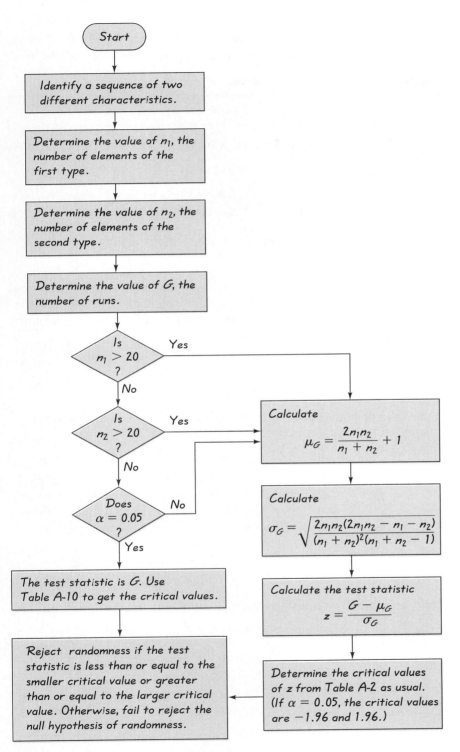

FIGURE 13-8 **Runs Test for Randomness**

For a test of the null hypothesis of randomness, the test statistic is G if the three conditions listed above are satisfied. Otherwise, the test statistic is z, calculated by using μ_G and σ_G.

Test Statistic for the Runs Test for Randomness

For $\alpha = 0.05$ and $n_1 \leq 20$ and $n_2 \leq 20$: G

For $\alpha \neq 0.05$ or $n_1 > 20$ or $n_2 > 20$: $z = \dfrac{G - \mu_G}{\sigma_G}$

Critical values: If the test statistic is G, critical values are found in Table A-10. If the test statistic is z, critical values are found in Table A-2 by using the same procedures introduced in Chapter 6.

The normal approximation (with test statistic z) is quite good. If the entire table of critical values (Table A-10) had been computed using this normal approximation, no critical value would be off by more than one unit.

We now illustrate the use of the runs test for randomness by presenting examples of complete tests of hypotheses.

▶ **EXAMPLE**

The president of an investment firm has observed that men and women have been hired in the following sequence:

$$\text{M} \quad \text{M} \quad \text{M} \quad \text{W} \quad \text{M} \quad \text{M} \quad \text{M} \quad \text{M} \quad \text{W} \quad \text{W} \quad \text{W} \quad \text{M}$$

At the 0.05 level of significance, test the personnel officer's claim that the sequence of men and women is random. (Note that we are not testing for a *bias* in favor of one sex over the other. There are 8 men and 4 women, but we are testing only for *randomness* in the way they appear in the given sequence.)

● **SOLUTION**

We follow the same hypothesis testing procedure outlined in Figure 7-4.

Steps 1, 2, and 3:   The null hypothesis is the claim of randomness, so

$$H_0: \text{The 8 men and 4 women have}$$
$$\text{been hired in a random sequence.}$$
$$H_1: \text{The sequence is not random.}$$

Step 4:   The significance level is $\alpha = 0.05$.

Step 5:   We use the runs test procedure in Figure 13-8 because we are testing for randomness.

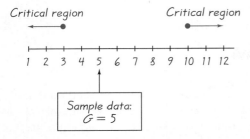

**FIGURE 13-9 Test Statistic and Critical Region for
Test of Randomness**

Step 6: Figure 13-8 indicates that to find the value of the test statistic, we must first find the values of n_1 and n_2 and the number of runs G.

$$n_1 = \text{number of men} = 8$$
$$n_2 = \text{number of women} = 4$$
$$G = \text{number of runs} = 5$$

Continuing with Figure 13-8, we answer no when asked if $n_1 > 20$ (because $n_1 = 8$), no when asked if $n_2 > 20$ (because $n_2 = 4$), and yes when asked if $\alpha = 0.05$. The test statistic is $G = 5$, and we refer to Table A-10 to find the critical values of 3 and 10.

Step 7: Figure 13-9 shows that the test statistic $G = 5$ does not fall in the critical region. We therefore fail to reject the null hypothesis that the given sequence of men and women is random.

Step 8: There is not sufficient evidence to warrant rejection of the claim that the sequence is random.

The next example will illustrate the same procedure when Table A-10 cannot be used and the normal approximation must be used instead.

▶ **EXAMPLE**

Refer to the sequence of 1s and 0s given in Figure 13-1. Recall that the sequence was designed to be a radio message that could be translated into a picture packed with information about us earthlings. If an intelligent extraterrestrial life form intercepted the message and used the runs test for randomness, would the sequence of 1s and 0s appear to be random and the message mistaken for "random noise"? Use a 0.05 significance level in testing the sequence.   *(continued)*

● **SOLUTION**

The null and alternative hypotheses are as follows:

H_0: The sequence is random. H_1: The sequence is not random.

The significance level is $\alpha = 0.05$; we are using the runs test for randomness. The test statistic is obtained by first finding the number of 1s, the number of 0s, and the number of runs. After painstakingly analyzing the sequence, we find that

$$n_1 = \text{number of 1s} = 249$$
$$n_2 = \text{number of 0s} = 1022$$
$$G = \text{number of runs} = 353$$

As we follow Figure 13-8, we answer yes to "Is $n_1 > 20$?" We therefore need to evaluate μ_G and σ_G before we can determine the test statistic. We get

$$\mu_G = \frac{2n_1 n_2}{n_1 + n_2} + 1 = \frac{2(249)(1022)}{249 + 1022} + 1 = 401.437$$

$$\sigma_G = \sqrt{\frac{(2n_1 n_2)(2n_1 n_2 - n_1 - n_2)}{(n_1 + n_2)^2 (n_1 + n_2 - 1)}}$$

$$= \sqrt{\frac{(2)(249)(1022)[2(249)(1022) - 249 - 1022]}{(249 + 1022)^2 (249 + 1022 - 1)}} = 11.223$$

We can now find the test statistic.

$$z = \frac{G - \mu_G}{\sigma_G} = \frac{353 - 401.437}{11.223} = -4.32$$

Because the significance level is $\alpha = 0.05$ and we have a two-tailed test, the critical values are $z = -1.96$ and $z = 1.96$. The test statistic of $z = -4.32$ does fall within the critical region, so we reject the null hypothesis of randomness. The given sequence does not appear to be random, which is good news for those who developed the radio message because it implies that the message would not be passed off as random noise.

Randomness Above and Below the Mean or Median

In each of the preceding examples, the data clearly fit into two categories, but we can also test for randomness in the way numerical data fluctuate above or below a mean or median. That is, in addition to analyzing sequences of nominal data of the type already discussed, we can also analyze sequences of data at the interval or ratio levels of measurement, as in the following example.

► **EXAMPLE**

An investor wants to analyze long-term trends in the stock market. The annual high points of the Dow Jones Industrial Average for a recent sequence of 10 years are given below.

  1287   1553   1956   2722   2184   2791   3000   3169   3413   3794

At the 0.05 significance level, test for randomness above and below the median.

● **SOLUTION**

For the given data, the median is 2756.5. Let B denote a value *below* the median of 2756.5, and let A represent a value *above* 2756.5. We can rewrite the given sequence as follows. (If a value is equal to the median, we simply delete it from the sequence.)

B	B	B	B	B	A	A	A	A	A
↓	↓	↓	↓	↓	↓	↓	↓	↓	↓
1287	1553	1956	2722	2184	2791	3000	3169	3413	3794

It is helpful to write the As and Bs directly above the numbers they represent. This makes checking easier and also reduces the chance of having the wrong number of letters. After finding the sequence of letters, we can proceed to apply the runs test of the null hypothesis that the sequence is random. We get

$$n_1 = \text{number of Bs} = 5$$
$$n_2 = \text{number of As} = 5$$
$$G = \text{number of runs} = 2$$

From Table A-10, the critical values of G are found to be 2 and 10. At the 0.05 level of significance, we reject the null hypothesis of randomness above and below the median if the number of runs is 2 or lower or 10 or more. Because $G = 2$, we reject the null hypothesis of randomness above and below the median. It isn't necessary to be a financial wizard to recognize that the yearly high levels of the Dow Jones Industrial Average appear to be following an upward trend. This example illustrates one way to recognize such trends, especially those that are not so obvious.

Random Walk and Stocks

In the book *A Random Walk Down Wall Street*, Burton Malkiel states, "When the term (random walk) is applied to the stock market, it means that short-run changes in stock prices cannot be predicted. Investment advisory services, earnings predictions, and complicated chart patterns are useless." He also writes that "taken to its logical extreme, it means that a blindfolded monkey throwing darts at a newspaper's financial pages could select a portfolio that would do just as well as one carefully selected by the experts." This suggests that investors could save money by no longer securing the expensive advice of financial consultants. They can simply select stocks in a way that is random.

We could also test for randomness above and below the *mean* by following the same procedure (after we have deleted from the sequence any values that are equal to the mean).

Economists use the runs test for randomness above and below the median in an attempt to identify trends or cycles. An upward economic trend would contain a predominance of Bs at the beginning and As at the end, so the number of runs would be small, as in the preceding example. A downward trend would have As dominating the beginning and Bs the end, with a low number of runs. A cyclical pattern would yield a sequence that systematically changes, so the number of runs would tend to be large.

13-7   Exercises A: Basic Skills and Concepts

In Exercises 1 and 2, use the given sequence to determine the values of n_1 and n_2, the number of runs, G, and the appropriate critical values from Table A-10. Use a 0.05 significance level.

1. A   A   A   A   A   A   A   B   B

2. A   B   B   B   A   B   B   B   B   B
A   B   A   A   A   A   A   B   A   A

In Exercises 3 and 4, use the given sequence to answer the following:
  a.   Find the median.
  b.   Let B represent a value below the median, and let A represent a value above the median. Rewrite the given numerical sequence as a sequence of As and Bs.
  c.   Find the values of n_1, n_2, and G.
  d.   Assuming a 0.05 level of significance, use Table A-10 to find the appropriate critical values of G.
  e.   What do you conclude about the randomness of the values above and below the median?

3. 3   8   7   7   9   12   10   16   20   18

4. 2   2   3   2   4   4   5   8   4   6   7   9   9   12

In Exercises 5–16, use the runs test to determine whether the given sequence is random. Use a significance level of $\alpha = 0.05$. (All data are listed in order by row.)

5. The Dallas Manufacturing Company uses a machine to produce surgical knives that must meet certain specifications. When a sample of knives were examined and judged to be defective (D) or acceptable (A), the following results were obtained.

D   A   A   A   A   D   A   A   A   A   A   D   A   A   A   A
D   A   A   A   A   D   A   A   A   A   A   D   A   A   A   A

6. A test for randomness of odd (O) and even (E) numbers on a roulette wheel yielded the results given in the following sequence.

O   E   O   O   O   O   E   E   O   O   O   E   O   O   O
O   O   O   E   E   E   O   O   O   E   O   O   O

7. The Flint Fabric Company supplies the major car manufacturers with upholstery fabric. The thicknesses of glue (in millimeters) applied to the upholstery are listed below for samples randomly selected once each half hour. Test for randomness above and below the median. The data are listed in order by rows.

0.17 0.18 0.16 0.19 0.17 0.18 0.16 0.17 0.16 0.14
0.15 0.12 0.09 0.08 0.06 0.08 0.07 0.06 0.07 0.08

8. The Apgar rating scale is used to rate the health of newborn babies. A sequence of births results in the following values. Test the claim that the ratings above and below the sample mean are random.

9 6 9 8 7 9 5 8 9 10 8 9 7 5 9 8 8

9. The numbers of housing starts in recent years in Orange County are as follows. Test the claim of randomness of those values above and below the *median*.

273 1067 856 971 1456 903 899 905
812 630 720 676 731 655 598 617

10. The gold reserves of the United States (in millions of fine troy ounces) for a recent 12-year sequence are given below (based on data from the International Monetary Fund). Test for randomness above (A) and below (B) the median. (The data are arranged in chronological order by rows.)

274.71 274.68 277.55 276.41 264.60 264.32
264.11 264.03 263.39 262.79 262.65 262.04

11. A nurse at the Millard Fillmore Elementary School records the grade levels of students who develop chicken pox; the sequential list is given here.

3 3 3 2 3 1 4 3 6 5 3 3 2 1 4
4 4 4 4 4 4 4 4 5 5 5 5 4 3 2
1 6 6 6 6 1 3 4 2 5 6 6 6 6 6

Consider children in grades 1, 2, and 3 to be younger and those in grades 4, 5, and 6 to be older. Test the claim that the sequence is random with respect to the categories of younger and older.

12. A teacher develops a true/false test with these answers:

T F F F F T T T T T T T F F F F T T T T
T T T F F F F F F F F F T T T T T T F
F F

Test the claim that the sequence of answers is random.

13. Test the claim that the sequence of World Series wins by American League and National League teams is random. Given below are recent results, with American and National League teams represented by A and N, respectively.

A N A N N N A A A A N A A A A N A N
N A A N N A A A A N A N N A A A A A
N A N A N A N A A A A A A N N A N
A N N A A N N N A N A N A N A A A N
N A A N N N N A A A N A N A N A A A

14. A sequence of dates is selected through a process that has the outward appearance of being random. Test the resulting sequence for randomness before and after the middle of the year.

Nov. 27	July 7	Aug. 3	Oct. 19	Dec. 19	Sept. 21	Apr. 1
Mar. 5	June 10	May 21	June 27	Jan. 5		

15. A *New York Times* article about the calculation of decimal places of π noted that "mathematicians are pretty sure that the digits of π are indistinguishable from any random sequence." Given below are the first 100 decimal places of π. Test for randomness of odd (O) and even (E) digits.

1	4	1	5	9	2	6	5	3	5	8	9	7	9	3	2	3	8	4	6
2	6	4	3	3	8	3	2	7	9	5	0	2	8	8	4	1	9	7	1
6	9	3	9	9	3	7	5	1	0	5	8	2	0	9	7	4	9	4	4
5	9	2	3	0	7	8	1	6	4	0	6	2	8	6	2	0	8	9	9
8	6	2	8	0	3	4	8	2	5	3	4	2	1	1	7	0	6	7	9

16. Use the 100 decimal digits for π given in Exercise 15. Test for randomness above (A) and below (B) the value of 4.5.

13-7   Exercises B: Beyond the Basics

17. If you use the elements A, A, B, B, what is the minimum number of possible runs that can be arranged? What is the maximum number of runs? Now refer to Table A-10 to find the critical G values for $n_1 = n_2 = 2$. What do you conclude about this case?

18. Let $z = 1.96$ and $n_1 = n_2 = 20$, and then compute μ_G and σ_G. Use those values in

$$z = \frac{G - \mu_G}{\sigma_G}$$

   to solve for G. What is the importance of this result? How does this result compare to the corresponding value found in Table A-10? How do you explain any discrepancy?

19. a. Using all of the elements A, A, A, B, B, B, B, B, B, list the 84 different possible sequences.
    b. Find the number of runs for each of the 84 sequences.
    c. Assuming that each sequence has a probability of 1/84, find $P(2$ runs$)$, $P(3$ runs$)$, $P(4$ runs$)$, and so on.
    d. Use the results of part c to establish your own critical G values in a two-tailed test with 0.025 in each tail.
    e. Compare your results to those given in Table A-10.
    f. Assuming that the 84 sequences from part a are all equally likely, use the results from part b to find the mean number of runs. Compare your result to the result obtained by using Formula 13-2.

20. If you use all of the elements A, A, A, B, B, B, B, B, B, B, B, B, it is possible to arrange 220 different sequences.

    a.   List all of those sequences having exactly 3 runs.

    b.   Using your result from part a, find P(3 runs).

    c.   Using your answer to part b, determine whether G = 3 should be in the critical region.

    d.   Find the lower critical value from Table A-10.

    e.   Find the lower critical value by using the normal approximation.

VOCABULARY LIST

Define and give an example of each term.

parametric tests	Mann-Whitney U test
nonparametric tests	Kruskal-Wallis test
distribution-free tests	H test
efficiency	rank correlation coefficient
rank	Spearman's rank correlation
sign test	coefficient
Wilcoxon signed-ranks test	runs test
Wilcoxon rank-sum test	run

REVIEW

In this chapter we examined six different nonparametric methods for analyzing sample data. Besides excluding involvement with population parameters such as the mean μ and standard deviation σ, nonparametric methods are not encumbered by many of the restrictions placed on parametric methods, such as the requirement that data come from normally distributed populations. Although nonparametric methods are generally not as efficient as their parametric counterparts, they can be used in a wider variety of circumstances and can accommodate more types of data. Also, the computations required in nonparametric tests are generally much simpler than the computations required in the corresponding parametric tests.

Table 13-7 lists the nonparametric tests presented in this chapter, along with their functions. The table also lists the corresponding parametric tests.

TABLE 13-7   Summary of Nonparametric Tests

Nonparametric Test	Function	Parametric Test
Sign test (Section 13-2)	Test for claimed value of average with one sample	t test or z test (Sections 7-2, 7-3, 7-4)
	Test for difference between two dependent samples	t test or z test (Section 8-2)
Wilcoxon signed-ranks test (Section 13-3)	Test for difference between two dependent samples	t test or z test (Section 8-2)
Wilcoxon rank-sum test (Section 13-4)	Test for difference between two independent samples	t test or z test (Sections 8-3, 8-5)
Kruskal-Wallis test (Section 13-5)	Test for more than two independent samples coming from identical populations	Analysis of variance (Sections 11-2, 11-3)
Rank correlation (Section 13-6)	Test for relationship between two variables	Linear correlation (Section 9-2)
Runs test (Section 13-7)	Test for randomness of sample data	(No parametric test)

Important Formulas

Test	Test Statistic	Critical Values
Sign test	x = number of times the less frequent sign occurs Test statistic when $n > 25$: $$z = \frac{(x + 0.5) - \left(\dfrac{n}{2}\right)}{\dfrac{\sqrt{n}}{2}}$$	If $n \leq 25$, see Table A-7. If $n > 25$, use the normal distribution (Table A-2).

Test	Test Statistic	Critical Values
Wilcoxon signed-ranks test for two dependent samples	T = smaller of the rank sums Test statistic when $n > 30$: $$z = \dfrac{T - \dfrac{n(n+1)}{4}}{\sqrt{\dfrac{n(n+1)(2n+1)}{24}}}$$	If $n \leq 30$, see Table A–8. If $n > 30$, use the normal distribution (Table A–2).
Wilcoxon rank-sum test for two independent samples	$$z = \dfrac{R - \mu_R}{\sigma_R}$$ where R = sum of ranks of the sample with size n_1 $$\mu_R = \dfrac{n_1(n_1 + n_2 + 1)}{2}$$ $$\sigma_R = \sqrt{\dfrac{n_1 n_2(n_1 + n_2 + 1)}{12}}$$	Normal distribution (Table A-2) (Requires that $n_1 > 10$ and $n_2 > 10$.)
Kruskal-Wallis test	$$H = \dfrac{12}{N(N+1)}\left(\dfrac{R_1^2}{n_1} + \dfrac{R_2^2}{n_2} + \cdots + \dfrac{R_k^2}{n_k}\right) - 3(N+1)$$ where N = total number of sample values R_k = sum of ranks for the kth sample k = number of samples	Chi-square (Table A-4) with $k - 1$ degrees of freedom (Requires that each sample have at least five values.)
Rank correlation	$$r_S = 1 - \dfrac{6\Sigma d^2}{n(n^2 - 1)}$$	If $n \leq 30$, use Table A-9. If $n > 30$, critical values are $$r_S = \dfrac{\pm z}{\sqrt{n - 1}}$$ where z is from the normal distribution.
Runs test for randomness	Test statistic when $n > 20$: $$z = \dfrac{G - \mu_G}{\sigma_G}$$ where $\mu_G = \dfrac{2n_1 n_2}{n_1 + n_2} + 1$ $$\sigma_G = \sqrt{\dfrac{(2n_1 n_2)(2n_1 n_2 - n_1 - n_2)}{(n_1 + n_2)^2(n_1 + n_2 - 1)}}$$	If $n_1 \leq 20$ and $n_2 \leq 20$, use Table A-10. If $n_1 > 20$ or $n_2 > 20$, use the normal distribution (Table A-2).

REVIEW EXERCISES

In Exercises 1–24, use the indicated test. If no particular test is specified, use the appropriate nonparametric test from this chapter.

1. A medical researcher at the National Institute for Health selected blood samples and tested for the presence (P) or absence (A) of an influenza virus. Test the claim that the following sequence of results (in order by row) is random.

    A A A A A A P P A A A A A
    A A A A P P P A A A A A

2. A biomedical researcher at the Medassist Pharmaceutical Company wanted to test the effectiveness of a synthetic antitoxin. Twelve randomly selected subjects were tested for resistance to bee stings. They were retested after receiving the antitoxin, with the results given below.

Before	18.2 21.6 23.5 22.9 16.3 19.2 21.6 21.8 20.3 19.5 18.9 20.3
After	18.4 20.3 21.5 20.2 17.6 18.5 21.7 22.3 19.4 18.6 20.1 19.7

    At the 0.05 level of significance, use the sign test to test the claim that the antitoxin is not effective and produces no change.

3. A ranking of randomly selected cities according to population density (people per square mile) and crime rate (crimes per 100,000 people) yielded the results given below (based on data from USA Today).

	Austin	Long Beach	San Diego	Detroit	Baltimore	Tampa	Boston
Population density	1	4	3	5	6	2	7
Crime rate	4	2	1	6	3	7	5

    At the 0.05 level of significance, use the rank correlation coefficient to determine whether there is a correlation.

4. A study was made of the response time of police cars after dispatching occurs. Sample results for 3 precincts are given below in seconds. At the 0.05 level of significance, test the claim that Precincts 1 and 2 have the same response times.

Precinct 1	Precinct 2
160 172 176 176 178 191	165 174 180 181 184 186
183 177 173 179 180 185	190 200 176 192 195 201

Precinct 3
162 175 177 179 187 195
210 215 216 220 222

5. Using the sample data given in Exercise 4, test the claim that the 3 precincts have the same response times.

6. The York School District plans to buy IQ tests to be administered to students entering high school. Two different companies have bid on the IQ testing contract, and school officials want to determine whether both tests produce the same results. Both tests were given to a sample of 9 randomly selected students, with the results given below.

Test A	100	111	93	92	99	85	117	110	98
Test B	106	112	95	90	107	100	126	105	110

At the 0.05 level of significance, use the Wilcoxon signed-ranks approach to determine whether both tests produce the same results.

7. Repeat Exercise 6, using the sign test.
8. To test the effect of smoking on pulse rate, a researcher at the Vermont Institute for Better Health compiled data consisting of pulse rates before and after smoking. The results are given below.

Before smoking	68	72	69	70	70	74	66	71
After smoking	69	76	68	73	72	76	66	71

At the 0.05 level of significance, use the sign test to test the claim that smoking does not affect pulse rate.

9. A police academy gives an entrance exam; sample results for applicants from two different counties follow. Use the Wilcoxon rank-sum test to test the claim that there is no difference between the scores from the two counties. Assume a significance level of 0.05.

Orange County	Westchester County
63 39 26 14 75 60	54 35 39 27 40 78
62 79 86 70 66	177 5 10 48 50 49

10. Final exam grades are listed below according to the order (by row) in which exams were completed.

45 50 92 87 79 89 93 75 76 74 76
73 65 68 69 70 78 60 60 60 55 100

At the 0.05 level of significance, test for randomness above and below the mean.

11. Several cities were ranked according to the number of hotel rooms and the amount of office space. The results are as follows.

	NY	Ch	SF	Ph	LA	At	Mi	KC	NO	Da	Ba	Bo	Se	Ho	SL
Office rank	1	3	2	7	6	10	14	15	11	9	12	4	8	5	13
Hotel rank	1	2	3	8	6	7	14	15	4	10	13	5	9	11	12

With a 0.05 significance level, use rank correlation to test for a relationship between office space and hotel rooms.

Treatment		
A	B	C
1	2	3
6	4	5
7	8	9
12	11	10
14	15	13

Newspaper	Radio	None
845	811	612
907	782	574
639	749	539
883	863	641
806	872	666

12. The Grange Fertilizer Company conducted an experiment of raising samples of corn under conditions identical except for the type of fertilizer used. The yields were obtained for 3 different fertilizers, and those values were ranked, with the results shown here. Does the type of fertilizer have an effect on the yield?

13. Randomly selected cars were tested for fuel consumption and then retested after an engine tune-up. The measures of fuel consumption follow.

Before tune-up	16	23	12	13	7	31	27	18	19	19	19	11	9	15
After tune-up	18	23	16	17	8	29	31	21	19	20	24	13	14	18

At the 0.05 significance level, use the Wilcoxon signed-ranks test to test the claim that the tune-up has no effect on fuel consumption.

14. Refer to the data given in Exercise 13. At the 0.05 level of significance, use rank correlation to test for a relationship between the before and after values.

15. Refer to the data given in Exercise 13. At the 0.05 level of significance, use the sign test to test the claim that the tune-up has no effect on fuel consumption.

16. The Newton Computer Store owner recorded the gross receipts for days randomly selected from periods during which she used only newspaper advertising, only radio advertising, or no advertising. The results are listed here. At the 0.05 level of significance, test the claim that the receipts are the same, regardless of advertising. Based on these results, does advertising seem to affect the gross receipts?

17. A consumer investigator obtained prices from mail order companies and computer stores. Listed below are the prices (in dollars) quoted for cartons of floppy disks from various manufacturers.

Mail order	Computer store
23.00 26.00 27.99 31.50	30.99 33.98 37.75 38.99
32.75 27.00 27.98 24.50	35.79 33.99 34.79 32.99
24.75 28.15 29.99 29.99	29.99 33.00 32.00

Use the Wilcoxon rank-sum test with a 0.05 level of significance to test the claim that there is no difference between mail order and store prices.

18. A pollster was hired to collect data from 30 randomly selected adults. As the data were turned in, the sexes of the interviewed subjects were noted and the sequence below was obtained.

M M M M M M M M F M M M M F F F
F F F F F F M M M M M M M M

At the 0.05 level of significance, test the claim that the sequence is random.

19. Three methods of instruction were used to train air traffic controllers. With Method A, an experienced controller was assigned to a trainee for practical field experience. With Method B, trainees were given extensive classroom instruction and then placed without supervision. Method C required a moderate amount of classroom training, and then several trainees were supervised on the job by an experienced instructor. A standardized test was used to measure levels of competency, and sample results are given to the right. At the 0.01 level of significance, test the claim that the three methods are equally effective.

Method		
A	B	C
195	187	193
198	210	212
223	222	215
240	238	231
251	256	252
		260
		267

20. The Medassist Pharmaceutical Company conducted a study to determine whether a drug affects eye movements. A standardized scale was developed, and the drug was administered to one group, while a control group was given a placebo that produces no effects. The eye movement ratings of subjects are as follows:

Drugged group	Control group
652 512 711 621 508 603 787	674 676 821 830 565 821 837 652 549
747 516 624 627 777 729	668 772 563 703 789 800 711 598

At the 0.01 level of significance, test the claim that the drug has no effect on eye movements. Use the Wilcoxon rank-sum test.

21. Two judges graded entries in a science fair. The grades for 8 randomly selected contestants are given below.

Judge A	6.3 7.2 6.6 8.5 9.7 7.0 7.3 8.8
Judge B	7.1 6.5 8.2 8.6 9.0 6.1 6.3 8.8

Use the data to test the claim that there is no difference between the judges' scoring. Use the Wilcoxon signed-ranks test at the 0.05 level of significance.

22. A researcher at the Boston Eye Clinic devised a test of depth perception. The test was given while the subject had one eye covered and then was repeated with the other eye covered. Results are given below for 8 randomly selected subjects.

Right eye	14.7 16.3 12.4 8.1 21.6 13.9 14.2 15.8
Left eye	15.2 16.7 12.6 10.4 24.1 17.2 11.9 18.4

At the 0.05 level of significance, test the claim that depth perception is the same for both eyes. Use the Wilcoxon signed-ranks test.

23. A telephone solicitor for *Time* magazine wrote a report in which she used N for unanswered calls and Y for answered calls. At the 0.05 level of significance, test her claim that the sample listed below is random.

N N N N N N N N Y Y N N N N N Y Y
Y Y Y Y N N N N N N N N N N N

24. For randomly selected states, the following table lists the per capita beer consumption (in gallons) and the per capita wine consumption (in gallons) (based on data from *Statistical Abstract of the United States*).

Beer	32.2 29.4 35.3 34.9 29.9 28.7 26.8 41.4
Wine	3.1 4.4 2.3 1.7 1.4 1.2 1.2 3.0

Using a 0.05 significance level, test the claim that there is no correlation between beer consumption and wine consumption.

FROM DATA TO DECISION: Was the Draft Lottery Random?

In 1970, a lottery was used to determine who would be drafted into the U.S. Army. The 366 dates in the year were placed in individual capsules. First, the 31 January capsules were placed in a box; then the 29 February capsules were added and the two months were mixed. Then the 31 March capsules were added and the three months were mixed. This process continued until all months were included. The first capsule selected was September 14, so men born on that date were drafted first. Listed below are the 366 dates in the order of selection.

a. Use the runs test to test the sequence for randomness above and below the median of 183.5.

b. Use the Kruskal-Wallis test to test the claim that the 12 months had priority numbers drawn from the same population.

c. Calculate the 12 monthly means. Then plot those 12 means on a graph. (The horizontal scale lists the 12 months, and the vertical scale ranges from 100 to 260.) Note any pattern suggesting that the original priority numbers were not randomly selected.

d. Based on the results from parts a, b, and c, decide whether this particular draft lottery was fair. Write a brief statement either supporting your position that it was fair or explaining why you believe that it was not fair. If you decided that this lottery was unfair, describe a process for selecting lottery numbers that would have been fair.

Jan:	305	159	251	215	101	224	306	199	194	325	329	221	318	238	017	121
	235	140	058	280	186	337	118	059	052	092	355	077	349	164	211	
Feb:	086	144	297	210	214	347	091	181	338	216	150	068	152	004	089	212
	189	292	025	302	363	290	057	236	179	365	205	299	285			
Mar:	108	029	267	275	293	139	122	213	317	323	136	300	259	354	169	166
	033	332	200	239	334	265	256	258	343	170	268	223	362	217	030	
Apr:	032	271	083	081	269	253	147	312	219	218	014	346	124	231	273	148
	260	090	336	345	062	316	252	002	351	340	074	262	191	208		
May:	330	298	040	276	364	155	035	321	197	065	037	133	295	178	130	055
	112	278	075	183	250	326	319	031	361	357	296	308	226	103	313	
Jun:	249	228	301	020	028	110	085	366	335	206	134	272	069	356	180	274
	073	341	104	360	060	247	109	358	137	022	064	222	353	209		
Jul:	093	350	115	279	188	327	050	013	277	284	248	015	042	331	322	120
	098	190	227	187	027	153	172	023	067	303	289	088	270	287	193	
Aug:	111	045	261	145	054	114	168	048	106	021	324	142	307	198	102	044
	154	141	311	344	291	339	116	036	286	245	352	167	061	333	011	
Sep:	225	161	049	232	082	006	008	184	263	071	158	242	175	001	113	207
	255	246	177	063	204	160	119	195	149	018	233	257	151	315		
Oct:	359	125	244	202	024	087	234	283	342	220	237	072	138	294	171	254
	288	005	241	192	243	117	201	196	176	007	264	094	229	038	079	
Nov:	019	034	348	266	310	076	051	097	080	282	046	066	126	127	131	107
	143	146	203	185	156	009	182	230	132	309	047	281	099	174		
Dec:	129	328	157	165	056	010	012	105	043	041	039	314	163	026	320	096
	304	128	240	135	070	053	162	095	084	173	078	123	016	003	100	

COMPUTER PROJECT

Use STATDISK, Minitab, or any other statistical software package to generate 100 numbers, and then use the runs test to test for randomness above and below the mean. Exit the program, restart it, and then generate another 100 random numbers. Use the rank correlation coefficient to test for a relationship between the original 100 numbers and the second sequence of 100 numbers. Use the Wilcoxon rank-sum test to test the claim that the two samples come from the same population. Now generate a third set of 100 numbers, and use the Kruskal-Wallis test to test the claim that the three samples come from the same population. Based on these results, does the computer's number generator appear to be random?

14 Epilogue

I   What Procedure Applies?

In attempting to implement a statistical analysis, one of the most difficult tasks is determining the specific procedure that is most appropriate. This text includes a wide variety of procedures that apply to many different circumstances. We generally begin by clearly identifying the questions that need to be answered and by evaluating the quality of the sampling procedure. We must also answer questions such as these: What is the level of measurement (nominal, ordinal, interval, ratio) of the data? Does the study involve one, two, or more populations? Is there a claim to be tested or a parameter to be estimated? What is the relevant parameter (mean, standard deviation, proportion)? Is the sample large ($n > 30$) or small? Is there reason to believe that the population is normally distributed? Figure 14-1 should help you find the relevant text section.

II   A Perspective

You have already studied at least several of the preceding chapters. It is quite natural for you to believe that you have not mastered the course to the extent necessary to become a serious user of statistics. To keep a proper perspective, you should recognize that no one expects a single introductory statistics course to transform you into an expert statistician. Many important topics have not even been discussed because they are too advanced for this introductory level. There are also some easier topics that we have not discussed for other reasons. It is important to know that professional help is available from those who have had much more extensive training and experience with statistical methods; your statistics course will help you open a dialogue with one of these statisticians.

Although you are not an expert statistician, you will be a better-educated person with improved job marketability if you know and understand some important statistical concepts. You should know and understand the basic concepts of probability and chance. You should know that in attempting to gain insight into a set of data, it is important to investigate measures of central tendency (such as mean and median), measures of dispersion (such as range and standard deviation), and the nature of the distribution (via a frequency table or graph). You should know and understand the importance of estimating population parameters (such as a mean, standard deviation, and proportion), as

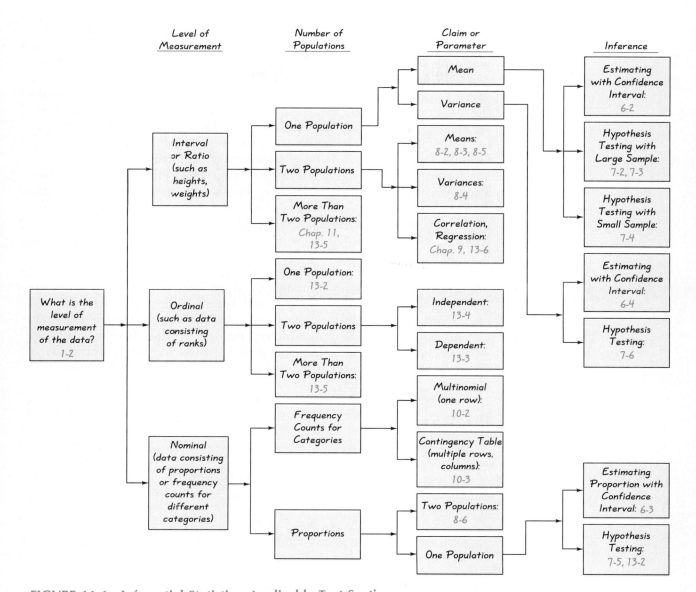

FIGURE 14-1   Inferential Statistics: Applicable Text Sections
Note: This figure applies to a fixed population. If the data are from a *process* that may change over time, construct a control chart (see Chapter 12) to determine whether the process is statistically stable. This figure applies to process data only if the process is statistically stable.

well as testing claims made about population parameters. You should realize that the nature and configuration of the data can have a dramatic effect on the particular statistical procedures that are used.

Throughout this text the importance of good sampling has been emphasized. You should recognize that a bad sample may be beyond repair by even the most expert statisticians, using the most sophisticated techniques. There are many mail, magazine, and telephone call-in surveys that allow respondents to be "self-selected." The results of such surveys are generally worthless when judged according to the criteria of sound statistical methodology. Keep this in mind when you are exposed to self-selected surveys, so that you don't let them affect your beliefs and decisions. You should also recognize, however, that many surveys and polls obtain very good results, even though the sample sizes might seem to be relatively small. Although many people refuse to believe it, a nationwide survey of only 1700 voters can provide good results if the sampling is carefully planned and executed.

At one time a person was considered educated if he or she could read, but our society has become highly complex and much more demanding. A modern education typically provides students with specific skills, such as the ability to read, write, understand the significance of the Renaissance, operate a computer, and do algebra. The larger picture involves several disciplines that use different approaches to a common goal—seeking the truth. The study of statistics helps us see the truth that is sometimes distorted by others or concealed by data that are disorganized or perhaps not yet collected. Understanding statistics is essential for a growing number of future employees, employers, and citizens. Congratulations on your success in the study of statistics!

Appendix A: Tables

TABLE A-1   Binomial Probabilities

n	x	.01	.05	.10	.20	.30	.40	.50	.60	.70	.80	.90	.95	.99	x
2	0	980	902	810	640	490	360	250	160	090	040	010	002	0+	0
	1	020	095	180	320	420	480	500	480	420	320	180	095	020	1
	2	0+	002	010	040	090	160	250	360	490	640	810	902	980	2
3	0	970	857	729	512	343	216	125	064	027	008	001	0+	0+	0
	1	029	135	243	384	441	432	375	288	189	096	027	007	0+	1
	2	0+	007	027	096	189	288	375	432	441	384	243	135	029	2
	3	0+	0+	001	008	027	064	125	216	343	512	729	857	970	3
4	0	961	815	656	410	240	130	062	026	008	002	0+	0+	0+	0
	1	039	171	292	410	412	346	250	154	076	026	004	0+	0+	1
	2	001	014	049	154	265	346	375	346	265	154	049	014	001	2
	3	0+	0+	004	026	076	154	250	346	412	410	292	171	039	3
	4	0+	0+	0+	002	008	026	062	130	240	410	656	815	961	4
5	0	951	774	590	328	168	078	031	010	002	0+	0+	0+	0+	0
	1	048	204	328	410	360	259	156	077	028	006	0+	0+	0+	1
	2	001	021	073	205	309	346	312	230	132	051	008	001	0+	2
	3	0+	001	008	051	132	230	312	346	309	205	073	021	001	3
	4	0+	0+	0+	006	028	077	156	259	360	410	328	204	048	4
	5	0+	0+	0+	0+	002	010	031	078	168	328	590	774	951	5
6	0	941	735	531	262	118	047	016	004	001	0+	0+	0+	0+	0
	1	057	232	354	393	303	187	094	037	010	002	0+	0+	0+	1
	2	001	031	098	246	324	311	234	138	060	015	001	0+	0+	2
	3	0+	002	015	082	185	276	312	276	185	082	015	002	0+	3
	4	0+	0+	001	015	060	138	234	311	324	246	098	031	001	4
	5	0+	0+	0+	002	010	037	094	187	303	393	354	232	057	5
	6	0+	0+	0+	0+	001	004	016	047	118	262	531	735	941	6
7	0	932	698	478	210	082	028	008	002	0+	0+	0+	0+	0+	0
	1	066	257	372	367	247	131	055	017	004	0+	0+	0+	0+	1
	2	002	041	124	275	318	261	164	077	025	004	0+	0+	0+	2
	3	0+	004	023	115	227	290	273	194	097	029	003	0+	0+	3
	4	0+	0+	003	029	097	194	273	290	227	115	023	004	0+	4
	5	0+	0+	0+	004	025	077	164	261	318	275	124	041	002	5
	6	0+	0+	0+	0+	004	017	055	131	247	367	372	257	066	6
	7	0+	0+	0+	0+	0+	002	008	028	082	210	478	698	932	7
8	0	923	663	430	168	058	017	004	001	0+	0+	0+	0+	0+	0
	1	075	279	383	336	198	090	031	008	001	0+	0+	0+	0+	1
	2	003	051	149	294	296	209	109	041	010	001	0+	0+	0+	2
	3	0+	005	033	147	254	279	219	124	047	009	0+	0+	0+	3
	4	0+	0+	005	046	136	232	273	232	136	046	005	0+	0+	4
	5	0+	0+	0+	009	047	124	219	279	254	147	033	005	0+	5
	6	0+	0+	0+	001	010	041	109	209	296	294	149	051	003	6
	7	0+	0+	0+	0+	001	008	031	090	198	336	383	279	075	7
	8	0+	0+	0+	0+	0+	001	004	017	058	168	430	663	923	8

NOTE: 0+ represents a positive probability less than 0.0005.

(continued)

TABLE A-1 Binomial Probabilities (continued)

n	x	.01	.05	.10	.20	.30	.40	.50	.60	.70	.80	.90	.95	.99	x
9	0	914	630	387	134	040	010	002	0+	0+	0+	0+	0+	0+	0
	1	083	299	387	302	156	060	018	004	0+	0+	0+	0+	0+	1
	2	003	063	172	302	267	161	070	021	004	0+	0+	0+	0+	2
	3	0+	008	045	176	267	251	164	074	021	003	0+	0+	0+	3
	4	0+	001	007	066	172	251	246	167	074	017	001	0+	0+	4
	5	0+	0+	001	017	074	167	246	251	172	066	007	001	0+	5
	6	0+	0+	0+	003	021	074	164	251	267	176	045	008	0+	6
	7	0+	0+	0+	0+	004	021	070	161	267	302	172	063	003	7
	8	0+	0+	0+	0+	0+	004	018	060	156	302	387	299	083	8
	9	0+	0+	0+	0+	0+	0+	002	010	040	134	387	630	914	9
10	0	904	599	349	107	028	006	001	0+	0+	0+	0+	0+	0+	0
	1	091	315	387	268	121	040	010	002	0+	0+	0+	0+	0+	1
	2	004	075	194	302	233	121	044	011	001	0+	0+	0+	0+	2
	3	0+	010	057	201	267	215	117	042	009	001	0+	0+	0+	3
	4	0+	001	011	088	200	251	205	111	037	006	0+	0+	0+	4
	5	0+	0+	001	026	103	201	246	201	103	026	001	0+	0+	5
	6	0+	0+	0+	006	037	111	205	251	200	088	011	001	0+	6
	7	0+	0+	0+	001	009	042	117	215	267	201	057	010	0+	7
	8	0+	0+	0+	0+	001	011	044	121	233	302	194	075	004	8
	9	0+	0+	0+	0+	0+	002	010	040	121	268	387	315	091	9
	10	0+	0+	0+	0+	0+	0+	001	006	028	107	349	599	904	10
11	0	895	569	314	086	020	004	0+	0+	0+	0+	0+	0+	0+	0
	1	099	329	384	236	093	027	005	001	0+	0+	0+	0+	0+	1
	2	005	087	213	295	200	089	027	005	001	0+	0+	0+	0+	2
	3	0+	014	071	221	257	177	081	023	004	0+	0+	0+	0+	3
	4	0+	001	016	111	220	236	161	070	017	002	0+	0+	0+	4
	5	0+	0+	002	039	132	221	226	147	057	010	0+	0+	0+	5
	6	0+	0+	0+	010	057	147	226	221	132	039	002	0+	0+	6
	7	0+	0+	0+	002	017	070	161	236	220	111	016	001	0+	7
	8	0+	0+	0+	0+	004	023	081	177	257	221	071	014	0+	8
	9	0+	0+	0+	0+	001	005	027	089	200	295	213	087	005	9
	10	0+	0+	0+	0+	0+	001	005	027	093	236	384	329	099	10
	11	0+	0+	0+	0+	0+	0+	0+	004	020	086	314	569	895	11
12	0	886	540	282	069	014	002	0+	0+	0+	0+	0+	0+	0+	0
	1	107	341	377	206	071	017	003	0+	0+	0+	0+	0+	0+	1
	2	006	099	230	283	168	064	016	002	0+	0+	0+	0+	0+	2
	3	0+	017	085	236	240	142	054	012	001	0+	0+	0+	0+	3
	4	0+	002	021	133	231	213	121	042	008	001	0+	0+	0+	4
	5	0+	0+	004	053	158	227	193	101	029	003	0+	0+	0+	5
	6	0+	0+	0+	016	079	177	226	177	079	016	0+	0+	0+	6
	7	0+	0+	0+	003	029	101	193	227	158	053	004	0+	0+	7
	8	0+	0+	0+	001	008	042	121	213	231	133	021	002	0+	8
	9	0+	0+	0+	0+	001	012	054	142	240	236	085	017	0+	9
	10	0+	0+	0+	0+	0+	002	016	064	168	283	230	099	006	10
	11	0+	0+	0+	0+	0+	0+	003	017	071	206	377	341	107	11
	12	0+	0+	0+	0+	0+	0+	0+	002	014	069	282	540	886	12

NOTE: 0+ represents a positive probability less than 0.0005.

(continued)

TABLE A-1 Binomial Probabilities *(continued)*

n	x	.01	.05	.10	.20	.30	.40	.50	.60	.70	.80	.90	.95	.99	x
13	0	878	513	254	055	010	001	0+	0+	0+	0+	0+	0+	0+	0
	1	115	351	367	179	054	011	002	0+	0+	0+	0+	0+	0+	1
	2	007	111	245	268	139	045	010	001	0+	0+	0+	0+	0+	2
	3	0+	021	100	246	218	111	035	006	001	0+	0+	0+	0+	3
	4	0+	003	028	154	234	184	087	024	003	0+	0+	0+	0+	4
	5	0+	0+	006	069	180	221	157	066	014	001	0+	0+	0+	5
	6	0+	0+	001	023	103	197	209	131	044	006	0+	0+	0+	6
	7	0+	0+	0+	006	044	131	209	197	103	023	001	0+	0+	7
	8	0+	0+	0+	001	014	066	157	221	180	069	006	0+	0+	8
	9	0+	0+	0+	0+	003	024	087	184	234	154	028	003	0+	9
	10	0+	0+	0+	0+	001	006	035	111	218	246	100	021	0+	10
	11	0+	0+	0+	0+	0+	001	010	045	139	268	245	111	007	11
	12	0+	0+	0+	0+	0+	0+	002	011	054	179	367	351	115	12
	13	0+	0+	0+	0+	0+	0+	0+	001	010	055	254	513	878	13
14	0	869	488	229	044	007	001	0+	0+	0+	0+	0+	0+	0+	0
	1	123	359	356	154	041	007	001	0+	0+	0+	0+	0+	0+	1
	2	008	123	257	250	113	032	006	001	0+	0+	0+	0+	0+	2
	3	0+	026	114	250	194	085	022	003	0+	0+	0+	0+	0+	3
	4	0+	004	035	172	229	155	061	014	001	0+	0+	0+	0+	4
	5	0+	0+	008	086	196	207	122	041	007	0+	0+	0+	0+	5
	6	0+	0+	001	032	126	207	183	092	023	002	0+	0+	0+	6
	7	0+	0+	0+	009	062	157	209	157	062	009	0+	0+	0+	7
	8	0+	0+	0+	002	023	092	183	207	126	032	001	0+	0+	8
	9	0+	0+	0+	0+	007	041	122	207	196	086	008	0+	0+	9
	10	0+	0+	0+	0+	001	014	061	155	229	172	035	004	0+	10
	11	0+	0+	0+	0+	0+	003	022	085	194	250	114	026	0+	11
	12	0+	0+	0+	0+	0+	001	006	032	113	250	257	123	008	12
	13	0+	0+	0+	0+	0+	0+	001	007	041	154	356	359	123	13
	14	0+	0+	0+	0+	0+	0+	0+	001	007	044	229	488	869	14
15	0	860	463	206	035	005	0+	0+	0+	0+	0+	0+	0+	0+	0
	1	130	366	343	132	031	005	0+	0+	0+	0+	0+	0+	0+	1
	2	009	135	267	231	092	022	003	0+	0+	0+	0+	0+	0+	2
	3	0−	031	129	250	170	063	014	002	0+	0+	0+	0+	0+	3
	4	0+	005	043	188	219	127	042	007	001	0+	0+	0+	0+	4
	5	0+	001	010	103	206	186	092	024	003	0+	0+	0+	0+	5
	6	0+	0+	002	043	147	207	153	061	012	001	0+	0+	0+	6
	7	0+	0+	0+	014	081	177	196	118	035	003	0+	0+	0+	7
	8	0+	0+	0+	003	035	118	196	177	081	014	0+	0+	0+	8
	9	0+	0+	0+	001	012	061	153	207	147	043	002	0+	0+	9
	10	0+	0+	0+	0+	003	024	092	186	206	103	010	001	0+	10
	11	0+	0+	0+	0+	001	007	042	127	219	188	043	005	0+	11
	12	0+	0+	0+	0+	0+	002	014	063	170	250	129	031	0+	12
	13	0+	0+	0+	0+	0+	0+	003	022	092	231	267	135	009	13
	14	0+	0+	0+	0+	0+	0+	0+	005	031	132	343	366	130	14
	15	0+	0+	0+	0+	0+	0+	0+	0+	005	035	206	463	860	15

NOTE: 0+ represents a positive probability less than 0.0005.

From Frederick C. Mosteller, Robert E. K. Rourke, and George B. Thomas, Jr., *Probability with Statistical Applications*, 2nd ed., © 1970 Addison-Wesley Publishing Co., Reading, MA. Reprinted with permission.

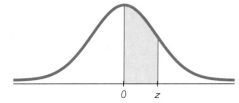

TABLE A-2   Standard Normal (z) Distribution

z	.00	.01	.02	.03	.04	.05	.06	.07	.08	.09
0.0	.0000	.0040	.0080	.0120	.0160	.0199	.0239	.0279	.0319	.0359
0.1	.0398	.0438	.0478	.0517	.0557	.0596	.0636	.0675	.0714	.0753
0.2	.0793	.0832	.0871	.0910	.0948	.0987	.1026	.1064	.1103	.1141
0.3	.1179	.1217	.1255	.1293	.1331	.1368	.1406	.1443	.1480	.1517
0.4	.1554	.1591	.1628	.1664	.1700	.1736	.1772	.1808	.1844	.1879
0.5	.1915	.1950	.1985	.2019	.2054	.2088	.2123	.2157	.2190	.2224
0.6	.2257	.2291	.2324	.2357	.2389	.2422	.2454	.2486	.2517	.2549
0.7	.2580	.2611	.2642	.2673	.2704	.2734	.2764	.2794	.2823	.2852
0.8	.2881	.2910	.2939	.2967	.2995	.3023	.3051	.3078	.3106	.3133
0.9	.3159	.3186	.3212	.3238	.3264	.3289	.3315	.3340	.3365	.3389
1.0	.3413	.3438	.3461	.3485	.3508	.3531	.3554	.3577	.3599	.3621
1.1	.3643	.3665	.3686	.3708	.3729	.3749	.3770	.3790	.3810	.3830
1.2	.3849	.3869	.3888	.3907	.3925	.3944	.3962	.3980	.3997	.4015
1.3	.4032	.4049	.4066	.4082	.4099	.4115	.4131	.4147	.4162	.4177
1.4	.4192	.4207	.4222	.4236	.4251	.4265	.4279	.4292	.4306	.4319
1.5	.4332	.4345	.4357	.4370	.4382	.4394	.4406	.4418	.4429	.4441
1.6	.4452	.4463	.4474	.4484	.4495 *	.4505	.4515	.4525	.4535	.4545
1.7	.4554	.4564	.4573	.4582	.4591	.4599	.4608	.4616	.4625	.4633
1.8	.4641	.4649	.4656	.4664	.4671	.4678	.4686	.4693	.4699	.4706
1.9	.4713	.4719	.4726	.4732	.4738	.4744	.4750	.4756	.4761	.4767
2.0	.4772	.4778	.4783	.4788	.4793	.4798	.4803	.4808	.4812	.4817
2.1	.4821	.4826	.4830	.4834	.4838	.4842	.4846	.4850	.4854	.4857
2.2	.4861	.4864	.4868	.4871	.4875	.4878	.4881	.4884	.4887	.4890
2.3	.4893	.4896	.4898	.4901	.4904	.4906	.4909	.4911	.4913	.4916
2.4	.4918	.4920	.4922	.4925	.4927	.4929	.4931	.4932	.4934	.4936
2.5	.4938	.4940	.4941	.4943	.4945	.4946	.4948	.4949 *	.4951	.4952
2.6	.4953	.4955	.4956	.4957	.4959	.4960	.4961	.4962	.4963	.4964
2.7	.4965	.4966	.4967	.4968	.4969	.4970	.4971	.4972	.4973	.4974
2.8	.4974	.4975	.4976	.4977	.4977	.4978	.4979	.4979	.4980	.4981
2.9	.4981	.4982	.4982	.4983	.4984	.4984	.4985	.4985	.4986	.4986
3.0	.4987	.4987	.4987	.4988	.4988	.4989	.4989	.4989	.4990	.4990

NOTE: For values of z above 3.09, use 0.4999 for the area.

*Use these common values that result from interpolation:

z score	Area
1.645	0.4500
2.575	0.4950

From Frederick C. Mosteller and Robert E. K. Rourke, *Sturdy Statistics,* 1973, Addison-Wesley Publishing Co., Reading, MA. Reprinted with permission of Frederick Mosteller.

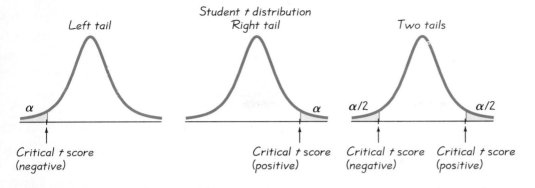

TABLE A-3 t Distribution

| | α | | | | | |
Degrees of freedom	.005 (one tail) .01 (two tails)	.01 (one tail) .02 (two tails)	.025 (one tail) .05 (two tails)	.05 (one tail) .10 (two tails)	.10 (one tail) .20 (two tails)	.25 (one tail) .50 (two tails)
1	63.657	31.821	12.706	6.314	3.078	1.000
2	9.925	6.965	4.303	2.920	1.886	.816
3	5.841	4.541	3.182	2.353	1.638	.765
4	4.604	3.747	2.776	2.132	1.533	.741
5	4.032	3.365	2.571	2.015	1.476	.727
6	3.707	3.143	2.447	1.943	1.440	.718
7	3.500	2.998	2.365	1.895	1.415	.711
8	3.355	2.896	2.306	1.860	1.397	.706
9	3.250	2.821	2.262	1.833	1.383	.703
10	3.169	2.764	2.228	1.812	1.372	.700
11	3.106	2.718	2.201	1.796	1.363	.697
12	3.054	2.681	2.179	1.782	1.356	.696
13	3.012	2.650	2.160	1.771	1.350	.694
14	2.977	2.625	2.145	1.761	1.345	.692
15	2.947	2.602	2.132	1.753	1.341	.691
16	2.921	2.584	2.120	1.746	1.337	.690
17	2.898	2.567	2.110	1.740	1.333	.689
18	2.878	2.552	2.101	1.734	1.330	.688
19	2.861	2.540	2.093	1.729	1.328	.688
20	2.845	2.528	2.086	1.725	1.325	.687
21	2.831	2.518	2.080	1.721	1.323	.686
22	2.819	2.508	2.074	1.717	1.321	.686
23	2.807	2.500	2.069	1.714	1.320	.685
24	2.797	2.492	2.064	1.711	1.318	.685
25	2.787	2.485	2.060	1.708	1.316	.684
26	2.779	2.479	2.056	1.706	1.315	.684
27	2.771	2.473	2.052	1.703	1.314	.684
28	2.763	2.467	2.048	1.701	1.313	.683
29	2.756	2.462	2.045	1.699	1.311	.683
Large (z)	2.575	2.327	1.960	1.645	1.282	.675

TABLE A-4   Chi-Square (χ^2) Distribution

Area to the Right of the Critical Value

Degrees of freedom	0.995	0.99	0.975	0.95	0.90	0.10	0.05	0.025	0.01	0.005
1	—	—	0.001	0.004	0.016	2.706	3.841	5.024	6.635	7.879
2	0.010	0.020	0.051	0.103	0.211	4.605	5.991	7.378	9.210	10.597
3	0.072	0.115	0.216	0.352	0.584	6.251	7.815	9.348	11.345	12.838
4	0.207	0.297	0.484	0.711	1.064	7.779	9.488	11.143	13.277	14.860
5	0.412	0.554	0.831	1.145	1.610	9.236	11.071	12.833	15.086	16.750
6	0.676	0.872	1.237	1.635	2.204	10.645	12.592	14.449	16.812	18.548
7	0.989	1.239	1.690	2.167	2.833	12.017	14.067	16.013	18.475	20.278
8	1.344	1.646	2.180	2.733	3.490	13.362	15.507	17.535	20.090	21.955
9	1.735	2.088	2.700	3.325	4.168	14.684	16.919	19.023	21.666	23.589
10	2.156	2.558	3.247	3.940	4.865	15.987	18.307	20.483	23.209	25.188
11	2.603	3.053	3.816	4.575	5.578	17.275	19.675	21.920	24.725	26.757
12	3.074	3.571	4.404	5.226	6.304	18.549	21.026	23.337	26.217	28.299
13	3.565	4.107	5.009	5.892	7.042	19.812	22.362	24.736	27.688	29.819
14	4.075	4.660	5.629	6.571	7.790	21.064	23.685	26.119	29.141	31.319
15	4.601	5.229	6.262	7.261	8.547	22.307	24.996	27.488	30.578	32.801
16	5.142	5.812	6.908	7.962	9.312	23.542	26.296	28.845	32.000	34.267
17	5.697	6.408	7.564	8.672	10.085	24.769	27.587	30.191	33.409	35.718
18	6.265	7.015	8.231	9.390	10.865	25.989	28.869	31.526	34.805	37.156
19	6.844	7.633	8.907	10.117	11.651	27.204	30.144	32.852	36.191	38.582
20	7.434	8.260	9.591	10.851	12.443	28.412	31.410	34.170	37.566	39.997
21	8.034	8.897	10.283	11.591	13.240	29.615	32.671	35.479	38.932	41.401
22	8.643	9.542	10.982	12.338	14.042	30.813	33.924	36.781	40.289	42.796
23	9.260	10.196	11.689	13.091	14.848	32.007	35.172	38.076	41.638	44.181
24	9.886	10.856	12.401	13.848	15.659	33.196	36.415	39.364	42.980	45.559
25	10.520	11.524	13.120	14.611	16.473	34.382	37.652	40.646	44.314	46.928
26	11.160	12.198	13.844	15.379	17.292	35.563	38.885	41.923	45.642	48.290
27	11.808	12.879	14.573	16.151	18.114	36.741	40.113	43.194	46.963	49.645
28	12.461	13.565	15.308	16.928	18.939	37.916	41.337	44.461	48.278	50.993
29	13.121	14.257	16.047	17.708	19.768	39.087	42.557	45.722	49.588	52.336
30	13.787	14.954	16.791	18.493	20.599	40.256	43.773	46.979	50.892	53.672
40	20.707	22.164	24.433	26.509	29.051	51.805	55.758	59.342	63.691	66.766
50	27.991	29.707	32.357	34.764	37.689	63.167	67.505	71.420	76.154	79.490
60	35.534	37.485	40.482	43.188	46.459	74.397	79.082	83.298	88.379	91.952
70	43.275	45.442	48.758	51.739	55.329	85.527	90.531	95.023	100.425	104.215
80	51.172	53.540	57.153	60.391	64.278	96.578	101.879	106.629	112.329	116.321
90	59.196	61.754	65.647	69.126	73.291	107.565	113.145	118.136	124.116	128.299
100	67.328	70.065	74.222	77.929	82.358	118.498	124.342	129.561	135.807	140.169

From Donald B. Owen, *Handbook of Statistical Tables*, © 1962 Addison-Wesley Publishing Co., Reading, MA. Reprinted with permission of the publisher.

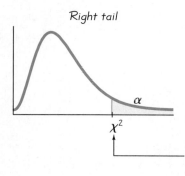

Right tail

x^2

To find this value, use the column with the area α given at the top of the table.

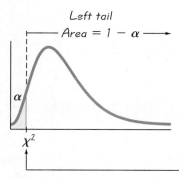

Left tail

Area $= 1 - \alpha$

α

x^2

To find this value, determine the area of the region to the right of this boundary (the unshaded area) and use the column with this value at the top. If the left tail has area α, use the column with the value of $1 - \alpha$ at the top of the table.

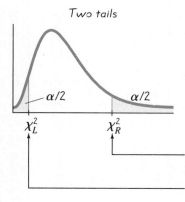

Two tails

$\alpha/2$ $\alpha/2$

x_L^2 x_R^2

To find this value, use the column with area $\alpha/2$ at the top of the table.

To find this value, use the column with area $1 - \alpha/2$ at the top of the table.

TABLE A-5   F Distribution (α = 0.01 in the right tail)

Numerator degrees of freedom (df₁)

df₂	1	2	3	4	5	6	7	8	9
1	4052.2	4999.5	5403.4	5624.6	5763.6	5859.0	5928.4	5981.1	6022.5
2	98.503	99.000	99.166	99.249	99.299	99.333	99.356	99.374	99.388
3	34.116	30.817	29.457	28.710	28.237	27.911	27.672	27.489	27.345
4	21.198	18.000	16.694	15.977	15.522	15.207	14.976	14.799	14.659
5	16.258	13.274	12.060	11.392	10.967	10.672	10.456	10.289	10.158
6	13.745	10.925	9.7795	9.1483	8.7459	8.4661	8.2600	8.1017	7.9761
7	12.246	9.5466	8.4513	7.8466	7.4604	7.1914	6.9928	6.8400	6.7188
8	11.259	8.6491	7.5910	7.0061	6.6318	6.3707	6.1776	6.0289	5.9106
9	10.561	8.0215	6.9919	6.4221	6.0569	5.8018	5.6129	5.4671	5.3511
10	10.044	7.5594	6.5523	5.9943	5.6363	5.3858	5.2001	5.0567	4.9424
11	9.6460	7.2057	6.2167	5.6683	5.3160	5.0692	4.8861	4.7445	4.6315
12	9.3302	6.9266	5.9525	5.4120	5.0643	4.8206	4.6395	4.4994	4.3875
13	9.0738	6.7010	5.7394	5.2053	4.8616	4.6204	4.4410	4.3021	4.1911
14	8.8616	6.5149	5.5639	5.0354	4.6950	4.4558	4.2779	4.1399	4.0297
15	8.6831	6.3589	5.4170	4.8932	4.5556	4.3183	4.1415	4.0045	3.8948
16	8.5310	6.2262	5.2922	4.7726	4.4374	4.2016	4.0259	3.8896	3.7804
17	8.3997	6.1121	5.1850	4.6690	4.3359	4.1015	3.9267	3.7910	3.6822
18	8.2854	6.0129	5.0919	4.5790	4.2479	4.0146	3.8406	3.7054	3.5971
19	8.1849	5.9259	5.0103	4.5003	4.1708	3.9386	3.7653	3.6305	3.5225
20	8.0960	5.8489	4.9382	4.4307	4.1027	3.8714	3.6987	3.5644	3.4567
21	8.0166	5.7804	4.8740	4.3688	4.0421	3.8117	3.6396	3.5056	3.3981
22	7.9454	5.7190	4.8166	4.3134	3.9880	3.7583	3.5867	3.4530	3.3458
23	7.8811	5.6637	4.7649	4.2636	3.9392	3.7102	3.5390	3.4057	3.2986
24	7.8229	5.6136	4.7181	4.2184	3.8951	3.6667	3.4959	3.3629	3.2560
25	7.7698	5.5680	4.6755	4.1774	3.8550	3.6272	3.4568	3.3239	3.2172
26	7.7213	5.5263	4.6366	4.1400	3.8183	3.5911	3.4210	3.2884	3.1818
27	7.6767	5.4881	4.6009	4.1056	3.7848	3.5580	3.3882	3.2558	3.1494
28	7.6356	5.4529	4.5681	4.0740	3.7539	3.5276	3.3581	3.2259	3.1195
29	7.5977	5.4204	4.5378	4.0449	3.7254	3.4995	3.3303	3.1982	3.0920
30	7.5625	5.3903	4.5097	4.0179	3.6990	3.4735	3.3045	3.1726	3.0665
40	7.3141	5.1785	4.3126	3.8283	3.5138	3.2910	3.1238	2.9930	2.8876
60	7.0771	4.9774	4.1259	3.6490	3.3389	3.1187	2.9530	2.8233	2.7185
120	6.8509	4.7865	3.9491	3.4795	3.1735	2.9559	2.7918	2.6629	2.5586
∞	6.6349	4.6052	3.7816	3.3192	3.0173	2.8020	2.6393	2.5113	2.4073

Denominator degrees of freedom (df₂)

From Maxine Merrington and Catherine M. Thompson, "Tables of Percentage Points of the Inverted Beta (F) Distribution," *Biometrika* 33 (1943): 80–84. Reproduced with permission of the Biometrika Trustees.

(continued)

TABLE A-5 *F* Distribution ($\alpha = 0.01$ in the right tail) *(continued)*

Numerator degrees of freedom (df_1)

df_2	10	12	15	20	24	30	40	60	120	∞
1	6055.8	6106.3	6157.3	6208.7	6234.6	6260.6	6286.8	6313.0	6339.4	6365.9
2	99.399	99.416	99.433	99.449	99.458	99.466	99.474	99.482	99.491	99.499
3	27.229	27.052	26.872	26.690	26.598	26.505	26.411	26.316	26.221	26.125
4	14.546	14.374	14.198	14.020	13.929	13.838	13.745	13.652	13.558	13.463
5	10.051	9.8883	9.7222	9.5526	9.4665	9.3793	9.2912	9.2020	9.1118	9.0204
6	7.8741	7.7183	7.5590	7.3958	7.3127	7.2285	7.1432	7.0567	6.9690	6.8800
7	6.6201	6.4691	6.3143	6.1554	6.0743	5.9920	5.9084	5.8236	5.7373	5.6495
8	5.8143	5.6667	5.5151	5.3591	5.2793	5.1981	5.1156	5.0316	4.9461	4.8588
9	5.2565	5.1114	4.9621	4.8080	4.7290	4.6486	4.5666	4.4831	4.3978	4.3105
10	4.8491	4.7059	4.5581	4.4054	4.3269	4.2469	4.1653	4.0819	3.9965	3.9090
11	4.5393	4.3974	4.2509	4.0990	4.0209	3.9411	3.8596	3.7761	3.6904	3.6024
12	4.2961	4.1553	4.0096	3.8584	3.7805	3.7008	3.6192	3.5355	3.4494	3.3608
13	4.1003	3.9603	3.8154	3.6646	3.5868	3.5070	3.4253	3.3413	3.2548	3.1654
14	3.9394	3.8001	3.6557	3.5052	3.4274	3.3476	3.2656	3.1813	3.0942	3.0040
15	3.8049	3.6662	3.5222	3.3719	3.2940	3.2141	3.1319	3.0471	2.9595	2.8684
16	3.6909	3.5527	3.4089	3.2587	3.1808	3.1007	3.0182	2.9330	2.8447	2.7528
17	3.5931	3.4552	3.3117	3.1615	3.0835	3.0032	2.9205	2.8348	2.7459	2.6530
18	3.5082	3.3706	3.2273	3.0771	2.9990	2.9185	2.8354	2.7493	2.6597	2.5660
19	3.4338	3.2965	3.1533	3.0031	2.9249	2.8442	2.7608	2.6742	2.5839	2.4893
20	3.3682	3.2311	3.0880	2.9377	2.8594	2.7785	2.6947	2.6077	2.5168	2.4212
21	3.3098	3.1730	3.0300	2.8796	2.8010	2.7200	2.6359	2.5484	2.4568	2.3603
22	3.2576	3.1209	2.9779	2.8274	2.7488	2.6675	2.5831	2.4951	2.4029	2.3055
23	3.2106	3.0740	2.9311	2.7805	2.7017	2.6202	2.5355	2.4471	2.3542	2.2558
24	3.1681	3.0316	2.8887	2.7380	2.6591	2.5773	2.4923	2.4035	2.3100	2.2107
25	3.1294	2.9931	2.8502	2.6993	2.6203	2.5383	2.4530	2.3637	2.2696	2.1694
26	3.0941	2.9578	2.8150	2.6640	2.5848	2.5026	2.4170	2.3273	2.2325	2.1315
27	3.0618	2.9256	2.7827	2.6316	2.5522	2.4699	2.3840	2.2938	2.1985	2.0965
28	3.0320	2.8959	2.7530	2.6017	2.5223	2.4397	2.3535	2.2629	2.1670	2.0642
29	3.0045	2.8685	2.7256	2.5742	2.4946	2.4118	2.3253	2.2344	2.1379	2.0342
30	2.9791	2.8431	2.7002	2.5487	2.4689	2.3860	2.2992	2.2079	2.1108	2.0062
40	2.8005	2.6648	2.5216	2.3689	2.2880	2.2034	2.1142	2.0194	1.9172	1.8047
60	2.6318	2.4961	2.3523	2.1978	2.1154	2.0285	1.9360	1.8363	1.7263	1.6006
120	2.4721	2.3363	2.1915	2.0346	1.9500	1.8600	1.7628	1.6557	1.5330	1.3805
∞	2.3209	2.1847	2.0385	1.8783	1.7908	1.6964	1.5923	1.4730	1.3246	1.0000

Denominator degrees of freedom (df_2)

From Maxine Merrington and Catherine M. Thompson, "Tables of Percentage Points of the Inverted Beta (*F*) Distribution," *Biometrika* 33 (1943): 80–84. Reproduced with permission of the Biometrika Trustees.

TABLE A-5   F Distribution ($\alpha = 0.025$ in the right tail)

0.025

Numerator degrees of freedom (df_1)

df_2	1	2	3	4	5	6	7	8	9
1	647.79	799.50	864.16	899.58	921.85	937.11	948.22	956.66	963.28
2	38.506	39.000	39.165	39.248	39.298	39.331	39.335	39.373	39.387
3	17.443	16.044	15.439	15.101	14.885	14.735	14.624	14.540	14.473
4	12.218	10.649	9.9792	9.6045	9.3645	9.1973	9.0741	8.9796	8.9047
5	10.007	8.4336	7.7636	7.3879	7.1464	6.9777	6.8531	6.7572	6.6811
6	8.8131	7.2599	6.5988	6.2272	5.9876	5.8198	5.6955	5.5996	5.5234
7	8.0727	6.5415	5.8898	5.5226	5.2852	5.1186	4.9949	4.8993	4.8232
8	7.5709	6.0595	5.4160	5.0526	4.8173	4.6517	4.5286	4.4333	4.3572
9	7.2093	5.7147	5.0781	4.7181	4.4844	4.3197	4.1970	4.1020	4.0260
10	6.9367	5.4564	4.8256	4.4683	4.2361	4.0721	3.9498	3.8549	3.7790
11	6.7241	5.2559	4.6300	4.2751	4.0440	3.8807	3.7586	3.6638	3.5879
12	6.5538	5.0959	4.4742	4.1212	3.8911	3.7283	3.6065	3.5118	3.4358
13	6.4143	4.9653	4.3472	3.9959	3.7667	3.6043	3.4827	3.3880	3.3120
14	6.2979	4.8567	4.2417	3.8919	3.6634	3.5014	3.3799	3.2853	3.2093
15	6.1995	4.7650	4.1528	3.8043	3.5764	3.4147	3.2934	3.1987	3.1227
16	6.1151	4.6867	4.0768	3.7294	3.5021	3.3406	3.2194	3.1248	3.0488
17	6.0420	4.6189	4.0112	3.6648	3.4379	3.2767	3.1556	3.0610	2.9849
18	5.9781	4.5597	3.9539	3.6083	3.3820	3.2209	3.0999	3.0053	2.9291
19	5.9216	4.5075	3.9034	3.5587	3.3327	3.1718	3.0509	2.9563	2.8801
20	5.8715	4.4613	3.8587	3.5147	3.2891	3.1283	3.0074	2.9128	2.8365
21	5.8266	4.4199	3.8188	3.4754	3.2501	3.0895	2.9686	2.8740	2.7977
22	5.7863	4.3828	3.7829	3.4401	3.2151	3.0546	2.9338	2.8392	2.7628
23	5.7498	4.3492	3.7505	3.4083	3.1835	3.0232	2.9023	2.8077	2.7313
24	5.7166	4.3187	3.7211	3.3794	3.1548	2.9946	2.8738	2.7791	2.7027
25	5.6864	4.2909	3.6943	3.3530	3.1287	2.9685	2.8478	2.7531	2.6766
26	5.6586	4.2655	3.6697	3.3289	3.1048	2.9447	2.8240	2.7293	2.6528
27	5.6331	4.2421	3.6472	3.3067	3.0828	2.9228	2.8021	2.7074	2.6309
28	5.6096	4.2205	3.6264	3.2863	3.0626	2.9027	2.7820	2.6872	2.6106
29	5.5878	4.2006	3.6072	3.2674	3.0438	2.8840	2.7633	2.6686	2.5919
30	5.5675	4.1821	3.5894	3.2499	3.0265	2.8667	2.7460	2.6513	2.5746
40	5.4239	4.0510	3.4633	3.1261	2.9037	2.7444	2.6238	2.5289	2.4519
60	5.2856	3.9253	3.3425	3.0077	2.7863	2.6274	2.5068	2.4117	2.3344
120	5.1523	3.8046	3.2269	2.8943	2.6740	2.5154	2.3948	2.2994	2.2217
∞	5.0239	3.6889	3.1161	2.7858	2.5665	2.4082	2.2875	2.1918	2.1136

Denominator degrees of freedom (df_2)

(continued)

TABLE A-5 F Distribution ($\alpha = 0.025$ in the right tail) (continued)

Numerator degrees of freedom (df_1)

	10	12	15	20	24	30	40	60	120	∞
1	968.63	976.71	984.87	993.10	997.25	1001.4	1005.6	1009.8	1014.0	1018.3
2	39.398	39.415	39.431	39.448	39.456	39.465	39.473	39.481	39.490	39.498
3	14.419	14.337	14.253	14.167	14.124	14.081	14.037	13.992	13.947	13.902
4	8.8439	8.7512	8.6565	8.5599	8.5109	8.4613	8.4111	8.3604	8.3092	8.2573
5	6.6192	6.5245	6.4277	6.3286	6.2780	6.2269	6.1750	6.1225	6.0693	6.0153
6	5.4613	5.3662	5.2687	5.1684	5.1172	5.0652	5.0125	4.9589	4.9044	4.8491
7	4.7611	4.6658	4.5678	4.4667	4.4150	4.3624	4.3089	4.2544	4.1989	4.1423
8	4.2951	4.1997	4.1012	3.9995	3.9472	3.8940	3.8398	3.7844	3.7279	3.6702
9	3.9639	3.8682	3.7694	3.6669	3.6142	3.5604	3.5055	3.4493	3.3918	3.3329
10	3.7168	3.6209	3.5217	3.4185	3.3654	3.3110	3.2554	3.1984	3.1399	3.0798
11	3.5257	3.4296	3.3299	3.2261	3.1725	3.1176	3.0613	3.0035	2.9441	2.8828
12	3.3736	3.2773	3.1772	3.0728	3.0187	2.9633	2.9063	2.8478	2.7874	2.7249
13	3.2497	3.1532	3.0527	2.9477	2.8932	2.8372	2.7797	2.7204	2.6590	2.5955
14	3.1469	3.0502	2.9493	2.8437	2.7888	2.7324	2.6742	2.6142	2.5519	2.4872
15	3.0602	2.9633	2.8621	2.7559	2.7006	2.6437	2.5850	2.5242	2.4611	2.3953
16	2.9862	2.8890	2.7875	2.6808	2.6252	2.5678	2.5085	2.4471	2.3831	2.3163
17	2.9222	2.8249	2.7230	2.6158	2.5598	2.5020	2.4422	2.3801	2.3153	2.2474
18	2.8664	2.7689	2.6667	2.5590	2.5027	2.4445	2.3842	2.3214	2.2558	2.1869
19	2.8172	2.7196	2.6171	2.5089	2.4523	2.3937	2.3329	2.2696	2.2032	2.1333
20	2.7737	2.6758	2.5731	2.4645	2.4076	2.3486	2.2873	2.2234	2.1562	2.0853
21	2.7348	2.6368	2.5338	2.4247	2.3675	2.3082	2.2465	2.1819	2.1141	2.0422
22	2.6998	2.6017	2.4984	2.3890	2.3315	2.2718	2.2097	2.1446	2.0760	2.0032
23	2.6682	2.5699	2.4665	2.3567	2.2989	2.2389	2.1763	2.1107	2.0415	1.9677
24	2.6396	2.5411	2.4374	2.3273	2.2693	2.2090	2.1460	2.0799	2.0099	1.9353
25	2.6135	2.5149	2.4110	2.3005	2.2422	2.1816	2.1183	2.0516	1.9811	1.9055
26	2.5896	2.4908	2.3867	2.2759	2.2174	2.1565	2.0928	2.0257	1.9545	1.8781
27	2.5676	2.4688	2.3644	2.2533	2.1946	2.1334	2.0693	2.0018	1.9299	1.8527
28	2.5473	2.4484	2.3438	2.2324	2.1735	2.1121	2.0477	1.9797	1.9072	1.8291
29	2.5286	2.4295	2.3248	2.2131	2.1540	2.0923	2.0276	1.9591	1.8861	1.8072
30	2.5112	2.4120	2.3072	2.1952	2.1359	2.0739	2.0089	1.9400	1.8664	1.7867
40	2.3882	2.2882	2.1819	2.0677	2.0069	1.9429	1.8752	1.8028	1.7242	1.6371
60	2.2702	2.1692'	2.0613	1.9445	1.8817	1.8152	1.7440	1.6668	1.5810	1.4821
120	2.1570	2.0548	1.9450	1.8249	1.7597	1.6899	1.6141	1.5299	1.4327	1.3104
∞	2.0483	1.9447	1.8326	1.7085	1.6402	1.5660	1.4835	1.3883	1.2684	1.0000

Denominator degrees of freedom (df_2)

From Maxine Merrington and Catherine M. Thompson, "Tables of Percentage Points of the Inverted Beta (F) Distribution," *Biometrika* 33 (1943): 80–84. Reprinted with permission of the Biometrika Trustees.

TABLE A-5   F Distribution ($\alpha = 0.05$ in the right tail)

Numerator degrees of freedom (df_1)

Denominator degrees of freedom (df_2)	1	2	3	4	5	6	7	8	9
1	161.45	199.50	215.71	224.58	230.16	233.99	236.77	238.88	240.54
2	18.513	19.000	19.164	19.247	19.296	19.330	19.353	19.371	19.385
3	10.128	9.5521	9.2766	9.1172	9.0135	8.9406	8.8867	8.8452	8.8123
4	7.7086	6.9443	6.5914	6.3882	6.2561	6.1631	6.0942	6.0410	5.9988
5	6.6079	5.7861	5.4095	5.1922	5.0503	4.9503	4.8759	4.8183	4.7725
6	5.9874	5.1433	4.7571	4.5337	4.3874	4.2839	4.2067	4.1468	4.0990
7	5.5914	4.7374	4.3468	4.1203	3.9715	3.8660	3.7870	3.7257	3.6767
8	5.3177	4.4590	4.0662	3.8379	3.6875	3.5806	3.5005	3.4381	3.3881
9	5.1174	4.2565	3.8625	3.6331	3.4817	3.3738	3.2927	3.2296	3.1789
10	4.9646	4.1028	3.7083	3.4780	3.3258	3.2172	3.1355	3.0717	3.0204
11	4.8443	3.9823	3.5874	3.3567	3.2039	3.0946	3.0123	2.9480	2.8962
12	4.7472	3.8853	3.4903	3.2592	3.1059	2.9961	2.9134	2.8486	2.7964
13	4.6672	3.8056	3.4105	3.1791	3.0254	2.9153	2.8321	2.7669	2.7144
14	4.6001	3.7389	3.3439	3.1122	2.9582	2.8477	2.7642	2.6987	2.6458
15	4.5431	3.6823	3.2874	3.0556	2.9013	2.7905	2.7066	2.6408	2.5876
16	4.4940	3.6337	3.2389	3.0069	2.8524	2.7413	2.6572	2.5911	2.5377
17	4.4513	3.5915	3.1968	2.9647	2.8100	2.6987	2.6143	2.5480	2.4943
18	4.4139	3.5546	3.1599	2.9277	2.7729	2.6613	2.5767	2.5102	2.4563
19	4.3807	3.5219	3.1274	2.8951	2.7401	2.6283	2.5435	2.4768	2.4227
20	4.3512	3.4928	3.0984	2.8661	2.7109	2.5990	2.5140	2.4471	2.3928
21	4.3248	3.4668	3.0725	2.8401	2.6848	2.5727	2.4876	2.4205	2.3660
22	4.3009	3.4434	3.0491	2.8167	2.6613	2.5491	2.4638	2.3965	2.3419
23	4.2793	3.4221	3.0280	2.7955	2.6400	2.5277	2.4422	2.3748	2.3201
24	4.2597	3.4028	3.0088	2.7763	2.6207	2.5082	2.4226	2.3551	2.3002
25	4.2417	3.3852	2.9912	2.7587	2.6030	2.4904	2.4047	2.3371	2.2821
26	4.2252	3.3690	2.9752	2.7426	2.5868	2.4741	2.3883	2.3205	2.2655
27	4.2100	3.3541	2.9604	2.7278	2.5719	2.4591	2.3732	2.3053	2.2501
28	4.1960	3.3404	2.9467	2.7141	2.5581	2.4453	2.3593	2.2913	2.2360
29	4.1830	3.3277	2.9340	2.7014	2.5454	2.4324	2.3463	2.2783	2.2229
30	4.1709	3.3158	2.9223	2.6896	2.5336	2.4205	2.3343	2.2662	2.2107
40	4.0847	3.2317	2.8387	2.6060	2.4495	2.3359	2.2490	2.1802	2.1240
60	4.0012	3.1504	2.7581	2.5252	2.3683	2.2541	2.1665	2.0970	2.0401
120	3.9201	3.0718	2.6802	2.4472	2.2899	2.1750	2.0868	2.0164	1.9588
∞	3.8415	2.9957	2.6049	2.3719	2.2141	2.0986	2.0096	1.9384	1.8799

(continued)

TABLE A-5 *F* Distribution (α = 0.05 in the right tail) *(continued)*

Numerator degrees of freedom (df$_1$)

df$_2$	10	12	15	20	24	30	40	60	120	∞
1	241.88	243.91	245.95	248.01	249.05	250.10	251.14	252.20	253.25	254.31
2	19.396	19.413	19.429	19.446	19.454	19.462	19.471	19.479	19.487	19.496
3	8.7855	8.7446	8.7029	8.6602	8.6385	8.6166	8.5944	8.5720	8.5494	8.5264
4	5.9644	5.9117	5.8578	5.8025	5.7744	5.7459	5.7170	5.6877	5.6581	5.6281
5	4.7351	4.6777	4.6188	4.5581	4.5272	4.4957	4.4638	4.4314	4.3985	4.3650
6	4.0600	3.9999	3.9381	3.8742	3.8415	3.8082	3.7743	3.7398	3.7047	3.6689
7	3.6365	3.5747	3.5107	3.4445	3.4105	3.3758	3.3404	3.3043	3.2674	3.2298
8	3.3472	3.2839	3.2184	3.1503	3.1152	3.0794	3.0428	3.0053	2.9669	2.9276
9	3.1373	3.0729	3.0061	2.9365	2.9005	2.8637	2.8259	2.7872	2.7475	2.7067
10	2.9782	2.9130	2.8450	2.7740	2.7372	2.6996	2.6609	2.6211	2.5801	2.5379
11	2.8536	2.7876	2.7186	2.6464	2.6090	2.5705	2.5309	2.4901	2.4480	2.4045
12	2.7534	2.6866	2.6169	2.5436	2.5055	2.4663	2.4259	2.3842	2.3410	2.2962
13	2.6710	2.6037	2.5331	2.4589	2.4202	2.3803	2.3392	2.2966	2.2524	2.2064
14	2.6022	2.5342	2.4630	2.3879	2.3487	2.3082	2.2664	2.2229	2.1778	2.1307
15	2.5437	2.4753	2.4034	2.3275	2.2878	2.2468	2.2043	2.1601	2.1141	2.0658
16	2.4935	2.4247	2.3522	2.2756	2.2354	2.1938	2.1507	2.1058	2.0589	2.0096
17	2.4499	2.3807	2.3077	2.2304	2.1898	2.1477	2.1040	2.0584	2.0107	1.9604
18	2.4117	2.3421	2.2686	2.1906	2.1497	2.1071	2.0629	2.0166	1.9681	1.9168
19	2.3779	2.3080	2.2341	2.1555	2.1141	2.0712	2.0264	1.9795	1.9302	1.8780
20	2.3479	2.2776	2.2033	2.1242	2.0825	2.0391	1.9938	1.9464	1.8963	1.8432
21	2.3210	2.2504	2.1757	2.0960	2.0540	2.0102	1.9645	1.9165	1.8657	1.8117
22	2.2967	2.2258	2.1508	2.0707	2.0283	1.9842	1.9380	1.8894	1.8380	1.7831
23	2.2747	2.2036	2.1282	2.0476	2.0050	1.9605	1.9139	1.8648	1.8128	1.7570
24	2.2547	2.1834	2.1077	2.0267	1.9838	1.9390	1.8920	1.8424	1.7896	1.7330
25	2.2365	2.1649	2.0889	2.0075	1.9643	1.9192	1.8718	1.8217	1.7684	1.7110
26	2.2197	2.1479	2.0716	1.9898	1.9464	1.9010	1.8533	1.8027	1.7488	1.6906
27	2.2043	2.1323	2.0558	1.9736	1.9299	1.8842	1.8361	1.7851	1.7306	1.6717
28	2.1900	2.1179	2.0411	1.9586	1.9147	1.8687	1.8203	1.7689	1.7138	1.6541
29	2.1768	2.1045	2.0275	1.9446	1.9005	1.8543	1.8055	1.7537	1.6981	1.6376
30	2.1646	2.0921	2.0148	1.9317	1.8874	1.8409	1.7918	1.7396	1.6835	1.6223
40	2.0772	2.0035	1.9245	1.8389	1.7929	1.7444	1.6928	1.6373	1.5766	1.5089
60	1.9926	1.9174	1.8364	1.7480	1.7001	1.6491	1.5943	1.5343	1.4673	1.3893
120	1.9105	1.8337	1.7505	1.6587	1.6084	1.5543	1.4952	1.4290	1.3519	1.2539
∞	1.8307	1.7522	1.6664	1.5705	1.5173	1.4591	1.3940	1.3180	1.2214	1.0000

Denominator degrees of freedom (df$_2$)

From Maxine Merrington and Catherine M. Thompson, "Tables of Percentage Points of the Inverted Beta *(F)* Distribution," *Biometrika* 33 (1943): 80–84. Reproduced with permission of the Biometrika Trustees.

TABLE A-6
Critical Values of the Pearson Correlation Coefficient r

n	$\alpha = .05$	$\alpha = .01$
4	.950	.999
5	.878	.959
6	.811	.917
7	.754	.875
8	.707	.834
9	.666	.798
10	.632	.765
11	.602	.735
12	.576	.708
13	.553	.684
14	.532	.661
15	.514	.641
16	.497	.623
17	.482	.606
18	.468	.590
19	.456	.575
20	.444	.561
25	.396	.505
30	.361	.463
35	.335	.430
40	.312	.402
45	.294	.378
50	.279	.361
60	.254	.330
70	.236	.305
80	.220	.286
90	.207	.269
100	.196	.256

NOTE: To test H_0: $\rho = 0$ against H_1: $\rho \neq 0$, reject H_0 if the absolute value of r is greater than the critical value in the table.

TABLE A-7   Critical Values for the Sign Test

n	.005 (one tail) .01 (two tails)	.01 (one tail) .02 (two tails)	.025 (one tail) .05 (two tails)	.05 (one tail) .10 (two tails)
1	*	*	*	*
2	*	*	*	*
3	*	*	*	*
4	*	*	*	*
5	*	*	*	0
6	*	*	0	0
7	*	0	0	0
8	0	0	0	1
9	0	0	1	1
10	0	0	1	1
11	0	1	1	2
12	1	1	2	2
13	1	1	2	3
14	1	2	2	3
15	2	2	3	3
16	2	2	3	4
17	2	3	4	4
18	3	3	4	5
19	3	4	4	5
20	3	4	5	5
21	4	4	5	6
22	4	5	5	6
23	4	5	6	7
24	5	5	6	7
25	5	6	7	7

NOTES:
1. * indicates that it is not possible to get a value in the critical region.
2. Reject the null hypothesis if the number of the less frequent sign (x) is less than or equal to the value in the table.
3. For values of n greater than 25, a normal approximation is used with

$$z = \frac{(x + 0.5) - \left(\frac{n}{2}\right)}{\frac{\sqrt{n}}{2}}$$

TABLE A-8   Critical Values of T for the Wilcoxon Signed-Rank Test

	α			
n	.005 (one tail) .01 (two tails)	.01 (one tail) .02 (two tails)	.025 (one tail) .05 (two tails)	.05 (one tail) .10 (two tails)
5	*	*	*	1
6	*	*	1	2
7	*	0	2	4
8	0	2	4	6
9	2	3	6	8
10	3	5	8	11
11	5	7	11	14
12	7	10	14	17
13	10	13	17	21
14	13	16	21	26
15	16	20	25	30
16	19	24	30	36
17	23	28	35	41
18	28	33	40	47
19	32	38	46	54
20	37	43	52	60
21	43	49	59	68
22	49	56	66	75
23	55	62	73	83
24	61	69	81	92
25	68	77	90	101
26	76	85	98	110
27	84	93	107	120
28	92	102	117	130
29	100	111	127	141
30	109	120	137	152

NOTES:
1. * indicates that it is not possible to get a value in the critical region.
2. Reject the null hypothesis if the test statistic T is less than or equal to the critical value found in this table. Fail to reject the null hypothesis if the test statistic T is greater than the critical value found in this table.

From *Some Rapid Approximate Statistical Procedures,* Copyright © 1949, 1964 Lederle Laboratories Division of American Cyanamid Company. Reprinted with the permission of the American Cyanamid Company.

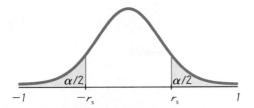

TABLE A-9   Critical Values of Spearman's Rank Correlation Coefficient r_s

n	$\alpha = 0.10$	$\alpha = 0.05$	$\alpha = 0.02$	$\alpha = 0.01$
5	.900	—	—	—
6	.829	.886	.943	—
7	.714	.786	.893	—
8	.643	.738	.833	.881
9	.600	.683	.783	.833
10	.564	.648	.745	.794
11	.523	.623	.736	.818
12	.497	.591	.703	.780
13	.475	.566	.673	.745
14	.457	.545	.646	.716
15	.441	.525	.623	.689
16	.425	.507	.601	.666
17	.412	.490	.582	.645
18	.399	.476	.564	.625
19	.388	.462	.549	.608
20	.377	.450	.534	.591
21	.368	.438	.521	.576
22	.359	.428	.508	.562
23	.351	.418	.496	.549
24	.343	.409	.485	.537
25	.336	.400	.475	.526
26	.329	.392	.465	.515
27	.323	.385	.456	.505
28	.317	.377	.448	.496
29	.311	.370	.440	.487
30	.305	.364	.432	.478

NOTE: For $n > 30$ use $r_s = \pm z/\sqrt{n-1}$, where z corresponds to the level of significance. For example, if $\alpha = 0.05$, then $z = 1.96$.

To test H_0: $\rho_s = 0$
against H_1: $\rho_s \neq 0$

From "Distribution of sums of squares of rank differences to small numbers of individuals," *The Annals of Mathematical Statistics*, Vol. 9, No. 2. Reprinted with permission of the Institute of Mathematical Statistics.

TABLE A-10   Critical Values for Number of Runs G

Value of n_2

Value of n_1	2	3	4	5	6	7	8	9	10	11	12	13	14	15	16	17	18	19	20
2	1	1	1	1	1	1	1	1	1	1	2	2	2	2	2	2	2	2	2
	6	6	6	6	6	6	6	6	6	6	6	6	6	6	6	6	6	6	6
3	1	1	1	1	2	2	2	2	2	2	2	2	2	3	3	3	3	3	3
	6	8	8	8	8	8	8	8	8	8	8	8	8	8	8	8	8	8	8
4	1	1	1	2	2	2	3	3	3	3	3	3	3	4	4	4	4	4	4
	6	8	9	9	9	10	10	10	10	10	10	10	10	10	10	10	10	10	10
5	1	1	2	2	3	3	3	3	3	4	4	4	4	4	4	4	5	5	5
	6	8	9	10	10	11	11	12	12	12	12	12	12	12	12	12	12	12	12
6	1	2	2	3	3	3	3	4	4	4	4	5	5	5	5	5	5	6	6
	6	8	9	10	11	12	12	13	13	13	13	14	14	14	14	14	14	14	14
7	1	2	2	3	3	3	4	4	5	5	5	5	5	6	6	6	6	6	6
	6	8	10	11	12	13	13	14	14	14	14	15	15	15	16	16	16	16	16
8	1	2	3	3	3	4	4	5	5	5	6	6	6	6	6	7	7	7	7
	6	8	10	11	12	13	14	14	15	15	16	16	16	16	17	17	17	17	17
9	1	2	3	3	4	4	5	5	5	6	6	6	7	7	7	7	8	8	8
	6	8	10	12	13	14	14	15	16	16	16	17	17	18	18	18	18	18	18
10	1	2	3	3	4	5	5	5	6	6	7	7	7	7	8	8	8	8	9
	6	8	10	12	13	14	15	16	16	17	17	18	18	18	19	19	19	20	20
11	1	2	3	4	4	5	5	6	6	7	7	7	8	8	8	9	9	9	9
	6	8	10	12	13	14	15	16	17	17	18	19	19	19	20	20	20	21	21
12	2	2	3	4	4	5	6	6	7	7	7	8	8	9	9	9	10	10	10
	6	8	10	12	13	14	16	16	17	18	19	19	20	20	21	21	21	22	22
13	2	2	3	4	5	5	6	6	7	7	8	8	9	9	9	10	10	10	10
	6	8	10	12	14	15	16	17	18	19	19	20	20	21	21	22	22	23	23
14	2	2	3	4	5	5	6	7	7	8	8	9	9	9	10	10	10	11	11
	6	8	10	12	14	15	16	17	18	19	20	20	21	22	22	23	23	23	24
15	2	3	3	4	5	6	6	7	7	8	8	9	9	10	10	11	11	11	12
	6	8	10	12	14	15	16	18	18	19	20	21	22	22	23	23	24	24	25
16	2	3	4	4	5	6	6	7	8	8	9	9	10	10	11	11	11	12	12
	6	8	10	12	14	16	17	18	19	20	21	21	22	23	23	24	25	25	25
17	2	3	4	4	5	6	7	7	8	9	9	10	10	11	11	11	12	12	13
	6	8	10	12	14	16	17	18	19	20	21	22	23	23	24	25	25	26	26
18	2	3	4	5	5	6	7	8	8	9	9	10	10	11	11	12	12	13	13
	6	8	10	12	14	16	17	18	19	20	21	22	23	24	25	25	26	26	27
19	2	3	4	5	6	6	7	8	8	9	10	10	11	11	12	12	13	13	13
	6	8	10	12	14	16	17	18	20	21	22	23	23	24	25	26	26	27	27
20	2	3	4	5	6	6	7	8	9	9	10	10	11	12	12	13	13	13	14
	6	8	10	12	14	16	17	18	20	21	22	23	24	25	25	26	27	27	28

NOTE:

1. The entries in this table are the critical G values, assuming a two-tailed test with a significance level of $\alpha = 0.05$.
2. The null hypothesis of randomness is rejected if the total number of runs G is less than or equal to the smaller entry or greater than or equal to the larger entry.

From "Tables for testing randomness of groupings in a sequence of alternatives," *The Annals of Mathematical Statistics*, Vol. 14, No. 1. Reprinted with permission of the Institute of Mathematical Statistics.

Appendix B: Data Sets

Data Set 1: Weights (in pounds) of Household Garbage for One Week

HHSIZE = household size
METAL = weight of discarded metals
PAPER = weight of discarded paper goods
PLAS = weight of discarded plastic goods
GLASS = weight of discarded glass products
FOOD = weight of discarded food items
YARD = weight of discarded yard waste
TEXT = weight of discarded textile goods
OTHER = weight of discarded goods not included in the above categories
TOTAL = total weight of discarded materials

 STATDISK: Variable names are HHSIZE, METAL, PAPER, PLAS, GLASS, FOOD, YARD, TEXT, OTHER, and TOTAL.

Minitab: Enter RETRIEVE 'GARBAGE' to get columns C1 through C10, which correspond to the 10 columns of data.

HOUSEHOLD	HHSIZE	METAL	PAPER	PLAS	GLASS	FOOD	YARD	TEXT	OTHER	TOTAL
1	2	1.09	2.41	0.27	0.86	1.04	0.38	0.05	4.66	10.76
2	3	1.04	7.57	1.41	3.46	3.68	0.00	0.46	2.34	19.96
3	3	2.57	9.55	2.19	4.52	4.43	0.24	0.50	3.60	27.60
4	6	3.02	8.82	2.83	4.92	2.98	0.63	2.26	12.65	38.11
5	4	1.50	8.72	2.19	6.31	6.30	0.15	0.55	2.18	27.90
6	2	2.10	6.96	1.81	2.49	1.46	4.58	0.36	2.14	21.90
7	1	1.93	6.83	0.85	0.51	8.82	0.07	0.60	2.22	21.83
8	5	3.57	11.42	3.05	5.81	9.62	4.76	0.21	10.83	49.27
9	6	2.32	16.08	3.42	1.96	4.41	0.13	0.81	4.14	33.27
10	4	1.89	6.38	2.10	17.67	2.73	3.86	0.66	0.25	35.54
11	4	3.26	13.05	2.93	3.21	9.31	0.70	0.37	11.61	44.44
12	7	3.99	11.36	2.44	4.94	3.59	13.45	4.25	1.15	45.17
13	3	2.04	15.09	2.17	3.10	5.36	0.74	0.42	4.15	33.07
14	5	0.99	2.80	1.41	1.39	1.47	0.82	0.44	1.03	10.35
15	6	2.96	6.44	2.00	5.21	7.06	6.14	0.20	14.43	44.44
16	2	1.50	5.86	0.93	2.03	2.52	1.37	0.27	9.65	24.13
17	4	2.43	11.08	2.97	1.74	1.75	14.70	0.39	2.54	37.60
18	4	2.97	12.43	2.04	3.99	5.64	0.22	2.47	9.20	38.96
19	3	1.42	6.05	0.65	6.26	1.93	0.00	0.86	0.00	17.17
20	3	3.60	13.61	2.13	3.52	6.46	0.00	0.96	1.32	31.60
21	2	4.48	6.98	0.63	2.01	6.72	2.00	0.11	0.18	23.11
22	2	1.36	14.33	1.53	2.21	5.76	0.58	0.17	1.62	27.56
23	4	2.11	13.31	4.69	0.25	9.72	0.02	0.46	0.40	30.96
24	1	0.41	3.27	0.15	0.09	0.16	0.00	0.00	0.00	4.08
25	4	2.02	6.67	1.45	6.85	5.52	0.00	0.68	0.03	23.22

(continued)

Data Set 1 (continued)

HOUSEHOLD	HHSIZE	METAL	PAPER	PLAS	GLASS	FOOD	YARD	TEXT	OTHER	TOTAL
26	6	3.27	17.65	2.68	2.33	11.92	0.83	0.28	4.03	42.99
27	11	4.95	12.73	3.53	5.45	4.68	0.00	0.67	19.89	51.90
28	3	1.00	9.83	1.49	2.04	4.76	0.42	0.54	0.12	20.20
29	4	1.55	16.39	2.31	4.98	7.85	2.04	0.20	1.48	36.80
30	3	1.41	6.33	0.92	3.54	2.90	3.85	0.03	0.04	19.02
31	2	1.05	9.19	0.89	1.06	2.87	0.33	0.01	0.03	15.43
32	2	1.31	9.41	0.80	2.70	5.09	0.64	0.05	0.71	20.71
33	2	2.50	9.45	0.72	1.14	3.17	0.00	0.02	0.01	17.01
34	4	2.35	12.32	2.66	12.24	2.40	7.87	4.73	0.78	45.35
35	6	3.69	20.12	4.37	5.67	13.20	0.00	1.15	1.17	49.37
36	2	3.61	7.72	0.92	2.43	2.07	0.68	0.63	0.00	18.06
37	2	1.49	6.16	1.40	4.02	4.00	0.30	0.04	0.00	17.41
38	2	1.36	7.98	1.45	6.45	4.27	0.02	0.12	2.02	23.67
39	2	1.73	9.64	1.68	1.89	1.87	0.01	1.73	0.58	19.13
40	2	0.94	8.08	1.53	1.78	8.13	0.36	0.12	0.05	20.99
41	3	1.33	10.99	1.44	2.93	3.51	0.00	0.39	0.59	21.18
42	3	2.62	13.11	1.44	1.82	4.21	4.73	0.64	0.49	29.06
43	2	1.25	3.26	1.36	2.89	3.34	2.69	0.00	0.16	14.95
44	2	0.26	1.65	0.38	0.99	0.77	0.34	0.04	0.00	4.43
45	3	4.41	10.00	1.74	1.93	1.14	0.92	0.08	4.60	24.82
46	6	3.22	8.96	2.35	3.61	1.45	0.00	0.09	1.12	20.80
47	4	1.86	9.46	2.30	2.53	6.54	0.00	0.65	2.45	25.79
48	4	1.76	5.88	1.14	3.76	0.92	1.12	0.00	0.04	14.62
49	3	2.83	8.26	2.88	1.32	5.14	5.60	0.35	2.03	28.41
50	3	2.74	12.45	2.13	2.64	4.59	1.07	0.41	1.14	27.17
51	10	4.63	10.58	5.28	12.33	2.94	0.12	2.94	15.65	54.47
52	3	1.70	5.87	1.48	1.79	1.42	0.00	0.27	0.59	13.12
53	6	3.29	8.78	3.36	3.99	10.44	0.90	1.71	13.30	45.77
54	5	1.22	11.03	2.83	4.44	3.00	4.30	1.95	6.02	34.79
55	4	3.20	12.29	2.87	9.25	5.91	1.32	1.87	0.55	37.26
56	7	3.09	20.58	2.96	4.02	16.81	0.47	1.52	2.13	51.58
57	5	2.58	12.56	1.61	1.38	5.01	0.00	0.21	1.46	24.81
58	4	1.67	9.92	1.58	1.59	9.96	0.13	0.20	1.13	26.18
59	2	0.85	3.45	1.15	0.85	3.89	0.00	0.02	1.04	11.25
60	4	1.52	9.09	1.28	8.87	4.83	0.00	0.95	1.61	28.15
61	2	1.37	3.69	0.58	3.64	1.78	0.08	0.00	0.00	11.14
62	2	1.32	2.61	0.74	3.03	3.37	0.17	0.00	0.46	11.70

Data provided by Masakazu Tani, the Garbage Project, University of Arizona.

Data Set 2: Body Temperatures (in degrees Fahrenheit) of Healthy Adults

This is the only Appendix B data set not available on disk.

SUBJECT	AGE	SEX	SMOKE	Temperature Day 1 8 AM	Temperature Day 1 12 AM	Temperature Day 2 8 AM	Temperature Day 2 12 AM
1	22	M	Y	98.0	98.0	98.0	98.6
2	23	M	Y	97.0	97.6	97.4	—
3	22	M	Y	98.6	98.8	97.8	98.6
4	19	M	N	97.4	98.0	97.0	98.0
5	18	M	N	98.2	98.8	97.0	98.0
6	20	M	Y	98.2	98.8	96.6	99.0
7	27	M	Y	98.2	97.6	97.0	98.4
8	19	M	Y	96.6	98.6	96.8	98.4
9	19	M	Y	97.4	98.6	96.6	98.4
10	24	M	N	97.4	98.8	96.6	98.4
11	35	M	Y	98.2	98.0	96.2	98.6
12	25	M	Y	97.4	98.2	97.6	98.6
13	25	M	N	97.8	98.0	98.6	98.8
14	35	M	Y	98.4	98.0	97.0	98.6
15	21	M	N	97.6	97.0	97.4	97.0
16	33	M	N	96.2	97.2	98.0	97.0
17	19	M	Y	98.0	98.2	97.6	98.8
18	24	M	Y	—	—	97.2	97.6
19	18	F	N	—	—	97.0	97.7
20	22	F	Y	—	—	98.0	98.8
21	20	M	Y	—	—	97.0	98.0
22	30	F	Y	—	—	96.4	98.0
23	29	M	N	—	—	96.1	98.3
24	18	M	Y	—	—	98.0	98.5
25	31	M	Y	—	98.1	96.8	97.3
26	28	F	Y	—	98.2	98.2	98.7
27	27	M	Y	—	98.5	97.8	97.4
28	21	M	Y	—	98.5	98.2	98.9
29	30	M	Y	—	99.0	97.8	98.6
30	27	M	N	—	98.0	99.0	99.5
31	32	M	Y	—	97.0	97.4	97.5
32	33	M	Y	—	97.3	97.4	97.3
33	23	M	Y	—	97.3	97.5	97.6
34	29	M	Y	—	98.1	97.8	98.2
35	25	M	Y	—	—	97.9	99.6
36	31	M	N	—	97.8	97.8	98.7

(continued)

Data Set 2 (continued)

SUBJECT	AGE	SEX	SMOKE	Temperature Day 1		Temperature Day 2	
				8 AM	12 AM	8 AM	12 AM
37	25	M	Y	—	99.0	98.3	99.4
38	28	M	N	—	97.6	98.0	98.2
39	30	M	Y	—	97.4	—	98.0
40	33	M	Y	—	98.0	—	98.6
41	28	M	Y	98.0	97.4	—	98.6
42	22	M	Y	98.8	98.0	—	97.2
43	21	F	Y	99.0	—	—	98.4
44	30	M	N	—	98.6	—	98.6
45	22	M	Y	—	98.6	—	98.2
46	22	F	N	98.0	98.4	—	98.0
47	20	M	Y	—	97.0	—	97.8
48	19	M	Y	—	—	—	98.0
49	33	M	N	—	98.4	—	98.4
50	31	M	Y	99.0	99.0	—	98.6
51	26	M	N	—	98.0	—	98.6
52	18	M	N	—	—	—	97.8
53	23	M	N	—	99.4	—	99.0
54	28	M	Y	—	—	—	96.5
55	19	M	Y	—	97.8	—	97.6
56	21	M	N	—	—	—	98.0
57	27	M	Y	—	98.2	—	96.9
58	29	M	Y	—	99.2	—	97.6
59	38	M	N	—	99.0	—	97.1
60	29	F	Y	—	97.7	—	97.9
61	22	M	Y	—	98.2	—	98.4
62	22	M	Y	—	98.2	—	97.3
63	26	M	Y	—	98.8	—	98.0
64	32	M	N	—	98.1	—	97.5
65	25	M	Y	—	98.5	—	97.6
66	21	F	N	—	97.2	—	98.2
67	25	M	Y	—	98.5	—	98.5
68	24	M	Y	—	99.2	97.0	98.8
69	25	M	Y	—	98.3	97.6	98.7
70	35	M	Y	—	98.7	97.5	97.8
71	23	F	Y	—	98.8	98.8	98.0
72	31	M	Y	—	98.6	98.4	97.1
73	28	M	Y	—	98.0	98.2	97.4
74	29	M	Y	—	99.1	97.7	99.4
75	26	M	Y	—	97.2	97.3	98.4
76	32	M	N	—	97.6	97.5	98.6
77	32	M	Y	—	97.9	97.1	98.4

(continued)

Data Set 2 (continued)

SUBJECT	AGE	SEX	SMOKE	Temperature Day 1 8 AM	Temperature Day 1 12 AM	Temperature Day 2 8 AM	Temperature Day 2 12 AM
78	21	F	Y	—	98.8	98.6	98.5
79	20	M	Y	—	98.6	98.6	98.6
80	24	F	Y	—	98.6	97.8	98.3
81	21	F	Y	—	99.3	98.7	98.7
82	28	M	Y	—	97.8	97.9	98.8
83	27	F	N	98.8	98.7	97.8	99.1
84	28	M	N	99.4	99.3	97.8	98.6
85	29	M	Y	98.8	97.8	97.6	97.9
86	19	M	N	97.7	98.4	96.8	98.8
87	24	M	Y	99.0	97.7	96.0	98.0
88	29	M	N	98.1	98.3	98.0	98.7
89	25	M	Y	98.7	97.7	97.0	98.5
90	27	M	N	97.5	97.1	97.4	98.9
91	25	M	Y	98.9	98.4	97.6	98.4
92	21	M	Y	98.4	98.6	97.6	98.6
93	19	M	Y	97.2	97.4	96.2	97.1
94	27	M	Y	—	—	96.2	97.9
95	32	M	N	98.8	96.7	98.1	98.8
96	24	M	Y	97.3	96.9	97.1	98.7
97	32	M	Y	98.7	98.4	98.2	97.6
98	19	F	Y	98.9	98.2	96.4	98.2
99	18	F	Y	99.2	98.6	96.9	99.2
100	27	M	N	—	97.0	—	97.8
101	34	M	Y	—	97.4	—	98.0
102	25	M	N	—	98.4	—	98.4
103	18	M	N	—	97.4	—	97.8
104	32	M	Y	—	96.8	—	98.4
105	31	M	Y	—	98.2	—	97.4
106	26	M	N	—	97.4	—	98.0
107	23	M	N	—	98.0	—	97.0

Data provided by Dr. Steven Wasserman, Dr. Philip Mackowiak, and Dr. Myron Levine of the University of Maryland.

Data Set 3: Survey of Persons Employed and Currently Working Full-Time

Age: Years

Sex: 1 = male, 2 = female

Married: 2 = married, 3 = widowed, 4 = divorced,
5 = separated, 6 = never married

Education: 21–28 = elementary grades 1–8 (2nd digit is grade)
31–34 = high school for 1–4 years (2nd digit is years)
41–45 = college for 1–5 years (2nd digit is years)

Weight: Pounds

Height: Inches

Size: Number of persons in the household

Cars: Number of cars (9 indicates blank)

Income: Family income in thousands of dollars

Region: 1 = Northeast, 2 = Midwest, 3 = South, 4 = West

Central City: 1 = in a central city that is a standard metropolitan statistical area
2 = in a standard metropolitan statistical area that is not a city
4 = not in a standard metropolitan statistical area

 STATDISK: Variable names are AGE, SEX, MARRIED, EDUC, WEIGHT, HEIGHT, SIZE, CARS, INCOME, REGION, CENTCITY.

Minitab: Enter RETRIEVE 'SURVEY' to get columns C1 through C11, which correspond to the 11 columns of data.

SUBJECT	AGE	SEX	MARRIED	EDUCATION	WEIGHT	HEIGHT	SIZE	CARS	INCOME	REGION	CENTRAL CITY
1	60	2	2	34	160	66.9	2	2	43.4	4	2
2	63	1	2	34	180	66.2	2	9	18.8	4	1
3	51	2	2	34	201	66.1	4	0	14.4	3	1
4	25	1	2	42	173	69.3	4	2	29.4	3	1
5	47	2	2	34	157	67.4	2	9	19.4	3	1
6	56	1	2	34	167	70.6	2	2	83.0	3	1
7	19	2	5	41	108	64.1	2	1	10.4	4	1
8	24	1	6	44	149	65.5	1	1	12.6	4	1
9	25	1	6	34	136	68.3	2	3	36.4	4	1
10	20	2	6	34	128	62.6	3	0	29.6	4	2
11	66	1	2	22	144	61.6	5	2	17.2	4	2
12	19	1	2	34	140	71.4	2	2	17.2	3	1

(continued)

Data Set 3 (continued)

SUBJECT	AGE	SEX	MARRIED	EDUCATION	WEIGHT	HEIGHT	SIZE	CARS	INCOME	REGION	CENTRAL CITY
13	48	2	2	45	192	66.7	2	3	67.0	3	1
14	52	1	2	33	163	65.7	5	3	33.0	3	1
15	27	1	2	42	161	67.3	2	2	37.4	2	4
16	61	1	2	34	155	68.1	2	2	106.4	2	4
17	35	1	2	34	157	71.2	4	2	41.2	2	4
18	59	1	2	28	182	69.0	2	2	15.8	4	4
19	21	1	2	34	194	68.4	2	2	26.4	2	1
20	41	1	2	34	176	70.0	7	5	39.0	4	4
21	33	2	4	34	133	66.8	3	1	19.4	4	4
22	28	2	2	34	135	62.8	2	2	67.2	2	1
23	26	1	6	43	172	73.1	1	1	13.6	2	2
24	29	1	2	34	148	65.5	4	1	21.4	2	1
25	20	2	6	34	147	70.0	1	1	7.4	3	4
26	61	1	2	34	194	70.6	2	2	36.6	3	4
27	21	1	2	34	167	69.3	3	2	27.6	3	4
28	23	2	6	34	136	65.9	7	1	11.2	3	4
29	30	1	2	34	144	65.4	3	1	16.2	3	4
30	46	1	2	23	217	71.6	3	3	9.4	3	4
31	64	1	2	45	183	70.8	2	2	6.4	3	4
32	25	1	2	34	135	69.8	3	2	18.8	3	4
33	38	1	2	32	155	72.2	5	1	42.4	3	4
34	40	1	2	33	235	72.0	5	3	16.0	3	4
35	65	1	2	28	152	64.1	3	2	46.4	1	1
36	25	2	2	34	211	64.3	2	2	47.0	1	2
37	60	1	2	32	148	68.8	2	1	23.6	1	2
38	20	1	6	41	156	71.8	3	2	43.2	1	2
39	21	1	6	41	146	67.3	3	3	32.4	1	2
40	22	2	6	42	138	63.6	5	3	33.6	1	2
41	48	1	2	45	171	67.7	6	3	49.0	1	2
42	54	1	2	42	200	72.0	2	1	49.2	1	2
43	18	2	6	34	186	64.6	7	2	7.8	1	4
44	47	2	2	34	176	61.6	5	2	29.8	1	4
45	28	1	2	44	184	72.1	4	1	56.8	1	2
46	66	1	2	31	165	68.5	1	2	17.2	1	1
47	29	1	4	34	146	71.3	3	3	45.0	1	1
48	30	1	2	34	163	70.0	4	2	48.2	1	2

(continued)

Data Set 3 (continued)

SUBJECT	AGE	SEX	MARRIED	EDUCATION	WEIGHT	HEIGHT	SIZE	CARS	INCOME	REGION	CENTRAL CITY
49	29	1	6	34	189	70.2	1	2	34.8	1	2
50	22	1	6	44	155	68.7	4	5	54.6	1	4
51	61	1	3	32	176	68.7	1	1	22.2	1	4
52	39	2	2	31	117	59.8	7	2	22.6	1	4
53	60	2	2	31	144	60.4	3	2	28.0	1	4
54	34	1	4	43	154	72.1	1	0	10.6	1	1
55	20	2	6	34	113	63.6	1	2	26.8	1	2
56	20	1	6	33	159	68.4	1	0	11.6	1	1
57	46	1	2	34	152	68.6	4	3	34.6	1	2
58	55	1	2	34	233	68.9	3	3	47.4	1	2
59	31	2	6	42	120	64.5	3	2	49.0	1	2
60	51	2	2	27	229	65.9	2	2	25.6	3	4
61	38	2	4	28	134	63.9	5	1	17.2	3	4
62	23	1	6	31	110	64.8	4	2	157.8	3	1
63	28	2	4	44	118	66.0	1	1	15.6	3	1
64	56	1	2	45	185	70.7	2	2	77.0	3	1
65	63	2	4	34	157	65.1	1	1	11.6	4	2
66	25	1	6	34	161	70.0	1	1	20.2	4	2
67	27	2	6	45	121	64.3	2	2	47.4	4	1
68	33	1	2	34	188	72.2	5	2	30.2	4	4
69	38	1	2	34	176	72.2	6	2	37.2	4	4
70	47	1	6	31	171	68.0	1	2	19.4	4	2
71	23	2	6	34	122	62.4	7	1	29.8	4	2
72	27	1	2	42	191	71.1	3	2	30.4	4	2
73	19	1	2	34	152	70.6	2	1	22.0	4	2
74	44	1	2	44	200	65.2	4	2	30.8	4	2
75	27	2	2	42	117	61.7	3	2	41.2	4	1
76	60	1	2	44	170	69.6	3	2	42.4	4	4
77	46	1	2	41	205	73.3	3	2	17.0	4	4
78	34	1	6	34	166	66.8	1	2	28.4	2	4
79	55	1	2	34	224	69.5	3	2	25.4	2	4
80	29	1	2	34	149	67.6	5	2	33.6	2	4
81	57	2	2	34	96	63.0	2	1	10.4	2	4
82	44	1	2	34	175	68.5	3	2	32.4	2	4
83	44	2	2	32	118	61.9	3	3	31.4	2	4
84	21	1	6	34	133	70.0	4	1	89.2	2	1
85	51	2	3	28	200	62.3	1	1	15.0	2	2
86	21	1	6	34	153	65.7	7	0	11.4	2	1
87	42	2	4	41	151	63.5	5	2	29.8	2	2

(continued)

Data Set 3 (continued)

SUBJECT	AGE	SEX	MARRIED	EDUCATION	WEIGHT	HEIGHT	SIZE	CARS	INCOME	REGION	CENTRAL CITY
88	51	1	2	33	228	70.2	5	2	42.6	2	1
89	30	1	4	45	173	69.6	1	1	26.0	2	4
90	55	1	2	31	200	72.4	5	4	75.0	2	4
91	42	1	2	41	216	74.9	4	2	40.0	2	4
92	44	1	2	45	185	69.6	3	2	59.4	2	4
93	46	2	2	34	136	64.6	4	8	31.4	3	4
94	51	1	2	34	196	67.2	4	3	62.4	3	4
95	37	1	2	34	137	66.7	3	2	28.8	3	4
96	69	1	2	28	137	67.1	2	2	9.2	3	4
97	56	2	2	34	125	64.6	2	2	30.0	3	4
98	39	1	2	34	176	72.1	5	3	33.0	3	4
99	63	1	2	26	193	67.1	2	3	31.8	3	4
100	64	1	2	34	196	68.8	2	2	29.0	3	4
101	62	2	3	34	147	64.9	1	1	33.8	3	4
102	21	1	6	34	164	72.0	10	2	18.8	3	2
103	26	2	2	41	116	64.1	2	2	31.8	3	2
104	31	2	2	33	143	64.1	4	2	80.0	3	2
105	49	2	2	32	119	66.6	2	2	26.4	4	4
106	41	1	2	42	162	72.0	5	2	21.0	4	4
107	27	1	2	43	174	67.9	2	2	22.6	4	4
108	22	2	2	44	118	59.8	2	3	23.6	4	4
109	29	2	2	34	124	62.1	3	2	33.0	3	4
110	40	1	2	34	220	70.7	4	2	49.2	3	4
111	43	1	4	32	148	69.4	2	1	55.8	2	4
112	74	1	3	26	192	71.3	2	1	3.6	2	4
113	22	1	6	34	166	68.1	3	2	48.8	2	4
114	34	1	2	28	146	70.7	4	1	16.6	2	4
115	24	1	2	42	133	66.8	3	2	21.4	2	4
116	61	1	2	32	147	71.5	2	2	30.0	2	2
117	24	1	2	34	173	68.7	4	2	29.6	2	2
118	31	2	2	41	133	61.7	3	2	34.6	2	2
119	50	1	2	44	178	72.0	3	3	66.2	2	1
120	37	1	4	34	176	70.6	3	2	29.4	2	2
121	28	1	6	34	158	67.6	1	3	40.4	2	2
122	30	1	2	45	153	69.7	3	1	21.2	1	2
123	42	1	2	34	191	70.7	5	2	27.4	1	4
124	40	1	2	34	261	71.1	5	2	45.0	1	4
125	24	2	2	34	92	61.3	2	1	21.2	1	4

Based on data from the U.S. National Center for Health Statistics.

Data Set 4: Domestic New Cars

STATDISK: Variable names are PRICE, CARWT, CARCITY, CARHWY, CARLEN, CARWIDTH, CARHT.

Minitab: Enter `RETRIEVE 'CARS'` to get columns C1 through C7, which correspond to the 7 columns of data.

Car	Car Type	Suggested List Price ($)	Weight (lb)	City (mi/gal)	Hwy (mi/gal)	Length (in.)	Width (in.)	Height (in.)
1	Buick Century	15965	2948	23	31	189.1	69.4	54.2
2	Buick Park Ave.	26040	3536	19	27	205.2	74.9	55.1
3	Buick Regal	19310	3472	19	27	194.8	72.5	54.5
4	Buick Skylark	15760	2782	20	29	189.1	67.5	53.2
5	Cadillac Allante	59975	3766	14	21	178.7	73.4	51.5
6	Cadillac Fleetwood	33990	4367	16	25	225.1	78.0	57.1
7	Chevrolet Beretta	12995	2649	21	30	183.4	68.2	56.2
8	Chevrolet Cavalier	9520	2526	23	31	182.3	66.3	52.0
9	Chevrolet Corsica	11395	2665	25	34	183.4	68.2	56.2
10	Chevrolet Lumina	18400	3374	21	29	199.3	71.7	53.3
11	Chrysler Le Baron	13999	2863	21	27	184.8	69.2	53.3
12	Chrysler Le Baron GTC	19185	3010	23	28	184.8	69.2	52.4
13	Dodge Daytona	10874	2779	21	28	179.0	69.3	51.8
14	Dodge Intrepid	15930	3315	20	28	201.7	74.4	56.3
15	Dodge Spirit	11941	2788	23	28	181.2	68.1	55.5
16	Eagle Vision	17387	3290	20	28	201.6	74.4	55.8
17	Ford Escort	10041	2360	25	33	170.9	66.7	52.7
18	Ford Mustang	13926	2775	17	24	179.6	68.3	52.1
19	Ford Probe	12845	2619	24	30	178.9	69.8	51.8
20	Ford Taurus	19989	3253	21	30	193.1	71.2	55.5
21	Geo Metro	8199	1650	46	50	147.4	62.7	52.4
22	Lincoln Continental	33328	3628	17	26	205.1	72.7	55.5
23	Mercury Grand Marquis	22609	3784	18	26	212.4	77.8	56.8
24	Mercury Topaz	10976	2602	22	27	177.0	68.3	52.9
25	Oldsmobile Achieva	14849	2717	22	29	187.9	67.2	53.1
26	Oldsmobile Cutlass Calais	14899	2992	19	26	194.4	69.5	54.1
27	Oldsmobile Cutlass Supreme	15795	3354	19	30	193.7	71.0	54.8
28	Oldsmobile 98	24999	3697	19	27	205.5	74.6	54.8
29	Plymouth Acclaim	11941	2784	23	28	181.2	68.1	55.5
30	Pontiac Grand Am	13924	2804	22	32	186.9	68.6	53.2
31	Pontiac Sunbird	9995	2823	20	28	176.3	67.6	52.5
32	Saturn SL	12820	2682	24	35	180.7	66.3	52.0

Data Set 5: Movie Ratings and Lengths

 STATDISK: Variable names are MOVSTAR and MOVLEN for the star ratings and movie lengths.

Minitab: Enter RETRIEVE 'MOVIES' to get columns C1 and C2, which correspond to the star ratings and movie lengths.

	Movie	Star Rating	Length (min)	Rating
1	After the Rehearsal	4.0	72	R
2	All of Me	3.5	93	PG
3	Amityville II	2.0	110	R
4	At Close Range	3.5	115	R
5	Best Little Whorehouse in Texas	2.0	114	R
6	Beverly Hills Cop II	1.0	105	R
7	Birdy	4.0	120	R
8	The Blues Brothers	3.0	133	R
9	Breakin' 2—Electric Boogaloo	3.0	94	PG
10	Bugsy Malone	3.5	94	G
11	Cannonball Run II	0.5	108	PG
12	Choose Me	3.5	106	R
13	The Class of 1984	3.5	93	R
14	Conan the Barbarian	3.0	129	R
15	Crocodile Dundee	2.0	98	PG-13
16	Dead of Winter	2.5	100	PG-13
17	Desert Hearts	2.5	96	R
18	The Dogs of War	3.0	102	R
19	Drive, He Said	3.0	90	R
20	The Elephant Man	2.0	123	PG
21	Extreme Prejudice	3.0	104	R
22	Father of the Bride	3.0	105	PG
23	Flashdance	1.5	96	R
24	Frances	3.5	139	R
25	The Gauntlet	3.0	111	R
26	Godzilla	1.0	91	PG
27	Gremlins	3.0	111	PG
28	Hardcore	4.0	106	R
29	Heat	2.0	101	R
30	Honkeytonk Man	3.0	123	PG
31	Infra-Man	2.5	92	PG
32	Jo Jo Dancer	3.0	97	R
33	The King of Marvin Gardens	3.0	104	R
34	Last House on the Left	3.5	82	R

(continued)

Data Set 5 (continued)

	Movie	Star Rating	Length (min)	Rating
35	The Little Drummer Girl	2.0	155	R
36	Love Letters	3.5	98	R
37	The Waterdance	3.5	107	R
38	Melvin and Howard	3.5	95	R
39	Mona Lisa	4.0	104	R
40	The Muppets Take Manhattan	3.0	94	G
41	Nashville	4.0	159	R
42	1984	3.5	117	R
43	Oh, God!	3.5	104	PG
44	Ordinary People	4.0	125	R
45	Peggy Sue Got Married	4.0	103	PG-13
46	Plenty	3.5	119	R
47	Pretty in Pink	3.0	96	PG-13
48	Q	2.5	92	R
49	Ran	4.0	160	R
50	The Right Stuff	4.0	193	PG
51	The Rose	3.0	134	R
52	Scenes from a Marriage	4.0	168	PG
53	Silent Movie	4.0	88	PG
54	A Soldier's Story	2.5	99	PG
55	Spaceballs	2.5	100	PG
56	Star Wars	4.0	121	PG
57	Stepping Out	2.0	108	PG
58	Superman	4.0	144	PG
59	Sweetie	3.5	100	R
60	Testament	4.0	90	PG

Data Set 6: Weights (in grams) of a Sample of M&M Candies

 STATDISK: Variable names are RED, ORANGE, YELLOW, BROWN, TAN, GREEN.

Minitab: Enter RETRIEVE 'M&M' to get columns C1 through C6, which correspond to the 6 columns of data.

RED	ORANGE	YELLOW	BROWN	TAN	GREEN
0.864	0.921	0.890	0.848	0.957	0.894
0.952	0.882	0.927	0.831	0.942	0.922
0.941	0.939	0.912	0.936	0.886	0.842
0.817	0.927	0.971	0.843	0.950	0.893
0.957	0.935	0.920	0.888	0.957	0.947
0.920	0.918	0.890	0.946	0.876	0.900
0.958	0.893	0.914	0.949	1.003	0.871
0.933	0.938	0.986	0.932	0.935	0.802
0.870	0.914	0.925	0.949	0.890	0.940
0.871		0.929	0.941	0.947	
0.921		0.892	0.924	0.899	
0.896		0.845	0.976		
0.913		0.958	0.946		
0.844		0.930	0.924		
0.931		0.922	0.972		
0.909		0.914	0.942		
0.876		0.912	0.889		
		0.936	0.905		
		0.822	0.936		
		0.861	1.027		
		0.912	0.922		
		0.935	0.824		
		0.905	0.888		
		0.931	0.992		
			0.978		
			0.963		
			0.966		
			0.961		
			0.955		
			0.816		

Data Set 7: Maryland Lottery Results

 STATDISK: Variable name for the 150 Pick Three numbers is PICK3.

Minitab: Enter RETRIEVE 'LOTTO' to get the 6 Lotto columns of C1, C2, . . . , C6. Enter RETRIEVE 'PICK' to get the 3 Pick Three columns of C1, C2, and C3.

	Maryland Lotto: 50 Consecutive Drawings						Maryland Pick Three: 50 Consecutive Drawings			
1	2	17	21	38	44	49	1	0	0	0
2	24	31	34	40	41	49	2	7	1	3
3	1	8	13	19	27	33	3	3	6	4
4	2	21	23	27	35	45	4	6	8	6
5	1	8	13	38	40	48	5	2	4	7
6	13	21	26	36	40	48	6	7	6	9
7	1	5	10	29	30	46	7	6	2	5
8	7	8	14	29	34	43	8	6	7	7
9	5	11	12	13	26	32	9	6	1	1
10	8	18	20	24	37	47	10	3	3	3
11	7	17	22	31	41	42	11	8	2	2
12	2	7	15	21	25	38	12	1	9	7
13	1	14	19	20	25	33	13	7	6	9
14	5	16	27	44	47	48	14	8	7	4
15	11	13	16	32	38	39	15	7	1	5
16	2	19	27	30	44	45	16	1	0	3
17	26	28	30	31	37	47	17	6	4	5
18	11	20	24	30	42	47	18	8	0	0
19	9	15	27	29	38	43	19	6	5	0
20	1	11	27	35	36	48	20	9	5	1
21	7	20	32	37	38	44	21	5	4	9
22	9	19	25	27	44	49	22	2	6	2
23	12	13	15	16	23	31	23	1	2	1
24	12	26	33	39	41	43	24	5	6	0
25	1	10	21	24	39	44	25	0	3	6
26	22	25	35	38	40	49	26	3	8	2
27	1	5	19	21	31	36	27	9	2	6
28	1	24	29	33	40	45	28	9	5	3
29	3	8	19	36	45	46	29	0	3	9
30	1	7	21	23	28	33	30	7	4	4
31	5	6	8	23	29	34	31	7	8	2
32	9	11	37	39	43	44	32	5	0	4
33	4	13	33	38	43	48	33	8	5	5
34	2	4	25	42	43	45	34	5	5	1

(continued)

Data Set 7 (continued)

	Maryland Lotto: 50 Consecutive Drawings							Maryland Pick Three: 50 Consecutive Drawings		
35	11	15	19	27	38	41	35	4	3	8
36	8	16	17	22	33	42	36	9	4	4
37	8	9	16	17	29	36	37	1	1	8
38	9	22	24	37	40	47	38	1	7	1
39	1	5	7	16	31	49	39	0	6	6
40	1	9	17	24	32	33	40	3	5	6
41	8	9	12	19	24	39	41	3	4	5
42	1	5	19	21	23	47	42	9	9	9
43	1	7	9	14	28	41	43	4	4	1
44	15	34	35	39	48	49	44	8	2	9
45	12	14	24	28	32	36	45	7	3	0
46	3	4	9	29	36	47	46	7	5	7
47	7	13	25	37	38	49	47	7	5	7
48	17	22	27	32	40	49	48	5	3	3
49	12	15	23	28	34	44	49	8	7	3
50	16	19	22	32	39	41	50	8	1	6

Data Set 8: Weights (in grams) of Quarters

STATDISK: Variable name is QUARTERS.

Minitab: Enter RETRIEVE 'QUARTERS' to get column C1, which contains the 50 weights.

5.60	5.63	5.58	5.56	5.66	5.58	5.57	5.59	5.67	5.61	5.84
5.73	5.53	5.58	5.52	5.65	5.57	5.71	5.59	5.53	5.63	5.68
5.62	5.60	5.53	5.58	5.60	5.58	5.59	5.66	5.73	5.59	5.63
5.66	5.67	5.60	5.74	5.57	5.62	5.73	5.60	5.60	5.57	5.71
5.62	5.72	5.57	5.70	5.60	5.49					

Data Set 9: Weights (in milligrams) of Bufferin Tablets

STATDISK: Variable name is ASPIRIN.

Minitab: Enter RETRIEVE 'ASPIRIN' to get column C1, which contains the 30 weights.

672.2	679.2	669.8	672.6	672.2	662.2	662.7	661.3	654.2
667.4	667.0	670.7	665.5	672.9	664.8	655.1	669.1	663.6
655.2	657.5	655.7	662.5	665.6	684.7	659.5	660.5	658.1
671.3	662.1	667.0						

Data Set 10: Ages (in years) of U.S. Commercial Aircraft

 STATDISK: Variable name is AIRCRAFT.

Minitab: Enter RETRIEVE 'AIRCRAFT' to get column C1, which contains the 40 ages.

3.2	22.6	23.1	16.9	0.4	6.6	12.5	22.8
26.3	8.1	13.6	17.0	21.3	15.2	18.7	11.5
4.9	5.3	5.8	20.6	23.1	24.7	3.6	12.4
27.3	22.5	3.9	7.0	16.2	24.1	0.1	2.1
7.7	10.5	23.4	0.7	15.8	6.3	11.9	16.8

Appendix C: Glossary

Acceptance sampling Sampling in which n items are randomly selected without replacement and an entire batch is rejected if the number of defects is at least a predetermined number

Addition rule Rule for determining the probability that, on a single trial, either event A occurs or event B occurs or they both occur

Adjusted coefficient of determination Multiple coefficient of determination R^2 modified to account for the number of variables and sample size

Alpha (α) Symbol used to represent the probability of a type I error (the significance level)

Alternative hypothesis Statement that is equivalent to the negation of the null hypothesis; denoted by H_1

Analysis of variance Method of analyzing population variance in order to make inferences about the population

ANOVA *See* analysis of variance.

Arithmetic mean Sum of a set of scores divided by the number of scores; usually referred to as the mean

Assignable variation Type of variation in a process that results from causes that can be identified

Attribute data Data that can be separated into different categories distinguished by a nonnumeric characteristic

Average Any one of several measures designed to reveal the central tendency of a collection of data

Beta (β) Symbol used to represent the probability of a type II error

Bimodal Having two modes

Binomial experiment Experiment with a fixed number of independent trials, where each outcome falls into exactly one of two categories

Binomial probability formula Expression used to calculate probabilities in a binomial experiment (see Formula 4-6 in Section 4-4)

Bivariate data Data arranged in pairs

Bivariate normal distribution Distribution of paired data in which, for any fixed value of one variable, the values of the other variable are normally distributed

Block In analysis of variance, a group of similar individuals

Boxplot Graphical representation of the spread of a set of data

Categorical data Data that can be separated into different categories distinguished by a nonnumeric characteristic

Cell Category used to separate qualitative (or attribute) data

Central limit theorem Theorem stating that sample means tend to be normally distributed

Centroid The point $(\overline{x}, \overline{y})$ determined from a collection of bivariate data

Chebyshev's theorem Theorem that uses standard deviation to provide information about the distribution of data (Section 2-5)

Chi-square distribution Continuous probability distribution with selected values (Table A-4)

Class boundaries Values obtained from a frequency table by increasing the upper class limits and decreasing the lower class limits by the same amount so that there are no gaps between consecutive classes

Classical approach to probability Approach in which the probability of an event is determined by dividing the number of ways the event can occur by the total number of possible outcomes

Classical method of testing hypotheses Method of testing hypotheses based on a comparison of the test statistic and critical values

Class marks Midpoints of the classes in a frequency table

Class width Difference between two consecutive lower class limits in a frequency table

Cluster sampling Sampling randomly selected sections of a population

Coefficient of determination Amount of the variation in y that is explained by the regression line

Combinations rule Rule for determining the number of different combinations of selected items

Complement of an event All outcomes in which the original event does not occur

Completely randomized design In analysis of variance, a design in which each element is given the same chance of belonging to the different categories or treatments

Compound event Combination of simple events

Conditional probability The probability of an event (say, B) given that another event (say, A) has already occurred; $P(A \text{ and } B)/P(A)$

Confidence coefficient *See* degree of confidence.

Confidence interval Range of values used to estimate a population parameter with a specific level of confidence; also called an interval estimate

Confidence interval limits Two numbers that are used as the high and low boundaries of a confidence interval

Contingency table   Table of observed frequencies where the rows correspond to one variable of classification and the columns correspond to another variable of classification; also called a two-way table

Continuity correction   Adjustment made when a discrete random variable is approximated by a continuous random variable (Section 5-5)

Continuous data   Data resulting from infinitely many possible values that can be associated with points on a continuous scale in such a way that there are no gaps or interruptions

Continuous random variable   Random variable with infinite values that can be associated with points on a continuous line interval

Control chart   Any one of several types of charts (Chapter 12) that can be used to depict some characteristic of a process in order to determine whether there is statistical stability

Control limit   Boundary used in a control chart for identifying unusual points

Convenience sampling   Sampling in which data are selected because they are readily available

Correlation   Statistical association between two variables

Correlation coefficient   Measurement of the strength of the relationship between two variables

Countable set   Set with either a finite number of values or values that can be made to correspond to the positive integers

Critical region   Area under a curve containing the values that lead to rejection of the null hypothesis

Critical value   Value separating the critical region from the values of the test statistic that would not lead to rejection of the null hypothesis

Cumulative frequency table   Frequency table in which each class and frequency represent cumulative data up to and including that class

Data   Numbers or information collected in an experiment

Deciles   The nine values that divide ranked data into ten groups, with 10% of the scores in each group

Degree of confidence   Probability that a population parameter is contained within a particular confidence interval; also called level of confidence

Degrees of freedom   Number of values that are free to vary after certain restrictions have been imposed on all values

Denominator degrees of freedom   Degrees of freedom corresponding to the denominator of the F test statistic

Dependent events   Events for which the occurrence of any one event affects the probabilities of the occurrences of the other events

Dependent sample   Sample whose values are related to the values in another sample

Descriptive statistics   Methods used to summarize the key characteristics of known population data

Deviation   Amount of difference between a score and the mean; often expressed as $x - \overline{x}$

Discrete data   Data resulting from either a finite number of possible values or a countable number of possible values

Discrete random variable   Random variable with either a finite number of values or a countable number of values

Distribution-free tests   Tests not requiring a particular distribution, such as the normal distribution. *See also* nonparametric tests.

Efficiency   Measure of the sensitivity of a nonparametric test in comparison to that of a corresponding parametric test

Empirical rule   Rule that uses standard deviation to provide information about data with a bell-shaped distribution (Section 2-5)

Estimate   Specific value or range of values used to approximate a population parameter

Estimator   Sample statistic (such as the sample mean $\overline{x}$) used to approximate a population parameter

Event   Result or outcome of an experiment

Expected frequency   Theoretical frequency for a cell of a contingency table or multinomial table

Expected value   For a discrete random variable, the average value of the outcomes

Experiment   Process that allows observations to be made

Explained deviation   For one pair of values in a collection of bivariate data, the difference between the predicted y value and the mean of the y values

Explained variation   Sum of the squares of the explained deviations for all pairs of bivariate data in a sample

Exploratory data analysis (EDA)   Branch of statistics emphasizing the investigation of data

Factor   In analysis of variance, a property or characteristic that allows us to distinguish different populations from one another

Factorial rule   Rule stating that n different items can be arranged $n!$ different ways

F **distribution**   Continuous probability distribution with selected values (Table A-5)

Finite population correction factor   Factor for correcting the standard error of the mean when a sample size exceeds 5% of the size of a finite population

Five-number summary   Minimum score, maximum score, median, and the two hinges of a set of data

Frequency polygon   Graphical representation of the distribution of data using connected straight-line segments

Frequency table   List of categories of scores along with their corresponding frequencies

Fundamental counting rule   Rule stating that, for a sequence of two events in which the first event can occur m ways and the second can occur n ways, the events together can occur a total of $m \cdot n$ ways

Goodness-of-fit test   Test of how well an observed frequency distribution fits a theoretical distribution

Hinge   Median value of the bottom (or top) half of a set of ranked data

Histogram   Graph of vertical bars representing the frequency distribution of a set of data

H test   *See* Kruskal-Wallis test.

Hypothesis   Statement or claim that a population characteristic is true

Hypothesis test   Method for testing claims made about populations; also called test of significance

Independent events   Events for which the occurrence of any one event does not affect the probabilities of the occurrences of the other events

Independent sample   Sample whose values are not related to the values in another sample

Inferential statistics   Methods involving the use of sample data to make generalizations or inferences about a population

Interaction   In two-way analysis of variance, the effect when one of the factors changes for different categories of the other factor

Interquartile range   Difference between the first and third quartiles

Interval   Level of measurement of data; characterizes data that can be arranged in order and for which differences between data values are meaningful

Interval estimate   *See* confidence interval.

Kruskal-Wallis test   Nonparametric hypothesis test used to compare three or more independent samples; also called an H test.

Least-squares property   Property stating that, for a regression line, the sum of the squares of the vertical deviations of the sample points from the line is the smallest sum possible

Left-tailed test   Hypothesis test in which the critical region is located in the extreme left area of the probability distribution

Level of confidence   *See* degree of confidence.

Linear correlation coefficient   Measure of the strength of the relationship between two variables; also called Pearson product moment correlation coefficient

Lower class limits   Smallest numbers that can actually belong to the different classes in a frequency table

Lower hinge   Median of the lower half of all scores (from the minimum score up to and including the original median); used in a boxplot

Mann-Whitney U test   Hypothesis test equivalent to the Wilcoxon rank-sum test for two independent samples

Marginal change   For variables related by a regression equation, the amount of change in the dependent variable when one of the independent variables changes by one unit and the other independent variables remain constant

Margin of error   Maximum likely (with the probability $1 - \alpha$) difference between the observed sample statistic and the true value of the population parameter; also called maximum error of estimate

Maximum error of estimate   *See* margin of error.

Mean   Sum of a set of scores divided by the number of scores

Mean deviation   Measure of dispersion equal to the sum of the deviations of each score from the mean, divided by the number of scores

Measure of central tendency   Value intended to indicate the center of the scores in a collection of data

Measure of dispersion   Any of several measures designed to reflect the amount of variability for a set of scores

Median   Middle value of a set of scores arranged in order of magnitude

Midquartile   One-half of the sum of the first and third quartiles

Midrange   One-half of the sum of the highest and lowest scores

Mode   Score that occurs most frequently

MS(error)   Mean square for error; used in analysis of variance

MS(total)   Mean square for the total variation; used in analysis of variance

MS(treatment)   Mean square for treatments; used in analysis of variance

Multimodal   Having more than two modes

Multinomial experiment   Experiment with a fixed number of independent trials, where each outcome falls into exactly one of several categories

Multiple coefficient of determination   Measure of how well a multiple regression equation fits the sample data

Multiple comparison procedures   Procedures for identifying which particular means are different, after it has been concluded that three or more means are not all equal

Multiple regression   Study of linear relationships among three or more variables

Multiple regression equation Equation that expresses a linear relationship between a dependent variable y and two or more independent variables $x_1, x_2, \cdots, x_k$

Multiplication rule Rule for determining the probability that event A will occur on one trial and event B will occur on a second trial

Mutually exclusive events Events that cannot occur simultaneously

Negatively skewed Skewed to the left

Nominal Level of measurement of data; characterizes data that consist of names, labels, or categories only

Nonparametric tests Statistical procedures, designed for testing hypotheses or estimating parameters, that are not based on population parameters and do not require many of the restrictions of parametric tests

Nonsampling errors Errors from external factors not related to sampling

Normal distribution Bell-shaped probability distribution described algebraically by Equation 5-1 in Section 5-1

np chart Control chart in which numbers of defects are plotted so that a process can be monitored

Null hypothesis Claim made about a population characteristic, usually involving the case of no difference; denoted by H_0

Numerator degrees of freedom Degrees of freedom corresponding to the numerator of the F test statistic

Numerical data Data consisting of numbers representing counts or measurements

Observed frequency Actual frequency count recorded in one cell of a contingency table or multinomial table

Odds against For an event A, odds obtained by finding $P(\overline{A})/P(A)$, usually expressed in the form of $a{:}b$, where a and b are integers having no common factors

Odds in favor For an event A, odds obtained by finding $P(A)/P(\overline{A})$, usually expressed as the ratio of two integers with no common factors

Ogive Graphical representation of a cumulative frequency table

One-way analysis of variance Analysis of variance involving data classified into groups according to a single criterion only; also called single factor analysis of variance

Ordinal Level of measurement of data; characterizes data that may be arranged in order, but for which differences between data values either cannot be determined or are meaningless

Parameter Measured characteristic of a population

Parametric tests Statistical procedures, based on population parameters, for testing hypotheses or estimating parameters

Pareto chart Bar graph for qualitative data, in which the bars are arranged in order according to frequencies

p chart Control chart used to monitor the proportion p for an attribute in a process

Pearson product moment correlation coefficient *See* linear correlation coefficient.

Percentiles The 99 values that divide ranked data into 100 groups, with 1% of the scores in each group

Permutations rule Rule for determining the number of different arrangements of selected items

Pie chart Graphical representation of data in the form of a circle containing wedges

Point estimate Single value that serves as an estimate of a population parameter

Poisson distribution Discrete probability distribution that applies to occurrences of an event over a specified interval of time, distance, area, volume, or some similar unit

Pooled estimate of p_1 and p_2 Probability obtained by combining the data from two sample proportions and dividing the total number of successes by the total number of observations

Population Complete and entire collection of elements to be studied

Positively skewed Skewed to the right

Power of a test Probability $(1 - \beta)$ of rejecting a false null hypothesis

Predicted value Value of one variable, derived from a regression equation by using given values for the other variables

Prediction interval Confidence interval estimate of a predicted value of y

Predictor variables Independent variables in a regression equation

Probability Measure of the likelihood that a given event will occur; expressed as a number between 0 and 1

Probability distribution Collection of values of a random variable along with their corresponding probabilities

Probability histogram Histogram with outcomes listed along the horizontal axis and probabilities listed along the vertical axis

Probability value *See* P-value.

Process data Data, arranged according to some time sequence, that measure a characteristic of goods or services resulting from some combination of equipment, people, materials, methods, and conditions

P-value Probability that a test statistic in a hypothesis test is at least as extreme as the one actually obtained

Qualitative data Data that can be separated into different categories distinguished by a nonnumeric characteristic

Quantitative data   Data consisting of numbers representing counts or measurements

Quartiles   The three values that divide ranked data into four groups, with approximately 25% of the scores in each group

Randomized block design   Design in which a measurement is obtained for each treatment on each of several individuals matched according to similar characteristics

Random sample   Sample selected in such a way that every member of the population has the same chance of being chosen

Random selection   Selection of sample elements in such a way that all elements available for selection have the same chance of being selected

Random variable   Variable (typically represented by x) that has a single numerical value (determined by chance) for each outcome of an experiment

Random variation   Type of variation in a process that is due to chance; the type of variation inherent in any process not capable of producing every good or service exactly the same way every time

Range   Measure of dispersion that is the difference between the highest and lowest scores

Range chart   Control chart, based on sample ranges, that is used to monitor variation in a process

Range rule of thumb   Rule stating that the range of a set of data is approximately four standard deviations ($4s$) wide

Rank   Numerical position of an item in a sample set arranged in order

Rank correlation coefficient   Measure of the strength of the relationship between two variables, based on the ranks of the values

Ratio   Level of measurement of data; characterizes data that can be arranged in order, for which differences between data values are meaningful, and for which there is an inherent zero starting point

R chart   *See* range chart.

Redundancy   Duplication of critical components so that their failure will not cause the failure of an entire system

Regression equation   Algebraic equation describing the relationship among variables

Regression line   Straight line that summarizes the relationship between two variables

Relative frequency approximation of probability   Estimated value of probability based on actual observations

Relative frequency histogram   Variation of the basic histogram in which frequencies are replaced by relative frequencies. *See also* relative frequency table.

Relative frequency table   Variation of the basic frequency table in which the frequency for each class is divided by the total of all frequencies

Right-tailed test   Hypothesis test in which the critical region is located in the extreme right area of the probability distribution

Rigorously controlled design   Design in which all factors are forced to be constant so that effects of extraneous factors are eliminated

Run   Sequence of data exhibiting the same characteristic; used in the runs test for randomness

Run chart   Sequential plot of individual data values over time, where one axis (usually the vertical axis) is used for the data values and the other axis (usually the horizontal axis) is used for the time sequence

Runs test   Nonparametric method used to test for randomness

Sample   Subset of a population

Sample space   Set of all possible outcomes or events in an experiment that cannot be further broken down

Sampling distribution of sample means   Distribution of the sample means that is obtained when we repeatedly draw samples of the same size from the same population

Sampling errors   Errors resulting from the sampling process itself

Scatter diagram   Graphical display of bivariate data

s chart   Control chart, based on sample standard deviations, that is used to monitor variation in a process

Semi-interquartile range   One-half of the difference between the first and third quartiles

Significance level   Probability of making a type I error when conducting a hypothesis test

Sign test   Nonparametric hypothesis test used to compare samples from two populations

Simple event   Experimental outcome that cannot be further broken down

Simulation   Process whose behavior is similar to that of some experiment, so similar results are produced

Single factor analysis of variance   *See* one-way analysis of variance.

Skewed   Not symmetric; extending more to one side than to the other

Slope   Measure of steepness of a straight line

Spearman's rank correlation coefficient   *See* rank correlation coefficient.

SS(error)   Sum of squares representing the variability that is assumed to be common to all the populations being considered; used in analysis of variance

SS(total)   Measure of the total variation (around $\bar{x}$) in all of the sample data combined; used in analysis of variance

SS(treatment)   Measure of the variation between the sample means; used in analysis of variance

Standard deviation   Measure of dispersion equal to the square root of the variance

Standard error of estimate   Measure of the spread of sample points about the regression line

Standard error of the mean   Standard deviation of all possible sample means $\overline{x}$

Standard normal distribution   Normal distribution with a mean of 0 and a standard deviation equal to 1

Standard score   Number of standard deviations that a given value is above or below the mean; also called a z score

Statistic   Measured characteristic of a sample

Statistically stable process   Process with only natural variation and no patterns, cycles, or unusual points

Statistical process control (SPC)   Use of statistical techniques such as control charts to analyze a process or its outputs so as to take appropriate actions to achieve and maintain a state of statistical control and improve the process capability

Statistics   Collection, organization, description, and analysis of data

Stem-and-leaf plot   Method of graphically sorting and arranging data to reveal the distribution

Stepwise regression   Process of using different combinations of variables until the best model has been obtained; used in multiple regression

Stratified sampling   Sampling in which samples are drawn from each stratum (class)

Student t distribution   See t distribution.

Subjective probability   Guess or estimate of a probability based on knowledge of relevant circumstances

Symmetric   Property of data for which the distribution can be divided into two halves that are approximately mirror images by drawing a vertical line through the middle

Systematic sampling   Sampling in which every kth element is selected

t distribution   Bell-shaped distribution usually associated with small sample experiments; also called the Student t distribution

10–90 percentile range   Difference between the 10th and 90th percentiles

Test of homogeneity   Test of the claim that different populations have the same proportion of a characteristic

Test of independence   Test of the null hypothesis that the row variable and column variable for a contingency table are not related

Test of significance   See hypothesis test.

Test statistic   Sample statistic based on the sample data; used in hypothesis testing

Total deviation   Sum of the explained deviation and unexplained deviation for a given pair of values in a collection of bivariate data

Total variation   Sum of the squares of the total deviation for all pairs of bivariate data in a sample

Traditional method of testing hypotheses   Method of testing hypotheses based on a comparison of the test statistic and critical values

Treatment   Property or characteristic that allows us to distinguish the different populations from one another; used in analysis of variance

Tree diagram   Graphical depiction of the different possible outcomes in a compound event

Two-tailed test   Hypothesis test in which the critical region is divided between the left and right extreme areas of the probability distribution

Two-way analysis of variance   Analysis of variance involving data classified according to two different factors

Two-way table   See contingency table.

Type I error   Mistake of rejecting the null hypothesis when it is true

Type II error   Mistake of failing to reject the null hypothesis when it is false

Unexplained deviation   For one pair of values in a collection of bivariate data, the difference between the y coordinate and the predicted value

Unexplained variation   Sum of the squares of the unexplained deviations for all pairs of bivariate data in a sample

Uniform distribution   Even distribution of values over the range of possibilities

Upper class limits   Largest numbers that can belong to the different classes in a frequency table

Upper hinge   Median of the upper half of all scores (from the original median up to the maximum score); used in a boxplot

Variance   Measure of dispersion found by using Formula 2-5 in Section 2-5

Variance between samples   In analysis of variance, the variation among the different samples

Variation due to error   See variation within samples.

Variation due to treatment   See variance between samples.

Variation within samples   In analysis of variance, the variation that is due to chance

Weighted mean   Mean of a collection of scores that have been assigned different degrees of importance

Wilcoxon rank-sum test   Nonparametric hypothesis test used to compare two independent samples

Wilcoxon signed-ranks test   Nonparametric hypothesis test used to compare two dependent samples

Within statistical control   *See* statistically stable process.

$\bar{x}$ **chart**   Control chart used to monitor the mean of a process

y-**intercept**   Point at which a straight line crosses the y-axis

z **score**   Number of standard deviations that a given score is above or below the mean

Appendix D: Bibliography

* An asterisk denotes a recommended reading. Other books are recommended as reference texts.

Beyer, W. 1991. *CRC Standard Probability and Statistics Tables and Formulae.* Boca Raton, Fla: CRC Press.

* Brook, R., and others, eds. 1986. *The Fascination of Statistics.* New York: Marcel Dekker.

* Campbell, S. 1974. *Flaws and Fallacies in Statistical Thinking.* Englewood Cliffs, N.J.: Prentice-Hall.

Cochran, W. 1982. *Contributions to Statistics.* New York: Wiley.

Conover, W. 1980. *Practical Nonparametric Statistics.* 2nd ed. New York: Wiley.

Devore, J., and R. Peck. 1986. *Statistics: The Exploration and Analysis of Data.* St. Paul, Minn.: West Publishing.

* Fairley, W., and F. Mosteller. 1977. *Statistics and Public Policy.* Reading, Mass.: Addison-Wesley.

Fisher, R. 1966. *The Design of Experiments.* 8th ed. New York: Hafner.

* Freedman, D., R. Pisani, R. Purves, and A. Adhikari. 1991. *Statistics.* 2nd ed. New York: Norton.

Freund, J. 1988. *Modern Elementary Statistics.* 7th ed. Englewood Cliffs, N.J.: Prentice-Hall.

* Gonick, L., and W. Smith. 1993. *The Cartoon Guide to Statistics.* New York: HarperCollins.

Hauser, P. 1975. *Social Statistics in Use.* New York: Russell Sage Foundation.

Heerman, E., and L. Braskam. 1970. *Readings in Statistics for the Behavioral Sciences.* Englewood Cliffs, N.J.: Prentice-Hall.

Hoaglin, D., F. Mosteller, and J. Tukey, eds. 1983. *Understanding Robust and Exploratory Data Analysis.* New York: Wiley.

* Hollander, M., and F. Proschan. 1984. *The Statistical Exorcist: Dispelling Statistics Anxiety.* New York: Marcel Dekker.

Hollander, M., and D. Wolfe. 1973. *Nonparametric Statistical Methods.* New York: Wiley.

* Holmes, C. 1990. *The Honest Truth About Lying with Statistics.* Springfield, IL: Charles C Thomas.

* Hooke, R. 1983. *How to Tell the Liars from the Statisticians.* New York: Dekker.

* Huff, D. 1954. *How to Lie with Statistics.* New York: Norton.

* Jaffe, A., and H. Spirer. 1987. *Misused Statistics.* New York: Marcel Dekker.

* Kimble, G. 1978. *How to Use (and Misuse) Statistics.* Englewood Cliffs, N.J.: Prentice-Hall.

King, R., and B. Julstrom. 1982. *Applied Statistics Using the Computer.* Sherman Oaks, Calif.: Alfred.

Kirk, R., ed. 1972. *Statistical Issues: A Reader for the Behavioral Sciences.* Belmont, Calif.: Brooks/Cole.

Kotz, S., and D. Stroup. 1983. *Educated Guessing—How to Cope in an Uncertain World.* New York: Marcel Dekker.

* Loyer, M. 1995. *Student Solutions Manual to Accompany Elementary Statistics.* 6th ed. Reading, Mass.: Addison-Wesley.

* Moore, D. 1991. *Statistics: Concepts and Controversies.* 3rd ed. San Francisco: Freeman.

Mosteller, F., and R. Rourke. 1973. *Sturdy Statistics, Nonparametrics and Order Statistics.* Reading, Mass.: Addison-Wesley.

Mosteller, F., R. Rourke, and G. Thomas, Jr. 1970. *Probability with Statistical Applications.* 2nd ed. Reading, Mass.: Addison-Wesley.

Neter, J., W. Wasserman, and M. Kutner. 1985. *Applied Linear Statistical Models.* Homewood, Ill.: Irwin.

Noether, G. 1976. *Elements of Nonparametric Statistics.* 2nd ed. New York: Wiley.

Ott, L., and W. Mendenhall. 1985. *Understanding Statistics.* 4th ed. Boston: Duxbury Press.

Owen, D. 1962. *Handbook of Statistical Tables.* Reading, Mass.: Addison-Wesley.

* Paulos, J. 1988. *Innumeracy: Mathematical Illiteracy and Its Consequences.* New York: Hill and Wang.

* Reichard, R. 1974. *The Figure Finaglers.* New York: McGraw-Hill.

* Reichmann, W. 1962. *Use and Abuse of Statistics.* New York: Oxford University Press.

* Runyon, R. 1977. *Winning with Statistics.* Reading, Mass.: Addison-Wesley.

Ryan, T., B. Joiner, and B. Ryan. 1985. *Minitab Student Handbook.* 2nd ed. Boston: Duxbury.

Schmid, C. 1983. *Statistical Graphics.* New York: Wiley.

Siegal, S. 1956. *Nonparametric Statistics for the Behavioral Sciences.* New York: McGraw-Hill.

Simon, J. 1992. *Resampling: The New Statistics.* Belmont, Calif.: Duxbury Press.

Smith, G. 1991. *Statistical Process Control and Quality Improvement.* New York: Macmillan.

* Stigler, S. 1986. *The History of Statistics.* Cambridge, Mass.: Harvard University Press.

* Tanur, J., ed. 1978. *Statistics: A Guide to the Unknown.* 2nd ed. San Francisco: Holden-Day.

Triola, M. 1995. *Minitab Student Laboratory Manual and Workbook.* 3rd ed. Reading, Mass.: Addison-Wesley.

Triola, M. 1995. *STATDISK Student Laboratory Manual and Workbook.* 4th ed. Reading, Mass.: Addison-Wesley.

Triola, M., and L. Franklin. 1994. *Business Statistics.* Reading, Mass.: Addison-Wesley.

Tukey, J. 1977. *Exploratory Data Analysis.* Reading, Mass.: Addison-Wesley.

Velleman, P., and D. Hoaglin. 1981. *Applications, Basics, and Computing of Exploratory Data Analysis.* Boston: Duxbury Press.

Wayne, D. 1978. *Applied Nonparametric Statistics.* Boston: Houghton Mifflin.

Weisberg, S. 1985. *Applied Linear Regression.* 2nd ed. New York: Wiley.

Wonnacott, R., and T. Wonnacott. 1985. *Introductory Statistics.* 4th ed. New York: Wiley.

Zeisel, H. 1968. *Say It with Figures.* 5th ed. New York: Harper & Row.

Appendix E: Answers to Odd-Numbered Exercises

Section 1-2

1. Discrete   3. Continuous   5. Continuous   7. Discrete
9. Ordinal   11. Nominal   13. Nominal   15. Ratio
17. Interval   19. a. Ratio   b. Interval (or possibly ordinal)
21. Differences between the years can be determined and are meaningful, but there is no inherent starting point since time did not begin in the year zero.

Section 1-3

1. The implication is that a study sponsored by the walnut industry is much more likely to reach conclusions favorable to that industry.
3. Because the graph does not start at the zero point, the differences are exaggerated.
5. No. It might be that lower grades make students nervous and cause them to smoke, or that the personality characteristics that make students prone to smoke also make them prone to achieve lower grades.
7. Not necessarily. Volvo does have a reputation for safe design, but Corvettes tend to attract disproportionately more reckless drivers.
9. Alumni with lower salaries would be less inclined to respond, so the reported salaries will tend to be disproportionately high. Also, those who cheat on their income tax forms might not want to reveal their true incomes. Others might want to retain privacy.
11. a. $320   b. $384   c. No
13. Part b is probably more accurate. It relates factual data without jumping to a broader conclusion which might not be justified on the basis of the available data.
15. The bars are not drawn in their proper proportions.
17. Because the groups consist of 20 subjects each, all percentages of success should be multiples of 5. The given percentages cannot be correct.
19. According to the *New York Times,* "It would have to remove all the plaque, remove it again and then remove it for a third time plus some more still."

Section 1-4

1. Systematic   3. Cluster   5. Convenience
7. Stratified   9. Cluster   11. Random
13. People often don't know the correct values, or they round off. Also, they often give responses that are wrong but create a more favorable impression.
15. a. An advantage of open questions is that they provide the subject and the interviewer with a much wider variety of responses; a disadvantage is that open questions can be very difficult to analyze.
    b. An advantage of closed questions is that they reduce the chance of misinterpretation of the topic; a disadvantage is that closed questions prevent the inclusion of valid responses the pollster might not have considered.
    c. Closed questions are easier to analyze with formal statistical procedures.

Section 1-5

1. MTB > SET C1
   DATA> 2 4 1 2 3 2 3 1
   DATA> ENDOFDATA
   MTB > PRINT C1
3. a. The sum of C1 and C2: 5.2 26.6 24.1 18.9 3.4 8.6 15.5 23.8
   b. No
5. The values in columns C1 through C6 should be displayed.
7. The number 12.345 should be displayed 25 times.
9. a. 2.25 is the mean (defined in Section 2-4).
   b. 4 is the maximum or highest score.
   c. 1 is the minimum or lowest score.   d. 18 is the sum of the scores.
   e. 48 is the sum of the squares of the scores.

Chapter 1: Review Exercises

1. a. Ordinal   b. Ratio   c. Ordinal   d. Nominal   e. Interval
3. They probably based their figure on the retail selling price, but they could have used cost, wholesale price, and so on. They might want to exaggerate in order to appear more effective.
5. a. Systematic   b. Random   c. Cluster   d. Stratified
   e. Convenience
7. Respondents often tend to round off to a nice even number like 50.

Section 2-2

1. Class width: 8. Class marks: 83.5, 91.5, 99.5, 107.5, 115.5. Class boundaries: 79.5, 87.5, 95.5, 103.5, 111.5, 119.5.
3. Class width: 5.0. Class marks: 18.65, 23.65, 28.65, 33.65, 38.65. Class boundaries: 16.15, 21.15, 26.15, 31.15, 36.15, 41.15.

5.

IQ	Relative Frequency
80–87	0.110
88–95	0.253
96–103	0.342
104–111	0.199
112–119	0.096

7.

Weight	Relative Frequency
16.2–21.1	0.296
21.2–26.1	0.278
26.2–31.1	0.222
31.2–36.1	0.148
36.2–41.1	0.056

9.

IQ	Cumulative Frequency
Less than 88	16
Less than 96	53
Less than 104	103
Less than 112	132
Less than 120	146

11.

Weight	Cumulative Frequency
Less than 21.2	16
Less than 26.2	31
Less than 31.2	43
Less than 36.2	51
Less than 41.2	54

13. 0.26–0.75   15. 1650–2049

17.

Weight	Frequency
0.26–0.75	2
0.76–1.25	9
1.26–1.75	16
1.76–2.25	8
2.26–2.75	8
2.76–3.25	7
3.26–3.75	7
3.76–4.25	1
4.26–4.75	3
4.76–5.25	1

19.

Weight	Frequency
1650–2049	1
2050–2449	1
2450–2849	14
2850–3249	4
3250–3649	8
3650–4049	3
4050–4449	1

21. The relative frequencies for men are

0.019   0.071   0.118   0.171   0.087   0.273   0.142   0.118

The relative frequencies for women are

0.010   0.072   0.173   0.265   0.042   0.279   0.060   0.100

23. The third guideline is clearly violated since the class width varies. The fifth guideline is also violated because the number of classes is not between 5 and 20.

Section 2-3

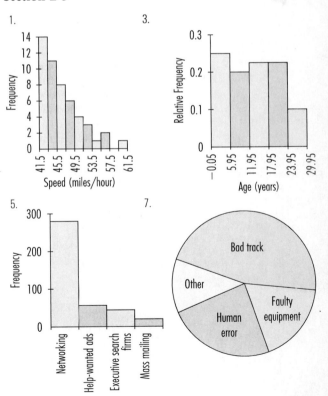

9. a.

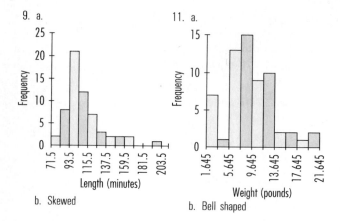

b. Skewed

11. a.

b. Bell shaped

13. a.

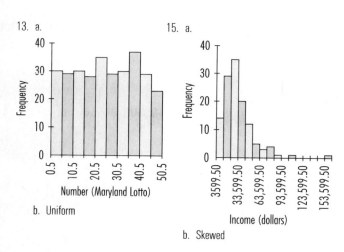

b. Uniform

15. a.

b. Skewed

17.

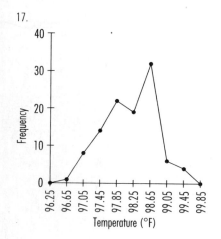

19.

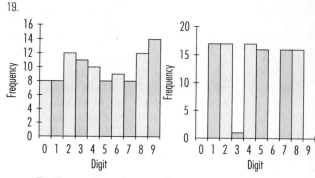

a. The histogram for π is more uniform.

b. π is irrational, whereas 22/7 is rational. The decimal form of π is nonrepeating, whereas the decimal form of 22/7 has the digits 142857 repeating.

21. a.

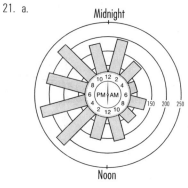

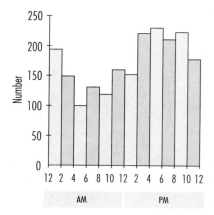

b. The circular bar chart is more effective because the bar chart hides the pattern by creating a break in the time sequence.

c. No, there aren't many cars being driven between 4:00 AM and 6:00 AM. You need to compare *rates* of fatal crashes.

Section 2-4

1. $\bar{x} = 71.5$; median = 72.0; mode = 77; midrange = 71.0
3. $\bar{x} = 0.187$; median = 0.170; mode = 0.16, 0.17; midrange = 0.205
5. $\bar{x} = 27.0$; median = 25.0; mode = 17, 18, 23, 25, 28; midrange = 32.5
7. $\bar{x} = 1.658$; median = 1.575; mode: none; midrange = 1.750
9. $\bar{x} = 667.63$; median = 668.60; mode = 672.2; midrange = 666.70
11. $\bar{x} = 3117.9$; median = 3122.5; mode: none; midrange = 3072.0
13. $\bar{x} = 98.13$; median = 98.20; mode = 97.4, 98.0, 98.2, 98.8; midrange = 97.80
15. $\bar{x} = 9.428$; median = 9.300; mode: none; midrange = 11.115
17. 46.7   19. 13.30
21. a. $\bar{x} = 193,000$; median = 206,000; mode = 236,000; midrange = 172,000
    b. $\bar{x} = 213,000$; median = 226,000; mode = 256,000; midrange = 192,000
    c. The mean, median, mode, and midrange also change by k.
    d. $\bar{x} = 193$; median = 206; mode = 236; midrange = 172
    e. The mean, median, mode, and midrange are also divided or multiplied by k.
23. a. No, the averages are the same.
    b. The data sets differ in the amount of variation.
25. No. For the scores 1, 2, 3, 4, 5, log $\bar{x}$ = log 3 = 0.477, but the mean of the logarithms of these scores is 0.416.
27. 1.092   29. a. 7.0   b. 7.1   c. 7.3
    In this case, the open-ended class doesn't have too much effect on the mean. The mean is likely to be around 7.1, give or take about 0.1.
31. a. 2.44   b. 2.37

Section 2-5

1. range = 12.0; $s^2 = 22.7$; $s = 4.8$
3. range = 0.170; $s^2 = 0.003$; $s = 0.051$
5. range = 37.0; $s^2 = 83.0$; $s = 9.1$
7. range = 2.240; $s^2 = 0.526$; $s = 0.725$
9. range = 25.00; $s^2 = 44.47$; $s = 6.67$
11. range = 1424.0; $s^2 = 245,855.9$; $s = 495.8$
13. a. 0.80   b. 0.76   15. a. 4.733   b. 4.168
17. 4.3   19. 3.04
21. a. range = 12.0; $s^2 = 22.7$; $s = 4.8$; same results
    b. same results
    c. range = 120.0; $s^2 = 2272.2$; $s = 47.7$; range and standard deviation are multiplied by 10, but the variance is multiplied by 10^2.

d. range = 6.0; $s^2 = 5.7$; $s = 2.4$; range and standard deviation are halved, but variance is 1/4 of original value.
23. Range is 675.0; variance is 44,748.3; standard deviation is 211.5. The outlier has a dramatic effect on these measures of dispersion.
25. a. 68%   b. 95%   c. 44.0, 68.0
27. Mean is 100.0, and standard deviation is 47.6.
29. $\bar{x} = 36.8$, $s = 0.34$
31. -0.97; no significant skewness

Section 2-6

1. a. 0.65   b. 0   c. 2.90   3. 0.55   5. a. 2.56   b. Yes
7. a. -1.50 b. No
9. The answer is b because $z = 0.60$ is greater than $z = 0.30$.
11. The answer is b since $z = 3.00$ is greater than $z = 2.00$ or 2.67.
13. 8   15. 92   17. 98.7   19. 98.6   21. 97.9   23. 98.6
25. 75   27. 35   29. 93   31. 123   33. 96   35. 141.5
37. a. 0.8   b. 98.2   c. 1.5   d. Yes; yes   e. No
39. 96.96, 99.44

Section 2-7

1. 406, 406, 407, 408, 410, 419, 419, 419, 419, 421, 423, 424, 426, 426, 430, 438, 438

3.
6	48
7	5679
8	000112334444569
9	0012233447

5.
5.4	9
5.5	2333
5.5	677777888889999
5.6	00000001222333
5.6	5666778
5.7	01123334
5.7	
5.8	4

7.   9.

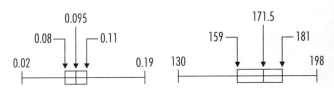

11.

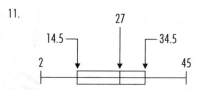

13. Lottery numbers should be uniformly distributed, but the boxplot does not appear to represent a uniform distribution.

15. a.

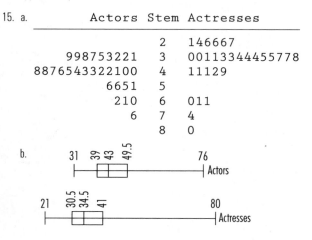

Actors	Stem	Actresses
	2	146667
998753221	3	00113344455778
8876543322100	4	11129
6651	5	
210	6	011
6	7	4
	8	0

b.

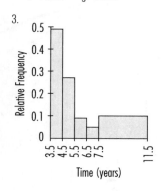

Actors

Actresses

c. Females who win Oscars tend to be younger than males.

Chapter 2: Review Exercises

1. a. 8.575   b. 7.745   c. None   d. 9.385   e. 10.570
   f. 14.759   g. 3.842

3.

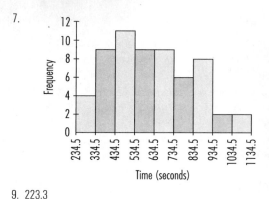

Time	Frequency
235–334	4
335–434	9
435–534	11
535–634	9
635–734	9
735–834	6
835–934	8
935–1034	2
1035–1134	2

7.

9. 223.3

11.
2	35 40 92
3	25 35 37 45 63 78 96 96
4	04 20 43 47 48 57 74 83 94 95
5	03 06 14 40 52 64 87
6	09 15 25 26 27 66 70 70 76 88 93
7	00 04 23 48 56 78 93 94
8	20 52 53 60 61 62 71
9	15 29 91
10	23 70
11	28

13. a. 16.0   b. 17.0   c. 8   d. 16.5   e. 17.0   f. 39.1
    g. 6.3

15. a. 20.8   b. 25.0   c. 29   d. 17.0
    e. 30.0   f. 107.0   g. 10.3

Section 3-2

1. 4/3, 1.001, -0.2, 2, $\sqrt{2}$

3. 0.320   5. 9/20   7. 0.361   9. 0.230
11. 0.395   13. 0.340   15. 0.0896   17. 0.050
19. 0.0460   21. b. 1/4 c. 1/2
23. b. 1/8 c. 1/8 d. 1/2   25. 0.600   27. 1
29. a. 37:1 b. 1:37

Section 3-3

1. a. No b. No c. Yes   3. a. 0.55 b. 0.487
5. 0.290   7. 0.580   9. 0.520   11. 0.471
13. 0.217   15. 0.824   17. 0.500   19. 0.490
21. 0.950   23. 0.530
25. a. 17/60 b. $P(A \text{ or } B) = 0.9$ c. $P(A \text{ or } B) < 0.9$
27. $P(A \text{ or } B) = P(A) + P(B) - 2P(A \text{ and } B)$

Section 3-4

1. a. Dependent b. Independent c. Dependent
3. 0.179   5. 0.000796   7. 0.00766   9. a. 0.0625 b. 0.130
11. 0.0000000206   13. 0.000000250   15. 0.00605   17. 0.480
19. 0.00455   21. 0.363   23. 0.262   25. 0.0728   27. 0.999
29. a. 0.750 b. 0.190 c. 0.290   31. a. 0.909 b. 0.364 c. 0.650
33. a. 0.431 b. 0.569
35. a. $1 - P(A) - P(B) + P(A \text{ and } B)$ b. $1 - P(A \text{ and } B)$ c. Different
37. 0.0192

Section 3-5

Note: Answers may vary because there are different ways to simulate the experiments.
1. Approx. 2   3. Approx. 10   5. Approx. 19.2
7. Approx. 5.2   9. a. 1.64 b. 0.98

Section 3-6

1. 5040   3. 4830   5. 720   7. 30   9. 120
11. 1326   13. $n!$ 15. 1   17. 6   19. 5040
21. 646,646   23. 10! = 3,628,800
25. a. 5040 b. 1/5040   27. 1,000,000,000
29. 43,680   31. 125,000 permutations   33. 18,720
35. a. 2002 b. 0.675
37. 1/13,983,816. You have a much better chance of being struck by lightning.
39. $6.20 \cdot 10^{22}$ 41. 120   43. 2,095,681,645,538 (about 2 trillion)
45. a. 2 b. $(n - 1)!$
47. Calculator: 3.0414093×10^{64}; approximation: 3.0363452×10^{64}
49. a. 1/70 b. $3.54 \cdot 10^{-19}$

Chapter 3: Review Exercises

1. a. 0.590 b. 0.410   3. a. 0.923 b. 0.0772
5. 0.860   7. 0.442   9. 1/3,838,380
11. a. 0.0225 b. 0.0214   13. 0.000000531
15. a. 0.08 b. 0.6 c. 0.8   17. 1/120
19. a. 1/16 b. 1/16 c. 1/8

Section 4-2

1. Probability distribution with $\mu = 5.9$, $\sigma^2 = 1.1$, $\sigma = 1.1$
3. Not a probability distribution; $\Sigma P(x) = 0.78 \neq 1$
5. Probability distribution with $\mu = 1.5$, $\sigma^2 = 0.8$, $\sigma = 0.9$
7. Probability distribution with $\mu = 2.8$, $\sigma^2 = 6.4$, $\sigma = 2.5$

9. Probability distribution with $\mu = 0.8$, $\sigma^2 = 0.5$, $\sigma = 0.7$
11. Probability distribution with $\mu = 2.0$, $\sigma^2 = 1.1$, $\sigma = 1.1$
13. $275   15. $97.50 per day   17. $-$106
19. $\mu = 1.5$, $\sigma^2 = 0.8$, $\sigma = 0.9$
21. $\mu = 1.7$, $\sigma^2 = 0.3$, $\sigma = 0.5$
23. a. Yes b. Yes c. No, $\Sigma P(x) > 1$
25. a. 31.25 b. 0.05 c. 1.25 d. 1.25
27. $\Sigma (x - \mu)^2 \cdot P(x) = \Sigma (x^2 - 2\mu x + \mu^2) \cdot P(x)$
    $= \Sigma x^2 \cdot P(x) - 2\mu \Sigma x \cdot P(x) + \mu^2 \Sigma P(x)$
    $= \Sigma x^2 \cdot P(x) - 2\mu \cdot \mu + \mu^2 \cdot 1 = \Sigma x^2 \cdot P(x) - \mu^2$

Section 4-3

1. a, c, d, e   3. 0.101   5. 0.349   7. 45/512
9. 15/64   11. 0.205   13. 0.000297   15. 0.264
17. 0.585   19. 0.189   21. 0.0338 (from Table A-1: 0.033)
23. 0.147   25. a. 0.364 b. 0.372
27. a. 0.225 b. 0.115   29. a. 0.105 b. 0.237 c. 0.113
31. 0.594   33. 0.0524   35. b. 0.000535   37. a. 0.317 b. 0.311

Section 4-4

1. $\mu = 24.5$, $\sigma^2 = 12.25$, $\sigma = 3.5$
3. $\mu = 404.4$, $\sigma^2 = 58.6$, $\sigma = 7.7$
5. $\mu = 25.0$, $\sigma^2 = 12.5$, $\sigma = 3.5$
7. $\mu = 8.8$, $\sigma^2 = 4.9$, $\sigma = 2.2$
9. $\mu = 94.0$, $\sigma^2 = 49.8$, $\sigma = 7.1$
11. $\mu = 4.4$, $\sigma = 1.7$
13. a. $\mu = 270.0$, $\sigma = 5.2$
    b. Yes, because 244 is more than 2 standard deviations away from 270.
15. a. $\mu = 8.3$, $\sigma^2 = 6.9$, $\sigma = 2.6$ b. Yes
17. Yes, it's more than 5 standard deviations above the mean of 48.96.
19. a. $\mu = 190.0$, $\sigma = 3.1$
    b. Yes, it's more than 3 standard deviations above the mean.
21. a. $\mu = 20.0$, $\sigma = 4.0$
    b. At least 3/4 of scores are between 12 and 28. c. $z = 2.50$

Chapter 4: Review Exercises

1. a. 0.103 b. 0.150 c. $\mu = 3.0$, $\sigma = 1.4$
3. 0.00000267
5. $P(4) = 0.227$, $P(5) = 0.060$, $\mu = 2.9$, $\sigma = 1.1$
7. a. $\mu = 350.0$, $\sigma^2 = 105.0$, $\sigma = 10.2$ b. 329.5, 370.5
    c. Yes, 375 is more than 2 standard deviations above the mean of 350.
9. a. 0.394 b. 0.00347 c. 0.000142 d. 0.0375

Section 5-2

1. 0.4   3. 0.9   5. 0.4772   7. 0.1469   9. 0.3085
11. 0.8621   13. 0.6826   15. 0.8702   17. 0.0792   19. 0.1362
21. 0.5   23. 0.0500   25. 0.0250   27. 0.9901   29. 1.28
31. 1.04   33. -2.33 35. $-1.645, 1.645$
37. a. 2.01 b. 1.23 c. 1.84 d. -1.04 e. -2.50
39. The heights at $x = 0, 1$ are 0.400 and 0.2434. Using a trapezoid, we approximate the area as 0.3217. The value from Table A-2 is 0.3413.

Section 5-3

1. 0.4641   3. 0.9452   5. 119.2   7. 87.4   9. 3.67%
11. 2.5, 16.3   13. 0.19%   15. 4.00   17. 41.68%   19. 0.3190
21. 2.28%   23. 0.4861   25. 78.8
27. 0.0038; either a very rare event has occurred or the husband is not the father.
29. 3.89%   31. a. 7.5% b. 200.46 c. 0.29 d. 2.94% e. Yes
33. A: 62.8 and above; B: above 55.2 and below 62.8; C: above 44.8 and below 55.2; D: above 37.2 and below 44.8; F: 37.2 and below
35. 29.6 min   37. ≈ 30.9

Section 5-4

1. a. 0.0793 b. 0.3413   3. 0.3026   5. 0.4699
7. 0.7734   9. 0.0668   11. 0.2766   13. 0.9768   15. 0.3566
17. a. 0.4681 b. 0.2033
   c. No, the mean of 440 could easily occur with an ineffective course.
19. a. 0.0011
   b. Yes, because it is so unlikely that a mean as high as 42.0 will occur by chance.
21. 0.0057   23. ≈ 0.1949 (using $s \approx 22.4$)

Section 5-5

1. a. 0.121 b. 0.1173
3. a. 0.677 b. Normal approximation not suitable
5. 0.0287   7. 0.1020   9. 0.0041   11. 0.6940   13. 0.0091
15. 0.5160   17. 0.0869   19. 0.2643   21. 0.0001; no
23. 0.2912; yes
25. 0.0075; something seems wrong with the sample.   27. 0.6368
29. a. 0.782 b. 0.782722294 c. 0.7852
   The binomial probability formula is most accurate.
31. 262   33. 0.1357

Chapter 5: Review Exercises

1. a. 0.0228 b. 0.0918 c. 0.9772 d. 0.8185 e. 0.0905
3. a. 0.3944 b. 0.3085 c. 0.8599 d. 0.6247 e. 0.2426

5. 0.7454   7. 0.3557
9. Using normal approximation: a. 0.0856 b. 0.8554

Section 6-2

1. a. 1.96   b. 2.33   c. 2.05   d. 2.093   e. 2.977
3. 20.6   5. 3.8625   7. $68.8 < \mu < 72.0$
9. $5.08 < \mu < 5.22$ 11. $1.79 < \mu < 3.01$ 13. 543
15. $5.597 < \mu < 5.647$; no, the quarters in circulation might be slightly lighter from wear.
17. 432   19. $23.37 < \mu < 31.51$; no
21. a. $40.0 < \mu < 43.6$ b. 7397   23. $7.29 < \mu < 9.37$
25. $8.391 < \mu < 10.466$ (using $\bar{x} = 9.428$, $s = 4.168$)
27. a. $104.4 < \mu < 117.9$ b. $99.8 < \mu < 122.2$
   c. Part a uses the normal distribution because the sample is large ($n = 35$), whereas part b uses the t distribution because the sample is small ($n = 23$) and σ is not known. Both confidence intervals are centered around the same approximate mean (111), but the interval for part a is narrower because of the larger sample size.
29. 0.5 (using $s = 7.2636$)   31. 147

Section 6-3

1. 0.0351   3. 0.0276   5. $0.208 < p < 0.292$
7. $0.553 < p < 0.654$ 9. 2944   11. $60.1\% < p < 65.9\%$
13. $0.619 < p < 0.661$ 15. 1393   17. $0.721 < p < 0.779$
19. $44.2\% < p < 51.8\%$ 21. 775   23. $0.243 < p < 0.311$
25. $70.3\% < p < 73.7\%$
27. $0.209 < p < 0.367$ (using $\hat{p} = 35/125$)
29. $45.9\% < p < 70.8\%$; yes   31. 9604   33. 89%   35. b. 305

Section 6-4

1. a. 97   b. 144.0   c. $\chi_L^2 = 13.844$, $\chi_R^2 = 41.923$
   d. $89.3 < \sigma^2 < 270.4$ e. $9.5 < \sigma < 16.4$
3. a. 141.7   b. $8.2 < \sigma < 21.7$ c. Yes
5. $1.42 < \sigma < 2.05$ 7. $0.99 < \sigma < 1.90$
9. $0.003 < \sigma^2 < 0.008$ 11. $35.2 < \sigma < 49.0$
13. $1.45 < \sigma < 3.03$ (using $s = 1.957$)
15. $3.567 < \sigma < 5.116$ (using $s = 4.168$)   17. 98%
19. $2.7 < \sigma < 2.9$

Chapter 6: Review Exercises

1. a. 17.6   b. $15.0 < \mu < 20.2$ 3. $2.92 < \sigma < 5.20$
5. $0.0696 < p < 0.110$ 7. 626   9. $36.5 < \mu < 44.9$

11. 423   13. 404   15. $21.5\% < p < 26.5\%$

17. $71.29 < \mu < 74.04$

Section 7-2

1. a. $\mu = 120$ b. $H_0: \mu = 120$ c. $H_1: \mu \neq 120$
   d. Two-tailed
   e. The error of rejecting the claim that the mean is 120 when it is equal to 120
   f. Failing to reject the claim that the mean is 120 when it is not equal to 120
   g. There is sufficient evidence to warrant rejection of the claim that the mean IQ of statistics instructors is equal to 120.
   h. There is not sufficient evidence to warrant rejection of the claim that the mean IQ of statistics instructors is equal to 120.

3. a. $\mu > 5$ b. $H_0: \mu \leq 5$ c. $H_1: \mu > 5$ d. Right-tailed
   e. The error of rejecting the claim that the mean is less than or equal to 5 years when it really is less than or equal to 5 years
   f. The error of failing to reject the claim that the mean is less than or equal to 5 years when it is actually greater than 5 years
   g. There is sufficient evidence to support the claim that the mean time is greater than 5 years.
   h. There is not sufficient evidence to support the claim that the mean time is greater than 5 years.

5. a. $\mu \geq 10$ b. $H_0: \mu \geq 10$ c. $H_1: \mu < 10$ d. Left-tailed
   e. The error of rejecting the claim that the mean is at least 10 years when it really is at least 10 years
   f. The error of failing to reject the claim that the mean is at least 10 years when it is actually less than 10 years
   g. There is sufficient evidence to warrant rejection of the claim that the mean is at least 10 years.
   h. There is not sufficient evidence to warrant rejection of the claim that the mean is at least 10 years.

7. ± 2.575 9. 2.33   11. -1.645

13. a. $\mu \leq 30$ b. $\mu \geq 5$ c. $\mu = 70$

15. By making $\alpha = 0$, we make it impossible to make a type I error because we *never* reject the null hypothesis, no matter how extreme the sample data might be.

Section 7-3

1. Test statistic: $z = 1.44$. Critical values: $z = \pm 2.575$. Fail to reject $H_0: \mu = 100$. There is not sufficient evidence to warrant rejection of the claim that the population mean is equal to 100. P-value $= 0.1498$.

3. Test statistic: $z = 3.46$. Critical values: $z = \pm 1.645$. Reject $H_0: \mu = 500$. There is sufficient evidence to warrant rejection of the claim that the mean is equal to 500. P-value $= 0.0002$.

5. Test statistic: $z = -2.84$. Critical value: $z = -2.33$. Reject $H_0: \mu \geq 12$. There is sufficient evidence to support the claim that the company is cheating consumers. P-value $= 0.0023$.

7. Test statistic: $z = 14.61$. Critical value: $z = 2.33$. Reject $H_0: \mu \leq 0.21$. There is sufficient evidence to support the claim that the mean is greater than 0.21. P-value $= 0.0001$.

9. Test statistic: $z = -38.82$. Critical value: $z = -2.33$. Reject $H_0: \mu \geq 1.39$. There is sufficient evidence to support the claim that the mean is less than 1.39 days. P-value $= 0.0001$.

11. Test statistic: $z = 1.37$. Critical value: $z = 1.645$. Fail to reject $H_0: \mu \leq 0$. There is not sufficient evidence to support the claim that the course is effective (with a mean increase greater than 0). P-value $= 0.0853$.

13. Test statistic: $z = 1.30$. Critical value: $z = 1.645$. Fail to reject $H_0: \mu \leq 51.0$. There is not sufficient evidence to support the claim that the mean is greater than 51.0. P-value $= 0.0968$.

15. Test statistic: $z = -3.12$. Critical value: $z = -1.645$. Reject $H_0: \mu \geq 24$ months. There is sufficient evidence to support the claim that the mean is less than 2 years. P-value $= 0.0001$.

17. Test statistic: $z = -10.02$. Critical value: $z = -1.645$. Reject $H_0: \mu \geq 7124$. There is sufficient evidence to support the claim that the mean is less than 7124 mi. P-value $= 0.0001$.

19. Test statistic: $z = 2.44$. Critical value: $z = 1.645$. Reject $H_0: \mu \leq 40,000$. There is sufficient evidence to support the claim that the mean is greater than \$40,000. P-value $= 0.0073$.

21. Test statistic: $z = -0.45$. Critical values: $z = \pm 1.96$. Fail to reject $H_0: \mu = 14$. There is not sufficient evidence to warrant rejection of the claim that the mean age is 14 years. P-value $= 0.6528$.

23. Test statistic: $z = -4.99$. Critical values: $z = \pm 2.575$. Reject $H_0: \mu = 5.670$. There is sufficient evidence to warrant rejection of the claim that the mean weight of quarters in circulation is 5.670 g. One possible explanation: Quarters in circulation become lighter through wear. P-value $= 0.0002$.

25. [Use $\sigma \approx (115 - 99)/2(0.67) = 11.9$, where 0.67 corresponds to an area of 0.25 in Table A-2.] Test statistic: $z = 1.62$. Critical values: $z = \pm 2.17$. Fail to reject $H_0: \mu = 105$. There is not sufficient evidence to warrant rejection of the claim that the population has a mean IQ score equal to 105.

27. \$24,696.

29. $\mu = 31.664$ oz

Section 7-4

1. a. ± 2.056 b. 1.337   c. -3.365

3. Test statistic: $t = 1.500$. Critical value: $t = 1.860$. Fail to reject $H_0: \mu \leq 10$. There is not sufficient evidence to warrant rejection of the claim that the mean is 10 or less.

5. Test statistic: $t = 2.014$. Critical values: $t = \pm 2.145$. Fail to reject H_0: $\mu = 75$. There is not sufficient evidence to warrant rejection of the claim that the mean equals 75.

7. Test statistic: $t = -3.214$. Critical values: $t = \pm 2.064$. Reject H_0: $\mu = 98.6$. There is sufficient evidence to warrant rejection of the claim that the mean body temperature of all healthy adults is equal to 98.6°F.

9. Test statistic: $t = 2.598$. Critical value: $t = 1.315$. Reject H_0: $\mu \leq 32$. There is sufficient evidence to support the claim that the mean fill amount is greater than 32 oz.

11. Test statistic: $t = 3.344$. Critical value: $t = 2.718$. Reject H_0: $\mu \leq \$1800$. There is sufficient evidence to support the claim that the mean exceeds $1800.

13. P-value < 0.005

15. Test statistic: $t = 1.826$. Critical values: $t = \pm 2.462$. Fail to reject H_0: $\mu = 20$. The pills appear to be acceptable.

17. Test statistic: $z = 2.10$. Critical value: $z = 1.96$. Reject H_0: $\mu \leq \$2000$. There is sufficient evidence to support the claim that the mean is greater than $2000. The monitoring system will be implemented.

19. P-value < 0.005

21. Test statistic: $t = -9.740$. Critical values: $t = \pm 2.080$. Reject H_0: $\mu = 243.5$. There is sufficient evidence to support the claim that the mean differs from 243.5. It appears that the VDT test should be revised.

23. Test statistic: $z = 0.80$. Critical value: $z = 1.28$. Fail to reject H_0: $\mu \leq 5$. There is not sufficient evidence to support the claim that the mean is greater than 5 years.

25. Test statistic: $t = 11.626$. Critical values: $t = \pm 2.045$. Reject H_0: $\mu = 650$. There is sufficient evidence to warrant rejection of the claim that the mean is equal to 650 mg.

27. $\bar{x} = 78.0500$, $s = 5.5958$. Test statistic: $t = 2.438$. Critical value: $t = 1.729$. Reject H_0: $\mu \leq 75$. There is sufficient evidence to support the claim that the class is above average.

29. $0.20 < P$-value < 0.5

31. $n = 15$, $\bar{x} = 188.133333$, $s = 33.204704$. Test statistic: $t = 1.415$. Critical values: $t = \pm 2.145$. Fail to reject H_0: $\mu = 176$ lb. There is not sufficient evidence to warrant rejection of the claim that the mean is equal to 176 lb.

33. With $z = 1.645$, the table and the approximation both result in $t = 1.833$.

Section 7-5

1. Test statistic: $z = 1.25$. Critical value: $z = 1.645$. Fail to reject H_0: $p \leq 0.04$. There is not sufficient evidence to warrant rejection of the claim that production is not out of control. No corrective action appears necessary. P-value $= 0.1056$.

3. Test statistic: $z = 1.48$. Critical values: $z = \pm 1.645$. Fail to reject H_0: $p = 0.71$. There is not sufficient evidence to warrant rejection of the claim that the actual percentage is 71%. P-value $= 0.1388$.

5. Test statistic: $z = 2.69$. Critical values: $z = \pm 2.33$. Reject H_0: $p = 0.2$. There is sufficient evidence to warrant rejection of the claim that computers are in 20% of all households. In fact, they appear to be in more than 20% of all households. P-value $= 0.0072$.

7. Test statistic: $z = 6.91$. Critical value: $z = 2.05$. Reject H_0: $p \leq 1/4$. There is sufficient evidence to support the claim that more than one-fourth of all white-collar criminals have attended college. P-value $= 0.0001$.

9. Test statistic: $z = 2.77$. Critical values: $z = \pm 2.33$. Reject H_0: $p = 1/8$. The Mendelian law does not appear to be working because the sample results differ significantly from the one-eighth proportion that was expected. P-value $= 0.0056$.

11. Test statistic: $z = 0.99$. Critical value: $z = 1.645$. Fail to reject H_0: $p \leq 0.08$. There is not sufficient evidence to support the claim that more than 8% of Seldane users experience drowsiness. P-value $= 0.1611$.

13. Test statistic: $z = -4.37$. Critical value: $z = -2.33$. Reject H_0: $p \geq 0.10$. There is sufficient evidence to support the claim that fewer than 10% of medical students prefer pediatrics. P-value $= 0.0001$.

15. Test statistic: $z = -1.75$. Critical value: $z = -1.645$ (assuming a 0.05 significance level). Reject H_0: $p \geq 0.07$. There is sufficient evidence to support the claim that the no-show rate is lower than 7%. The new system does appear to be effective in reducing no-shows. P-value $= 0.0401$.

17. Test statistic: $z = 1.33$. Critical value: $z = 1.645$. Fail to reject H_0: $p \leq 0.5$. There is not sufficient evidence to support the claim that most students don't know what the Holocaust is. P-value $= 0.0918$.

19. Test statistic: $z = -14.00$. Critical value: $z = -2.575$. Reject H_0: $p \geq 0.5$. There is sufficient evidence to reject the claim that at least half of all adults eat their fruitcakes. Fruitcake producers should consider changes. P-value $= 0.0001$.

21. a. Test statistic: $z = -0.75$. Critical values: $z = \pm 1.96$. Fail to reject H_0: $p = 0.2$. There is not sufficient evidence to warrant rejection of the claim that 20% of M&Ms are red.

    b. P-value $= 0.4532$, which exceeds the significance level of $\alpha = 0.05$. Therefore, form the same conclusions as in part a.

    c. $0.0964 < p < 0.244$; because the confidence interval limits contain 0.20, fail to reject the claim that 20% of M&Ms are red.

23. From the table of binomial probabilities with $n = 15$ and $p = 0.1$, we get $P(0) = 0.206$. If p is really 0.1, then there is a good chance

(0.206) that none of the 15 residents will believe that the mayor is doing a good job. Because that result could easily occur by chance, there is not sufficient evidence to reject the reporter's claim. We would reject the claim only if the probability were found to be less than 0.05.

25. 0.0023

Section 7-6

1. a. 8.907, 32.852   b. 10.117
   c. Interpolated values: 52.116, 99.665
3. Test statistic: $\chi^2 = 114.586$. Critical values: $\chi^2 = 57.153, 106.629$. Reject H_0: $\sigma = 43.7$ ft. There is sufficient evidence to warrant rejection of the claim that the standard deviation is equal to 43.7 ft. The new method appears to be worse than the old method.
5. Test statistic: $\chi^2 = 24.576$. Critical value: $\chi^2 = 43.188$. Reject H_0: $\sigma \geq 0.75$. There is sufficient evidence to support the claim that the standard deviation is lower than it was in the past.
7. Test statistic: $\chi^2 = 44.800$. Critical value: $\chi^2 = 51.739$. Reject H_0: $\sigma^2 \geq 0.0225$. There is sufficient evidence to support the claim that the new machine produces less variance. Its purchase should be considered.
9. Test statistic: $\chi^2 = 69.135$. Critical value: $\chi^2 = 67.505$. Reject H_0: $\sigma \leq 19.7$. There is sufficient evidence to support the claim that women have a larger standard deviation.
11. Test statistic: $\chi^2 = 62.016$. Critical values: $\chi^2 = 32.357, 71.420$. Fail to reject H_0: $\sigma = 2.4$. There is not sufficient evidence to warrant rejection of the claim that the standard deviation is equal to 2.4 in.
13. Test statistic: $\chi^2 = 73.676$. Critical values: $\chi^2 = 40.482, 83.298$. Fail to reject H_0: $\sigma = 3$. There is not sufficient evidence to warrant rejection of the claim that the standard deviation is 3 lb.
15. $n = 12$, $\bar{x} = 33.050$, $s = 1.129$. Test statistic: $\chi^2 = 3.505$. Critical value: $\chi^2 = 3.816$. Reject H_0: $\sigma \geq 2.0$. There is sufficient evidence to support the claim that the standard deviation is less than 2.0 mg.
17. The sample standard deviation is $s = 0.657$. Test statistic: $\chi^2 = 29.311$. Critical values: $\chi^2 = 6.262, 27.488$ (assuming a 0.05 significance level). Reject H_1: $\sigma = 0.470$ kg. There is sufficient evidence to warrant rejection of the claim that the standard deviation is equal to 0.470 kg.
19. a. $0.01 < P\text{-value} < 0.025$ b. $0.005 < P\text{-value} < 0.01$
   c. $P\text{-value} < 0.01$
21. a. Estimated values: 73.772, 129.070; Table A-4 values: 74.222, 129.561
   b. 116.643, 184.199
23. Approximately 0.10

Chapter 7: Review Exercises

1. a. $z = -1.645$ b. $z = 2.33$ c. $t = -3.106, 3.106$
   d. $\chi^2 = 10.856$ e. $\chi^2 = 16.047, 45.722$
3. a. H_0: $\mu \geq 20.0$ b. Left-tailed
   c. Rejecting the claim that the mean is at least 20.0 min when it really is at least 20.0 min
   d. Failing to reject the claim that the mean is at least 20.0 min when it really is less than 20.0 min
   e. 0.01
5. Test statistic: $z = 3.14$. Critical value: $z = 2.33$. Reject H_0: $p \leq 0.5$. There is sufficient evidence to support the claim that more than 50% of gun owners favor stricter gun laws.
7. Test statistic: $t = -1.457$. Critical values: $t = \pm 2.045$. Fail to reject H_0: $\mu = 3393$. There is not sufficient evidence to warrant rejection of the claim that boys and girls have the same mean birth weight.
9. Test statistic: $z = -2.73$. Critical value: $z = -2.33$. Reject H_0: $\mu \geq 5.00$. There is sufficient evidence to support the claim that the mean radiation dosage is below 5.00 milliroentgens.
11. Test statistic: $z = 13.50$. Critical value: $z = 1.96$. Reject H_0: $\mu \leq 55.0$. There is sufficient evidence to support the claim that the mean is above 55.0 mi/h.
13. Test statistic: $\chi^2 = 77.906$. Critical value: $\chi^2 = 74.397$. Reject H_0: $\sigma^2 \leq 6410$. There is sufficient evidence to support the counselor's claim that the current group has more varied aptitudes.
15. Test statistic: $z = -0.25$. Critical values: $z = \pm 1.645$. Fail to reject H_0: $p = 0.98$. There is not sufficient evidence to warrant rejection of the claim that the recognition rate is equal to 98%.
17. Test statistic: $\chi^2 = 9.900$. Critical value: $\chi^2 = 6.571$. Fail to reject H_0: $\sigma \geq 10.0$. There is not sufficient evidence to support the claim that the standard deviation is less than 10.0.
19. Test statistic: $z = -0.36$. Critical value: $z = -1.28$. Fail to reject H_0: $p \geq 0.05$. There is not sufficient evidence to support the claim that the burglary rate is now less than 5%. $P\text{-value} = 0.3594$.

Section 8-2

1. a. 1.3636   b. 1.5015   c. 3.012   d. ± 2.228
3. $0.4 < \mu_d < 2.4$
5. Test statistic: $t = 3.822$. Critical values: $t = \pm 2.365$. Reject H_0: $\mu_d = 0$. There is sufficient evidence to warrant rejection of the claim that the differences between the commercials have a mean of 0. The humorous commercial seems to be better.
7. $3.9 < \mu_d < 19.9$
9. Test statistic: $t = 3.036$. Critical value: $t = 1.895$. Reject H_0: $\mu_d \leq 0$. There is sufficient evidence to support the claim that the after scores

are lower than the before scores. It appears that hypnosis does have an effect on pain, as measured by the sensory scores.

11. $-23.2 < \mu_d < 4.0$

13. Test statistic: $t = 2.301$. Critical values: $t = \pm 2.262$. Reject H_0: $\mu_1 = \mu_2$. There is sufficient evidence to warrant rejection of the claim that both weights have the same mean. The training appears to reduce weight.

15. Test statistic: $t = -2.840$. Critical values: $t = \pm 2.228$. Reject H_0: $\mu_1 = \mu_2$. There is sufficient evidence to warrant rejection of the claim that the morning and night body temperatures are the same.

17. a. P-value < 0.01 b. $0.005 < P$-value < 0.01

19. 0.025

Section 8-3

1. Test statistic: $z = -1.67$. Critical values: $z = \pm 1.96$. Fail to reject H_0: $\mu_1 = \mu_2$. There is not sufficient evidence to warrant rejection of the claim that the two population means are equal.

3. $-10.0 < \mu_1 - \mu_2 < 0.8$

5. Test statistic: $z = 1.57$. Critical value: $z = 2.33$. Fail to reject H_0: $\mu_1 \le \mu_2$. There is not sufficient evidence to support the claim that the mean family income is greater for males.

7. $8.4 < \mu_1 - \mu_2 < 15.6$

9. Test statistic: $z = 2.64$. Critical values: $z = \pm 1.645$. Reject H_0: $\mu_1 = \mu_2$. There is sufficient evidence to warrant rejection of the claim that both airlines have the same mean. Because the means are significantly different, salary might well be an important factor in selecting an airline for which to work.

11. Test statistic: $z = 0.81$. Critical values: $z = \pm 1.96$. Fail to reject H_0: $\mu_1 = \mu_2$. There is not sufficient evidence to warrant rejection of the claim that the two teachers produce students with the same mean score.

13. $27 < \mu_1 - \mu_2 < 121$

15. Test statistic: $z = -4.93$. Critical values: $z = \pm 1.96$. Reject H_0: $\mu_1 = \mu_2$. There is sufficient evidence to warrant rejection of the claim that both ambulance services have the same mean response time. Weston seems preferable.

17. 18–24: $n = 37$, $\bar{x} = 98.224324$, $s = 0.659716$. 25 and older: $n = 56$, $\bar{x} = 98.057143$, $s = 0.635283$

a. Test statistic: $z = 1.21$. Critical values: $z = \pm 1.96$. Fail to reject H_0: $\mu_1 = \mu_2$. There is not sufficient evidence to warrant rejection of the claim that both age groups have the same mean body temperature.

b. P-value $= 0.2262$. Because the P-value is not less than the significance level of $\alpha = 0.05$, conclusions are the same as in part a.

c. $-0.10 < \mu_1 - \mu_2 < 0.44$; because the interval contains 0, the conclusions are the same as in part a.

19. a. 50/3   b. 2/3   c. 52/3

Section 8-4

1. Test statistic: $F = 2.0000$. Critical value: $F = 4.0260$. Fail to reject H_0: $\sigma_1^2 = \sigma_2^2$. There is not sufficient evidence to warrant rejection of the claim that the variances are equal.

3. Test statistic: $F = 1.4038$. Critical value: $F = 1.6664$. Fail to reject H_0: $\sigma_1^2 \le \sigma_2^2$. There is not sufficient evidence to support the claim that population A has a larger variance.

5. Test statistic: $F = 1.4246$. Critical value: $F = 2.1540$. Fail to reject H_0: $\sigma_1^2 = \sigma_2^2$. There is not sufficient evidence to warrant rejection of the claim that the two production methods yield batteries with the same variance. Therefore, there is not sufficient evidence to justify use of the experimental method.

7. Test statistic: $F = 1.1352$. Critical value of F is between 3.5257 and 3.4296. Fail to reject H_0: $\sigma_1^2 = \sigma_2^2$. There is not sufficient evidence to warrant rejection of the claim that both time periods have the same variance.

9. Test statistic: $F = 1.0722$. Critical value: $F = 1.7444$. Fail to reject H_0: $\sigma_1 = \sigma_2$. There is not sufficient evidence to warrant rejection of the claim that the standard deviations for American and TWA are the same.

11. Test statistic: $F = 4.0179$. Critical value: $F = 4.9621$. Fail to reject H_0: $\sigma_1^2 = \sigma_2^2$. There is not sufficient evidence to warrant rejection of the claim that both processes have the same variance.

13. Test statistic: $F = 1.3478$. Critical value: $F = 2.6171$. Fail to reject H_0: $\sigma_1^2 = \sigma_2^2$. There is not sufficient evidence to warrant rejection of the claim that both groups come from populations with the same variance.

15. Test statistic: $F = 1.2478$. Critical value: $F = 3.0502$. Fail to reject H_0: $\sigma_1 = \sigma_2$. There is not sufficient evidence to warrant rejection of the claim that both groups come from populations with the same standard deviation.

17. Test statistic: $F = 2.6045$. Critical value: $F = 1.9838$. Reject H_0: $\sigma_1^2 = \sigma_2^2$. There is sufficient evidence to support the claim that the first professor's grading exhibits greater variance. The student should choose the second professor because the lower variance should result in a score closer to the mean. That is, his score from the second professor should be higher than his score from the first professor.

19. a. $F_L = 0.2484$, $F_R = 4.0260$ b. $F_L = 0.2315$, $F_R = 5.5234$
    c. $F_L = 0.1810$, $F_R = 4.3197$ d. $F_L = 0.3071$, $F_R = 4.7290$
    e. $F_L = 0.2115$, $F_R = 3.2560$

21. a. No   b. No   c. No

Section 8-5

1. F-test results: Test statistic is $F = 2.0000$. Critical value: $F = 4.0260$. Fail to reject H_0: $\sigma_1^2 = \sigma_2^2$. Test of means: Test statistic is $t = 5.477$.

Critical values: $t = \pm 2.101$. Reject H_0: $\mu_1 = \mu_2$. There is sufficient evidence to warrant rejection of the claim that the two population means are equal.

3. $9 < \mu_1 - \mu_2 < 21$

5. *F*-test results: Test statistic is $F = 1.4246$. Critical value: $F = 2.1540$. Fail to reject H_0: $\sigma_1^2 = \sigma_2^2$. Test of means: Test statistic is $t = 2.618$. Critical values: $t = \pm 1.96$. Reject H_0: $\mu_1 = \mu_2$. There is sufficient evidence to warrant rejection of the claim that the two production methods yield batteries with the same mean. The traditional method seems better.

7. *F*-test results: Test statistic is $F = 1.1352$. Critical *F* value is between 3.5257 and 3.4296. Fail to reject H_0: $\sigma_1^2 = \sigma_2^2$. Test of means: Test statistic is $t = -1.045$. Critical values: $t = \pm 2.074$. Fail to reject H_0: $\mu_1 = \mu_2$. There is not sufficient evidence to warrant rejection of the claim that the two time periods have the same mean.

9. *F*-test results: Test statistic is $F = 4.0179$. Critical value: $F = 4.9621$. Fail to reject H_0: $\sigma_1^2 = \sigma_2^2$. Test of means: Test statistic is $t = 8.820$. Critical values: $t = \pm 2.492$. Reject H_0: $\mu_1 = \mu_2$. There is sufficient evidence to warrant rejection of equality of the two means.

11. $0.07 < \mu_1 - \mu_2 < 0.27$

13. *F*-test results: Test statistic is $F = 1.0000$, so fail to reject H_0: $\sigma_1^2 = \sigma_2^2$. Test of means: Test statistic is $t = -3.075$. Critical values: $t = \pm 2.878$. Reject H_0: $\mu_1 = \mu_2$. There is sufficient evidence to warrant rejection of equality of the two population means. It appears that obsessive-compulsive disorders have a biological basis.

15. $-17.16 < \mu_1 - \mu_2 < 260.40$

17. *F*-test results: Test statistic is $F = 3.0785$. Critical value of *F* is between 2.8442 and 2.7608. Reject H_0: $\sigma_1^2 = \sigma_2^2$. Test of means: Test statistic is $t = 1.900$. Critical values: $t = \pm 2.540$. Fail to reject H_0: $\mu_1 = \mu_2$. There is not sufficient evidence to warrant rejection of the claim that both divisions have the same mean.

19. Red: $n = 17$, $\bar{x} = 0.90430$, $s = 0.04144$. Brown: $n = 30$, $\bar{x} = 0.92563$, $s = 0.05210$. (Assume a significance level of $\alpha = 0.05$.) *F*-test results: Test statistic is $F = 1.5807$. Critical value: $F = 2.5678$. Fail to reject H_0: $\sigma_1^2 = \sigma_2^2$. Test of means: Test statistic is $t = -1.447$. Critical values: $t = \pm 1.96$. Fail to reject H_0: $\mu_1 = \mu_2$. There is not sufficient evidence to warrant rejection of the claim that red M&Ms and brown M&Ms have the same mean weight. No corrective action appears necessary at this time.

21. df $= 50$, but the result will be the same as in Exercise 11.

23. a. No, assume equal variances. b. s^2

Section 8-6

1. a. $160/300 = 0.533$ b. -2.05 c. ± 1.96 d. 0.0404

3. Test statistic: $z = -1.92$. Critical values: $z = \pm 1.96$. Fail to reject

H_0: $p_1 = p_2$. There is not sufficient evidence to warrant rejection of the claim that there is no difference between the proportions of Democrats and Republicans who believe that the government should regulate airline prices. Based on the evidence, the consultant should not consider different approaches for Democrats and Republicans.

5. Test statistic: $z = 1.07$. Critical value: $z = 1.645$. Fail to reject H_0: $p_1 \geq p_2$. There is not sufficient evidence to support the claim that the rate of severe injuries is lower for children wearing seat belts.

7. $-0.271 < p_1 - p_2 < -0.089$; the differences are significant.

9. Test statistic: $z = 12.86$. Critical value: $z = 2.575$. Reject H_0: $p_1 \leq p_2$. There is sufficient evidence to support the claim that vinyl gloves have a larger leak rate than latex gloves.

11. Test statistic: $z = 4.46$. Critical values: $z = \pm 2.575$. Reject H_0: $p_1 = p_2$. There is sufficient evidence to warrant rejection of the claim that the central city refusal rate is the same as the refusal rate in other areas.

13. Test statistic: $z = 1.61$. Critical values: $z = \pm 1.96$. Fail to reject H_0: $p_1 = p_2$. There is not sufficient evidence to warrant rejection of the claim that four-year public and private colleges have the same freshman dropout rate.

15. Test statistic: $z = 0.10$. Critical values: $z = \pm 1.96$. Fail to reject H_0: $p_1 = p_2$. There is not sufficient evidence to warrant rejection of the claim that the proportion of males with at least some college education is equal to the proportion of females with at least some college education.

17. Test statistic: $z = -2.40$. Critical values: $z = \pm 1.96$. Reject H_0: $p_1 = p_2$. There is sufficient evidence to warrant rejection of the claim that the California percentage exceeds the New York percentage by an amount equal to 25%.

19. 2135

Chapter 8: Review Exercises

1. a. Test statistic: $z = -13.23$. Critical values: $z = \pm 1.96$. Reject H_0: $\mu_1 = \mu_2$. There is sufficient evidence to warrant rejection of the claim that the means are equal.
   b. $-80.5 < \mu_1 - \mu_2 < -59.7$

3. Test statistic: $t = 2.951$. Critical value: $t = 1.860$: Reject H_0: $\mu_1 \leq \mu_2$. There is sufficient evidence to support Minton's claim that the new spark plugs have a higher mean.

5. Test statistic: $z = 23.67$. Critical value: $z = 2.33$. Reject H_0: $p_1 \leq p_2$. There is sufficient evidence to support the claim that Queens County has a lower DWI conviction rate.

7. Test statistic: $z = 0.78$. Critical values: $z = \pm 2.33$. Fail to reject H_0: $\mu_1 = \mu_2$. There is not sufficient evidence to warrant rejection of the claim that the mean score for women is equal to the mean score for men.

9. Test statistic: $z = 0.96$. Critical value: $z = 2.33$. Fail to reject

$H_0: \mu_1 \leq \mu_2$. There is not sufficient evidence to support the claim that the 10-day program results in scores with a lower mean.

11. Test statistic: $z = 3.67$. Critical value: $z = 1.645$. Reject $H_0: p_1 \leq p_2$. There is sufficient evidence to support the claim that the proportion of correct answers by prepared students is greater.

13. a. Test statistic: $t = -23.00$. Critical value: $t = -2.365$. Reject $H_0: \mu_1 \geq \mu_2$. There is sufficient evidence to support the claim that the lesson was effective.
    b. $-3.2 < \mu_d < -2.6$ (or $2.6 < \mu_d < 3.2$)

15. F-test results: Test statistic is $F = 1.1061$. Critical value: $F = 2.5598$. Fail to reject $H_0: \sigma_1^2 = \sigma_2^2$. Test of means: Test statistic is $t = 2.738$. Critical values: $t = \pm 1.96$. Reject $H_0: \mu_1 = \mu_2$. There is sufficient evidence to warrant rejection of the claim that the two systems have the same mean.

Section 9-2

1. a. Significant linear correlation b. No significant linear correlation
   c. Significant linear correlation
3. b. $n = 4$, $\Sigma x = 7$, $\Sigma x^2 = 15$, $(\Sigma x)^2 = 49$, $\Sigma xy = 20$
   c. $r = -0.191$
5. Test statistic: $r = 0.630$. Critical values: $r = \pm 0.707$. Fail to reject the claim of no linear correlation. There does not appear to be a linear correlation between the weight of discarded paper and household size.
7. Test statistic: $r = 0.761$. Critical values: $r = \pm 0.632$. Reject the claim of no linear correlation. There does appear to be a linear correlation between the total weight of discarded garbage and household size.
9. Test statistic: $r = -0.069$. Critical values: $r = \pm 0.707$. Fail to reject the claim of no linear correlation. There does not appear to be a linear correlation between age and blood alcohol concentration for convicted DWI jail inmates.
11. Test statistic: $r = 0.564$. Critical values: $r = \pm 0.632$. Fail to reject the claim of no linear correlation. There does not appear to be a linear correlation between the score on the algebra placement test and the calculus achievement score.
13. Test statistic: $r = 0.845$. Critical values: $r = \pm 0.707$. Reject the claim of no linear correlation. There does appear to be a linear correlation between hydrocarbon emissions and carbon monoxide emissions.
15. Test statistic: $r = -0.926$. Critical values: $r = \pm 0.707$. Reject the claim of no linear correlation. There does appear to be a linear correlation between temperature and snow depth.
17. Test statistic: $r = -0.102$. Critical values: $r = \pm 0.396$. Fail to reject the claim of no linear correlation. For persons who never married, there does not appear to be a correlation between height and income.

19. Test statistic: $r = 0.888$. Critical values: $r = \pm 0.361$. Reject the claim of no linear correlation. There does appear to be a linear correlation between weights of cars and their lengths.
21. The conclusion implies that there is a significant linear correlation, but the correlation is actually close to 0.
23. Although there is no *linear* correlation, the two variables may be related in some other *nonlinear* way.
25. a. 0.972 b. 0.905 c. 0.999 d. 0.992 e. -0.984
    Case c yields largest r.
27. In attempting to calculate r, we get a denominator of 0, so a real value of r does not exist. However, it should be obvious that the value of x is not at all related to the value of y.
29. r changes from -0.926 to -0.631. The effect of an extreme value can be minimal or severe, depending on the other data.
31. With $r = 0.963$ and $n = 11$, it is reasonable to conclude that there is a significant linear correlation. This section of the parabola can be approximated by a straight line.

Section 9-3

1. $\hat{y} = 5.97 - 1.09x$ 3. $\hat{y} = 3.64 - 0.364x$
5. $\hat{y} = 0.152 + 0.398x$ 7. $\hat{y} = 0.183 + 0.119x$
9. $\hat{y} = 0.214 - 0.000182x$ 11. $\hat{y} = 52.7 + 1.02x$
13. $\hat{y} = -8.22 + 30.1x$; 14.4 g/m
15. $\hat{y} = -6.50 - 0.438x$; 19.8 cm   17. $\hat{y} = 106 - 1.05x$
19. $\hat{y} = -3130 + 32.6x$
21. a. 110.0 b. 25.0 c. 110.0 d. 25.0 e. 110.0
23. a. 6 b. 17 c. -13 d. 6 e. 6
25. $\hat{y} = -182 + 0.000351x$; $\hat{y} = -182 + 0.351x$. The slope is multiplied by 1000 and the y-intercept doesn't change. If each y entry is divided by 1000, the slope and y-intercept are both divided by 1000.
27. $\hat{y} = -5.66 + 0.505x$
29. In $\hat{y} = \hat{\beta}_0 + \hat{\beta}_1 x$, replace $\bar{x}$ by $\Sigma x/n$ and replace $\hat{\beta}_0$ and $\hat{\beta}_1$ by their expressions as given in Section 9-3, and show that $\hat{y} = \Sigma y/n = \bar{y}$.

Section 9-4

1. 0.111; 11.1%   3. 0.640; 64.0%
5. a. 154.8 b. 0 c. 154.8 d. 1 e. 0
7. a. 144.651 b. 57.9979 c. 202.649 d. 0.713801
   e. 3.10907
9. a. 14
   b. Because the standard error of estimate is 0, $E = 0$ and there is no "interval" estimate.
11. a. 14.4 b. $5.67 < y < 23.1$ 13. $-0.611 < y < 4.67$
15. $0.752 < y < 4.79$

17. $-1.377 < \beta_0 < 2.47$; $0.534 < \beta_1 < 2.43$

19. a. $(n - 2) s_e^2$ b. $\dfrac{r^2 \cdot \text{(unexplained variation)}}{1 - r^2}$ c. $r = -0.949$

Section 9-5

1. 85 3. 75 5. $\hat{y} = 0.92 - 0.244x_2 + 1.75x_4 - 0.073x_6$

7. No, because the P-value of 0.116 indicates that the equation lacks overall significance

9. a. $\hat{y} = 0.804 - 0.363x_2 + 1.76x_4$ b. 0.040 c. 0.723
   d. 0.613 e. Yes

11. $\hat{y} = 2.17 + 2.44x + 0.464x^2$. Because $R^2 = 1$, the parabola fits perfectly.

Chapter 9: Review Exercises

1. a. 0.316 b. ± 0.707 c. No significant linear correlation
   d. $\hat{y} = 24.4 + 0.456x$

3. a. 0.967 b. ± 0.666 c. Significant linear correlation
   d. $\hat{y} = -101 + 0.643x$

5. a. Test statistic: $r = 0.999$. Critical values: $r = \pm 0.666$. Reject the claim of no linear correlation. There does appear to be a linear correlation between minutes before midnight on the doomsday clock and the index of savings.
   b. $\hat{y} = 5.92 + 0.636x_1$ c. 9.10

7. $8.81 < y < 9.38$

Section 10-2

1. Test statistic: $\chi^2 = 156.500$. Critical value: $\chi^2 = 21.666$. Reject the claim that the last digits occur with the same frequency.

3. Test statistic: $\chi^2 = 7.417$. Critical value: $\chi^2 = 12.592$. Fail to reject the claim that fatal crashes occur on the different days of the week with equal frequency. The weekend frequencies seem to be higher, but there isn't sufficient evidence to support that theory.

5. Test statistic: $\chi^2 = 23.431$. Critical value: $\chi^2 = 9.488$. Reject the claim that the accidents are distributed as claimed. Rejection of the claim does not provide help in correcting the problem.

7. Test statistic: $\chi^2 = 1.954$. Critical value: $\chi^2 = 12.592$. Fail to reject the claim that the given percentages are correct.

9. Test statistic: $\chi^2 = 4.200$. Critical value: $\chi^2 = 16.919$. Fail to reject the claim that the digits are uniformly distributed.

11. Test statistic: $\chi^2 = 27.332$. Critical value: $\chi^2 = 13.277$. Reject the claim that Mendelian principles hold.

13. Test statistic: $\chi^2 = 1.550$. Critical value: $\chi^2 = 11.071$. There is not sufficient evidence to warrant rejection of the claim that the colors are distributed as claimed.

15. Test statistic: $\chi^2 = 21.467$. Critical value: $\chi^2 = 7.815$. Reject the claim that the movies are distributed evenly among the 4 categories.

17. $0.10 < P$-value < 0.90

19. a.
$$\chi^2 = \dfrac{\left(f_1 - \dfrac{f_1 + f_2}{2}\right)^2}{\dfrac{f_1 + f_2}{2}} + \dfrac{\left(f_2 - \dfrac{f_1 + f_2}{2}\right)^2}{\dfrac{f_1 + f_2}{2}} = \dfrac{(f_1 - f_2)^2}{f_1 + f_2}$$

b.
$$z^2 = \dfrac{\left(\dfrac{f_1}{f_1 + f_2} - 0.5\right)^2}{\dfrac{1/4}{f_1 + f_2}} = \dfrac{(f_1 - f_2)^2}{f_1 + f_2}$$

21. Combining time slots 1 and 2, 5 and 6, 7 and 8, and 9 and 10, we get a test statistic of $\chi^2 = 4.118$ and a critical value of $\chi^2 = 11.071$ (assuming that $\alpha = 0.05$). Fail to reject the claim that observed and expected frequencies are compatible.

Section 10-3

1. Test statistic: $\chi^2 = 119.330$. Critical value: $\chi^2 = 5.991$. Reject the claim that the type of crime is independent of whether the criminal is a stranger.

3. Test statistic: $\chi^2 = 1.174$. Critical value: $\chi^2 = 3.841$. Fail to reject the claim that the treatment is independent of the reaction.

5. Test statistic: $\chi^2 = 12.321$. Critical value: $\chi^2 = 3.841$. Reject the claim that success and group are independent.

7. Test statistic: $\chi^2 = 10.708$. Critical value: $\chi^2 = 3.841$. Reject the claim that wearing a helmet is independent of whether the injuries included facial injuries. Based on the results, it appears that wearing a helmet does seem to be effective in helping to prevent facial injuries in a crash, but this conclusion is not substantiated by the statistical methods that were used.

9. Test statistic: $\chi^2 = 1.505$. Critical value: $\chi^2 = 3.841$. Fail to reject the claim that the accident rate is independent of the use of cellular phones. Based on the results, it appears that the use of cellular phones does not affect driving safety, but this conclusion is not substantiated by the statistical methods that were used.

11. Test statistic: $\chi^2 = 0.515$. Critical value: $\chi^2 = 7.815$. Fail to reject the claim that the original death certificates and the medical panel have the same proportions of deaths in the different categories; that is, they appear to agree.

13. Test statistic: $\chi^2 = 0.615$. Critical value: $\chi^2 = 7.815$. Fail to reject the claim that smoking is independent of age group.

15. Test statistic: $\chi^2 = 13.143$. Critical value: $\chi^2 = 12.017$. Reject the claim that the time of day of fatal accidents is the same in New York City as in all other New York State locations.

17. P-value < 0.005

19. The test statistic is multiplied by the same constant, but the critical value doesn't change.

Chapter 10: Review Exercises

1. Test statistic: $\chi^2 = 6.780$. Critical value: $\chi^2 = 12.592$. Fail to reject the claim that gunfire death rates are the same for the different days of the week.

3. Test statistic: $\chi^2 = 22.600$. Critical value: $\chi^2 = 19.675$. Reject the claim that months were selected with equal frequencies. This result does not support extrasensory perception; we need data showing the frequency of correct selections.

5. Test statistic: $\chi^2 = 53.451$. Critical value: $\chi^2 = 12.592$. Reject the claim that the county is independent of the type of conviction.

7. Test statistic: $\chi^2 = 4.556$. Critical value: $\chi^2 = 5.991$. Fail to reject the claim that teachers have the same accident rate.

Section 11-2

1. Test statistic: $F = 3.6926$. Critical value: $F = 3.8853$. Fail to reject the claim of equal means.

3. Test statistic: $F = 103.1651$. Critical value: $F = 2.6049$. Reject the claim of equal means. Paper seems to be a particularly large part of the problem (although this cannot be concluded using the statistical methods in Section 11-2).

5. Test statistic: $F = 1.5430$. Critical value: $F = 2.6060$. Fail to reject the claim of equal means. The sociologist cannot proceed because the means appear to be equal.

7. Test statistic: $F = 1.6500$. Critical value: $F = 3.0556$. Fail to reject the claim of equal means. The choice of car model does not seem to affect gas mileage.

9. Test statistic: $F = 0.1587$. Critical value: $F = 3.3541$. Fail to reject the claim of equal means. The zones appear to have the same mean selling price.

11. Test statistic: $F = 5.0793$. Critical value: $F = 3.3541$. Reject the claim of equal means. Zone 1 appears to have larger lots (although this cannot be concluded using the statistical methods of Section 11-2).

13. a. With test statistic $z = -1.31$ and critical values $z = \pm 1.96$, fail to reject the claim that $\mu_1 = \mu_2$.
    b. With test statistic $z = -1.39$ and critical values $z = \pm 1.96$, fail to reject the claim that $\mu_2 = \mu_3$.
    c. With test statistic $z = -2.66$ and critical values $z = \pm 1.96$, reject the claim that $\mu_1 = \mu_3$.

d. With test statistic $F = 3.6313$ and critical value $F = 3.0000$ (approx.), reject the claim of equal means.
   e. It is better to use the approach of part d.

15. a. The test statistic does not change.
    b. The test statistic is multiplied by the square of the constant.
    c. 0

Section 11-3

1. Test statistic: $F = 15.8142$. Critical value: $F = 3.6823$. Reject the claim of equal means. It appears that high school performance does affect success in the calculus course.

3. Test statistic: $F = 11.6743$. Critical value: $F = 4.6755$. Reject the claim of equal means. Programmer B seems to make the lowest number of errors (although this cannot be concluded by using the methods of Section 11-3).

5. Test statistic: $F = 2.9493$. Critical value: $F = 2.6060$. Reject the claim of equal means. The different laboratories do not appear to have the same mean.

7. Test statistic: $F = 1.5081$. Critical value: $F = 2.2899$ (the critical value is approximate and is based on $\alpha = 0.05$). Fail to reject the claim of equal means. It appears that no corrective action is necessary.

9. SS(treatments) = 2.17, df(treatments) = 2, df(error) = 16, df(total) = 18, MS(treatments) = 1.085, MS(error) = 7.036, F = 0.1542.

11. The SS and MS values all increase dramatically. The test statistic changes from 11.6743 to 1.3502.

Section 11-4

1. a. 1263   b. 695   c. 350   d. 14

3. Test statistic: $F = 0.5036$. Critical value: $F = 4.0662$, assuming that the significance level is $\alpha = 0.05$. It appears that star rating does not have an effect on movie length.

5. Test statistic: $F = 0$. Critical value: $F = 10.128$. It appears that MPAA rating does not have an effect on movie length.

7. Test statistic: $F = 17.5851$. Critical value: $F = 4.7571$. It appears that the typist does have an effect on the time.

9. Transposing the table does not change the results or conclusions.

11. The SS and MS values are multiplied by 100, but the values of the test statistics do not change.

Chapter 11:   Review Exercises

1. Test statistic: $F = 31.1111$. Critical value: $F = 3.2389$. Reject the claim of equal means.

3. Test statistic: $F = 46.9000$. Critical value: $F = 3.7389$. Reject the claim of equal means.

5. Test statistic: $F = 0.0531$. Critical value: $F = 5.1433$. Fuel consumption does not appear to be affected by an interaction between transmission type and engine size.
7. Test statistic: $F = 3.5664$. Critical value: $F = 5.9874$. The type of transmission does not appear to have an effect on fuel consumption.

Section 12-2

1. Process appears to be within statistical control.
3. Process appears to be out of statistical control because there is a point lying beyond the lower control limit. The process should be corrected because too little epoxy cement hardener is being put into some containers.
5. Process appears to be out of statistical control because of a shift up in values of the sample range. Also, there are 8 points all lying below the center line, and there are 8 consecutive points all lying above the center line.
7. Process appears to be within statistical control.
9. Process appears to be within statistical control and no corrective action seems necessary. If the $\bar{x}$ chart were to have a downward trend, the process would be out of statistical control, but it is actually *improving* because delivery times are decreasing. In this case, the cause of the decrease should be investigated to ensure that the reduced times continue.
11. Process appears to be out of statistical control because there is a shift up in values, there are 8 consecutive points all lying below the center line, and there are points beyond the upper control limit.
13. a. The s chart and the R chart are similar, and in this case they lead to the same conclusions.
    b. The $\bar{x}$ chart based on sample standard deviations is similar to the $\bar{x}$ chart based on sample ranges, and in this case they lead to the same conclusions.

Section 12-3

1. Process appears to be within statistical control.
3. Process appears to be out of statistical control because there is a pattern of an upward trend and there is a point that lies beyond the upper control limit.
5. Process appears to be out of statistical control because the 6th and 7th points lie beyond the upper control limit. If temporary employees were hired on the 6th and 7th days, that could easily explain the jump in defective filled orders. The company should consider staggering vacations so that the negative effect of temporary employees can be reduced.
7. Process appears to be out of statistical control because there is a point that lies beyond the upper control limit.

9. a. $0.0158 < p < 0.0217$
   b. Test statistic: $z = 7.87$. Critical value: $z = 1.645$. Reject H_0: $p \le 0.01$. There is sufficient evidence to warrant rejection of the claim that the rate of defects is 1% or less.
11. a. LCL = 0; center line is at 0.0500; UCL = 0.1154.
    b. LCL = 0.01225; center line is at 0.0500; UCL = 0.08775.
    c. As the sample size n increases, the control lines get closer. The advantage is that smaller shifts in the process are detected. A disadvantage is that more items must be sampled. The chart in part b would be better for detecting a change from 5% to 10%.

Chapter 12:   Review Exercises

1. Process appears to be within statistical control.
3. Process appears to be out of statistical control because there is a point beyond the upper control limit. The process appears to require corrective action.
5. Process appears to be out of statistical control because there seems to be a cyclical pattern. Also, there are 8 consecutive points all lying below the center line. (The 16th value is 0.12, which is below the center line value of 0.1208.) Corrective action should be taken to improve the process. Examination of the control chart should show that there appears to be a problem with excessively higher defects for shift B, so that is where the investigation should begin.

Section 13-2

1. The test statistic $x = 1$ is not less than or equal to the critical value of 0. Fail to reject the claim that there is no difference between the reactions to the humorous and serious commercials. Neither commercial seems to be better.
3. The test statistic $x = 1$ is less than or equal to the critical value of 1. Reject the null hypothesis of no effect. The drug appears to lower blood pressure.
5. The test statistic $x = 1$ is less than or equal to the critical value of 1. Reject the null hypothesis that the after scores are equal to or less than the before scores. The sample evidence supports the claim that the scores are higher after the course.
7. The test statistic $x = 1$ is less than or equal to the critical value of 1. Reject the null hypothesis of no difference. The tests appear to be different, and further research should be funded.
9. The test statistic $x = 0$ is less than or equal to the critical value of 1. Reject the claim of no difference.
11. The test statistic $x = 3$ is not less than or equal to the critical value of 1. Fail to reject the null hypothesis that Covariant is favored by at most half of all dentists. The sample data do not provide a reasonable basis for making the claim that most dentists favor Covariant toothpaste.

13. Convert $x = 36$ to the test statistic $z = -2.26$. The critical value is $z = -1.645$. Reject the null hypothesis that the training is ineffective. It appears to be effective.

15. Convert $x = 22$ to the test statistic $z = -1.71$. The critical values are $z = \pm 2.575$. Fail to reject the null hypothesis of no change. It appears that the diet produces no significant change in cholesterol levels.

17. Convert $x = 16$ to the test statistic $z = -1.11$. The critical values are $z = \pm 1.96$. Fail to reject the null hypothesis that both pistols are equally effective. Based on the available sample evidence, the less expensive pistol should be bought.

19. Convert $x = 10$ to the test statistic $z = -3.90$. The critical values are $z = \pm 1.96$, assuming a significance level of $\alpha = 0.05$. Reject the null hypothesis that the quarters have a median equal to 5.670 g.

21. With $k = 3$, we get $6 < M < 17$.

23. 78

Section 13-3

1. $T = 1$, $n = 8$. The test statistic $T = 1$ is not less than or equal to the critical value of 0, so fail to reject the null hypothesis that both samples come from the same population distribution.

3. $T = 2$, $n = 9$. The test statistic $T = 2$ is less than or equal to the critical value of 6, so reject the null hypothesis that both samples come from the same population distribution.

5. $T = 4$, $n = 9$. The test statistic $T = 4$ is not less than or equal to the critical value of 2, so fail to reject the null hypothesis that both samples come from the same population distribution.

7. $T = 3.5$, $n = 10$. The test statistic $T = 3.5$ is not less than or equal to the critical value of 3, so fail to reject the null hypothesis that both samples come from the same population distribution.

9. $T = 0$, $n = 10$. The test statistic $T = 0$ is less than or equal to the critical value of 8, so reject the null hypothesis that both samples come from the same population distribution.

11. a. $T = 51.5$, $n = 36$. The test statistic $z = -4.42$ is less than or equal to the critical value of $z = -2.575$, so reject the null hypothesis that both systems require the same times. (It appears that the scanner system is faster.)

   b. $T = 51.5$, $n = 36$. The test statistic is $z = -4.42$ and the critical value is $z = -2.33$, so reject the null hypothesis that the distribution of scanner times is the same as (or to the right of) the distribution of times for the manual system. There is sufficient evidence to support the claim that the distribution of scanner times is shifted to the left of the distribution of times for the manual system. It appears that the scanner times are lower.

13. 1954

Section 13-4

1. $\mu_R = 232.5$, $\sigma_R = 24.109$, $R = 285$, $z = 2.18$. Test statistic: $z = 2.18$. Critical values: $z = \pm 1.96$. Reject the null hypothesis that the salaries of teachers are the same in California and Maryland. The California salaries seem to be higher.

3. $\mu_R = 150$, $\sigma_R = 17.321$, $R = 96.5$, $z = -3.09$. Test statistic: $z = -3.09$. Critical values: $z = \pm 2.575$. Reject the null hypothesis that obsessive-compulsive patients and healthy persons have the same volumes of the right cordate. However, we cannot conclude that obsessive-compulsive disorders have a biological basis; that is, we cannot conclude that a physical change causes the disorder or vice versa. Such conclusions cannot be made on the basis of a statistical test.

5. $\mu_R = 162$, $\sigma_R = 19.442$, $R = 89.5$, $z = -3.73$. Test statistic: $z = -3.73$. Critical values: $z = \pm 1.96$. Reject the null hypothesis that beer and liquor drinkers have the same BAC levels. It appears that liquor drinkers are more dangerous.

7. $\mu_R = 370$, $\sigma_R = 31.411$, $R = 254$, $z = -3.69$. Test statistic: $z = -3.69$. Critical values: $z = \pm 1.96$. Reject the null hypothesis of equal scores.

9. $\mu_R = 224$, $\sigma_R = 20.265$, $R = 254$, $z = 1.48$. Test statistic: $z = 1.48$. Critical values: $z = \pm 1.96$. Fail to reject the null hypothesis that both come from populations with the same scores.

11. $\mu_R = 357$, $\sigma_R = 37.789$, $R = 336.5$, $z = -0.54$. Test statistic: $z = -0.54$. Critical values: $z = \pm 1.96$ (assuming a 0.05 level of significance). Fail to reject the null hypothesis that the weights of red M&Ms and yellow M&Ms are the same.

13. a. $\mu_R = 232.5$, $\sigma_R = 24.11$, $R = 120$, $z = -4.67$. Test statistic: $z = -4.67$. Critical values: $z = \pm 1.96$. Reject the null hypothesis that both groups come from the same population.

   b. $\mu_R = 232.5$, $\sigma_R = 24.11$, $R = 225$, $z = -0.31$. Test statistic: $z = -0.31$. Critical values: $z = \pm 1.96$. Fail to reject the null hypothesis that both groups come from the same population.

   c. There is an obvious difference in part a because the low ranks are all found in sample 1, whereas the high ranks are in sample 2. In contrast, the ranks in part b are evenly distributed. These observations are reflected in the test statistics of $z = -4.67$ for part a and $z = -0.31$ for part b.

   d. R changes to 345, but μ_R, σ_R, the test statistic z, and the conclusion remain the same.

   e. R changes to 149 and the test statistic changes to $z = -3.46$, but μ_R, σ_R, and the conclusion remain the same.

15. a. A B A B 4
      A B B A 5
      B B A A 7
      B A A B 5
      B A B A 6

b.

R	p
3	1/6
4	1/6
5	2/6
6	1/6
7	1/6

c. No, the most extreme rank distribution has a probability of at least 1/6 and we can never get into a critical region with a probability of 0.10 or less.

Section 13-5

1. Test statistic: $H = 2.180$. Critical value: $\chi^2 = 5.991$. Fail to reject the null hypothesis that the 3 age-group populations of body temperatures are identical.

3. Test statistic: $H = 2.480$. Critical value: $\chi^2 = 5.991$ (assuming a significance level of 0.05). Fail to reject the null hypothesis that the samples come from identical populations. No one supplier seems to be superior to the others.

5. Test statistic: $H = 9.129$. Critical value: $\chi^2 = 7.779$. Reject the null hypothesis that the samples come from identical populations.

7. Test statistic: $H = 1.589$. Critical value: $\chi^2 = 5.991$. Fail to reject the null hypothesis that the living areas are the same in all 3 zones.

9. Test statistic: $H = 1.732$. Critical value: $\chi^2 = 5.991$. Fail to reject the null hypothesis that the taxes are the same in all 3 zones.

11. Test statistic: $H = 16.083$. Critical value: $\chi^2 = 7.815$. Reject the null hypothesis that the different weight categories have the same fuel consumption levels. It appears that fuel consumption is different for the different weight categories.

13. a. $H = \dfrac{1}{1176}(R_1^2 + R_2^2 + \cdots + R_8^2) - 147$ b. No change

    c. No change

15. Dividing H by the correction factor results in $3.489 \div 0.99914$, which is rounded off to 3.492.

Section 13-6

1. a. ± 0.450 b. ± 0.280 c. ± 0.373 3. -0.486

5. $r_s = 0.261$. Critical values: $r_s = \pm 0.648$. No significant correlation.

7. $r_s = 0.571$. Critical values: $r_s = \pm 0.738$. No significant correlation.

9. $r_s = 0.715$. Critical values: $r_s = \pm 0.591$. Significant correlation.

11. $r_s = 0.869$. Critical values: $r_s = \pm 0.738$. Significant correlation.

13. $r_s = 0.006$. Critical values: $r_s = \pm 0.738$. No significant correlation.

15. $r_s = 0.494$. Critical values: $r_s = \pm 0.648$. No significant correlation.

17. $r_s = 0.911$. Critical values: $r_s = \pm 0.738$. Significant correlation.

19. $r_s = 0.292$. Critical values: $r_s = \pm 0.738$. No significant correlation.

21. $r_s = 0.017$. Critical values: $r_s = \pm 0.490$. No significant correlation.

23. $r_s = 0.772$. Critical values: $r_s = \pm 0.352$. Significant correlation.

25. a. 123   132   213   231   312   321

   b. 1   0.5   0.5 -0.5 -0.5 -1

   c. 1/6 or 0.167

27. a. The rank correlation coefficient, because the trend is not linear

   b. The sign changes.   c. Both will be the same.

   d. Test statistic: $r_s = 0.855$. Critical value: $r_s = 0.564$. Reject H_0: $\rho_s \leq 0$. There is sufficient evidence to support the claim that there is a positive correlation.

Section 13-7

1. $n_1 = 7$, $n_2 = 2$, $G = 2$; critical values: 1, 6.

3. a. Median is 9.5.   c. $n_1 = 5$, $n_2 = 5$, $G = 2$.

   d. Critical values are 2, 10.   e. Reject randomness.

5. $n_1 = 6$, $n_2 = 24$, $G = 12$, $\mu_G = 10.6$, $\sigma_G = 1.6873$. Test statistic: $z = 0.83$. Critical values: $z = \pm 1.96$. Fail to reject randomness.

7. Median is 14.5, $n_1 = 10$, $n_2 = 10$, $G = 4$. Critical values are 6, 16. Reject randomness.

9. Median is 834. $n_1 = 8$, $n_2 = 8$, $G = 2$. Critical values are 4, 14. Reject randomness. (There appears to be a downward trend.)

11. $n_1 = 17$, $n_2 = 28$, $G = 12$, $\mu_G = 22.16$, $\sigma_G = 3.11$. Test statistic: $z = -3.27$. Critical values: $z = \pm 1.96$. Reject randomness.

13. $n_1 = 53$, $n_2 = 37$, $G = 49$, $\mu_G = 44.58$, $\sigma_G = 4.566$. Test statistic: $z = 0.97$. Critical values: $z = \pm 1.96$. Fail to reject randomness.

15. $n_1 = 49$, $n_2 = 51$, $G = 43$, $\mu_G = 50.98$, $\sigma_G = 4.9727$. Test statistic: $z = -1.60$. Critical values: $z = \pm 1.96$. Fail to reject randomness.

17. Minimum is 2, maximum is 4. Critical values of 1 and 6 can never be realized, so the null hypothesis of randomness can never be rejected.

19. b. The 84 sequences yield 2 runs of 2, 7 runs of 3, 20 runs of 4, 25 runs of 5, 20 runs of 6, and 10 runs of 7.

   c. $P(2 \text{ runs}) = 2/84$, $P(3 \text{ runs}) = 7/84$, $P(4 \text{ runs}) = 20/84$, $P(5 \text{ runs}) = 25/84$, $P(6 \text{ runs}) = 20/84$, and $P(7 \text{ runs}) = 10/84$.

   d. From this we conclude that the G values of 3, 4, 5, 6, 7 can easily occur by chance, whereas $G = 2$ is unlikely since $P(2 \text{ runs})$ is less than 0.025. The lower critical G value is therefore 2.

   e. Critical value of $G = 2$ agrees with Table A-10. The table lists 8 as the upper critical value, but it is impossible to get 8 runs using the given elements.

   f. You get the same result of 5.

Chapter 13:   Review Exercises

1. $n_1 = 20$, $n_2 = 5$, $G = 5$. Critical values are 5, 12. Reject randomness.

3. $r_s = 0$. Critical values: $r_s = \pm 0.786$. No significant correlation.

5. Test statistic: $H = 6.181$. Critical value: $\chi^2 = 5.991$ (assuming a 0.05 significance level). Reject the null hypothesis that the 3 precincts have the same response times.

7. The test statistic $x = 2$ is not less than or equal to the critical value of 1. Fail to reject the null hypothesis that both tests produce the same results.

9. $\mu_R = 137.50$, $\sigma_R = 17.26$, $R = 177.5$, $z = 2.32$. Test statistic: $z = 2.32$. Critical values: $z = \pm 1.96$. Reject the null hypothesis of no difference.

11. $r_s = 0.818$. Critical values: $r_s = \pm 0.525$. Significant correlation.

13. $T = 4$, $n = 12$. The test statistic of $T = 4$ is less than or equal to the critical value of 14, so reject the null hypothesis that the tune-up has no effect on fuel consumption.

15. Discard the two 0s to get 11 negative signs and 1 positive sign, so $x = 1$, which is less than the critical value of 2 found from Table A-7. Reject the null hypothesis of no effect.

17. $\mu_R = 144$, $\sigma_R = 16.25$, $R = 84$, $z = -3.69$. Test statistic: $z = -3.69$. Critical values: $z = \pm 1.96$. Reject the null hypothesis of no difference between mail order and store prices.

19. Test statistic: $H = 0.775$. Critical value: $\chi^2 = 9.210$. Fail to reject the null hypothesis of equally effective methods.

21. $T = 12$, $n = 7$. The test statistic of $T = 12$ is greater than the critical value of 2, so fail to reject the null hypothesis that there is no difference between the scores of Judge A and Judge B.

23. $n_1 = 25$, $n_2 = 8$, $G = 5$, $\mu_G = 13.12$, $\sigma_G = 2.05$. Test statistic: $z = -3.96$. Critical values: $z = \pm 1.96$. Reject randomness.

Index of Minitab Commands

Command	Use	Page
AOVONEWAY	Analysis of variance (one-way)	583, 596, 600, 617
BINOMIAL	Binomial distribution	202, 646
BOXPLOT	Generates boxplot	108
CHISQUARE	Chi-square test statistic	558
CORRELATION	Linear correlation coefficient	480, 502, 503, 700
DELETE	Deletes a score	26
DESCRIBE	Gives basic statistics	97, 280, 402
ENDOFDATA	Terminates entry of data	24, 25, 334, 415, 518, 596
GICHART	Runs chart of individual values	622
GPCHART	Control chart for attributes	640
GRCHART	Control chart for variation	629
GXBARCHART	Control chart for means	633
HISTOGRAM	Provides histogram	41, 47, 159, 280
HLINES	Control chart for horizontal lines	622, 623, 629, 630, 633
ICHART	Runs chart of individual values	623
INSERT	Inserts a score	26
INTEGERS	Uniform distribution of integers	159, 280
KRUSKAL-WALLIS	Kruskal-Wallis test	687
LET	Assigns data to columns	26, 159, 280, 415, 424, 673
MANN-WHITNEY	Mann-Whitney test (equivalent to Wilcoxon rank-sum test)	679
NAME	Gives name to column of scores	518
NOPAPER	Discontinues printing	25
NORMAL	Normal distribution	402, 424, 616
PDF	Calculates probability for a distribution	202
PLOT	Scatter diagram	476, 480, 502
POOLED	Combines variances	445, 448
PRINT	Displays data	25, 334, 616, 646
RANDOM	Generates random numbers	159, 280, 334, 402, 424, 616, 646
RANK	Ranks data	700
RBAR	R chart using sample ranges	630
READ	Enters data	24, 415, 502, 503, 518, 558, 606, 608
REGRESSION	Linear regression equation	502, 503, 519
RETRIEVE	Retrieves or recalls stored data	25
RMEAN	Calculates row means	334
SAVE	Saves or stores data	25, 31

Command	Use	Page
SET	Enters data	24, 31, 41, 334, 362, 502, 596
SORT	Sorts data	334
STACK	Combines data in one column	687
STEM-AND-LEAF	Stem-and-leaf plot	102
STEST	Sign test	658
STOP	Exits the Minitab program	25
SUBSCRIPTS	Group identifiers	687
TINTERVAL	Confidence interval (uses Student t distribution)	298, 414
TTEST	Hypothesis test using t distribution	362, 414, 415
TWOSAMPLE	Hypothesis test of equal means	424, 448
TWOWAY	Two-way analysis of variance	602, 608, 610, 611, 615
WTEST	Wilcoxon signed-ranks test	673

Index

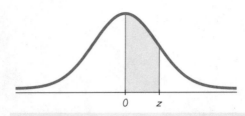

op for closer value

TABLE A-2 Standard Normal (z) Distribution

z	.00	.01	.02	.03	.04	.05	.06	.07	.08	.09
0.0	.0000	.0040	.0080	.0120	.0160	.0199	.0239	.0279	.0319	.0359
0.1	.0398	.0438	.0478	.0517	.0557	.0596	.0636	.0675	.0714	.0753
0.2	.0793	.0832	.0871	.0910	.0948	.0987	.1026	.1064	.1103	.1141
0.3	.1179	.1217	.1255	.1293	.1331	.1368	.1406	.1443	.1480	.1517
0.4	.1554	.1591	.1628	.1664	.1700	.1736	.1772	.1808	.1844	.1879
0.5	.1915	.1950	.1985	.2019	.2054	.2088	.2123	.2157	.2190	.2224
0.6	.2257	.2291	.2324	.2357	.2389	.2422	.2454	.2486	.2517	.2549
0.7	.2580	.2611	.2642	.2673	.2704	.2734	.2764	.2794	.2823	.2852
0.8	.2881	.2910	.2939	.2967	.2995	.3023	.3051	.3078	.3106	.3133
0.9	.3159	.3186	.3212	.3238	.3264	.3289	.3315	.3340	.3365	.3389
1.0	.3413	.3438	.3461	.3485	.3508	.3531	.3554	.3577	.3599	.3621
1.1	.3643	.3665	.3686	.3708	.3729	.3749	.3770	.3790	.3810	.3830
1.2	.3849	.3869	.3888	.3907	.3925	.3944	.3962	.3980	.3997	.4015
1.3	.4032	.4049	.4066	.4082	.4099	.4115	.4131	.4147	.4162	.4177
1.4	.4192	.4207	.4222	.4236	.4251	.4265	.4279	.4292	.4306	.4319
1.5	.4332	.4345	.4357	.4370	.4382	.4394	.4406	.4418	.4429	.4441
1.6	.4452	.4463	.4474	.4484	.4495	* .4505	.4515	.4525	.4535	.4545
1.7	.4554	.4564	.4573	.4582	.4591	.4599	.4608	.4616	.4625	.4633
1.8	.4641	.4649	.4656	.4664	.4671	.4678	.4686	.4693	.4699	.4706
1.9	.4713	.4719	.4726	.4732	.4738	.4744	.4750	.4756	.4761	.4767
2.0	.4772	.4778	.4783	.4788	.4793	.4798	.4803	.4808	.4812	.4817
2.1	.4821	.4826	.4830	.4834	.4838	.4842	.4846	.4850	.4854	.4857
2.2	.4861	.4864	.4868	.4871	.4875	.4878	.4881	.4884	.4887	.4890
2.3	.4893	.4896	.4898	.4901	.4904	.4906	.4909	.4911	.4913	.4916
2.4	.4918	.4920	.4922	.4925	.4927	.4929	.4931	.4932	.4934	.4936
2.5	.4938	.4940	.4941	.4943	.4945	.4946	.4948	.4949	* .4951	.4952
2.6	.4953	.4955	.4956	.4957	.4959	.4960	.4961	.4962	.4963	.4964
2.7	.4965	.4966	.4967	.4968	.4969	.4970	.4971	.4972	.4973	.4974
2.8	.4974	.4975	.4976	.4977	.4977	.4978	.4979	.4979	.4980	.4981
2.9	.4981	.4982	.4982	.4983	.4984	.4984	.4985	.4985	.4986	.4986
3.0	.4987	.4987	.4987	.4988	.4988	.4989	.4989	.4989	.4990	.4990

NOTE: For values of z above 3.09, use 0.4999 for the area.

*Use these common values that result from interpolation:

z score	Area
1.645	0.4500
2.575	0.4950

Frederick Mosteller and Robert E. K. Rourke, *Sturdy Statistics*, 1973, Addison-Wesley Publishing Co., Reading, MA.
Reprinted with permission of Frederick Mosteller.